AF361659

Molecular Psychiatry

Molecular Psychiatry

Special Issue Editors

Theo Rein
Gabriel R. Fries

MDPI • Basel • Beijing • Wuhan • Barcelona • Belgrade • Manchester • Tokyo • Cluj • Tianjin

Special Issue Editors
Theo Rein
Max Planck Institute of Psychiatry
Germany

Gabriel R. Fries
The University of Texas Health Science Center at Houston
USA

Editorial Office
MDPI
St. Alban-Anlage 66
4052 Basel, Switzerland

This is a reprint of articles from the Special Issue published online in the open access journal *International Journal of Molecular Sciences* (ISSN 1422-0067) (available at: https://www.mdpi.com/journal/ijms/special_issues/molecular_psychiatry).

For citation purposes, cite each article independently as indicated on the article page online and as indicated below:

LastName, A.A.; LastName, B.B.; LastName, C.C. Article Title. *Journal Name* **Year**, *Article Number*, Page Range.

ISBN 978-3-03936-120-5 (Hbk)
ISBN 978-3-03936-121-2 (PDF)

Contents

About the Special Issue Editors

Theo Rein University studies in chemistry and biochemistry in Stuttgart and Munich. Diplom (equivalent of M-Sc.) in chemistry and Dr. rer. nat. (equivalent of PhD) in biochemistry from the Ludwig Maximilian University in Munich. Postdoctoral studies at the Roche Institute of Molecular Biology, Nutley, NJ, USA and at the National Institute of Child Health and Human Development in Bethesda, Maryland, USA. From 1998 to 2013, head of the chaperone research group, and from 2004 to 2013, Research Coordinator at the Max-Planck-Institute of Psychiatry, Munich, Germany. Since 2002, Lecturer in Biochemistry, Medical Faculty, Ludwig Maximilian University, Munich, Germany, and since 2014 head of the project group molecular signaling pathways in depression, Max-Planck-Institute of Psychiatry, Munich, Germany.

Gabriel R. Fries B.Sc. in Biomedical Sciences from the Federal University of Rio Grande do Sul, Brazil. M.Sc. and Ph.D. in Biochemistry from the Federal University of Rio Grande do Sul, Brazil. Graduate research fellowship at the Max Planck Institute of Psychiatry in Munich, Germany (2013–2014). Postdoctoral fellowship at the University of Texas Health Science Center at Houston, TX, USA (2015–2018). Instructor (2018–2019) and currently Assistant Professor (2019 - present) in the Faillace Department of Psychiatry and Behavioral Sciences at the University of Texas Health Science Center at Houston, TX, USA.

Preface to "Molecular Psychiatry"

The concept that psychiatric diseases have a molecular basis is now widely accepted. Nevertheless, the complexity of these diseases poses a particular challenge for the translational efforts to integrate disease manifestation, behavior, neuronal circuits, and molecular pathways to form a complete theory with significant clinical implications. The Ebook "Molecular Psychiatry" of the International Journal of Molecular Sciences features exciting examples of these efforts (9 reviews and 11 original articles), encompassing studies on molecular mechanisms, animal models, biomarkers, advanced methodology, drug development and responsiveness, as well as genetics and epigenetics.

Theo Rein, Gabriel R. Fries
Special Issue Editors

International Journal of
Molecular Sciences

Editorial

Molecular Psychiatry: Trends and Study Examples

Theo Rein [1],* and Gabriel R. Fries [2],*

[1] Translational Psychiatry Program, Department of Psychiatry and Behavioral Sciences, McGovern Medical School, The University of Texas Health Science Center at Houston, Houston, TX 77054, USA

[2] Department of Translational Science in Psychiatry, Max Planck Institute of Psychiatry, 80804 Munich, Germany

* Correspondence: theorein@psych.mpg.de (T.R.); Gabriel.R.Fries@uth.tmc.edu (G.R.F.)

Received: 2 December 2019; Accepted: 9 January 2020; Published: 10 January 2020

In contrast to about 20–30 years ago, the concept that psychiatric diseases have a molecular basis is now widely accepted. Nevertheless, the complexity of these diseases poses a particular challenge for the translational efforts to integrate disease manifestation, behavior, neuronal circuits, and molecular pathways to form a complete theory with significant clinical implications. The Special Issue "Molecular Psychiatry" of the International Journal of Molecular Sciences presents exciting examples of these efforts (9 reviews and 11 original articles), encompassing studies on molecular mechanisms, animal models, biomarkers, advanced methodology, drug development and responsiveness, as well as genetics and epigenetics.

1. Molecular Mechanisms

Basic science mechanisms are a vital part of the search for the biological basis of psychiatric disorders, providing molecular hints that can later be tested as biomarkers or as targets for the development of new medications. Several manuscripts published in this Special Issue describe interesting mechanisms that may underlie the biology of these disorders.

The role of corticosteroid receptors in psychiatric diseases has been recognized for a long time, in part as executors of the stress response that is pivotal in a number of diseases. The review of Baker et al. illustrates this while focusing on the molecular mechanisms regulating steroid receptor activity [1]. The authors summarize our current knowledge on the control of glucocorticoid receptor (GR) activity by the heat shock protein (Hsp) 90 based chaperone system, with a focus on the established stress factor and co-chaperone FK506 binding protein (FKBP) 51. The link to the stress response and circadian rhythm is outlined and the potential for chaperone-targeting therapeutics is discussed.

The article by Kretzschmar and colleagues focusses on the molecular effects of the stress- and GR-inducible protein Downregulated in renal cell carcinoma 1 (DRR1) in organizing the actin cytoskeleton [2]. DRR1 has been associated with several brain disorders and is described as a resilience factor. The authors demonstrate that DRR1 affects actin dynamics through several mechanisms that likely impact neuronal function, as well as stress physiology and pathophysiology.

The article by van Weert et al. provides intriguing mechanistic insight into the regulation of genes that can be target of either GR or the mineralocorticoid receptor (MR) [3]. It is established for a long time that the balanced activity of these two corticosteroid receptors is pivotal for stress coping and mental health. The authors provide evidence suggesting that the basic helix-loop-helix transcription factor NeuroD facilitates MR binding to gene regulatory elements, questioning competition for DNA binding as a mechanism of MR- over GR-specific binding.

Strategies to decode the molecular mechanisms of fast-acting antidepressants are the focus of the review by Herzog and colleagues, with emphasis on gender-specific aspects [4]. The authors survey the literature documenting the need for elucidating gender-specific mechanisms and provide the current state of the art to propose a framework for experiments in rodents to tackle the issue of gender difference in treatment response.

In the search for molecular mechanisms of the protein Disrupted in Schizophrenia 1 (DISC1), whose gene translocation frequently is found correlated with cases of schizophrenia, bipolar disorder, and major depression, Ramos et al. followed an unbiased proteomic approach [5]. The analysis of the proteome of primary neurons in which DISC1 was knocked down provided evidence that DISC1 has a role in both neurodevelopment and synaptic function.

In light of the accumulating evidence for the association between chronic inflammation and major depressive disorder (MDD), Milenkovic et al. review the role of chemokines in MDD. Chemokines are known as small cytokines impacting the induction of chemotaxis, the migration of leukocytes and macrophages, and the propagation of inflammation. The authors conclude that these cytokines could serve as peripheral markers of psychiatric disorders, or even targets for novel treatment strategies in depression [6].

Finally, the article by Ambrée et al. examining mechanistic aspects of the T cell response in stress-induced depression-like behavior [7] is outlined in the next section on animal models in more detail.

2. Animal Models

The application of basic science findings to the clinics is an important and particularly hard part of translational science. One of the commonly used approaches to aid in this bench-to-bed translation is the use and study of so-called 'animal models' that can test specific mechanisms in vivo prior to the use of human subjects. This Special Issue includes two interesting articles applying such models.

Ambrée and colleagues used social defeat stress (SDS) as a model for stress-induced depression-like behavior to shed light on the T cell phenotype associated with susceptibility and resilience to depression [7]. The authors grouped SDS-exposed mice into susceptible and resilient and found significantly increased numbers of interleukin-17 producing CD4+ and CD8+ T cells in the spleen of susceptible mice. Harnessing the power of genetic intervention, mice with a conditional deletion of PPARγ in CD4+ cells were analyzed; PPARγ is an inhibitor of Th17 development and its deletion thus enhances Th17 differentiation. However, this genetic manipulation did not change susceptibility to SDS. Thus, while SDS promotes Th17 cell and suppresses Treg cell differentiation predominantly in susceptible mice, the effects in immune responses after stress exposure remain to be elucidated [7].

Another rodent model for susceptibility and resilience to stress-induced depression-like behavior is represented by the Roman High-Avoidance (RHA) and the Roman Low-Avoidance (RLA) rats, which were used by Serra et al. to investigate the effects of forced swimming on factors of neuronal plasticity [8]. This stressor elicited changes in the expression of brain-derived neurotrophic factor (BDNF), its receptor trkB, and the Polysialilated-Neural Cell Adhesion Molecule in distinct regions of the brain. These changes pronouncedly differed between the two rat lines, consistent with a role of BDNF/trkB signaling and neuroplasticity in susceptibility and resistance to stress-induced depression [8]. Animal models are also outlined in the above-mentioned review, where Herzog et al. suggest procedures for experiments in rodents to investigate gender differences in treatment response in depression [4].

3. Biomarkers

As is the case in other medical fields, such as oncology and cardiology, biomarkers may provide valuable proxy information in psychiatric patients, potentially assisting in measures of prognosis, treatment response, diagnosis, and progression. Psychiatric disorders pose a particular challenge due to the tissue specificity of many currently investigated biomarkers; i.e., not all blood-based measures directly represent changes in the brain, and the study of the correlations between the periphery and the central nervous system is a rapidly evolving field. This Special Issue includes five manuscripts focused on the challenges of identifying clinically and biologically relevant biomarkers for psychiatric disorders.

Recent findings in the field of Psychiatry have suggested a potentially key role for epigenetic mechanisms in determining not only risk and resilience in patients and vulnerable subjects but also in modulating one's response to a given treatment. The paper by Goud Alladi and colleagues explored through a systematic review the evidence of DNA methylation mechanisms involved in the clinical treatment response in serious mental illness, specifically bipolar disorder, schizophrenia, and major depressive disorder [9]. The authors of this interesting study emphasize the potential clinical use of such markers in predicting whether a patient will respond or not to a medication, which is a highly anticipated approach in the emerging field of personalized medicine.

Among promising biomarkers, significant effort has also been made in the study of inflammatory mediators to predict diagnosis, prognosis, and treatment. In fact, most psychiatric disorders have been shown to present immune dysfunctions, as measured by cytokines and inflammatory mediators, and such molecular phenotypes are thought to at least partly mediate the higher cardiovascular and metabolic disturbances seen in psychiatric patients. Baghai and colleagues, for instance, performed a large study to investigate the combined influence of major depressive disorder and cardiovascular disorders on immune mediators [10]. Interestingly, their findings not only suggest that higher inflammatory biomarkers underlie the risk for cardiovascular disease in depression patients but also that higher levels of these molecules are associated with better clinical outcome and faster remission in patients.

This Special Issue also includes the study by Mühle and colleagues [11], which investigated another type of peripheral biomarkers in alcohol-dependent patients: the acid sphingomyelinase, an enzyme that breaks sphingomyelin into ceramide and thereby changes the composition of plasma membranes. As hypothesized, the authors found clinically relevant changes in the enzyme levels in patients, which also correlated with several other biochemical markers of dependence and health.

Another interesting approach has been taken by Rotter and colleagues [12], which measured the expression levels of alpha-synuclein (a protein known to be associated with Lewy bodies in Parkinson's disease) in the peripheral blood of patients with major depressive disorder as an attempt to understand the high comorbidity between depression and Parkinson's disease. Accordingly, their interesting findings were suggestive of an increase in alpha-synuclein levels in depressed patients, with a positive correlation between the severity of depression and the levels of this biomarker.

Finally, Steiger and Pawlowski discussed in a review paper how biomarkers do not necessarily entail laboratory measures using biospecimens [13]. The authors discuss the use of sleep electroencephalogram (EEG) as a biomarker of impaired sleep in the context of depression, and comprehensively review many applications and biological underpinning of this marker in the depressed population.

4. Advanced Methods

One of the challenges of the field of molecular psychiatry is related to methodological limitations. Many commonly used techniques are particularly limited and have not been successful in answering the rapidly evolving questions of the field. Two manuscripts in this Special Issue are focused on innovative approaches that can revolutionize the field, including the use and proper analysis of Big Data through machine learning methods and the development and application of induced pluripotent stem cells (iPSC) from psychiatric patients.

The study by Cao and colleagues tested eight machine learning algorithms in the same transcriptome-wide expression datasets of patients with schizophrenia and controls to identify reproducible biological signatures of disease [14]. This type of study to identify the most robust and effective algorithms can have significant impacts in the field since integrative analysis of complex datasets is becoming the cornerstone of biological psychiatry studies.

Novel methods are also being developed at the bench side, including the establishment of patient-derived iPSCs. These cells can provide an effective tool for the study of complex diseases by allowing the establishment of cellular models accounting for the patient's genetic background while removing the effect of outside environmental influences. The review by Hoffman and colleagues [15] provides an insightful overview of the use of these cell models to study the neurobiological basis of childhood-onset schizophrenia, discussing advantages, limitations, and challenges of the field and the use of these modern technologies. Overall, the discussion also applies to other psychiatric diagnoses which may benefit from the use of patient-derived cell models as opposed to commonly used animal primary cells or cell lines.

5. Drugs/Antidepressant Response

One of the major limitations of currently available psychiatric medications stems from the long duration of treatment and the need for several days of treatment until the detection of proper medication response and efficacy. Due to the heterogeneity of patients and the lack of proper biomarkers of treatment response, more often than not the choice of treatment is made in a trial-and-error mode by the clinician, potentially delaying symptom resolution. Several authors aim at understanding the specific mechanisms of actions of existing drugs with the ultimate goal of identifying ways to predict treatment response in patients. Three papers in this Special Issue provide an overview of this area of investigation, with a particular focus on major depression and schizophrenia.

The abovementioned review by Herzog et al. focusses on gender-specific aspects of antidepressant response [4]. The article by Ising and colleagues presents the analysis of FKBP5 gene polymorphism and of RNA and protein levels of its gene product FKBP51 in peripheral blood in 297 inpatients treated for acute depression according to doctor's choice [16]. Pronounced reduction of FKBP5 gene and FKBP51 protein expression was observed in patients responding to antidepressant treatment, while non-responders had increased levels. The FKBP5 genotype moderated this effect [16]. This study significantly contributes to the complexity of the link of the stress factor FKBP51 to antidepressant responsiveness.

The review by Kondej and colleagues discusses established and novel drug targets for schizophrenia, in particular, the concept of multi-target drugs [17]. This is an increasingly important aspect in drug discovery, also considering the limited success of single-target drugs in polygenic diseases with complex pathomechanisms.

6. Genetics and Epigenetics

Family studies have convincingly proven that psychiatric disorders run in families and have a strong genetic basis. Notwithstanding, the high heterogeneity of psychiatric diagnoses has significantly hindered the discovery of their molecular genetics, with few genome-wide association studies (GWAS) suggesting the need for extremely large samples and robust statistical methods. In addition, the 'heritability gap' (i.e., the gap between the heritability detected in family studies and that detected by GWAS) has suggested an important role for the environment in modulating genetic mechanisms, which has led to the hypothesis that epigenetic mechanisms may be especially important in these disorders. Three manuscripts in this Special Issue go deeper into this topic and highlight approaches and findings related to the genetics and epigenetics of psychiatric disorders.

Lesiewska and colleagues undertook an experiment to investigate the association between affective temperament traits and polymorphisms in dopaminergic genes in a sample of obese subjects [18]. Among other findings, they found interesting associations between specific temperament dimensions with a polymorphism in the catechol-O-methyltransferase (*COMT*) gene, which supports the hypothesis of a key role for dopaminergic genes in determining temperament expression in obese individuals.

The influence of genetic markers in psychiatric traits is also subject of the review from Ohi and colleagues [19], which explored the genetic determinants of general cognitive function and how they overlap with schizophrenia. Lower general cognitive function has been repeatedly associated with a higher risk for schizophrenia, and a genetic risk may likely underlie this association. More specifically, the authors discussed risk *loci* identified by GWAS studies of both schizophrenia and cognitive function, the polygenic nature of both conditions, and recent evidence showing how their genetic determinants are particularly similar.

Finally, in light of evidence suggesting an important role for the serotonergic system in many physiological and pathological conditions, including psychiatric disorders, the paper by Rebholz and colleagues reviewed biological findings regarding the serotonin 4 receptor (5-HT4R) [20] in brain regions and how its genetic regulation and gene expression changes can modulate reward and executive function and potentially give rise to mood changes in vulnerable subjects. The authors also extend their discussion to explore methodological advancements that will be needed for a better understanding of this receptor in brain function, including the development of more suitable genetic mouse models.

Overall, this rich collection of studies provides a diversified portfolio of the several approaches that can be used for tackling the biology of psychiatric disorders, reinforcing the importance of using multiple lines of converging evidence for their study. While not exhaustive and comprehensive, these studies are effective in identifying current limitations of the field and provide the reader with an enlightening overview of the directions and future of translational research in molecular psychiatry.

Conflicts of Interest: The authors declare no conflict of interest.

References

1. Baker, J.D.; Ozsan, I.; Rodriguez, O.S.; Gulick, D.; Blair, L.J. Hsp90 Heterocomplexes Regulate Steroid Hormone Receptors: From Stress Response to Psychiatric Disease. *Int. J. Mol. Sci.* **2018**, *20*, 79. [CrossRef] [PubMed]
2. Kretzschmar, A.; Schulke, J.P.; Masana, M.; Durre, K.; Muller, M.B.; Bausch, A.R.; Rein, T. The Stress-Inducible Protein DRR1 Exerts Distinct Effects on Actin Dynamics. *Int. J. Mol. Sci.* **2018**, *19*, 3993. [CrossRef] [PubMed]
3. van Weert, L.T.C.M.; Buurstede, J.C.; Sips, H.C.M.; Mol, I.M.; Puri, T.; Damsteegt, R.; Roozendaal, B.; Sarabdjitsingh, R.A.; Meijer, O.C. Mechanistic Insights in NeuroD Potentiation of Mineralocorticoid Receptor Signaling. *Int. J. Mol. Sci.* **2019**, *20*, 1575. [CrossRef] [PubMed]
4. Herzog, D.P.; Wegener, G.; Lieb, K.; Muller, M.B.; Treccani, G. Decoding the Mechanism of Action of Rapid-Acting Antidepressant Treatment Strategies: Does Gender Matter? *Int. J. Mol. Sci.* **2019**, *20*, 949. [CrossRef] [PubMed]
5. Ramos, A.; Rodríguez-Seoane, C.; Rosa, I.; Gorroño-Etxebarria, I.; Alonso, J.; Veiga, S.; Korth, C.; Kypta, R.M.; García, Á.; Requena, J.R. Proteomic Studies Reveal Disrupted in Schizophrenia 1 as a Player in Both Neurodevelopment and Synaptic Function. *Int. J. Mol. Sci.* **2019**, *20*, 119.
6. Milenkovic, V.M.; Stanton, E.H.; Nothdurfter, C.; Rupprecht, R.; Wetzel, C.H. The Role of Chemokines in the Pathophysiology of Major Depressive Disorder. *Int. J. Mol. Sci.* **2019**, *20*, 2283. [CrossRef] [PubMed]
7. Ambrée, O.; Ruland, C.; Zwanzger, P.; Klotz, L.; Baune, B.T.; Arolt, V.; Scheu, S.; Alferink, J. Social Defeat Modulates T Helper Cell Percentages in Stress Susceptible and Resilient Mice. *Int. J. Mol. Sci.* **2019**, *20*, 3512. [CrossRef] [PubMed]
8. Serra, M.P.; Poddighe, L.; Boi, M.; Sanna, F.; Piludu, M.A.; Sanna, F.; Corda, M.G.; Giorgi, O.; Quartu, M. Effect of Acute Stress on the Expression of BDNF, trkB, and PSA-NCAM in the Hippocampus of the Roman Rats: A Genetic Model of Vulnerability/Resistance to Stress-Induced Depression. *Int. J. Mol. Sci.* **2018**, *19*, 3745. [CrossRef] [PubMed]
9. Alladi, C.G.; Etain, B.; Bellivier, F.; Marie-Claire, C. DNA Methylation as a Biomarker of Treatment Response Variability in Serious Mental Illnesses: A Systematic Review Focused on Bipolar Disorder, Schizophrenia, and Major Depressive Disorder. *Int. J. Mol. Sci.* **2018**, *19*, 3026. [CrossRef] [PubMed]
10. Baghai, T.C.; Varallo-Bedarida, G.; Born, C.; Hafner, S.; Schule, C.; Eser, D.; Zill, P.; Manook, A.; Weigl, J.; Jooyandeh, S.; et al. Classical Risk Factors and Inflammatory Biomarkers: One of the Missing Biological Links between Cardiovascular Disease and Major Depressive Disorder. *Int. J. Mol. Sci.* **2018**, *19*, 1740. [CrossRef] [PubMed]
11. Mühle, C.; Weinland, C.; Gulbins, E.; Lenz, B.; Kornhuber, J. Peripheral Acid Sphingomyelinase Activity Is Associated with Biomarkers and Phenotypes of Alcohol Use and Dependence in Patients and Healthy Controls. *Int. J. Mol. Sci.* **2018**, *19*, 4028. [CrossRef] [PubMed]
12. Rotter, A.; Lenz, B.; Pitsch, R.; Richter-Schmidinger, T.; Kornhuber, J.; Rhein, C. Alpha-Synuclein RNA Expression is Increased in Major Depression. *Int. J. Mol. Sci.* **2019**, *20*, 2029. [CrossRef] [PubMed]
13. Steiger, A.; Pawlowski, M. Depression and Sleep. *Int. J. Mol. Sci.* **2019**, *20*, 607. [CrossRef] [PubMed]

14. Cao, H.; Meyer-Lindenberg, A.; Schwarz, E. Comparative Evaluation of Machine Learning Strategies for Analyzing Big Data in Psychiatry. *Int. J. Mol. Sci.* **2018**, *19*, 3387. [CrossRef] [PubMed]

15. Hoffmann, A.; Ziller, M.; Spengler, D. Childhood-Onset Schizophrenia: Insights from Induced Pluripotent Stem Cells. *Int. J. Mol. Sci.* **2018**, *19*, 3829. [CrossRef] [PubMed]

16. Ising, M.; Maccarrone, G.; Bruckl, T.; Scheuer, S.; Hennings, J.; Holsboer, F.; Turck, C.W.; Uhr, M.; Lucae, S. FKBP5 Gene Expression Predicts Antidepressant Treatment Outcome in Depression. *Int. J. Mol. Sci.* **2019**, *20*, 485. [CrossRef] [PubMed]

17. Kondej, M.; Stepnicki, P.; Kaczor, A.A. Multi-Target Approach for Drug Discovery against Schizophrenia. *Int. J. Mol. Sci.* **2018**, *19*, 3105. [CrossRef] [PubMed]

18. Lesiewska, N.; Borkowska, A.; Junik, R.; Kaminska, A.; Pulkowska-Ulfig, J.; Tretyn, A.; Bielinski, M. The Association Between Affective Temperament Traits and Dopamine Genes in Obese Population. *Int. J. Mol. Sci.* **2019**, *20*, 1847. [CrossRef] [PubMed]

19. Ohi, K.; Sumiyoshi, C.; Fujino, H.; Yasuda, Y.; Yamamori, H.; Fujimoto, M.; Shiino, T.; Sumiyoshi, T.; Hashimoto, R. Genetic Overlap between General Cognitive Function and Schizophrenia: A Review of Cognitive GWASs. *Int. J. Mol. Sci.* **2018**, *19*, 3822. [CrossRef] [PubMed]

20. Rebholz, H.; Friedman, E.; Castello, J. Alterations of Expression of the Serotonin 5-HT4 Receptor in Brain Disorders. *Int. J. Mol. Sci.* **2018**, *19*, 3581. [CrossRef] [PubMed]

International Journal of
Molecular Sciences

Review

Hsp90 Heterocomplexes Regulate Steroid Hormone Receptors: From Stress Response to Psychiatric Disease

Jeremy D. Baker, Ilayda Ozsan, Santiago Rodriguez Ospina, Danielle Gulick and Laura J. Blair *

USF Health Byrd Institute, Morsani College of Medicine, Department of Molecular Medicine, University of South Florida, 4001 East Fowler Ave, Tampa, FL 33613, USA; jeremyb@health.usf.edu (J.D.B.); iozsan@health.usf.edu (I.O.); santiago3@health.usf.edu (S.R.O.); dgulick@health.usf.edu (D.G.)
* Correspondence: lblair@health.usf.edu; Tel.: +1-813-396-0639

Received: 17 November 2018; Accepted: 17 December 2018; Published: 25 December 2018

Abstract: The hypothalamus-pituitary-adrenal (HPA) axis directly controls the stress response. Dysregulation of this neuroendocrine system is a common feature among psychiatric disorders. Steroid hormone receptors, like glucocorticoid receptor (GR), function as transcription factors of a diverse set of genes upon activation. This activity is regulated by molecular chaperone heterocomplexes. Much is known about the structure and function of these GR/heterocomplexes. There is strong evidence suggesting altered regulation of steroid receptor hormones by chaperones, particularly the 51 kDa FK506-binding protein (FKBP51), may work with environmental factors to increase susceptibility to various psychiatric illnesses including post-traumatic stress disorder (PTSD), major depressive disorder (MDD), and anxiety. This review highlights the regulation of steroid receptor dynamics by the 90 kDa heat shock protein (Hsp90)/cochaperone heterocomplexes with an in depth look at how the structural regulation and imbalances in cochaperones can cause functional effects on GR activity. Links between the stress response and circadian systems and the development of novel chaperone-targeting therapeutics are also discussed.

Keywords: Hsp90; GR; stress response; steroid hormones; molecular chaperones; psychiatric disease; circadian rhythms; FKBP51; FKBP52; CyP40; PP5

---

1. Introduction

The HPA, or hypothalamic-pituitary-adrenal, axis is a critical neuroendocrine system controlling numerous processes including autonomic functions (e.g., digestion), the immune response, metabolic activity, and, importantly, the stress response [1–4]. The axis consists of three major glands, for which it is named: the hypothalamus, the pituitary, and the adrenal or suprarenal gland. When stressors are encountered, including both physical insult and psychological stress, the HPA axis is activated [5–7]. The paraventricular nucleus of the hypothalamus releases corticotropin releasing hormone (CRH), which activates the anterior lobe of the pituitary gland causing the release of adrenocorticotropic hormone (ACTH). ACTH production stimulates the release of the glucocorticoid hormone, cortisol (CORT), from the zona fasciculata of the adrenal cortex. Then, circulating CORT inhibits the release of CRH and ACTH through a negative feedback loop ending the HPA activated stress response (Figure 1) [8–10]. The stress response is adaptive, but dysregulation can occur after long-term stress or other insults that can result imbalanced serum CORT [11].

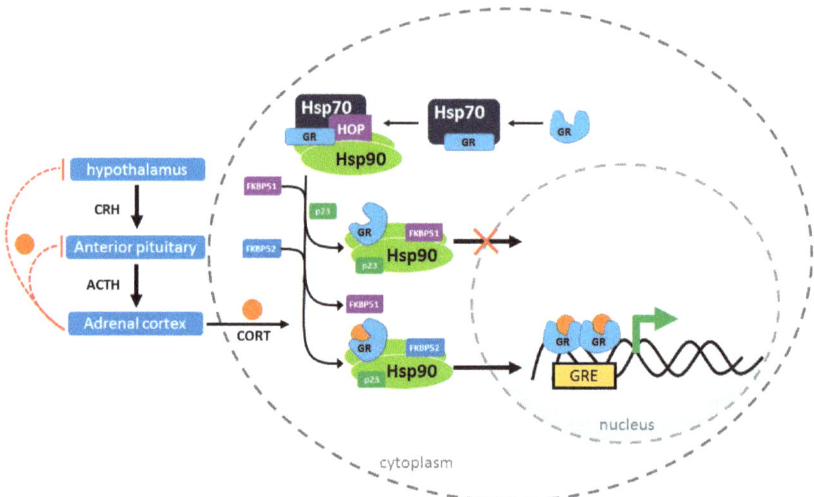

Figure 1. Schematic of glucocorticoid receptor (GR) transactivation in response to cortisol (CORT). After stress, the hypothalamus releases corticotropin releasing hormone (CRH) stimulating the anterior pituitary to release adrenocorticotropic hormone (ACTH). ACTH stimulates CORT release from the adrenal cortex which crosses the plasma membrane of the cell. Through negative feedback, CORT inhibits hormone release from the hypothalamus and anterior pituitary glands. Inside the cell, Hsp70 binds to and unfolds GR in the cytosol. HOP recruits GR:Hsp70 to Hsp90. Cochaperones (including FKBP51 and FKBP52) bind to Hsp90 as HOP is released. p23 binds and stabilizes the GR:Hsp90 heterocomplex. FKBP51 inhibits nuclear transactivation of GR, while FKBP52 and other copchaperones may promote translocation. Subsequently, GR binds CORT, dimerizes and translocates to the nucleus binding to glucocorticoid response elements (GREs).

CORT has broad physiological impacts and modulates behavior, memory, cognition, metabolism, development, inflammation, gluconeogenesis, and circadian rhythmicity [12,13]. Here, we will focus on the regulation of CORT by chaperone heterocomplexes and discuss how this relates to disruption of the normal physiological stress response. It has been well-described that short-term memory formation is intimately tied to the actions of CORT and epinephrine, particularly in response to traumatic emotional events. In the brain, CORT acts directly on the amygdala, an emotional hub, and regulates neural connections to the hippocampus, which is required for memory formation [14–16]. Dysregulation of CORT levels, whether positive or negative, can impair memory consolidation [17]. Memory retrieval can also be negatively impacted by the levels of CORT [18].

Importantly, altered CORT levels have been linked to psychiatric disorders including major depressive disorder (MDD) [19,20], general anxiety [21,22], bipolar disorder [23,24], and post-traumatic stress disorder (PTSD) [14] as well as substance use disorder [25]. CORT influences brain activity by binding glucocorticoid receptor (GR) and, to a lesser affinity, mineralocorticoid receptor (MR). Upon activation, GR homodimerizes and translocates to the nucleus where it regulates the transcription of GR-responsive genes [26].

GR translocation is tightly regulated by a well-characterized chaperone ensemble [27–29]. At the center of this complex, the 90 kDa heat shock protein (Hsp90) collaborates with cochaperones, including two FK506-binding proteins, FKBP51 and FKBP52, cyclophilin 40 (CyP40), and protein phosphatase 5 (PP5) to control GR transactivation, affecting both sensitivity to CORT and nuclear translocation [30,31]. Notably, FKBP51 affects GR transactivation in a dissimilar manner to the other cochaperones. Increased FKBP51 slows GR nuclear translocation, at least in part, through impairing the interaction between GR heterocomplexes and dynein, resulting in reduced GR activity. FKBP52, as well as CyP40 and PP5, have been shown to promote GR activity, which may be a combination of increased dynein binding as well as through displacing the inhibitory effects of FKBP51 from the Hsp90-heterocomplex, since these cochaperones bind Hsp90 at the same site (Figure 1) [32–37]. This review discusses chaperone involvement in GR physiology and the impact of chaperone imbalances that may lower resilience against psychiatric disorders.

2. Chaperones in GR Signaling

The cellular stress response is highly conserved within eukaryotes [38,39], granting the ability to rapidly cope with adverse physiological insults. This process, controlled by molecular chaperones, is integral to maintaining homeostasis in all cells. A key stress response chaperone, Hsp90, is highly abundant throughout mammalian cells and uses ATP to interact with numerous substrates [40,41]. Hsp90 interacts with up to 10% of all proteins and is involved in nearly every cellular homeostatic process and is essential for signaling pathways, including GR activation [42–46]. Hsp90 functions as a homodimer [47]. Hsp90 consists of an ATP-binding domain at the amino-terminus separated by a flexible linker from a middle domain, which is important for client binding. Conformation of Hsp90 determines activity and function as well as regulates substrate binding [48]. Hsp90 adopts an open conformation in the absence of ATP and proceeds to a closed state when it is ATP-bound [49]. In addition, cytosolic Hsp90 contains a carboxy-terminal MEEVD (met-glu-glu-val-asp) motif that regulates the interaction with a host of cochaperones [50]. This allows for direct binding of cochaperones containing a tetratricopeptide repeat (TPR) domain, including FKBP51, FKBP52, and CyP40.

Hsp90 is regulated transcriptionally by heat shock factor 1 (HSF1), posttranslationally by modifications including phosphorylation and acetylation [50], and functionally by a diverse set of cochaperones [51]. Although the main binding of FKBP51, FKBP52, CyP40, and PP5 is through a conserved TPR domain, cochaperones have been shown to bind to Hsp90 in all three domains [50]. These cochaperones compete for Hsp90 binding; however, simultaneous binding of more than one cochaperone has also been demonstrated [52,53]. Cochaperones influence Hsp90 substrate recognition, alter Hsp90 ATPase activity and conformational dynamics, and have the ability to interact with Hsp90 substrates directly [54,55].

Hsp90 Heterocomplex

It has been demonstrated, in vitro, that an Hsp90 heterocomplex is required for GR maturation [31,56]. In a stepwise process, GR interacts first with Hsp70, and then is passed to Hsp90 via Hsp70-Hsp90 organizing protein (HOP) [46,57,58]. HOP is dislodged from this Hsp90-GR complex upon Hsp90 binding ATP and subsequent association of cochaperones, FKBP51, FKBP52, CyP40, or PP5 [59,60]. p23 preferentially associates with ATP-bound Hsp90 through the N-terminus and middle domains stabilizing the complex in a conformation with high affinity for CORT [30,61]. Upon CORT binding, GR dimerizes and translocates into the nucleus where it regulates the transcription of GR-responsive genes.

With the development of new technologies to evaluate the detailed structure of large, multiprotein complexes, we are continually learning more about how these proteins interact. This information is important to understand how changes in structure are tightly linked to functional effects. Recently, cryoelectron microscopy has further clarified the interactions between Hsp70 and Hsp90 that are required for GR maturation and function [46]. It was shown that the coupling of the ATP-dependent chaperone cycles of Hsp70 and Hsp90 may be required for GR maturation. Using recombinant proteins and the ligand binding domain of GR (GRLBD), it was shown that GRLBD is first unfolded by Hsp70 and inactivated. Hsp90 reverses this step resulting in folded, aggregation resistant, and functional GRLBD. This could explain the necessity of Hsp90 in GR maturation.

Remarkably, during the hand-off of GR from Hsp70 to Hsp90, the substrate binding domains of Hsp70 and client binding domain of Hsp90 align. HOP recruits GRLBD:Hsp70 to Hsp90 resulting in an intermediary complex of GRLBD:Hsp70:Hsp90:HOP in which GRLBD is physically in contact with both Hsp70 and Hsp90 simultaneously [46]. This conformation allows for an interaction between the ATP domains of Hsp90 and Hsp70 and subsequent coupling of ATP hydrolysis. A key finding is that Hsp90 ATP hydrolysis is required for client loading from Hsp70, however blocking ATPase activity does not perturb the GRLBD:Hsp70:Hsp90:HOP intermediary complex. This is important because it helps to explain how inhibition of Hsp90 ATPase activity results in Hsp70-mediated degradation, as client loading is blocked. It was recently described that the intermediary complex actually contains two Hsp70 molecules, where one Hsp70 delivers GR to Hsp90, while the other supports the HOP interaction [62]. Overall, the Hsp70-Hsp90 system maintains GR in a competent high-affinity state for CORT, allowing for response to changing CORT levels. This provides an explanation for the necessity of this cycle for GR activity. Further studies are needed to evaluate how cochaperones affect this heterocomplex to functionally regulate GR.

Rearrangement of Hsp90 by cochaperones, in addition to ATP-mediated effects [50], directly impacts client binding [30]. For example, recent nuclear magnetic resonance spectroscopy studies have solved the structure of the Hsp90/FKBP51 [48]. FKBP51 binds between the two dimers of Hsp90 and upon binding stabilizes Hsp90 in the open conformation, reducing ATP hydrolysis activity. Other TPR-containing cochaperones may share the same Hsp90 binding site, but potentially have disparate effects on Hsp90 conformation. Furthermore, it has been found that GR preferentially binds Hsp90 when in the closed ATP-bound state, where both a TPR-containing protein and HOP are associated [30]. Given this, Hsp90/FKBP51 association may disrupt Hsp90/GR binding.

The relationship of cochaperones to CORT levels is complex as cochaperones may have diverse effects on GR and CORT physiology. FKBP51 is upregulated by GR activity, which directly promotes the transcription of *FKBP5*, the gene that encodes FKBP51 [63]. FKBP51, then, negatively inhibits GR activity. The proposed mechanism of action is a short, negative feedback loop whereby FKBP51, in an Hsp90-dependent mechanism, decreases the binding of CORT to GR leading to CORT resistance [64]. *FKBP4*, the gene that encodes the highly similar FKBP51 homolog, FKBP52, may have an opposing effect on GR activity; however, conflicting evidence suggests FKBP52 may not alter GR nuclear transactivation [32,35,36,65,66]. Another TPR-containing cochaperone, CyP40, can also regulate GR through an Hsp90 heterocomplex. It has also been reported that CyP40 may facilitate the export of CORT from the nucleus [65]. CyP40, like FKBP52 and PP5, interacts with dynein, which is a cytoskeletal motor protein that can regulate GR transport [67]. Evidence suggests that PP5 binds to the Hsp90/GR complex at an intermediate step in CORT activation, following the binding of FKBP51 during the basal state [68,69]. PP5 has also been shown to dephosphorylate GR, which can alter GR activity [70]. Thus, PP5 may regulate GR activity through two distinct, but linked mechanisms.

Interestingly, it has been shown that about half of GR within the cell is in complex with Hsp90 and FKBP51 or FKBP52 [68]. About one third of GR is in complex with Hsp90 and PP5, and only a fraction of GR has been found in an Hsp90/CyP40 complex. However, since cochaperones compete to bind Hsp90, this normal distribution of GR/Hsp90 heterocomplexes may become imbalanced with alterations of any TPR-containing proteins, potentially disrupting GR regulation and the stress response.

3. Chaperones Implicated in Psychiatric Disorders

Cochaperone variants and altered expression levels have been linked to psychiatric illness [71,72]. Moreover, cochaperone dysregulation may also contribute to circadian desynchrony, which is exacerbated by and implicated in the etiology of mood disorders. Candidate studies have identified common single nucleotide polymorphisms (SNPs) in the gene that encodes FKBP51, *FKBP5*, that interact synergistically with environmental factors to increase susceptibility to develop PTSD, MDD, anxiety, and bipolar disorder [73–75]. Some of these *FKBP5* SNPs result in increased FKBP51 levels following stress [76]. Increased FKBP51 reduces GR sensitivity, prolonging the HPA-mediated stress response and resulting in increased circulating CORT [77]. We and others have shown that mice lacking *Fkbp5* (*Fkbp5*$^{-/-}$ mice) are resilient to depressive-like behavior following stress [78,79]. These mice have reduced levels of serum CORT following restraint stress or exogenous CORT administration, suggesting that these stress-resilient phenotypes may be linked to CORT regulation by FKBP51.

FKBP52, CyP40 and PP5 have not been currently linked to psychiatric disease, but since each of these cochaperones play an important role in GR regulation through Hsp90, it is possible that dysregulation could increase risk. Work in vitro and in vivo has started to reveal the physiological implications of imbalances in these cochaperones. A recent study in mice with reduced levels of FKBP52 revealed that FKBP52 may not have an apparent role in regulating anxiety-like behaviors or recognition memory, including fear conditioning, but rather may be more important in motor coordination [80]. This corroborated previous work that did not detect any GR-related physiological changes in FKBP52 knockout mice [66]. However, since FKBP52 competes with FKBP51 to bind Hsp90, it has been suggested that reduced FKBP52 levels could regulate the stress response by increasing the Hsp90/FKBP51 interaction [63]. Recent work identified a role for CyP40 in the regulation of amygdala-mediated fear extinction [81]. An NIH-led study identified CyP40 as being enriched in the basolateral amygdala of normal mice, however, CyP40 expression is reduced in mice that demonstrate impairments in extinction learning. This group also showed that CyP40 colocalizes with GR in these mice and that the extinction regulating effects can be blocked by a GR antagonist. This work may provide a role for CyP40 in PTSD. A specific role in stress-related phenotypes has been not yet been described for PP5. However, GR-specific effects may be difficult to identify, since PP5 primarily functions as a protein phosphatase.

HPA hyperactivity has been extensively shown for depressed individuals. GR dysfunction contributes to this hyperactivity [82–84]. There is evidence that the use of antidepressants inhibits transcription of genes with glucocorticoid responsive elements (GREs). Further, antidepressants increase GR expression in patients. It has been shown that antidepressants have activity beyond targeting monoamine transporters, since they also upregulate autophagy markers [85]. FKBP51 enhances the activity of some antidepressants, which may be a result of priming autophagy pathways. Further, FKBP51 inhibition improved stress coping behavior in a mouse model of depression co-treated with a selective serotonin reuptake inhibitor (SSRI), while decreasing anti-anxiety effects of the SSRI [86]. In further support of a role for FKBP51 in modulating antidepressant effects, it has been shown that *FKBP5* SNPs affect antidepressant response [87,88]. It should also be noted however, that a separate study in geriatric depressed patients showed no role for *FKBP5* SNPs in antidepressant efficacy prediction [89]. Additionally, it has been shown that depression may be linked to epigenetic changes through methylation or differential gene expression. Some antidepressants may mediate epigenetic changes; however, FKBP51 may affect this activity. High FKBP51 expression reduces the activity of DNA methyltransferase, DNMT1, by modulating its activating kinases. This reduction in DNMT1 activity reduces DNA methylation, which broadly impacts expression of stress-induced genes and alters antidepressant activity [90].

4. Stress Response and Circadian Rhythmicity

Physiological stress is frequently associated with circadian disruption, and the regulation of the stress response system is one mechanism by which circadian rhythms are altered [91]. Thus, it is not surprising that disruptions in the circadian rhythms are a common symptom across stress-related psychiatric disorders [92]. Similar to the stress response system, the circadian clock is a well-conserved mechanism that allows organisms to adapt to their environment. This homeostasis occurs not only in individual tissues, but in the coordination within and between systems. Interestingly, there are multiple interactions between the HPA axis and the circadian system, including circadian control over the daily cycling of CORT release [93] and stress reactivity [94]. There is also a growing body of literature suggesting that CORT synchronizes the rhythms of peripheral clocks [95], and even modulates that rhythms of central nervous system clocks outside of the mast oscillator in the suprachiasmatic nucleus [96]. Furthermore, it is likely that GR activity is an important link between these two systems. As already described, CORT promotes GR activation, transcriptionally regulating many genes that contain GREs. This includes core clock genes, *PERIOD 1* and *PERIOD 2*, and the accessory clock genes, *REV-ERBα* and *RORA*. The latter genes are essential for the normal activity of the positive arm of the clock via BMAL1 transcription factor activity [97]. Knockdown of either the positive (BMAL1) or negative (PERIOD genes) arm of the clock disrupts circadian rhythmicity at both the molecular and physiological levels [98–100]. This suggests that the regulation of GR by the Hsp90 heterocomplex can directly impact the expression of clock genes and circadian rhythmicity at a cellular and organismal level, as summarized in Figure 2. This is particularly interesting for FKBP51, since the *FKBP5* gene also contains GREs [101], so the levels of FKBP51 are increased by GR activation. Since FKBP51 works in a short, negative feedback loop with GR, increased FKBP51 levels may directly regulate the activation of the negative feedback loop of the clock FKBP51 levels can be affected not only by SNPs, but they also dramatically increase with age [102,103]. Sleep architecture and quality decline with age, largely due to impairments in circadian rhythmicity [104]. Thus, strategies aimed at restoring CORT homeostasis or depleting FKBP51 could be beneficial for both acute and chronic stress-related co-morbidities, like circadian rhythm sleep disorders. In support of this, mice lacking *Fkbp5* demonstrated increased wake times and protection from stress-induced sleep disruption [105]. More work needs to be done to fully understand the connection between these essential processes, but it is possible that finding treatments that restore normal stress response or circadian rhythmicity may be beneficial for both systems.

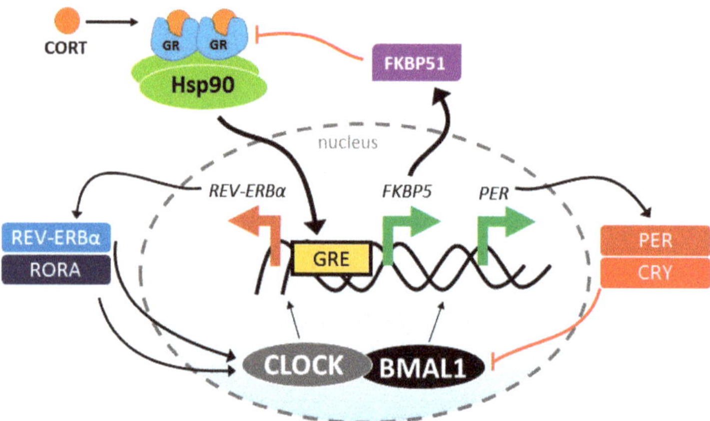

Figure 2. Schematic of the feedback between the molecular clock and stress response systems. Stress produces CORT, which binds to the GR/Hsp90 heterocomplex. GR forms a homodimer and translocate to the nucleus where it binds the glucocorticoid response elements in the promoter region. This leads to increased FKBP5/FKBP51, which slows GR activity, and increased PER expression, a component of the negative arm of the circadian clock; at the same time, REV-ERBα, a positive arm protein is down regulated.

5. Therapeutic Progress

It is clear that selective inhibitory molecules are not only needed for therapeutic interventions, but also as tools to investigate the complex interplay at work in diverse chaperone heterocomplexes. Although Hsp90 inhibitors are available, because of the vast network of Hsp90-interactions, numerous on- and off-target effects are a primary concern [106]. Geldanamycin (GA), a long-used Hsp90 inhibitor, causes cytotoxicity through the production of reactive oxygen species, which can result in hepato- and ocular toxicity [107–109]. Additionally, the GA backbone binds to ion channels of the mitochondrial membrane resulting in increased Ca^{2+} levels [110]. Furthermore, because the N-terminal of Hsp90 includes a conserved fold for binding ATP, inhibitors targeting this region can inadvertently inhibit other important ATP-binding proteins. For example, radicicol has been shown to inhibit the activity of a Type II DNA topoisomerase [111]. In addition, N-terminal Hsp90 ATP inhibitors activate the stress response, which upregulates Hsp70, Hsp40, and Hsp27 along with other pro-survival factors [112]. With this in mind, targeting Hsp90 cochaperones may be a promising alternative therapeutic approach.

The principal tool in determining FKBP51 and FKBP52 functional activity has been FK506 (also tacrolimus), a potent immunosuppressant. FK506 binds FKBP51 and FKBP52, disrupting signaling events mediated by the calcium-dependent serine/threonine protein phosphatase, calcineurin (CaN/PP2B). FK506 will non-specifically bind to FKBPs. Therefore, specific inhibitors that work through an alternate mechanism are needed to discriminate between the FKBPs, especially the highly homologous FKBP51 and FKBP52.

Using induced-fit modeling, highly selective inhibitors of FKBP51, SAFit 1 and SAFit2, have now been generated [113]. SAFit1 showed protection from FKBP51-mediated neurite outgrowth suppression in primary neurons, while SAFit2 treatment in mice led to antidepressant-like effects [113]. Additionally, microRNA-511 (miR-511), a non-coding RNA molecule, was shown to silence *FKBP5* post-transcriptionally [114]. miR-511 suppressed CORT-induced upregulation of FKBP51 and promote neurite outgrowth in primary neurons. miR-511 may be a promising therapeutic candidate for suppressing FKBP51. Recently, benztropine was shown to restore GR activity in the presence of high FKBP51 and interact with FKBP51, but not FKBP52 [77].

Selective inhibitors for FKBP52, CyP40, and PP5 have not been reported. Structural insights may help guide strategies to target these chaperones. For example, the proline-rich loop extending over the FK1 catalytic domain of FKBP52 has been described and may be targetable, since this domain is suggested to have a role GR regulation [115]. However, inhibiting FKBP52 may lead to reduced fertility, as this has been found in mice lacking this protein [116]. Additional studies are still needed to better understand the regions on each cochaperone that are most important for regulating GR activity. At the same time, further development of selective inhibitors will both benefit from and aid these studies.

6. Conclusions

Although there has been substantial research investigating the role of GR in regulating the stress response, recent work has advanced our understanding of both the structural and functional regulation of GR by Hsp90 heterocomplexes. There is a growing body of evidence describing the functional effects of cochaperones on GR activity. Still there is much to learn about how these cochaperones regulate GR signaling in vivo and how the intracellular feedback loops and HPA axis are interconnected. Even less well known are the structural effects of cochaperones on GR-Hsp90 heterocomplexes. More work needs to be done to elucidate the effects of cochaperones on GR structure, function, and feedback regulation.

The stress response has now been suggested to be linked to circadian rhythms. However, a detailed investigation at the molecular level has yet to be done, despite apparent overlaps in regulation through GR activity. Additional studies are needed to start to understand how stress response and circadian rhythms are molecularly linked and how this link impacts the susceptibility and severity of psychiatric disorders.

Author Contributions: Writing—original draft preparation, J.D.B, I.O., L.J.B.; writing—review and editing, I.O., S.R.O., D.G., L.J.B.; funding acquisition, L.J.B.

Funding: This research was funded by NIMH, grant number R01 MH103848.

Conflicts of Interest: The authors declare no conflict of interest.

Abbreviations

ACTH	adrenocorticotropic hormone
ATP	adenosine triphosphate
CORT	cortisol
CRH	corticotropin-releasing hormone
CaN/PP2B	calcineurin
CyP40	cyclophilin 40
FKBP	FK506-binding protein
GR	glucocorticoid receptor
GRE	glucocorticoid response element
HPA	hypothalamic-pituitary-adrenal
Hsp90	heat shock protein 90
LBD	ligand-binding domain
MDD	major depressive disorder
MEEVD	binding motif in Hsp90 that binds TPR domain
MR	mineralocorticoid receptors
N2A	Neuro2A, a mouse neuroblastoma cell line
PTSD	post-traumatic stress disorder
SNP	single nucleotide polymorphism
TPR	tetratricopeptide repeat

References

1. Holsboer, F. The corticosteroid receptor hypothesis of depression. *Neuropsychopharmacol. Off. Publ. Am. Coll. Neuropsychopharmacol.* **2000**, *23*, 477–501. [CrossRef]
2. Eisenlohr-Moul, T.A.; Miller, A.B.; Giletta, M.; Hastings, P.D.; Rudolph, K.D.; Nock, M.K.; Prinstein, M.J. HPA axis response and psychosocial stress as interactive predictors of suicidal ideation and behavior in adolescent females: A multilevel diathesis-stress framework. *Neuropsychopharmacol. Off. Publ. Am. Coll. Neuropsychopharmacol.* **2018**, *43*, 2564–2571. [CrossRef] [PubMed]
3. Daskalakis, N.P.; Lehrner, A.; Yehuda, R. Endocrine Aspects of Post-traumatic Stress Disorder and Implications for Diagnosis and Treatment. *Endocrinol. Metab. Clin. N. Am.* **2013**, *42*, 503–513. [CrossRef] [PubMed]
4. McEwen, B.S. Protective and damaging effects of stress mediators. *N. Engl. J. Med.* **1998**, *338*, 171–179. [CrossRef] [PubMed]
5. Stephens, M.A.C.; Wand, G. Stress and the HPA axis: Role of glucocorticoids in alcohol dependence. *Alcohol. Res. Curr. Rev.* **2012**, *34*, 468–483.
6. Rose, A.K.; Shaw, S.G.; Prendergast, M.A.; Little, H.J. The importance of glucocorticoids in alcohol dependence and neurotoxicity. *Alcohol. Clin. Exp. Res.* **2010**, *34*, 2011–2028. [CrossRef] [PubMed]
7. Smith, S.M.; Vale, W.W. The role of the hypothalamic-pituitary-adrenal axis in neuroendocrine responses to stress. *Dialogues Clin. Neurosci.* **2006**, *8*, 383–395.
8. Keller-Wood, M. Hypothalamic-Pituitary—Adrenal Axis-Feedback Control. *Compr. Physiol.* **2015**, *5*, 1161–1182. [CrossRef]
9. Zhe, D.; Fang, H.; Yuxiu, S. Expressions of hippocampal mineralocorticoid receptor (MR) and glucocorticoid receptor (GR) in the single-prolonged stress-rats. *Acta Histochem. Cytochem.* **2008**, *41*, 89–95. [CrossRef]
10. Herman, J.P.; Patel, P.D.; Akil, H.; Watson, S.J. Localization and regulation of glucocorticoid and mineralocorticoid receptor messenger RNAs in the hippocampal formation of the rat. *Mol. Endocrinol. (Baltimore MD.)* **1989**, *3*, 1886–1894. [CrossRef]
11. Qin, D.-D.; Rizak, J.; Feng, X.-L.; Yang, S.-C.; Lü, L.-B.; Pan, L.; Yin, Y.; Hu, X.-T. Prolonged secretion of cortisol as a possible mechanism underlying stress and depressive behaviour. *Sci. Rep.* **2016**, *6*, 30187. [CrossRef] [PubMed]
12. Hannibal, K.E.; Bishop, M.D. Chronic stress, cortisol dysfunction, and pain: A psychoneuroendocrine rationale for stress management in pain rehabilitation. *Phys. Ther.* **2014**, *94*, 1816–1825. [CrossRef] [PubMed]

13. Fries, G.R.; Vasconcelos-Moreno, M.P.; Gubert, C.; dos Santos, B.T.M.Q.; Sartori, J.; Eisele, B.; Ferrari, P.; Fijtman, A.; Rüegg, J.; Gassen, N.C.; et al. Hypothalamic-pituitary-adrenal axis dysfunction and illness progression in bipolar disorder. *Int. J. Neuropsychopharmacol.* **2014**, *18*, pyu043. [CrossRef] [PubMed]

14. Meewisse, M.L.; Reitsma, J.B.; de Vries, G.J.; Gersons, B.P.; Olff, M. Cortisol and post-traumatic stress disorder in adults: Systematic review and meta-analysis. *Br. J. Psychiatry J. Ment. Sci.* **2007**, *191*, 387–392. [CrossRef] [PubMed]

15. Tatomir, A.; Micu, C.; Crivii, C. The impact of stress and glucocorticoids on memory. *Clujul Med. (1957)* **2014**, *87*, 3–6. [CrossRef] [PubMed]

16. Roozendaal, B.; McEwen, B.S.; Chattarji, S. Stress, memory and the amygdala. *Nat. Rev. Neurosci.* **2009**, *10*, 423–433. [CrossRef] [PubMed]

17. de Quervain, D.J.; Roozendaal, B.; McGaugh, J.L. Stress and glucocorticoids impair retrieval of long-term spatial memory. *Nature* **1998**, *394*, 787–790. [CrossRef] [PubMed]

18. Wolf, O.T.; Kuhlmann, S.; Buss, C.; Hellhammer, D.H.; Kirschbaum, C. Cortisol and memory retrieval in humans: Influence of emotional valence. *Ann. N. Y. Acad. Sci.* **2004**, *1032*, 195–197. [CrossRef] [PubMed]

19. Dedovic, K.; Ngiam, J. The cortisol awakening response and major depression: Examining the evidence. *Neuropsychiatr. Dis. Treat.* **2015**, *11*, 1181–1189. [CrossRef]

20. Dougherty, L.R.; Klein, D.N.; Olino, T.M.; Dyson, M.; Rose, S. Increased waking salivary cortisol and depression risk in preschoolers: The role of maternal history of melancholic depression and early child temperament. *J. Child Psychol. Psychiatry Allied Discip.* **2009**, *50*, 1495–1503. [CrossRef]

21. Otte, C.; Hart, S.; Neylan, T.C.; Marmar, C.R.; Yaffe, K.; Mohr, D.C. A meta-analysis of cortisol response to challenge in human aging: Importance of gender. *Psychoneuroendocrinology* **2005**, *30*, 80–91. [CrossRef] [PubMed]

22. Chaudieu, I.; Beluche, I.; Norton, J.; Boulenger, J.P.; Ritchie, K.; Ancelin, M.L. Abnormal reactions to environmental stress in elderly persons with anxiety disorders: Evidence from a population study of diurnal cortisol changes. *J. Affect. Disord.* **2008**, *106*, 307–313. [CrossRef] [PubMed]

23. Cervantes, P.; Gelber, S.; Kin, F.N.; Nair, V.N.; Schwartz, G. Circadian secretion of cortisol in bipolar disorder. *J. Psychiatry Neurosci. JPN* **2001**, *26*, 411–416. [PubMed]

24. Joyce, P.R.; Donald, R.A.; Elder, P.A. Individual differences in plasma cortisol changes during mania and depression. *J. Affect. Disord.* **1987**, *12*, 1–5. [CrossRef]

25. Wand, G. The influence of stress on the transition from drug use to addiction. *Alcohol. Res. Health J. Natl. Inst. Alcohol Abuse Alcohol.* **2008**, *31*, 119–136.

26. Meijsing, S.H. Mechanisms of Glucocorticoid-Regulated Gene Transcription. *Adv. Exp. Med. Biol.* **2015**, *872*, 59–81. [CrossRef]

27. Echeverría, P.C.; Mazaira, G.; Erlejman, A.; Gomez-Sanchez, C.; Piwien Pilipuk, G.; Galigniana, M.D. Nuclear import of the glucocorticoid receptor-hsp90 complex through the nuclear pore complex is mediated by its interaction with Nup62 and importin beta. *Mol. Cell. Biol.* **2009**, *29*, 4788–4797. [CrossRef]

28. Pratt, W.B.; Galigniana, M.D.; Morishima, Y.; Murphy, P.J. Role of molecular chaperones in steroid receptor action. *Essays Biochem.* **2004**, *40*, 41–58. [CrossRef]

29. Noguchi, T.; Makino, S.; Matsumoto, R.; Nakayama, S.; Nishiyama, M.; Terada, Y.; Hashimoto, K. Regulation of glucocorticoid receptor transcription and nuclear translocation during single and repeated immobilization stress. *Endocrinology* **2010**, *151*, 4344–4355. [CrossRef]

30. Lorenz, O.R.; Freiburger, L.; Rutz, D.A.; Krause, M.; Zierer, B.K.; Alvira, S.; Cuellar, J.; Valpuesta, J.M.; Madl, T.; Sattler, M.; et al. Modulation of the Hsp90 chaperone cycle by a stringent client protein. *Mol. Cell* **2014**, *53*, 941–953. [CrossRef]

31. Picard, D.; Khursheed, B.; Garabedian, M.J.; Fortin, M.G.; Lindquist, S.; Yamamoto, K.R. Reduced levels of hsp90 compromise steroid receptor action in vivo. *Nature* **1990**, *348*, 166. [CrossRef] [PubMed]

32. Wochnik, G.M.; Ruegg, J.; Abel, G.A.; Schmidt, U.; Holsboer, F.; Rein, T. FK506-binding proteins 51 and 52 differentially regulate dynein interaction and nuclear translocation of the glucocorticoid receptor in mammalian cells. *J. Biol. Chem.* **2005**, *280*, 4609–4616. [CrossRef] [PubMed]

33. Galigniana, M.D.; Radanyi, C.; Renoir, J.M.; Housley, P.R.; Pratt, W.B. Evidence that the peptidylprolyl isomerase domain of the hsp90-binding immunophilin FKBP52 is involved in both dynein interaction and glucocorticoid receptor movement to the nucleus. *J. Biol. Chem.* **2001**, *276*, 14884–14889. [CrossRef] [PubMed]

34. Storer, C.L.; Dickey, C.A.; Galigniana, M.D.; Rein, T.; Cox, M.B. FKBP51 and FKBP52 in signaling and disease. *Trends Endocrinol. Metab. TEM* **2011**, *22*, 481–490. [CrossRef] [PubMed]

35. Riggs, D.L.; Roberts, P.J.; Chirillo, S.C.; Cheung-Flynn, J.; Prapapanich, V.; Ratajczak, T.; Gaber, R.; Picard, D.; Smith, D.F. The Hsp90-binding peptidylprolyl isomerase FKBP52 potentiates glucocorticoid signaling in vivo. *EMBO J.* **2003**, *22*, 1158–1167. [CrossRef] [PubMed]

36. Riggs, D.L.; Cox, M.B.; Tardif, H.L.; Hessling, M.; Buchner, J.; Smith, D.F. Noncatalytic role of the FKBP52 peptidyl-prolyl isomerase domain in the regulation of steroid hormone signaling. *Mol. Cell. Biol.* **2007**, *27*, 8658–8669. [CrossRef] [PubMed]

37. Banerjee, A.; Periyasamy, S.; Wolf, I.M.; Hinds, T.D., Jr.; Yong, W.; Shou, W.; Sanchez, E.R. Control of glucocorticoid and progesterone receptor subcellular localization by the ligand-binding domain is mediated by distinct interactions with tetratricopeptide repeat proteins. *Biochemistry* **2008**, *47*, 10471–10480. [CrossRef]

38. Johnson, J.L. Evolution and function of diverse Hsp90 homologs and cochaperone proteins. *Biochim. Biophys. Acta (BBA) Mol. Cell Res.* **2012**, *1823*, 607–613. [CrossRef]

39. Liu, X.D.; Liu, P.C.; Santoro, N.; Thiele, D.J. Conservation of a stress response: Human heat shock transcription factors functionally substitute for yeast HSF. *EMBO J.* **1997**, *16*, 6466–6477. [CrossRef]

40. Picard, D. Heat-shock protein 90, a chaperone for folding and regulation. *Cell. Mol. Life Sci. CMLS* **2002**, *59*, 1640–1648. [CrossRef]

41. Zhao, R.; Davey, M.; Hsu, Y.-C.; Kaplanek, P.; Tong, A.; Parsons, A.B.; Krogan, N.; Cagney, G.; Mai, D.; Greenblatt, J.; et al. Navigating the Chaperone Network: An Integrative Map of Physical and Genetic Interactions Mediated by the Hsp90 Chaperone. *Cell* **2005**, *120*, 715–727. [CrossRef] [PubMed]

42. Wu, Z.; Gholami, A.M.; Kuster, B. Systematic identification of the HSP90 candidate regulated proteome. *Mol. Cell. Proteomics MCP* **2012**, *11*, M111.016675. [CrossRef] [PubMed]

43. Mailhos, C.; Howard, M.K.; Latchman, D.S. Heat shock proteins hsp90 and hsp70 protect neuronal cells from thermal stress but not from programmed cell death. *J. Neurochem.* **1994**, *63*, 1787–1795. [CrossRef] [PubMed]

44. Gallo, L.I.; Lagadari, M.; Piwien-Pilipuk, G.; Galigniana, M.D. The 90-kDa heat-shock protein (Hsp90)-binding immunophilin FKBP51 is a mitochondrial protein that translocates to the nucleus to protect cells against oxidative stress. *J. Biol. Chem.* **2011**, *286*, 30152–30160. [CrossRef] [PubMed]

45. Swaminathan, S. Protein quality control: Knowing when to fold. *Nat. Cell Biol.* **2005**, *7*, 647. [CrossRef]

46. Kirschke, E.; Goswami, D.; Southworth, D.; Griffin, P.R.; Agard, D.A. Glucocorticoid Receptor Function Regulated by Coordinated Action of the Hsp90 and Hsp70 Chaperone Cycles. *Cell* **2014**, *157*, 1685–1697. [CrossRef] [PubMed]

47. Li, J.; Buchner, J. Structure, function and regulation of the hsp90 machinery. *Biomed. J.* **2013**, *36*, 106–117. [CrossRef]

48. Oroz, J.; Chang, B.J.; Wysoczanski, P.; Lee, C.-T.; Pérez-Lara, Á.; Chakraborty, P.; Hofele, R.V.; Baker, J.D.; Blair, L.J.; Biernat, J.; Urlaub, H.; et al. Structure and pro-toxic mechanism of the human Hsp90/PPIase/Tau complex. *Nat. Commun.* **2018**, *9*, 4532. [CrossRef]

49. Hellenkamp, B.; Wortmann, P.; Kandzia, F.; Zacharias, M.; Hugel, T. Multidomain structure and correlated dynamics determined by self-consistent FRET networks. *Nat. Methods* **2017**, *14*, 174–180. [CrossRef]

50. Schopf, F.H.; Biebl, M.M.; Buchner, J. The HSP90 chaperone machinery. *Nat. Rev. Mol. Cell Biol.* **2017**, *18*, 345–360. [CrossRef]

51. Mayer, M.P.; Le Breton, L. Hsp90: Breaking the symmetry. *Mol. Cell* **2015**, *58*, 8–20. [CrossRef] [PubMed]

52. Hildenbrand, Z.L.; Molugu, S.K.; Herrera, N.; Ramirez, C.; Xiao, C.; Bernal, R.A. Hsp90 can accommodate the simultaneous binding of the FKBP52 and HOP proteins. *Oncotarget* **2011**, *2*, 43–58. [CrossRef] [PubMed]

53. Harst, A.; Lin, H.; Obermann, W.M.J. Aha1 competes with Hop, p50 and p23 for binding to the molecular chaperone Hsp90 and contributes to kinase and hormone receptor activation. *Biochem. J.* **2005**, *387 Pt 3*, 789–796. [CrossRef] [PubMed]

54. Zuehlke, A.; Johnson, J.L. Hsp90 and co-chaperones twist the functions of diverse client proteins. *Biopolymers* **2010**, *93*, 211–217. [CrossRef] [PubMed]

55. Sahasrabudhe, P.; Rohrberg, J.; Biebl, M.M.; Rutz, D.A.; Buchner, J. The Plasticity of the Hsp90 Co-chaperone System. *Mol. Cell* **2017**, *67*, 947–961. [CrossRef] [PubMed]

56. Dittmar, K.D.; Hutchison, K.A.; Owens-Grillo, J.K.; Pratt, W.B. Reconstitution of the steroid receptor.hsp90 heterocomplex assembly system of rabbit reticulocyte lysate. *J. Biol. Chem.* **1996**, *271*, 12833–12839. [CrossRef]

57. Röhl, A.; Wengler, D.; Madl, T.; Lagleder, S.; Tippel, F.; Herrmann, M.; Hendrix, J.; Richter, K.; Hack, G.; Schmid, A.B.; et al. Hsp90 regulates the dynamics of its cochaperone Sti1 and the transfer of Hsp70 between modules. *Nat. Commun.* **2015**, *6*, 6655. [CrossRef] [PubMed]

58. Chrousos, G.P.; Kino, T. Glucocorticoid signaling in the cell. Expanding clinical implications to complex human behavioral and somatic disorders. *Ann. N. Y. Acad. Sci.* **2009**, *1179*, 153–166. [CrossRef] [PubMed]

59. Grenert, J.P.; Johnson, B.D.; Toft, D.O. The importance of ATP binding and hydrolysis by hsp90 in formation and function of protein heterocomplexes. *J. Biol. Chem.* **1999**, *274*, 17525–17533. [CrossRef] [PubMed]

60. Echeverria, P.C.; Picard, D. Molecular chaperones, essential partners of steroid hormone receptors for activity and mobility. *Biochim. Biophys. Acta* **2010**, *1803*, 641–649. [CrossRef]

61. Morishima, Y.; Kanelakis, K.C.; Murphy, P.J.; Lowe, E.R.; Jenkins, G.J.; Osawa, Y.; Sunahara, R.K.; Pratt, W.B. The hsp90 cochaperone p23 is the limiting component of the multiprotein hsp90/hsp70-based chaperone system in vivo where it acts to stabilize the client protein: Hsp90 complex. *J. Biol. Chem.* **2003**, *278*, 48754–48763. [CrossRef] [PubMed]

62. Blair, L.J.; Genest, O.; Mollapour, M. The multiple facets of the Hsp90 machine. *Nat. Struct. Mol. Biol.* in press, **2018**. [CrossRef]

63. Tatro, E.T.; Everall, I.P.; Kaul, M.; Achim, C.L. Modulation of glucocorticoid receptor nuclear translocation in neurons by immunophilins FKBP51 and FKBP52: Implications for major depressive disorder. *Brain Res.* **2009**, *1286*, 1–12. [CrossRef] [PubMed]

64. Zannas, A.S.; Wiechmann, T.; Gassen, N.C.; Binder, E.B. Gene-Stress-Epigenetic Regulation of FKBP5: Clinical and Translational Implications. *Neuropsychopharmacol. Off. Publ. Am. Coll. Neuropsychopharmacol.* **2016**, *41*, 261–274. [CrossRef] [PubMed]

65. Davies, T.H.; Ning, Y.M.; Sanchez, E.R. Differential control of glucocorticoid receptor hormone-binding function by tetratricopeptide repeat (TPR) proteins and the immunosuppressive ligand FK506. *Biochemistry* **2005**, *44*, 2030–2038. [CrossRef] [PubMed]

66. Wolf, I.M.; Periyasamy, S.; Hinds, T., Jr.; Yong, W.; Shou, W.; Sanchez, E.R. Targeted ablation reveals a novel role of FKBP52 in gene-specific regulation of glucocorticoid receptor transcriptional activity. *J. Steroid Biochem. Mol. Biol.* **2009**, *113*, 36–45. [CrossRef] [PubMed]

67. Galigniana, M.D.; Harrell, J.M.; Murphy, P.J.; Chinkers, M.; Radanyi, C.; Renoir, J.M.; Zhang, M.; Pratt, W.B. Binding of hsp90-associated immunophilins to cytoplasmic dynein: Direct binding and in vivo evidence that the peptidylprolyl isomerase domain is a dynein interaction domain. *Biochemistry* **2002**, *41*, 13602–13610. [CrossRef]

68. Silverstein, A.M.; Galigniana, M.D.; Chen, M.S.; Owens-Grillo, J.K.; Chinkers, M.; Pratt, W.B. Protein phosphatase 5 is a major component of glucocorticoid receptor.hsp90 complexes with properties of an FK506-binding immunophilin. *J. Biol. Chem.* **1997**, *272*, 16224–16230. [CrossRef]

69. Golden, T.; Swingle, M.; Honkanen, R.E. The role of serine/threonine protein phosphatase type 5 (PP5) in the regulation of stress-induced signaling networks and cancer. *Cancer Metastasis Rev.* **2008**, *27*, 169–178. [CrossRef]

70. Wang, Z.; Chen, W.; Kono, E.; Dang, T.; Garabedian, M.J. Modulation of glucocorticoid receptor phosphorylation and transcriptional activity by a C-terminal-associated protein phosphatase. *Mol. Endocrinol.* **2007**, *21*, 625–634. [CrossRef]

71. O'Leary, J.C., 3rd; Zhang, B.; Koren, J., 3rd; Blair, L.; Dickey, C.A. The role of FKBP5 in mood disorders: Action of FKBP5 on steroid hormone receptors leads to questions about its evolutionary importance. *CNS Neurol. Disord. Drug Targets* **2013**, *12*, 1157–1162.

72. Jaaskelainen, T.; Makkonen, H.; Palvimo, J.J. Steroid up-regulation of FKBP51 and its role in hormone signaling. *Curr. Opin. Pharmacol.* **2011**, *11*, 326–331. [CrossRef] [PubMed]

73. Binder, E.B. The role of FKBP5, a co-chaperone of the glucocorticoid receptor in the pathogenesis and therapy of affective and anxiety disorders. *Psychoneuroendocrinology* **2009**, *34* (Suppl. 1), S186–S195. [CrossRef] [PubMed]

74. Criado-Marrero, M.; Rein, T.; Binder, E.B.; Porter, J.T.; Koren, J.; Blair, L.J. Hsp90 and FKBP51: Complex regulators of psychiatric diseases. *Philos. Trans. R. Soc. B Biol.Sci.* **2018**, *373*. [CrossRef] [PubMed]

75. Xie, P.; Kranzler, H.R.; Poling, J.; Stein, M.B.; Anton, R.F.; Farrer, L.A.; Gelernter, J. Interaction of FKBP5 with childhood adversity on risk for post-traumatic stress disorder. *Neuropsychopharmacol. Off. Publ. Am. Coll. Neuropsychopharmacol.* **2010**, *35*, 1684–1692. [CrossRef]

76. Klengel, T.; Mehta, D.; Anacker, C.; Rex-Haffner, M.; Pruessner, J.C.; Pariante, C.M.; Pace, T.W.; Mercer, K.B.; Mayberg, H.S.; Bradley, B.; et al. Allele-specific FKBP5 DNA demethylation mediates gene-childhood trauma interactions. *Nat. Neurosci.* **2013**, *16*, 33–41. [CrossRef] [PubMed]

77. Sabbagh, J.J.; Cordova, R.A.; Zheng, D.; Criado-Marrero, M.; Lemus, A.; Li, P.; Baker, J.D.; Nordhues, B.A.; Darling, A.L.; Martinez-Licha, C.; et al. Targeting the FKBP51/GR/Hsp90 Complex to Identify Functionally Relevant Treatments for Depression and PTSD. *ACS Chem. Biol.* **2018**, *13*, 2288–2299. [CrossRef]

78. O'Leary, J.C., 3rd; Dharia, S.; Blair, L.J.; Brady, S.; Johnson, A.G.; Peters, M.; Cheung-Flynn, J.; Cox, M.B.; de Erausquin, G.; Weeber, E.J.; et al. A new anti-depressive strategy for the elderly: Ablation of FKBP5/FKBP51. *PLoS ONE* **2011**, *6*, e24840. [CrossRef]

79. Touma, C.; Gassen, N.C.; Herrmann, L.; Cheung-Flynn, J.; Bull, D.R.; Ionescu, I.A.; Heinzmann, J.M.; Knapman, A.; Siebertz, A.; Depping, A.M.; et al. FK506 binding protein 5 shapes stress responsiveness: Modulation of neuroendocrine reactivity and coping behavior. *Biol. Psychiatry* **2011**, *70*, 928–936. [CrossRef]

80. Young, M.J.; Geiszler, P.C.; Pardon, M.C. A novel role for the immunophilin FKBP52 in motor coordination. *Behav. Brain Res.* **2016**, *313*, 97–110. [CrossRef] [PubMed]

81. Gunduz-Cinar, O.; Brockway, E.; Lederle, L.; Wilcox, T.; Halladay, L.R.; Ding, Y.; Oh, H.; Busch, E.F.; Kaugars, K.; Flynn, S.; et al. Identification of a novel gene regulating amygdala-mediated fear extinction. *Mol. Psychiatry* **2018**. [CrossRef] [PubMed]

82. Vreeburg, S.A.; Hoogendijk, W.G.; van Pelt, J.; Derijk, R.H.; Verhagen, J.C.; van Dyck, R.; Smit, J.H.; Zitman, F.G.; Penninx, B.W. Major depressive disorder and hypothalamic-pituitary-adrenal axis activity: Results from a large cohort study. *Arch. Gen. Psychiatry* **2009**, *66*, 617–626. [CrossRef]

83. Varghese, F.P.; Brown, E.S. The Hypothalamic-Pituitary-Adrenal Axis in Major Depressive Disorder: A Brief Primer for Primary Care Physicians. *Primary Care Companion J. Clin. Psychiatry* **2001**, *3*, 151–155. [CrossRef]

84. Pariante, C.M.; Lightman, S.L. The HPA axis in major depression: Classical theories and new developments. *Trends Neurosci.* **2008**, *31*, 464–468. [CrossRef] [PubMed]

85. Gassen, N.C.; Hartmann, J.; Zschocke, J.; Stepan, J.; Hafner, K.; Zellner, A.; Kirmeier, T.; Kollmannsberger, L.; Wagner, K.V.; Dedic, N.; et al. Association of FKBP51 with priming of autophagy pathways and mediation of antidepressant treatment response: Evidence in cells, mice, and humans. *PLoS Med.* **2014**, *11*, e1001755. [CrossRef] [PubMed]

86. Pöhlmann, M.L.; Häusl, A.S.; Harbich, D.; Balsevich, G.; Engelhardt, C.; Feng, X.; Breitsamer, M.; Hausch, F.; Winter, G.; Schmidt, M.V. Pharmacological Modulation of the Psychiatric Risk Factor FKBP51 Alters Efficiency of Common Antidepressant Drugs. *Front. Behav. Neurosci.* **2018**, *12*, 262. [CrossRef]

87. Binder, E.B.; Salyakina, D.; Lichtner, P.; Wochnik, G.M.; Ising, M.; Pütz, B.; Papiol, S.; Seaman, S.; Lucae, S.; Kohli, M.A.; et al. Polymorphisms in FKBP5 are associated with increased recurrence of depressive episodes and rapid response to antidepressant treatment. *Nat. Genet.* **2004**, *36*, 1319. [CrossRef]

88. Lekman, M.; Laje, G.; Charney, D.; Rush, A.J.; Wilson, A.F.; Sorant, A.J.M.; Lipsky, R.; Wisniewski, S.R.; Manji, H.; McMahon, F.J.; et al. The FKBP5-gene in depression and treatment response—An association study in the Sequenced Treatment Alternatives to Relieve Depression (STAR*D) Cohort. *Biol. Psychiatry* **2008**, *63*, 1103–1110. [CrossRef]

89. Sarginson, J.E.; Lazzeroni, L.C.; Ryan, H.S.; Schatzberg, A.F.; Murphy, G.M., Jr. FKBP5 polymorphisms and antidepressant response in geriatric depression. *Ame. J. Med. Genet. Part B Neuropsychiatr. Genet.* **2010**, *153B*, 554–560. [CrossRef]

90. Gassen, N.C.; Fries, G.R.; Zannas, A.S.; Hartmann, J.; Zschocke, J.; Hafner, K.; Carrillo-Roa, T.; Steinbacher, J.; Preissinger, S.N.; Hoeijmakers, L.; et al. Chaperoning epigenetics: FKBP51 decreases the activity of DNMT1 and mediates epigenetic effects of the antidepressant paroxetine. *Sci. Signal.* **2015**, *8*, ra119. [CrossRef]

91. Koch, C.E.; Leinweber, B.; Drengberg, B.C.; Blaum, C.; Oster, H. Interaction between circadian rhythms and stress. *Neurobiol. Stress* **2017**, *6*, 57–67. [CrossRef] [PubMed]

92. Pilz, L.K.; Carissimi, A.; Oliveira, M.A.B.; Francisco, A.P.; Fabris, R.C.; Medeiros, M.S.; Scop, M.; Frey, B.N.; Adan, A.; Hidalgo, M.P. Rhythmicity of Mood Symptoms in Individuals at Risk for Psychiatric Disorders. *Sci. Rep.* **2018**, *8*, 11402. [CrossRef] [PubMed]

93. Debono, M.; Ghobadi, C.; Rostami-Hodjegan, A.; Huatan, H.; Campbell, M.J.; Newell-Price, J.; Darzy, K.; Merke, D.P.; Arlt, W.; Ross, R.J. Modified-release hydrocortisone to provide circadian cortisol profiles. *J. Clin. Endocrinol. Metab.* **2009**, *94*, 1548–1554. [CrossRef] [PubMed]

94. Chabot, C.C.; Taylor, D.H. Daily rhythmicity of the rat acoustic startle response. *Physiol. Behav.* **1992**, *51*, 885–889. [CrossRef]

95. Dickmeis, T. Glucocorticoids and the circadian clock. *J. Endocrinol.* **2009**, *200*, 3–22. [CrossRef] [PubMed]

96. Woodruff, E.R.; Chun, L.E.; Hinds, L.R.; Spencer, R.L. Diurnal Corticosterone Presence and Phase Modulate Clock Gene Expression in the Male Rat Prefrontal Cortex. *Endocrinology* **2016**, *157*, 1522–1534. [CrossRef] [PubMed]

97. Guillaumond, F.; Dardente, H.; Giguere, V.; Cermakian, N. Differential control of Bmal1 circadian transcription by REV-ERB and ROR nuclear receptors. *J. Biol. Rhythms* **2005**, *20*, 391–403. [CrossRef] [PubMed]

98. Zheng, B.; Larkin, D.W.; Albrecht, U.; Sun, Z.S.; Sage, M.; Eichele, G.; Lee, C.C.; Bradley, A. The mPer2 gene encodes a functional component of the mammalian circadian clock. *Nature* **1999**, *400*, 169–173. [CrossRef]

99. Bunger, M.K.; Wilsbacher, L.D.; Moran, S.M.; Clendenin, C.; Radcliffe, L.A.; Hogenesch, J.B.; Simon, M.C.; Takahashi, J.S.; Bradfield, C.A. Mop3 is an essential component of the master circadian pacemaker in mammals. *Cell* **2000**, *103*, 1009–1017. [CrossRef]

100. Zheng, B.; Albrecht, U.; Kaasik, K.; Sage, M.; Lu, W.; Vaishnav, S.; Li, Q.; Sun, Z.S.; Eichele, G.; Bradley, A.; et al. Nonredundant roles of the mPer1 and mPer2 genes in the mammalian circadian clock. *Cell* **2001**, *105*, 683–694. [CrossRef]

101. Scharf, S.H.; Liebl, C.; Binder, E.B.; Schmidt, M.V.; Müller, M.B. Expression and Regulation of the Fkbp5 Gene in the Adult Mouse Brain. *PLoS ONE* **2011**, *6*, e16883. [CrossRef] [PubMed]

102. Blair, L.J.; Nordhues, B.A.; Hill, S.E.; Scaglione, K.M.; O'Leary, J.C., III; Fontaine, S.N.; Breydo, L.; Zhang, B.; Li, P.; Wang, L.; Cotman, C.; et al. Accelerated neurodegeneration through chaperone-mediated oligomerization of tau. *J. Clin. Investig.* **2013**, *123*, 4158–4169. [CrossRef] [PubMed]

103. Sabbagh, J.J.; O'Leary, J.C., 3rd; Blair, L.J.; Klengel, T.; Nordhues, B.A.; Fontaine, S.N.; Binder, E.B.; Dickey, C.A. Age-associated epigenetic upregulation of the FKBP5 gene selectively impairs stress resiliency. *PLoS ONE* **2014**, *9*, e107241. [CrossRef] [PubMed]

104. Myers, B.L.; Badia, P. Changes in circadian rhythms and sleep quality with aging: Mechanisms and interventions. *Neurosci. Biobehav. Rev.* **1995**, *19*, 553–571. [CrossRef]

105. Albu, S.; Romanowski, C.P.; Letizia Curzi, M.; Jakubcakova, V.; Flachskamm, C.; Gassen, N.C.; Hartmann, J.; Schmidt, M.V.; Schmidt, U.; Rein, T.; et al. Deficiency of FK506-binding protein (FKBP) 51 alters sleep architecture and recovery sleep responses to stress in mice. *J. Sleep Res.* **2014**, *23*, 176–185. [CrossRef] [PubMed]

106. Yuno, A.; Lee, M.J.; Lee, S.; Tomita, Y.; Rekhtman, D.; Moore, B.; Trepel, J.B. Clinical Evaluation and Biomarker Profiling of Hsp90 Inhibitors. *Methods Mol. Biol.* **2018**, *1709*, 423–441. [CrossRef] [PubMed]

107. Samuni, Y.; Ishii, H.; Hyodo, F.; Samuni, U.; Krishna, M.C.; Goldstein, S.; Mitchell, J.B. Reactive oxygen species mediate hepatotoxicity induced by the Hsp90 inhibitor geldanamycin and its analogs. *Free Radic. Biol. Med.* **2010**, *48*, 1559–1563. [CrossRef]

108. Amin, K.; Ip, C.; Jimenez, L.; Tyson, C.; Behrsing, H. In vitro detection of differential and cell-specific hepatobiliary toxicity induced by geldanamycin and 17-allylaminogeldanamycin using dog liver slices. *Toxicol. Sci. Off. J. Soc. Toxicol.* **2005**, *87*, 442–450. [CrossRef]

109. Rajan, A.; Kelly, R.J.; Trepel, J.B.; Kim, Y.S.; Alarcon, S.V.; Kummar, S.; Gutierrez, M.; Crandon, S.; Zein, W.M.; Jain, L.; et al. A phase I study of PF-04929113 (SNX-5422), an orally bioavailable heat shock protein 90 inhibitor, in patients with refractory solid tumor malignancies and lymphomas. *Clin. Cancer Res.* **2011**, *17*, 6831–6839. [CrossRef]

110. Xie, Q.; Wondergem, R.; Shen, Y.; Cavey, G.; Ke, J.; Thompson, R.; Bradley, R.; Daugherty-Holtrop, J.; Xu, Y.; Chen, E.; et al. Benzoquinone ansamycin 17AAG binds to mitochondrial voltage-dependent anion channel and inhibits cell invasion. *Proc. Natl. Acad. Sci. USA* **2011**, *108*, 4105–4110. [CrossRef]

111. Gadelle, D.; Bocs, C.; Graille, M.; Forterre, P. Inhibition of archaeal growth and DNA topoisomerase VI activities by the Hsp90 inhibitor radicicol. *Nucleic Acids Res.* **2005**, *33*, 2310–2317. [CrossRef] [PubMed]

112. Kijima, T.; Prince, T.L.; Tigue, M.L.; Yim, K.H.; Schwartz, H.; Beebe, K.; Lee, S.; Budzynski, M.A.; Williams, H.; Trepel, J.B.; et al. HSP90 inhibitors disrupt a transient HSP90-HSF1 interaction and identify a noncanonical model of HSP90-mediated HSF1 regulation. *Sci. Rep.* **2018**, *8*, 6976. [CrossRef] [PubMed]

113. Gaali, S.; Kirschner, A.; Cuboni, S.; Hartmann, J.; Kozany, C.; Balsevich, G.; Namendorf, C.; Fernandez-Vizarra, P.; Sippel, C.; Zannas, A.S.; et al. Selective inhibitors of the FK506-binding protein 51 by induced fit. *Nat. Chem. Biol.* **2015**, *11*, 33–37. [CrossRef] [PubMed]
114. Zheng, D.; Sabbagh, J.J.; Blair, L.J.; Darling, A.L.; Wen, X.; Dickey, C.A. MicroRNA-511 Binds to FKBP5 mRNA, Which Encodes a Chaperone Protein, and Regulates Neuronal Differentiation. *J. Biol. Chem.* **2016**, *291*, 17897–17906. [CrossRef]
115. Guy, N.C.; Garcia, Y.A.; Cox, M.B. Therapeutic Targeting of the FKBP52 Co-Chaperone in Steroid Hormone Receptor-Regulated Physiology and Disease. *Curr. Mol. Pharmacol.* **2015**, *9*, 109–125. [CrossRef] [PubMed]
116. Cheung-Flynn, J.; Prapapanich, V.; Cox, M.B.; Riggs, D.L.; Suarez-Quian, C.; Smith, D.F. Physiological role for the cochaperone FKBP52 in androgen receptor signaling. *Mol. Endocrinol.* **2005**, *19*, 1654–1666. [CrossRef]

International Journal of
Molecular Sciences

Article

The Stress-Inducible Protein DRR1 Exerts Distinct Effects on Actin Dynamics

Anja Kretzschmar [1], Jan-Philip Schülke [1], Mercè Masana [1,2,3], Katharina Dürre [4],
Marianne B. Müller [1,2], Andreas R. Bausch [4] and Theo Rein [1,*]

[1] Max Planck Institute of Psychiatry, Kraepelinstraße 2-10, 80805 München, Germany;
    anja.kretzschmar@mytum.de (A.K.); jan.schuelke@gmail.com (J.-P.S.); mmasana@ub.edu (M.M.);
    marianne.mueller@unimedizin-mainz.de (M.B.M.)
[2] Department of Psychiatry and Psychotherapy & Focus Program Translational Neuroscience, Johannes
    Gutenberg Universität Medical Center, 55131 Mainz, Germany
[3] Department of Biomedical Sciences, Faculty of Medicine and Health Sciences, University of Barcelona,
    IDIBAPS, CIBERNED, Casanova, 143, 08036 Barcelona, Spain
[4] Lehrstuhl für Biophysik E27, Technische Universität München, 85748 Garching, Germany;
    katharina.duerre@tum.de (K.D.); abausch@mytum.de (A.R.B.)
* Correspondence: theorein@psych.mpg.de; Tel.: +49-(0)89-30622-531

Received: 29 October 2018; Accepted: 10 December 2018; Published: 11 December 2018

Abstract: Cytoskeletal dynamics are pivotal to memory, learning, and stress physiology, and thus psychiatric diseases. Downregulated in renal cell carcinoma 1 (DRR1) protein was characterized as the link between stress, actin dynamics, neuronal function, and cognition. To elucidate the underlying molecular mechanisms, we undertook a domain analysis of DRR1 and probed the effects on actin binding, polymerization, and bundling, as well as on actin-dependent cellular processes. Methods: DRR1 domains were cloned and expressed as recombinant proteins to perform in vitro analysis of actin dynamics (binding, bundling, polymerization, and nucleation). Cellular actin-dependent processes were analyzed in transfected HeLa cells with fluorescence recovery after photobleaching (FRAP) and confocal microscopy. Results: DRR1 features an actin binding site at each terminus, separated by a coiled coil domain. DRR1 enhances actin bundling, the cellular F-actin content, and serum response factor (SRF)-dependent transcription, while it diminishes actin filament elongation, cell spreading, and actin treadmilling. We also provide evidence for a nucleation effect of DRR1. Blocking of pointed end elongation by addition of profilin indicates DRR1 as a novel barbed end capping factor. Conclusions: DRR1 impacts actin dynamics in several ways with implications for cytoskeletal dynamics in stress physiology and pathophysiology.

Keywords: stress physiology; cytoskeleton; actin dynamics; DRR1; TU3A; FAM107A

---

1. Introduction

Stress is a risk factor for several pathologies, including mental disorders such as psychiatric diseases [1,2]. Underlying mental disorders are alterations in the pattern of synaptic structure and activity, which has been repeatedly shown to be impacted by stress [2,3]. Actin, as the most prominent cytoskeletal component at the synapse, plays a major role in synaptic transmission by regulating synaptic shape, neurotransmitter vesicle release, and post-synaptic receptor distribution [4]. Actin dynamics and rearrangements of actin filaments are crucial during structural and functional alterations of neurons in response to stress shaping synaptic plasticity and behavior [5]. More specifically, acute and chronic stress have been shown to dramatically impact on numerous processes, including neuronal architecture, network dynamics, synaptic efficacy, and dendritic spine shape [2,6]. Further, dynamics of dendritic spines have been implicated in both memory formation and the development of psychiatric or neurological disorders [7]. Dysregulation of synaptic actin dynamics has been proposed as a convergent mechanism of mental disorders [8]. Therefore, investigating how specific actin binding proteins modulate actin dynamics is essential to understanding cell physiology and disease pathophysiology. Since the actin cytoskeleton exerts a major modulatory function in a plethora of additional cellular processes such as morphogenesis, motility or endocytosis, deciphering the processes contributing to dynamic actin cytoskeleton rearrangements is relevant to understanding several human pathologies [9].

The variety of actin-dependent processes is accomplished by its highly dynamic structure: globular actin (G-actin) polymerizes to filamentous actin (F-actin), while this polymerization reaction and the organization of actin filaments to higher-order actin-structures is orchestrated by numerous actin binding proteins. Adenosine triphosphate (ATP)-bound actin monomers are added at the barbed (+) end of the filament, ATP is then hydrolyzed by actin along the filament leading to its destabilization and depolymerization at the opposite, pointed (−) end. Thereby, actin filaments undergo a constant turnover of monomers called treadmilling [10].

The rate-limiting step of filament polymerization, the formation of actin dimers and trimers, is enhanced by nucleating factors like formins [11,12]. In contrast, the Arp2/3 complex generates new actin filaments by nucleation from existing filaments [13]. Elongation is terminated by capping proteins that bind to the barbed ends and inhibit the addition of further actin monomers, thereby limiting the length of the filament [14,15].

While sheet-like structures necessary for lamellipodial protrusions of the cells are created by Arp2/3 and crosslinkers like filamin, finger-like filopodia are arranged by thick actin bundles crosslinked, e.g., by fascin or α-actinin [16–18]. The cellular G-/F-actin equilibrium further changes the intracellular processes, for example, the transcription factor serum response factor (SRF) [19]. SRF-responsive genes, in turn, encode regulators of the actin and microtubule cytoskeleton, cell growth, and motility, adhesion, extracellular matrix synthesis and processing, and transcription [20].

Previously, we have identified a novel stress-induced protein enhancing cognition and social behavior, primarily localizing to actin-rich structures like stress fibers, membrane ruffles, and synapses [21–23]. This protein had initially been described as a tumor suppressor and, thus, had been termed downregulated in renal cell carcinoma gene 1 (DRR1). It is also known as Tohoku University cDNA clone A on chromosome 3 (TU3A) or Family with sequence similarity 107, member A (FAM107A) [24,25]. DRR1 is downregulated in various cancer cell lines, including renal cell, ovarian, cervical, laryngeal, gastric, prostate, liver, lymph, and non-small cell lung cancer and is associated with the progression of neuroblastoma, meningioma and malignant glioma [26–36]. On the other hand, DRR1 is highly expressed in outer radial glial cells [37] and in the invasive component of glioblastoma [38,39]. Lately, DRR1 has been associated to several brain disorders. Gene expression analyses indicated altered expression of DRR1 in neurodegenerative diseases as well as in bipolar disorder, autism spectrum disorder, and schizophrenia, presumably indicating an aberrant adaptation to chronic stress [40–46].

DRR1 shows basal expression in several brain regions and is strongly upregulated in mouse models of stress, as well as by dexamethasone in the hippocampus [23,47–49]. Its virus-mediated upregulation—aiming at mimicking stress-induced DRR1 increase—in the Cornu Ammonis region 3 (CA3) hippocampal region and the lateral septum increased hippocampus-dependent memory and social behavior, respectively [22,23]. Recently, cognitive impairment was measured 4 h after social defeat stress, when DRR1 protein levels were not increased yet, but not after 8 h, when DRR1 protein levels were found increased [50]. However, viral-mediated overexpression of DRR1 was not able to prevent the cognitive impairments 4 h after social defeat [50]. These findings suggest DRR1 to act as an adaptation factor that contributes to the molecular machinery counterbalancing aversive stress effects, but cannot act in a preventive manner. On the molecular and cellular levels, it was found to directly interact with β-actin and inhibit neurite outgrowth [23].

The link between stress and actin dynamics appears to be a critical component of the general adaptation mechanism [5,22,23]. However, up to now, a more detailed mechanistic understanding of DRR1's action on actin is lacking. Given the relevance of DRR1 not only during the stress response, but also in brain disorders and tumor development and progression, we aimed at elucidating its molecular mechanism and its significance in actin-dependent cell function. We found that DRR1 impacts actin dynamics in an intriguing multifaceted fashion by bundling, capping and nucleating filaments, altogether leading to stabilization of F-actin.

2. Results

2.1. DRR1 Features an Actin Binding Site at Each Terminus

Murine DRR1 is a highly conserved protein with 144 amino acids containing the "conserved domain of unknown function 1151". Secondary structure prediction in DRR1 indicates a predominantly helical protein with three helices and a coiled coil motif from amino acids 66 to 93 within the central helical region. Coiled coil motifs are abundant in the eukaryotic proteome and frequently involved in protein-protein and protein-DNA interactions [51,52]. Based on this structure prediction, DRR1 was divided into three domains: an N-terminal domain, a middle domain, and a C-terminal domain. Truncation mutants were generated accordingly to map the functions of DRR1 on actin dynamics in vitro and actin-dependent cellular processes (Figure 1A).

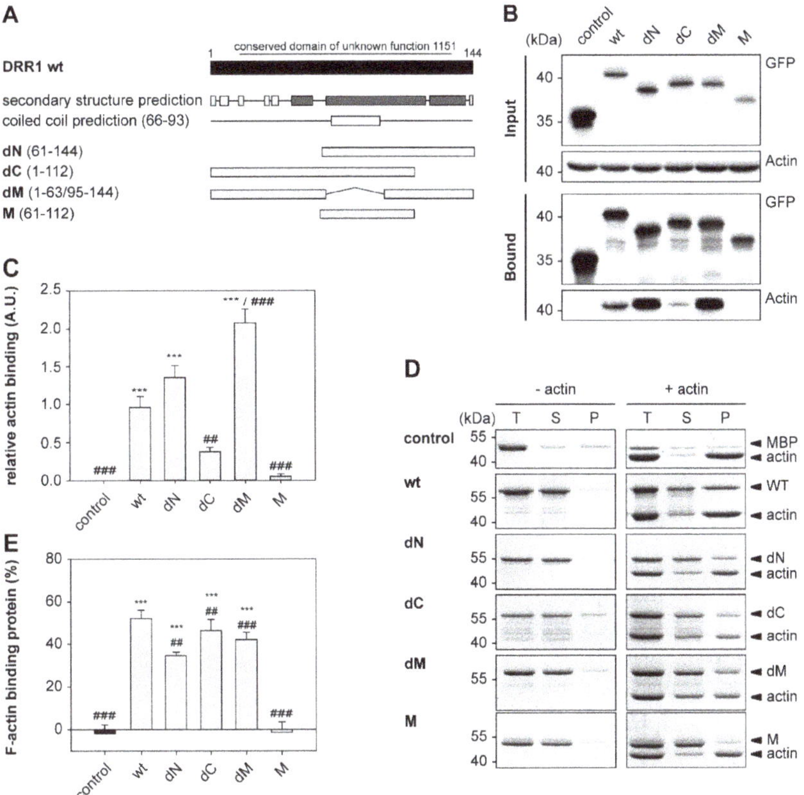

Figure 1. Downregulated in renal cell carcinoma 1 (DRR1) features an actin binding site at each terminus. (**A**) Domain structure of DRR1 wt and mutants. DRR1 harbors a conserved domain of unknown function from amino acid 16–133. Secondary structure prediction was performed with the

"Predict Protein Server" (dark: helix, light: loop; https://www.predictprotein.org/, accessed on 31 July 2012). Coiled coil prediction performed with "Coils" (http://embnet.vital-it.ch/software/COILS_form.html; accessed no 10 December 2018); (**B**) Co-immunoprecipitation of actin with DRR1 wt and mutants fused to Enhanced Green Fluorescent Protein EGFP overexpressed in Human embryonic kidney 293 cells (HEK)-293 cells using Green Fluorescent Protein (GFP)-Trap® beads. Control was performed with EGFP alone. Lysate and eluate samples were analyzed by SDS-PAGE and Western blot. A representative Western blot is shown; (**C**) Quantification of Co-immunoprecipitation (n = 8, dN and M n = 7); (**D**) Co-sedimentation of recombinant wt and mutant DRR1 protein with preformed F-actin by ultracentrifugation. Coomassie-stained sodium dodecyl sulfate (SDS) – polyacrylamide gel electrophoresis (PAGE) with total (T), supernatant (S) and pellet (P) fractions are shown; (**E**) Quantification of co-sedimented protein (n = 3). Bars represent means + SEM. **/## $p < 0.01$, ***/### $p < 0.001$ in comparison to control/wt DRR1 (only significant differences are marked). Statistical analysis was performed with one-way analysis of variants (ANOVA) and Bonferroni post hoc.

Actin binding of wild-type (wt) and mutant DRR1 was verified by co-immunoprecipitation (CoIP) from cellular extracts and by co-sedimentation of purified recombinant DRR1 proteins with F-actin. For immunoprecipitation, enhanced green fluorescent protein (EGFP)-tagged DRR1 proteins were ectopically expressed in Human embryonic kidney 293 cells (HEK)-293 cells. CoIP revealed actin binding of wt and all mutants except for the middle domain M. Quantification of relative actin binding in the CoIP revealed significant binding for DRR1 wt, dN, and dM, while it was not significant for dC and M. However, some actin binding could still be detected for dC in the Western blot. This is consistent with the presence of an actin binding site at both the N- and the C-terminus (Figure 1B,C).

For the co-sedimentation assays, purified G-actin from rabbit skeletal muscle was polymerized and then incubated with purified wt and mutant DRR1 proteins (tagged with maltose binding protein (MBP)) followed by high speed centrifugation and analysis of the total (T), supernatant (S), and pellet (P) fractions by Sodium dodecyl sulfate polyacrylamide gel electrophoresis (SDS-PAGE) and Coomassie staining (Figure 1D). Largely consistent with the results from the CoIP, there was significant binding to F-actin in the co-sedimentation assay for DRR1 wt and the mutants dN, dC and dM. The mutant M showed no binding (Figure 1D,E). In comparison to wt, deletion of the N, M and C domain modulated actin binding, which appeared more pronounced in the coprecipitation experiment (Figure 1 C,E), while deletion of both N and C domain completely abolished it.

2.2. DRR1 Enhances Actin Bundling Via Its Two Actin Binding Regions and Potentially through Homo-Dimerization

To visualize the DRR1-induced alterations in F-actin networks, in vitro actin networks were polymerized in the presence or absence of recombinant DRR1 (purified via the MBP-tag) until equilibrium and then imaged in a confocal microscope. While the networks of the control (MBP added) showed a purely filamentous network lacking distinguishable higher order structures, the addition of DRR1 resulted in strong bundle formation with a completely bundled network in a concentration-dependent manner (Figure 2A), consistent with our previous results [23]. At a DRR1:actin ratio (R) of 0.1, DRR1 already generates clear actin bundling. At the highest DRR1:actin ratio tested of R = 0.5, the whole network appears as bundles without distinguishable single filaments.

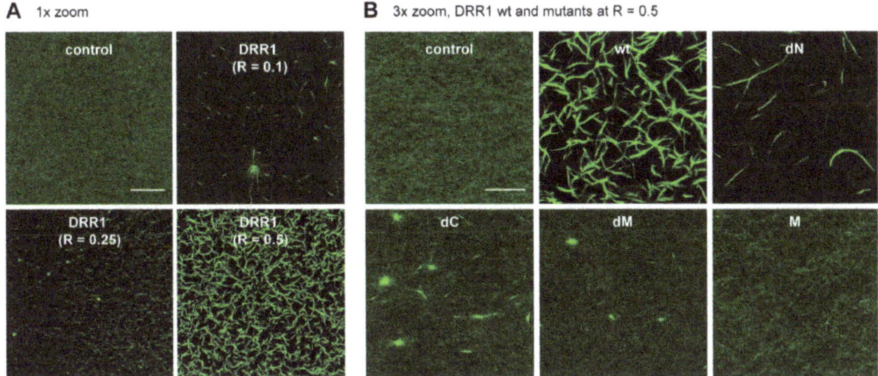

Figure 2. DRR1 enhances actin bundling via its two actin binding regions and potentially through homo-dimerization. (**A**) DRR1 enhances bundling of F-actin in a concentration-dependent manner. Z-stacks from actin networks polymerized at room temperature (RT) for >2 h in the presence of DRR1 and visualized with phalloidin-488. Scale bar denotes 50 µm. In all panels, "R" refers to the molar ratio of recombinant protein: actin protein; (**B**) Z-stacks from actin networks polymerized at RT for > 2 h in the presence of DRR1 proteins (R = 0.5) and visualized with phalloidin-488. Control = MBP added, because the DRR1 proteins are MBP-tagged. Scale bar denotes 20 µm.

To dissect the bundling mechanism of DRR1, actin networks polymerized in the presence of each mutant at R = 0.5 were also visualized in the confocal microscope (Figure 2B). Despite of retaining both actin binding regions, addition of the mutant dM only led to amorphous bundle "aggregates", but no proper actin bundling. This suggests that the central region is necessary as a spacer for accurate positioning of the two actin binding regions for proper bundle formation. In contrast, the mutants dN and dC both impacted on actin networks by producing bundle-like structures, although they both harbor one actin binding region only. While dC showed bundle-like aggregates, dN generated actin bundling comparable to wt DRR1 at a lower concentration (compare Figure 2B dN R = 0.5 and Figure 2A wt R = 0.1). These findings could be explained by homo-dimerization of the mutants dC and dN through the putative coiled coil interaction motif (compare with Figure 1). The mutant M exhibited no effect on the actin networks and appeared similar to the control. This is consistent with the lack of F-actin binding observed in Figure 1.

In an effort to provide experimental evidence for the dimerization of DRR1, we expressed untagged wt and mutant DRR1 in HEK293 cells and probed for dimerization using the crosslinker 1,4-Bismaleimidobutane (BMB) that reacts with sulfhydryl groups (cysteines) in close proximity. Western blot analysis revealed signals at the dimer positions of wt DRR1, dC and dN, but not for dM and M (Figure A1). Thus, as predicted by the secondary structure analysis (Figure 1), DRR1 dimerizes very likely through the middle domain that features the only cysteine (position 94) available for crosslinking by BMB. However, it appears that actin binding strongly promotes dimerization, because the M-domain alone did not produce any sign of dimerization, probably because it is too dispersed throughout the cell. It also should be noted that on formal grounds, these data do not exclude the possibility that the N- and the C-domain each dimerize independently.

2.3. DRR1 Bundling Diminishes Cellular Actin Treadmilling

The strong bundling effect of DRR1 on actin filaments in vitro could lead to a stabilization of F-actin as well as reduced actin treadmilling in cells. To test this hypothesis, fluorescence recovery after photobleaching (FRAP) of Green Fluorescent Protein (GFP)-labeled actin was measured in HeLa cells co-expressing untagged DRR1 (or empty vector as control) for 24 h. Time-lapse images were acquired with a confocal microscope during 5 min (five frames were recorded pre-bleach). After 5 min, the recovery of fluorescence in control cells reached about 75% of the bleached fluorescence. In DRR1 wt overexpressing cells, however, the recovery reached about 55%, indicating a higher immobile fraction of actin and, thus, a reduced actin treadmilling rate. None of the deletion mutants analyzed changed FRAP, except for the mutant dN, which exerted a similar effect as wt DRR1 (Figure 3).

The images of Figure 3 display nuclear localization of Actin, consistent with other reports [53]; DRR1 also has been reported to be both in the nuclear and cytosolic compartment [25,47,54]. We performed biochemical fractionation of HEK-293 cells ectopically expressing wt or mutant DRR1 to analyze the nuclear and cytosolic fraction by Western blotting (Figure A2). The efficiency of fractionation was monitored by probing the membranes for the cytosolic kinase AKT and the nuclear histone H4 (acetylated). All DRR1 forms could be detected both in the cytosol and in the nucleus (Figure A2).

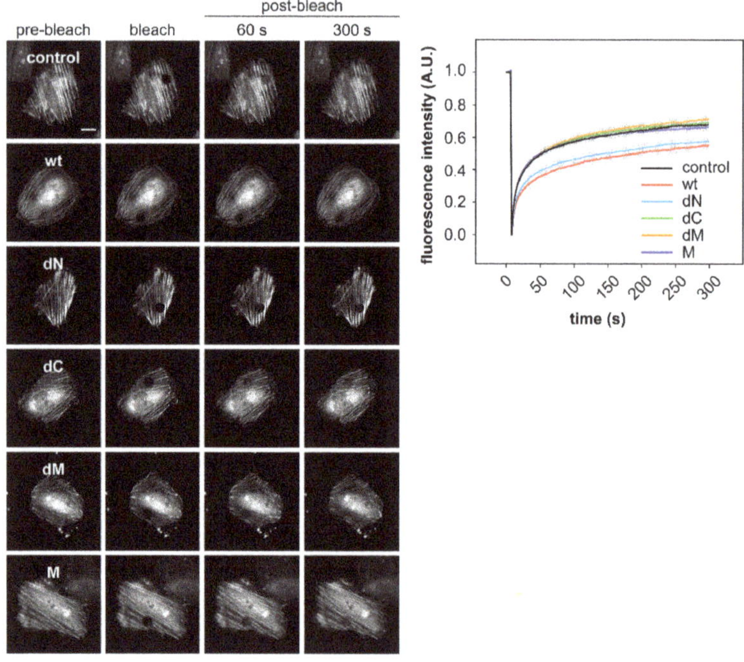

Figure 3. DRR1 bundling effect diminishes cellular actin treadmilling. DRR1 wt and the mutant dN—but none of the other mutants–slow down actin treadmilling in HeLa cells. Fluorescence recovery after photobleaching (FRAP) in HeLa cells co-transfected with plasmids expressing GFP-actin and untagged DRR1 wt, dN, dC, dM, and M was recorded. Representative cells are shown. Quantification was performed in ImageJ (25–30 cells from 2–3 independent experiments). Scale bar denotes 20 μm. Movies of FRAP experiments are available on request.

2.4. DRR1 Reduces Actin Filament Elongation but Increases Nucleation

To explore the effects of DRR1 on actin beyond the previously described bundling ability of DRR1 [23], we examined actin polymerization with pyrene-actin in the presence of DRR1 (Figure 4A,B) as well as single filament elongation and nucleation using total internal reflection fluorescence (TIRF) microscopy (Figure 4C).

As a first step to analyze actin polymerization in the presence of DRR1, pyrene-labeled actin, which shows enhanced fluorescence upon polymerization, was polymerized in the presence of increasing concentrations of DRR1 wt. At a DRR1:actin ratio of R = 0.5 and 1, the polymerization reaction was strongly inhibited in comparison to the control (Figure 4A). The only mutant to have an effect comparable to DRR1 was dM (Figure 4B).

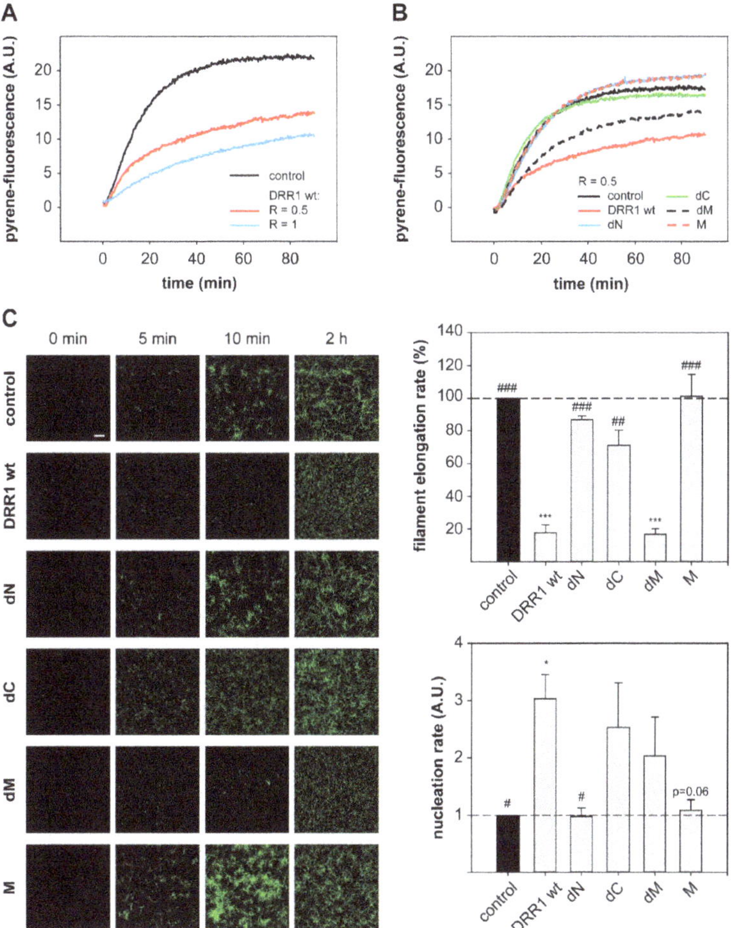

Figure 4. DRR1 reduces actin filament elongation but increases nucleation. (**A,B**) DRR1 and the mutant dM exert an inhibitory effect on in vitro polymerization of pyrene-actin. 20% pyrene-labeled actin (4 μM) was polymerized in the presence of wt (**A,B**) and mutant (**B**) DRR1 proteins (purified via the MBP-tag) as indicated. Increase in fluorescence of pyrene-actin during polymerization was monitored in 5 s intervals for 90 min; (**C**) Single filament elongation of actin is strongly reduced by DRR1 and the mutant dM. Actin (c = 0.5 μM, 10% labeled with ATTO-488) was polymerized in the presence of DRR1 proteins or MBP as control (R = 0.5) and visualized by TIRF microscopy for 10 min with 3 s intervals starting 2 min after the beginning of the reaction. An endpoint image was taken at 2 h of polymerization. Scale bar denotes 10 μm for all images. Bars indicating the filament elongation rate and the nucleation rate represent means + SEM of three independent experiments. $*/^{\#}\ p < 0.05$, $**/^{\#\#}\ p < 0.01$, $***/^{\#\#\#}\ p < 0.001$ in comparison to control/wt DRR1 (only significant differences are marked; $p = 0.06$ refers to the comparison of M to wt DRR1). Statistical analysis was performed with one-way ANOVA and Bonferroni post hoc. Movies of single filament elongation experiments are available on request.

In order to verify the slowdown of polymerization by DRR1, the polymerization reaction of fluorescently labeled G-actin was monitored using TIRF. The overall slowdown of actin polymerization by DRR1 observed in the pyrene-assay was well reproduced in the TIRF polymerization. DRR1 significantly slowed down actin polymerization to less than 20% control (Figure 4C). The mutant dM was the only mutant to retain the inhibitory effect of wt DRR1 on filament elongation, even though it had previously not shown an effect on bundling. The mutant dC showed a mild reduction of filament elongation, which was not significant. This finding indicates that the reduction of single filament elongation by DRR1 is independent of actin bundling, but both actin binding sites are necessary to affect elongation, since mutants lacking one or both actin binding sites displayed no significant effect.

Intriguingly, the visualization of single filament elongation revealed more but shorter filaments in the presence of DRR1 versus the control. Thus, the number of new filaments per time frame was quantified and the slope of the resulting plot was determined as read-out of the nucleation rate. DRR1 moderately enhanced the filament nucleation rate up to three-fold above the control at a molar ratio of DRR1:actin of 0.5. The mutants dC and dM both showed a trend towards increased nucleation versus the control, although not statistically significant (Figure 4C).

2.5. In the Presence of Profilin, DRR1, and the Mutants dM and dC Block Elongation More Effectively, Suggesting DRR1 as a Novel Barbed End Capping Factor

Different actin binding factors interact in the cell to control overall actin dynamics. To reconstitute more complex conditions in vitro, we analyzed the effects of wt DRR1 and mutants on actin polymerization in the presence of profilin, a well-described blocker of pointed end polymerization. Thus, we assessed elongation exclusively from the barbed end.

Addition of profilin strongly enhanced DRR1's inhibitory effect on actin polymerization supporting the notion of DRR1 as a novel capping protein at the barbed end. In the pyrene-assay, an inhibitory effect of DRR1 on barbed end polymerization was already detectable at a DRR1:actin ratio of R = 0.01. At R = 0.1, polymerization was almost completely blocked (Figure 5A). At R = 0.5, the mutant dM most noticeably reduced actin polymerization, but also dC slowed down the polymerization, while all other mutants lacked strong effects (Figure 5B).

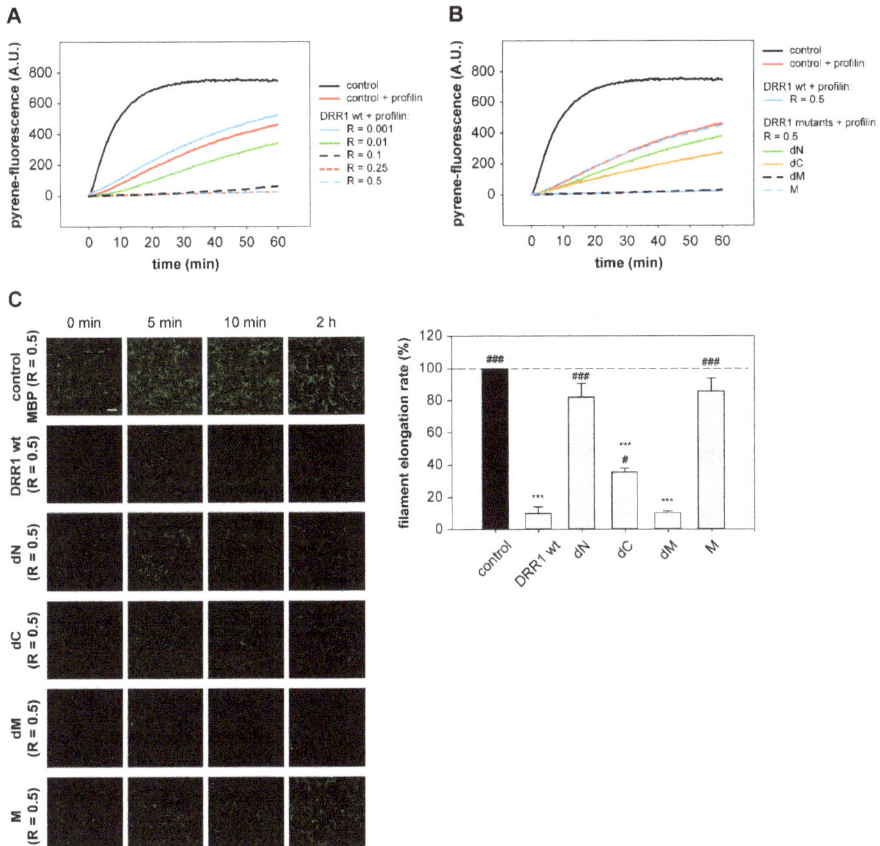

Figure 5. In the presence of profilin, DRR1 and the mutants dM and dC block elongation more effectively, suggesting DRR1 as a novel barbed end capping factor. (**A,B**) Pyrene-actin polymerization is blocked by DRR1 and the mutant dM at R = 0.5 in the presence of profilin (12 μM). 20% pyrene-labeled actin (4 μM) was polymerized in the presence of wt (**A,B**) and mutant (**B**) DRR1 proteins (purified via the MBP tag) as indicated. An increase in fluorescence of pyrene-actin during polymerization was monitored in 5 s intervals for 60 min; (**C**) Visualization of actin in vitro polymerization by TIRF microscopy (c = 0.5 μM, 10% labeled with ATTO-488) in the presence of profilin (1.5 μM). Actin was polymerized in the presence of DRR1 proteins for 10 min with 3 s intervals imaging starting 2 min after the beginning of the reaction. An endpoint image was taken at 2 h of polymerization. Scale bar denotes 10 μm for all images. Bars indicating the filament elongation rate represent means + SEM of three independent experiments. */# $p < 0.05$, ***/### $p < 0.001$ in comparison to control/wt DRR1 (only significant differences are marked). Statistical analysis was performed with one-way ANOVA and Bonferroni post hoc. Movies of single filament elongation experiments are available on request.

Similar results were obtained with polymerization of actin and single filament analysis using TIRF microscopy. At R = 0.5, DRR1 wt displayed a pronounced barbed end capping activity in the presence of profilin by reducing the filament elongation rate to about 10% of the control. The same effect was reproduced for dM. In addition, dC which had only a mild effect in the absence of profilin, reached significant capping activity in its presence reducing the filament elongation rate to around 36% of the control. Nevertheless, this capping activity of dC was less pronounced than for DRR1 wt (Figure 5C).

2.6. DRR1 Modulates Actin-Dependent Processes in Cells

To further analyze the cellular consequences of DRR1-induced changes of actin dynamics, we evaluated known actin-dependent processes such as cell spreading and activity of the transcription factor serum response factor (SRF). Cell spreading is relevant in many cellular functions, such as migration or wound healing. Spreading of HeLa cells ectopically expressing EGFP-DRR1 wt or mutants, was analyzed by replating on a fibronectin-coated surface and fixation after 30 min of spreading; F-actin was stained with phalloidin.

DRR1 wt strongly reduced spreading of HeLa cells: while (EGFP transfected) control cells showed a mean size of about 700 μm^2, DRR1 wt expressing cells had a mean cell size below 500 μm^2 (Figure 6A). In addition, control cells expressing EGFP showed extension of filopodial protrusions after 30 min of spreading, while DRR1 wt-expressing cells were still round–shaped, lacking any protrusions. In these freshly-seeded cells, DRR1 wt colocalized with F-actin at the cortex area of the cells, where the filaments' barbed ends are oriented [55]; this is consistent with DRR1's capping activity at the barbed ends, thereby inhibiting extension of protrusions during cell spreading. Evaluation of the deletion mutants of DRR1 revealed that all mutants except M also inhibited cell spreading. This indicates that either capping or bundling by DRR1 is sufficient to reduce cell spreading.

Cell imaging revealed that DRR1 colocalizes with F-actin (Figure 3 and [23]). As expected from the binding analyses (Figure 1), the mutants dN, dC, and dM also exhibit colocalization with actin filaments (Figure A3). Among them, dC, whose actin binding did not reach significance in the co-immunoprecipitation experiment (Figure 1C), exhibited the lowest correlation coefficient. M showed no colocalisation with F-actin (Figure A3). The overall cellular content of F-actin was increased by wt DRR1, dN, and dM, while dC and M had no significant effect (Figure A4).

The equilibrium between G- and F-actin has further repercussions for intracellular processes, for example activation of the transcription factor serum response factor (SRF). With decreasing levels of G-actin, the SRF cofactor MAL detaches from G-actin, translocates to the nucleus and activates SRF [19]. We employed SRF reporter gene assays as a G-actin sensor to monitor the effects of wt and mutant DRR1. The nucleator formin mDia lacking its autoinhibitory "DAD" region was used as a positive control for SRF activation [56].

DRR1 wt increased SRF activity about 10-fold in serum-stimulated cells, similar to the effect of mDia. In the absence of serum, the stimulation was still about eight-fold above the serum-stimulated control sample and again comparable to mDia, indicating strong SRF activation by DRR1 independently of serum. The mutant dN also significantly enhanced SRF activity, while the mutants dC, and M showed no effect. While dM showed a minor increase in SRF activity, its stimulation did not reach significance (Figure 6B). This data indicate that DRR1 expression levels modulate SRF-dependent gene expression through modulation of the equilibrium between G- and F-actin

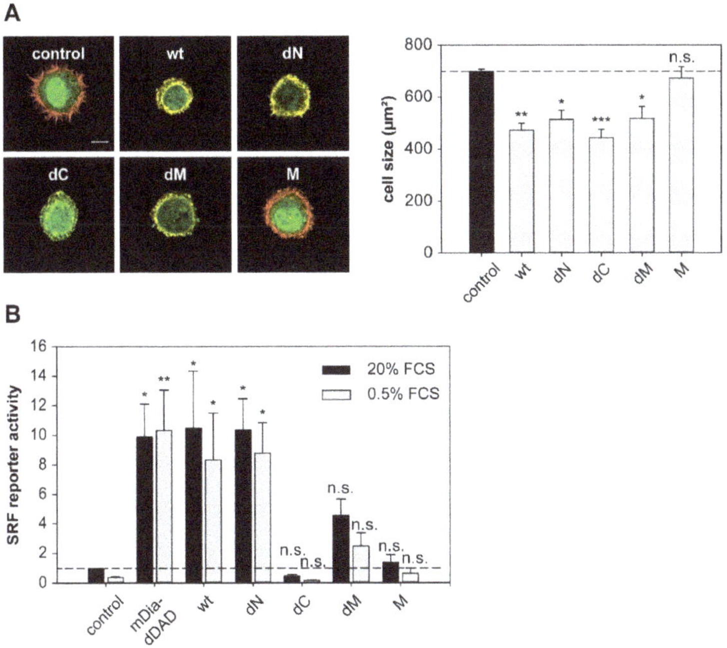

Figure 6. DRR1 modulates actin-dependent processes in cells. (**A**) DRR1 wt and the mutants dN, dC, and dM inhibit spreading of HeLa cells. Cells were transfected with constructs expressing EGFP-DRR1 wt or mutants (control: EGFP), cultivated for 24 h and re-plated on fibronectin-coated coverslips. After 30 min, cells were fixed and F-actin was stained with phalloidin. Representative cells are displayed (green: EGFP or EGFP-DRR1; red: F-actin). Scale bar denotes 20 μm. Bars represent mean cell sizes + SEM of four independent experiments (50–200 cells in each experiment). * $p < 0.05$, ** $p < 0.01$, *** $p < 0.001$ in comparison to control. Statistical analysis was performed with one-way ANOVA and Bonferroni post hoc; (**B**) DRR1 overexpression leads to a strong activation of the serum response factor (SRF) independently of serum, indicating a stabilization of cellular F-actin by DRR1 bundling and capping effects. SRF reporter gene assays in HEK-293 cells show 8–10 fold enhanced SRF activity after overexpression of DRR1 wt or dN with and without serum. Cells were transfected with the SRF reporter 3DA.luc, the gaussia luciferase control vector and the indicated plasmids or vector control. Serum stimulation or withdrawal was for 16–20 h. Luciferase activity is shown as the fold-increase of serum-stimulation over control samples. Bars represent means + SEM of five independent experiments. * $p < 0.05$, ** $p < 0.01$, n.s. = not significant in comparison to control. Statistical analysis was performed with one-way ANOVA and Bonferroni post hoc.

3. Discussion

Actin binding proteins orchestrate the temporal and spatial remodeling of the actin cytoskeleton in cells as the structural basis for several cellular functions [57]. This highly dynamic process also responds to specific stimuli and, thus, conveys the ability to adapt to new environmental demands. Here, we present a domain and functional analysis of the stress-induced protein DRR1 with respect to its action on actin dynamics.

Our findings extend the characterization of DRR1 as actin bundler [23] and add it to the list of actin cappers. While the contact points of DRR1 on actin filaments remain unknown, capping might be achieved in two ways by the binding of at least one of the two actin binding domains of DRR1 described here close to the barbed end of the filament: either by inducing a conformational change at the outmost actin unit or by sterically interfering with the addition of the next actin molecule to the extending filament. Deletion of the N-terminal domain leads to complete loss of capping activity, while deletion of the C-terminus retains a somewhat lower capping activity. Therefore, we hypothesize that the N-terminal binding domain of DRR1 is required for capping. In addition, the second actin binding domain appears to contribute to capping possibly by stabilizing the interaction with actin. According to our model, the C-terminal deletion mutant dC is able to form dimers yielding two actin binding sites and, thus, enhances binding affinity to actin. Thus, the deletion mutant dC exerts some capping activity.

Most capping proteins appear to exert their activity at nanomolar concentrations [58], similar to the concentrations used here for DRR1. The ratio of DRR1 to actin of 1:10 at which significant capping was observed is also close to the range of other capping proteins, for example gelsolin [59]. However, the dominant actin capper in the cell, called "capping protein", displays a very high binding affinity and is effective at ratios as low as 1:1000 [60]. Thus, the other capping proteins, like DRR1 and gelsolin, may have more specialized roles. In general, both actin interaction domains and capping mechanisms are not conserved. Nevertheless, the mode of action of DRR1 in capping at the barbed end might be similar to the mechanism proposed for Twinfilin [61,62] and the Gelsolin protein family [63–65]. These proteins feature multiple actin binding sites and contact the actin filament both at the barbed end and at the side of the filament.

We cannot exclude that the nucleation effect of DRR1 observed here in the in vitro assays could be secondary to the capping activity, i.e., due to the extended availability of non-polymerized actin when polymerization is diminished. For capping protein, a concentration-dependent nucleation activity has been reported: it inhibits elongation of actin already at low concentration by blocking the barbed end, while at higher concentrations it enhances nucleation by mimicking a non-dissociable actin dimer [14]. At this stage, it is likely that DRR1 may act as a nucleation factor by pulling together actin monomers or by stabilizing short oligomers, which appears possible in particular with dimerized DRR1 (compare also graphical abstract).

Efficient bundling of actin by DRR1 requires both binding sites of DRR1 as deletion of one of the domains severely compromises actin bundling. Reduced bundling was also observed for the deletion of the middle domain, suggesting that proper spacing of the two actin binding domains is required, possibly in conjunction with dimerization of the full length protein. Similarly, the residual bundling activity of each of the terminal deletion mutants might be attributed to their dimerization. Even though the middle domain does not bind to actin, our data do not allow excluding the possibility that it actively contributes to bundling.

Visually, with respect to bundle thickness and length, and mesh size of the bundled network, bundled actin networks with DRR1 and α-actinin, respectively, look similar. Furthermore, actin:DRR1 networks at a ratio of 1:2 are comparable to α-actinin:actin networks of 1:1 ([66] and this work), suggesting DRR1's bundling activity to be at least as strong as the respective effect of α-actinin. Other actin bundling proteins inhibit in vitro actin depolymerization similarly to DRR1 in this study [67], and DRR1 itself has been shown to reduce dilution-induced actin depolymerization at ratios of DRR1:actin of 0.7 [23].

It should be noted that all experiments with recombinant DRR1 in this study had to be performed with a large (MBP) tag at the wt and mutant DRR1 proteins. Even though the cellular effects of DRR1 with a smaller (GFP) tag or with no tag reflect the in vitro results, and even though we distinguish the DRR1 effects from the effects of MBP alone, we cannot exclude the possibility that he MBP tag influenced the experimental outcome.

In cells, actin dynamics is shaped by the concerted action of several actin binding proteins. Although the effects observed in vitro with purified compounds may not always reliably predict the outcome in the cell, the effects on actin dynamics found in cells expressing wt DRR1 and its mutants were largely congruent with the in vitro results (summaries in Tables 1 and 2). For example, the mutants that exerted proper bundle formation, i.e., DRR1 wt and dN, were the only ones to reduce actin treadmilling in the FRAP experiment, suggesting that it is mainly the bundling activity that leads to stabilization of F-actin in cells. We noted though, that the effect of dN in the FRAP assay is comparable to that of DRR1 wt, while the effect on bundling is not as strong (Figure 2). Thus, we cannot exclude the possibility that additional actions of dN contribute to the overall effect in the FRAP assay.

Both bundling and capping effects appear to contribute to activation of cellular SRF, since serum-independent SRF activation was observed for DRR1, dN, and dM. Meanwhile, bundling and inhibition of filament polymerization seem to be largely independent effects: dN generated bundles but had no effect on filament elongation, whereas dM had formed no proper bundles, but strong inhibition of filament elongation similar to the wild-type.

Table 1. Schematic overview of DRR1's molecular effects on actin dynamics.

	F-Actin Binding	Bundling	Filament Elongation	Nucleation	Capping
DRR1 wt	+ + +	+ + +	− − −	+ +	+ + +
dN	+	+ +	0	0	0
dC	+ +	+	− (n.s.)	+ (n.s.)	+
dM	+ +	+	− − −	+ (n.s.)	+ + +
M	0	0	0	0	0

Summary of the data presented in Figures 1, 2, 4 and 5. + enhancement, − decrease, 0 no effect, n.s. not significant.

Table 2. Schematic overview of DRR1 effects on actin-dependent cellular processes.

DRR1	Colocalization F-Actin	Cellular F-Actin	Actin Treadmillling	Cell Spreading	SRF Activation
wt	+++	+++	− − −	− − −	+ + +
dN	++	+	− − −	−	+ +
dC	+	0	0	− − −	0
dM	++	++	0	−	+
M	0	0	0	0	0

Summary of the data presented in Figures 3 and 6. + enhancement, − decrease, 0 no effect.

In general, cell spreading is known to be a complex process influenced by several parameters, including substrate stiffness and density and actin polymerization [68–70]. Early spreading was proposed to depend on the mechanical properties of the cell, and the actin cortex in particular [71]. In the cell spreading assay, mutants that showed bundling or capping activity (or both) exhibited an inhibitory effect. Presumably, bundling by DRR1 inhibits the early phases of spreading by increasing cell stiffness while capping likely interferes with extension of the lamella at the later stages of spreading. These results might explain the reduced spine density found in hippocampal neurons overexpressing DRR1 in rodents, and with the reduced neurite growth found in cultured Neuro2a cells [23]. The here-observed effect on cell spreading is unlikely to be a result of the function of DRR1 as tumor suppressor. Upon ectopic overexpression, no effect on cell viability or induction of apoptosis could be observed [23]. Furthermore, in the cell spreading assay only newly attached cells are followed 30 min after seeding. Conversely, we cannot exclude that impairment of cell spreading contributes to the tumor suppressive action of DRR1. However, another tumor suppressor, p14ARF, has been demonstrated to enhance cell spreading, reflecting its dual role in tumor suppression and apoptosis protection [72].

The recovery rate of GFP-fluorescence upon photobleaching reflects the actin turnover or treadmilling rate, as the free diffusion of monomeric G-actin is much faster [73–75]. It was somewhat surprising that the mutant lacking the middle domain, which displayed significant capping activity in vitro, did not affect actin turnover. Of the tested mutants only full length and the mutant dN showed an effect. It is, thus, possible that bundling of actin is causing the decrease in actin turnover. This was observed also for the actin bundling protein Ca^{2+}/calmodulin-dependent protein kinase IIb (CaMKIIb), which reduced actin turnover in dendrites but did not directly impact on polymerization and depolymerization kinetics of actin [76]. Since actin treadmilling was shown to consume about half of the ATP pool in neurons [77], one of DRR1's physiological roles upon stress could be saving ATP that might be required for the proper adaptive reaction to stress.

Increased levels of G-actin not only impact SRF, but have also been reported to reduce glucocorticoid receptor (GR)-dependent transcription, possibly through inducing the GR inhibitor c-jun [78]. Accordingly, the F-actin depolymerization factor cofilin 1 has been found to inhibit GR activity. Thus, it is possible that DRR1 enhances GR activity under certain conditions (constituting a feed-forward mechanism), which may furthermore be cell type-dependent because DRR1 is not only expressed in neurons, but also in other cell types, such as glial cells and various tissues [37,41,79].

Several studies proposed a role of actin dynamics and remodeling in psychiatric disorders such as depression, based on case-control comparisons and animal models [80–83]. A recent proteome study revealed increased levels of F-actin-capping protein subunit beta (CAPZB) in platelets from patients suffering from major depression in comparison to healthy controls [84]. Other studies reported changes of actin regulatory proteins by antidepressants and mood stabilizers [85,86] mutations in genes of the regulatory network of the actin cytoskeleton appear to be enriched in treatment-resistant major depression [87]. Pathway-based methods to genetic data have been suggested to blend biological information with the power of –omics approaches [83]; we propose that the stress- and glucocorticoid-regulated DRR1 [22,23,47,88] should be included when analyzing the role of the actin cytoskeleton in physiology and pathology, particularly in stress-related processes. Furthermore, since actin regulatory factors work in concert, future biochemical investigation of DRR1 should include the combination with additional actin binding proteins, as this study now firmly established DRR1 as an actin-regulatory protein.

4. Materials and Methods

Several of the methods outlined in the following are also described in the Ph.D. thesis of Anja Kretzschmar [89].

4.1. Plasmids

Plasmids for transfection in cell culture were cloned downstream of the Cytomegalovirus (CMV) promoter of the vector pRK5-SV40-MCS. DRR1 mutants were generated by PCR mutagenesis from murine DRR1 wild-type (wt) construct in pRK5. Cloning of the murine DRR1 wt construct was previously described in [23]. The nucleotide sequences of all constructs were confirmed after cloning by Sanger sequencing. For expression of DRR1 proteins N-terminally fused to EGFP or MBP, inserts of DRR1 wt and mutants were subcloned into the vector pEGFP-C1 (Clontech, Saint-Germain-en-Laye, France) or pMAL-CR1 (New England Biolabs, Ipswich, MA, USA), respectively. Details of the cloning strategies and primer sequences are available on request.

4.2. Cell Culture and Transfection

HeLa and HEK-293 cells were cultured in Dulbecco's Modified Eagle Medium (DMEM, Life Technologies, Carlsbad, CA, USA) containing 10% fetal bovine serum, 1% sodium pyruvate, and 100 U/mL penicillin and streptomycin at 37 °C in a 5% CO_2 atmosphere. A confluent 10 cm dish of HEK-293 cells was transfected by electroporation with 15 µg plasmid and cultured for two days until conduction of the experiments. HeLa cells were transfected using TurboFect (Thermo Scientific, Waltham, MA, USA) according to the manufacturer's instructions and incubated for 24 h after transfection.

4.3. SDS-PAGE, Colloidal Coomassie Staining, and Immunoblot

Samples were separated on 10, 12, or 15% poly-acrylamide gels with 3.2% stacking gels and stained with colloidal coomassie brilliant blue G (Sigma-Aldrich, St. Louis, MO, USA) or electrophoretically transferred onto nitrocellulose membranes (GE Healthcare, Chalfont St Giles,

UK). Immunodetection was performed by blocking the membrane with 5% non-fat milk in Tris-buffered saline, supplemented with 0.05% Tween (TBS-T, Sigma-Aldrich) for 1 h at room temperature, and then incubated with primary antibody overnight at 4 °C. The blots were washed and probed with the respective horseradish peroxidase- or fluorophor-conjugated secondary antibody for 3 h at room temperature. All antibodies were diluted in TBS-T with 2% milk powder. The immunoreaction was visualized with ECL detection reagent (Millipore, Darmstadt, Germany) or by fluorescence. The following antibodies were used: rabbit-anti-DRR1 (1:2000, Biogenes, Berlin, Germany, as described in [23]), goat-anti-actin (I-19, 1:2000, Santa Cruz Biotechnology, Dallas, TX, USA), mouse-anti-GFP (B-2, 1:2000, Santa Cruz Biotechnology), rabbit-anti-AKT (1:1000, Cell Signaling, Frankfurt, Germany), rabbit-acetyl-H4 (1:4000, Upstate, Schwalbach, Germany), donkey-anti-rabbit-HRP (1:10,000, Cell Signaling, Cambridge, UK), donkey-anti-goat-HRP (1:10,000, Santa Cruz, Heidelberg, Germany), and Alexa Fluor 488-donkey-anti-mouse (1:5000, Life Technologies). Determination of the relative optical density and quantification of band intensities were performed using the ImageLab 4.1 Software (Bio-Rad, Munich, Germany).

4.4. Protein Expression and Purification

Recombinant DRR1 proteins were expressed and purified as maltose binding protein (MBP) fusion proteins in order to enhance stability and solubility. In our hands, various efforts to purify DRR1 without a tag [90] revealed insufficient stability of DRR1 [23]. We observed that with only a small tag this protein was prone to aggregation at high concentrations and required some urea (1M) and sodium dodecyl sulfate (SDS) (0.1%) to keep it in solution. Similarly, DRR1 turned out to be unstable when the MBP-tag was cleaved off. Therefore, control conditions with buffer only and with MBP only were included in all experiments with recombinant DRR1 proteins. Since there were no detectable differences in the results between buffer and MBP conditions (see Appendix A Figure A5), the latter is shown in all figures as control. Proteins were expressed in *Escherichia coli* BL21(DE3)pLysS bacteria (Life Technologies) induced by 0.3 mM isopropyl-beta-D-1-thiogalactopyranoside (IPTG) for 2 h at 37 °C. The bacterial pellets were lysed by the freeze-thaw method in a dry-ice ethanol bath and then re-suspended in lysis buffer (binding buffer supplemented with protease inhibitor cocktail, 1 mg/mL lysozyme, 0.1 mM Phenylmethane sulfonyl fluoride (PMSF) and 1 mM Dithiothreitol (DTT)), incubated on ice for 1 h, and sonicated. The lysates were cleared by centrifugation at $48,400 \times g$ for 1 h at 4 °C (Beckmann Avanti J-25, Krefeld, Germany) and then filtered through a 0.22 μm syringe filter. The ÄKTA purifier system (General Electrics Healthcare) was used for protein purification with affinity chromatography (MBPTrap HP, 1 mL, GE Healthcare) and gel filtration (Superdex200 10/300 GL, GE Healthcare) as a second step. All buffers used were first filtered through a 0.22 μm filter and then degassed. Bacterial lysates were loaded on equilibrated MBPTrap columns after clearing and filtration at a flow rate of 0.5 mL/min with 15 mL binding buffer (20 mM Tris-HCl pH 7.4, 200 mM NaCl, 1 mM Ethylenediaminetetraacetic acid (EDTA), 1 mM DTT), washed with 5 mL binding buffer, and eluted with 10 mL elution buffer (20 mM Tris-HCl pH 7.4, 200 mM NaCl, 1 mM EDTA, 10 mM maltose, 1 mM DTT). Samples containing recombinant protein as controlled by SDS-PAGE and Coomassie staining were pooled and concentrated with Vivaspin 2, MWCO 30 kDa, columns (GE Healthcare). The buffer was changed to Superdex running buffer with (20 mM Tris-HCl pH 7.4, 150 mM NaCl, 1 mM DTT). A Superdex200 10/300 GL column (GE Healthcare) was used for gel filtration at a flow rate of 0.5 mL/min. Collected samples were loaded and analyzed with SDS-PAGE, with subsequent pooling of samples containing recombinant protein. Protein concentration was measured with UV absorbance at 280 nm and with colloidal Coomassie stained SDS-PAGE with a protein standard and densitometry.

Recombinant Profilin2a from mouse tagged with glutathione s-transferase (GST) was expressed from *E. coli* and purified with 2–4 mL glutathion sepharose 4B resin (GE Healthcare) in disposable columns. Binding and elution was performed in 50 mM Tris-HCl pH 7.0, 150 nM NaCl, 1 mM EDTA, 1 mM DTT. Elution of Profilin was performed by cleaving off the tag overnight at 4 °C with PreScission

Protease (GE Healthcare). Protein concentration was determined by UV absorbance at 280 nm. Dialysis was performed against 20 mM Tris-HCl pH 7.0, 150 mM NaCl, 1 mM EGTA, 1 mM DTT.

All proteins were aliquoted and frozen in liquid nitrogen. Fresh aliquots of recombinant DRR1 or Profilin protein were used for all experiments.

4.5. Co-Immunoprecipitation

For co-immunoprecipitation (CoIP), HEK-293 cells transfected with plasmids expressing EGFP-fusion proteins were lysed with 200 µL ice-cold lysis buffer (10 mM Tris-HCl pH 7.5, 150 mM NaCl, 0.5 mM EDTA, 0.5% NP-40, 1:100 protease inhibitor cocktail P2714 from Sigma-Aldrich). The extract was incubated for 1 h on ice, diluted with 700 µL wash buffer (10 mM Tris-HCl pH 7.5, 150 mM NaCl, 0.5 mM EDTA, 1:100 protease inhibitor cocktail), and centrifuged for 10 min at 13,000 rpm to remove cell debris. Lysates were incubated with 25 µL GFP-Trap pre-equilibrated agarose beads (ChromoTek, Planegg-Martinsried, Germany) for 1 h at 4 °C. The beads were washed two times with 1 mL wash buffer and samples were eluted by incubation for 10 min at 95 °C in 50 µL 1× Laemmli sample buffer (1% SDS, 8% glycerol, 32 mM Tris-HCl pH 6.8, 5% mercaptoethanol, bromophenol blue). Fifteen microliters (15 µL) of each input/elution sample were loaded on gels for detection of Co-IP signals, and 5 µL were loaded for detection of EGFP-fusion-proteins. To calculate relative actin binding, IP and CoIP bands were revealed using an enhanced chemiluminiscence system (Millipore), detected with the Chemidoc system (BioRad, Munich, Germany), and quantified by densitometry (using the ImageLab 4.1 Software from Bio-Rad) with background signal, corresponding to areas of the membrane without signal, subtracted. Next, the corrected grey density value of the co-precipitated actin was referred to its corresponding value of the precipitated DRR1 protein and the actin/DRR1 ratio was defined as "actin binding". To be able to compare the values between different experiments and blots, the "average actin binding" of all DRR1 proteins (wt, dN, dC, dM, and M) for each experiment was calculated and "actin binding" of each mutant was normalized to this average actin binding of the respective experiment. Therefore, "1.0" at the y-axis of Figure 1C denotes the (arbitrary value of) average actin binding of the DRR1 proteins and carries no further meaning.

4.6. Dual Luciferase Reporter Assay

For SRF reporter gene assays, Simian virus 40 promoter-driven non-secretory Gaussia luciferase expression vector [91] (10 ng per well in a 96-well-plate) was co-transfected in HEK-293 cells with SRF reporter plasmid (25 ng) to correct for transfection efficiency and the respective test plasmids (150 ng per well). The SRF reporter 3DA.luc and mDia1-dDAD plasmids were kind gifts from Robert Grosse (Universität Marburg, Germany). SRF activity was stimulated with 20% FBS or inhibited (0.5% FBS) for 16–18 h twenty-four hours after transfection. Cell lysis was performed with 50 µL passive lysis buffer (0.2% Triton X-100, 100 mM K_2HPO_4/KH_2PO_4 pH 7.8) for 30 min at room temperature. Activity of the firefly luciferase was measured in white microtiter plates in a luminometer ((TriStar LB941 Luminometer, Berthold Technologies, Bad Wildbad, Germany)) by adding 50 µL Firefly substrate solution (3 mM MgCl2, 2.4 mM ATP, 120 mM D-Luciferin) to 10 µL lysate. Then, 50 µL of Gaussia substrate solution (1.1 M NaCl, 2.2 mM Na_2EDTA, 0.22 M K_2HPO_4/KH_2PO_4 pH 5.1, 0.44 mg/mL bovine serum albumin (BSA), Coelenterazine 3 µg/mL) was added to the same well to quench the firefly reaction and measure Gaussia luminescence with a 5 s delay. Firefly luminescence was corrected with Gaussia values to calculate Firefly activity data. The SRF activity in the serum-stimulated control with pRK5 was set to 1 in order to compare different experiments.

4.7. Chemical Crosslinking

HEK-293 cells were transfected with wt and mutant DRR1 expressing plasmids by electroporation. After cultivation for two days, cells were detached, washed, and then incubated in conjugation buffer (PBS with 1 mM EDTA) with either 200 µM crosslinker BMB (1,4-bismaleimidobutane, a crosslinker with a spacer arm length of 10.9 Å generating chemical bonds between sulfhydryl groups) or DMSO as control at 4 °C on a shaker for 2 h. Samples were quenched by incubation for 30 min at 4 °C in quenching buffer (10 mM DTT in PBS). Finally, protein extracts were prepared by centrifuging the cells and resuspension in SDS-lysis (20 mM Tris-HCl pH 7.4, 3.3% sucrose, 0.66% SDS, 1:100 protease inhibitor cocktail), short sonication and heating to 95 °C for 5 min. Protein concentration was determined with the bicinchoninic acid (BCA) method. A total of 5–10 μg protein were loaded on SDS-PAGE for Western blot analysis.

4.8. Subcellular Fractionation

HEK-293 cells were transfected with wt or mutant DRR1 expressing plasmids; after two days of cultivation, cells were trypsinized, washed with PBS, and resuspended in 250 μL hypotonic lysis buffer (10 mM HEPES pH 7.9, 10 mM KCl, 0.5 mM EDTA, 0.1% NP-40, 10% glycerol, 1 mM DTT, and 1:100 protease inhibitor cocktail (freshly added)) per 10 cm dish. After 10 min on ice and brief vortexing, disruption of the outer cell membrane was analyzed in the microscope. Centrifugation was at 6500 rpm for 30 s at 4 °C), followed by transfer of the supernatant (containing the cytosolic proteins) to a fresh tube. The pellet was washed three times with 500 μL hypotonic lysis buffer and the nuclei were lysed by incubating in 200 μL SDS-lysis buffer ($1\times$ diluted from $3\times$ which is: 62.5 mM Tris-HCl pH 7.4, 10% sucrose, 2% SDS, 1:100 protease inhibitor cocktail (freshly added)) 5 min at 95 °C and short sonication. After centrifugation, the supernatant was transferred to a fresh tube. Protein concentration of cytosolic and nuclear fractions was determined by BCA and the samples were analyzed by SDS-PAGE and Western blot. About 7–10 μg cytosolic fraction and the same volume of the corresponding nuclear fraction were loaded.

4.9. HeLa Cell Spreading

HeLa cells were transfected with DRR1 in pEGFP (using TurboFect), harvested on the next day and replated on fibronectin-coated (50 μg/mL) 12 mm round coverslips in 24-well plates (Merck Millipore, Darmstadt, Germany). Cell spreading was stopped after 30 min of spreading at 37 °C by fixing cells with 4% paraformaldehyde for 20 min at room temperature. The actin cytoskeleton was stained with Alexa Fluore 594-phalloidin and stained cells were placed on glass slides with a drop of Prolong Gold Antifade Medium (Life Technologies). A laser scanning microscope was used for analysis ($10\times/0.40$ NA or $40\times/1.15$ NA objective, LSM FV-1000, Olympus, Shinjuku, Japan). Up to 10 fields were randomly selected with 50–100 cells each and analyzed with ImageJ software. The images were scaled, the phalloidin channel was thresholded (lower threshold level 250, upper threshold level 4095 for a 16-bit image) and adjacent cells were separated by the "Watershed" algorithm (controlling manually for correct cell separation). The thresholded phalloidin channel was used determine cell size in the original phallidin channel and to measure the mean gray value per cell in the EGFP channel using the "Analyze Particles" algorithm. Cells with a mean gray value > 500 in the EGFP channel were determined to be transfected. Cells with a mean gray value of > 500 were defined as transfected. In order to compare different conditions, the mean area of transfected cell was normalized with the mean area of untransfected cells in the same condition.

4.10. Fluorescence Recovery after Photobleaching (FRAP)

FRAP was analyzed in HeLa cells plated on 50 μg/mL fibronectin-coated 35 mm glass dishes transfected with TurboFect. Per glass dish, 1 μg GFP-actin was cotransfected with 3 μg (unlabeled) DRR1 constructs in pRK5. After 24 h, the medium was changed to fresh medium. Time-lapse image frames with 2 s intervals for 5 min were acquired in the confocal microscope ($20\times/0.8$ NA objective, $5\times$ zoom, C.A. 200 μm, LSM FV-1000, 2% laser power). Prior to bleaching, five frames were recorded. At an image resolution of 320×320 px, bleaching was performed with the circular "TurboTool" for 1000 ms at 100% 488-laser power. This led to a bleach of the GFP-fluorescence of about 80%. ImageJ software was used to quantify FRAP. First, the mean gray value measured in the bleached area was divided through an equally-sized arbitrary non-bleached area within the same cell in each frame to correct for acquisition photobleaching and possible laser fluctuations. In order to normalize each fluorescence recovery curve to compare different cells and conditions, the mean gray value of the bleached area immediately after bleaching (C(t0)) was set to 0, while the pre-bleached value (C(pre)) was set to one. The mean gray value of each time point was, thus, normalized using the following formula described in [92]: N(t) = [C(t) − C(t0)]/[C(pre) − C(t0)]. The average gray values in the time lapse imaging was finally averaged across different cells and experiments and plotted as depicted. "N" refers to the number of cells analyzed from 2–3 independent experiments.

4.11. Cellular Stainings

HeLa cells were seeded on 50 µg/mL fibronectin-coated glass coverslips (Merck Millipore), placed in 24-well-plates and transfected using TurboFect as described above. For immunofluorescence, cells were fixed 24 h after transfection with 4% paraformaldehyde in PBS for 20 min at room temperature. Cells were permeabilized with 0.1% Triton X-100 for 10 min and blocked with 10% goat serum for 1 h (both in PBS at room temperature). Primary antibodies were diluted 1:200 in 0.1% Triton X-100/PBS and incubated overnight at 4 °C. Secondary antibodies were diluted 1:500 in 0.1% Triton X-100/PBS and incubated for 3 h at room temperature. The following antibodies were used: rabbit-anti-DRR1 (Biogenes, Berlin, Germany), Alexa Fluor 647-goat-anti-rabbit (Life Technologies). For staining of actin filaments, cells were incubated with 165 nM AlexaFluor 594, or 546 phalloidin (Life Technologies) in PBS for 20 min at room temperature. Nuclei were counterstained with 1 µg/mL DAPI in PBS for 15 min at room temperature.

4.12. Colocalization Analysis

Images for colocalization analysis were taken with the 40×/1.15 NA objective, 3× zoom, and a pinhole of 200 µm. Colocalization analysis was performed in ImageJ with the plugin "Coloc 2" using individual cells as ROI and a point spread function of 4.25. Pearson's correlation coefficient (PCC) R and Costes *p* value were calculated using 100 Costes randomizations. For each condition, 5–15 randomly-selected cells from two independent experiments were analyzed.

4.13. Quantification of Mean Cellular F-actin Content

Quantification of mean cellular F-actin content was performed in ImageJ. Shortly, the images were scaled, the phalloidin channel was thresholded (lower threshold level 150, upper threshold level 4095) and adjacent cells were separated by the "Watershed" algorithm. Correct cell separation was double-checked manually. By means of the thresholded phalloidin channel, the mean grey value of individual cells in both original channels was measured. Cells with a mean gray value of >500 were defined as transfected. For each condition, the mean gray value of F-actin in transfected cells was normalized to the untransfected cells in order to compare different samples.

4.14. Actin Preparation

G-actin was obtained from rabbit skeletal muscle actin and labeled with pyrene, as described previously [67,93]. All in vitro actin experiments except the F-actin co-sedimentation were performed in the lab of Andreas R. Bausch. In all experiments, "R" refers to the molar ratio of recombinant protein and actin. Experiments were performed under reducing conditions with 1 mM DTT.

4.15. Pyrene-Actin Polymerization Assay

Actin polymerization was monitored by the increase in fluorescence of 20% pyrenyl-actin at 407 nm (excitation at 365 nm) in a fluorescence spectrometer (Jasco FP-8500, Gross-Umstadt, Germany). The final concentration of actin in the reaction was 4 µM. DRR1 proteins were added to G-actin in a constant volume and polymerization was induced by the addition of 1:10 volume of 10× F-buffer

(250 mM Tris-HCl pH 7.4, 250 mM KCl, 40 mM $MgCl_2$, 10 mM EGTA, 10 mM ATP, 10 mM DTT). The polymerization was monitored for 1 h at 21 °C with a cycle interval of 5.5 s.

4.16. Actin-Filament Elongation and Nucleation Assay

For visualization of single filament polymerization samples containing 1× F-buffer and recombinant proteins (in a constant volume) were prepared. Polymerization was induced by the addition of G-actin (0.5 or 1 µM final concentration). The sample was then immediately pipetted into a flow chamber consisting of two high precision coverslips (60 × 24 mm and 20 × 20 mm, Carl Roth, Karlsruhe, Germany) separated by vacuum grease and placed in a TIRF or confocal microscope (TIRF: Leica DMI6000B, 100×/1.47 NA oil immersion objective, confocal: 63×/1.4 NA oil immersion objective, 5× optical zoom, Leica TSC SP5, Solms, Germany). Samples prepared with 10% actin-ATTO488 were visualized in a TIRF microscope, samples with Alexa Fluor 488-phalloidin were visualized in the confocal microscope.

To avoid unspecific surface interactions casein was added to the samples in 0.15 mg/mL. The larger coverslips were previously cleaned with a plasma cleaner (40–50 s at 4–6 mbar) and N-ethylmaleimide-modified heavy meromyosin (NEM-HMM, 2.7 µg/mL diluted in F-buffer) was bound to the surface to keep actin filaments close to the surface during live visualization. The chambers were washed with 1× F-buffer prior to applying the sample. The time between the addition of actin to the sample and initiation of the visualization was 2 min. Time-lapse images of polymerization were acquired for 10 min every 3 s.

The image analysis for single filament elongation rate was performed with ImageJ. In all images the background was subtracted and, subsequently, brightness and contrast were adjusted if necessary for ideal visualization. Shortly thereafter, a segmented line was drawn along the filament and plotted time versus filament length (i.e., fluorescence intensity) with the plugin "multiple kymograph". The slope of this linear graph corresponds to the filament elongation rate at the barbed end and was as follows for the MBP control: about 5.6 actin monomers per second in the TIRF assay (0.5 µM actin), and 11.5 actin monomers per second in the confocal assay (1 µM actin), based on the published value of 0.0027 µm/actin monomer in actin filaments [94]. The filament elongation rate of the MBP control was set to 100% in order to compare different experiments. The polymerization speed strongly depended on the actin preparation. This is in accordance to values described in the literature for ATP-actin at similar buffer conditions [95]. Barbed and pointed ends could be easily distinguished in the ImageJ graph in the control, as the elongation rate is much lower at the pointed end. For samples containing DRR1 this was not possible due to the strong inhibition of elongation. Thus, profilin (R = 3) was added to the samples to block pointed end elongation. Ten filaments from three independent experiments were measured for each condition.

For nucleation analysis (both in TIRF and confocal assays), filaments in 4–8 frames with 30 s intervals were counted manually from 3–6 independent experiments, respectively. The number of filaments was plotted versus time of polymerization and the slope of the resulting linear graph was defined as relative nucleation rate. The MBP control was set to 1. MBP exerted no significant effect on polymerization or nucleation in comparison to buffer conditions (see Appendix A Figure A5).

4.17. Reconstituted Actin Networks

Samples with 4 µM actin and 2 µM recombinant DRR1 proteins (R = 0.5, wt also in R = 0.1 and 0.25) at a constant volume were prepared for in vitro actin networks and Alexa Fluor 488-phalloidin was added to visualized actin (0.08 µM, Life Technologies). Furthermore, casein was added to the samples in 0.15 mg/mL to avoid unspecific surface interactions. After induction of polymerization by addition of 1:10 volume of 10× F-buffer, the samples were placed into a flow chamber. The chamber was sealed with vacuum grease and samples were polymerized at room temperature for 1.5–2 h protected from light. At equilibrium of polymerization, networks were visualized using a confocal microscope (63×/1.4 NA oil immersion objective, 1× or 3× optical zoom, Leica TSC SP5). Z-stacks of each sample were taken with a 10 µm depth and slices with 0.38 µm step size and maximum projections of the staces were generated with ImageJ software. In addition, the background of this maximum projections was subtracted and brightness and contrast were automatically adjusted.

4.18. F-Actin Co-Sedimentation Assay

Polymerization of 1 µM G-Actin was induced by the addition of 1:10 volume of $10\times$ F-buffer for 1 h at room temperature. Freshly thawed recombinant DRR1 proteins were added to the F-actin samples at 0.5 µM final concentration in a constant volume and incubation on ice was performed for 30 min to allow proteins to bind to F-actin. The samples were centrifuged at $150.000\times g$ for 1 h at 21 °C (Beckmann LB-70M). An aliquot was taken prior to centrifugation (representing "T" = total protein) and after centrifugation supernatant and pellet samples were collected ("S" = supernatant, "P" = pellet). Supernatant samples contain G-actin and non-sedimented protein, while F-actin and F-actin-binding proteins are found in the pellet. All samples were loaded on SDS-PAGE, stained with colloidal Coomassie and subsequently analyzed by densitometry scanning (ChemiDoc Imaging System, Bio-Rad Laboratories, Munich, Germany). Since the pellet fractions are easily contaminated by residual supernatant proteins, the fraction of co-sedimented protein was calculated by the subtraction of the amount in the supernatant fraction from the amount of total protein determined on the same gel as the more reliable method. To determine the background precipitation of each protein, the proteins were each centrifuged alone (i.e., without the presence of F-actin) as a control. The amount of pelleted protein in these samples ($^{-\text{Actin}}$) was subtracted from the co-sedimented protein amount in the samples with F-actin ($^{+\text{Actin}}$). This value was then related to (i.e., divided by) the total and is presented as % in Figure 1E. Thus, the amount of DRR1 (wt and mutants) and MBP control co-sedimenting with F-actin was determined by ($[(T^{+\text{Actin}} - S^{+\text{Actin}}) - (T^{-\text{Actin}} - S^{-\text{Actin}})]/T^{+\text{Actin}}) \times 100$. This method was highly reproducible and consistent.

Control experiments were performed with ddH$_2$O, buffer and MBP at R = 0.5, respectively. No differences were detected in the samples containing MBP vs. buffer and MBP controls are shown in all panels. Both buffer and MBP sample, however, seemed to slightly slow down actin polymerization in comparison to ddH$_2$O samples (see Appendix A, Figure A5).

4.19. Statistical Analysis

The commercially available program SigmaPlot 14.0 (Erkrath, Germany) was used for statistical analysis. A one-way ANOVA (analysis of variants) was performed followed by Bonferroni *post hoc* analysis for multi-group comparisons. The level of significance was set at $p < 0.05$. All data are presented as mean $\pm$ SEM. Grubb's test (with alpha = 0.05) was run to identify significant outliers.

5. Conclusions

This study characterizes the tumor suppressor and stress-regulated protein DRR1 as actin binding protein that affects several aspects of actin dynamics such as nucleation, elongation, capping and bundling of F-Actin. We are only beginning to understand the cellular and physiological implications. Through induction of DRR1 stress changes the cellular make-up of actin dynamic regulators. The interplay of DRR1 with the array of other actin binding proteins needs further exploration, as well as the cellular effects that are expected to extend beyond G-actin dependent transcription and cell spreading. Future experiments should also refine the analysis of the structural and functional changes of neuronal networks that might contribute to stress-related psychiatric diseases.

Author Contributions: Conceptualization: A.K., J.-P.S., T.R.; data curation: A.K.; formal analysis: A.K.; funding acquisition: A.R.B., T.R.; investigation: A.K.; methodology: A.K., K.D.; project administration: A.K., T.R.; resources: K.D., A.R.B.; supervision: J.-P.S.; T.R.; validation: A.K.; visualization: A.K.; writing—original draft preparation: A.K., T.R.; writing—review and editing: A.K. J.-P.S., M.M., K.D. M.B.M., A.R.B, T.R.

Funding: This research received no external funding.

Acknowledgments: We are indebted to Christine Wurm, Carina Pelzl, Simone Köhler, and Kurt Schmoller (all Technical University Munich) for technical help, Angela Oberhofer (Technical University Munich, Germany) for introduction to TIRF microscopy, and Robert Grosse and Dominique T. Brandt (University of Marburg, Germany) for providing SRF reporter plasmids.

Conflicts of Interest: The authors declare no conflict of interest.

Abbreviations

ANOVA	Analysis of variants
ATP	Adenosine triphosphate
BCA	Bicinchoninic acid
BMB	1,4-Bismaleimidobutane
BSA	Bovine serum albumin
CA3	Cornu Ammonis region 3 (hippocampal region)
CaMKIIb	Ca2+/calmodulin-dependent protein kinase IIb
CMV	Cytomegalovirus
CoIP	Co-immunoprecipitation
DRR1	Down-regulated in renal cell carcinoma 1 (also known as FAM107A or TU3A)
DTT	Dithiothreitol
EDTA	Ethylenediaminetetraacetic acid
EGFP	Enhanced green fluorescent protein
F-actin	Filamentous actin
FAM107A	Family with sequence similarity 107, member A
G-actin	Globular actin
GFP	Green fluorescent protein
HEK-293	Human embryonic kidney 293 cells
MBP	Maltose binding protein
PCC	Pearson's correlation coefficient
PMSF	Phenylmethane sulfonyl fluoride
R	Molar ratio DRR1 (wt or mutants):actin
RT	Room temperature
SDS	Sodium dodecyl sulfate
SDS-PAGE	Sodium dodecyl sulfate polyacrylamide gel electrophoresis
SRF	Serum response factor
TIRF	Total internal reflection fluorescence
TU3A	Tohoku University cDNA clone A on chromosome 3
wt	Wild-type

Appendix A

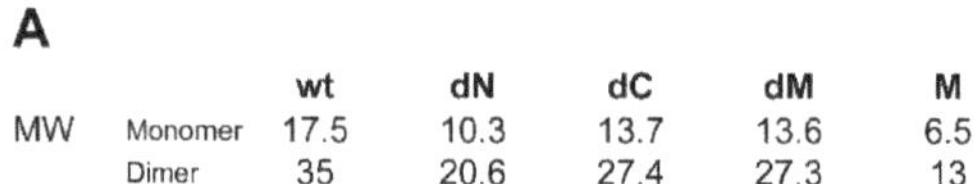

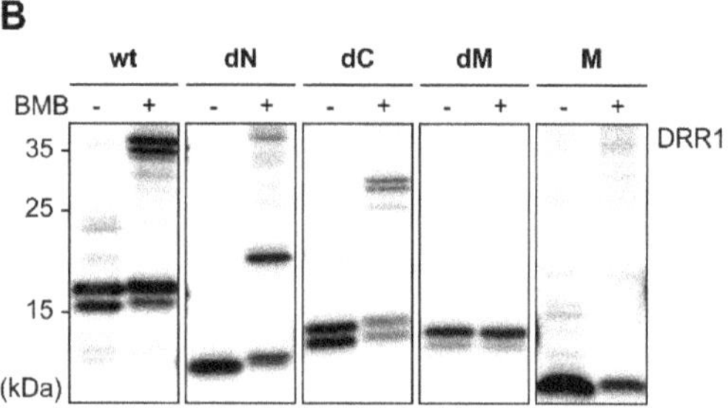

Figure A1. Crosslinking with BMB of DRR1 proteins from HEK-293 cell extracts suggests potential dimerization of DRR1 wt, dN and dC. (**A**) Expected molecular weight (MW) of DRR1 monomers and dimers in kDa; (**B**) HEK-293 cells transfected with wt or mutant DRR1 (no tag) were incubated with the cysteine-cysteine crosslinker BMB (+) or DMSO (−) as a control. Lysates were analyzed by SDS-PAGE and Western blot. Representative Western blots are shown.

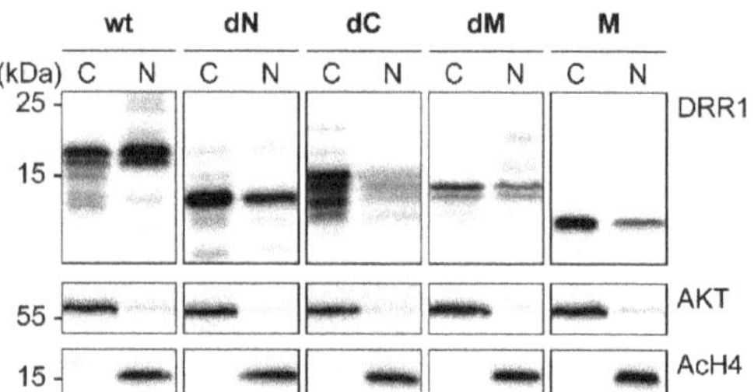

Figure A2. DRR1 wt and mutants display nuclear localization. Cytosolic and nuclear fractions of HEK-293 cells expressing ectopic DRR1 wt or mutants were analyzed by SDS-PAGE and Western blot. Efficiency of fractionation into cytosol (C) and nucleus (N) was confirmed with antibodies against the cytosolic kinase AKT and the nuclear acetyl-histone H4. Representative Western blots are shown.

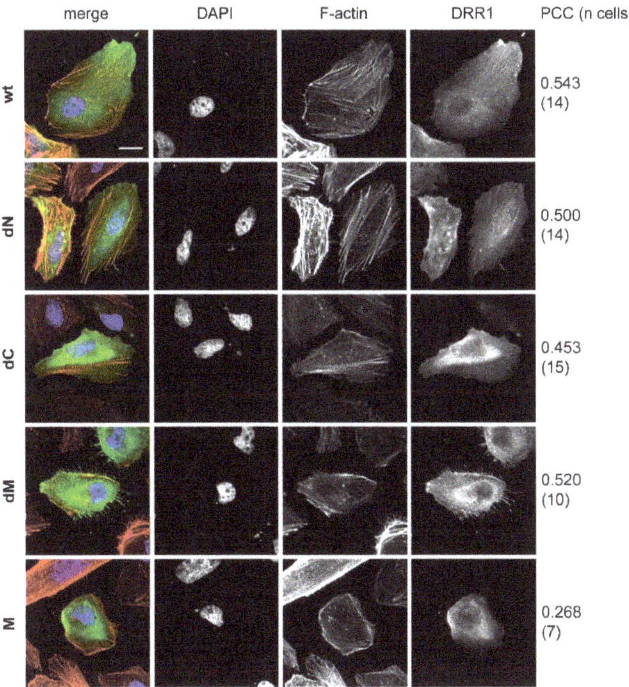

Figure A3. DRR1 wt and the mutants dN, dC and dM co-localize with F-actin in HeLa cells, particularly along stress fibers. HeLa cells were transfected with DRR1 plasmids or vector control and fixed 24 h after transfection. Nuclei were counterstained with DAPI (blue), F-actin with phalloidin (red) and DRR1 with a specific antibody (green). Representative cells are shown. Scale bar denotes 20 µm. Pearson's correlation coefficient (PCC) and the number of cells used for colocalization analysis (n, from two independent experiments) are shown. DRR1 wt and the mutants dN, dC, and dM exhibit colocalization with PCCs above 0.4, which generally can be regarded as a strong positive correlation.

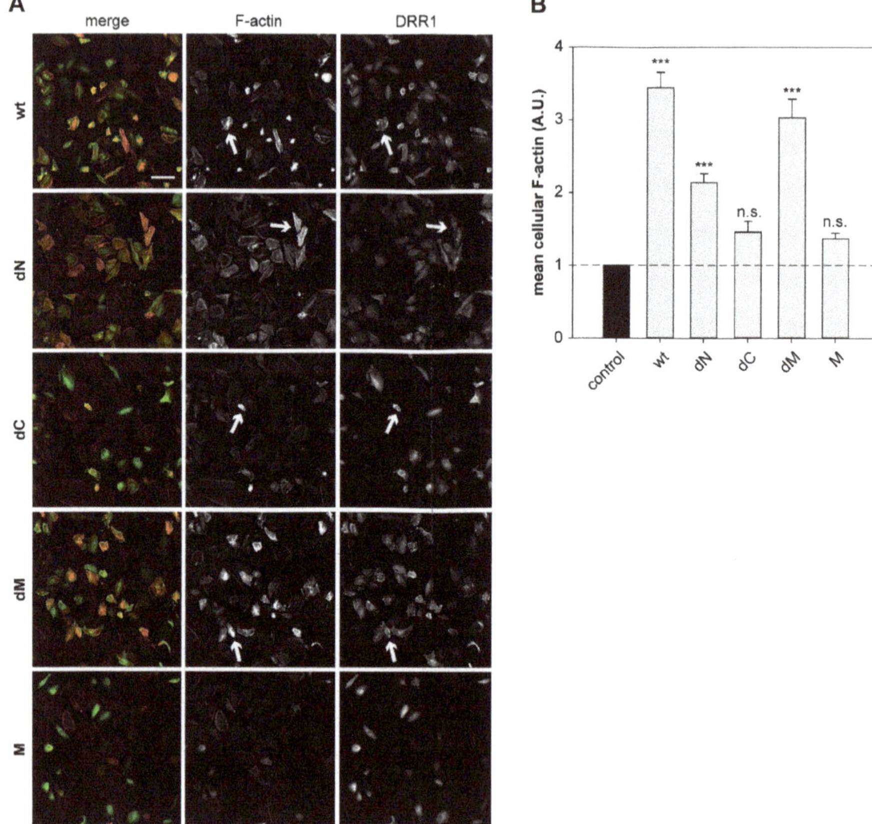

Figure A4. DRR1 modulates F-actin content in cultured cells. (**A**) HeLa cells were transfected with DRR1 plasmids and fixed 24 h after transfection. F-actin was stained with phalloidin (red) and DRR1 with a specific antibody (green). Scale bar denotes 100 μm. White arrows indicate exemplary cells with increased F-actin correlating with high DRR1 wt or mutant expression. (**B**) Mean F-actin per cell is increased upon overexpression of DRR1 wt and the mutants dN and dM. Quantification of mean cellular F-actin from HeLa cells described in A was performed and values of transfected cells were normalized to untransfected cells in the same image (number of cells in one experiment, control $n = 200$, DRR1 wt $n = 102$, dN $n = 93$, dC $n = 38$, dM $n = 54$, M $n = 52$). Bars represent means + SEM of five independent experiments. ** $p < 0.01$ in comparison to control. Statistical analysis was performed with one-way ANOVA and Bonferroni post hoc.

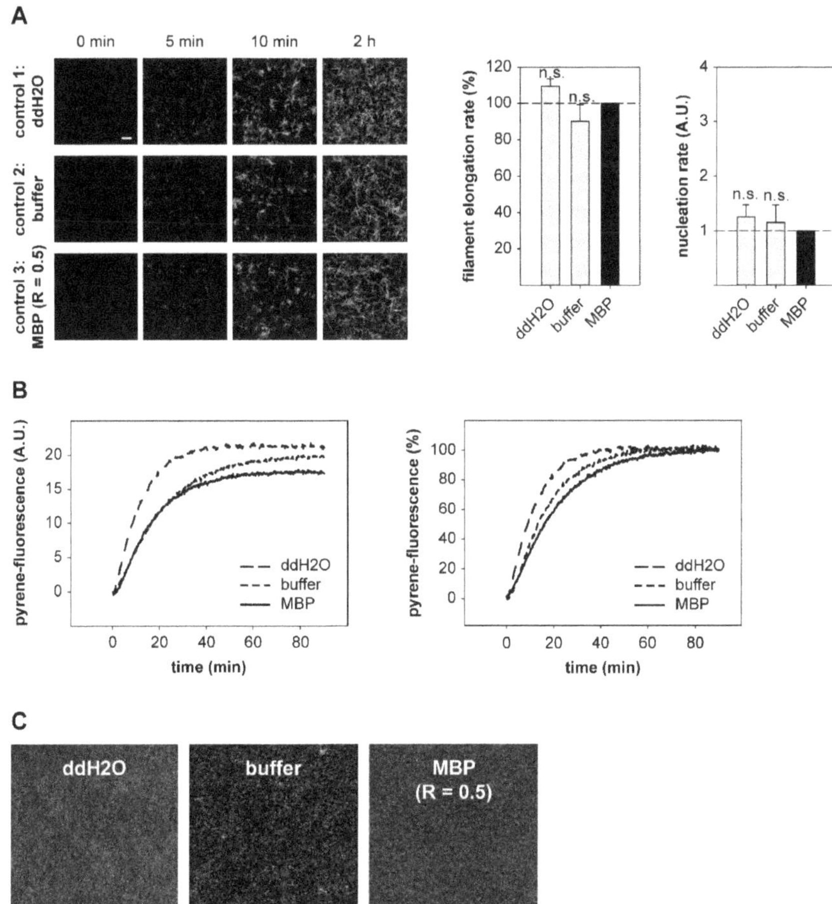

Figure A5. Overview of the controls used in all in vitro actin assays: ddH_2O, buffer, and purified MBP protein. In all controls ddH_2O and buffer were added in a constant volume, and MBP was added at the same ratio as DRR1 at the highest concentration, i.e., at R = MBP:actin = 0.5. (**A**) In vitro polymerization of actin visualized by TIRF microscopy for 10 min with 3 s intervals starting 2 min after the beginning of the reaction. An endpoint image was taken at 2 h of polymerization. Scale bar denotes 10 μm. Quantification of filament elongation rate normalized to the control (*n* = 3). Nucleation rate defined as the slope of number of filaments per frame quantified in 3–5 frames (with 30 s intervals) normalized to the control (*n* = 3). Statistical analysis was performed with one-way ANOVA and Bonferroni post hoc. n.s. = not significant in comparison to control (MBP). (**B**) Pyrene-polymerization assay. Actin filaments (5 μM, 20% pyrenyl-labeled) were polymerized and the increase in fluorescence of pyrene-actin during polymerization was monitored for 60 min. Representative curves are shown (**left**: raw background-corrected data, **right**: data normalized to endpoint). The buffer slightly slows down actin polymerization in comparison to the control with ddH_2O, while there appears to be only marginal variation between the control with MBP (R = 0.5) and the buffer. (**C**) Actin networks (4 μM) were polymerized at RT for > 2 h and visualized with phalloidin-488. Images were taken in a confocal microscope (63× /1.4 NA objective, 10 μm z-stacks). Scale bar denotes 50 μm.

References

1. McEwen, B.S.; Bowles, N.P.; Gray, J.D.; Hill, M.N.; Hunter, R.G.; Karatsoreos, I.N.; Nasca, C. Mechanisms of stress in the brain. *Nat. Neurosci.* **2015**, *18*, 1353–1363. [CrossRef] [PubMed]
2. De Kloet, E.R.; Joels, M.; Holsboer, F. Stress and the brain: From adaptation to disease. *Nat. Rev. Neurosci.* **2005**, *6*, 463–475. [CrossRef] [PubMed]
3. Fenster, R.J.; Lebois, L.A.M.; Ressler, K.J.; Suh, J. Brain circuit dysfunction in post-traumatic stress disorder: From mouse to man. *Nat. Rev. Neurosci.* **2018**, *19*, 535–551. [CrossRef] [PubMed]
4. Cingolani, L.A.; Goda, Y. Actin in action: The interplay between the actin cytoskeleton and synaptic efficacy. *Nat. Rev. Neurosci.* **2008**, *9*, 344–356. [CrossRef] [PubMed]
5. Van der Kooij, M.A.; Masana, M.; Rust, M.B.; Müller, M.B. The stressed cytoskeleton: How actin dynamics can shape stress-related consequences on synaptic plasticity and complex behavior. *Neurosci. Biobehav. Rev.* **2016**, *62*, 69–75. [CrossRef] [PubMed]
6. Fuchs, E.; Flügge, G. Stress, glucocorticoids and structural plasticity of the hippocampus. *Neurosci. Biobehav. Rev.* **1998**, *23*, 295–300. [CrossRef]
7. Kasai, H.; Fukuda, M.; Watanabe, S.; Hayashi-Takagi, A.; Noguchi, J. Structural dynamics of dendritic spines in memory and cognition. *Trends Neurosci.* **2010**, *33*, 121–129. [CrossRef]
8. Yan, Z.; Kim, E.; Datta, D.; Lewis, D.A.; Soderling, S.H. Synaptic Actin Dysregulation, a Convergent Mechanism of Mental Disorders? *J. Neurosci.* **2016**, *36*, 11411–11417. [CrossRef]
9. Davidson, A.J.; Wood, W. Unravelling the Actin Cytoskeleton: A New Competitive Edge? *Trends Cell Biol.* **2016**, *26*, 569–576. [CrossRef]
10. Kuhn, J.R.; Pollard, T.D. Real-time measurements of actin filament polymerization by total internal reflection fluorescence microscopy. *Biophys. J.* **2005**, *88*, 1387–1402. [CrossRef]
11. Harris, E.S.; Higgs, H.N. Biochemical analysis of mammalian formin effects on actin dynamics. *Methods Enzymol.* **2006**, *406*, 190–214. [PubMed]
12. Dominguez, R. Structural insights into de novo actin polymerization. *Curr. Opin. Struct. Biol.* **2010**, *20*, 217–225. [CrossRef] [PubMed]
13. Mullins, R.D.; Heuser, J.A.; Pollard, T.D. The interaction of Arp2/3 complex with actin: Nucleation, high affinity pointed end capping, and formation of branching networks of filaments. *Proc. Natl. Acad. Sci. USA* **1998**, *95*, 6181–6186. [CrossRef] [PubMed]
14. Cooper, J.A.; Pollard, T.D. Effect of capping protein on the kinetics of actin polymerization. *Biochemistry* **1985**, *24*, 793–799. [CrossRef] [PubMed]
15. Edwards, M.; Zwolak, A.; Schafer, D.A.; Sept, D.; Dominguez, R.; Cooper, J.A. Capping protein regulators fine-tune actin assembly dynamics. *Nat. Rev. Mol. Cell Biol.* **2014**, *15*, 677–689. [CrossRef] [PubMed]
16. Jayo, A.; Parsons, M. Fascin: A key regulator of cytoskeletal dynamics. *Int. J. Biochem. Cell Biol.* **2010**, *42*, 1614–1617. [CrossRef]
17. Nakamura, F.; Stossel, T.P.; Hartwig, J.H. The filamins: Organizers of cell structure and function. *Cell Adh. Migr.* **2011**, *5*, 160–169. [CrossRef]
18. Sjöblom, B.; Salmazo, A.; Djinovic-Carugo, K. Alpha-actinin structure and regulation. *Cell Mol. Life Sci.* **2008**, *65*, 2688–2701. [CrossRef]
19. Posern, G.; Treisman, R. Actin' together: Serum response factor, its cofactors and the link to signal transduction. *Trends Cell Biol.* **2006**, *16*, 588–596. [CrossRef]
20. Esnault, C.; Stewart, A.; Gualdrini, F.; East, P.; Horswell, S.; Matthews, N.; Treisman, R. Rho-actin signaling to the MRTF coactivators dominates the immediate transcriptional response to serum in fibroblasts. *Genes Dev.* **2014**, *28*, 943–958. [CrossRef]
21. Liebl, C.; Panhuysen, M.; Putz, B.; Trumbach, D.; Wurst, W.; Deussing, J.M.; Muller, M.B.; Schmidt, M.V. Gene expression profiling following maternal deprivation: Involvement of the brain Renin-Angiotensin system. *Front. Mol. Neurosci.* **2009**, *2*, 1. [CrossRef] [PubMed]
22. Masana, M.; Su, Y.A.; Liebl, C.; Wang, X.D.; Jansen, L.; Westerholz, S.; Wagner, K.V.; Labermaier, C.; Scharf, S.H.; Santarelli, S.; et al. The stress-inducible actin-interacting protein DRR1 shapes social behavior. *Psychoneuroendocrinology* **2014**, *48*, 98–110. [CrossRef] [PubMed]

23. Schmidt, M.V.; Schülke, J.P.; Liebl, C.; Stiess, M.; Avrabos, C.; Bock, J.; Wochnik, G.M.; Davies, H.A.; Zimmermann, N.; Scharf, S.H.; et al. Tumor suppressor down-regulated in renal cell carcinoma 1 (DRR1) is a stress-induced actin bundling factor that modulates synaptic efficacy and cognition. *Proc. Natl. Acad. Sci. USA* **2011**, *108*, 17213–17218. [CrossRef] [PubMed]

24. Yamato, T.; Orikasa, K.; Fukushige, S.; Orikasa, S.; Horii, A. Isolation and characterization of the novel gene, TU3A, in a commonly deleted region on 3p14.3–>p14.2 in renal cell carcinoma. *Cytogenet. Cell Genet.* **1999**, *87*, 291–295. [CrossRef] [PubMed]

25. Wang, L.; Darling, J.; Zhang, J.S.; Liu, W.; Qian, J.; Bostwick, D.; Hartmann, L.; Jenkins, R.; Bardenhauer, W.; Schutte, J.; et al. Loss of expression of the DRR 1 gene at chromosomal segment 3p21.1 in renal cell carcinoma. *Genes Chromosom. Cancer* **2000**, *27*, 1–10. [CrossRef]

26. Udali, S.; Guarini, P.; Ruzzenente, A.; Ferrarini, A.; Guglielmi, A.; Lotto, V.; Tononi, P.; Pattini, P.; Moruzzi, S.; Campagnaro, T.; et al. DNA methylation and gene expression profiles show novel regulatory pathways in hepatocellular carcinoma. *Clin. Epigenetics* **2015**, *7*, 43. [CrossRef]

27. Kiwerska, K.; Szaumkessel, M.; Paczkowska, J.; Bodnar, M.; Byzia, E.; Kowal, E.; Kostrzewska-Poczekaj, M.; Janiszewska, J.; Bednarek, K.; Jarmuz-Szymczak, M.; et al. Combined deletion and DNA methylation result in silencing of FAM107A gene in laryngeal tumors. *Sci. Rep.* **2017**, *7*, 5386. [CrossRef]

28. Lawrie, A.; Han, S.; Sud, A.; Hosking, F.; Cezard, T.; Turner, D.; Clark, C.; Murray, G.I.; Culligan, D.J.; Houlston, R.S.; et al. Combined linkage and association analysis of classical Hodgkin lymphoma. *Oncotarget* **2018**, *9*, 20377–20385. [CrossRef]

29. Pastuszak-Lewandoska, D.; Czarnecka, K.H.; Migdalska-Sek, M.; Nawrot, E.; Domanska, D.; Kiszalkiewicz, J.; Kordiak, J.; Antczak, A.; Gorski, P.; Brzezianska-Lasota, E. Decreased FAM107A Expression in Patients with Non-small Cell Lung Cancer. *Adv. Exp. Med. Biol.* **2015**, *852*, 39–48.

30. Mu, P.; Akashi, T.; Lu, F.; Kishida, S.; Kadomatsu, K. A novel nuclear complex of DRR1, F-actin and COMMD1 involved in NF-kappaB degradation and cell growth suppression in neuroblastoma. *Oncogene* **2017**, *36*, 5745–5756. [CrossRef]

31. Asano, Y.; Kishida, S.; Mu, P.; Sakamoto, K.; Murohara, T.; Kadomatsu, K. DRR1 is expressed in the developing nervous system and downregulated during neuroblastoma carcinogenesis. *Biochem. Biophys. Res. Commun.* **2010**, *394*, 829–835. [CrossRef] [PubMed]

32. Fèvre-Montange, M.; Champier, J.; Durand, A.; Wierinckx, A.; Honnorat, J.; Guyotat, J.; Jouvet, A. Microarray gene expression profiling in meningiomas: Differential expression according to grade or histopathological subtype. *Int. J. Oncol.* **2009**, *35*, 1395–1407. [CrossRef] [PubMed]

33. Kholodnyuk, I.D.; Kozireva, S.; Kost-Alimova, M.; Kashuba, V.; Klein, G.; Imreh, S. Down regulation of 3p genes, LTF, SLC38A3 and DRR1, upon growth of human chromosome 3-mouse fibrosarcoma hybrids in severe combined immunodeficiency mice. *Int. J. Cancer* **2006**, *119*, 99–107. [CrossRef] [PubMed]

34. Liu, Q.; Zhao, X.Y.; Bai, R.Z.; Liang, S.F.; Nie, C.L.; Yuan, Z.; Wang, C.T.; Wu, Y.; Chen, L.J.; Wei, Y.Q. Induction of tumor inhibition and apoptosis by a candidate tumor suppressor gene DRR1 on 3p21.1. *Oncol. Rep.* **2009**, *22*, 1069–1075. [PubMed]

35. Vanaja, D.K.; Ballman, K.V.; Morlan, B.W.; Cheville, J.C.; Neumann, R.M.; Lieber, M.M.; Tindall, D.J.; Young, C.Y. PDLIM4 repression by hypermethylation as a potential biomarker for prostate cancer. *Clin. Cancer Res.* **2006**, *12*, 1128–1136. [CrossRef]

36. Van den, B.J.; Wolter, M.; Blaschke, B.; Knobbe, C.B.; Reifenberger, G. Identification of novel genes associated with astrocytoma progression using suppression subtractive hybridization and real-time reverse transcription-polymerase chain reaction. *Int. J. Cancer* **2006**, *119*, 2330–2338. [CrossRef] [PubMed]

37. Pollen, A.A.; Nowakowski, T.J.; Chen, J.; Retallack, H.; Sandoval-Espinosa, C.; Nicholas, C.R.; Shuga, J.; Liu, S.J.; Oldham, M.C.; Diaz, A.; et al. Molecular identity of human outer radial glia during cortical development. *Cell* **2015**, *163*, 55–67. [CrossRef]

38. Ma, Y.S.; Wu, Z.J.; Bai, R.Z.; Dong, H.; Xie, B.X.; Wu, X.H.; Hang, X.S.; Liu, A.N.; Jiang, X.H.; Wang, G.R.; et al. DRR1 promotes glioblastoma cell invasion and epithelial-mesenchymal transition via regulating AKT activation. *Cancer Lett.* **2018**, *423*, 86–94. [CrossRef]

39. Le, P.U.; Angers-Loustau, A.; de Oliveira, R.M.; Ajlan, A.; Brassard, C.L.; Dudley, A.; Brent, H.; Siu, V.; Trinh, G.; Molenkamp, G.; et al. DRR drives brain cancer invasion by regulating cytoskeletal-focal adhesion dynamics. *Oncogene* **2010**, *29*, 4636–4647. [CrossRef]

40. Quartier, A.; Chatrousse, L.; Redin, C.; Keime, C.; Haumesser, N.; Maglott-Roth, A.; Brino, L.; Le, G.S.; Benchoua, A.; Mandel, J.L.; et al. Genes and Pathways Regulated by Androgens in Human Neural Cells, Potential Candidates for the Male Excess in Autism Spectrum Disorder. *Biol. Psychiatry* **2018**, *84*, 239–252. [CrossRef]

41. Wan, B.; Feng, P.; Guan, Z.; Sheng, L.; Liu, Z.; Hua, Y. A severe mouse model of spinal muscular atrophy develops early systemic inflammation. *Hum. Mol. Genet.* **2018**, *27*. [CrossRef] [PubMed]

42. Li, M.D.; Burns, T.C.; Morgan, A.A.; Khatri, P. Integrated multi-cohort transcriptional meta-analysis of neurodegenerative diseases. *Acta Neuropathol. Commun.* **2014**, *2*, 93. [CrossRef] [PubMed]

43. Shao, L.; Vawter, M.P. Shared gene expression alterations in schizophrenia and bipolar disorder. *Biol. Psychiatry* **2008**, *64*, 89–97. [CrossRef] [PubMed]

44. Ranno, E.; D'Antoni, S.; Spatuzza, M.; Berretta, A.; Laureanti, F.; Bonaccorso, C.M.; Pellitteri, R.; Longone, P.; Spalloni, A.; Iyer, A.M.; et al. Endothelin-1 is over-expressed in amyotrophic lateral sclerosis and induces motor neuron cell death. *Neurobiol. Dis.* **2014**, *65*, 160–171. [CrossRef] [PubMed]

45. Murray, L.M.; Lee, S.; Baumer, D.; Parson, S.H.; Talbot, K.; Gillingwater, T.H. Pre-symptomatic development of lower motor neuron connectivity in a mouse model of severe spinal muscular atrophy. *Hum. Mol. Genet.* **2010**, *19*, 420–433. [CrossRef] [PubMed]

46. Shin, J.; Salameh, J.S.; Richter, J.D. Impaired neurodevelopment by the low complexity domain of CPEB4 reveals a convergent pathway with neurodegeneration. *Sci. Rep.* **2016**, *6*, 29395. [CrossRef] [PubMed]

47. Masana, M.; Westerholz, S.; Kretzschmar, A.; Treccani, G.; Liebl, C.; Santarelli, S.; Dournes, C.; Popoli, M.; Schmidt, M.V.; Rein, T.; et al. Expression and glucocorticoid-dependent regulation of the stress-inducible protein DRR1 in the mouse adult brain. *Brain Struct. Funct.* **2018**, *223*, 4039–4052. [CrossRef]

48. Masana, M.; Jukic, M.M.; Kretzschmar, A.; Wagner, K.V.; Westerholz, S.; Schmidt, M.V.; Rein, T.; Brodski, C.; Muller, M.B. Deciphering the spatio-temporal expression and stress regulation of Fam107B, the paralog of the resilience-promoting protein DRR1 in the mouse brain. *Neuroscience* **2015**, *290*, 147–158. [CrossRef]

49. Stankiewicz, A.M.; Goscik, J.; Swiergiel, A.H.; Majewska, A.; Wieczorek, M.; Juszczak, G.R.; Lisowski, P. Social stress increases expression of hemoglobin genes in mouse prefrontal cortex. *BMC Neurosci.* **2014**, *15*, 130. [CrossRef]

50. Jene, T.; Gassen, N.C.; Opitz, V.; Endres, K.; Muller, M.B.; van der Kooij, M.A. Temporal profiling of an acute stress-induced behavioral phenotype in mice and role of hippocampal DRR1. *Psychoneuroendocrinology* **2018**, *91*, 149–158. [CrossRef]

51. Lupas, A.N.; Gruber, M. The structure of alpha-helical coiled coils. *Adv. Protein Chem.* **2005**, *70*, 37–78. [PubMed]

52. Liu, J.; Rost, B. Comparing function and structure between entire proteomes. *Protein Sci.* **2001**, *10*, 1970–1979. [CrossRef] [PubMed]

53. Falahzadeh, K.; Banaei-Esfahani, A.; Shahhoseini, M. The potential roles of actin in the nucleus. *Cell J.* **2015**, *17*, 7–14. [PubMed]

54. Zhao, X.Y.; Liang, S.F.; Yao, S.H.; Ma, F.X.; Hu, Z.G.; Yan, F.; Yuan, Z.; Ruan, X.Z.; Yang, H.S.; Zhou, Q.; et al. Identification and preliminary function study of Xenopus laevis DRR1 gene. *Biochem. Biophys. Res. Commun.* **2007**, *361*, 74–78. [CrossRef]

55. Siton-Mendelson, O.; Bernheim-Groswasser, A. Functional Actin Networks under Construction: The Cooperative Action of Actin Nucleation and Elongation Factors. *Trends Biochem. Sci.* **2017**, *42*, 414–430. [CrossRef] [PubMed]

56. Li, F.; Higgs, H.N. The mouse Formin mDia1 is a potent actin nucleation factor regulated by autoinhibition. *Curr. Biol.* **2003**, *13*, 1335–1340. [CrossRef]

57. Dominguez, R.; Holmes, K.C. Actin structure and function. *Annu. Rev. Biophys.* **2011**, *40*, 169–186. [CrossRef]

58. Hertzog, M.; Milanesi, F.; Hazelwood, L.; Disanza, A.; Liu, H.; Perlade, E.; Malabarba, M.G.; Pasqualato, S.; Maiolica, A.; Confalonieri, S.; et al. Molecular basis for the dual function of Eps8 on actin dynamics: Bundling and capping. *PLoS Biol.* **2010**, *8*, e1000387. [CrossRef]

59. Gremm, D.; Wegner, A. Gelsolin as a calcium-regulated actin filament-capping protein. *Eur. J. Biochem.* **2000**, *267*, 4339–4345. [CrossRef]

60. Schirenbeck, A.; Bretschneider, T.; Arasada, R.; Schleicher, M.; Faix, J. The Diaphanous-related formin dDia2 is required for the formation and maintenance of filopodia. *Nat. Cell Biol.* **2005**, *7*, 619–625. [CrossRef]

61. Paavilainen, V.O.; Oksanen, E.; Goldman, A.; Lappalainen, P. Structure of the actin-depolymerizing factor homology domain in complex with actin. *J. Cell Biol.* **2008**, *182*, 51–59. [CrossRef]

62. Paavilainen, V.O.; Hellman, M.; Helfer, E.; Bovellan, M.; Annila, A.; Carlier, M.F.; Permi, P.; Lappalainen, P. Structural basis and evolutionary origin of actin filament capping by twinfilin. *Proc. Natl. Acad. Sci. USA* **2007**, *104*, 3113–3118. [CrossRef] [PubMed]

63. Way, M.; Pope, B.; Weeds, A.G. Evidence for functional homology in the F-actin binding domains of gelsolin and alpha-actinin: Implications for the requirements of severing and capping. *J. Cell Biol.* **1992**, *119*, 835–842. [CrossRef] [PubMed]

64. Robinson, R.C.; Mejillano, M.; Le, V.P.; Burtnick, L.D.; Yin, H.L.; Choe, S. Domain movement in gelsolin: A calcium-activated switch. *Science* **1999**, *286*, 1939–1942. [CrossRef] [PubMed]

65. McGough, A.; Chiu, W.; Way, M. Determination of the gelsolin binding site on F-actin: Implications for severing and capping. *Biophys. J.* **1998**, *74*, 764–772. [CrossRef]

66. Falzone, T.T.; Lenz, M.; Kovar, D.R.; Gardel, M.L. Assembly kinetics determine the architecture of alpha-actinin crosslinked F-actin networks. *Nat. Commun.* **2012**, *3*, 861. [CrossRef] [PubMed]

67. Schmoller, K.M.; Semmrich, C.; Bausch, A.R. Slow down of actin depolymerization by cross-linking molecules. *J. Struct. Biol.* **2011**, *173*, 350–357. [CrossRef]

68. Chamaraux, F.; Fache, S.; Bruckert, F.; Fourcade, B. Kinetics of cell spreading. *Phys. Rev. Lett.* **2005**, *94*, 158102. [CrossRef]

69. Li, J.; Han, D.; Zhao, Y.P. Kinetic behaviour of the cells touching substrate: The interfacial stiffness guides cell spreading. *Sci. Rep.* **2014**, *4*, 3910. [CrossRef]

70. Yauch, R.L.; Felsenfeld, D.P.; Kraeft, S.K.; Chen, L.B.; Sheetz, M.P.; Hemler, M.E. Mutational evidence for control of cell adhesion through integrin diffusion/clustering, independent of ligand binding. *J. Exp. Med.* **1997**, *186*, 1347–1355. [CrossRef]

71. Cuvelier, D.; Théry, M.; Chu, Y.S.; Dufour, S.; Thiery, J.P.; Bornens, M.; Nassoy, P.; Mahadevan, L. The universal dynamics of cell spreading. *Curr. Biol.* **2007**, *17*, 694–699. [CrossRef]

72. Vivo, M.; Fontana, R.; Ranieri, M.; Capasso, G.; Angrisano, T.; Pollice, A.; Calabro, V.; La, M.G. p14ARF interacts with the focal adhesion kinase and protects cells from anoikis. *Oncogene* **2017**, *36*, 4913–4928. [CrossRef] [PubMed]

73. Frost, N.A.; Shroff, H.; Kong, H.; Betzig, E.; Blanpied, T.A. Single-molecule discrimination of discrete perisynaptic and distributed sites of actin filament assembly within dendritic spines. *Neuron* **2010**, *67*, 86–99. [CrossRef] [PubMed]

74. Honkura, N.; Matsuzaki, M.; Noguchi, J.; Ellis-Davies, G.C.; Kasai, H. The subspine organization of actin fibers regulates the structure and plasticity of dendritic spines. *Neuron* **2008**, *57*, 719–729. [CrossRef] [PubMed]

75. Star, E.N.; Kwiatkowski, D.J.; Murthy, V.N. Rapid turnover of actin in dendritic spines and its regulation by activity. *Nat. Neurosci.* **2002**, *5*, 239–246. [CrossRef] [PubMed]

76. Okamoto, K.; Narayanan, R.; Lee, S.H.; Murata, K.; Hayashi, Y. The role of CaMKII as an F-actin-bundling protein crucial for maintenance of dendritic spine structure. *Proc. Natl. Acad. Sci. USA* **2007**, *104*, 6418–6423. [CrossRef] [PubMed]

77. Bernstein, B.W.; Bamburg, J.R. Actin-ATP hydrolysis is a major energy drain for neurons. *J. Neurosci.* **2003**, *23*, 1–6. [CrossRef] [PubMed]

78. Rüegg, J.; Holsboer, F.; Turck, C.; Rein, T. Cofilin 1 is revealed as an inhibitor of glucocorticoid receptor by analysis of hormone-resistant cells 2195. *Mol. Cell Biol.* **2004**, *24*, 9371–9382. [CrossRef]

79. Cahoy, J.D.; Emery, B.; Kaushal, A.; Foo, L.C.; Zamanian, J.L.; Christopherson, K.S.; Xing, Y.; Lubischer, J.L.; Krieg, P.A.; Krupenko, S.A.; et al. A transcriptome database for astrocytes, neurons, and oligodendrocytes: A new resource for understanding brain development and function. *J. Neurosci.* **2008**, *28*, 264–278. [CrossRef]

80. Durand, C.M.; Perroy, J.; Loll, F.; Perrais, D.; Fagni, L.; Bourgeron, T.; Montcouquiol, M.; Sans, N. SHANK3 mutations identified in autism lead to modification of dendritic spine morphology via an actin-dependent mechanism. *Mol. Psychiatry* **2012**, *17*, 71–84. [CrossRef]

81. Nakatani, N.; Ohnishi, T.; Iwamoto, K.; Watanabe, A.; Iwayama, Y.; Yamashita, S.; Ishitsuka, Y.; Moriyama, K.; Nakajima, M.; Tatebayashi, Y.; et al. Expression analysis of actin-related genes as an underlying mechanism for mood disorders. *Biochem. Biophys. Res. Commun.* **2007**, *352*, 780–786. [CrossRef] [PubMed]

82. Nakatani, N.; Aburatani, H.; Nishimura, K.; Semba, J.; Yoshikawa, T. Comprehensive expression analysis of a rat depression model. *Pharmacogenomics J.* **2004**, *4*, 114–126. [CrossRef] [PubMed]

83. Wang, K.; Li, M.; Bucan, M. Pathway-based approaches for analysis of genomewide association studies. *Am. J. Hum. Genet.* **2007**, *81*, 1278–1283. [CrossRef]

84. Huang, T.L.; Sung, M.L.; Chen, T.Y. 2D-DIGE proteome analysis on the platelet proteins of patients with major depression. *Proteome Sci.* **2014**, *12*. [CrossRef]

85. Calabrese, B.; Halpain, S. Lithium prevents aberrant NMDA-induced F-actin reorganization in neurons. *Neuroreport* **2014**, *25*, 1331–1337. [CrossRef] [PubMed]

86. Piubelli, C.; Vighini, M.; Mathe, A.A.; Domenici, E.; Carboni, L. Escitalopram affects cytoskeleton and synaptic plasticity pathways in a rat gene-environment interaction model of depression as revealed by proteomics. Part II: Environmental challenge. *Int. J. Neuropsychopharmacol.* **2011**, *14*, 834–855. [CrossRef]

87. O'Dushlaine, C.; Ripke, S.; Ruderfer, D.M.; Hamilton, S.P.; Fava, M.; Iosifescu, D.V.; Kohane, I.S.; Churchill, S.E.; Castro, V.M.; Clements, C.C.; et al. Rare copy number variation in treatment-resistant major depressive disorder. *Biol. Psychiatry* **2014**, *76*, 536–541. [CrossRef]

88. Frijters, R.; Fleuren, W.; Toonen, E.J.; Tuckermann, J.P.; Reichardt, H.M.; van der, M.H.; van, E.A.; van Lierop, M.J.; Dokter, W.; de, V.J.; et al. Prednisolone-induced differential gene expression in mouse liver carrying wild type or a dimerization-defective glucocorticoid receptor. *BMC Genomics* **2010**, *11*, 359. [CrossRef]

89. Kretzschmar, A. The Modulation of Actin Dynamics by the Stress-Induced Protein DRR1 and Antidepressants. Ph.D. Thesis, Ludmig Maximilians University Munich, Munich, Germany, 2015.

90. Zhao, X.Y.; Li, H.X.; Liang, S.F.; Yuan, Z.; Yan, F.; Ruan, X.Z.; You, J.; Xiong, S.Q.; Tang, M.H.; Wei, Y.Q. Soluble expression of human DRR1 (down-regulated in renal cell carcinoma 1) in Escherichia coli and preparation of its polyclonal antibodies. *Biotechnol. Appl. Biochem.* **2008**, *49*, 17–23. [CrossRef]

91. Schülke, J.P.; Wochnik, G.M.; Lang-Rollin, I.; Gassen, N.C.; Knapp, R.T.; Berning, B.; Yassouridis, A.; Rein, T. Differential impact of tetratricopeptide repeat proteins on the steroid hormone receptors. *PLoS ONE* **2010**, *5*, e11717. [CrossRef] [PubMed]

92. Rino, J.; Martin, R.M.; Carvalho, T.; Carmo-Fonseca, M. Imaging dynamic interactions between spliceosomal proteins and pre-mRNA in living cells. *Methods* **2014**, *65*, 359–366. [CrossRef] [PubMed]

93. Cooper, J.A.; Walker, S.B.; Pollard, T.D. Pyrene actin: Documentation of the validity of a sensitive assay for actin polymerization. *J. Muscle Res. Cell Motil.* **1983**, *4*, 253–262. [CrossRef]

94. Liu, X.; Pollack, G.H. Stepwise sliding of single actin and Myosin filaments. *Biophys. J.* **2004**, *86*, 353–358. [CrossRef]

95. Pollard, T.D. Rate constants for the reactions of ATP- and ADP-actin with the ends of actin filaments. *J. Cell Biol.* **1986**, *103*, 2747–2754. [CrossRef] [PubMed]

International Journal of
Molecular Sciences

Article

Mechanistic Insights in NeuroD Potentiation of Mineralocorticoid Receptor Signaling

Lisa T. C. M. van Weert [1,2,3], Jacobus C. Buurstede [1], Hetty C. M. Sips [1], Isabel M. Mol [1], Tanvi Puri [1], Ruth Damsteegt [4], Benno Roozendaal [2,3], R. Angela Sarabdjitsingh [4] and Onno C. Meijer [1,*]

[1] Department of Medicine, Division of Endocrinology, Leiden University Medical Center, 2333 ZA Leiden, The Netherlands; L.T.C.M.van_Weert@lumc.nl (L.T.C.M.v.W.); J.C.Buurstede@lumc.nl (J.C.B.); H.C.M.Sips@lumc.nl (H.C.M.S.); I.M.Mol@lumc.nl (I.M.M.); tanvi-04@hotmail.com (T.P.)

[2] Department of Cognitive Neuroscience, Radboudumc, 6525 GA Nijmegen, The Netherlands; Benno.Roozendaal@radboudumc.nl

[3] Donders Institute for Brain, Cognition and Behaviour, Radboud University, 6525 EN Nijmegen, The Netherlands

[4] Department of Translational Neuroscience, UMC Utrecht Brain Center, University Medical Center Utrecht, 3584 CG Utrecht, The Netherlands; R.Damsteegt@umcutrecht.nl (R.D.); r.a.sarabdjitsingh@gmail.com (R.A.S.)

* Correspondence: O.C.Meijer@lumc.nl

Received: 1 March 2019; Accepted: 25 March 2019; Published: 29 March 2019

Abstract: Mineralocorticoid receptor (MR)-mediated signaling in the brain has been suggested as a protective factor in the development of psychopathology, in particular mood disorders. We recently identified genomic loci at which either MR or the closely related glucocorticoid receptor (GR) binds selectively, and found members of the NeuroD transcription factor family to be specifically associated with MR-bound DNA in the rat hippocampus. We show here using forebrain-specific MR knockout mice that GR binding to MR/GR joint target loci is not affected in any major way in the absence of MR. Neurod2 binding was also independent of MR binding. Moreover, functional comparison with MyoD family members indicates that it is the chromatin remodeling aspect of NeuroD, rather than its direct stimulation of transcription, that is responsible for potentiation of MR-mediated transcription. These findings suggest that NeuroD acts in a permissive way to enhance MR-mediated transcription, and they argue against competition for DNA binding as a mechanism of MR- over GR-specific binding.

Keywords: basic-helix-loop-helix; brain; coactivator; glucocorticoids; hippocampus; mineralocorticoid receptor knockout; stress; transcription biology

---

1. Introduction

The mineralocorticoid receptor (MR) regulates stress coping and has gained significant attention in the field of psychopathology. In general higher brain MR expression levels or MR activity parallel improved cognition and reduced anxiety [1]. An MR gain-of-function variant is associated with optimism and provides a decreased risk for depression in females [2]. One single nucleotide polymorphism (SNP) that is part of this haplotype affected the cortisol-awakening response only in those subjects using antidepressants [3]. Furthermore, administration of an MR agonist as a supplement to antidepressant therapy led to faster treatment response [4], and MR activation alone could improve cognitive function in young depressed patients [5]. In contrast, chronic stimulation of the highly related glucocorticoid receptor (GR) predisposes to stress-related disorders [6], and GR antagonism seems of benefit in psychotic depression [7]. A study combining standard dexamethasone (GR activation) for leukemia treatment with add-on cortisol (concurrent MR activation), shows that MR activity is important for neuronal processes such as sleep cycle and mood regulation [8]. Therefore, it is of great relevance to characterize and enable selective modulation of MR-mediated effects, serving a potential antidepressant approach.

Being part of the nuclear receptor family, MR and GR function as ligand-activated transcription factors, binding the glucocorticoid response element (GRE) at the DNA to mediate transcriptional changes. Even though the two receptors share their ligand cortisol/corticosterone (albeit with a different affinity) and recognize the same motif, receptor-specific binding loci exist as demonstrated in the rat hippocampus [9]. This suggests that other factors might be necessary to guide MR/GR-specific binding and subsequent transcriptional effects. Indeed, we found that binding sites for NeuroD factors were present selectively near MR-bound loci, and confirmed Neurod2 binding near MR-bound but not GR-bound GREs [9]. Furthermore NeuroD factors were able to potentiate glucocorticoid-mediated signaling in an in vitro setting, although MR/GR specificity was not recapitulated in reporter assays [9].

NeuroD proteins belong to the basic-helix-loop-helix (bHLH) family of transcription factors, and regulate neuronal differentiation. Related MyoD factors are expressed in the muscle, where they induce myogenesis. The bHLH transcription factors bind to E-boxes, which have the sequence CANNTG [10]. Specificity is obtained via the middle two nucleotides, with CAGATG known to be a NeuroD-specific binding site, whereas CAGCTG is a shared site that is bound by both MyoD and NeuroD [11]. The previously found interaction between NeuroD and glucocorticoid signaling was based on the presence of the NeuroD-specific motif [9]. As the MyoD proteins are better understood in terms of functional domains [12], we also examined transcriptional modulation by bHLH factors at the MyoD/NeuroD shared motif to unravel the interaction between NeuroD and MR here.

The current study aimed to provide mechanistic insights in the NeuroD potentiation of MR signaling, and how MR over GR specificity is achieved. We selected the protein Neurod2 as a representative of the NeuroD family [9]. We first questioned whether GR binding would be affected by MR absence, and if Neurod2 binding would be dependent on MR presence. Therefore, we assessed GR and Neurod2 binding at previously identified MR targets [9] in the hippocampus of forebrain-specific MR knockout mice (fbMRKO). Subsequently using various E-box binders in a reporter assay, we further explored the mechanism by which NeuroD can enhance glucocorticoid signaling. Our data show that at MR target loci both GR and Neurod2 binding seem independent of MR binding, and it is likely the chromatin remodeling effect of NeuroD is responsible for the transcriptional potentiation.

2. Results

2.1. DNA Binding Assessed by Chromatin Immunoprecipitation

In order to define the mechanism behind the NeuroD potentiation of glucocorticoid signaling in more detail, we first tested whether MR binding to its hippocampal DNA targets affects local GR and Neurod2 binding. Although family members Neurod1, Neurod2 and Neurod6 are all expressed in the adult mouse hippocampus and are able to bind the same NeuroD binding site [9], we focus here on Neurod2. GR and Neurod2 occupancy of MR-binding loci was measured by chromatin immunoprecipitation coupled with quantitative polymerase chain reaction (ChIP-qPCR) on hippocampus of wild-type (WT) and forebrain-specific mineralocorticoid receptor knockout (fbMRKO) mice. The fbMRKO mice show ablated hippocampal MR mRNA levels [13], which is accompanied by efficient knockdown of MR protein (Bonapersona et al., in preparation). Plasma corticosterone of all animals was over 140 ng/mL, ensuring ligand occupancy of both MR and GR [14]. No difference in corticosterone plasma levels was observed between the two genotypes, with an average of 363 ± 30 ng/mL for WT mice and 313 ± 44 ng/mL for fbMRKO mice (Figure 1A).

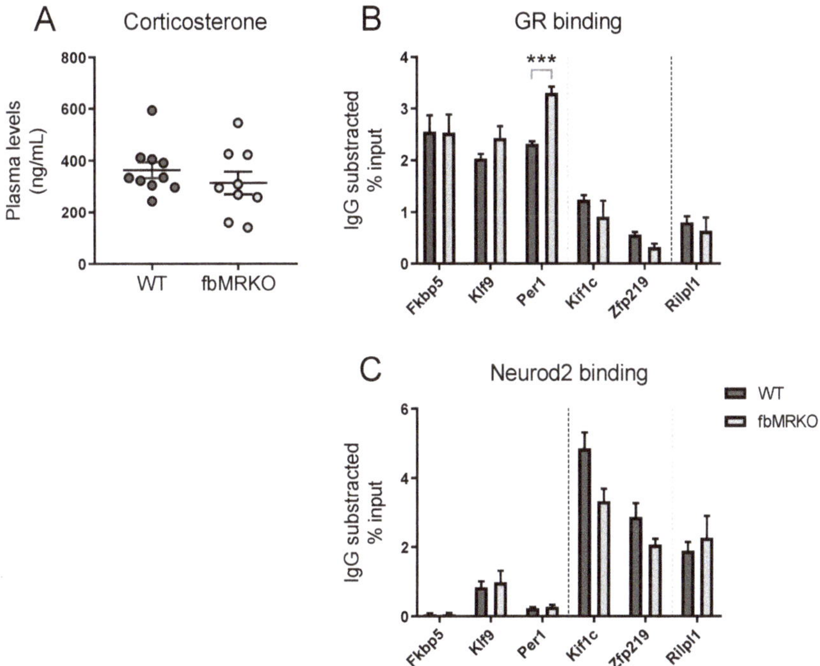

Figure 1. (**A**) Corticosterone levels of wild-type (WT) and forebrain-specific mineralocorticoid receptor (MR) knockout (fbMRKO) mice. In these mice chromatin immunoprecipitation coupled with quantitative polymerase chain reaction (ChIP-qPCR) measurements for (**B**) glucocorticoid receptor (GR) and (**C**) Neurod2 were performed. For each gene, the corresponding immunoglobulin G (IgG) background signal is subtracted from detected binding levels, expressed as the percentage of immunoprecipitated DNA. The binding sites near *Fkbp5*, *Klf9*, *Per1*, *Kif1c* and *Zfp219* are joint MR/GR loci, while *Rilpl1* has been identified as an MR-specific target [9] (separated by the right dotted line). Genes are further sorted based on the absence (*Fkbp5*, *Klf9*, *Per1*) or presence (*Kif1c*, *Zfp219*, *Rilpl1*) of a NeuroD binding sequence near the MR binding site (separated by the left dotted line). *** *p* < 0.001

2.1.1. MR Effect on GR Binding

We aimed to investigate if the joint binding of MR and NeuroD on the DNA is related to competition for GR binding at the same locus. GR binding was confirmed in WT mice for classical glucocorticoid target genes *Fkbp5* and *Per1* (Figure 1B), which are occupied by both MR and GR [15]. Other MR-GR overlapping loci near the *Klf9* [16] and *Kif1c* [9] genes showed evident GR binding. Previously identified MR-specific target *Rilpl1* [9] showed low GR signal, to the same extent as MR-GR overlapping target *Zfp219* [9]. GR binding levels were similar in the fbMRKO mice for most of the genes measured, suggesting that GR binding is not dependent on MR binding at these target loci. Only the GR binding at *Per1* was slightly enhanced in MR absence (*p* = 0.00055), which might point to a compensatory mechanism at this specific binding site. However, in general GR binding does not seem to compensate for the lack of MR binding in fbMRKO mice.

2.1.2. MR Effect on Neurod2 Binding

Next, we addressed the question of whether the association between MR and NeuroD factors that we observed previously implies that Neurod2 binding at these loci depends on the presence of MR. We measured Neurod2 binding at the same loci as for GR binding. No Neurod2 binding motif was detected in the ChIP-identified MR-GR overlapping binding sequences near *Fkbp5*, *Klf9* and *Per1*. For the *Kif1c* and *Zfp219* associated MR-GR overlapping loci a directed motif search [9] did reveal a Neurod2 binding motif. Neurod2 binding was indeed observed for *Kif1c*, *Zfp219* and to a lesser extent in *Klf9*, and for MR-specific *Rilpl1* as observed before [9] (Figure 1C). Those genes with relatively low GR binding showed higher Neurod2 binding and vice versa, supporting the earlier finding that Neurod2 seems to interact preferentially with MR [9]. The fbMRKO mice demonstrated unchanged Neurod2 binding levels, indicating the presence of MR is not crucial for Neurod2 binding. For *Kif1c* there might be an interaction, as the Neurod2 signal seems to be lower in fbMRKO compared to WT animals, but this difference does not statistically hold after multiple comparison correction ($p = 0.23$). Overall, these data show that Neurod2 binding to MR-associated loci is independent of MR binding.

2.2. Structure–Function Relationship

We continued unraveling the mechanism behind the NeuroD potentiation of glucocorticoid signaling by exploring which coactivation property of the NeuroD protein is responsible for the transcriptional potentiating effects. While the structure–function relationship of the NeuroD family is not known in detail, much more is known about the related bHLH family of MyoD proteins [12]. We therefore used the myogenic regulatory factors MyoD and Myf5 as tools to study the effect of bHLH factors in the potentiation of glucocorticoid signaling. Where MyoD can induce both histone acetylation at H4 (chromatin remodeling) and in addition recruit RNA polymerase II (direct activation mediated by the transcriptional activation domain), Myf5 is only able to induce H4 acetylation as a manner to enhance transcription [12]. NeuroD family members have been shown to affect both chromatin accessibility and direct transcriptional activation [11,17], although these functions have not been assigned to a specific part of the protein. Comparing the myogenic variants will enable us to dissect the process important for the potentiation of glucocorticoid signaling.

2.2.1. Transcriptional Potentiation by MyoD

We started by exploring whether MyoD is able to show a similar coactivation effect for MR/GR-mediated signaling as Neurod2 did in our reporter assay. Despite the in vivo binding selectivity of Neurod2 with MR (and not GR), Neurod2 exhibits coactivation of MR but also GR transcriptional activity in vitro [9]. MyoD and NeuroD have both unique and common response elements [11]. Our original reporter construct that is based on in vivo MR ChIP-sequencing binding sites [9], harbors the NeuroD-specific CAGATG along a GRE. In a first experiment we tested the effect of Neurod2, MyoD and a chimeric MyoD protein with its bHLH domain substituted by that of Neurod2 (MyoD(ND2bHLH)) in the concentrations of 1–3–10 ng/well (Figure 2). Both a cofactor ($F_{2,24} = 356.3$ for MR; $F_{2,24} = 708.3$ for GR, both $p < 0.000001$) and concentration ($F_{3,24} = 247.6$ for MR; $F_{3,24} = 489.0$ for GR, both $p < 0.000001$) effect, plus an interaction ($F_{6,24} = 71.0$ for MR; $F_{6,24} = 159.2$ for GR, both $p < 0.000001$) were observed.

We confirmed Neurod2 could potentiate glucocorticoid signaling for both MR and GR (Figure 2A,B). The observed Neurod2 effect was receptor-mediated, as in absence or with lower amounts of nuclear receptor expression vector Neurod2 did not enhance the glucocorticoid-dependent transcriptional increase (Figure S1). We showed that also MyoD can potentiate MR- and GR-mediated transcriptional activity, once brought to the DNA. Coactivation by MyoD itself is minimal with a slightly higher fold induction in the upper tested dose compared to control cells without cofactor ($p = 0.0062$ for MR; $p = 0.0019$ for GR), but can be enhanced to an extent similar to Neurod2 by swopping the MyoD DNA-binding domain (DBD) with that of Neurod2 as demonstrated using the MyoD(ND2bHLH) chimera (Figure 2). In its highest tested dose the chimera could even potentiate glucocorticoid signaling to a superior extent. Of note, the chimera showed a clear dose-dependent increase in potentiation over the concentration range tested. These findings indicate the Neurod2 DBD is required for coactivation, and the DNA sequence rather than the bHLH protein function drives specificity.

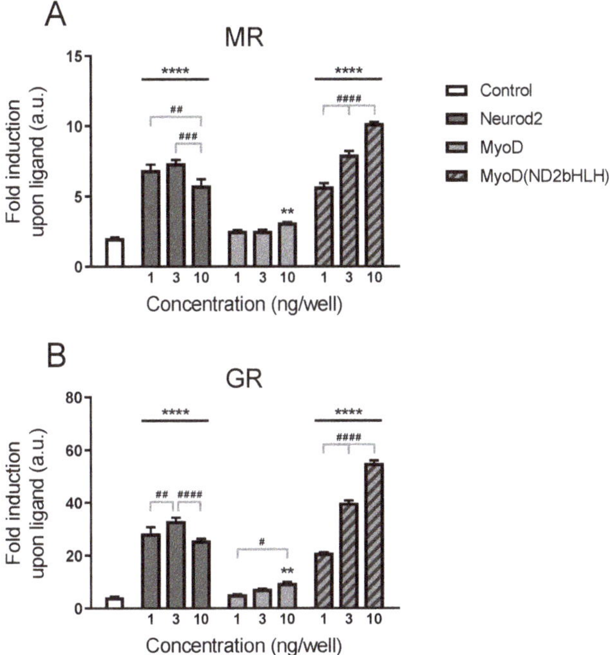

Figure 2. Specificity of NeuroD coactivation at the previously identified binding motif (CAGATG) for (**A**) MR and (**B**) GR. HEK293 cells were transfected with GRE-At_GA luciferase construct, MR or GR (10 ng/well), various amounts of Neurod2, MyoD or the MyoD/Neurod2 chimera (MyoD(ND2bHLH)) (1–3–10 ng/well), and stimulated with corticosterone (10^{-7} M). Data are presented as luciferase activity fold induction upon corticosterone treatment. a.u. = arbitrary unit; ** $p < 0.01$, **** $p < 0.0001$ compared to control condition; # $p < 0.05$, ## $p < 0.01$, ### $p < 0.001$, #### $p < 0.0001$ for within group comparisons

2.2.2. Activation Domain Not Crucial for Potentiation

Finally we tested several bHLH factors for their coactivation ability in our reporter assay to examine the contribution of different protein domains. In order to have a fair comparison of all variants, we ensured a similar binding affinity of NeuroD and MyoD by further studying a reporter construct containing the shared CAGCTG motif [11]. At this reporter Neurod2 and MyoD could potentiate MR signaling to the same extent (Figure 3A), while for GR-mediated transcription the MyoD potentiation was somewhat lower than by Neurod2 ($p = 0.000003$, Figure 3B). MyoD lacking its activation domain (MyoDΔN) demonstrated a less strong potentiation of GR-mediated signaling compared to full length MyoD ($p = 0.0012$), as did family member Myf5 ($p = 0.0035$), but both MyoDΔN ($p = 0.047$) and Myf5 ($p = 0.016$) still showed a significantly higher transcriptional effect upon corticosterone treatment than the control condition without overexpression (Figure 3B).

The effect of the bHLH proteins on MR transactivation was more modest. Interestingly, the MyoDΔN and Myf5 coactivating potential for MR-mediated signaling was not different from Neurod2 and MyoD (Figure 3A). However, MyoDΔN did not reach significance in corticosterone induction compared to control cells (Figure 3A). Although potentiation of GR transcriptional activity by bHLH factors seems thus partly dependent on their activation domain, these data suggest that the coactivation of MR signaling by Neurod2 postulated to happen in vivo [9] is likely mediated via chromatin remodeling rather than direct transcriptional activation.

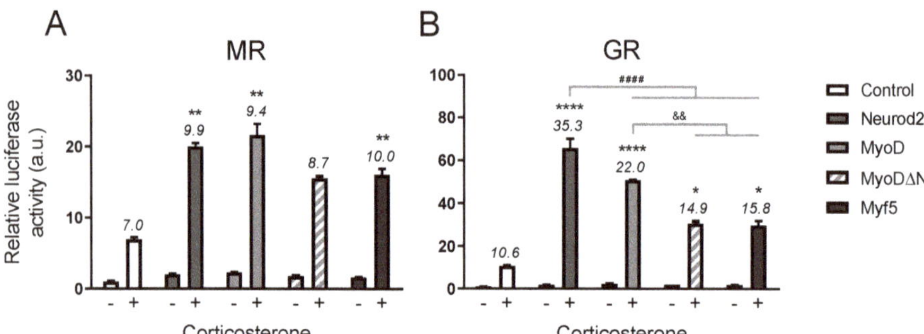

Figure 3. Modulation by NeuroD and MyoD variants at the shared binding motif (CAGCTG) for (**A**) MR- and (**B**) GR-mediated transcription. HEK293 cells were transfected with GRE-At_GC luciferase construct, MR or GR (10 ng/well), and Neurod2, MyoD, MyoDΔN or Myf5 (10 ng/well), and stimulated with corticosterone (10^{-7} M). Luciferase activity of nonstimulated control cells was normalized to 1. Numbers represent fold induction upon corticosterone treatment. a.u. = arbitrary unit; * $p < 0.05$, ** $p < 0.01$, **** $p < 0.0001$ compared to control condition; #### $p < 0.0001$ compared to Neurod2 condition; && $p < 0.01$ compared to MyoD condition

3. Discussion

This study further elucidates the mechanism behind NeuroD potentiation of brain MR signaling. First transcription factor DNA binding was assessed by ChIP-qPCR in mice lacking MR in (amongst other brain regions) their hippocampus. Both GR and Neurod2 binding were not altered in these fbMRKO mice compared to control mice, except for an enhanced GR signal at the *Per1* promoter in absence of MR. Subsequently bHLH factors of the NeuroD and MyoD families were used to study coactivator effects in an MR/GR-driven reporter assay. Those factors lacking (MyoDΔN) or with diminished (Myf5) activator function were able to potentiate the glucocorticoid-stimulated transcriptional activation as well as Neurod2 and MyoD in case of MR-dependent transcription, suggesting coactivation of MR signaling by Neurod2 does not require its activation domain.

3.1. Effects on DNA Binding

Because MR and GR can bind the same DNA sequences, GREs, the absence of MR might affect genomic binding by GR. Competition between MR and GR at a specific locus does not seem to play a major role, as there was no overall enhanced GR binding in the fbMRKO mice at the sites we examined, even though hippocampal GR expression is upregulated in these animals [13,18]. Only in the case of *Per1*, higher GR occupancy levels were observed at the promoter region in the absence of MR. At this locus it has been demonstrated that besides homodimerization, MR and GR can combine to form heterodimers [15]. However, we cannot distinguish between these two binding modes in our measurements. The increased GR binding could reflect a compensatory mechanism to maintain a required degree of *Per1* expression and is in agreement with the fact that basal *Per1* mRNA levels were not altered in fbMRKO mice [13]. Rather than competition, data on joint occupancy suggest there can be synergism between two transcription factors binding the same site, via a process called 'assisted loading'. For concurrent stimulation of the GR and estrogen receptor (ER; where ER is altered to also recognize the GRE), GR activation could enhance ER binding at the same locus [19]. In the present study, GR binding is not significantly diminished when MR is lacking, suggesting such assisted loading is not applicable for MR-GR joint loci here. In our measurements of whole hippocampus we should acknowledge that we work under the assumption that all studied cells have (similar amounts of) MR and GR, but effects on DNA binding could be diluted as MR/GR expression is not homogeneous throughout the hippocampal regions and in the various cell types present [20]. Single cell analysis will offer a solution to study transcription biology in a cell-type specific manner [21]. Nevertheless, our data indicate that GR binding is predominantly independent from the presence of MR in the hippocampus.

In the same setting we studied if Neurod2 binding was affected by absence of MR. No differences in Neurod2 signal at the MR target loci were observed in MR deficient mice, which implies that NeuroD facilitates MR binding in a unidirectional manner. We cannot exclude the possibility that Neurod2 binding is affected by or dependent on changes in stress hormone levels, since this was not studied here. The presence of another collaborative transcription factor (nuclear factor-1) found near preaccessible GR-bound loci was independent of corticosterone treatment or exposure to restraint stress [22]. As discussed below, our reporter assay data suggest that the potentiation of MR signaling by NeuroD is likely mediated via chromatin accessibility.

3.2. Mechanism of Glucocorticoid Signaling Potentiation

Unfortunately, the NeuroD activation domain is not well documented/distinguished, but MyoD family members do have well described domains [12]. We first tested whether MyoD was able to potentiate glucocorticoid signaling at a reporter construct containing a GRE and NeuroD-specific E-box (CAGATG). When the MyoD DBD was adapted to that of Neurod2 in order to bind this motif efficiently, MyoD could coactivate glucocorticoid-mediated signaling to a similar (or even superior) extent as Neurod2. This is in line with findings by Fong et al., showing that MyoD could be redirected to NeuroD target sites through replacement of its bHLH domain by the analogue sequence of Neurod2 and, hereby, could activate part of the neuronal differentiation program [23]. The same group has demonstrated that NeuroD and MyoD can bind and drive transcription at the E-box that is specific for the other bHLH factor, but have a strong preference for their specific motifs [11,23]. This explains why unmodified MyoD showed a slight transcriptional potentiation on the NeuroD-specific binding site at its highest concentration tested. In concordance with the DBD being decisive in converting MyoD into a neurogenic factor [23], the specificity of the interaction between NeuroD/MyoD and MR/GR in our data is also determined by the ability of the factor to bind the DNA rather than a protein-specific functionality. Interactions between bHLH transcription factors and steroid receptors can be speculated to be generic but have cell/tissue-type dependent mechanisms. For instance, bHLH proteins DEC1/DEC2 (differentiated embryo chondrocyte) were found to corepress liver retinoid X receptors [24]. Likewise, E47 can modulate hepatic glucocorticoid action by promoting GR occupancy of metabolic target loci [25]. Of relevance in the testis, Pod-1 (also: transcription factor 21) could diminish transactivation by the androgen receptor [26].

For unbiased comparisons we proceeded our experiments with a reporter construct containing the shared E-box (CAGCTG), which is bound with similar affinity by both Neurod2 and MyoD [23]. Coregulators can modulate transcription by affecting chromatin accessibility and/or recruitment and stabilization of the transcriptional machinery [27]. To distinguish between these two modes, we made use of a truncated version of MyoD lacking its activation domain (responsible for direct recruitment), and the myogenic Myf5 that has a weak activation domain (and, therefore, relies mainly on its chromatin remodeling ability) compared to MyoD [12]. All MyoD variants were able to coactivate the GRE-driven reporter. Strikingly, while potentiation of GR signaling was partly dependent on the bHLH activation domain, coactivation of MR signaling was almost unaffected when using the factors with diminished direct transcriptional activation. Extrapolating these findings to the NeuroD family, the chromatin remodeling aspect of NeuroD thus seems sufficient for effective potentiation of MR-mediated signaling. This is in accordance with the pioneer function of family member Neurod1 demonstrated in a ChIP-sequencing experiment on developing neurons [17]. Of note, during neurogenesis occupancy of the Neurod2-specific motif was linked to gene expression effects, while the shared motif related mostly to chromatin modifications [11]. Despite the fact that transient systems might be considered to have an undefined chromatin context, it has been shown that exogenous plasmids do interact with endogenous histone proteins [28,29] and can serve as a proper model to study effects mediated via chromatin accessibility as observed here.

3.3. MR Selective Signaling and Future Implications

A number of issues have remained unaddressed. In the current study we have been looking at only a subset of Neurod2 sites, and mainly focused on targets bound by both MR and GR. It would be of interest to study genome-wide effects and observe if MR-specific sites become GR-bound in the absence of MR. We also have to point out that we have not assessed in vivo which NeuroD factor(s) is/are responsible for potentiation of MR signaling, as we only measured and detected Neurod2 binding at MR-bound sites [9]. The basis for MR over GR specificity in full chromatin is not known, but the fact that bHLH chromatin remodeling plays a more important role in case of MR-mediated reporter activation is in line with the fact that we could correlate MR and Neurod2 binding in vivo [9]. Besides, those MR target genes with relatively low GR signal had high Neurod2 binding in our current ChIP data. A study by Pooley et al. found that 17% of GR-bound loci contained a NeuroD binding site in their vicinity [22]. These are likely MR/GR joint sites comparable to those studied here, some of which do show an E-box and could be co-bound by Neurod2. MyoD family inhibitor domain-containing protein (MDFIC) has been found to bind the hinge region of unliganded GR, is capable of regulating GR phosphorylation and can by this means define the receptor transcriptome [30]. This interaction might play a role in the MR/GR binding selectivity near Neurod2-bound sites, as our earlier studies suggested that proteins in the nuclear receptor complex might account for the MR preference [9]. One promising approach to further elucidate the MR over GR specificity would be to have ChIP experiments followed-up by proteomics [31].

The question emerges what the NeuroD potentiation of MR signaling implicates for stress processing and stress-related disorders. Increased *Neurod2* expression levels were detected in the ventromedial prefrontal cortex of men with major depressive disorder compared to healthy control subjects [32]. In a mouse model of chronic social defeat paradigm, overexpression of Neurod2 in the ventral hippocampus reduces, while overexpression in the nucleus accumbens increases social interaction time [33]. Antidepressant agomelatine could normalize the rise in hippocampal *Neurod1* expression of mice that underwent chronic mild stress [34]. Furthermore, fish in touristic zones were shown to express higher levels of *Neurod1* and the MR gene *Nr3c2* relative to fish at control sites [35]. Together these observations strongly suggest a functional and context-dependent link between NeuroD and stress regulation. How this might depend on MR or influence MR function remains to be investigated. Further research is needed focusing on the in vivo specificity of the interaction between MR and NeuroD, and directionality in the highly adaptable stress system. MR activation is considered a promising strategy to promote stress resilience [1]. It would be of great interest to test if SNPs in the MR gene can affect NeuroD potentiation. In conclusion, we show that GR and Neurod2 binding at MR target loci is not dependent on MR presence and that Neurod2 potentiation of MR signaling is likely mediated via chromatin remodeling. We summarize the findings of this study in Figure 4. Future studies will have to point out how the interaction between Neurod2 and MR might be exploited to modulate MR-specific effects in the brain and affect associated behavior.

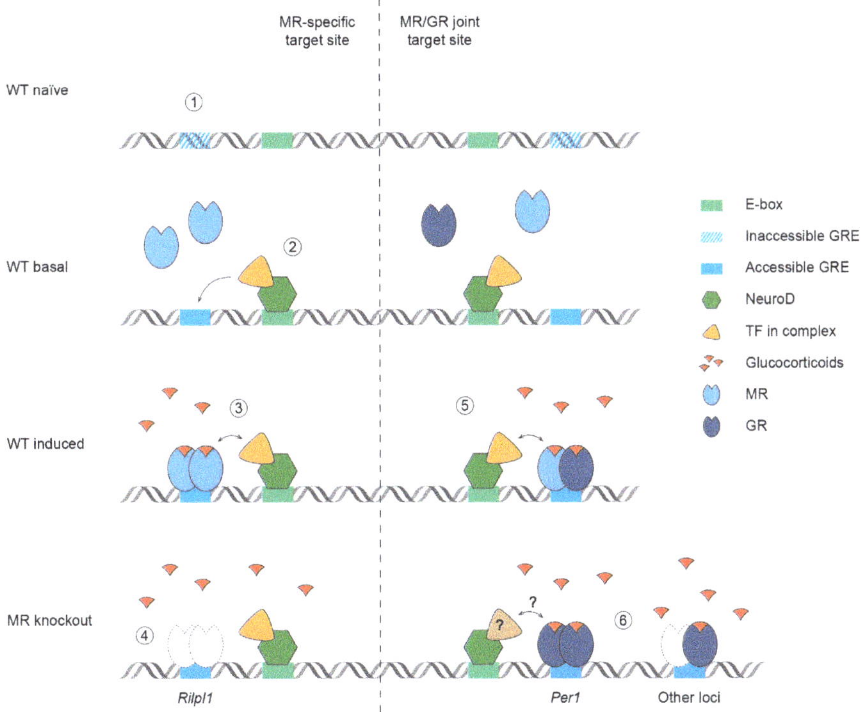

Figure 4. Summary of the interaction between hippocampal MR and NeuroD. Glucocorticoid response elements (GREs) previously inaccessible (**1**) could be rendered accessible by chromatin remodeling (one-way arrow) induced by NeuroD (**2**) binding at a nearby E-box (the NeuroD-specific sequence CAGATG). Upon ligand availability MR can bind an accessible GRE (**3**) in order to modulate transcriptional activity of its target genes. This interaction between NeuroD and MR (two-way arrow) is likely mediated via additional TF(s) in the transcriptional complex [9]. In forebrain MR knockout mice (**4**) GR is not compensating for the lack of MR binding at the MR-specific *Rilpl1* site. Also at several MR/GR joint target sites (**5**) NeuroD occupancy is observed in the vicinity. Of note, we cannot discriminate between the binding of homo- and heterodimers in the present study. In absence of MR (**6**) GR binding is increased at the *Per1* promoter, while for the other tested loci GR binding levels are unaltered. For sites that become GR-specific due to MR knockout, interactions with NeuroD remain to be explored, and other TF(s) might be involved (**?**). MR = mineralocorticoid receptor, GR = glucocorticoid receptor, GRE = glucocorticoid response element, TF = transcription factor, WT = wild type

4. Materials and Methods

4.1. Animals

Male homozygous forebrain-specific MR knockout (MR$^{flox/flox_Cre}$, fbMRKO, n = 9) and littermate flox heterozygous control mice (MR$^{flox/wt_wt}$, n = 10) [18] aged 10-19 weeks, were housed on a 12-h light/12-h dark reversed cycle (lights off at 8:00AM). Mice were group-housed with fbMRKOs and controls combined, and a total of four mice per cage. Each mouse was individually transferred to a novel cage 45 min before harvesting the tissue, in order to ensure GR binding for ChIP analysis. Mice were sacrificed by cervical dislocation around the time of their endogenous corticosterone peak, between 9:00AM–11:30AM. Genotypes were equally distributed over the sacrifice window to prevent an effect by time of the day. Trunk blood was collected, and hippocampal hemispheres were freshly dissected, snap-frozen in liquid nitrogen and stored at $-80\ ^\circ$C for later analysis. The experiment was performed according to the European Commission Council Directive 2010/63/EU and the Dutch law on animal experiments and approved by the animal ethical committee from Utrecht University (authorization number 2014.I.08.057, approval date: 9 October 2014).

4.2. Plasma Corticosterone

Trunk blood was centrifuged for 10 min at $7000 \times g$, after which plasma was transferred to new tubes and stored at $-20\ ^\circ$C for later analysis. Corticosterone levels were determined using an Enzyme ImmunoAssay, according to the manufacturer's instruction (Immunodiagnostic Systems, Boldon, UK).

4.3. ChIP-qPCR

To assess GR and Neurod2 binding at MR-bound loci, we performed ChIP-qPCR on hippocampal tissue as described previously [9]. Briefly, two fixated hippocampal hemispheres of the same animal were pooled and used for a single ChIP sample (500 µL) to measure GR binding ($n = 4$–5) with 6 µg of anti-GR antibody H-300 (sc-8992X, Santa Cruz, Dallas, TX, USA) or Neurod2 binding ($n = 4$) with 6 µg of anti-Neurod2 antibody (ab109406, Abcam, Cambridge, UK). Hippocampi were allocated for either GR or Neurod2 detection, with tissue from each group of co-housed mice divided over the two transcription factors. A ChIP using 6 µg of control IgG antibody (ab37415, Abcam) was taken along for background measurements, on a mixed hippocampal chromatin sample per genotype and transcription factor. This was followed by qPCR on undiluted Chelex-isolated (200 µL) ChIP samples, using the primers listed in Table 1.

Table 1. Primer sequences used for qPCR on mouse hippocampal ChIP samples. Primers target a mineralocorticoid receptor binding site near the listed gene.

Gene	Full Name	Forward & Reverse (5′ > 3′)	Product Length (bp)
Fkbp5	FK506 binding protein 5	TGCCAGCCACATTCAGAACA TCAAGTGAGTCTGGTCACTGC	122
Kif1c	Kinesin family member 1C	GCTGGGGTGTACACAGATGG TGACTAGCCAGAGCAGTATGTC	156
Klf9	Kruppel-like factor 9	ATCTAGGGCAGTTTGTTCAA GGCAGGTTCATCTGAGGACA	96
Per1	Period circadian clock 1	GGAGGCGCCAAGGCTGAGTG CGGCCAGCGCACTAGGGAAC	73
Rilpl1	Rab interacting lysosomal protein-like 1	CAGGCAGATGCCAGGCT CCCATGCCTGTTCCTCTAGT	106
Zfp219	Zinc finger protein 219	AGTCCATCACATTCTGTTGCTTTC TAGTCAGCTATGACCATGCAGT	131

4.4. Reporter Assays

For mechanistic insights into the role of NeuroD factors on MR/GR-driven promoter activity, we performed luciferase reporter as described previously [9]. In short, HEK293 cells were transfected using FuGENE (Promega, Leiden, The Netherlands) with luciferase construct (GRE-At, 30 ng/well), expression vector for either MR or GR (10 ng/well), with or without NeuroD/MyoD cofactor (10 ng/well), and Renilla (1 ng/well) for normalization. To exclude glucocorticoid effects from the medium we used charcoal-stripped fetal bovine serum (Sigma-Aldrich, Zwijndrecht, The Netherlands) during the experiments. After 24 h stimulation of the cells with 10^{-7} M corticosterone (Sigma) reporter protein levels were measured using the Dual Luciferase Reporter Assay System according to the manufacturer's instruction (Promega).

4.5. Plasmids

Transcriptional activity was assessed at a GRE-driven promoter combined with either the NeuroD-specific (CA<u>GA</u>TG) or the MyoD/NeuroD-shared (CA<u>GC</u>TG) motif. The GRE and NeuroD binding site-containing vector (GRE-At_GA) was constructed before (GRE-At-pGL4 [9]). For the generation of the GRE-At_GC luciferase construct, we exploited mutagenesis targeting the NeuroD binding site (GA > GC) using a QuikChange II Site-Directed Mutagenesis Kit (Agilent Technologies, Santa Clara, CA, USA). PAGE-purified mutagenic primers were: 5′-CTCGAGGATGGCAG<u>C</u>TGGAGCTAAGAACAGAA-3′ and 5′-TTCTGTTCTTAG CTCCA<u>G</u>CTGCCATCCTCGAG-3′. For MR and GR expression we used the 6RMR and 6RGR-based plasmids [36]. Expression vectors (all pCS2) for Neurod2, MyoD, a chimera of MyoD with the DNA-binding domain of Neurod2 (MyoD(ND2bHLH)), MyoD lacking the N-terminal domain (MyoDΔN) and Myf5 were kindly provided by Dr. Tapscott [12,23].

4.6. Statistics

On the ChIP data we ran unpaired *t*-tests with the Holm–Sidak multiple comparison correction. For the reporter assays we performed statistics on the fold induction by ligand (calculated for each corticosterone-treated sample as signal in the presence of hormone divided by the average signal from the same condition in absence of hormone). The first reporter experiment (different cofactors at various concentrations) was analyzed by two-way analysis of variance (ANOVA); the second reporter experiment (different cofactors) was analyzed by one-way ANOVA, both followed by Tukey's post-hoc tests. All data are presented as mean ± standard error of the mean.

Supplementary Materials: Supplementary materials can be found at http://www.mdpi.com/1422-0067/20/7/1575/s1.

Author Contributions: Conceptualization, L.T.C.M.v.W. and O.C.M.; Formal analysis, L.T.C.M.v.W.; Funding acquisition, B.R. and O.C.M.; Investigation, L.T.C.M.v.W., J.C.B., H.C.M.S., I.M.M., T.P., and R.D.; Project administration, L.T.C.M.v.W. and R.A.S.; Resources, R.A.S.; Supervision, L.T.C.M.v.W. and O.C.M.; Validation, H.C.M.S.; Visualization, L.T.C.M.v.W.; Writing—original draft, L.T.C.M.v.W. and O.C.M.; Writing—review and editing, J.C.B., B.R. and R.A.S.

Funding: This research was supported by the Netherlands Organisation for Scientific Research (NWO) ALW, grant number 823.02.002.

Acknowledgments: We thank Trea Streefland for technical assistance and Stephen Tapscott for providing plasmids.

Conflicts of Interest: The authors declare no conflict of interest. The funder had no role in the design of the study; in the collection, analyses, or interpretation of data; in the writing of the manuscript; or in the decision to publish the results.

References

1. De Kloet, E.R.; Otte, C.; Kumsta, R.; Kok, L.; Hillegers, M.H.; Hasselmann, H.; Kliegel, D.; Joels, M. Stress and Depression: A Crucial Role of the Mineralocorticoid Receptor. *J. Neuroendocrinol.* **2016**, *28*. [CrossRef] [PubMed]
2. Klok, M.D.; Giltay, E.J.; Van der Does, A.J.; Geleijnse, J.M.; Antypa, N.; Penninx, B.W.; de Geus, E.J.; Willemsen, G.; Boomsma, D.I.; van Leeuwen, N.; et al. A common and functional mineralocorticoid receptor haplotype enhances optimism and protects against depression in females. *Transl. Psychiatry* **2011**, *1*, e62. [CrossRef]
3. Klok, M.D.; Vreeburg, S.A.; Penninx, B.W.; Zitman, F.G.; de Kloet, E.R.; DeRijk, R.H. Common functional mineralocorticoid receptor polymorphisms modulate the cortisol awakening response: Interaction with SSRIs. *Psychoneuroendocrinology* **2011**, *36*, 484–494. [CrossRef] [PubMed]
4. Otte, C.; Hinkelmann, K.; Moritz, S.; Yassouridis, A.; Jahn, H.; Wiedemann, K.; Kellner, M. Modulation of the mineralocorticoid receptor as add-on treatment in depression: A randomized, double-blind, placebo-controlled proof-of-concept study. *J. Psychiatr. Res.* **2010**, *44*, 339–346. [CrossRef]

5. Otte, C.; Wingenfeld, K.; Kuehl, L.K.; Kaczmarczyk, M.; Richter, S.; Quante, A.; Regen, F.; Bajbouj, M.; Zimmermann-Viehoff, F.; Wiedemann, K.; et al. Mineralocorticoid receptor stimulation improves cognitive function and decreases cortisol secretion in depressed patients and healthy individuals. *Neuropsychopharmacology* **2015**, *40*, 386–393. [CrossRef] [PubMed]

6. Judd, L.L.; Schettler, P.J.; Brown, E.S.; Wolkowitz, O.M.; Sternberg, E.M.; Bender, B.G.; Bulloch, K.; Cidlowski, J.A.; de Kloet, E.R.; Fardet, L.; et al. Adverse consequences of glucocorticoid medication: Psychological, cognitive, and behavioral effects. *Am. J. Psychiatry* **2014**, *171*, 1045–1051. [CrossRef]

7. DeBattista, C.; Belanoff, J.; Glass, S.; Khan, A.; Horne, R.L.; Blasey, C.; Carpenter, L.L.; Alva, G. Mifepristone versus placebo in the treatment of psychosis in patients with psychotic major depression. *Biol. Psychiatry* **2006**, *60*, 1343–1349. [CrossRef] [PubMed]

8. Warris, L.T.; van den Heuvel-Eibrink, M.M.; Aarsen, F.K.; Pluijm, S.M.; Bierings, M.B.; van den Bos, C.; Zwaan, C.M.; Thygesen, H.H.; Tissing, W.J.; Veening, M.A.; et al. Hydrocortisone as an Intervention for Dexamethasone-Induced Adverse Effects in Pediatric Patients with Acute Lymphoblastic Leukemia: Results of a Double-Blind, Randomized Controlled Trial. *J. Clin. Oncol.* **2016**, *34*, 2287–2293. [CrossRef]

9. Van Weert, L.T.C.M.; Buurstede, J.C.; Mahfouz, A.; Braakhuis, P.S.M.; Polman, J.A.E.; Sips, H.C.M.; Roozendaal, B.; Balog, J.; de Kloet, E.R.; Datson, N.A.; et al. NeuroD Factors Discriminate Mineralocorticoid From Glucocorticoid Receptor DNA Binding in the Male Rat Brain. *Endocrinology* **2017**, *158*, 1511–1522. [CrossRef]

10. Murre, C. Helix-loop-helix proteins and the advent of cellular diversity: 30 years of discovery. *Genes Dev.* **2019**, *33*, 6–25. [CrossRef] [PubMed]

11. Fong, A.P.; Yao, Z.; Zhong, J.W.; Cao, Y.; Ruzzo, W.L.; Gentleman, R.C.; Tapscott, S.J. Genetic and epigenetic determinants of neurogenesis and myogenesis. *Dev. Cell* **2012**, *22*, 721–735. [CrossRef]

12. Conerly, M.L.; Yao, Z.; Zhong, J.W.; Groudine, M.; Tapscott, S.J. Distinct Activities of Myf5 and MyoD Indicate Separate Roles in Skeletal Muscle Lineage Specification and Differentiation. *Dev. Cell* **2016**, *36*, 375–385. [CrossRef]

13. Van Weert, L.T.C.M.; Buurstede, J.C.; Sips, H.C.M.; Vettorazzi, S.; Mol, I.M.; Hartmann, J.; Prekovic, S.; Zwart, W.; Schmidt, M.V.; Roozendaal, B.; et al. Identification of mineralocorticoid receptor target genes in the mouse hippocampus. *J. Neuroendocrinol.* **2019**. submitted.

14. Reul, J.M.; de Kloet, E.R. Two receptor systems for corticosterone in rat brain: Microdistribution and differential occupation. *Endocrinology* **1985**, *117*, 2505–2511. [CrossRef] [PubMed]

15. Mifsud, K.R.; Reul, J.M. Acute stress enhances heterodimerization and binding of corticosteroid receptors at glucocorticoid target genes in the hippocampus. *Proc. Natl. Acad. Sci. USA* **2016**, *113*, 11336–11341. [CrossRef] [PubMed]

16. Polman, J.A.; Welten, J.E.; Bosch, D.S.; de Jonge, R.T.; Balog, J.; van der Maarel, S.M.; de Kloet, E.R.; Datson, N.A. A genome-wide signature of glucocorticoid receptor binding in neuronal PC12 cells. *BMC Neurosci.* **2012**, *13*, 118. [CrossRef] [PubMed]

17. Pataskar, A.; Jung, J.; Smialowski, P.; Noack, F.; Calegari, F.; Straub, T.; Tiwari, V.K. NeuroD1 reprograms chromatin and transcription factor landscapes to induce the neuronal program. *EMBO J.* **2016**, *35*, 24–45. [CrossRef]

18. Berger, S.; Wolfer, D.P.; Selbach, O.; Alter, H.; Erdmann, G.; Reichardt, H.M.; Chepkova, A.N.; Welzl, H.; Haas, H.L.; Lipp, H.P.; et al. Loss of the limbic mineralocorticoid receptor impairs behavioral plasticity. *Proc. Natl. Acad. Sci. USA* **2006**, *103*, 195–200. [CrossRef] [PubMed]

19. Voss, T.C.; Schiltz, R.L.; Sung, M.H.; Yen, P.M.; Stamatoyannopoulos, J.A.; Biddie, S.C.; Johnson, T.A.; Miranda, T.B.; John, S.; Hager, G.L. Dynamic exchange at regulatory elements during chromatin remodeling underlies assisted loading mechanism. *Cell* **2011**, *146*, 544–554. [CrossRef] [PubMed]

20. De Kloet, E.R.; Joels, M.; Holsboer, F. Stress and the brain: From adaptation to disease. *Nat. Rev. Neurosci.* **2005**, *6*, 463–475. [CrossRef]

21. Hodge, R.D.; Bakken, T.E.; Miller, J.A.; Smith, K.A.; Barkan, E.R.; Graybuck, L.T.; Close, J.L.; Long, B.; Penn, O.; Yao, Z.; et al. Conserved cell types with divergent features between human and mouse cortex. *bioRxiv* **2018**, 384826. [CrossRef]

22. Pooley, J.R.; Flynn, B.P.; Grontved, L.; Baek, S.; Guertin, M.J.; Kershaw, Y.M.; Birnie, M.T.; Pellatt, A.; Rivers, C.A.; Schiltz, R.L.; et al. Genome-Wide Identification of Basic Helix-Loop-Helix and NF-1 Motifs Underlying GR Binding Sites in Male Rat Hippocampus. *Endocrinology* **2017**, *158*, 1486–1501. [CrossRef]

23. Fong, A.P.; Yao, Z.; Zhong, J.W.; Johnson, N.M.; Farr, G.H., 3rd; Maves, L.; Tapscott, S.J. Conversion of MyoD to a neurogenic factor: Binding site specificity determines lineage. *Cell Rep.* **2015**, *10*, 1937–1946. [CrossRef] [PubMed]

24. Cho, Y.; Noshiro, M.; Choi, M.; Morita, K.; Kawamoto, T.; Fujimoto, K.; Kato, Y.; Makishima, M. The basic helix-loop-helix proteins differentiated embryo chondrocyte (DEC) 1 and DEC2 function as corepressors of retinoid X receptors. *Mol. Pharmacol.* **2009**, *76*, 1360–1369. [CrossRef]

25. Hemmer, M.C.; Wierer, M.; Schachtrup, K.; Downes, M.; Hubner, N.; Evans, R.M.; Uhlenhaut, N.H. E47 modulates hepatic glucocorticoid action. *Nat. Commun.* **2019**, *10*, 306. [CrossRef]

26. Hong, C.Y.; Gong, E.Y.; Kim, K.; Suh, J.H.; Ko, H.M.; Lee, H.J.; Choi, H.S.; Lee, K. Modulation of the expression and transactivation of androgen receptor by the basic helix-loop-helix transcription factor Pod-1 through recruitment of histone deacetylase 1. *Mol. Endocrinol.* **2005**, *19*, 2245–2257. [CrossRef]

27. Tetel, M.J.; Auger, A.P.; Charlier, T.D. Who's in charge? Nuclear receptor coactivator and corepressor function in brain and behavior. *Front. Neuroendocrinol.* **2009**, *30*, 328–342. [CrossRef] [PubMed]

28. Christensen, M.D.; Nitiyanandan, R.; Meraji, S.; Daer, R.; Godeshala, S.; Goklany, S.; Haynes, K.; Rege, K. An inhibitor screen identifies histone-modifying enzymes as mediators of polymer-mediated transgene expression from plasmid DNA. *J. Control. Release* **2018**, *286*, 210–223. [CrossRef]

29. Ochiai, H.; Fujimuro, M.; Yokosawa, H.; Harashima, H.; Kamiya, H. Transient activation of transgene expression by hydrodynamics-based injection may cause rapid decrease in plasmid DNA expression. *Gene Ther.* **2007**, *14*, 1152–1159. [CrossRef]

30. Oakley, R.H.; Busillo, J.M.; Cidlowski, J.A. Cross-talk between the glucocorticoid receptor and MyoD family inhibitor domain-containing protein provides a new mechanism for generating tissue-specific responses to glucocorticoids. *J. Biol. Chem.* **2017**, *292*, 5825–5844. [CrossRef] [PubMed]

31. Rafiee, M.R.; Girardot, C.; Sigismondo, G.; Krijgsveld, J. Expanding the Circuitry of Pluripotency by Selective Isolation of Chromatin-Associated Proteins. *Mol. Cell* **2016**, *64*, 624–635. [CrossRef] [PubMed]

32. Labonte, B.; Engmann, O.; Purushothaman, I.; Menard, C.; Wang, J.; Tan, C.; Scarpa, J.R.; Moy, G.; Loh, Y.E.; Cahill, M.; et al. Sex-specific transcriptional signatures in human depression. *Nat. Med.* **2017**, *23*, 1102–1111. [CrossRef] [PubMed]

33. Bagot, R.C.; Cates, H.M.; Purushothaman, I.; Lorsch, Z.S.; Walker, D.M.; Wang, J.; Huang, X.; Schluter, O.M.; Maze, I.; Pena, C.J.; et al. Circuit-wide Transcriptional Profiling Reveals Brain Region-Specific Gene Networks Regulating Depression Susceptibility. *Neuron* **2016**, *90*, 969–983. [CrossRef] [PubMed]

34. Boulle, F.; Massart, R.; Stragier, E.; Paizanis, E.; Zaidan, L.; Marday, S.; Gabriel, C.; Mocaer, E.; Mongeau, R.; Lanfumey, L. Hippocampal and behavioral dysfunctions in a mouse model of environmental stress: Normalization by agomelatine. *Transl. Psychiatry* **2014**, *4*, e485. [CrossRef] [PubMed]

35. Geffroy, B.; Sadoul, B.; Bouchareb, A.; Prigent, S.; Bourdineaud, J.P.; Gonzalez-Rey, M.; Morais, R.N.; Mela, M.; Nobre Carvalho, L.; Bessa, E. Nature-Based Tourism Elicits a Phenotypic Shift in the Coping Abilities of Fish. *Front. Physiol.* **2018**, *9*, 13. [CrossRef] [PubMed]

36. Pearce, D.; Yamamoto, K.R. Mineralocorticoid and glucocorticoid receptor activities distinguished by nonreceptor factors at a composite response element. *Science* **1993**, *259*, 1161–1165. [CrossRef] [PubMed]

International Journal of
Molecular Sciences

Review

Decoding the Mechanism of Action of Rapid-Acting Antidepressant Treatment Strategies: Does Gender Matter?

David P. Herzog [1,2], Gregers Wegener [3], Klaus Lieb [1,2], Marianne B. Müller [1,2,*,†] and Giulia Treccani [1,2,3,*,†]

[1] Department of Psychiatry and Psychotherapy, Johannes Gutenberg University Medical Center Mainz, Untere Zahlbacher Straße 8, 55131 Mainz, Germany; daherzog@uni-mainz.de (D.P.H.); klaus.lieb@unimedizin-mainz.de (K.L.)

[2] Focus Program Translational Neurosciences, Johannes Gutenberg University Medical Center Mainz, Langenbeckstraße 1, 55131 Mainz, Germany

[3] Translational Neuropsychiatry Unit, Department of Clinical Medicine, Aarhus University, Skovagervej 2, 8240 Risskov, Denmark; wegener@clin.au.dk

* Correspondence: marianne.mueller@uni-mainz.de (M.B.M.); gtreccan@uni-mainz.de (G.T.)

† Shared last authorship.

Received: 14 January 2019; Accepted: 19 February 2019; Published: 22 February 2019

Abstract: Gender differences play a pivotal role in the pathophysiology and treatment of major depressive disorder. This is strongly supported by a mean 2:1 female-male ratio of depression consistently observed throughout studies in developed nations. Considering the urgent need to tailor individualized treatment strategies to fight depression more efficiently, a more precise understanding of gender-specific aspects in the pathophysiology and treatment of depressive disorders is fundamental. However, current treatment guidelines almost entirely neglect gender as a potentially relevant factor. Similarly, the vast majority of animal experiments analysing antidepressant treatment in rodent models exclusively uses male animals and does not consider gender-specific effects. Based on the growing interest in innovative and rapid-acting treatment approaches in depression, such as the administration of ketamine, its metabolites or electroconvulsive therapy, this review article summarizes the evidence supporting the importance of gender in modulating response to rapid acting antidepressant treatment. We provide an overview on the current state of knowledge and propose a framework for rodent experiments to ultimately decode gender-dependent differences in molecular and behavioural mechanisms involved in shaping treatment response.

Keywords: gender; sex difference; depression; antidepressant; rapid-acting; antidepressant; Ketamine; endocrinology; (2R,6R)-Hydroxynorketamine; electroconvulsive therapy

---

1. Introduction

1.1. Does Gender Matter in Major Depressive Disorder (MDD)?

MDD poses a serious threat to global mental health. It is the second leading cause of disability worldwide [1], with more than 300 million people affected [2]. The lifetime prevalence for MDD is 2–3 times higher in women compared to men [3], however current diagnosis and treatment guidelines for MDD around the world (e.g., [4]) do not really consider gender differences. Apart from general recommendations regarding sex hormone dysfunction, premenstrual and menopausal hormonal changes that can be causal factors contributing to the development of depression-like syndromes, the diagnosis of MDD relies on still unisex criteria (ICD-10 or DSM-5). Interestingly, the incidence of new onset of depression cases drops tremendously after menopause [5]. These women are susceptible to the same extent to MDD as men, thus highlighting a crucial role of sexual hormones in the pathophysiological mechanisms of depression.

Oestrogens influence synaptic plasticity, neurotransmission, neurodegeneration and cognitive function [6]. In addition, several studies revealed a neuroprotective role of progesterone in rodent and in humans [7]. Therefore, the underlying pathways driven by hormones and other molecular players are in the spotlight of gender-specific MDD research, with the intent of possibly bridging the gap between gender-specific disease parameters (i.e., higher symptom severity [8–10], an earlier onset of disease [9,10] and an increased duration of depressive episodes [9,10]) and underlying neurobiological and molecular changes.

1.2. Genetics, Epigenetics and Hormones: Powerful Players Shaping Gender-Specificity of MDD

The sexually dimorphic anatomy and function of the brain is strongly influenced by a broad variety of parameters, such as hormonal status, variance in body fat, liver metabolism, the transcription machinery, gene accessibility via epigenetic modifications and others [11–14]. Moreover, the heritability of MDD is higher in women than in men implicating an increased genetic vulnerability [15].

A longitudinal study by Bundy et al. showed that 198 genes were differentially expressed in the hippocampus between male and female mice across developmental stages [16]. The older the animals were, the more differentially expressed genes were found between male and female mice, indicating the importance and the increase over time in sexual dimorphism of the genome. Indeed, the difference of the transcriptome profile is bigger between female rodents with high and low hormonal states than between female and male rodents [17]. A recent translational study from the Nestler group focused on sex differences of transcriptome profiles comparing gender in both humans and mice [18]. In both species, the authors revealed that there is only a limited overlap of regulated genes between males and females in several brain regions [18]. These data highlighted that the transcriptome profiles are gender-specific both in depressed patients and in animal models resembling some of the features of depression. In depressed patients, only 5–10% of genes were shared between women and men [18], while in the animals were 20% [18]. Finally, by means of a translational approach, the researchers were able to identify and validate gender-specific candidate genes for depression, namely *Dusp6* for female and *Emx1* for male gender [18].

In the complex symphony of genetic risk load and environmental factors of depression, epigenetics plays a pivotal role in the gene-environmental interaction [19]. Epigenetics (i.e., all mechanisms to modify chromatin accessibility) affects brain functions as well as disease conditions [20]. In 2015, Nugent and colleagues reported that in male rodents the enzyme DNA methyltransferase 3a was required to keep a male phenotype [21]. Manipulation of this process led to feminization of the brain, thus displaying the importance of epigenetic mechanisms for gender determination of the organism. A recent review by Marija Kundakovic highlighted the need to include both sexes in research studies of depression and at the same time to consider the hormonal state of female to study brain function in healthy and disease conditions [11].

The most relevant risk factor of MDD is stress. Stress lowers the threshold for an organism to develop mental diseases like MDD. Stress can be similarly used in animal models to study mental illnesses, as it is a highly conserved and evolutionary important factor across species. Response to stress is highly gender-specific. Female rodents have a greater sensitivity to stress with a more prominent stress response, a longer period of recovery and a more active hypothalamic–pituitary–adrenal (HPA) axis [22,23]. During the stress response, the induction of immediate early gene *Fos* after acute stress is greater in female rodents [24], especially in the hippocampus, where immediate early gene expression is reported to be more pronounced [25]. At the morphological level, stress has different impact on gender. Indeed, chronic stress reduced dendritic complexity and spine remodelling in hippocampal pyramidal cells of male rodents but not in females [26,27]. Moreover, social deprivation experiments using single housing of rodents over a long period of time revealed more anxious and anhedonia-like phenotypes in both male and female rodents, although the effect was more pronounced in males [28].

For a more in-depth and recent review of gender specificity of behavioural tests in terms of depression, depression treatment and stress see [12].

The discovery of the HPA axis together with the detection of glucocorticoid receptors found in hypothalamus and hippocampus [29], led to the conclusion that hormones and the brain were strongly connected and influenced by each other. This is supported by a plethora of interesting findings (Figure 1): Hormonal alterations during the menstrual cycle affect females across species. Elevated oestrogen levels mediate an enhanced hippocampal spine density [30] and plasticity [31]. Hagemann et al. revealed that during ovulation women had a significant increase in grey matter volume [32]. Finally, the end of the menstrual cycle and by that, the decline of sexual hormones oestrogen and progesterone may induce depression-like symptoms [33]. Sex differences in the stress-induced remodelling of dendrites and synapses in brain regions such as the hippocampus or the prefrontal cortex first emerge in puberty [34], thus highlighting the link between female susceptibility to stress related disorders and hormone effects, mainly oestrogen and progesterone [14]. For a review article summarizing the hormonal impact on the female brain see [13].

To support the evidence of the importance of a sexual dimorphic brain, several studies revealed a crosstalk between sexual hormones and key neurotransmitters altered in depression. Amin and colleagues showed that in rodents, brain serotonin pharmacodynamics and –kinetics are affected by oestrogens [35]. Serotonin, the major player in the monoaminergic hypothesis depression, is also influenced by gender; human studies revealed that women had higher levels of serotonin, serotonin metabolites [36] and serotonin transporters [37]. This might also explain why a lack of serotonin levels has a stronger effect in women. Finally, brain-derived neurotrophic factor (BDNF), a key mediator of neurogenesis and neuronal survival altered in depression and inversely associated with depressive symptoms [38] expression, synthesis and function is influenced by oestradiol, whereas BDNF itself is a downstream mediator activated by oestradiol signalling in the hippocampus [39].

1.3. Gender-Specific Differences in MDD Therapy

In line with the lack of gender-specific recommendations for the diagnosis of MDD, there is also a neglect of gender in depression therapy. During antidepressant treatment, women generally show higher response rates than men [40]. What is the evidence for gender-specificity of antidepressant treatment? A meta-analysis covering 30 randomized controlled trials found no gender-specificity of tricyclic antidepressant drugs (TCAs) like imipramine [41], whereas other studies reported a superior response rate of males receiving TCAs (e.g., [42,43]). It was also shown that females after menopause benefit from TCA treatment compared to pre-menopausal women, similarly like males [42,43]. Selective serotonin reuptake inhibitors (SSRI) like fluoxetine or citalopram, commonly used antidepressant agents, induced a response only in 50% of SSRI-treated patients while 70% of the same patients do not have full remission after 12 weeks of treatment with SSRIs [44]. There are numerous studies reporting that SSRIs have a higher efficacy in women [42,43,45,46]. In addition, SSRI and hormone replacement therapy was reported to be beneficial for women after menopause [47,48]. In contrast, a meta-analysis from Cuijpers et al. did not detect sex differences in the treatment with either SSRIs or TCAs [49]. The inconsistencies within those reports are not easy to be explained. One reason might be highly different inclusion and exclusion criteria of clinical trials. These might lead to different patient stratification, thus creating a huge bias and complicating the interpretation of the results. Another possible reason might be that psychiatric nosology and diagnosing is a highly artificial process. Two persons diagnosed with depression may express completely different symptoms, however in clinical trials, they might be put together as namely "depressed patient." A recent National Institute of Mental Health (NIMH) initiative called the Research Domain Criteria (RDoC), introduces a transdiagnostic approach suited to specify symptoms and phenotypes at individual level [50].

For a detailed review about the evidence of gender-specificity of *classical* antidepressant drug treatment see [14].

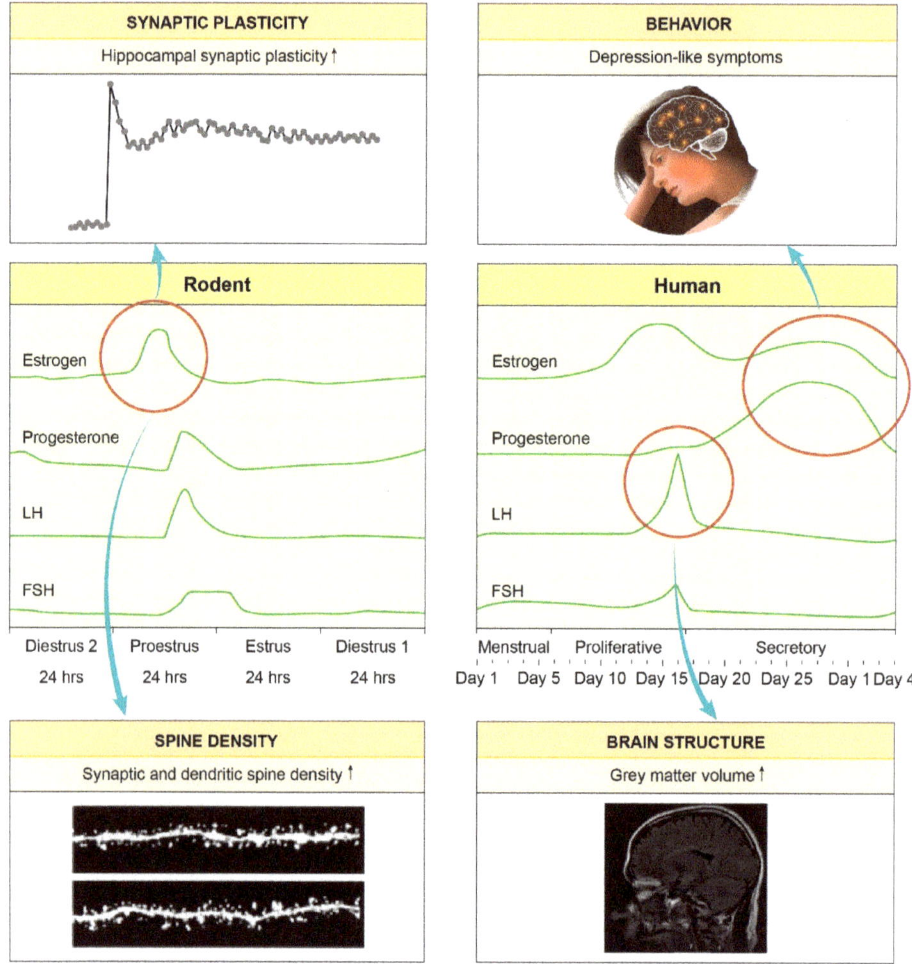

Figure 1. Highlighting the hormonal impact on the female brain: changes throughout the menstrual cycle. Rodents and women share a similar menstrual cycle pattern. Alterations in hormone levels are known to play an important role in neurobiology and mental health. The increased levels of oestrogen in the rodent proestrus phase have been reported to guide an increase in both hippocampal spine density [30] and plasticity [31]. In women, a significant increase in grey matter volume was found around ovulation [32]. The decrease of circulating hormones at the end of the menstrual cycle may induce depression-like symptoms [33].

Additional factors have been reported to play a role in gender differences in antidepressant efficacy. Different pharmacokinetics of TCAs [51,52] could explain why women report more drug related side effects and why women seem to prefer medication with SSRIs. Women showed a superior adherence to continuous antidepressant drug treatment [53].

1.4. Why We Need Antidepressant Treatment Approaches with a Rapid Onset of Action

After initiating antidepressant drug treatment with classical antidepressant compounds (i.e., TCAs or SSRIs) there is a latency lasting up to four weeks before a detectable treatment response becomes evident. A possible loss of antidepressant treatment efficacy in long term-treated patients, the danger of manic switches and several prominent side effects are further limitations of antidepressant treatment [54,55]. In addition, the large heterogeneity of antidepressant treatment response and the lack of biomarkers to monitor or stratify disease state, to facilitate diagnostic decisions or to predict treatment success already early turn antidepressant drug treatment into a large "trial and error" game [56]. To overcome the current limitations in neuropsychopharmacology, increasing attention is given to biomarker research by using truly translational research projects, combining animal and human cohorts in the same study design. The accessibility of the central nervous system for the validation of potent drug candidate structures and the possibility of biomarker studies in human and animal blood offer a unique and meaningful platform to really contribute to the development of novel antidepressant agents and improve MDD therapy [57]. Another promising approach is the emergence of rapid-acting antidepressant treatments, which may possess a distinct mechanism of action than the conventional antidepressants, promote faster recovery and thereby overcome the high socio-economic costs of long-lasting depression courses. Among those, electroconvulsive therapy, ketamine and other compounds showing similar properties, cause higher response rate also in treatment resistant, heavily depressed patients - for some patients even within hours. In the following paragraphs we will highlight the major findings of this class of therapy and discuss the implication of gender in their mechanism of action.

2. Gender-Specific Differences in the Molecular Mechanisms of Rapid-Acting Antidepressant Drugs

2.1. Ketamine: Evidence of Gender-Specific Differences in the Effect on the Brain

Recently, the discovery of drugs with rapid acting antidepressant efficacy has built the basis for novel treatment strategies in MDD. Among those, ketamine, a non-competitive NMDA receptor antagonist, originally introduced as a dissociative anaesthetic, revealed to induce a rapid (within hours) and sustained (up to 1 week) antidepressant effect in treatment-resistant [58,59]. Furthermore, its rapid action on suicide ideation significantly improves the management of acute MDD treatment [60,61].

Ketamine has been shown to exert antidepressant-like effects in different animal models of depression by activation of the mammalian target of rapamycin (mTOR), by increasing the expression of synaptic proteins and increasing synaptogenesis in the prefrontal cortex [62]. The mechanism by which ketamine produces fast antidepressant-like effects depends on the rapid synthesis of brain-derived neurotrophic factor [63] and on dendritic release/translation of BDNF.

Preliminary clinical observations suggest gender differences to play a role in ketamine metabolism and clearance in a dose dependent manner. However, only recently a more comprehensive characterization of the gender differences in the effects of ketamine has been done. Female C57BL/6J stress-naïve mice were more sensitive to the rapid and sustained antidepressant-like effects of ketamine in the forced swim test (FST) and responsive to lower doses of ketamine [64]. Moreover, the female mice responded earlier to a single injection of ketamine than the male mice.

A possible explanation may lie in sex differences of ketamine pharmacokinetics. Female rats exhibited greater concentrations of ketamine and norketamine over the first 30 min following treatment in both brain and plasma, due to slower clearance rates and longer half-lives. Gender differences can influence the metabolism of ketamine and therefore the amount of ketamine and norketamine reaching brain areas [65].

In another study, instead, oestrogens augmented the effect of ketamine and its metabolites (2R,6R)-HNK and (2S,6S)-HNK via induction of the CYP2A6 and CYP2B6 enzymes responsible for ketamine's biotransformation into its active metabolites [66].

The pharmacodynamics of ketamine is also modified in a gender specific way with a possible synergism with sexual hormone. Female enhanced sensitivity to ketamine during proestrus was mediated via oestradiol activation of oestrogen receptor (ER)alpha and ERbeta, leading to greater activation of synaptic plasticity related kinases within prefrontal cortex and hippocampus (Table 1) [67].

Different molecular pathways are activated by ketamine in a gender specific manner. Indeed male chronic isolated rats showed an increased spine density after ketamine treatment in medial prefrontal cortex via restoration of synapsin 1, PSD95 and GluR1 levels while those proteins were not altered in female rats after ketamine treatment [28].

However, the literature reported as well studies showing no effects of gender in the effects of ketamine. Indeed, a recent work revealed no effects of gender in the acute and chronic effects of ketamine in ICR mice [68].

2.2. Rapid Antidepressant-like Effects with Less Side Effects? Emerging Data on Ketamine Metabolites

Recently, the major ketamine metabolites have been drawn into focus of international research studies. In 2016, Zanos and colleagues showed that the ketamine metabolite (2R,6R)-hydroxynorketamine (HNK) produced antidepressant-like effects in mice similarly to those of ketamine but without the ketamine-associated side effects [69]. They were the first to propose a NMDA receptor-independent mechanism for the antidepressant-like effects of ketamine and its metabolites, indicating an important role of AMPA receptors [69]. Although the relevance of HNK and the question of AMPA receptor involvement still remains under debate ([70,71] and Table 2), HNK is a very interesting compound offering the great potential of rapid antidepressant-like effects without strong side effects.

Chou and colleagues applied the learned helplessness paradigm in male and female rats as a depression model in rodents [72]. They could show that HNK rapidly rescued depression-like conditions assessed by using the FST and the sucrose preference test (SPT) [72]. In addition, they did not detect a difference between female and male rats with respect to HNK effects and no effect at all using (2S,6S)-hydroxynorketamine [72]. Yamaguchi and colleagues recently reported that in male mice using the lipopolysaccharide (LPS) inflammation model of depression the metabolism from R-ketamine to HNK is not exclusively essential for the antidepressant-like effects of ketamine in rodents [70]. They validated this finding using cytochrome p 450 inhibitors. By combining a microdialysis experiment with the FST, HNK was found both in plasma and in brain as an antidepressant-like acting metabolite [73]. At the morphological level, HNK produces an increase of structural plasticity in murine and human dopaminergic neurons [74]. Similarly, Collo et al. reported that in human dopaminergic neurons HNK produced effects on dendritic outgrowth similar to those seen with ketamine [75] and Yao et al. showed that ketamine and HNK effects on AMPA-receptors and synapse alterations in the murine mesolimbic system are strongly aligned [76].

On the other hand, some publications show a lack of response to HNK in rodents. Shirayama et al. reported that they could not find an effect of HNK in a rat model of learned helplessness [77] using the Conditioned Avoidance Test. On top of that, Yang and colleagues were only able to find a very weak effect of HNK in an LPS model of depression in mice and did not find an effect in a chronic social defeat model of depression [78] using the FST, the Tail Suspension Test and the SPT.

To our knowledge, there is only one paper [72] focusing on the gender aspect of HNK treatment. In this study, no difference of the antidepressant-like effects of HNK could be observed. It is important to fill this gap, especially because a more pronounced effect of ketamine in females has been repeatedly reported both in rodents and humans.

Table 1. In vitro and in vivo experiments using ketamine treatment.

Publication	Test Subject	Study Design	Antidepressant-Like Effect	Molecular Mechanism
Franceschelli et al., 2015 [64]	Male and female C57/BL6J mice	KET in naïve and CMS animals: female and male mice (FST)	KET effect: Female mice > male mice	Effects on excitatory amino acids (glutamate and aspartate), serotoninergic activity.
Saland et al., 2018 [65]	Male and female Sprague-Dawley rats	KET metabolism and distribution		↑ level of KET and NK in both brain and plasma
Ho et al., 2018 [66]	Human iPSC-derived astrocytes	Oestrogen + KET in vitro	Oestrogens augmented the effect of KET	↑ level of AMPA receptor subunit and ERα. Oestrogens: ↑ level of CYP2A6 and CYP2B6.
Dossat et al., 2018 [67]	Male and female C57/BL6J mice	Oestrogen and Progesterone receptor agonist and KET (FST)	Female in proestrus + KET: sensitive to lower dose.	Proestrus female ↑ p-Akt and p-CaMKIIα.
Sarkar et al. 2016 [28]	Male and female Sprague-Dawley rats	KET and social isolation stress (behaviour and synaptic protein level)	IS: male depression like behaviour at 8 weeks while female at 11 weeks. KET rescued the phenotype.	Decline in spine density and synaptic proteins reversed by KET only in male but not female

We list relevant, ketamine-associated publications with significant impact in the field. *KET* ketamine, *CMS* chronic mild stress, *FST* forced swim test, *NK* norketamine, *iPSC* induced Pluripotent Stem Cells, *AMPA* α-amino-3-hydroxy-5-methyl-4-isoxazolepropionic acid receptor, *CYP* Cytochrom P 450 Enzyme, *IS* Isolation Stress.

Table 2. In vitro and in vivo experiments using HNK treatment.

Publication	Test Subject	Study Design	Antidepressant-Like Effect	Molecular Mechanism
Zanos et al. 2016 [69]	Male and female C57/BL6J mice	KET: female and male mice (FST) Ketamine metabolites: male mice (CSD, FST, ST)	KET: Female mice > male mice HNK: HNK > (2S,6S)-hydroxynorketamine HNK: lacks ketamine-related side effects	HNK-effects independent of NMDAR-signalling by AMPAR-signalling
Yamaguchi et al. 2018 [70]	Male C57/BL6 mice	LPS with (R)-ketamine and HNK (FST, TST)	(R)-ketamine > HNK Blocking CYP: (R)-ketamine effects ↑	(R)-ketamine and HNK in plasma, brain, CSF
Chou et al. 2018 [72]	Male and female Sprague-Dawley rats	LH with ketamine metabolites (FST, SPT)	HNK: Male ≈ Female (2S,6S)-hydroxynorketamine: no effect	HNK: enhancement of AMPAR-signalling in vlPAG

Table 2. *Cont.*

Publication	Test Subject	Study Design	Antidepressant-Like Effect	Molecular Mechanism
Pham et al. 2018 [73]	Male BALB/cJ mice	Local (mPFC) and systemic injection of KET and HNK (FST)	Local injection: HNK ≈ KET Systemic injection: HNK ≈ KET	HNK+KET: extracellular 5-hydroxytryptamine (mPFC) ↑, extracellular glutamate (mPFC) ↑ KET: extracellular GABA ↑
Cavalleri et al. 2018 [74]	Murine and human DA neurons	KET and HNK in vitro	-	HNK+KET: structural plasticity ↑ (arborization ↑, soma size ↑)
Collo et al. 2018 [75]	Human DA neurons (PSCs)	KET and HNK in vitro	-	HNK+KET: structural plasticity ↑ (dendrite length ↑ and number ↑)
Yao et al. 2017 [76]	Male C57/BL6 mice	KET and HNK tested ex vivo with electrophysiology	-	HNK+KET: lasting modulation of AMPAR and synaptic plasticity (NAc+VTA), potentiation ↓ and depression ↑ of GA synapses (NAc+VTA-DA neurons)
Shirayama et al. 2018 [77]	Male Sprague-Dawley rats	LH with ketamine metabolites (CAT)	KET: antidepressant-like effect HNK: no effect	-
Yang et al. 2017 [78]	Male C57/BL6 mice	LH and CSD with KET and HNK (FST, TST, SPT)	KET: antidepressant-like effect HNK: no effect	-

We list relevant, HNK-associated publications with significant impact in the field. *HNK* (2R,6R)-hydroxynorketamine, *KET* (R,S)-ketamine, *FST* Forced Swim Test, *CSD* Chronic Social Defeat Model of Depression, *ST* Sociability Test, *NMDAR* N-methyl-D-aspartate receptor, *AMPAR* α-amino-3-hydroxy-5-methyl-4-isoxazolepropionic acid receptor, *LPS* Lipopolysaccharide Model of Depression, *TST* Tail Suspension Test, *CYP* Cytochrom P 450 Enzyme, *CSF* Cerebrospinal Fluid, *LH* Learned Helplessness Model of Depression, *SPT* Sucrose Preference Test, *vlPAG* ventrolateral Periaqueductal grey, *mPFC* medial Prefrontal Cortex, *GABA* gamma-aminobutyric acid, *DA* dopaminergic, *PSC* pluripotent stem cell, *NAc* Nucleus Accumbens, *VTA* Ventral Tegmental Area, *GA* glutamatergic, *CAT* Conditioned Avoidance Test.

2.3. Other Rapid-Acting Antidepressant Agents

Rapid-acting antidepressant agents share some key neurobiological pathways, which probably mediate their antidepressant-like effects [79]. By altering glutamate transmission, they enhance mTOR signalling, which leads to increased BDNF levels, a process strongly connected to enhanced synaptic activity and plasticity in the prefrontal cortex (PFC). Besides ketamine and its metabolites, there are several substances and molecules known, which fit into this mechanistic framework and at the same time have shown antidepressant-like effects (Table 3). For recent in-depth reviews see [71,80].

Table 3. Other rapid-acting antidepressant agents.

Agent	Molecular Target	Reference
Scopolamine	M1/2-antagonist, AMPAR↑, mTOR↑	[81,82]
GLYX-13	Partial agonist and modulator of NMDAR, AMPAR↑	[83,84]
MGS0039, LY3020371	mGlu2/3 antagonists, AMPAR↑	[85–87]
L-655,708, MRK-016	NAM of α5-GABA$_A$-R, cortex & HC-specific	[88,89]
Cannabidiol	5-HT$_{1A}$-R↑, CB1↑, vmPFC	[90–93]
Psychedelics (LSD, DOI, DMT, MDMA)	TrkB→mTOR↑+BDNF↑, 5-HT$_{2A}$-R↑, PFC	[94]

We list several other important compounds, which have been shown to provide antidepressant-like effects similar to ketamine and its metabolites. *M1/2* Muscarinic acetylcholine Receptor Type 1 and 2, *AMPAR* α-amino-3-hydroxy-5-methyl-4-isoxazolepropionic acid receptor, *mTOR* mammalian target of rapamycin, *mGlu2/3* Metabotropic glutamate receptor type 2 and 3, *NAM* negative allosteric modulator, *α5-GABA$_A$-R* Alpha 5 subunit of the gamma-aminobutyric acid type A receptor, *HC* hippocampus, *NMDAR* N-methyl-D-aspartate receptor, *5-HT$_{1A}$-R* 5-Hydroxy tryptophan receptor type 1A, *CB1* Cannabinoid receptor type 1, *vmPFC* ventromedial prefrontal cortex, *LSD* lysergic acid diethylamide, *DOI* (2,5)-dimethoxy-4-iodoamphetamine, *DMT* N,N-dimethyltryptamine, *MDMA* (3,4)-methylenedioxymethamphetamine, *TrkB* Tyrosine receptor kinase B, *BDNF* Brain-derived neurotrophic factor, *5-HT$_{2A}$-R* 5-Hydroxy tryptophan receptor type 2A, *PFC* Prefrontal cortex.

In line with the glutamate hypothesis of depression and depression treatment, pharmacological research focuses on the role of NMDA and AMPA receptors. Based on the evidence showing that the NMDA receptor antagonist ketamine produced sustained antidepressant effects in both humans and rodents, other ways to manipulate glutamate transmission were subsequently examined. Scopolamine, a muscarinic acetylcholine receptor (mAchR) antagonist, produced rapid, antidepressant effects even in treatment resistant depressed patients [81]. Rodent studies revealed that type 1 and 2 mAchRs are exclusively responsible for these effects [82], which led to enhanced AMPA receptor signalling and mTOR activation.

GLYX-13 which is also known as Rapastinel—targets the outer surface of the NMDA receptor without occupying its ion pore and acts as a partial agonist [71,80]. That is probably why a single injection of GLYX-13 can decrease depression scores in patients [84], without inducing ketamine-related side effects. The antidepressant effect appeared after two hours and persisted for at least 7 days. This compound was effective in phase 2 clinical trials [83]. In addition, the direct and specific manipulation of type 2 and 3 metabotropic glutamate receptors emerged as an interesting approach. By enhancing AMPA receptor signalling, antagonists of these receptors produced antidepressant-like effects in rodents, without ketamine-related side effects [85–87]. GABA$_A$ receptor signalling reduces glutamate transmission, which explains why GABA receptor antagonists and modulators are also important in the search for rapid-acting antidepressant drugs. Negative allosteric modulation of the alpha 5 subunit of the type A GABA receptor provides cortex- and hippocampus specific [95] antidepressant-like effects [88] in an animal study by Fischell et al. These drugs have also been shown to lack the ketamine-related side effects [89].

Cannabidiol (CBD) is the non-psychotomimetic compound of *Cannabis sativa*. Anxiolytic [96] and antidepressant-like effects [97] have been reported in the literature. More recently, it was shown that type 1A serotonin receptor (5-HT$_{1A}$) is important for the effects of cannabidiol. In a study with rats [91], cannabidiol was injected bilaterally into ventromedial parts of the PFC. Acute antidepressant-like effects were observed, which were dependent on 5-HT$_{1A}$ and cannabinoid receptor CB1 signalling. Sales et al. have shown that i.p. injection of cannabidiol also led to an acute (30 min) antidepressant response in rodents [90]. They were able to observe an increase in synaptic plasticity in the PFC, accompanied by elevated BDNF levels in PFC and hippocampus. The antidepressant-like effects could be prevented by blocking the mTOR pathway. Two studies with CBD found that CBD in the saccharin preference test showed a pro-hedonic effect of CBD in male and female Wistar Kyoto rats (WKY) and decreased immobility in the FST in male Flinders Sensitive Line rats (FSL) and male and female WKY but not female FSL [92,93].

Inducing structural and functional plasticity in the PFC is a key process of antidepressant action. It has been shown that psychedelics such as lysergic acid diethylamide (LSD) and others (see Table 3) are also able to exert rapid antidepressant-like effects in humans and rodents. Recently, Ly et al. showed that LSD and alike agents produced a robust increase in neuritogenesis and spinogenesis in vitro and in vivo [94]. They were able to show these effects spanning a range from drosophila larvae, zebrafish embryos and rats, leading to the conclusion that it is a process highly conserved during evolution. Finally, they could show that mTOR and type 2A serotonin receptor play crucial roles in inducing these morphological and molecular changes.

3. Non-Pharmacological, Rapid-Acting Treatment for MDD: Electroconvulsive Therapy

3.1. Molecular Pathways Shaping the Effect of Stimulating the Brain

Brain stimulation techniques represent an important part of last-line treatment options for MDD therapy. They are usually considered for patients with severe depression and treatment-resistant MDD courses. By applying electricity, all brain stimulation approaches interfere with activation and/or inactivation of certain brain circuits, brain regions and molecular pathways. Electroconvulsive therapy (ECT) is one of the oldest treatment paradigms in psychiatry, which is effective and approved for treatment-resistant depression, depression with psychotic symptoms, mania, catatonia and treatment-resistant schizophrenia [98]. Under anaesthesia and muscle relaxation, it induces a generalized seizure [99,100]. Some patients report an improved depressive symptomatology score already after their first ECT session within hours after 2-3 sessions reliable and sustained antidepressant effects occur even in chronic, severely ill patients and after 8 bilateral sessions the average patient shows full remission of symptoms [101].

Neuroinflammation, reduced levels of monoaminergic neurotransmitters and altered HPA axis activity have been identified as key mechanisms in depression pathophysiology [102]. Similar to pharmacological treatment with antidepressant drugs [103], long-term ECT reduces the activation of the immune system [104]. Specifically, Yrondi et al. described an immediate immuno-inflammatory response after the first ECT session, which was reversed after long-term ECT [105]. By normalizing the brain immuno-inflammatory state with ECT, biomarkers linked to antidepressant-like effects—like serum BDNF—are upregulated [106]. ECT also has a strong effect on modulating the HPA axis activity, with the neuroendocrine system being closely linked to depression and antidepressant response. By means of neuroendocrine challenge tests (dexamethasone suppression test and the combined dexamethasone/corticotropin releasing hormone test), a reduced hormonal response can be found after long-term ECT [107,108], indicating a normalization of HPA axis activity. Along with these findings, there are several studies reporting a long-term decrease in stress hormone cortisol levels following ECT (e.g., [109]).

ECT increased several monoamines after long-term treatment in patients [110]. In addition, animal studies showed that ECT enhanced neurogenesis [111] and neuroplasticity [112], processes which both are known to play important roles in antidepressant-like effects. Neuroimaging studies showed that the hippocampus, the amygdala, prefrontal cortex, anterior cingulate cortex and basal ganglia are main target brain regions of ECT [113]. Amygdala and hippocampus volumes increased after ECT [113], together with an improved functional connectivity [113]. Benson-Martin and colleagues provided an overview of the genetic pathways involved in the mechanisms of action of ECT treatment [114]. Finally, de Jong and colleagues and others reported that ECT resulted in a robust impact on epigenetic mechanisms [115,116].

3.2. Clinical Efficacy and the Role of Gender

Interestingly, although ECT is a very old technique, there is not much evidence for gender-specific aspects of ECT treatment reported in literature. A retrospective comparison from 2005 analysed patients with MDD, bipolar disorder and schizophrenia and found that women with MDD and schizophrenia received their first ECT session earlier meaning they had less previous antidepressant drug trials than men [117]. In addition, the authors of this study reported that ECT was more effective in women with schizophrenia compared to male schizophrenic patients [117]. In a study including ECT cases conducted in a hospital in Turkey, the authors did not find a difference between men and women in response to ECT across all tested diagnoses [118]. A study by Bousman et al. reported that the beneficial effect of a catechol-O-methyltransferase polymorphism on ECT response can be exclusively found in male patients [119]. In summary, there is very little knowledge about the role of gender in ECT and more studies are needed to fill this lack of knowledge.

4. Conclusions

There is a considerable amount of evidence collected from both animal and human studies, highlighting the central role of gender in depression pathophysiology and treatment. A better understanding of the molecular mechanism activated by hormones both in health and psychiatric disorders combined with a precise knowledge of the pharmacological interaction between hormones and antidepressant drugs, would be suited to redefine treatment guideline and possible identify molecular targets relevant for drug discovery and for gender personalized therapy.

Author Contributions: D.H. and G.T. wrote the first draft of this paper. D.H., G.W., K.L., M.M. and G.T. contributed to writing and discussing the paper and approved its final version.

Funding: Please add: M.B.M. and K.L. are supported by the German Research Foundation (DFG) within the Collaborative Research Centre 1193 (CRC1193, https://crc1193.de/) and by the Boehringer Ingelheim Foundation. G.T. is supported by a 2014 NARSAD Young Investigator Grant from the Brain & Behaviour Research Foundation and the Danish Council for Independent Research grant number DFF-5053-00103. G.W. is supported by the Independent Research Fund Denmark (grant 8020-00310B), Aarhus University Research Foundation (AU-IDEAS initiative (eMOOD) and EU Horizon 2020 (ExEDE).

Acknowledgments: D.P.H. is supported by the Mainz Research School of Translational Biomedicine (TransMed) with a MD-PhD fellowship.

Conflicts of Interest: D.H., G.T., M.B.M., K.L. report no conflict of interest. G.W. declares having received research support/lecture/consultancy fees from H. Lundbeck A/S, Servier SA, Astra Zeneca AB, Eli Lilly A/S, Sun Pharma Pty Ltd., Pfizer Inc., Shire A/S, HB Pharma A/S, Arla Foods A.m.b.A., Alkermes Inc, Johnson & Johnson Inc. and Mundipharma International Ltd.

References

1. Patel, V.; Chisholm, D.; Parikh, R.; Charlson, F.J.; Degenhardt, L.; Dua, T.; Ferrari, A.J.; Hyman, S.; Laxminarayan, R.; Levin, C.; et al. Addressing the burden of mental, neurological and substance use disorders: Key messages from Disease Control Priorities. *Lancet* **2016**, *387*, 1672–1685. [CrossRef]
2. WHO. World Health Organization Factsheet: Depression. Available online: www.who.int/news-room/fact-sheets/detail/depression (accessed on 10 October 2018).
3. Kessler, R.C.; Berglund, P.; Demler, O.; Jin, R.; Merikangas, K.R.; Walters, E.E. Lifetime prevalence and age-of-onset distributions of DSM-IV disorders in the National Comorbidity Survey Replication. *Arch. Gen. Psychiatry* **2005**, *62*, 593–602. [CrossRef] [PubMed]

4. NICE. Depression in Adults: Recognition and Management—National Institute for Health and Care Excellence Guidelines [CG90]. Available online: www.nice.org.uk/guidance/cg90 (accessed on 2 January 2019).

5. Freeman, E.W.; Sammel, M.D.; Boorman, D.W.; Zhang, R. Longitudinal pattern of depressive symptoms around natural menopause. *JAMA Psychiatry* **2014**, *71*, 36–43. [CrossRef] [PubMed]

6. Gillies, G.E.; McArthur, S. Estrogen actions in the brain and the basis for differential action in men and women: A case for sex-specific medicines. *Pharmacol. Rev.* **2010**, *62*, 155–198. [CrossRef] [PubMed]

7. Wei, J.; Xiao, G.M. The neuroprotective effects of progesterone on traumatic brain injury: Current status and future prospects. *Acta Pharmacol. Sin.* **2013**, *34*, 1485–1490. [CrossRef] [PubMed]

8. Kessler, R.C.; McGonagle, K.A.; Swartz, M.; Blazer, D.G.; Nelson, C.B. Sex and depression in the National Comorbidity Survey. I: Lifetime prevalence, chronicity and recurrence. *J. Affect. Disord.* **1993**, *29*, 85–96. [CrossRef]

9. Kornstein, S.G.; Schatzberg, A.F.; Thase, M.E.; Yonkers, K.A.; McCullough, J.P.; Keitner, G.I.; Gelenberg, A.J.; Ryan, C.E.; Hess, A.L.; Harrison, W.; et al. Gender differences in chronic major and double depression. *J. Affect. Disord.* **2000**, *60*, 1–11. [CrossRef]

10. Marcus, S.M.; Young, E.A.; Kerber, K.B.; Kornstein, S.; Farabaugh, A.H.; Mitchell, J.; Wisniewski, S.R.; Balasubramani, G.K.; Trivedi, M.H.; Rush, A.J. Gender differences in depression: findings from the STAR*D study. *J. Affect. Disord.* **2005**, *87*, 141–150. [CrossRef] [PubMed]

11. Kundakovic, M. Sex-Specific Epigenetics: Implications for Environmental Studies of Brain and Behavior. *Curr. Environ. Health Rep.* **2017**, *4*, 385–391. [CrossRef]

12. LeGates, T.A.; Kvarta, M.D.; Thompson, S.M. Sex differences in antidepressant efficacy. *Neuropsychopharmacology* **2018**. [CrossRef]

13. Marrocco, J.; McEwen, B.S. Sex in the brain: Hormones and sex differences. *Dial. Clin. Neurosci.* **2016**, *18*, 373–383.

14. Sramek, J.J.; Murphy, M.F.; Cutler, N.R. Sex differences in the psychopharmacological treatment of depression. *Dial. Clin. Neurosci.* **2016**, *18*, 447–457.

15. Kendler, K.S.; Thornton, L.M.; Prescott, C.A. Gender differences in the rates of exposure to stressful life events and sensitivity to their depressogenic effects. *Am. J. Psychiatry* **2001**, *158*, 587–593. [CrossRef] [PubMed]

16. Bundy, J.L.; Vied, C.; Nowakowski, R.S. Sex differences in the molecular signature of the developing mouse hippocampus. *BMC Genom.* **2017**, *18*, 237. [CrossRef] [PubMed]

17. Duclot, F.; Kabbaj, M. The estrous cycle surpasses sex differences in regulating the transcriptome in the rat medial prefrontal cortex and reveals an underlying role of early growth response 1. *Genome Biol.* **2015**, *16*, 256. [CrossRef] [PubMed]

18. Labonte, B.; Engmann, O.; Purushothaman, I.; Menard, C.; Wang, J.; Tan, C.; Scarpa, J.R.; Moy, G.; Loh, Y.E.; Cahill, M.; et al. Sex-specific transcriptional signatures in human depression. *Nat. Med.* **2017**, *23*, 1102–1111. [CrossRef] [PubMed]

19. Binder, E.B. Dissecting the molecular mechanisms of gene x environment interactions: Implications for diagnosis and treatment of stress-related psychiatric disorders. *Eur. J. Psychotraumatol.* **2017**, *8*, 1412745. [CrossRef] [PubMed]

20. Nestler, E.J. Epigenetic mechanisms of depression. *JAMA Psychiatry* **2014**, *71*, 454–456. [CrossRef] [PubMed]

21. Nugent, B.M.; Wright, C.L.; Shetty, A.C.; Hodes, G.E.; Lenz, K.M.; Mahurkar, A.; Russo, S.J.; Devine, S.E.; McCarthy, M.M. Brain feminization requires active repression of masculinization via DNA methylation. *Nat. Neurosci.* **2015**, *18*, 690–697. [CrossRef]

22. Bale, T.L. Stress sensitivity and the development of affective disorders. *Horm. Behav.* **2006**, *50*, 529–533. [CrossRef]

23. Fernandez-Guasti, A.; Fiedler, J.L.; Herrera, L.; Handa, R.J. Sex, stress and mood disorders: At the intersection of adrenal and gonadal hormones. *Horm. Metab. Res.* **2012**, *44*, 607–618. [CrossRef] [PubMed]

24. Babb, J.A.; Masini, C.V.; Day, H.E.; Campeau, S. Stressor-specific effects of sex on HPA axis hormones and activation of stress-related neurocircuitry. *Stress (Amsterdam, Netherlands)* **2013**, *16*, 664–677. [CrossRef] [PubMed]

25. Bohacek, J.; Manuella, F.; Roszkowski, M.; Mansuy, I.M. Hippocampal gene expression induced by cold swim stress depends on sex and handling. *Psychoneuroendocrinology* **2015**, *52*, 1–12. [CrossRef]

26. Galea, L.A.; McEwen, B.S.; Tanapat, P.; Deak, T.; Spencer, R.L.; Dhabhar, F.S. Sex differences in dendritic atrophy of CA3 pyramidal neurons in response to chronic restraint stress. *Neuroscience* **1997**, *81*, 689–697. [CrossRef]

27. McLaughlin, K.J.; Baran, S.E.; Wright, R.L.; Conrad, C.D. Chronic stress enhances spatial memory in ovariectomized female rats despite CA3 dendritic retraction: Possible involvement of CA1 neurons. *Neuroscience* **2005**, *135*, 1045–1054. [CrossRef] [PubMed]

28. Sarkar, A.; Kabbaj, M. Sex Differences in Effects of Ketamine on Behavior, Spine Density and Synaptic Proteins in Socially Isolated Rats. *Biol. Psychiatry* **2016**, *80*, 448–456. [CrossRef] [PubMed]

29. McEwen, B.S.; De Kloet, E.R.; Rostene, W. Adrenal steroid receptors and actions in the nervous system. *Physiol. Rev.* **1986**, *66*, 1121–1188. [CrossRef] [PubMed]

30. Woolley, C.S.; McEwen, B.S. Estradiol mediates fluctuation in hippocampal synapse density during the estrous cycle in the adult rat. *J. Neurosci.* **1992**, *12*, 2549–2554. [CrossRef] [PubMed]

31. Warren, S.G.; Humphreys, A.G.; Juraska, J.M.; Greenough, W.T. LTP varies across the estrous cycle: Enhanced synaptic plasticity in proestrus rats. *Brain Res.* **1995**, *703*, 26–30. [CrossRef]

32. Hagemann, G.; Ugur, T.; Schleussner, E.; Mentzel, H.J.; Fitzek, C.; Witte, O.W.; Gaser, C. Changes in brain size during the menstrual cycle. *PLoS ONE* **2011**, *6*, e14655. [CrossRef]

33. Epperson, C.N.; Steiner, M.; Hartlage, S.A.; Eriksson, E.; Schmidt, P.J.; Jones, I.; Yonkers, K.A. Premenstrual dysphoric disorder: Evidence for a new category for DSM-5. *Am. J. Psychiatry* **2012**, *169*, 465–475. [CrossRef] [PubMed]

34. Eiland, L.; Ramroop, J.; Hill, M.N.; Manley, J.; McEwen, B.S. Chronic juvenile stress produces corticolimbic dendritic architectural remodeling and modulates emotional behavior in male and female rats. *Psychoneuroendocrinology* **2012**, *37*, 39–47. [CrossRef] [PubMed]

35. Amin, Z.; Canli, T.; Epperson, C.N. Effect of estrogen-serotonin interactions on mood and cognition. *Behav. Cogn. Neurosci. Rev.* **2005**, *4*, 43–58. [CrossRef] [PubMed]

36. Young, S.N.; Gauthier, S.; Anderson, G.M.; Purdy, W.C. Tryptophan, 5-hydroxyindoleacetic acid and indoleacetic acid in human cerebrospinal fluid: Interrelationships and the influence of age, sex, epilepsy and anticonvulsant drugs. *J. Neurol. Neurosurg. Psychiatry* **1980**, *43*, 438–445. [CrossRef] [PubMed]

37. Staley, J.K.; Krishnan-Sarin, S.; Zoghbi, S.; Tamagnan, G.; Fujita, M.; Seibyl, J.P.; Maciejewski, P.K.; O'Malley, S.; Innis, R.B. Sex differences in [123I]beta-CIT SPECT measures of dopamine and serotonin transporter availability in healthy smokers and nonsmokers. *Synapse* **2001**, *41*, 275–284. [CrossRef] [PubMed]

38. Lee, B.H.; Kim, Y.K. The roles of BDNF in the pathophysiology of major depression and in antidepressant treatment. *Psychiatry Investig.* **2010**, *7*, 231–235. [CrossRef]

39. McEwen, B.S.; Akama, K.T.; Spencer-Segal, J.L.; Milner, T.A.; Waters, E.M. Estrogen effects on the brain: Actions beyond the hypothalamus via novel mechanisms. *Behav. Neurosci.* **2012**, *126*, 4–16. [CrossRef]

40. Keers, R.; Aitchison, K.J. Gender differences in antidepressant drug response. *Int. Rev. Psychiatry* **2010**, *22*, 485–500. [CrossRef]

41. Wohlfarth, T.; Storosum, J.G.; Elferink, A.J.; van Zwieten, B.J.; Fouwels, A.; van den Brink, W. Response to tricyclic antidepressants: Independent of gender? *Am. J. Psychiatry* **2004**, *161*, 370–372. [CrossRef]

42. Khan, A.; Brodhead, A.E.; Schwartz, K.A.; Kolts, R.L.; Brown, W.A. Sex differences in antidepressant response in recent antidepressant clinical trials. *J. Clin. Psychopharmacol.* **2005**, *25*, 318–324. [CrossRef]

43. Kornstein, S.G.; Schatzberg, A.F.; Thase, M.E.; Yonkers, K.A.; McCullough, J.P.; Keitner, G.I.; Gelenberg, A.J.; Davis, S.M.; Harrison, W.M.; Keller, M.B. Gender differences in treatment response to sertraline versus imipramine in chronic depression. *Am. J. Psychiatry* **2000**, *157*, 1445–1452. [CrossRef] [PubMed]

44. Gaynes, B.N.; Warden, D.; Trivedi, M.H.; Wisniewski, S.R.; Fava, M.; Rush, A.J. What did STAR*D teach us? Results from a large-scale, practical, clinical trial for patients with depression. *Psychiatr. Serv.* **2009**, *60*, 1439–1445. [CrossRef] [PubMed]

45. Berlanga, C.; Flores-Ramos, M. Different gender response to serotonergic and noradrenergic antidepressants. A comparative study of the efficacy of citalopram and reboxetine. *J. Affect. Disord.* **2006**, *95*, 119–123. [CrossRef] [PubMed]

46. Haykal, R.F.; Akiskal, H.S. The long-term outcome of dysthymia in private practice: Clinical features, temperament and the art of management. *J. Clin. Psychiatry* **1999**, *60*, 508–518. [CrossRef] [PubMed]

47. Schneider, L.S.; Small, G.W.; Hamilton, S.H.; Bystritsky, A.; Nemeroff, C.B.; Meyers, B.S. Estrogen replacement and response to fluoxetine in a multicenter geriatric depression trial. Fluoxetine Collaborative Study Group. *Am. J. Geriatr. Psychiatry* **1997**, *5*, 97–106. [CrossRef] [PubMed]

48. Thase, M.E.; Entsuah, R.; Cantillon, M.; Kornstein, S.G. Relative antidepressant efficacy of venlafaxine and SSRIs: Sex-age interactions. *J. Womens Health (Larchmt)* **2005**, *14*, 609–616. [CrossRef] [PubMed]

49. Cuijpers, P.; Weitz, E.; Twisk, J.; Kuehner, C.; Cristea, I.; David, D.; DeRubeis, R.J.; Dimidjian, S.; Dunlop, B.W.; Faramarzi, M.; et al. Gender as predictor and moderator of outcome in cognitive behavior therapy and pharmacotherapy for adult depression: An "individual patient data" meta-analysis. *Depress. Anxiety* **2014**, *31*, 941–951. [CrossRef]

50. NIMH. RDoC Matrix. Available online: https://www.nimh.nih.gov/research-priorities/rdoc/constructs/rdoc-matrix.shtml (accessed on 13 February 2019).

51. Gex-Fabry, M.; Balant-Gorgia, A.E.; Balant, L.P.; Garrone, G. Clomipramine metabolism. Model-based analysis of variability factors from drug monitoring data. *Clin. Pharmacokinet.* **1990**, *19*, 241–255. [CrossRef]

52. Preskorn, S.H.; Mac, D.S. Plasma levels of amitriptyline: Effect of age and sex. *J. Clin. Psychiatry* **1985**, *46*, 276–277.

53. Degli Esposti, L.; Piccinni, C.; Sangiorgi, D.; Fagiolini, A.; Buda, S. Patterns of antidepressant use in Italy: Therapy duration, adherence and switching. *Clin. Drug Investig.* **2015**, *35*, 735–742. [CrossRef]

54. Fornaro, M.; Anastasia, A.; Monaco, F.; Novello, S.; Fusco, A.; Iasevoli, F.; De Berardis, D.; Veronese, N.; Solmi, M.; de Bartolomeis, A. Clinical and psychopathological features associated with treatment-emergent mania in bipolar-II depressed outpatients exposed to antidepressants. *J. Affect. Disord.* **2018**, *234*, 131–138. [CrossRef]

55. Fornaro, M.; Anastasia, A.; Novello, S.; Fusco, A.; Pariano, R.; De Berardis, D.; Solmi, M.; Veronese, N.; Stubbs, B.; Vieta, E.; et al. The emergence of loss of efficacy during antidepressant drug treatment for major depressive disorder: An integrative review of evidence, mechanisms and clinical implications. *Pharmacol. Res.* **2019**, *139*, 494–502. [CrossRef] [PubMed]

56. Labermaier, C.; Masana, M.; Muller, M.B. Biomarkers predicting antidepressant treatment response: How can we advance the field? *Dis. Mark.* **2013**, *35*, 23–31. [CrossRef] [PubMed]

57. Herzog, D.P.; Beckmann, H.; Lieb, K.; Ryu, S.; Müller, M.B. Understanding and Predicting Antidepressant Response: Using Animal Models to Move Toward Precision Psychiatry. *Front. Psychiatry* **2018**, *9*, 512. [CrossRef]

58. Xu, Y.; Hackett, M.; Carter, G.; Loo, C.; Galvez, V.; Glozier, N.; Glue, P.; Lapidus, K.; McGirr, A.; Somogyi, A.A.; et al. Effects of Low-Dose and Very Low-Dose Ketamine among Patients with Major Depression: A Systematic Review and Meta-Analysis. *Int. J. Neuropsychopharmacol.* **2016**, *19*, pyv124. [CrossRef] [PubMed]

59. Zarate, C.A., Jr.; Singh, J.B.; Carlson, P.J.; Brutsche, N.E.; Ameli, R.; Luckenbaugh, D.A.; Charney, D.S.; Manji, H.K. A randomized trial of an N-methyl-D-aspartate antagonist in treatment-resistant major depression. *Arch. Gen. Psychiatry* **2006**, *63*, 856–864. [CrossRef] [PubMed]

60. De Berardis, D.; Fornaro, M.; Valchera, A.; Cavuto, M.; Perna, G.; Di Nicola, M.; Serafini, G.; Carano, A.; Pompili, M.; Vellante, F.; et al. Eradicating Suicide at Its Roots: Preclinical Bases and Clinical Evidence of the Efficacy of Ketamine in the Treatment of Suicidal Behaviors. *Int. J. Mol. Sci.* **2018**, *19*, 2888. [CrossRef] [PubMed]

61. Tomasetti, C.; Iasevoli, F.; Buonaguro, E.F.; De Berardis, D.; Fornaro, M.; Fiengo, A.L.; Martinotti, G.; Orsolini, L.; Valchera, A.; Di Giannantonio, M.; et al. Treating the Synapse in Major Psychiatric Disorders: The Role of Postsynaptic Density Network in Dopamine-Glutamate Interplay and Psychopharmacologic Drugs Molecular Actions. *Int. J. Mol. Sci.* **2017**, *18*, 135. [CrossRef] [PubMed]

62. Li, N.; Lee, B.; Liu, R.J.; Banasr, M.; Dwyer, J.M.; Iwata, M.; Li, X.Y.; Aghajanian, G.; Duman, R.S. mTOR-dependent synapse formation underlies the rapid antidepressant effects of NMDA antagonists. *Science* **2010**, *329*, 959–964. [CrossRef]

63. Autry, A.E.; Adachi, M.; Nosyreva, E.; Na, E.S.; Los, M.F.; Cheng, P.F.; Kavalali, E.T.; Monteggia, L.M. NMDA receptor blockade at rest triggers rapid behavioural antidepressant responses. *Nature* **2011**, *475*, 91–95. [CrossRef]

64. Franceschelli, A.; Sens, J.; Herchick, S.; Thelen, C.; Pitychoutis, P.M. Sex differences in the rapid and the sustained antidepressant-like effects of ketamine in stress-naive and "depressed" mice exposed to chronic mild stress. *Neuroscience* **2015**, *290*, 49–60. [CrossRef] [PubMed]

65. Saland, S.K.; Kabbaj, M. Sex Differences in the Pharmacokinetics of Low-dose Ketamine in Plasma and Brain of Male and Female Rats. *J. Pharmacol. Exp. Ther.* **2018**, *367*, 393–404. [CrossRef] [PubMed]

66. Ho, M.F.; Correia, C.; Ingle, J.N.; Kaddurah-Daouk, R.; Wang, L.; Kaufmann, S.H.; Weinshilboum, R.M. Ketamine and ketamine metabolites as novel estrogen receptor ligands: Induction of cytochrome P450 and AMPA glutamate receptor gene expression. *Biochem. Pharmacol.* **2018**, *152*, 279–292. [CrossRef] [PubMed]

67. Dossat, A.M.; Wright, K.N.; Strong, C.E.; Kabbaj, M. Behavioral and biochemical sensitivity to low doses of ketamine: Influence of estrous cycle in C57BL/6 mice. *Neuropharmacology* **2018**, *130*, 30–41. [CrossRef]

68. Kara, N.Z.; Agam, G.; Anderson, G.W.; Zitron, N.; Einat, H. Lack of effect of chronic ketamine administration on depression like behavior and frontal cortex autophagy in female and male ICR mice. *Behav. Brain Res.* **2017**, *317*, 576–580. [CrossRef] [PubMed]

69. Zanos, P.; Moaddel, R.; Morris, P.J.; Georgiou, P.; Fischell, J.; Elmer, G.I.; Alkondon, M.; Yuan, P.; Pribut, H.J.; Singh, N.S.; et al. NMDAR inhibition-independent antidepressant actions of ketamine metabolites. *Nature* **2016**, *533*, 481–486. [CrossRef] [PubMed]

70. Yamaguchi, J.I.; Toki, H.; Qu, Y.; Yang, C.; Koike, H.; Hashimoto, K.; Mizuno-Yasuhira, A.; Chaki, S. (2R,6R)-Hydroxynorketamine is not essential for the antidepressant actions of (R)-ketamine in mice. *Neuropsychopharmacology* **2018**, *43*, 1900–1907. [CrossRef]

71. Zanos, P.; Thompson, S.M.; Duman, R.S.; Zarate, C.A., Jr.; Gould, T.D. Convergent Mechanisms Underlying Rapid Antidepressant Action. *CNS Drugs* **2018**, *32*, 197–227. [CrossRef]

72. Chou, D.; Peng, H.Y.; Lin, T.B.; Lai, C.Y.; Hsieh, M.C.; Wen, Y.C.; Lee, A.S.; Wang, H.H.; Yang, P.S.; Chen, G.D.; et al. (2R,6R)-hydroxynorketamine rescues chronic stress-induced depression-like behavior through its actions in the midbrain periaqueductal gray. *Neuropharmacology* **2018**, *139*, 1–12. [CrossRef]

73. Pham, T.H.; Defaix, C.; Xu, X.; Deng, S.X.; Fabresse, N.; Alvarez, J.C.; Landry, D.W.; Brachman, R.A.; Denny, C.A.; Gardier, A.M. Common Neurotransmission Recruited in (R,S)-Ketamine and (2R,6R)-Hydroxynorketamine-Induced Sustained Antidepressant-like Effects. *Biol. Psychiatry* **2018**, *84*, e3–e6. [CrossRef]

74. Cavalleri, L.; Merlo Pich, E.; Millan, M.J.; Chiamulera, C.; Kunath, T.; Spano, P.F.; Collo, G. Ketamine enhances structural plasticity in mouse mesencephalic and human iPSC-derived dopaminergic neurons via AMPAR-driven BDNF and mTOR signaling. *Mol. Psychiatry* **2018**, *23*, 812–823. [CrossRef] [PubMed]

75. Collo, G.; Cavalleri, L.; Chiamulera, C.; Merlo Pich, E. (2R,6R)-Hydroxynorketamine promotes dendrite outgrowth in human inducible pluripotent stem cell-derived neurons through AMPA receptor with timing and exposure compatible with ketamine infusion pharmacokinetics in humans. *Neuroreport* **2018**, *29*, 1425–1430. [CrossRef] [PubMed]

76. Yao, N.; Skiteva, O.; Zhang, X.; Svenningsson, P.; Chergui, K. Ketamine and its metabolite (2R,6R)-hydroxynorketamine induce lasting alterations in glutamatergic synaptic plasticity in the mesolimbic circuit. *Mol. Psychiatry* **2017**, *23*, 2066–2077. [CrossRef] [PubMed]

77. Shirayama, Y.; Hashimoto, K. Lack of Antidepressant Effects of (2R,6R)-Hydroxynorketamine in a Rat Learned Helplessness Model: Comparison with (R)-Ketamine. *Int. J. Neuropsychopharmacol.* **2018**, *21*, 84–88. [CrossRef] [PubMed]

78. Yang, C.; Qu, Y.; Abe, M.; Nozawa, D.; Chaki, S.; Hashimoto, K. (R)-Ketamine Shows Greater Potency and Longer Lasting Antidepressant Effects Than Its Metabolite (2R,6R)-Hydroxynorketamine. *Biol. Psychiatry* **2017**, *82*, e43–e44. [CrossRef] [PubMed]

79. Thomas, A.M.; Duman, R.S. Novel rapid-acting antidepressants: Molecular and cellular signaling mechanisms. *Neuronal Signal.* **2017**, *1*. [CrossRef] [PubMed]

80. Witkin, J.M.; Knutson, D.E.; Rodriguez, G.J.; Shi, S. Rapid-Acting Antidepressants. *Curr. Pharm. Des.* **2018**, *24*, 2556–2563. [CrossRef]

81. Furey, M.L.; Drevets, W.C. Antidepressant efficacy of the antimuscarinic drug scopolamine: A randomized, placebo-controlled clinical trial. *Arch. Gen. Psychiatry* **2006**, *63*, 1121–1129. [CrossRef]

82. Griffiths, R.; Richards, W.; Johnson, M.; McCann, U.; Jesse, R. Mystical-type experiences occasioned by psilocybin mediate the attribution of personal meaning and spiritual significance 14 months later. *J. Psychopharmacol.* **2008**, *22*, 621–632. [CrossRef]

83. Moskal, J.R.; Burgdorf, J.S.; Stanton, P.K.; Kroes, R.A.; Disterhoft, J.F.; Burch, R.M.; Khan, M.A. The Development of Rapastinel (Formerly GLYX-13); A Rapid Acting and Long Lasting Antidepressant. *Curr. Neuropharmacol.* **2017**, *15*, 47–56. [CrossRef]

84. Preskorn, S.; Macaluso, M.; Mehra, D.O.; Zammit, G.; Moskal, J.R.; Burch, R.M.; Group, G.-C.S. Randomized proof of concept trial of GLYX-13, an N-methyl-D-aspartate receptor glycine site partial agonist, in major depressive disorder nonresponsive to a previous antidepressant agent. *J. Psychiatr. Pract.* **2015**, *21*, 140–149. [CrossRef] [PubMed]

85. Chaki, S.; Fukumoto, K. mGlu receptors as potential targets for novel antidepressants. *Curr. Opin. Pharmacol.* **2018**, *38*, 24–30. [CrossRef] [PubMed]

86. Chappell, M.D.; Li, R.; Smith, S.C.; Dressman, B.A.; Tromiczak, E.G.; Tripp, A.E.; Blanco, M.J.; Vetman, T.; Quimby, S.J.; Matt, J.; et al. Discovery of (1S,2R,3S,4S,5R,6R)-2-Amino-3-[(3,4-difluorophenyl)sulfanylmethyl]-4-hydroxy-bicy clo[3.1.0]hexane-2,6-dicarboxylic Acid Hydrochloride (LY3020371.HCl): A Potent, Metabotropic Glutamate 2/3 Receptor Antagonist with Antidepressant-Like Activity. *J. Med. Chem.* **2016**, *59*, 10974–10993. [CrossRef] [PubMed]

87. Witkin, J.M.; Ornstein, P.L.; Mitch, C.H.; Li, R.; Smith, S.C.; Heinz, B.A.; Wang, X.S.; Xiang, C.; Carter, J.H.; Anderson, W.H.; et al. In vitro pharmacological and rat pharmacokinetic characterization of LY3020371, a potent and selective mGlu2/3 receptor antagonist. *Neuropharmacology* **2017**, *115*, 100–114. [CrossRef] [PubMed]

88. Fischell, J.; Van Dyke, A.M.; Kvarta, M.D.; LeGates, T.A.; Thompson, S.M. Rapid Antidepressant Action and Restoration of Excitatory Synaptic Strength After Chronic Stress by Negative Modulators of Alpha5-Containing GABAA Receptors. *Neuropsychopharmacology* **2015**, *40*, 2499–2509. [CrossRef] [PubMed]

89. Zanos, P.; Nelson, M.E.; Highland, J.N.; Krimmel, S.R.; Georgiou, P.; Gould, T.D.; Thompson, S.M. A Negative Allosteric Modulator for alpha5 Subunit-Containing GABA Receptors Exerts a Rapid and Persistent Antidepressant-Like Action without the Side Effects of the NMDA Receptor Antagonist Ketamine in Mice. *eNeuro* **2017**, *4*. [CrossRef] [PubMed]

90. Sales, A.J.; Fogaca, M.V.; Sartim, A.G.; Pereira, V.S.; Wegener, G.; Guimaraes, F.S.; Joca, S.R.L. Cannabidiol Induces Rapid and Sustained Antidepressant-Like Effects Through Increased BDNF Signaling and Synaptogenesis in the Prefrontal Cortex. *Mol. Neurobiol.* **2018**. [CrossRef]

91. Sartim, A.G.; Guimaraes, F.S.; Joca, S.R. Antidepressant-like effect of cannabidiol injection into the ventral medial prefrontal cortex-Possible involvement of 5-HT1A and CB1 receptors. *Behav. Brain Res.* **2016**, *303*, 218–227. [CrossRef]

92. Shbiro, L.; Hen-Shoval, D.; Hazut, N.; Rapps, K.; Dar, S.; Zalsman, G.; Mechoulam, R.; Weller, A.; Shoval, G. Effects of cannabidiol in males and females in two different rat models of depression. *Physiol. Behav.* **2018**, *201*, 59–63. [CrossRef]

93. Shoval, G.; Shbiro, L.; Hershkovitz, L.; Hazut, N.; Zalsman, G.; Mechoulam, R.; Weller, A. Prohedonic Effect of Cannabidiol in a Rat Model of Depression. *Neuropsychobiology* **2016**, *73*, 123–129. [CrossRef]

94. Ly, C.; Greb, A.C.; Cameron, L.P.; Wong, J.M.; Barragan, E.V.; Wilson, P.C.; Burbach, K.F.; Soltanzadeh Zarandi, S.; Sood, A.; Paddy, M.R.; et al. Psychedelics Promote Structural and Functional Neural Plasticity. *Cell Rep.* **2018**, *23*, 3170–3182. [CrossRef]

95. Sur, C.; Fresu, L.; Howell, O.; McKernan, R.M.; Atack, J.R. Autoradiographic localization of alpha5 subunit-containing GABAA receptors in rat brain. *Brain Res.* **1999**, *822*, 265–270. [CrossRef]

96. Guimaraes, F.S.; Chiaretti, T.M.; Graeff, F.G.; Zuardi, A.W. Antianxiety effect of cannabidiol in the elevated plus-maze. *Psychopharmacology* **1990**, *100*, 558–559. [CrossRef] [PubMed]

97. Zanelati, T.V.; Biojone, C.; Moreira, F.A.; Guimaraes, F.S.; Joca, S.R. Antidepressant-like effects of cannabidiol in mice: Possible involvement of 5-HT1A receptors. *Br. J. Pharmacol.* **2010**, *159*, 122–128. [CrossRef] [PubMed]

98. Husain, S.S.; Kevan, I.M.; Linnell, R.; Scott, A.I. Electroconvulsive therapy in depressive illness that has not responded to drug treatment. *J. Affect. Disord.* **2004**, *83*, 121–126. [CrossRef] [PubMed]

99. Hoy, K.E.; Fitzgerald, P.B. Brain stimulation in psychiatry and its effects on cognition. *Nat. Rev. Neurol.* **2010**, *6*, 267–275. [CrossRef] [PubMed]

100. NIMH. National Institute of Mental Health Information: Brain Stimulation Therapies. Available online: https://www.nimh.nih.gov/health/topics/brain-stimulation-therapies/brain-stimulation-therapies.shtml (accessed on 29 October 2018).

101. The UK ECT Review Group. Efficacy and safety of electroconvulsive therapy in depressive disorders: A systematic review and meta-analysis. *Lancet* **2003**, *361*, 799–808. [CrossRef]

102. Miller, A.H.; Maletic, V.; Raison, C.L. Inflammation and its discontents: The role of cytokines in the pathophysiology of major depression. *Biol. Psychiatry* **2009**, *65*, 732–741. [CrossRef] [PubMed]

103. Nazimek, K.; Strobel, S.; Bryniarski, P.; Kozlowski, M.; Filipczak-Bryniarska, I.; Bryniarski, K. The role of macrophages in anti-inflammatory activity of antidepressant drugs. *Immunobiology* **2017**, *222*, 823–830. [CrossRef] [PubMed]

104. Guloksuz, S.; Rutten, B.P.; Arts, B.; van Os, J.; Kenis, G. The immune system and electroconvulsive therapy for depression. *J. ECT* **2014**, *30*, 132–137. [CrossRef]

105. Yrondi, A.; Sporer, M.; Peran, P.; Schmitt, L.; Arbus, C.; Sauvaget, A. Electroconvulsive therapy, depression, the immune system and inflammation: A systematic review. *Brain Stimul.* **2018**, *11*, 29–51. [CrossRef]

106. Bouckaert, F.; Dols, A.; Emsell, L.; De Winter, F.L.; Vansteelandt, K.; Claes, L.; Sunaert, S.; Stek, M.; Sienaert, P.; Vandenbulcke, M. Relationship Between Hippocampal Volume, Serum BDNF and Depression Severity Following Electroconvulsive Therapy in Late-Life Depression. *Neuropsychopharmacology* **2016**, *41*, 2741–2748. [CrossRef]

107. Albala, A.A.; Greden, J.F.; Tarika, J.; Carroll, B.J. Changes in serial dexamethasone suppression tests among unipolar depressive receiving electroconvulsive treatment. *Biol. Psychiatry* **1981**, *16*, 551–560.

108. Yuuki, N.; Ida, I.; Oshima, A.; Kumano, H.; Takahashi, K.; Fukuda, M.; Oriuchi, N.; Endo, K.; Matsuda, H.; Mikuni, M. HPA axis normalization, estimated by DEX/CRH test but less alteration on cerebral glucose metabolism in depressed patients receiving ECT after medication treatment failures. *Acta Psychiatr. Scand.* **2005**, *112*, 257–265. [CrossRef] [PubMed]

109. Dored, G.; Stefansson, S.; d'Elia, G.; Kagedal, B.; Karlberg, E.; Ekman, R. Corticotropin, cortisol and beta-endorphin responses to the human corticotropin-releasing hormone during melancholia and after unilateral electroconvulsive therapy. *Acta Psychiatr. Scand.* **1990**, *82*, 204–209. [CrossRef] [PubMed]

110. Saijo, T.; Takano, A.; Suhara, T.; Arakawa, R.; Okumura, M.; Ichimiya, T.; Ito, H.; Okubo, Y. Electroconvulsive therapy decreases dopamine D(2)receptor binding in the anterior cingulate in patients with depression: A controlled study using positron emission tomography with radioligand [(1)(1)C]FLB 457. *J. Clin. Psychiatry* **2010**, *71*, 793–799. [CrossRef] [PubMed]

111. Madsen, T.M.; Treschow, A.; Bengzon, J.; Bolwig, T.G.; Lindvall, O.; Tingstrom, A. Increased neurogenesis in a model of electroconvulsive therapy. *Biol. Psychiatry* **2000**, *47*, 1043–1049. [CrossRef]

112. Chen, F.; Madsen, T.M.; Wegener, G.; Nyengaard, J.R. Repeated electroconvulsive seizures increase the total number of synapses in adult male rat hippocampus. *Eur. Neuropsychopharmacol.* **2009**, *19*, 329–338. [CrossRef]

113. Yrondi, A.; Peran, P.; Sauvaget, A.; Schmitt, L.; Arbus, C. Structural-functional brain changes in depressed patients during and after electroconvulsive therapy. *Acta Neuropsychiatr.* **2018**, *30*, 17–28. [CrossRef]

114. Benson-Martin, J.J.; Stein, D.J.; Baldwin, D.S.; Domschke, K. Genetic mechanisms of electroconvulsive therapy response in depression. *Hum. Psychopharmacol.* **2016**, *31*, 247–251. [CrossRef]

115. de Jong, J.O.; Arts, B.; Boks, M.P.; Sienaert, P.; van den Hove, D.L.; Kenis, G.; van Os, J.; Rutten, B.P. Epigenetic effects of electroconvulsive seizures. *J. ECT* **2014**, *30*, 152–159. [CrossRef] [PubMed]

116. Pusalkar, M.; Ghosh, S.; Jaggar, M.; Husain, B.F.; Galande, S.; Vaidya, V.A. Acute and Chronic Electroconvulsive Seizures (ECS) Differentially Regulate the Expression of Epigenetic Machinery in the Adult Rat Hippocampus. *Int. J. Neuropsychopharmacol.* **2016**, *19*, pyw040. [CrossRef] [PubMed]

117. Bloch, Y.; Ratzoni, G.; Sobol, D.; Mendlovic, S.; Gal, G.; Levkovitz, Y. Gender differences in electroconvulsive therapy: A retrospective chart review. *J. Affect. Disord.* **2005**, *84*, 99–102. [CrossRef] [PubMed]

118. Bolu, A.; Ozselek, S.; Akarsu, S.; Alper, M.; Balikci, A. Is There a Role of Gender in Electroconvulsive Therapy Response? *Klinik Psikofarmakoloji Bülteni-Bull. Clin. Psychopharmacol.* **2016**, *26*, 32–38. [CrossRef]

119. Bousman, C.A.; Katalinic, N.; Martin, D.M.; Smith, D.J.; Ingram, A.; Dowling, N.; Ng, C.; Loo, C.K. Effects of COMT, DRD2, BDNF and APOE Genotypic Variation on Treatment Efficacy and Cognitive Side Effects of Electroconvulsive Therapy. *J. ECT* **2015**, *31*, 129–135. [CrossRef] [PubMed]

International Journal of
Molecular Sciences

Article

Proteomic Studies Reveal Disrupted in Schizophrenia 1 as a Player in Both Neurodevelopment and Synaptic Function

Adriana Ramos [1,2,*,†], Carmen Rodríguez-Seoane [1,†], Isaac Rosa [1,3], Irantzu Gorroño-Etxebarria [4], Jana Alonso [5], Sonia Veiga [1], Carsten Korth [6], Robert M. Kypta [4,7], Ángel García [1,3] and Jesús R. Requena [1,8]

[1] CIMUS Biomedical Research Institute, University of Santiago de Compostela-IDIS,
     15782 Santiago de Compostela, Spain; C.Rodriguez@ed.ac.uk (C.R.-S.); isaacrbenito@gmail.com (I.R.);
     sonia.veiga@usc.es (S.V.); angel.garcia@usc.es (Á.G.); jesus.requena@usc.es (J.R.R.)
[2] Department of Psychiatry and Behavioral Sciences, Johns Hopkins University, Baltimore, MD 21287, USA
[3] Department of Pharmacology, University of Santiago de Compostela, 15782 Santiago de Compostela, Spain
[4] Cell Biology and Stem Cells Unit, CIC-bioGUNE, Parque Tecnologico de Bizkaia, 48160 Derio, Spain;
     igorrono@cicbiogune.es (I.G.-E.); rkypta@cicbiogune.es (R.M.K.)
[5] Proteomics Unit, IDIS, 15706 Santiago de Compostela, Spain; janaalonsol@hotmail.com
[6] Department of Neuropathology, Heinrich Heine University, Medical School, 40225 Düsseldorf, Germany;
     ckorth@uni-duesseldorf.de
[7] Department of Surgery and Cancer, Imperial College London, London W12 OUQ, UK
[8] Department of Medical Sciences, University of Santiago de Compostela,
     15782 Santiago de Compostela, Spain
*   Correspondence: Adriana.Ramos@jhmi.edu; Tel.: +1-410-9551617
†   These authors contributed equally to this work.

Received: 17 November 2018; Accepted: 24 December 2018; Published: 29 December 2018

Abstract: A balanced chromosomal translocation disrupting DISC1 (Disrupted in Schizophrenia 1) gene has been linked to psychiatric diseases, such as major depression, bipolar disorder and schizophrenia. Since the discovery of this translocation, many studies have focused on understating the role of the truncated isoform of DISC1, hypothesizing that the gain of function of this protein could be behind the neurobiology of mental conditions, but not so many studies have focused in the mechanisms impaired due to its loss of function. For that reason, we performed an analysis on the cellular proteome of primary neurons in which DISC1 was knocked down with the goal of identifying relevant pathways directly affected by DISC1 loss of function. Using an unbiased proteomic approach, we found that the expression of 31 proteins related to neurodevelopment (e.g., CRMP-2, stathmin) and synaptic function (e.g., MUNC-18, NCS-1) is altered by DISC1 in primary mouse neurons. Hence, this study reinforces the idea that DISC1 is a unifying regulator of both neurodevelopment and synaptic function, thereby providing a link between these two key anatomical and cellular circuitries.

Keywords: DISC1; neurodevelopment; synapse; CRMP-2; proteomics

---

1. Introduction

The Disrupted in Schizophrenia 1 (DISC1) gene was found mutated when studying a chromosomal translocation t(1;11)(q42.1;q14.3) in a Scottish family; this translocation correlated with cases of schizophrenia, bipolar disorder and major depression [1,2]. Further studies also found that the truncation of this gene in an American family segregated with cases of schizophrenia [3].

Since the discovery of this translocation, many groups have invested their efforts in understanding the role of DISC1 protein, with the hope of revealing new mechanisms that could explain the neurobiology behind mental disease. Therefore, DISC1 was proposed to be involved in diverse processes such as neurogenesis [4,5], synapse regulation [6–10], neurite outgrowth [6,11,12], and neural migration and proliferation [13–15]. Also, yeast two hybrid experiments [16] and other molecular studies have revealed several important interacting partners of DISC1 including GSK3β [5], PDE4B [17], Rac1 [8], Girdin [18] or TNIK [9] among others. Thus, DISC1 might act as a molecular scaffold, providing cohesion and coordination among different biological events in the brain [19].

To acquire a deeper understanding of the mechanisms of action of DISC1, several proteomic analyses have been conducted to specifically address the role of the truncated isoform of DISC1 on the cellular proteome of neural cells [20,21]. In this study, we decided to specifically address the role of DISC1 loss of function, for that we carried out an unbiased proteomic analysis in DISC1-silenced neurons.

We report that DISC1 alters the expression of many relevant proteins related to neurodevelopment and synaptic function, reinforcing the idea that DISC1 is a key molecular link bridging neurodevelopmental functions with the regulation of synaptic formation and neurosignaling processes.

2. Results

2.1. Proteomic Analysis

Cell extracts from control and DISC1 knockdown murine primary neurons (Figure S1) were subjected to proteomic analysis. Four bidimensional gels for the silenced condition versus four of the control condition were analyzed. 3474 identical spots per gel were detected (Figure S2) and 75 of them were found differentially expressed with a fold change ≥2 and p value < 0.05 (Table S1). 68 of these spots were identified using mass spectrometry, corresponding to 48 unique proteins (Table 1). The functions of these proteins were mainly related to neurodevelopmental processes or synaptic function (Table 1, Figure S3). Particularly, 19 of them were related to neurodevelopmental processes (Table 1) and other 19 unique proteins were related to synaptic function (Table 1). Of note, 7 of these proteins have shared functions (Table 1, Figure S3). Therefore, these results suggest that DISC1 plays an important role linking these two processes.

Remarkably, some of the identified proteins have previously been described as DISC1 binding partners, it is the case of 14-3-3 proteins [12] and LIS1 [22], while CRMP-2 has been identified as a possible DISC1 interactor [16]. However, to the best of our knowledge, this is the first time that DISC1 has been found to also alter their expression. As well, we could identify some of the proteins as substrates of similar enzymes; this is the case of stathmin, CRMP-2, and MAP1B. These proteins are known to be phosphorylated by GSK3β to exert their functions.

Table 1. Proteins involved in neurodevelopment or synaptic function identified through proteomic analysis of primary neurons [1].

Function	Protein	Fold Change	*p* Value
Neurite outgrowth or neural migration	Dihydropyrimidinase-related protein 5 (CRMP-5)	2.59	3.066×10^{-5}
	Dihydropyrimidinase-related protein 3 (CRMP-3)	3.12, 2.23	1.469×10^{-4}, 2.894×10^{-4}
	Dihydropyrimidinase-relatedprotein 2 (CRMP-2)	2.21, 2.03	2.457×10^{-4}, 0.059
	Dihydropyrimidinase-related protein 1 (CRMP-1)	2.10	9.180×10^{-5}
	Tubulin alpha-1A chain (TBA1A)	2.01,2.99,2.13	0.0043, 1.326×10^{-4}, 4.067×10^{-4}
	Tubulin beta-2B chain (TBB2B)	Inf, 2.42	0.0065, 0.0156
	Microtubule-associated protein (MAP1B)	2.04, 2.20,3.03	3.661×10^{-7}, 2.894×10^{-4}, 2.717×10^{-5}
	14-3-3 protein epsilon (14-3-3ε)	2.67, 3.19	6.713×10^{-5}, 1.854×10^{-4}
	14-3-3 protein zeta/delta (14-3-3θ/Δ)	8.42, 2.63, 3.26, 3.88, 3.89, 6.81, 3.82, 6.20	3.028×10^{-6}, 2.334×10^{-4}, 1.579×10^{-4} 9.307×10^{-4}, 0.021, 2.080×10^{-5}, 0.0022, 2.572×10^{-6}
	14-3-3 protein gamma (14-3-3γ)	3.24, 2.30	6.104×10^{-5}, 0.0084
	Platelet-activating factor acetylhydrolase IB (Lis-1)	2.48	0.0049
	Stathmin (STMN)	2.12, 4.64	8.206×10^{-4}, 1.021×10^{-4}
	Syntaxin-7 (STX7)	2.13	0.0022
	Tropomyosin alpha-3 chain (TPM3)	2.88	0.0115
	Actin, cytoplasmic 2 (ACTG)	4.94, 2.95	4.398×10^{-5}, 1.081×10^{-5}
	Cadherin-13 (CAD13)	2.31	0.0108
	Calreticulin (CALR)	2.60	0.0088
	Septin-5 (SEPT5)	2.15	0.0034
	Apolipoprotein A-I (APOA1)	2.41	9.215×10^{-5}
	Dynamin 1 (DYN1)	4.33	1.440×10^{-4}
Dynamin 1 (DYN1)	Dihydropyrimidinase-related protein 5 (CRMP-5)	2.59	3.066×10^{-5}
	Dihydropyrimidinase-relatedprotein 2 (CRMP-2)	2.21, 2.03	2.457×10^{-4}, 0.059
	Microtubule-associated protein (MAP1B)	2.04, 2.20, 3.03	3.661×10^{-7}, 2.894×10^{-4}, 2.717×10^{-5}
	Transitional endoplasmic Reticulum ATPase (TERA)	2.21	5.537×10^{-4}
	Stathmin (STMN)	2.12, 4.64	8.206×10^{-4}, 1.021×10^{-4}
	Syntaxin-binding protein 1 (STXB1)	3.43	0.0010
	Syntaxin-7 (STX7)	2.13	0.0022
	Ras-related protein Rab-1A (RAB1A)	2.01	0.0023
	Ras-related protein Rab-2A (RAB2A)	2.43	4.164×10^{-4}
	Ras-related protein Rab-11B (RB11B)	3.25	0.0305
	Ras-related protein Rab-18 (RAB18)	3.23	5.527×10^{-4}
	Cadherin-13 (CAD13)	2.31	0.0108
	Rho GDP-dissociation inhibitor 2 (GDIR2)	2.28	1.061×10^{-4}
	Phosphatidylethanolamine-binding protein 1 (HCNP)	3.84, 6.49	4.081×10^{-4}, 2.377×10^{-4}
	Calreticulin (CALR)	2.60	0.0088
	Adaptin ear-binding coat-associated protein 1 (NECP1)	2.51	6.028×10^{-5}
	Neuronal calcium sensor 1 (NCS1)	2.24	9.215×10^{-5}
	Dynamin 1 (DYN1)	4.33	1.440×10^{-4}

[1] All the proteins had a fold change > 2 and *p* value < 0.05. Fold change in red indicates that the protein is overexpressed in DISC1 silenced cells, while fold change in black indicates a downregulation in DISC1 silenced cells.

2.2. Ingenuity Pathway

To identify common molecular pathways regulated by DISC1 in our sample set we used the Ingenuity Pathways Analysis (IPA) software. The 5 top canonical pathways involved in our analysis are represented in Table 2. It is interesting that CRMP (collapsin response mediator protein) family was highlighted in the analysis as part of the Semaphorin signaling in neurons, since this signaling cascade is known to play an important role in neuronal differentiation and axonal growth [23,24]. Previous studies also concluded that the overexpression of the truncated isoform of DISC1 leads to dysregulation of Semaphorin signaling [20]. This could be a corroborative evidence for the fact that DISC1 expression has to be tightly and precisely regulated in a small window and that both, above and below that window you have dysregulation of similar signaling pathways.

Table 2. Ingenuity top canonical pathways.

Name	p Value	Proteins
14-3-3 mediated signaling	4.99×10^{-7}	TUBA1A, 14-3-3G, TUBB2B, PDIA3,1 4-3-3E, 14-3-3Z
Semaphorin signaling in neurons	5.28×10^{-6}	CRMP3, CRMP1, CRMP2, CRMP5
Remodeling of epithelial adherent junctions	1.52×10^{-5}	DNM1L, TUBA1A, ACTG1, TUBB2B
Cell cycle: G2/M DNA damage checkpoint regulation	1.75×10^{-4}	14-3-3G, 14-3-3E, 14-3-3Z
PI3K/AKT signaling	1.87×10^{-4}	14-3-3G, 14-3-3E, HSP90AA1, 14-3-3Z

The top molecular and cellular functions identified by IPA are represented in Table 3. The analysis particularly highlighted proteins involved in neurite outgrowth and branching of neurons.

Table 3. Ingenuity Top 10 molecular and cellular functions.

Name	p Value	Proteins
Outgrowth of cells	3.94×10^{-8}	DNM1L, TUBA1A, HBA1/HBA2, CRMP3, MAP1B, SET, PDIA3, CRMP2, 14-3-3G, HSP90AA1, CRMP5
Patterning of dendrites	9.56×10^{-8}	CRMP1, CRMP2, GDA
Outgrowth of neurites	1.94×10^{-7}	DNM1L, TUBA1A, HBA1/HBA2, DPYSL3, MAP1B, SET, PDIA3, CRMP2, 14-3-3Z, CRMP5
Branching of neurons	2.53×10^{-7}	DNM1L, HNRNPK, CRMP3, MAP1B, PDIA3, CRMP1, CRMP2, CRMP5, GDA
Organization of cytoplasm	7.08×10^{-7}	CDH13, RAB2A, HNRNPK, CRMP1, CRMP2, CRMP5, STMN1, CALR, TPM3, DNM1L, ACTG1, PEX5, CRMP3, MAP1B, RAB1A, PDIA3, HSP90AA1, GDA
Fibrogenesis	8.53×10^{-7}	CALR, CDH13, TPM3, ACTG1, CRMP3, MAP1B, APOA1, CRMP2, GDA, STMN1
Endocytosis	1.39×10^{-6}	CALR, CDH13, HNRNPK, MAP1B, RAB1A, APOA1, CRMP2, VCP, HSP90AA1, NECAP1
Neuritogenesis	2.09×10^{-6}	DNM1L, HNRNPK, CRMP3, MAP1B, PDIA3, CRMP1, CRMP2, HSP90AA1, CRMP5, GDA, STMN1
Branching of neurites	2.50×10^{-6}	DNM1L, HNRNPK, MAP1B, PDIA3, CRMP1, CRMP2, CRMP5, GDA
Microtubule dynamics	3.39×10^{-6}	CDH13, RAB2A, HNRNPK, CRMP1, CRMP2, CRMP5, STMN1, TPM3, DNM1L, ACTG1, CRMP3, MAP1B, PDIA3, HSP90AA1, GDA

2.3. DISC1 Alters the Expression of Neurodevelopmental Related Proteins

Considering the results obtained by IPA analysis we focused on the collapsin response mediator proteins (CRMPs) to perform our validations. These proteins constitute a family of five homologous cytosolic proteins (CRMP-1-5) involved in microtubule regulation. All of them are phosphorylated and highly expressed in the developing and adult nervous system where they play important roles in neuronal development and maturation [25]. Six spots corresponding to CRMP-5, CRMP-3, CRMP-2 and CRMP-1 were differentially expressed in silenced vs. control cells (Table 1) in our study; in all cases the proteins were upregulated in DISC1 silenced cells.

Particularly, CRMP-2 has been described as a candidate gene for susceptibility to schizophrenia [26] and was found upregulated in a proteomic study performed with brain samples from patients with bipolar disorder, schizophrenia and major depression [27]. We showed differential expression of multiple CRMP2 isoforms upon DISC1 silencing (Figure 1) in primary neurons. The existence of different isoforms of CRMP2 has been highlighted in several studies [28,29]. Here, CRMP2 was detected as three isoforms (labelled 1 to 3). Isoforms 1 and 2, most likely corresponding to CRMP2A and CRMP2B [28] were found to be downregulated in DISC1-silenced cells, while isoform 3 was upregulated. A similar pattern was observed using antibodies that recognize CRMP-2 phosphorylated at Thr-514 (Figure 1). Therefore, isoform 3 most likely corresponds to the spot that was differentially expressed in our proteomic analysis.

Some studies described this isoform as a calpain-associated degradation product [30,31], while others highlight its role in neurite outgrowth inhibition [32]. If this is the case, it suggests that DISC1 silencing leads to increased expression of CRMP-2 and, as a result, inhibition of neurite outgrowth. Of note, Septin-5, a protein that directly interacts with CRMP-2, was also found differentially expressed in our study (Table 1).

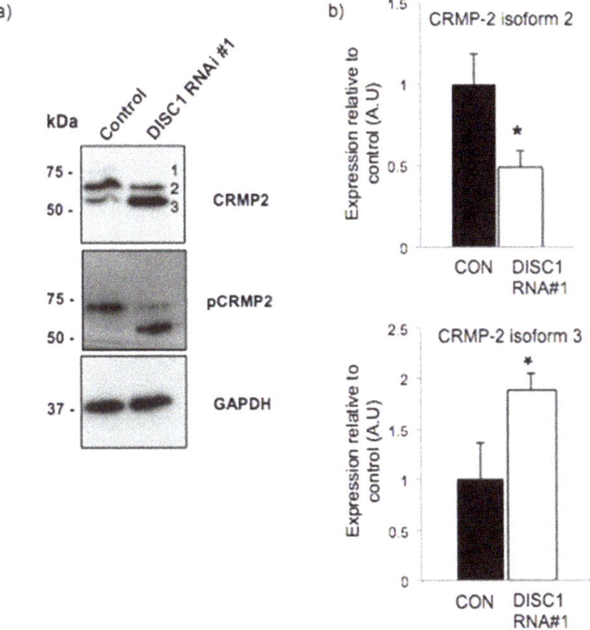

Figure 1. DISC1 differentially affects CRMP2 isoform levels. (**a**) Western blot of CRMP2 and pCRMP2 proteins. The total content of CRMP2 falls in DISC1 silenced cells, and the smallest one, thought to be a cleavage product, rises. The three isoforms are indicated (1–3). (**b**) Densitometric analysis of CRMP2 bands 2 and 3 ($n = 4$, * $p < 0.05$).

2.4. DISC1 Alters the Expression of Synaptic Function Related Proteins

We also consider of great relevance that endocytosis was highlighted under the top molecular and cellular functions in our IPA analysis (Table 3). Endocytosis and exocytosis are crucial processes for neurotransmission [33] and regulated by SNARE and SM proteins (Sec1/Munc18-like proteins) [34]. In particular, syntaxin-7 (member of the SNARE complex present on plasma membrane) and syntaxin binding protein (STXBP, also known as MUNC18) were found upregulated in DISC1-silenced cells

(Table 1). Other proteins that regulate the exocytic processes responsible for neuronal communication are Rab proteins [35], which catalyze SNARE complex assembly [36]. In this study four different Rab proteins were found differentially expressed in DISC1-silenced cells (Table 1).

2.5. DISC1 Silenced SH-SY5Y Cells Show Impaired Neurite Outgrowth

To further test that silencing of DISC1 results in disruption of neural development, we performed a morphological study in SH-SY5Y cells in which DISC1 was silenced [37]. The absence of DISC1 in this cell line resulted in morphological changes (Figure 2). Thus, upon retinoic acid-induced differentiation, DISC1-silenced cells exhibited fewer and shorter neurites (Figure 2, Figure S4).

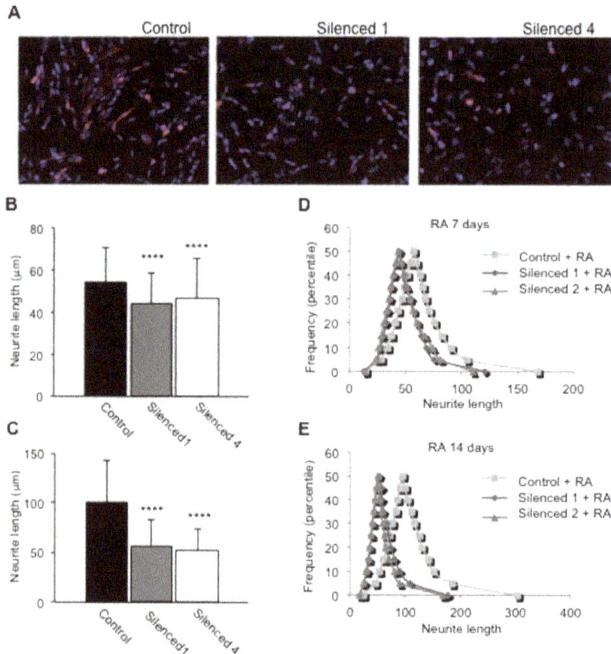

Figure 2. DISC1-silenced cells show morphological impairment in neurite outgrowth assays. Cells were treated with retinoic acid (RA) for 7 and 14 days and neurite length was measured using Image J. (**A**) Fluorescence images of SH-SY5Y cells expressing control and DISC1 shRNAs treated with RA for 7 days and immunostained for βIII-tubulin (red); nuclei were stained using DAPI (blue). (**B,C**) Average neurite length ± SD; (**** $p < 0.0001$, significantly different between control and DISC1-silenced cells, $n > 200$ for each cell line). (**D,E**) Frequency (percentile) of cells according to neurite length at 7 days (**D**) and 14 days (**E**) for each cell population; $p < 0.0001$ control vs. silenced 1 at 7 and 14 days, $p < 0.0001$ control vs. silenced 4 at 7 days, $p < 0.001$ control vs. silenced 4 at 14 days (Mann–Whitney U test).

3. Discussion

We have taken advantage of a well-established murine primary neuron DISC1 knock-down experimental system [8,14,37] to carry out an unbiased proteomic analysis and thus, identify proteins which have their expression affected by DISC1.

The results of our analysis highlight the importance of DISC1 both in neurodevelopment and synaptic regulation. Both functions have been already ascribed to DISC1; however, this study describes new important routes to explore, as the effect DISC1 silencing on the expression of CRMP family of proteins. This could be a powerful mechanism to further investigate considering the relevance this family of proteins has in the neurobiology of mental disease [27,38–40].

Furthermore, DISC1 knockdown resulted in a neurite outgrowth deficit in RA-treated SH-SY5Y cells. Previous studies have reported an impaired neurite outgrowth in cell models that overexpress mutant isoforms of DISC1 [11,41] and an increase of neurite outgrowth was seen in PC12 cells that overexpress DISC1 [42]. Therefore, our study reinforces the idea that the loss of function of DISC1 is critical for proper regulation of neurite outgrowth. In this direction, other studies have previously shown DISC1 silencing affected neurite outgrowth using PC12 cells [14]. We have to consider neurite outgrowth in PC12 cells is a result of two processes, neural differentiation and subsequent neurite extension, so the effects of silencing may be interpreted as measuring an effect on either/both processes. In contrast, SH-SY5Y cells are already neuronal and forming neurites, so we could compare neurite length and the effect is specific to neurite outgrowth.

At the same time, we have found that several proteins that participate in synaptic membrane trafficking and synapse formation are altered in DISC1 silenced neurons, such as syntaxin 7, MUNC-18, cadherin-13, and Rab proteins (Table 1), but we cannot conclude whether trafficking is up- or downregulated in our system. Previous studies have shown that DISC1 enhances the transport of synaptic vesicles, therefore we could expect that knocking down DISC1 expression produced an attenuated vesicle transport in primary cortical neurons [43].

Summarizing, our study shows that DISC1 works as an important modulator of proteins that are directly involved both in neurodevelopment and in adult synaptic regulation, representing a unifier factor of two seemingly different categories.

4. Materials and Methods

4.1. Antibodies

Commercial antibodies specific for the following proteins were used: CRMP-2, p(Thr514)CRMP-2, Stathmin, p(Ser38)Stathmin (1:1000; Cell Signaling Technology, Danvers, MA, USA); tubulin, GAPDH (1:5000; Sigma-Aldrich, St. Louis, MO, USA); the human DISC1-specific antibody 14F2 has been previously described [44]; the mouse DISC1-specific antibody D27 was a kind gift from Merck (Kenilworth, NJ, USA). Goat anti-rabbit (1:2000; Dako Cytomation, Glosstrup, Denmark), sheep anti-mouse (1:5000; GE Healthcare Amersham Bioscience, Uppsala, Sweden) and donkey anti-goat (1:2000; Santa Cruz Biotechnologies, Santa Cruz, CA, USA) were used as secondary antibodies.

4.2. Cell Culture

SH-SY5Y neuroblastoma cells (European Collection of Cell Cultures, Salisbury, UK) were maintained in 1:1 Earle's Balanced Salt Solution (EBSS)- F12HAM (Sigma Aldrich) with 15% fetal bovine serum (FBS) (Gibco, Life Technologies, Gaithesburg, MD, USA), 1% Glutamine (Gln) (Sigma Aldrich), 1% non-essential amino acids (NEAA) (Sigma Aldrich), and 1% Penicillin-Streptomycin (P/S) (Invitrogen). 293FT cells (Invitrogen) were maintained in Dulbecco's Modified Eagle's Medium (DMEM) (Sigma Aldrich) with 10% FBS, 1% sodium pyruvate (Sigma Aldrich), 1% NEAA, 1% Gln, and 1% P/S.

Murine cortex and hippocampal primary neurons were prepared from 14–15 days embryos (see below ethical statement). Pregnant dams were killed by cervical dislocation in accordance with institutional guidelines for care and use of animals. The embryos were maintained and dissected in PBS Ca/Mg (Invitrogen) supplemented with 33 mM glucose. Pooled tissue was mechanically dissociated, treated with trypsin (Invitrogen) and DNaseI (Roche Applied Science, Mannheim, Germany) and resuspended in Neurobasal medium (Invitrogen) supplemented with 50X B27 (Invitrogen), 0.55g/100mL glucose (Sigma Aldrich), 42 mg/100 mL sodium bicarbonate (Sigma Aldrich), 1% P/S and 1% glutamine. The cells were plated on poly-D-lysine (Sigma Aldrich) coated Petri dishes. Cultures were maintained in serum free medium at 37 °C in 95% air/5% CO_2.

4.3. Ethics Statement

Animal experiments were carried out in accordance with the European Union Council Directive 86/609/EEC, and were approved by the University of Santiago de Compostela Ethics Committee (protocol 15005AE/12/FUN 01/PAT 05/JRR2, 5 January 2012).

4.4. DISC1 Silencing

For DISC1 knock-down in murine primary neurons, we chose a validated shRNA construct developed by Akira Sawa's group (DISC1 RNAi #1) that has been shown to specifically decrease the amount of DISC1 in cortical neural cell cultures [8,14,37]. The commercial pLK0.1-puro non-mammalian shRNA control construct from Sigma Aldrich (reference: SHC002) was used as a scramble control. Lentiviruses were produced by calcium phosphate triple co-transfection of shRNA (see Table S3 and Figure S1 in Supporting Information), VSVG and ΔR8.9 constructs into 293FT packaging cells. Virus-containing medium was collected 48 h after transfection, and added (10 mL of lentiviral solution/3 × 10^6 neurons) to the medium of primary neurons at 7 DIV. The medium was changed 24 h after infection, and incubation continued for 72 h.

In SH-SY5Y cells, DISC1 was silenced using commercial Mission® shRNA lentiviral transduction particles (Sigma Aldrich, reference NM_018662) containing two alternative PLKO.1-Puro-CMV shRNA plasmids (Table S2 in Supporting Information). Mission® pLKO.1-puro non-mammalian shRNA particles (reference: SHC002V) were used as control. Stable cell lines were generated for any of these constructs after selection with puromycin as previously described [37].

4.5. Sample Preparation for Proteomic Studies

Cells (confluent 100 mm plates) were washed twice with cold PBS and solubilized in lysis buffer (20 mM HEPES, 2 mM EGTA, 1 mM DTT, 1 mM sodium orthovanadate, 1% Triton X-100, 10% Glycerol, 2 µM leupeptin, 400 µM PMSF, 50 µM β-glycerophosphate, 100 µg/mL Trasylol). The cells were scraped on ice for 10 min, incubated on ice for 30 min with periodic vortexing every 5 min and centrifuged for 20 min at 14,000 g, 4 °C. The supernatant was saved and the pellet discarded. The protein content was determined using the BCA protein assay kit (Pierce Chemical). Proteins were precipitated with 60% trichloroacetic acid (TCA) in acetone. After 2–3 acetone washes, proteins were dissolved in 500 µL of 2D sample buffer (5 M urea, 2 M thiourea, 2 mM tributyl-phosphine, 65 mM DTT, 65 mM CHAPS, 0.15 M NDSB-256, 1 mM sodium vanadate, 0.1 mM sodium fluoride, and 1 mM benzamidine). Ampholytes (Servalyte 4–7) were added to the sample to a final concentration of 1.6% (*v/v*).

4.6. Proteomic Studies

The primary neuron cell lysates were subjected to two-dimensional gel electrophoresis (2-DE). Protein quantitation was performed with the Coomassie plus protein reagent (Thermo Scientific, Asheville, NC). Five hundred micrograms of protein were loaded onto each gel to allow detection of low abundance proteins. Four gels per study group (DISC1 knock-down and control) were compared. Immobilized pH gradient (IPG) strips (4–7, 24 cm, GE Healthcare, Uppsala, Sweden) were rehydrated in the sample, and isoelectric focusing (IEF) was performed in a Multiphor (GE Healthcare) for 85 kVh at 17 °C. Following focusing, the IPG strips were immediately equilibrated for 15 min in 4 M urea, 2 M thiourea, 130 mM DTT, 50 mM Tris pH 6.8, 2% *w/v* SDS, 30% *v/v* glycerol. Later, the strips were placed for 15 min in the same buffer, in which DTT was replaced by 4.5% iodoacetamide (Sigma Aldrich). The IPG strips were placed on top of the second dimension gels and embedded with 0.5% melted agarose. Proteins were separated in the second dimension by SDS-polyacrylamide gel electrophoresis (PAGE) on 10% gels at run conditions of 10 °C, 20 mA per gel for 1 h, followed by 40 mA per gel for 4 h by using an Ettan Dalt 6 system (GE Healthcare). Following electrophoresis, gels were fixed in 10% methanol/7% acetic acid for 1 h, and stained overnight with Sypro Ruby fluorescent dye (Lonza,

Switzerland). After staining, gels were washed for 1 h in 10% methanol/7% acetic acid, and scanned in a Typhoon 9410 (GE Healthcare).

4.7. Differential Image Analysis

Image analysis was performed with the Ludesi REDFIN 3 Solo software (Ludesi, Malmö, Sweden). The integrated intensity of each of the spots was measured, and the background corrected and normalized. Differential expression of proteins was defined on the basis of $\geq$2-fold change between group averages and $p < 0.05$.

4.8. Mass Spectrometric Analysis

Spots of interest were carefully excised and subjected to in-gel digestion with trypsin [45]. Tryptic digests were analyzed using a 4800 MALDI-TOF/TOF analyzer (Applied Biosystems). Dried peptides were dissolved in 4 μL of 0.5% formic acid. Equal volumes (0.5 μL) of peptide and matrix solution, consisting of 3 mg alpha-cyano-4-hydroxycinnamic acid (α-CHCA) dissolved in 1 mL of 50% acetonitrile in 0.1% trifluoroacetic acid, were deposited using the thin layer method, onto a 384 Opti-TOF MALDI plate (Applied Biosystems). MS spectra were acquired in reflectron positive-ion mode with a Nd:YAG, 355 nm wavelength laser, averaging 1000 laser shots and using at least three trypsin autolysis peaks as internal calibration. All MS/MS spectra were performed by selecting the precursors with a relative resolution of 300 (FWHM) and metastable suppression. Automated analysis of mass data was achieved by using the 4000 Series Explorer Software V3.5. MS and MS/MS spectra data were combined through the GPS Explorer Software v3.6. Database search was performed with the Mascot v2.1 search tool (Matrix Science, London, UK) screening SwissProt (release 56.0). Searches were restricted to mouse taxonomy allowing carbamidomethyl cysteine as a fixed modification and oxidized methionine as potential variable modification. Both the precursor mass tolerance and the MS/MS tolerance were set at 30 ppm and 0.35 Da, respectively, allowing 1 missed tryptic cleavage site. All spectra and database results were manually inspected in detail using the above software. Protein scores greater than 56 were accepted as statistically significant ($p < 0.05$), considering positive the identification when protein score CI (confidence interval) was above 98%. In case of MS/MS spectra, total ion score CI was above 95%.

4.9. SDS-PAGE and Western Blotting

A total of 50 μg of protein was mixed with Laemmli sample buffer (BioRad), heated at 100 °C for 10 min, spun, and the supernatant loaded on a 7.5% SDS-PAGE gel. Samples were subjected to electrophoresis and transferred to polyvinylidenedifluoride (PVDF) membranes (Millipore, Bedford, MA, USA). The conditions of the electrophoresis were 200 V, 1 h. Electrophoresis was performed using a Mini-PROTEAN 3 cell electrophoresis system (BioRad). The transfer was performed in a Trans-blot SD semi-dry transfer cell (BioRad) using the following conditions: 0.8 mA/cm^2, 90 min. The PVDF membranes were blocked in 5% non-fat milk in PBS-0.1% Tween solution overnight at 4 °C, then 4 washes of 5 min with PBS-0.1% Tween20 were performed, and the membrane was incubated with the primary antiserum (in 5% BSA in PBS-0.1%Tween20) for 1 h at room temperature, washed again and incubated with the peroxidase-conjugated secondary antibody (in PBS-0.1% Tween20), and subjected to 4 washes of 5 min each with PBS-0.1% Tween20. Finally the membrane was incubated with the chemiluminescence solution Luminata Forte Western HRP substrate (Merck Millipore). To develop the membranes Hypercassette (GE Healthcare) and Amersham Hyperfilm ECL (GE Healthcare) were used.

4.10. Ingenuity Pathway

Ingenuity Pathway Analysis software (Ingenuity Systems, CA, USA) was used to investigate interactions between all the 48 identified proteins. Interactive pathways were generated to observe potential direct and indirect relations among the differentially expressed proteins. To test the enriched pathways we consider as settings direct and indirect relationships that were experimentally observed.

4.11. Neurite Outgrowth Assays

Stable SH-SY5Y cell lines generated using TRCN0000118997 (Silenced 1), TRCN0000119000 (Silenced 4) and non-target shRNAs were cultured for 7 and 14 days in medium containing 10 μM retinoic acid (RA) (Sigma Aldrich). To analyze neurite outgrowth, images of live cells were taken under a microscope and processed using Image J software (http://rsb.info.nih.gov/ij). Cells with and without neurites longer than two cell bodies were counted in photomicrographs of the differentiated control and DISC1-silenced cells.

4.12. Immunocytochemistry of SH-SY5Y Cells

Retinoic acid-treated cells were fixed in paraformaldehyde and immunostained for β3-tubulin and nuclei were visualized using DAPI, as previously described by the authors of [46].

4.13. Statistical Analysis

One-way ANOVA was employed in the proteomic analysis to determine statistically significant differences between groups of samples. For each spot ID, ANOVA p-value was calculated using the quantified and normalized spots volumes for the matched spot in each of the images. Differential expression of proteins was defined on the basis of $\geq$2-fold change between group averages and $p < 0.05$.

In the neurite outgrowth assay, three fields of up to 100 cells were analyzed for each condition and the experiment was performed twice. Statistical analysis was performed using a non-parametric unpaired Mann-Whitney U-test (two-tailed); results were considered significant with $p < 0.05$.

5. Conclusions

This study shows DISC1 disrupts the expression of a number of proteins involved in neurodevelopment and synaptic function. Thus, DISC1 acts as a key modulator of two mechanisms that have been critically implicated in the development of mental disease.

Supplementary Materials: Supplementary materials can be found at http://www.mdpi.com/1422-0067/20/1/119/s1.

Author Contributions: Conceptualization, A.R., C.R.-S., R.M.K., A.G., and J.R.R.; methodology, A.R., C.R.-S., R.M.K., A.G., and J.R.R.; software, A.R. and A.G.; validation, A.R., C.R.S.; formal analysis, A.R., C.R.-S., I.G.-E. and J.A.; investigation, A.R., C.R.-S., I.R., I.G.-E., J.A., S.V.; resources, C.K., J.A., R.M.K., A.G. and J.R.R.; data curation, A.R., C.R.-S., J.A., R.M.K. and J.R.R.; writing—original draft preparation, A.R., C.R.-S. and J.R.R.; writing—review and editing, C.K., R.M.K. and A.G.; visualization, A.R. and I.G.-E.; supervision, A.G., R.M.K. and J.R.R.; project administration, J.R.R.; funding acquisition, C.K., R.M.K., A.G., I.G.-E. and J.R.R.

Funding: This research was funded by ERANET-NEURON, grant DISCover (National funding institution grants: ISCIII PI09/2688 and BMBF 01EW1003, respectively, to J.R.R. and C.K.), and grants from the DFG (Ko1679/3-1), NARSAD 2013 Independent Investigator Award #20350 and EU-FP MC-ITN "IN-SENS" #607616) (C.K.); The Spanish Ministry of Science and Innovation (SAF2011-30494 and BFU2017-86692-P, partially funded by European Union regional funds (FRDER)); the Departments of Industry, Tourism and Trade (Etortek) and Innovation Technology of the Government of the Autonomous Community of the Basque Country (R.M.K. and I.G-E.); and the Spanish Ministry of Economy and Competitiveness (SAF2013-45014-R, A.G., SAF2011-30494 (R.M.K. and I.G-E.)).

Conflicts of Interest: The authors declare no conflict of interest.

Abbreviations

2-DE	Two-dimensional electrophoresis
CRMP-2	Collapsing response mediator protein 2
DISC1	Disrupted in Schizophrenia 1
MAP1B	Microtubule associated binding protein 1
MUNC18	Mammalian uncoordinated-18
MALDI	Matrix-assisted laser desorption ionization
MS	Mass Spectrometry

NCS-1	Neural calcium sensor 1
SM	Sec1/Munc18-like proteins
SNARE	(Soluble NSF Attachment Protein) Receptor
WB	Western blot

References

1. St. Clair, D.; Blackwood, D.; Muir, W.; Walker, M.; Carothers, A.; Spowart, G.; Gosden, C.; Evans, H.J. Association within a family of a balanced autosomal translocation with major mental illness. *Lancet* **1990**, *336*, 13–16. [CrossRef]
2. Millar, J.K. Disruption of two novel genes by a translocation co-segregating with schizophrenia. *Hum. Mol. Genet.* **2000**, *9*, 1415–1423. [CrossRef] [PubMed]
3. Sachs, N.A.; Sawa, A.; Holmes, S.E.; Ross, C.A.; DeLisi, L.E.; Margolis, R.L. A frameshift mutation in Disrupted in Schizophrenia 1 in an American family with schizophrenia and schizoaffective disorder. *Mol. Psychiatry* **2005**, *10*, 758–764. [CrossRef] [PubMed]
4. Kim, J.Y.; Liu, C.Y.; Zhang, F.; Duan, X.; Wen, Z.; Song, J.; Feighery, E.; Lu, B.; Rujescu, D.; St Clair, D.; et al. Interplay between DISC1 and GABA signaling regulates neurogenesis in mice and risk for schizophrenia. *Cell* **2012**, *148*, 1051–1064. [CrossRef] [PubMed]
5. Mao, Y.; Ge, X.; Frank, C.L.; Madison, J.M.; Koehler, A.N.; Doud, M.K.; Tassa, C.; Berry, E.M.; Soda, T.; Singh, K.K.; et al. Disrupted in schizophrenia 1 regulates neuronal progenitor proliferation via modulation of GSK3beta/beta-catenin signaling. *Cell* **2009**, *136*, 1017–1031. [CrossRef] [PubMed]
6. Lepagnol-Bestel, A.M.; Kvajo, M.; Karayiorgou, M.; Simonneau, M.; Gogos, J.A. A Disc1 mutation differentially affects neurites and spines in hippocampal and cortical neurons. *Mol. Cell. Neurosci.* **2013**, *54*, 84–92. [CrossRef] [PubMed]
7. Zhou, M.; Li, W.; Huang, S.; Song, J.; Kim, J.Y.; Tian, X.; Kang, E.; Sano, Y.; Liu, C.; Balaji, J.; et al. MTOR Inhibition Ameliorates Cognitive and Affective Deficits Caused by Disc1 Knockdown in Adult-Born Dentate Granule Neurons. *Neuron* **2013**, *77*, 647–654. [CrossRef]
8. Hayashi-Takagi, A.; Takaki, M.; Graziane, N.; Seshadri, S.; Murdoch, H.; Dunlop, A.J.; Makino, Y.; Seshadri, A.J.; Ishizuka, K.; Srivastava, D.P.; et al. Disrupted-in-Schizophrenia 1 (DISC1) regulates spines of the glutamate synapse via Rac1. *Nat. Neurosci.* **2010**, *13*, 327–332. [CrossRef]
9. Wang, Q.; Charych, E.I.; Pulito, V.L.; Lee, J.B.; Graziane, N.M.; Crozier, R.A.; Revilla-Sanchez, R.; Kelly, M.P.; Dunlop, A.J.; Murdoch, H.; et al. The psychiatric disease risk factors DISC1 and TNIK interact to regulate synapse composition and function. *Mol. Psychiatry* **2011**, *16*, 1006–1023. [CrossRef]
10. Tsuboi, D.; Kuroda, K.; Tanaka, M.; Namba, T.; Iizuka, Y.; Taya, S.; Shinoda, T.; Hikita, T.; Muraoka, S.; Iizuka, M.; et al. Disrupted-in-schizophrenia 1 regulates transport of ITPR1 mRNA for synaptic plasticity. *Nat. Neurosci.* **2015**, *18*, 698–707. [CrossRef]
11. Wen, Z.; Nguyen, H.N.; Guo, Z.; Lalli, M.A.; Wang, X.; Su, Y.; Kim, N.S.; Yoon, K.J.; Shin, J.; Zhang, C.; et al. Synaptic dysregulation in a human iPS cell model of mental disorders. *Nature* **2014**, *515*, 414–418. [CrossRef] [PubMed]
12. Ozeki, Y.; Tomoda, T.; Kleiderlein, J.; Kamiya, A.; Bord, L.; Fujii, K.; Okawa, M.; Yamada, N.; Hatten, M.E.; Snyder, S.H.; et al. Disrupted-in-Schizophrenia-1 (DISC-1): Mutant truncation prevents binding to NudE-like (NUDEL) and inhibits neurite outgrowth. *Proc. Natl. Acad. Sci. USA* **2003**, *100*, 289–294. [CrossRef] [PubMed]
13. Namba, T.; Ming, G.-L.; Song, H.; Waga, C.; Enomoto, A.; Kaibuchi, K.; Kohsaka, S.; Uchino, S. NMDA receptor regulates migration of newly generated neurons in the adult hippocampus via Disrupted-In-Schizophrenia 1 (DISC1). *J. Neurochem.* **2011**, *118*, 34–44. [CrossRef] [PubMed]
14. Kamiya, A.; Kubo, K.; Tomoda, T.; Takaki, M.; Youn, R.; Ozeki, Y.; Sawamura, N.; Park, U.; Kudo, C.; Okawa, M.; et al. A schizophrenia-associated mutation of DISC1 perturbs cerebral cortex development. *Nat. Cell Biol.* **2005**, *7*, 1167–1178. [CrossRef] [PubMed]
15. Ishizuka, K.; Kamiya, A.; Oh, E.C.; Kanki, H.; Seshadri, S.; Robinson, J.F.; Murdoch, H.; Dunlop, A.J.; Kubo, K.I.; Furukori, K.; et al. DISC1-dependent switch from progenitor proliferation to migration in the developing cortex. *Nature* **2011**, *473*, 92–96. [CrossRef] [PubMed]

16. Camargo, L.M.; Collura, V.; Rain, J.-C.; Mizuguchi, K.; Hermjakob, H.; Kerrien, S.; Bonnert, T.P.; Whiting, P.J.; Brandon, N.J. Disrupted in Schizophrenia 1 Interactome: Evidence for the close connectivity of risk genes and a potential synaptic basis for schizophrenia. *Mol. Psychiatry* **2007**, *12*, 74–86. [CrossRef] [PubMed]

17. Millar, J.K.; Pickard, B.S.; Mackie, S.; James, R.; Christie, S.; Buchanan, S.R.; Malloy, M.P.; Chubb, J.E.; Huston, E.; Baillie, G.S.; et al. DISC1 and PDE4B are interacting genetic factors in schizophrenia that regulate cAMP signaling. *Science* **2005**, *310*, 1187–1191. [CrossRef]

18. Enomoto, A.; Asai, N.; Namba, T.; Wang, Y.; Kato, T.; Tanaka, M.; Tatsumi, H.; Taya, S.; Tsuboi, D.; Kuroda, K.; et al. Roles of Disrupted-In-Schizophrenia 1-Interacting Protein Girdin in Postnatal Development of the Dentate Gyrus. *Neuron* **2009**, *63*, 774–787. [CrossRef]

19. Brandon, N.J.; Sawa, A. Linking neurodevelopmental and synaptic theories of mental illness through DISC1. *Nat. Rev. Neurosci.* **2011**, *12*, 707–722. [CrossRef]

20. Sialana, F.J.; Wang, A.-L.; Fazari, B.; Kristofova, M.; Smidak, R.; Trossbach, S.V.; Korth, C.; Huston, J.P.; de Souza Silva, M.A.; Lubec, G. Quantitative Proteomics of Synaptosomal Fractions in a Rat Overexpressing Human DISC1 Gene Indicates Profound Synaptic Dysregulation in the Dorsal Striatum. *Front. Mol. Neurosci.* **2018**, *11*, 26. [CrossRef]

21. Xia, M.; Broek, J.A.C.; Jouroukhin, Y.; Schoenfelder, J.; Abazyan, S.; Jaaro-Peled, H.; Sawa, A.; Bahn, S.; Pletnikov, M. Cell Type-Specific Effects of Mutant DISC1: A Proteomics Study. *Mol. Neuropsychiatry* **2016**, *2*, 28–36. [CrossRef] [PubMed]

22. Taya, S.; Shinoda, T.; Tsuboi, D.; Asaki, J.; Nagai, K.; Hikita, T.; Kuroda, S.; Kuroda, K.; Shimizu, M.; Hirotsune, S.; et al. DISC1 regulates the transport of the NUDEL/LIS1/14-3-3epsilon complex through kinesin-1. *J. Neurosci.* **2007**, *27*, 15–26. [CrossRef] [PubMed]

23. Nagai, J.; Baba, R.; Ohshima, T. CRMPs Function in Neurons and Glial Cells: Potential Therapeutic Targets for Neurodegenerative Diseases and CNS Injury. *Mol. Neurobiol.* **2017**, *54*, 4243–4256. [CrossRef] [PubMed]

24. Schmidt, E.F.; Strittmatter, S.M. The CRMP family of proteins and their role in Sema3A signaling. In *Semaphorins: Receptor and Intracellular Signaling Mechanisms*; Advances in Experimental Medicine and Biology; Springer: New York, NY, USA, 2007.

25. Yamashita, N.; Goshima, Y. Collapsin response mediator proteins regulate neuronal development and plasticity by switching their phosphorylation status. *Mol. Neurobiol.* **2012**, *45*, 234–246. [CrossRef] [PubMed]

26. Nakata, K.; Ujike, H.; Sakai, A.; Takaki, M.; Imamura, T.; Tanaka, Y.; Kuroda, S. The Human Dihydropyrimidinase-Related Protein 2 Gene on Chromosome 8p21 Is Associated with Paranoid-Type Schizophrenia. *Biol. Psychiatry* **2003**, *53*, 571–576. [CrossRef]

27. Johnston-Wilson, N.L.; Sims, C.D.; Hofmann, J.P.; Anderson, L.; Shore, A.D.; Torrey, E.F.; Yolken, R.H. Disease-specific alterations in frontal cortex brain proteins in schizophrenia, bipolar disorder, and major depressive disorder. The Stanley Neuropathology Consortium. *Mol. Psychiatry* **2000**, *5*, 142–149. [CrossRef] [PubMed]

28. Hensley, K.; Venkova, K.; Christov, A.; Gunning, W.; Park, J. Collapsin response mediator protein-2: An emerging pathologic feature and therapeutic target for neurodisease indications. *Mol. Neurobiol.* **2011**, *43*, 180–191. [CrossRef]

29. Wakatsuki, S.; Saitoh, F.; Araki, T. ZNRF1 promotes Wallerian degeneration by degrading AKT to induce GSK3B-dependent CRMP2 phosphorylation. *Nat. Cell Biol.* **2011**, *13*, 1415–1423. [CrossRef]

30. Zhang, Z.; Ottens, A.K.; Sadasivan, S.; Kobeissy, F.H.; Fang, T.; Hayes, R.L.; Wang, K.K. Calpain-mediated collapsin response mediator protein-1, -2, and -4 proteolysis after neurotoxic and traumatic brain injury. *J. Neurotrauma* **2007**, *24*, 460–472. [CrossRef]

31. Zhang, J.-N.; Michel, U.; Lenz, C.; Friedel, C.C.; Köster, S.; d'Hedouville, Z.; Tönges, L.; Urlaub, H.; Bähr, M.; Lingor, P.; et al. Calpain-mediated cleavage of collapsin response mediator protein-2 drives acute axonal degeneration. *Sci. Rep.* **2016**, *6*, 37050. [CrossRef]

32. Rogemond, V.; Auger, C.; Giraudon, P.; Becchi, M.; Auvergnon, N.; Belin, M.F.; Honnorat, J.; Moradi-Améli, M. Processing and nuclear localization of CRMP2 during brain development induce neurite outgrowth inhibition. *J. Biol. Chem.* **2008**, *283*, 14751–14761. [CrossRef] [PubMed]

33. Alabi, A.A.; Tsien, R.W. Perspectives on Kiss-and-Run: Role in Exocytosis, Endocytosis, and Neurotransmission. *Annu. Rev. Physiol.* **2013**, *75*, 393–422. [CrossRef] [PubMed]

34. Dulubova, I.; Khvotchev, M.; Liu, S.; Huryeva, I.; Su, T.C. Munc18-1 binds directly to the neuronal SNARE complex. *Proc. Natl. Acad. Sci. USA* **2007**, *104*, 2697–2702. [CrossRef]

35. von Mollard, G.F.; Stahl, B.; Li, C.; Südhof, T.C.; Jahn, R. Rab proteins in regulated exocytosis. *Trends Biochem. Sci.* **1994**, *19*, 164–168. [CrossRef]
36. Søgaard, M.; Tani, K.; Ye, R.R.; Geromanos, S.; Tempst, P.; Kirchhausen, T.; Rothman, J.E.; Söllner, T. A rab protein is required for the assembly of SNARE complexes in the docking of transport vesicles. *Cell* **1994**, *78*, 937–948. [CrossRef]
37. Ramos, A.; Rodríguez-Seoane, C.; Rosa, I.; Trossbach, S.V.; Ortega-Alonso, A.; Tomppo, L.; Ekelund, J.; Veijola, J.; Järvelin, M.R.; Alonso, J.; et al. Neuropeptide precursor VGF is genetically associated with social anhedonia and underrepresented in the brain of major mental illness: Its downregulation by DISC1. *Hum. Mol. Genet.* **2014**, *23*, 5859–5865. [CrossRef]
38. Niwa, M.; Cash-Padgett, T.; Kubo, K.; Saito, A.; Ishii, K.; Sumitomo, A.; Taniguchi, Y.; Ishizuka, K.; Jaaro-Peled, H.; Tomoda, T.; et al. DISC1 a key molecular lead in psychiatry and neurodevelopment: No-More Disrupted-in-Schizophrenia. *Mol. Psychiatry* **2016**, *21*, 1488–1489. [CrossRef]
39. Bader, V.; Tomppo, L.; Trossbach, S.V.; Bradshaw, N.J.; Prikulis, I.; Rutger Leliveld, S.; Lin, C.Y.; Ishizuka, K.; Sawa, A.; Ramos, A.; et al. Proteomic, genomic and translational approaches identify CRMP1 for a role in schizophrenia and its underlying traits. *Hum. Mol. Genet.* **2012**, *21*, 4406–4418. [CrossRef]
40. McLean, C.K.; Narayan, S.; Lin, S.Y.; Rai, N.; Chung, Y.; Hipolito, M.M.S.; Cascella, N.G.; Nurnberger, J.I.; Ishizuka, K.; Sawa, A.S.; et al. Lithium-associated transcriptional regulation of CRMP1 in patient-derived olfactory neurons and symptom changes in bipolar disorder. *Transl. Psychiatry* **2018**, *8*, 81. [CrossRef]
41. Pletnikov, M.V.; Ayhan, Y.; Nikolskaia, O.; Xu, Y.; Ovanesov, M.V.; Huang, H.; Mori, S.; Moran, T.H.; Ross, C.A. Inducible expression of mutant human DISC1 in mice is associated with brain and behavioral abnormalities reminiscent of schizophrenia. *Mol. Psychiatry* **2008**, *13*, 173–186. [CrossRef]
42. Miyoshi, K.; Honda, A.; Baba, K.; Taniguchi, M.; Oono, K.; Fujita, T.; Kuroda, S.; Katayama, T.; Tohyama, M. Disrupted-In-Schizophrenia 1, a candidate gene for schizophrenia, participates in neurite outgrowth. *Mol. Psychiatry* **2003**, *8*, 685–694. [CrossRef] [PubMed]
43. Flores, R.; Hirota, Y.; Armstrong, B.; Sawa, A.; Tomoda, T. DISC1 regulates synaptic vesicle transport via a lithium-sensitive pathway. *Neurosci. Res.* **2011**, *71*, 71–77. [CrossRef] [PubMed]
44. Ottis, P.; Bader, V.; Trossbach, S.V.; Kretzschmar, H.; Michel, M.; Leliveld, S.R.; Korth, C. Convergence of two independent mental disease genes on the protein level: Recruitment of dysbindin to cell-invasive disrupted-in-schizophrenia 1 aggresomes. *Biol. Psychiatry* **2011**, *70*, 604–610. [CrossRef] [PubMed]
45. Shevchenko, A.; Wilm, M.; Vorm, O.; Mann, M. Mass spectrometric sequencing of proteins silver-stained polyacrylamide gels. *Anal. Chem.* **1996**, *68*, 850–858. [CrossRef] [PubMed]
46. Castaño, Z.; Gordon-Weeks, P.R.; Kypta, R.M. The neuron-specific isoform of glycogen synthase kinase-3beta is required for axon growth. *J. Neurochem.* **2010**, *113*, 117–130. [CrossRef] [PubMed]

International Journal of
Molecular Sciences

Review

The Role of Chemokines in the Pathophysiology of Major Depressive Disorder

Vladimir M. Milenkovic *, Evan H. Stanton, Caroline Nothdurfter, Rainer Rupprecht and Christian H. Wetzel

Department of Psychiatry and Psychotherapy, Molecular Neurosciences, University of Regensburg, D-93053 Regensburg, Germany; evan.h.stanton@gmail.com (E.H.S.); Caroline.Nothdurfter@medbo.de (C.N.); Rainer.Rupprecht@medbo.de (R.R.); Christian.Wetzel@klinik.uni-regensburg.de (C.H.W.)
* Correspondence: vladimir.milenkovic@ukr.de; Tel.: +49-941-944-8955

Received: 31 March 2019; Accepted: 8 May 2019; Published: 9 May 2019

Abstract: Major depressive disorder (MDD) is a debilitating condition, whose high prevalence and multisymptomatic nature set its standing as a leading contributor to global disability. To better understand this psychiatric disease, various pathophysiological mechanisms have been proposed, including changes in monoaminergic neurotransmission, imbalance of excitatory and inhibitory signaling in the brain, hyperactivity of the hypothalamic-pituitary-adrenal (HPA) axis, and abnormalities in normal neurogenesis. While previous findings led to a deeper understanding of the disease, the pathogenesis of MDD has not yet been elucidated. Accumulating evidence has confirmed the association between chronic inflammation and MDD, which is manifested by increased levels of the C-reactive protein, as well as pro-inflammatory cytokines, such as Interleukin 1 beta, Interleukin 6, and the Tumor necrosis factor alpha. Furthermore, recent findings have implicated a related family of cytokines with chemotactic properties, known collectively as chemokines, in many neuroimmune processes relevant to psychiatric disorders. Chemokines are small (8–12 kDa) chemotactic cytokines, which are known to play roles in direct chemotaxis induction, leukocyte and macrophage migration, and inflammatory response propagation. The inflammatory chemokines possess the ability to induce migration of immune cells to the infection site, whereas their homeostatic chemokine counterparts are responsible for recruiting cells for their repair and maintenance. To further support the role of chemokines as central elements to healthy bodily function, recent studies suggest that these proteins demonstrate novel, brain-specific mechanisms including the modulation of neuroendocrine functions, chemotaxis, cell adhesion, and neuroinflammation. Elevated levels of chemokines in patient-derived serum have been detected in individuals diagnosed with major depressive disorder, bipolar disorder, and schizophrenia. Furthermore, despite the considerable heterogeneity of experimental samples and methodologies, existing biomarker studies have clearly demonstrated the important role of chemokines in the pathophysiology of psychiatric disorders. The purpose of this review is to summarize the data from contemporary experimental and clinical studies, and to evaluate available evidence for the role of chemokines in the central nervous system (CNS) under physiological and pathophysiological conditions. In light of recent results, chemokines could be considered as possible peripheral markers of psychiatric disorders, and/or targets for treating depressive disorders.

Keywords: major depressive disorder; chemokines; neuroinflammation

---

1. Introduction

Major depressive disorder (MDD) is a highly prevalent condition, and is the third leading cause of disability worldwide [1]. Despite the availability of numerous anti-depressive treatments, 30% of patients diagnosed with MDD fail to respond to anti-depressant therapy, or show only a partial response [2,3]. Bipolar disorder, which is characterized by recurrent depressive and manic episodes, is difficult to diagnose [4], and is often misdiagnosed as MDD, particularly during a depressive episode [5]. Diagnostic and Statistical Manual of Mental Disorders (DSM-5) criteria for unipolar and bipolar depression are the same during a major depressive episode [6]. Therefore, there is a need for novel biomarkers, which could distinguish between these two conditions [5]. This inadequate response to treatment reflects an incomplete understanding of the actual pathogenesis of depression, which was initially linked to changes in monoaminergic transmission [7,8]. Subsequent hypotheses include the disturbance of excitatory and inhibitory signaling in the brain [9,10], hyperactivity of the hypothalamic-pituitary-adrenal (HPA) axis [11,12], and hindrance upon the healthy progression of neurogenesis [13,14]. However, increasingly compelling lines of evidence indicate a role of nearly or completely asymptomatic subclinical systemic inflammation in the pathophysiology of MDD [15–26]. While using the reassessment of immune privilege in the central nervous system [27,28] as a foundation, complex interactions between the immune system and the brain began to emerge. The immune system regulates key aspects of brain development, neurogenesis, central nervous system (CNS) homeostasis, mood, and behavior [29–35]. As such, perturbations of the neuroimmune functions have been implicated in a number of psychiatric disorders, including MDD [36–39], bipolar disorder [40,41], schizophrenia [42–45], and autism [46,47].

Recent advances in neuroscience have linked chemotactic cytokines (chemokines) to neurobiological processes relevant to psychiatric disorders, such as synaptic transmission and plasticity, neurogenesis, and neuron-glia communication [48–51]. The disruption of any of these functions, by activation of the inflammatory response system, could be central for the pathogenesis of MDD. Impaired CXCL12/CXCR4 signaling is implicated in abnormal development, proliferation, and migration of neural progenitor cells [52,53], which is suggestive of their essential roles in mammalian neurogenesis. Furthermore, the dysregulation of various chemokines, which modulate neuronal activity by means of inducing signal transduction [54,55] and Ca^{2+} mobilization [56,57], could also be involved in pathophysiological processes leading to MDD. To add to the wide breadth of chemokine functionality, these ligands and their receptors, which are widely expressed in the CNS [58–62], coordinate immune cell recruitment and their subsequent migration to sites of inflammation. Therefore, this links peripheral and central inflammation. This phenomenon can be observed in the quantitative increase of chemokine concentrations within the serum of patients with MDD, relative to homeostatic levels. Moreover, this discrepancy is associated with the onset and progression of depression in humans [63].

To further investigate the potential connection between chemokines and depression, chemokine receptor knockout mice (CCR6 and CCR7) were created and observed to display behavioral phenotypes similar to psychiatric disorders, including MDD [64].

Altogether, these data provide evidence of the involvement of chemokines in processes underlying MDD. In this work, we will examine the role of chemokines in healthy and depressed states, as well as summarize to the best of our knowledge evidence to date for the possible role of chemokines in the pathogenesis of MDD.

2. Chemokine Superfamily

The chemokine superfamily contains a large number of ligands and receptors, which are classified into four sub-families (CXC, CC, C, and CX3C) [65], according to the number and spacing of their two N-terminal, disulfide bonding participating cysteine residues. Chemokines are small (8–12 kDa) heparin binding proteins, structurally related to cytokines that can induce directed chemotaxis of immune cells. However, chemokines are additionally involved in the regulation of migration of immune cells [66,67], blood-brain barrier (BBB) permeability [68], and synaptic pruning processes [69]. In addition to their structural criteria, chemokines can be subdivided into inflammatory chemokines, which are upregulated under inflammatory conditions, homeostatic chemokines that are responsible for maintaining homeostasis, and chemokines, which exhibit dual functionality [70].

The chemokine superfamily has expanded rapidly after the initial identification of secreted platelet factor 4 (PF4/CXCL4) [71] in 1977. Subsequent studies have identified more than 50 chemokines, as well as 20 chemokine receptors [72]. The majority of human chemokine genes are clustered on chromosomes 4 and 17. CXC chemokines can be found at chromosomal location 4q12-21, whereas most of the CC chemokines are located at 17q11-21 [73]. This suggests a rapid evolution by repeated gene duplications [74]. All chemokines share a very similar tertiary structure [75], including a highly flexible N-terminal domain and a long rigid loop, which are essential for interacting with their respective receptors [76], and a C-terminal α-helix. Typically, a given chemokine can bind to more than one receptor (Table 1) and, correspondingly, a number of different chemokines can be recognized by the same receptor [65]. Chemokines are secreted in response to inflammatory cytokines, and they selectively recruit monocytes, lymphocytes, and neutrophil-inducing chemotaxis by activating G-protein-coupled receptors (GPCRs) [77].

Table 1. Chemokines and their known receptors. Chemokine receptors, which belong to the superfamily of GPCRs, can bind to multiple chemokines, and certain chemokines can similarly bind to more than one receptor. Adapted from Zlotnik and Yoshie 2012 [65].

Subfamily	Chemokine	Synonyms	Receptors
CXC	CXCL1	Growth-related oncogene α (GROα)	CXCR1/CXCR2
	CXCL2	Growth-related oncogene β (GROβ)	CXCR2
	CXCL3	Growth-related oncogene γ (GROγ)	CXCR2
	CXCL4	Platelet factor 4 (PF-4)	CXCR3-B
	CXCL5	Epithelial cell-derived neutrophil-activating factor 78 (ENA-78)	CXCR2
	CXCL6	Granulocyte chemoattractant protein (GCP-2)	CXCR1/CXCR2
	CXCL7	Neutrophil-activating protein (NAP-2)	CXCR1/CXCR2
	CXCL8	Interleukin-8 (IL-8)	CXCR1/CXCR2
	CXCL9	Monokine induced by γ-interferon (MIG)	CXCR3
	CXCL10	γ-interferon-inducible protein 10 (IP-10)	CXCR3
	CXCL11	Interferon-inducible T cell α-Chemoattractant (I-TAC)	CXCR3
	CXCL12	Stromal cell-derived factor 1 (SDF-1)	CXCR4
	CXCL13	B cell-activating chemokine 1 (BCA-1)	CXCR5
	CXCL14	Breast and kidney chemokine (BRAK)	CXCR4
	CXCL15	Lungkine	-
	CXCL16	Scavenger receptor for phosphatidylserine and oxidized lipoprotein (SR-POX)	CXCR6
	CXCL17	dendritic cell-attracting and monocyte-attracting chemokine-like protein (DMC)	CXCR8
CC	CCL1	I-309	CCR8
	CCL2	Monocyte chemoattractant protein 1 (MCP-1)	CCR2/CCR9/CCR11
	CCL3	Macrophage inflammatory protein 1α (MIP-1α)	CCR1/CCR5/CCR9
	CCL4	Macrophage inflammatory protein 1β (MIP-1β)	CCR1/CCR5/CCR9
	CCL5	Regulated on activation of normal T cell-expressed and secreted (RANTES) entities	CCR1/CCR3/CCR4/CCR5
	CCL7	Monocyte chemoattractant protein 3 (MCP-3)	CCR1/CCR2/CCR3
	CCL8	Monocyte chemoattractant protein 2 (MCP-2)	CCR2/CCR9/CCR11
	CCL11	Eosinophil chemotactic protein (Eotaxin-1)	CCR2/CCR3/CCR5

Table 1. *Cont.*

Subfamily	Chemokine	Synonyms	Receptors
	CCL13	Monocyte chemoattractant protein 4 (MCP-4)	CCR2/CCR3/CCR5
	CCL14	Hemofiltrate CC chemokine (HCC1)	CCR1/CCR5
	CCL15	Leukotactin-1, macrophage inflammatory protein 5 (MIP-5)	-
	CCL16	Liver-expressed chemokine (LEC), monotactin-1 (MTN-1)	CCR1/CCR2/CCR5/CCR8
	CCL17	Thymus and activation-related chemokine (TARC)	CCR4
	CCL18	Macrophage inflammatory protein 4 (MIP-4)	CCR8
	CCL19	Epstein–Barr virus-induced receptor ligand chemokine (ELC)	CCR7
	CCL20	Liver-related and activation-related chemokine (LARC)	CCR6
	CCL21	Secondary lymphoid tissue chemokine (SCL)	CCR7
	CCL22	Macrophage-derived chemokine (MDC)	CCR4
	CCL23	Macrophage inflammatory protein 3 (MIP-3)	CCR1
	CCL24	Eosinophil chemotactic protein 2 (Eotaxin-2)	CCR3
	CCL25	Thymus lymphoma cell-stimulating factor (TECK)	CCR9
	CCL26	Macrophage inflammatory protein 4-α (MIP-4-α)	CCR3
	CCL27	Cutaneous T cell-attracting chemokine (CTACK)	CCR10
	CCL28	Mucosae-associated epithelial chemokine (MEC)	CCR10
C	XCL1	Lymphotactin-α	XCR1
	XCL2	Lymphotactin-β	XCR1
CX3C	CX3CL1	Fractalkine	CX3CR1

3. Chemokines and Chemokine Receptors in the Brain

Chemokines and their receptors are broadly expressed in the CNS in both physiological and pathophysiological states [58–60,78]. The glia cells (astrocytes, oligodendrocytes, and microglia), and neuronal cells constitutively express several chemokines, including CCL2, CCL3, CCL19, CCL21, CXCL10, and CX3CL1 [58,78–80], as well as others, which can be upregulated in response to pathological conditions. Endothelial cells of the BBB may, under severe inflammatory conditions, likewise produce several chemokines such as CCL2 [68], CCL4 and CCL5 [81], which bind CCR1, CCR2, and CCR5 [82] chemokine receptors that are expressed by circulating mononuclear cells.

In addition to their traditional role in immune surveillance and immune cell chemotaxis, chemokines and chemokine receptors residing in the brain are also involved in the homeostatic maintenance of the CNS through either autocrine or paracrine activity [83]. Different expression patterns of various chemokines during embryonic and postnatal development is suggestive of their essential role for typical brain development. For example, CXCL12 and its receptors CXCR4/CXCR7 are involved in the proliferation and migration of neural progenitor cells (NPC). They are distinctively expressed in both the developing and the adult brain [61,84]. On the other hand, the CX3CL1 chemokine (fractalkine) and its receptor CX3CR1, which are constitutively present in the CNS, act to modulate inflammatory responses of microglia by suppressing its neurotoxicity [85] by reducing levels of the tumor necrosis factor α (TNF-α) and nitric oxide (NO) [86]. Other chemokines such as CXCL1 and CXCL8 exert neuro-modulatory effect on the synapsis of cerebellar neurons [87].

Consequently, the chemokine system, which plays an important role in neurogenesis, neuron-glia communication, synaptic transmission, and plasticity under physiological and pathophysiological conditions, might participate in the pathogenesis of depression. Evidence in support of this claim is that alterations to all of the previously mentioned processes are consistently implicated in various psychiatric disorders including MDD [11,88].

4. Regulation of Neurogenesis and Neuronal Plasticity by Chemokines

The process of neurogenesis, by which new neurons are continuously generated in discrete brain regions of many vertebrate species including humans, is particularly prominent in the dentate gyrus of the hippocampus [89,90]. Initial studies in patients with recurrent major depression, which have shown stress-induced loss of the hippocampal volume, suggested association of hippocampal atrophy with depression [91,92]. Furthermore, the decrease in hippocampal volume was correlated with the total duration of the depressive episodes [93]. Further studies have established a link between reduced adult hippocampal neurogenesis with the pathophysiology of several psychiatric disorders, including anxiety and depression [78,90,94,95]. Therapeutic interventions, such as electro-convulsive and anti-depressive therapy [96,97], are, on the other hand, able to promote recovery from depression, in part by enhancing hippocampal neurogenesis.

Chemokines play an important role in the regulation of neuronal development and plasticity, proliferation, migration, and neural progenitor cell (NPC) differentiation [98,99]. Because of the significant redundancy in chemokine receptor-ligand interactions, most of the chemokine or chemokine receptor knockout animals are viable and show no apparent neural phenotype [100]. The only exception to this is the knockout mice from either CXCL12 or its receptor CXCR4, which are not viable and exhibit cerebellum malformation. This is suggestive of their essential role in the migration of the NPCs [101]. NPCs derived from the hippocampus and the subventricular zone (SVZ) express various chemokine receptors on their surface [102], which are important for the regulation of proliferation and differentiation of these cells. The CX3CL1 chemokine, which is abundantly present on mature neurons and astrocytes, and its receptor CX3CR1 that is mostly expressed on microglia cells [103], are additionally involved in the regulation of neurogenesis and neuroplasticity. The CX3CL1 chemokine regulates microglial synaptic pruning of mature neurons [104], modulates several neurotransmitter systems [105], and regulates the activation state of microglia [85]. Therefore, this influences the development and plasticity of the CNS. Exogenous application of the CX3CL1 chemokine further enhanced in vivo neurogenesis in aged rats by modulating the microglia phenotype [106]. Other chemokines such as CCL2, CCL21, and CXCL9, promote neuronal differentiation, whereas CCL2, CXCL1, and CXCL9 favor oligodendrocyte differentiation [107]. Further support for the association of adult hippocampal neurogenesis and MDD arise from the studies, which demonstrated that various chronic anti-depressive treatments stimulate hippocampal neurogenesis [108]. However, recent evidence suggests that the alterations in adult hippocampal neurogenesis are not solely responsible for the development of depression [109].

Altogether, chemokines play a significant role in both neurogenesis and neuronal plasticity, which are essential for proper brain functioning, and any disturbance in any of these functions could lead to a depressed state.

5. Chemokines and Neurotransmission in the Adult CNS

Chemokines and their respective receptors, which are constitutively expressed in glial cells and neurons [59,110–112], are responsible for homeostatic maintenance of the developed brain. Recent data suggest that chemokines present a unique class of neurotransmitters and neuromodulators that regulate cell survival and synaptic transmission [103,113]. For example, patch-clamp experiments performed in Purkinje neurons demonstrated an increase in spontaneous GABAergic activity upon the application of CXCL12 [56]. Application of CXCL12 in rat hypothalamic slices similarly caused an increase of GABA release from melanin-concentrating hormone neurons [114]. According to subcellular studies, chemokines are detected in presynaptic nerve terminals, where they co-localize with various neurotransmitters, and are released ensuing membrane depolarization [115–117]. CX3CL1, which co-localize with serotonin in neurons of the dorsal raphe nucleus, may indirectly inhibit serotonin neurotransmission by upregulating the sensitivity of serotonin dorsal raphe nucleus neurons to GABA inputs [50]. Furthermore, results from electrophysiological studies suggest that CCL2, CCL5, CCL22, CXCL12, CXCL8, and CX3CL1 chemokines can modulate the electrical activity in cortical, cerebral, hippocampal, and hypothalamic neurons [59,105,118–121].

Overall, the data presented in this case suggest a significant role of chemokines in neurotransmission and modulation of neurotransmitter release, which are increasingly being implicated in the pathogenesis of MDD.

6. Pre-Clinical Evidence Linking Changes in the Chemokine Network to Depressive Behavior

Animal models of psychiatric disease are a potent tool to investigate possible causes and treatments for human diseases. However, they face a number of challenges given the lack of objective diagnostic tests, biomarkers, and low predictive power [122]. Early animal-utilizing studies of depression investigated stress-response paradigms [123], and would often involve the subjugation of models to mild, unpredictable stressors that were either acute or chronic in application. The response, which is reasoned to be analogous to stress-induced depression in humans, involves the dysregulation of the hypothalamic-pituitary-adrenal (HPA) axis, as well as the neuroendocrine and neurotransmitter systems [124]. Findings indicate that immobilization and painful stress experiments demonstrated increased expression of CXCL1 chemokine in various regions of the CNS [125,126]. On the other hand, in a mouse model of depressive behavior based on chronic variable stress, no significant differences in expression of the CCL2 chemokine in hippocampus were found [127]. In prenatally stressed rats, as a further animal model of depressive behavior, levels of CCL2, and CXCL12 chemokines were upregulated in the hippocampus and prefrontal cortex, which is suggestive of excessive microglial activation [128]. Moreover, chronic anti-depressant treatment has been shown to revert those changes [129].

An alternative attempt to model depression-like behavior in animals involves inducing sickness-like behavior by administering inflammatory cytokines or lipopolysaccharide (LPS), which mimic the depressive symptoms induced by treatment of human patients with interferons [130]. CXCL1, CXCL10, and CCL5 were up-regulated in mice in which the depressive-like behavior was induced by application of Interferon α [131]. Rats treated with CXCL1 chemokines have shown a dose-dependent reduction in both spontaneous open field activity and burrowing behavior [132]. Peripheral administration of LPS have further induced the expression of CXCL1 and CCL2 in the prefrontal cortex, hypothalamus, and plasma of rats exposed to chronic, intermittent, cold stress [133]. Animals lacking CX3CR1 receptors experienced an increased duration of sickness-like behavior on the tail suspension test after peripheral LPS challenge [134], which additionally implicates the role of the chemokine system in sickness-like behavior. Virus-induced sickness-like behavior can additionally cause impaired learning and cognitive dysfunction by mechanisms that remain poorly understood. However, recent studies performed in mice have suggested a key role of an innate immune system of the brain in mediating the behavioral effects of viral infection [135,136]. Virus associated activation of a subpopulation of circulating monocytes expressing the CX3CR1 receptor causes release of TNF-α, which induces dendritic spine loss and motor learning impairment [135]. The exact mechanism by which monocytes modulate synaptic activity is not known, but there is evidence to suggest it is microglia-independent [137]. Brain endothelial cells, which serve as a natural barrier to interferon-induced sickness behavior, could also play an important role for the communication between the central nervous and immune systems [136].

Stress has been shown to play an important role in the etiology of neuropsychiatric diseases, including depression [138], and a number of animal studies have identified that exposure to stress greatly increases the risk of developing depression [139]. However, most of the stressors applied were artificial, and, thus, are not a representational model of stress exposure in humans, which is mostly social in nature [140]. Lately, alternative animal models of depression have begun to focus on psychosocial stress, particularly on a paradigm based on social defeat [141]. Repeated social defeat

(RSD) in mice causes an exposure-dependent increase of CXCL1 and CXCL2 levels in the brain, which is indicative of higher leukocyte recruitment in the brain vasculature [142]. Animals repeatedly exposed to social defeat show decreased volume and cell proliferation in the hippocampus and prefrontal cortex, which can be reverted by an anti-depressant treatment [123], bearing similar resemblance to the human studies. Altogether, various animal models of depressive-like behavior have provided evidence for the involvement of a chemokine network in the pathophysiology of major depression.

7. Involvement of Chemokines in the Pathophysiology of MDD—Clinical Studies

Several studies in humans and animal models have linked elevated levels of chemokines with the depressive behavioral symptoms, particularly increased levels of circulating inflammatory chemokines. The majority of the published clinical studies included the CCL2 and CXCL8 chemokines, and were based on the detection of the chemokine expression in blood or cerebrospinal fluid [48,73]. CCL2, which belongs to the group of the inflammatory chemokines, has been implicated in the chemotactic migration of peripheral monocytes to the brain [143]. Significantly higher concentrations of CCL2 in the serum of depressed patients compared to controls were described in numerous studies [23,144,145]. Moreover, antidepressant drug treatment effectively reduced peripheral levels of the CCL2 chemokine [146]. Although a considerable number of publications, including recent meta-analyses [15,48,63,147], have reported an increased CCL2 expression in patients diagnosed with MDD, studies involving MDD patients with suicidal ideation have surprisingly shown unchanged or reduced levels of the chemokine [148,149]. Considering a similar correlative elevation of CCL2 levels reported in patients diagnosed with bipolar disorder [150], more research is needed in order to effectively use elevated serum CCL2 levels as a marker of MDD.

CXCL8 levels in blood samples from a total of 40 studies involving 3788 participants were significantly elevated in depressed subjects when compared to controls reported in a recent comprehensive meta-analysis performed by Leighton et al. [15]. However, these results were obtained only after exclusion of a subgroup with physical illness. Significant differences that were observed in CXCL8 chemokine levels only after restricting the analysis to healthy subjects, suggest that inflammatory changes of underlying physical disease could mask the changes in chemokine levels in depressed patients [15]. Plasma levels of CCL3, also known as macrophage inflammatory protein-1α (MIP-1α), were similarly increased in depressed patients compared to healthy control subjects [15,23,151,152]. A significant increase of blood levels were also shown for further chemokines including CCL11, CXCL4, and CXCL7 [15]. Inflammation can also play an important role in the etiology of bipolar disorder, which has been suggested by several studies [153–155], in which patients with bipolar disorder showed increased levels of CCL11 and CXCL10 in the plasma. On the other hand, plasma levels of another chemokine from the CC group, CCL4, decreased in depressed patients in several studies [15,148,156]. Many other chemokines examined, such as CCL5, CCL7, CXCL9, and CXCL10, showed no significant differences [15].

During depressive episodes, biochemical measurements indicate an increased level of the microglia-enriched protein, translocator protein 18 kDa (TSPO), which is elucidated by the correlative increase of binding by TSPO-specific ligands [157]. It is still a matter of debate whether an increased TSPO ligand binding in depression is due to the proliferation of microglial cells or infiltration of circulating macrophages, which also express high amounts of TSPO protein through the blood-brain-barrier (BBB). Our recent published data show higher levels of CCL22, macrophage-derived chemokine (MDC) in the blood of the MDD patients who responded to anti-depressive therapy [158]. Therefore, this suggests that chemotaxis and infiltration of monocytes, as well as recruiting T-helper 2 cells (Th2) and T-regulatory cells through the BBB, could play a significant role in the pathophysiology of MDD. A link between macrophages and depression was initially proposed in 1991 [159], where excessive activity of macrophages has been suggested as a key factor in the etiology of this illness. Recent studies in various models of CNS injury and neurodegenerative diseases have highlighted the essential role of infiltrating monocyte-derived macrophages for the CNS repair process by resolving inflammation [160].

Typical pharmacological treatment of MDD can also decrease peripheral inflammation, as demonstrated by the reduction in levels of CCL2 [146]. However, other approaches are necessary in order to improve treatment outcome. Targeting immune-related pathways, which are altered in MDD and in bipolar disorder, could constitute a novel therapeutic mechanism for the treatment of both MDD and BD [161,162]. CCL11, which has been associated with many psychiatric disorders and its CCR3 receptor, may have represented attractive targets for treating both MDD and BD [163]. Moreover, the use of nonsteroidal anti-inflammatory drugs, including celecoxib, as an adjunctive treatment in MDD patients, and minocycline demonstrate a significant anti-depressive effect [164,165]. Even electroconvulsive therapy, which is one of the most effective treatment options for treatment-resistant depression, modulates peripheral immune activation [166]. In order to provide an accurate diagnosis, and to monitor treatment response in MDD and BD patients, novel biomarkers are urgently needed. Biobanks with well-defined phenotype of MDD and BD patients [167–169] were established with a goal to expedite development of novel diagnostic and therapeutic compounds.

According to the available clinical studies reviewed in this work, it is clear that chemokines play an important role in regulating neurobiological processes relevant to psychiatric disorders, and that dysregulation of various chemokines could play an important role in the pathophysiology of MDD.

8. Conclusions

Elucidating the neurobiological basis of depression and the development of more effective pharmacological treatments are the principal challenges, and one of the main goals of modern medicine. Less than a third of MDD patients adequately respond to the initial antidepressant treatment, and over 35% of depressed patients fail to respond to different antidepressants altogether [170]. Considering that the majority of commonly prescribed anti-depressants act primarily by increasing or modulating monoamine neurotransmission [171], there is a need for novel therapeutic agents. Identification of specific biomarkers of depression, which could be used to predict a response to anti-depressive drugs, and develop new treatment options would help reduce the burden of depression.

An increasing body of evidence, reviewed in this study, demonstrates an important role for chemokines in the biology of depression. However, the majority of the studies were performed on peripheral blood samples, and had a cross-sectional design. In order to fully comprehend the changes that occur in depression, longitudinal studies with treated MDD patients will be necessary. An additional limitation of the majority of human studies published thus far is that the patho-physiological changes detected in the periphery might not reliably indicate changes in the CNS. Furthermore, many of the investigations utilized a small subset of chemokines, which limits our total understanding of inflammatory processes in vivo.

In summary, the data reviewed in this manuscript demonstrates the important role of chemokines in pathophysiology of MDD. Chemokines and their receptors, which are widely expressed in the CNS, could become novel diagnostic markers or therapeutic targets for MDD. However, additional research in larger populations, which should also include longitudinal studies, is necessary.

9. Methods

We performed literature searches through Pubmed and Google Scholar databases for articles published before September 2018. The search terms (chemokines OR cytokines OR neuroinflammation OR inflammation) AND (Depression OR Depressive Disorder OR Major Depressive Disorder) were used. Obtained references were additionally inspected and all relevant publications were included.

Author Contributions: All authors contributed to the writing of the manuscript.

Acknowledgments: We are grateful to colleagues from our department for thoughtful discussions.

Conflicts of Interest: The authors declare no conflict of interest.

References

1. Disease, G.B.D.; Injury, I.; Prevalence, C. Global, regional, and national incidence, prevalence, and years lived with disability for 310 diseases and injuries, 1990–2015: A systematic analysis for the Global Burden of Disease Study 2015. *Lancet* **2016**, *388*, 1545–1602. [CrossRef]
2. Al-Harbi, K.S. Treatment-resistant depression: Therapeutic trends, challenges, and future directions. *Patient Prefer. Adher.* **2012**, *6*, 369–388. [CrossRef] [PubMed]
3. Amsterdam, J.D.; Maislin, G.; Potter, L. Fluoxetine efficacy in treatment resistant depression. *Prog. Neuro-Psychopharmacol. Biol. Psychiatry* **1994**, *18*, 243–261. [CrossRef]
4. Hirschfeld, R.M.; Lewis, L.; Vornik, L.A. Perceptions and impact of bipolar disorder: How far have we really come? Results of the national depressive and manic-depressive association 2000 survey of individuals with bipolar disorder. *J. Clin. Psychiatry* **2003**, *64*, 161–174. [CrossRef] [PubMed]
5. Han, K.M.; De Berardis, D.; Fornaro, M.; Kim, Y.K. Differentiating between bipolar and unipolar depression in functional and structural MRI studies. *Prog. Neuro-Psychopharmacol. Biol. Psychiatry* **2019**, *91*, 20–27. [CrossRef]
6. Grande, I.; Berk, M.; Birmaher, B.; Vieta, E. Bipolar disorder. *Lancet* **2016**, *387*, 1561–1572. [CrossRef]
7. Schildkraut, J.J. The catecholamine hypothesis of affective disorders: A review of supporting evidence. *Am. J. Psychiatry* **1965**, *122*, 509–522. [CrossRef]
8. Turner, W.J.; Merlis, S. A Clinical Trial of Pargyline and Dopa in Psychotic Subjects. *Dis. Nerv. Syst.* **1964**, *25*, 538–541. [PubMed]
9. Hashimoto, K. Emerging role of glutamate in the pathophysiology of major depressive disorder. *Brain Res. Rev.* **2009**, *61*, 105–123. [CrossRef]
10. Hashimoto, K.; Sawa, A.; Iyo, M. Increased levels of glutamate in brains from patients with mood disorders. *Biol. Psychiatry* **2007**, *62*, 1310–1316. [CrossRef]
11. Pariante, C.M.; Lightman, S.L. The HPA axis in major depression: Classical theories and new developments. *Trends Neurosci.* **2008**, *31*, 464–468. [CrossRef] [PubMed]
12. Holsboer, F.; Von Bardeleben, U.; Gerken, A.; Stalla, G.K.; Muller, O.A. Blunted corticotropin and normal cortisol response to human corticotropin-releasing factor in depression. *New Engl. J. Med.* **1984**, *311*, 1127. [CrossRef]
13. Duman, R.S. Role of neurotrophic factors in the etiology and treatment of mood disorders. *NeuroMol. Med.* **2004**, *5*, 11–25. [CrossRef]
14. Shimizu, E.; Hashimoto, K.; Okamura, N.; Koike, K.; Komatsu, N.; Kumakiri, C.; Nakazato, M.; Watanabe, H.; Shinoda, N.; Okada, S.; et al. Alterations of serum levels of brain-derived neurotrophic factor (BDNF) in depressed patients with or without antidepressants. *Biol. Psychiatry* **2003**, *54*, 70–75. [CrossRef]
15. Leighton, S.P.; Nerurkar, L.; Krishnadas, R.; Johnman, C.; Graham, G.J.; Cavanagh, J. Chemokines in depression in health and in inflammatory illness: A systematic review and meta-analysis. *Mol. Psychiatry* **2017**. [CrossRef] [PubMed]
16. Misiak, B.; Beszlej, J.A.; Kotowicz, K.; Szewczuk-Boguslawska, M.; Samochowiec, J.; Kucharska-Mazur, J.; Frydecka, D. Cytokine alterations and cognitive impairment in major depressive disorder: From putative mechanisms to novel treatment targets. *Prog. Neuro-Psychopharmacol. Biol. Psychiatry* **2018**, *80*, 177–188. [CrossRef] [PubMed]
17. Dowlati, Y.; Herrmann, N.; Swardfager, W.; Liu, H.; Sham, L.; Reim, E.K.; Lanctot, K.L. A meta-analysis of cytokines in major depression. *Biol. Psychiatry* **2010**, *67*, 446–457. [CrossRef] [PubMed]
18. Dantzer, R.; O'Connor, J.C.; Freund, G.G.; Johnson, R.W.; Kelley, K.W. From inflammation to sickness and depression: When the immune system subjugates the brain. *Nat. Rev. Neurosci.* **2008**, *9*, 46–56. [CrossRef]
19. Maes, M. Evidence for an immune response in major depression: A review and hypothesis. *Prog. Neuro-Psychopharmacol. Biol. Psychiatry* **1995**, *19*, 11–38. [CrossRef]
20. Strawbridge, R.; Arnone, D.; Danese, A.; Papadopoulos, A.; Herane Vives, A.; Cleare, A.J. Inflammation and clinical response to treatment in depression: A meta-analysis. *Eur. Neuropsychopharmacol. J. Eur. Coll. Neuropsychopharmacol.* **2015**, *25*, 1532–1543. [CrossRef]
21. O'Brien, S.M.; Scully, P.; Fitzgerald, P.; Scott, L.V.; Dinan, T.G. Plasma cytokine profiles in depressed patients who fail to respond to selective serotonin reuptake inhibitor therapy. *J. Psychiatr. Res.* **2007**, *41*, 326–331. [CrossRef]

22. Sasayama, D.; Hattori, K.; Wakabayashi, C.; Teraishi, T.; Hori, H.; Ota, M.; Yoshida, S.; Arima, K.; Higuchi, T.; Amano, N.; et al. Increased cerebrospinal fluid interleukin-6 levels in patients with schizophrenia and those with major depressive disorder. *J. Psychiatr. Res.* **2013**, *47*, 401–406. [CrossRef]

23. Simon, N.M.; McNamara, K.; Chow, C.W.; Maser, R.S.; Papakostas, G.I.; Pollack, M.H.; Nierenberg, A.A.; Fava, M.; Wong, K.K. A detailed examination of cytokine abnormalities in Major Depressive Disorder. *Eur. Neuropsychopharmacol. J. Eur. Coll. Neuropsychopharmacol.* **2008**, *18*, 230–233. [CrossRef]

24. van den Biggelaar, A.H.; Gussekloo, J.; de Craen, A.J.; Frolich, M.; Stek, M.L.; van der Mast, R.C.; Westendorp, R.G. Inflammation and interleukin-1 signaling network contribute to depressive symptoms but not cognitive decline in old age. *Exp. Gerontol.* **2007**, *42*, 693–701. [CrossRef]

25. Maes, M.; Stevens, W.; DeClerck, L.; Bridts, C.; Peeters, D.; Schotte, C.; Cosyns, P. Immune disorders in depression: Higher T helper/T suppressor-cytotoxic cell ratio. *Acta Psychiatr. Scand.* **1992**, *86*, 423–431. [CrossRef]

26. Eller, T.; Vasar, V.; Shlik, J.; Maron, E. Pro-inflammatory cytokines and treatment response to escitalopram in major depressive disorder. *Prog. Neuro-Psychopharmacol. Biol. Psychiatry* **2008**, *32*, 445–450. [CrossRef]

27. Galea, I.; Bechmann, I.; Perry, V.H. What is immune privilege (not)? *Trends Immunol.* **2007**, *28*, 12–18. [CrossRef]

28. Matyszak, M.K.; Perry, V.H. Demyelination in the central nervous system following a delayed-type hypersensitivity response to bacillus Calmette-Guerin. *Neuroscience* **1995**, *64*, 967–977. [CrossRef]

29. de Miranda, A.S.; Zhang, C.J.; Katsumoto, A.; Teixeira, A.L. Hippocampal adult neurogenesis: Does the immune system matter? *J. Neurol. Sci.* **2017**, *372*, 482–495. [CrossRef]

30. Marques-Deak, A.; Cizza, G.; Sternberg, E. Brain-immune interactions and disease susceptibility. *Mol. Psychiatry* **2005**, *10*, 239–250. [CrossRef]

31. Fung, T.C.; Olson, C.A.; Hsiao, E.Y. Interactions between the microbiota, immune and nervous systems in health and disease. *Nat. Neurosci.* **2017**, *20*, 145–155. [CrossRef]

32. Brenhouse, H.C.; Schwarz, J.M. Immunoadolescence: Neuroimmune development and adolescent behavior. *Neurosci. Biobehav. Rev.* **2016**, *70*, 288–299. [CrossRef]

33. Vukovic, J.; Colditz, M.J.; Blackmore, D.G.; Ruitenberg, M.J.; Bartlett, P.F. Microglia modulate hippocampal neural precursor activity in response to exercise and aging. *J. Neurosci. Off. J. Soc. Neurosci.* **2012**, *32*, 6435–6443. [CrossRef]

34. Zheng, P.; Zeng, B.; Zhou, C.; Liu, M.; Fang, Z.; Xu, X.; Zeng, L.; Chen, J.; Fan, S.; Du, X.; et al. Gut microbiome remodeling induces depressive-like behaviors through a pathway mediated by the host's metabolism. *Mol. Psychiatry* **2016**, *21*, 786–796. [CrossRef]

35. do Prado, C.H.; Narahari, T.; Holland, F.H.; Lee, H.N.; Murthy, S.K.; Brenhouse, H.C. Effects of early adolescent environmental enrichment on cognitive dysfunction, prefrontal cortex development, and inflammatory cytokines after early life stress. *Dev. Psychobiol.* **2016**, *58*, 482–491. [CrossRef]

36. Muller, N.; Schwarz, M.J. The immune-mediated alteration of serotonin and glutamate: Towards an integrated view of depression. *Mol. Psychiatry* **2007**, *12*, 988–1000. [CrossRef]

37. Miller, A.H.; Raison, C.L. The role of inflammation in depression: From evolutionary imperative to modern treatment target. *Nat. Rev. Immunol.* **2016**, *16*, 22–34. [CrossRef]

38. Reichenberg, A.; Yirmiya, R.; Schuld, A.; Kraus, T.; Haack, M.; Morag, A.; Pollmacher, T. Cytokine-associated emotional and cognitive disturbances in humans. *Arch. Gen. Psychiatry* **2001**, *58*, 445–452. [CrossRef]

39. Mostafavi, S.; Battle, A.; Zhu, X.; Potash, J.B.; Weissman, M.M.; Shi, J.; Beckman, K.; Haudenschild, C.; McCormick, C.; Mei, R.; et al. Type I interferon signaling genes in recurrent major depression: Increased expression detected by whole-blood RNA sequencing. *Mol. Psychiatry* **2014**, *19*, 1267–1274. [CrossRef]

40. Watkins, C.C.; Sawa, A.; Pomper, M.G. Glia and immune cell signaling in bipolar disorder: Insights from neuropharmacology and molecular imaging to clinical application. *Transl. Psychiatry* **2014**, *4*, e350. [CrossRef]

41. Munkholm, K.; Vinberg, M.; Vedel Kessing, L. Cytokines in bipolar disorder: A systematic review and meta-analysis. *J. Affect. Disord.* **2013**, *144*, 16–27. [CrossRef]

42. Horvath, S.; Mirnics, K. Immune system disturbances in schizophrenia. *Biol. Psychiatry* **2014**, *75*, 316–323. [CrossRef]

43. Khandaker, G.M.; Cousins, L.; Deakin, J.; Lennox, B.R.; Yolken, R.; Jones, P.B. Inflammation and immunity in schizophrenia: Implications for pathophysiology and treatment. *Lancet Psychiatry* **2015**, *2*, 258–270. [CrossRef]

44. Song, X.Q.; Lv, L.X.; Li, W.Q.; Hao, Y.H.; Zhao, J.P. The interaction of nuclear factor-kappa B and cytokines is associated with schizophrenia. *Biol. Psychiatry* **2009**, *65*, 481–488. [CrossRef]

45. Khandaker, G.M.; Zimbron, J.; Dalman, C.; Lewis, G.; Jones, P.B. Childhood infection and adult schizophrenia: A meta-analysis of population-based studies. *Schizophrenia Res.* **2012**, *139*, 161–168. [CrossRef]

46. Estes, M.L.; McAllister, A.K. Immune mediators in the brain and peripheral tissues in autism spectrum disorder. *Nat. Rev. Neurosci.* **2015**, *16*, 469–486. [CrossRef]

47. Li, X.; Chauhan, A.; Sheikh, A.M.; Patil, S.; Chauhan, V.; Li, X.M.; Ji, L.; Brown, T.; Malik, M. Elevated immune response in the brain of autistic patients. *J. Neuroimmunol.* **2009**, *207*, 111–116. [CrossRef]

48. Stuart, M.J.; Baune, B.T. Chemokines and chemokine receptors in mood disorders, schizophrenia, and cognitive impairment: A systematic review of biomarker studies. *Neurosci. Biobehav. Rev.* **2014**, *42*, 93–115. [CrossRef]

49. de Jong, E.K.; Vinet, J.; Stanulovic, V.S.; Meijer, M.; Wesseling, E.; Sjollema, K.; Boddeke, H.W.; Biber, K. Expression, transport, and axonal sorting of neuronal CCL21 in large dense-core vesicles. *FASEB J.* **2008**, *22*, 4136–4145. [CrossRef]

50. Heinisch, S.; Kirby, L.G. Fractalkine/CX3CL1 enhances GABA synaptic activity at serotonin neurons in the rat dorsal raphe nucleus. *Neuroscience* **2009**, *164*, 1210–1223. [CrossRef]

51. Pujol, F.; Kitabgi, P.; Boudin, H. The chemokine SDF-1 differentially regulates axonal elongation and branching in hippocampal neurons. *J. Cell Sci.* **2005**, *118*, 1071–1080. [CrossRef]

52. Zou, Y.R.; Kottmann, A.H.; Kuroda, M.; Taniuchi, I.; Littman, D.R. Function of the chemokine receptor CXCR4 in haematopoiesis and in cerebellar development. *Nature* **1998**, *393*, 595–599. [CrossRef]

53. Peng, H.; Wu, Y.; Duan, Z.; Ciborowski, P.; Zheng, J.C. Proteolytic processing of SDF-1alpha by matrix metalloproteinase-2 impairs CXCR4 signaling and reduces neural progenitor cell migration. *Protein Cell* **2012**, *3*, 875–882. [CrossRef]

54. Oh, S.B.; Cho, C.; Miller, R.J. Electrophysiological analysis of neuronal chemokine receptors. *Methods* **2003**, *29*, 335–344. [CrossRef]

55. Ragozzino, D. CXC chemokine receptors in the central nervous system: Role in cerebellar neuromodulation and development. *J. NeuroVirol.* **2002**, *8*, 559–572. [CrossRef]

56. Limatola, C.; Giovannelli, A.; Maggi, L.; Ragozzino, D.; Castellani, L.; Ciotti, M.T.; Vacca, F.; Mercanti, D.; Santoni, A.; Eusebi, F. SDF-1alpha-mediated modulation of synaptic transmission in rat cerebellum. *Eur. J. Neurosci.* **2000**, *12*, 2497–2504. [CrossRef]

57. Qin, X.; Wan, Y.; Wang, X. CCL2 and CXCL1 trigger calcitonin gene-related peptide release by exciting primary nociceptive neurons. *J. Neurosci. Res.* **2005**, *82*, 51–62. [CrossRef]

58. Jaerve, A.; Muller, H.W. Chemokines in CNS injury and repair. *Cell Tissue Res.* **2012**, *349*, 229–248. [CrossRef]

59. Rostene, W.; Dansereau, M.A.; Godefroy, D.; Van Steenwinckel, J.; Reaux-Le Goazigo, A.; Melik-Parsadaniantz, S.; Apartis, E.; Hunot, S.; Beaudet, N.; Sarret, P. Neurochemokines: A menage a trois providing new insights on the functions of chemokines in the central nervous system. *J. Neurochem.* **2011**, *118*, 680–694. [CrossRef]

60. Banisadr, G.; Fontanges, P.; Haour, F.; Kitabgi, P.; Rostene, W.; Melik Parsadaniantz, S. Neuroanatomical distribution of CXCR4 in adult rat brain and its localization in cholinergic and dopaminergic neurons. *Eur. J. Neurosci.* **2002**, *16*, 1661–1671. [CrossRef]

61. Schonemeier, B.; Kolodziej, A.; Schulz, S.; Jacobs, S.; Hoellt, V.; Stumm, R. Regional and cellular localization of the CXCl12/SDF-1 chemokine receptor CXCR7 in the developing and adult rat brain. *J. Comp. Neurol.* **2008**, *510*, 207–220. [CrossRef] [PubMed]

62. Gosselin, R.D.; Varela, C.; Banisadr, G.; Mechighel, P.; Rostene, W.; Kitabgi, P.; Melik-Parsadaniantz, S. Constitutive expression of CCR2 chemokine receptor and inhibition by MCP-1/CCL2 of GABA-induced currents in spinal cord neurones. *J. Neurochem.* **2005**, *95*, 1023–1034. [CrossRef] [PubMed]

63. Eyre, H.A.; Air, T.; Pradhan, A.; Johnston, J.; Lavretsky, H.; Stuart, M.J.; Baune, B.T. A meta-analysis of chemokines in major depression. *Prog. Neuro-Psychopharmacol. Biol. Psychiatry* **2016**, *68*, 1–8. [CrossRef] [PubMed]

64. Jaehne, E.J.; Baune, B.T. Effects of chemokine receptor signalling on cognition-like, emotion-like and sociability behaviours of CCR6 and CCR7 knockout mice. *Behav. Brain Res.* **2014**, *261*, 31–39. [CrossRef] [PubMed]

65. Zlotnik, A.; Yoshie, O. The chemokine superfamily revisited. *Immunity* **2012**, *36*, 705–716. [CrossRef] [PubMed]

66. Eugenin, E.A.; Dyer, G.; Calderon, T.M.; Berman, J.W. HIV-1 tat protein induces a migratory phenotype in human fetal microglia by a CCL2 (MCP-1)-dependent mechanism: Possible role in NeuroAIDS. *Glia* **2005**, *49*, 501–510. [CrossRef]

67. Biber, K.; Vinet, J.; Boddeke, H.W. Neuron-microglia signaling: Chemokines as versatile messengers. *J. Neuroimmunol.* **2008**, *198*, 69–74. [CrossRef]

68. Dimitrijevic, O.B.; Stamatovic, S.M.; Keep, R.F.; Andjelkovic, A.V. Effects of the chemokine CCL2 on blood-brain barrier permeability during ischemia-reperfusion injury. *J. Cereb. Blood Flow Metab.* **2006**, *26*, 797–810. [CrossRef]

69. Kettenmann, H.; Kirchhoff, F.; Verkhratsky, A. Microglia: New roles for the synaptic stripper. *Neuron* **2013**, *77*, 10–18. [CrossRef]

70. Le Thuc, O.; Blondeau, N.; Nahon, J.L.; Rovere, C. The complex contribution of chemokines to neuroinflammation: Switching from beneficial to detrimental effects. *Ann. N. Y. Acad. Sci.* **2015**, *1351*, 127–140. [CrossRef]

71. Wu, V.Y.; Walz, D.A.; McCoy, L.E. Purification and characterization of human and bovine platelet factor 4. *Prep. Biochem.* **1977**, *7*, 479–493. [CrossRef] [PubMed]

72. Zlotnik, A.; Yoshie, O. Chemokines: A new classification system and their role in immunity. *Immunity* **2000**, *12*, 121–127. [CrossRef]

73. Slusarczyk, J.; Trojan, E.; Chwastek, J.; Glombik, K.; Basta-Kaim, A. A Potential Contribution of Chemokine Network Dysfunction to the Depressive Disorders. *Curr. Neuropharmacol.* **2016**, *14*, 705–720. [CrossRef]

74. Nomiyama, H.; Osada, N.; Yoshie, O. The evolution of mammalian chemokine genes. *Cytokine Growth Factor Rev.* **2010**, *21*, 253–262. [CrossRef] [PubMed]

75. Allen, S.J.; Crown, S.E.; Handel, T.M. Chemokine: Receptor structure, interactions, and antagonism. *Annu. Rev. Immunol.* **2007**, *25*, 787–820. [CrossRef]

76. Blanpain, C.; Buser, R.; Power, C.A.; Edgerton, M.; Buchanan, C.; Mack, M.; Simmons, G.; Clapham, P.R.; Parmentier, M.; Proudfoot, A.E. A chimeric MIP-1alpha/RANTES protein demonstrates the use of different regions of the RANTES protein to bind and activate its receptors. *J. Leukoc. Biol.* **2001**, *69*, 977–985. [PubMed]

77. Deshmane, S.L.; Kremlev, S.; Amini, S.; Sawaya, B.E. Monocyte chemoattractant protein-1 (MCP-1): An overview. *J. Interferon Cytokine Res.* **2009**, *29*, 313–326. [CrossRef]

78. Stuart, M.J.; Singhal, G.; Baune, B.T. Systematic Review of the Neurobiological Relevance of Chemokines to Psychiatric Disorders. *Front. Cell. Neurosci.* **2015**, *9*, 357. [CrossRef]

79. Che, X.; Ye, W.; Panga, L.; Wu, D.C.; Yang, G.Y. Monocyte chemoattractant protein-1 expressed in neurons and astrocytes during focal ischemia in mice. *Brain Res.* **2001**, *902*, 171–177. [CrossRef]

80. Biber, K.; Zuurman, M.W.; Dijkstra, I.M.; Boddeke, H.W. Chemokines in the brain: Neuroimmunology and beyond. *Curr. Opin. Pharmacol.* **2002**, *2*, 63–68. [CrossRef]

81. Quandt, J.; Dorovini-Zis, K. The beta chemokines CCL4 and CCL5 enhance adhesion of specific CD4+ T cell subsets to human brain endothelial cells. *J. Neuropathol. Exp. Neurol.* **2004**, *63*, 350–362. [CrossRef]

82. Szczucinski, A.; Losy, J. Chemokines and chemokine receptors in multiple sclerosis. Potential targets for new therapies. *Acta Neurol. Scand.* **2007**, *115*, 137–146. [CrossRef]

83. Cardona, A.E.; Li, M.; Liu, L.; Savarin, C.; Ransohoff, R.M. Chemokines in and out of the central nervous system: Much more than chemotaxis and inflammation. *J. Leukocyte Biol.* **2008**, *84*, 587–594. [CrossRef]

84. Sanchez-Alcaniz, J.A.; Haege, S.; Mueller, W.; Pla, R.; Mackay, F.; Schulz, S.; Lopez-Bendito, G.; Stumm, R.; Marin, O. Cxcr7 controls neuronal migration by regulating chemokine responsiveness. *Neuron* **2011**, *69*, 77–90. [CrossRef]

85. Cardona, A.E.; Pioro, E.P.; Sasse, M.E.; Kostenko, V.; Cardona, S.M.; Dijkstra, I.M.; Huang, D.; Kidd, G.; Dombrowski, S.; Dutta, R.; et al. Control of microglial neurotoxicity by the fractalkine receptor. *Nat. Neurosci.* **2006**, *9*, 917–924. [CrossRef]

86. Mattison, H.A.; Nie, H.; Gao, H.; Zhou, H.; Hong, J.S.; Zhang, J. Suppressed pro-inflammatory response of microglia in CX3CR1 knockout mice. *J. Neuroimmunol.* **2013**, *257*, 110–115. [CrossRef]

87. Giovannelli, A.; Limatola, C.; Ragozzino, D.; Mileo, A.M.; Ruggieri, A.; Ciotti, M.T.; Mercanti, D.; Santoni, A.; Eusebi, F. CXC chemokines interleukin-8 (IL-8) and growth-related gene product alpha (GROalpha) modulate Purkinje neuron activity in mouse cerebellum. *J. Neuroimmunol.* **1998**, *92*, 122–132. [CrossRef]

88. Schoenfeld, T.J.; Cameron, H.A. Adult neurogenesis and mental illness. *Neuropsychopharmacology* **2015**, *40*, 113–128. [CrossRef]

89. Schmidt-Hieber, C.; Jonas, P.; Bischofberger, J. Enhanced synaptic plasticity in newly generated granule cells of the adult hippocampus. *Nature* **2004**, *429*, 184–187. [CrossRef]

90. Jacobs, B.L.; van Praag, H.; Gage, F.H. Adult brain neurogenesis and psychiatry: A novel theory of depression. *Mol. Psychiatry* **2000**, *5*, 262–269. [CrossRef]

91. Sheline, Y.I.; Wang, P.W.; Gado, M.H.; Csernansky, J.G.; Vannier, M.W. Hippocampal atrophy in recurrent major depression. *Proc. Natl. Acad. Sci. USA* **1996**, *93*, 3908–3913. [CrossRef]

92. Bremner, J.D.; Narayan, M.; Anderson, E.R.; Staib, L.H.; Miller, H.L.; Charney, D.S. Hippocampal volume reduction in major depression. *Am. J. Psychiatry* **2000**, *157*, 115–118. [CrossRef]

93. Sheline, Y.I.; Sanghavi, M.; Mintun, M.A.; Gado, M.H. Depression duration but not age predicts hippocampal volume loss in medically healthy women with recurrent major depression. *J. Neurosci.* **1999**, *19*, 5034–5043. [CrossRef]

94. Eyre, H.; Baune, B.T. Neuroplastic changes in depression: A role for the immune system. *Psychoneuroendocrinology* **2012**, *37*, 1397–1416. [CrossRef]

95. Eisch, A.J.; Petrik, D. Depression and hippocampal neurogenesis: A road to remission? *Science* **2012**, *338*, 72–75. [CrossRef]

96. Smitha, J.S.; Roopa, R.; Sagar, B.K.; Kutty, B.M.; Andrade, C. Images in electroconvulsive therapy: ECS dose-dependently increases cell proliferation in the subgranular region of the rat hippocampus. *J. ECT* **2014**, *30*, 193–194. [CrossRef]

97. Moylan, S.; Maes, M.; Wray, N.R.; Berk, M. The neuroprogressive nature of major depressive disorder: Pathways to disease evolution and resistance, and therapeutic implications. *Mol. Psychiatry* **2013**, *18*, 595–606. [CrossRef]

98. Tran, P.B.; Banisadr, G.; Ren, D.; Chenn, A.; Miller, R.J. Chemokine receptor expression by neural progenitor cells in neurogenic regions of mouse brain. *J. Comp. Neurol.* **2007**, *500*, 1007–1033. [CrossRef]

99. Miller, R.J.; Rostene, W.; Apartis, E.; Banisadr, G.; Biber, K.; Milligan, E.D.; White, F.A.; Zhang, J. Chemokine action in the nervous system. *J. Neurosci.* **2008**, *28*, 11792–11795. [CrossRef]

100. Bajetto, A.; Bonavia, R.; Barbero, S.; Florio, T.; Schettini, G. Chemokines and their receptors in the central nervous system. *Front. Neuroendocrinol.* **2001**, *22*, 147–184. [CrossRef]

101. Li, M.; Ransohoff, R.M. Multiple roles of chemokine CXCL12 in the central nervous system: A migration from immunology to neurobiology. *Prog. Neurobiol.* **2008**, *84*, 116–131. [CrossRef]

102. Flynn, G.; Maru, S.; Loughlin, J.; Romero, I.A.; Male, D. Regulation of chemokine receptor expression in human microglia and astrocytes. *J. Neuroimmunol.* **2003**, *136*, 84–93. [CrossRef]

103. Reaux-Le Goazigo, A.; Van Steenwinckel, J.; Rostene, W.; Melik Parsadaniantz, S. Current status of chemokines in the adult CNS. *Prog. Neurobiol.* **2013**, *104*, 67–92. [CrossRef]

104. Paolicelli, R.C.; Bolasco, G.; Pagani, F.; Maggi, L.; Scianni, M.; Panzanelli, P.; Giustetto, M.; Ferreira, T.A.; Guiducci, E.; Dumas, L.; et al. Synaptic pruning by microglia is necessary for normal brain development. *Science* **2011**, *333*, 1456–1458. [CrossRef]

105. Piccinin, S.; Di Angelantonio, S.; Piccioni, A.; Volpini, R.; Cristalli, G.; Fredholm, B.B.; Limatola, C.; Eusebi, F.; Ragozzino, D. CX3CL1-induced modulation at CA1 synapses reveals multiple mechanisms of EPSC modulation involving adenosine receptor subtypes. *J. Neuroimmunol.* **2010**, *224*, 85–92. [CrossRef]

106. Bachstetter, A.D.; Morganti, J.M.; Jernberg, J.; Schlunk, A.; Mitchell, S.H.; Brewster, K.W.; Hudson, C.E.; Cole, M.J.; Harrison, J.K.; Bickford, P.C.; et al. Fractalkine and CX 3 CR1 regulate hippocampal neurogenesis in adult and aged rats. *Neurobiol. Aging* **2011**, *32*, 2030–2044. [CrossRef]

107. Turbic, A.; Leong, S.Y.; Turnley, A.M. Chemokines and inflammatory mediators interact to regulate adult murine neural precursor cell proliferation, survival and differentiation. *PLoS ONE* **2011**, *6*, e25406. [CrossRef]

108. Santarelli, L.; Saxe, M.; Gross, C.; Surget, A.; Battaglia, F.; Dulawa, S.; Weisstaub, N.; Lee, J.; Duman, R.; Arancio, O.; et al. Requirement of hippocampal neurogenesis for the behavioral effects of antidepressants. *Science* **2003**, *301*, 805–809. [CrossRef]

109. Hanson, N.D.; Owens, M.J.; Nemeroff, C.B. Depression, antidepressants, and neurogenesis: A critical reappraisal. *Neuropsychopharmacology* **2011**, *36*, 2589–2602. [CrossRef]

110. Banisadr, G.; Dicou, E.; Berbar, T.; Rostene, W.; Lombet, A.; Haour, F. Characterization and visualization of [125I] stromal cell-derived factor-1alpha binding to CXCR4 receptors in rat brain and human neuroblastoma cells. *J. Neuroimmunol.* **2000**, *110*, 151–160. [CrossRef]

111. Coughlan, C.M.; McManus, C.M.; Sharron, M.; Gao, Z.; Murphy, D.; Jaffer, S.; Choe, W.; Chen, W.; Hesselgesser, J.; Gaylord, H.; et al. Expression of multiple functional chemokine receptors and monocyte chemoattractant protein-1 in human neurons. *Neuroscience* **2000**, *97*, 591–600. [CrossRef]

112. Meucci, O.; Fatatis, A.; Simen, A.A.; Bushell, T.J.; Gray, P.W.; Miller, R.J. Chemokines regulate hippocampal neuronal signaling and gp120 neurotoxicity. *Proc. Natl. Acad. Sci. USA* **1998**, *95*, 14500–14505. [CrossRef]

113. Riek-Burchardt, M.; Kolodziej, A.; Henrich-Noack, P.; Reymann, K.G.; Hollt, V.; Stumm, R. Differential regulation of CXCL12 and PACAP mRNA expression after focal and global ischemia. *Neuropharmacology* **2010**, *58*, 199–207. [CrossRef]

114. Guyon, A.; Banisadr, G.; Rovere, C.; Cervantes, A.; Kitabgi, P.; Melik-Parsadaniantz, S.; Nahon, J.L. Complex effects of stromal cell-derived factor-1 alpha on melanin-concentrating hormone neuron excitability. *Eur. J. Neurosci.* **2005**, *21*, 701–710. [CrossRef]

115. Van Steenwinckel, J.; Reaux-Le Goazigo, A.; Pommier, B.; Mauborgne, A.; Dansereau, M.A.; Kitabgi, P.; Sarret, P.; Pohl, M.; Melik Parsadaniantz, S. CCL2 released from neuronal synaptic vesicles in the spinal cord is a major mediator of local inflammation and pain after peripheral nerve injury. *J. Neurosci.* **2011**, *31*, 5865–5875. [CrossRef]

116. Dansereau, M.A.; Gosselin, R.D.; Pohl, M.; Pommier, B.; Mechighel, P.; Mauborgne, A.; Rostene, W.; Kitabgi, P.; Beaudet, N.; Sarret, P.; et al. Spinal CCL2 pronociceptive action is no longer effective in CCR2 receptor antagonist-treated rats. *J. Neurochem.* **2008**, *106*, 757–769. [CrossRef]

117. Rostene, W.; Kitabgi, P.; Parsadaniantz, S.M. Chemokines: A new class of neuromodulator? *Nat. Rev. Neurosci.* **2007**, *8*, 895–903. [CrossRef]

118. Guyon, A.; Nahon, J.L. Multiple actions of the chemokine stromal cell-derived factor-1alpha on neuronal activity. *J. Mol. Endocrinol.* **2007**, *38*, 365–376. [CrossRef]

119. Lax, P.; Limatola, C.; Fucile, S.; Trettel, F.; Di Bartolomeo, S.; Renzi, M.; Ragozzino, D.; Eusebi, F. Chemokine receptor CXCR2 regulates the functional properties of AMPA-type glutamate receptor GluR1 in HEK cells. *J. Neuroimmunol.* **2002**, *129*, 66–73. [CrossRef]

120. Ragozzino, D.; Renzi, M.; Giovannelli, A.; Eusebi, F. Stimulation of chemokine CXC receptor 4 induces synaptic depression of evoked parallel fibers inputs onto Purkinje neurons in mouse cerebellum. *J. Neuroimmunol.* **2002**, *127*, 30–36. [CrossRef]

121. Sciaccaluga, M.; Fioretti, B.; Catacuzzeno, L.; Pagani, F.; Bertollini, C.; Rosito, M.; Catalano, M.; D'Alessandro, G.; Santoro, A.; Cantore, G.; et al. CXCL12-induced glioblastoma cell migration requires intermediate conductance Ca2+-activated K+ channel activity. *Am. J. Physiol.-Cell Physiol.* **2010**, *299*, C175–C184. [CrossRef]

122. Nestler, E.J.; Hyman, S.E. Animal models of neuropsychiatric disorders. *Nat. Neurosci.* **2010**, *13*, 1161–1169. [CrossRef]

123. Hollis, F.; Kabbaj, M. Social defeat as an animal model for depression. *ILAR J.* **2014**, *55*, 221–232. [CrossRef]

124. Baune, B. Conceptual challenges of a tentative model of stress-induced depression. *PLoS ONE* **2009**, *4*, e4266. [CrossRef]

125. Sakamoto, Y.; Koike, K.; Kiyama, H.; Konishi, K.; Watanabe, K.; Tsurufuji, S.; Bicknell, R.J.; Hirota, K.; Miyake, A. A stress-sensitive chemokinergic neuronal pathway in the hypothalamo-pituitary system. *Neuroscience* **1996**, *75*, 133–142. [CrossRef]

126. Matsumoto, K.; Koike, K.; Miyake, A.; Watanabe, K.; Konishi, K.; Kiyama, H. Noxious stimulation enhances release of cytokine-induced neutrophil chemoattractant from hypothalamic neurosecretory cells. *Neurosci. Res.* **1997**, *27*, 181–184. [CrossRef]

127. Tagliari, B.; Tagliari, A.P.; Schmitz, F.; da Cunha, A.A.; Dalmaz, C.; Wyse, A.T. Chronic variable stress alters inflammatory and cholinergic parameters in hippocampus of rats. *Neurochem. Res.* **2011**, *36*, 487–493. [CrossRef]

128. Slusarczyk, J.; Trojan, E.; Glombik, K.; Budziszewska, B.; Kubera, M.; Lason, W.; Popiolek-Barczyk, K.; Mika, J.; Wedzony, K.; Basta-Kaim, A. Prenatal stress is a vulnerability factor for altered morphology and biological activity of microglia cells. *Front. Cell. Neurosci.* **2015**, *9*, 82. [CrossRef]

129. Trojan, E.; Slusarczyk, J.; Chamera, K.; Kotarska, K.; Glombik, K.; Kubera, M.; Basta-Kaim, A. The Modulatory Properties of Chronic Antidepressant Drugs Treatment on the Brain Chemokine - Chemokine Receptor Network: A Molecular Study in an Animal Model of Depression. *Front. Pharmacol.* **2017**, *8*, 779. [CrossRef]

130. Dantzer, R.; O'Connor, J.C.; Lawson, M.A.; Kelley, K.W. Inflammation-associated depression: From serotonin to kynurenine. *Psychoneuroendocrinology* **2011**, *36*, 426–436. [CrossRef]

131. Hoyo-Becerra, C.; Liu, Z.; Yao, J.; Kaltwasser, B.; Gerken, G.; Hermann, D.M.; Schlaak, J.F. Rapid Regulation of Depression-Associated Genes in a New Mouse Model Mimicking Interferon-alpha-Related Depression in Hepatitis C Virus Infection. *Mol. Neurobiol.* **2015**, *52*, 318–329. [CrossRef]

132. Campbell, S.J.; Meier, U.; Mardiguian, S.; Jiang, Y.; Littleton, E.T.; Bristow, A.; Relton, J.; Connor, T.J.; Anthony, D.C. Sickness behaviour is induced by a peripheral CXC-chemokine also expressed in multiple sclerosis and EAE. *Brain Behav. Immun.* **2010**, *24*, 738–746. [CrossRef]

133. Girotti, M.; Donegan, J.J.; Morilak, D.A. Chronic intermittent cold stress sensitizes neuro-immune reactivity in the rat brain. *Psychoneuroendocrinology* **2011**, *36*, 1164–1174. [CrossRef]

134. Corona, A.W.; Huang, Y.; O'Connor, J.C.; Dantzer, R.; Kelley, K.W.; Popovich, P.G.; Godbout, J.P. Fractalkine receptor (CX3CR1) deficiency sensitizes mice to the behavioral changes induced by lipopolysaccharide. *J. Neuroinflamm.* **2010**, *7*, 93. [CrossRef]

135. Garre, J.M.; Silva, H.M.; Lafaille, J.J.; Yang, G. CX3CR1(+) monocytes modulate learning and learning-dependent dendritic spine remodeling via TNF-alpha. *Nat. Med.* **2017**, *23*, 714–722. [CrossRef]

136. Blank, T.; Detje, C.N.; Spiess, A.; Hagemeyer, N.; Brendecke, S.M.; Wolfart, J.; Staszewski, O.; Zoller, T.; Papageorgiou, I.; Schneider, J.; et al. Brain Endothelial- and Epithelial-Specific Interferon Receptor Chain 1 Drives Virus-Induced Sickness Behavior and Cognitive Impairment. *Immunity* **2016**, *44*, 901–912. [CrossRef]

137. Priller, J.; Bottcher, C. Patrolling monocytes sense peripheral infection and induce cytokine-mediated neuronal dysfunction. *Nat. Med.* **2017**, *23*, 659–661. [CrossRef]

138. Barden, N. Implication of the hypothalamic-pituitary-adrenal axis in the physiopathology of depression. *J. Psychiatry Neurosci.* **2004**, *29*, 185–193.

139. Lupien, S.J. Brains under stress. *Can. J. Psychiat.-Rev. Can. Psychiat.* **2009**, *54*, 4–5. [CrossRef]

140. Almeida, D.M.; Wethington, E.; Kessler, R.C. The daily inventory of stressful events: An interview-based approach for measuring daily stressors. *Assessment* **2002**, *9*, 41–55. [CrossRef]

141. Stein, D.J.; Vasconcelos, M.F.; Albrechet-Souza, L.; Cereser, K.M.M.; de Almeida, R.M.M. Microglial Over-Activation by Social Defeat Stress Contributes to Anxiety- and Depressive-Like Behaviors. *Front. Behav. Neurosci.* **2017**, *11*, 207. [CrossRef]

142. Sawicki, C.M.; McKim, D.B.; Wohleb, E.S.; Jarrett, B.L.; Reader, B.F.; Norden, D.M.; Godbout, J.P.; Sheridan, J.F. Social defeat promotes a reactive endothelium in a brain region-dependent manner with increased expression of key adhesion molecules, selectins and chemokines associated with the recruitment of myeloid cells to the brain. *Neuroscience* **2015**, *302*, 151–164. [CrossRef]

143. Ge, S.; Song, L.; Serwanski, D.R.; Kuziel, W.A.; Pachter, J.S. Transcellular transport of CCL2 across brain microvascular endothelial cells. *J. Neurochem.* **2008**, *104*, 1219–1232. [CrossRef]

144. Sutcigil, L.; Oktenli, C.; Musabak, U.; Bozkurt, A.; Cansever, A.; Uzun, O.; Sanisoglu, S.Y.; Yesilova, Z.; Ozmenler, N.; Ozsahin, A.; et al. Pro- and anti-inflammatory cytokine balance in major depression: Effect of sertraline therapy. *Clin. Dev. Immunol.* **2007**, *2007*, 76396. [CrossRef]

145. Piletz, J.E.; Halaris, A.; Iqbal, O.; Hoppensteadt, D.; Fareed, J.; Zhu, H.; Sinacore, J.; Devane, C.L. Pro-inflammatory biomakers in depression: Treatment with venlafaxine. *World J. Biol. Psychiatry* **2009**, *10*, 313–323. [CrossRef]

146. Kohler, C.A.; Freitas, T.H.; Stubbs, B.; Maes, M.; Solmi, M.; Veronese, N.; de Andrade, N.Q.; Morris, G.; Fernandes, B.S.; Brunoni, A.R.; et al. Peripheral Alterations in Cytokine and Chemokine Levels After Antidepressant Drug Treatment for Major Depressive Disorder: Systematic Review and Meta-Analysis. *Mol. Neurobiol.* **2017**. [CrossRef]

147. Kohler, C.A.; Freitas, T.H.; Maes, M.; de Andrade, N.Q.; Liu, C.S.; Fernandes, B.S.; Stubbs, B.; Solmi, M.; Veronese, N.; Herrmann, N.; et al. Peripheral cytokine and chemokine alterations in depression: A meta-analysis of 82 studies. *Acta Psychiatr. Scand.* **2017**, *135*, 373–387. [CrossRef]

148. Lehto, S.M.; Niskanen, L.; Herzig, K.H.; Tolmunen, T.; Huotari, A.; Viinamaki, H.; Koivumaa-Honkanen, H.; Honkalampi, K.; Ruotsalainen, H.; Hintikka, J. Serum chemokine levels in major depressive disorder. *Psychoneuroendocrinology* **2010**, *35*, 226–232. [CrossRef]

149. Black, C.; Miller, B.J. Meta-Analysis of Cytokines and Chemokines in Suicidality: Distinguishing Suicidal Versus Nonsuicidal Patients. *Biol. Psychiatry* **2015**, *78*, 28–37. [CrossRef]

150. Drexhage, R.C.; Hoogenboezem, T.H.; Versnel, M.A.; Berghout, A.; Nolen, W.A.; Drexhage, H.A. The activation of monocyte and T cell networks in patients with bipolar disorder. *Brain Behav. Immun.* **2011**, *25*, 1206–1213. [CrossRef]

151. Dahl, J.; Ormstad, H.; Aass, H.C.; Malt, U.F.; Bendz, L.T.; Sandvik, L.; Brundin, L.; Andreassen, O.A. The plasma levels of various cytokines are increased during ongoing depression and are reduced to normal levels after recovery. *Psychoneuroendocrinology* **2014**, *45*, 77–86. [CrossRef]

152. Fontenelle, L.F.; Barbosa, I.G.; Luna, J.V.; de Sousa, L.P.; Abreu, M.N.; Teixeira, A.L. A cytokine study of adult patients with obsessive-compulsive disorder. *Compr. Psychiatry* **2012**, *53*, 797–804. [CrossRef]

153. Reus, G.Z.; Fries, G.R.; Stertz, L.; Badawy, M.; Passos, I.C.; Barichello, T.; Kapczinski, F.; Quevedo, J. The role of inflammation and microglial activation in the pathophysiology of psychiatric disorders. *Neuroscience* **2015**, *300*, 141–154. [CrossRef]

154. Barbosa, I.G.; Nogueira, C.R.; Rocha, N.P.; Queiroz, A.L.; Vago, J.P.; Tavares, L.P.; Assis, F.; Fagundes, C.T.; Huguet, R.B.; Bauer, M.E.; et al. Altered intracellular signaling cascades in peripheral blood mononuclear cells from BD patients. *J. Psychiatr. Res.* **2013**, *47*, 1949–1954. [CrossRef]

155. Barbosa, I.G.; Rocha, N.P.; Bauer, M.E.; de Miranda, A.S.; Huguet, R.B.; Reis, H.J.; Zunszain, P.A.; Horowitz, M.A.; Pariante, C.M.; Teixeira, A.L. Chemokines in bipolar disorder: Trait or state? *Eur. Arch. Psych. Clin. Neurosci.* **2013**, *263*, 159–165. [CrossRef]

156. Einvik, G.; Vistnes, M.; Hrubos-Strom, H.; Randby, A.; Namtvedt, S.K.; Nordhus, I.H.; Somers, V.K.; Dammen, T.; Omland, T. Circulating cytokine concentrations are not associated with major depressive disorder in a community-based cohort. *Gen. Hosp. Psychiatry* **2012**, *34*, 262–267. [CrossRef]

157. Setiawan, E.; Wilson, A.A.; Mizrahi, R.; Rusjan, P.M.; Miler, L.; Rajkowska, G.; Suridjan, I.; Kennedy, J.L.; Rekkas, P.V.; Houle, S.; et al. Role of translocator protein density, a marker of neuroinflammation, in the brain during major depressive episodes. *JAMA psychiatry* **2015**, *72*, 268–275. [CrossRef]

158. Milenkovic, V.M.; Sarubin, N.; Hilbert, S.; Baghai, T.C.; Stoffler, F.; Lima-Ojeda, J.M.; Manook, A.; Almeqbaali, K.; Wetzel, C.H.; Rupprecht, R.; et al. Macrophage-Derived Chemokine: A Putative Marker of Pharmacological Therapy Response in Major Depression? *Neuroimmunomodulation* **2017**, *24*, 106–112. [CrossRef]

159. Smith, R.S. The macrophage theory of depression. *Med. Hypotheses* **1991**, *35*, 298–306. [CrossRef]

160. Wattananit, S.; Tornero, D.; Graubardt, N.; Memanishvili, T.; Monni, E.; Tatarishvili, J.; Miskinyte, G.; Ge, R.; Ahlenius, H.; Lindvall, O.; et al. Monocyte-Derived Macrophages Contribute to Spontaneous Long-Term Functional Recovery after Stroke in Mice. *J. Neurosci.* **2016**, *36*, 4182–4195. [CrossRef]

161. Wohleb, E.S.; McKim, D.B.; Sheridan, J.F.; Godbout, J.P. Monocyte trafficking to the brain with stress and inflammation: A novel axis of immune-to-brain communication that influences mood and behavior. *Front. Neurosci.* **2014**, *8*, 447. [CrossRef] [PubMed]

162. Kappelmann, N.; Lewis, G.; Dantzer, R.; Jones, P.B.; Khandaker, G.M. Antidepressant activity of anti-cytokine treatment: A systematic review and meta-analysis of clinical trials of chronic inflammatory conditions. *Mol. Psychiatry* **2018**, *23*, 335–343. [CrossRef]

163. Teixeira, A.L.; Gama, C.S.; Rocha, N.P.; Teixeira, M.M. Revisiting the Role of Eotaxin-1/CCL11 in Psychiatric Disorders. *Front. Psychiatry* **2018**, *9*, 241. [CrossRef] [PubMed]

164. Na, K.S.; Lee, K.J.; Lee, J.S.; Cho, Y.S.; Jung, H.Y. Efficacy of adjunctive celecoxib treatment for patients with major depressive disorder: A meta-analysis. *Prog. Neuro-Psychopharmacol. Biol. Psychiatry* **2014**, *48*, 79–85. [CrossRef]

165. Husain, M.I.; Strawbridge, R.; Stokes, P.R.; Young, A.H. Anti-inflammatory treatments for mood disorders: Systematic review and meta-analysis. *J. Psychopharmacol.* **2017**, *31*, 1137–1148. [CrossRef]

166. Guloksuz, S.; Rutten, B.P.; Arts, B.; van Os, J.; Kenis, G. The immune system and electroconvulsive therapy for depression. *J. ECT* **2014**, *30*, 132–137. [CrossRef]

167. Teixeira, A.L.; Colpo, G.D.; Fries, G.R.; Bauer, I.E.; Selvaraj, S. Biomarkers for bipolar disorder: Current status and challenges ahead. *Expert Rev. Neurother.* **2019**, *19*, 67–81. [CrossRef] [PubMed]

168. Frye, M.A.; McElroy, S.L.; Fuentes, M.; Sutor, B.; Schak, K.M.; Galardy, C.W.; Palmer, B.A.; Prieto, M.L.; Kung, S.; Sola, C.L.; et al. Development of a bipolar disorder biobank: Differential phenotyping for subsequent biomarker analyses. *Int. J. Bipolar Disord.* **2015**, *3*, 30. [CrossRef]

169. Howard, D.M.; Adams, M.J.; Shirali, M.; Clarke, T.K.; Marioni, R.E.; Davies, G.; Coleman, J.R.I.; Alloza, C.; Shen, X.; Barbu, M.C.; et al. Genome-wide association study of depression phenotypes in UK Biobank identifies variants in excitatory synaptic pathways. *Nat. Commun.* **2018**, *9*, 1470. [CrossRef]

170. Trivedi, M.H.; Fava, M.; Wisniewski, S.R.; Thase, M.E.; Quitkin, F.; Warden, D.; Ritz, L.; Nierenberg, A.A.; Lebowitz, B.D.; Biggs, M.M.; et al. Medication augmentation after the failure of SSRIs for depression. *N. Engl. J. Med.* **2006**, *354*, 1243–1252. [CrossRef]

171. Ball, S.; Classi, P.; Dennehy, E.B. What happens next?: A claims database study of second-line pharmacotherapy in patients with major depressive disorder (MDD) who initiate selective serotonin reuptake inhibitor (SSRI) treatment. *Ann. Gen. Psychiatry* **2014**, *13*, 8. [CrossRef]

International Journal of
Molecular Sciences

Article

Social Defeat Modulates T Helper Cell Percentages in Stress Susceptible and Resilient Mice

Oliver Ambrée [1,2,*], Christina Ruland [1], Peter Zwanzger [3,4], Luisa Klotz [5], Bernhard T Baune [1,6,7], Volker Arolt [1], Stefanie Scheu [8,†] and Judith Alferink [1,9,*,†]

[1]   Department of Psychiatry, University of Münster, Münster 48149, Germany
[2]   Department of Behavioural Biology, University of Osnabrück, 49076 Osnabrück, Germany
[3]   Kbo-Inn-Salzach-Klinikum, 83512 Wasserburg am Inn, Germany
[4]   Department of Psychiatry and Psychotherapy, Ludwig-Maximilians-Universität München, 80336 Munich, Germany
[5]   Department of Neurology, University of Münster, Münster 49149, Germany
[6]   Department of Psychiatry, Melbourne Medical School, The University of Melbourne, Parkville, VIC 3010, Australia
[7]   The Florey Institute of Neuroscience and Mental Health, The University of Melbourne, Parkville, VIC 3010, Australia
[8]   Institute of Medical Microbiology and Hospital Hygiene, University of Düsseldorf, 40225 Düsseldorf, Germany
[9]   Cluster of Excellence EXC 1003, Cells in Motion, University of Münster, 48149 Münster, Germany
*   Correspondence: oliver.ambree@uos.de (O.A.); judith.alferink@ukmuenster.de (J.A.); Tel.: +49-541-969-2830 (O.A.); +49-251-835-7209 (J.A.)
†   These authors contributed equally to this work.

Received: 14 June 2019; Accepted: 16 July 2019; Published: 17 July 2019

Abstract: Altered adaptive immunity involving T lymphocytes has been found in depressed patients and in stress-induced depression-like behavior in animal models. Peripheral T cells play important roles in homeostasis and function of the central nervous system and thus modulate behavior. However, the T cell phenotype and function associated with susceptibility and resilience to depression remain largely unknown. Here, we characterized splenic T cells in susceptible and resilient mice after 10 days of social defeat stress (SDS). We found equally decreased T cell frequencies and comparably altered expression levels of genes associated with T helper (Th) cell function in resilient and susceptible mice. Interleukin (IL)-17 producing $CD4^+$ and $CD8^+$ T cell numbers in the spleen were significantly increased in susceptible mice. These animals further exhibited significantly reduced numbers of regulatory T cells (T_{reg}) and decreased gene expression levels of TGF-β. Mice with enhanced Th17 differentiation induced by conditional deletion of PPARγ in $CD4^+$ cells (CD4-PPARγ^{KO}), an inhibitor of Th17 development, were equally susceptible to SDS when compared to CD4-PPARγ^{WT} controls. These data indicate that enhanced Th17 differentiation alone does not alter stress vulnerability. Thus, SDS promotes Th17 cell and suppresses T_{reg} cell differentiation predominantly in susceptible mice with yet unknown effects in immune responses after stress exposure.

Keywords: social defeat; Immune response; T cells; susceptibility; resilience; major depression; T_{reg} cells; Th17 cells; behavior; PPARγ

1. Introduction

Stressful life events have been shown to result in long-term alterations of the immune system [1–3] and to increase the risk for major depressive disorder (MDD) [4,5]. Multiple studies have demonstrated a chronic mild inflammation characterized by increased levels of acute phase proteins, pro-inflammatory cytokines, and chemokines in depressed patients and stress-exposed individuals [6–10]. However, growing evidence also supports a role for the adaptive immune response and its cellular components, in particular T cells, in the pathophysiology of MDD [11,12] and in depression-like behaviors in rodents [13]. T cells have been shown to play an important role in neural plasticity and maintenance of CNS function [14–16]. Thus, alterations in the T cell compartment affect microglia function and adult neurogenesis that are involved in stress responses and MDD [17–19].

In patients with MDD and individuals exposed to stress, lower numbers of circulating T cells, as well as altered T cell responses, have been found by meta-analytic approaches [20,21]. Recent studies suggested that CD4$^+$ T helper (Th)1, Th17 and T regulatory (T$_{reg}$) cells are involved in the pathophysiology of MDD [13,22]. Interleukin-17 (IL-17)-producing Th17 cells exhibit potent inflammatory activity and have been functionally implicated in neuroinflammation and CNS autoimmunity [23]. On the other hand, T$_{reg}$ cells play a key role in immune tolerance and downregulation of Th17 responses and exert inhibitory functions on immune effector cells and pro-inflammatory responses [24]. Individuals with MDD have been shown to exhibit altered percentages in circulating Th17 and T$_{reg}$ cells. With regard to Th17 cells, different studies reported increased as well as decreased percentages of circulating Th17 cells in patients with MDD [25,26], while T$_{reg}$ cells were mainly found to be decreased in the peripheral blood [26–28]. In summary, these findings point towards an imbalance of Th17 and T$_{reg}$ cell populations in MDD.

Also in rodent models, stress-induced depression-like behavior has been shown to be associated with alterations of adaptive immune responses [29]. For example, various CD4$^+$ Th cell subsets have been implicated in stress induced depression-like behavior: Percentages of Th17 cells were found to be elevated in brains of mice exhibiting learned helplessness and after chronic restraint stress [30]. Furthermore, adoptive transfer of Th17 cells increased depression-like behavior after foot shock stress while depletion of Th17 cells reduced the acquisition of learned helplessness [30]. However, the view that Th17 cells exclusively exhibit pathogenic actions has been challenged by studies pointing toward a potential beneficial role of Th17 cells in depression-like behavior and MDD. In a rodent model of depression-like behavior following chronic unpredictable mild stress, a decrease of Th17 cell percentages and an increase in percentages of T$_{reg}$ cells was demonstrated [31]. In addition, Th17 cells have also been reported to promote adult hippocampal neurogenesis [16] which is usually associated with antidepressive effects [32,33]. A study in humans also pointed toward a potential beneficial role of Th17 cells in MDD by maintaining the functional and structural integrity of the brain [34]. Taken together, these findings in animals and humans suggest that circulating Th cells may contribute to stress responses and the development of MDD.

Recently, the nuclear receptor peroxisome proliferator-activated receptor gamma (PPARγ) has been identified as a key negative regulator of human and mouse Th17 differentiation and has been shown to suppress CNS autoimmunity [35,36]. In rodents, it has been demonstrated that PPARγ-agonists reduce depression-like behavior [37–39]. Furthermore, in MDD, PPARγ-agonists promoted enhanced remission [40,41]. However, whether PPARγ-mediated antidepressant effects are due to altered Th17 differentiation has not been investigated.

An important factor in understanding the consequences of stress on the organism, is a sound knowledge of the individual immune variations associated with stress vulnerability. It has been shown before that an early increase in plasma IL-6 levels predicts susceptibility to social defeat [42]. In addition, our earlier findings demonstrate that specific alterations in innate immune cells occur in monocytes and dendritic cells in susceptible mice that develop depression-like behavior after exposure to chronic social defeat [43]. However, the implication of T cells in stress susceptibility and resilience in this model remains undefined. In this study, we characterized T cell responses associated with stress vulnerability to social defeat by assessing expression levels of T cell differentiation and effector genes and numbers of cytokine-producing T cells in socially defeated animals. In addition, we examined the effect of increased IL-17 producing CD4$^+$ T cells on stress vulnerability in socially defeated CD4$^+$ T cell-specific PPARγ knockout mice. Our data identified a specific pattern of T cell responses associated with social defeat stress and point toward an involvement of the adaptive immune system as cellular contributor to brain homeostasis relevant for MDD and the physiological stress response.

2. Results

2.1. Susceptible Mice Show Social Avoidance after Social Defeat Stress

To study T cell responses associated with susceptibility and resilience to prolonged stress, we utilized repeated social defeat in mice as a paradigm for social stress [43,44]. For this, C57BL/6J mice were subjected to repeated social defeats over 10 days. After ten days of exposure to dominant conspecifics and repeated social defeat, we assessed social interaction behavior to determine susceptible and resilient individuals. In analogy to our previous study [43], susceptible mice had a significantly lower interaction ratio than control and resilient animals, whose interaction ratio was comparable to that of control animals (Figure 1A, C vs S: $p < 0.001$; S vs. R: $p < 0.001$, see Table S1 for details of statistics). In addition, the behavior of control and resilient mice clearly differed with regard to the time spent in the interaction zone that was reduced in susceptible mice (Figure 1B, C vs S: $p < 0.001$; S vs. R: $p < 0.001$). In line with that, the time spent in the corner zone was increased in susceptible animals when compared to resilient animals and undefeated controls (Figure 1C, C vs S: $p < 0.001$; S vs. R: $p < 0.001$).

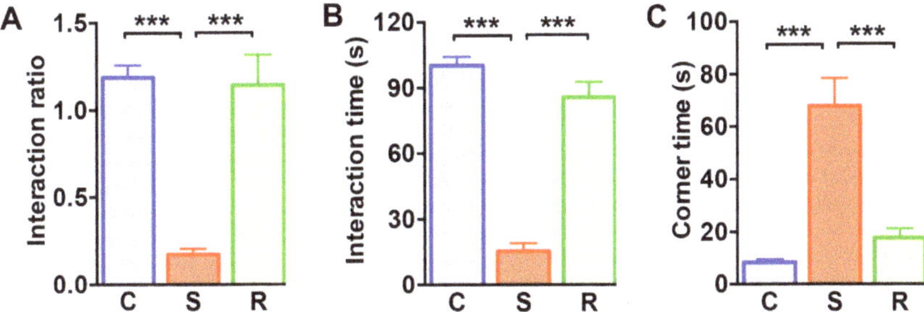

Figure 1. Social interaction test. (**A**) Interaction ratio, (**B**) the time spent in the interaction zone during the social interaction trial, and (**C**) the time spent in the corners on the opposite site of the interaction enclosure. Bar graphs represent mean + SEM. C: control, S: susceptible, R: resilient. $n_C = 15$, $n_S = 16$, $n_R = 11$. ***: $p < 0.001$ (Bonferroni post hoc).

2.2. Expression Levels of Molecules Associated with T Cell Differentiation and Function Were Reduced after Social Defeat

We next determined expression of genes associated with T cell differentiation and function in the spleen of resilient and susceptible mice after social defeat and controls. A panel of genes that were differentially expressed after social defeat, was selected based on the results of a qPCR-based gene array on 84 genes encoding pro- and anti-inflammatory cytokines and chemokines (Table S2). Granulocyte-macrophage colony-stimulating factor (GM-CSF; also designated as colony stimulating factor 2, CSF2) represents a pro-inflammatory mediator for T cell function and myeloid cell responses during tissue inflammation [45,46]. We found lower levels of *Csf2* mRNA in socially defeated mice when compared to non-defeated controls, independent of the susceptible or resilient phenotype of defeated animals (Figure 2A, C vs. S: $p = 0.003$, C vs. R: $p = 0.026$). The expression levels of genes encoding the Th1 differentiation cytokine interleukin (IL)-12 and interferon (IFN)-γ were reduced in the spleen of susceptible and resilient mice after social defeat when compared to non-defeated controls (Figure 2B,C, *Il12a*: C vs. S: $p < 0.001$, C vs. R: $p = 0.006$; *Ifng*: C vs. S: $p < 0.001$, C vs. R: $p = 0.001$). In addition, mRNA levels of *Il27*, the gene encoding the pleiotropic cytokine IL-27, an inhibitor of Th17 development [47], were reduced in susceptible and resilient mice compared to controls (Figure 2D, C vs. S: $p = 0.001$, C vs. R: $p = 0.003$). In accordance, expression of *Il17f* mRNA encoding the pro-inflammatory cytokine IL-17F tended to be increased in defeated mice (Figure 2E, $p = 0.067$). Thus, expression levels of genes associated with Th cell functions are modulated after social defeat stress similarly in susceptible and resilient animals.

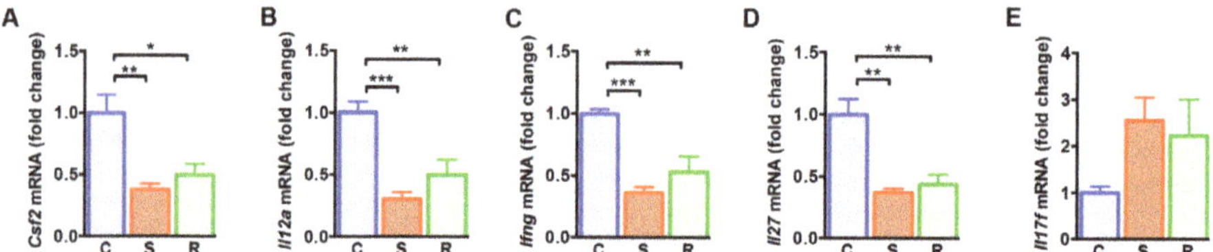

Figure 2. Gene expression analysis of splenocytes from control, susceptible and resilient mice after 10 days of social defeat. (**A**) mRNA expression of *Csf2*, (**B**) *Il12a*, (**C**) *Ifng*, (**D**) *Il27*, and (**E**) *Il17f*. Expression levels were normalized to the mean expression of housekeeping genes *Gapdh* and *Hsp90ab1*. Fold changes were calculated relative to control mice. Bar graphs represent mean + SEM. C: control, S: susceptible, R: resilient. $n_C = 6$, $n_S = 6$, $n_R = 4$. *: $p < 0.05$, **: $p < 0.01$, ***: $p < 0.001$ (Bonferroni post hoc).

2.3. Reduced Percentages of T Lymphocytes in Susceptible and Resilient Mice after Social Defeat Stress

We next examined splenocytes and T cell subsets in the spleen of these animals by flow cytometry. After 10 days of social defeat, the numbers of splenic mononuclear cells were increased in susceptible mice when compared to control animals (Figure 3A, $p = 0.019$). Percentages of splenic $\alpha\beta$ T cells were markedly reduced in defeated mice when compared to non-defeated controls (Figure 3B) resulting in equivalent $\alpha\beta$ T cell numbers in the spleen of these animals (Figure 3C). Frequencies of CD4$^+$ and CD8$^+$ cells among $\alpha\beta$ T cells and absolute numbers of these subsets were comparable in all groups excluding that social defeat differentially affected homeostasis of these T cell subsets (Figure 3D,E, CD4 %: $p = 0.551$; #: $p = 0.092$; CD8 %: $p = 0.099$; #: $p = 0.042$). Thus, social defeat stress reduces percentages of T cells in the spleen independent of the behavioral outcome with regard to susceptibility or resilience.

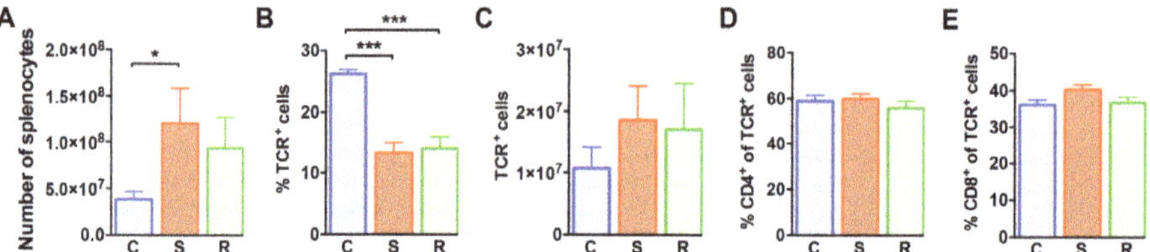

Figure 3. Numbers of splenocytes and $\alpha\beta$ T cells in mice after social defeat and control animals. (**A**) Absolute numbers of splenocytes in control, susceptible and resilient mice. (**B**) Percentages and (**C**) numbers of TCR$^+$ T cells, (**D**) percentages of CD4$^+$ cells among TCR$^+$ T cells, and (**E**) percentages of CD8$^+$ cells among TCR$^+$ T cells in the spleen as determined by flow cytometry. Bar graphs represent mean + SEM. C: control, S: susceptible, R: resilient. $n_C = 10$, $n_S = 11$, $n_R = 9$. *: $p < 0.05$, ***: $p < 0.001$ (Bonferroni post hoc).

2.4. Increased Numbers of IL-17 Producing T Cells after Social Defeat Stress

To study whether T cell functions were differentially affected in susceptible versus resilient mice following social defeat, we studied the cytokine producing capacity of T cells in these animals. For this, interferon (IFN)-γ and IL-17 production by splenic CD4$^+$ and CD8$^+$ T cells was determined after 10 days of social defeat by flow cytometry. Percentages of CD4$^+$ T cells producing IFN-γ but not IL-17 were decreased in susceptible mice after social defeat when compared to controls ($p = 0.019$); however, absolute numbers were comparable between all groups (Figure 4B). In contrast, IL-17$^+$ IFN-γ^- CD4$^+$ T cell proportions and numbers were elevated in susceptible animals after social defeat (Figure 4C,

%: $p = 0.006$; #: $p = 0.032$). Percentages and absolute numbers of CD4$^+$ T cells co-expressing IFN-γ and IL-17 were rather low and similarly distributed in all three groups. The entire population of IL-17 producing CD4$^+$ T cells, comprising IL-17$^+$ IFN-γ^+ and IL-17$^+$ IFN-γ^- cells, showed similar effects as the IL-17$^+$ IFN-γ^- population. Susceptible mice again presented increased proportions and numbers compared to controls (%: $p = 0.002$; #: $p = 0.015$). Percentages and numbers of CD8$^+$ T cells producing IFN-γ did not differ between the groups (Table S1, Figure 4D). However, we found enhanced percentages and absolute numbers of IL-17 expressing CD8$^+$ T cells in the spleen of susceptible mice after social defeat compared to controls (Table S1, %: $p = 0.006$; #: Figure 4E, $p = 0.036$). Thus, social defeat affects CD4$^+$ and CD8$^+$ T cells producing IL-17 in susceptible animals.

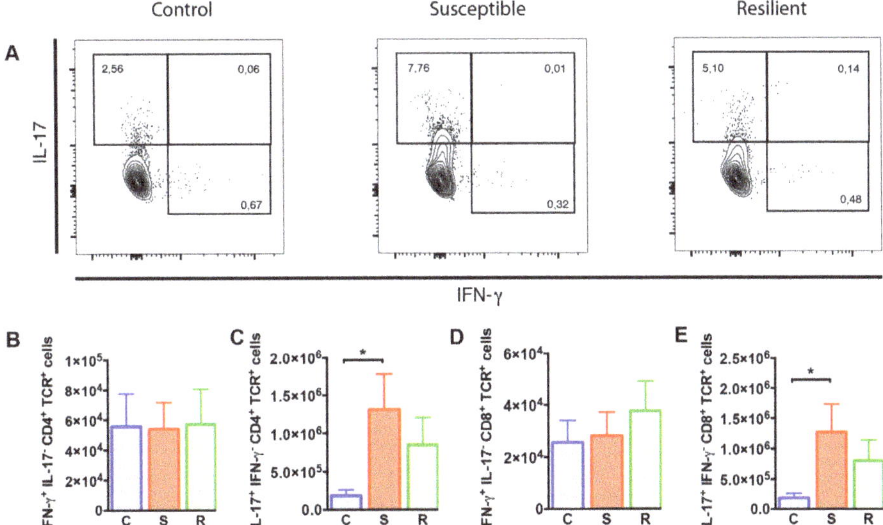

Figure 4. Cytokine expression by CD4$^+$ and CD8$^+$ T cells from the spleen of mice after social defeat. Representative contour plots showing expression of (**A**) IFN-γ and IL-17 in CD4$^+$ T cells from the spleen as determined by flow cytometry. (**B**) Numbers of IFN-γ and (**C**) IL-17 producing CD4$^+$ T cells. (**D**) Numbers of IFN-γ cells and (**E**) IL-17 producing CD8$^+$ T cells. Bar graphs represent mean + SEM. C: control, S: susceptible, R: resilient. $n_C = 9$, $n_S = 10$, $n_R = 9$. *: $p < 0.05$ (Bonferroni post hoc).

2.5. Reduced Numbers of Regulatory T Cells after Social Defeat

We next investigated whether distinct behavioral changes after social defeat affect the immunoregulatory T cell compartment. While percentages of CD4$^+$ T cells expressing the IL-2 receptor α-chain (CD25) were equivalent in all groups of mice (Table S1), absolute numbers of CD25$^+$ CD4$^+$ T cells were significantly reduced in susceptible animals (Figure 5A, $p = 0.003$). FoxP3 expressing CD4$^+$ T cells specifically linked to immune regulation were markedly reduced in numbers in susceptible mice when compared to controls (Figure 5B, $p = 0.009$). In resilient animals, a similar trend was observed ($p = 0.070$). Splenic FoxP3$^+$ T$_{reg}$ cell percentages were not affected by social defeat (Table S1). Furthermore, splenic mRNA levels of transforming growth factor β (*Tgfb*) encoding the immunomodulatory TGF β which is crucial for T$_{reg}$ cell-mediated suppression in vivo were significantly reduced in susceptible mice compared to control animals (Figure 5C, C vs. S: $p < 0.001$, S vs. R: $p = 0.053$). Together these findings point toward an altered immunoregulatory status in defeated mice.

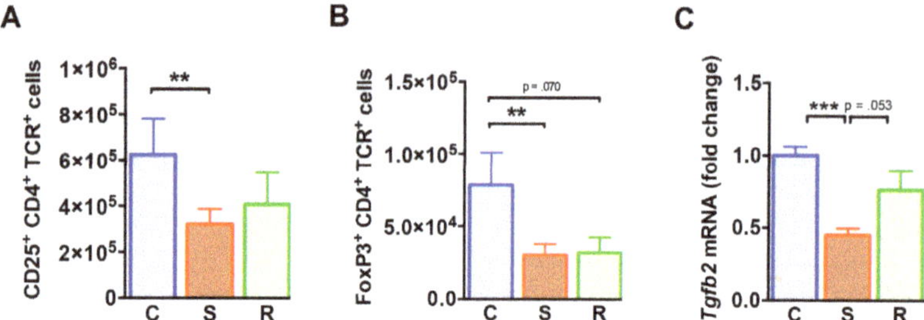

Figure 5. T regulatory cells. (**A**) Numbers of CD25$^+$ CD4$^+$ T cells. (**B**) Absolute numbers of FoxP3$^+$ CD4$^+$ T cells. (**C**) mRNA expression of *Tgfb2* in splenocytes. Expression levels were normalized to the mean expression of housekeeping genes *Gapdh* and *Hsp90ab1*. Fold changes were calculated relative to control mice. Bar graphs represent mean + SEM. C: control, S: susceptible, R: resilient. $n_C = 15$, $n_S = 16$, $n_R = 11$. **: $p < 0.01$, ***: $p < 0.001$ (Bonferroni post hoc).

2.6. Enhancement of Th17 Differentiation Did Not Alter Behavioral Responses to Social Defeat

Finally, we investigated whether the increase in IL-17 producing CD4$^+$ T cell percentages observed in defeated mice is sufficient to alter behavioral responses to social defeat stress. We therefore utilized mice with CD4-specific knockout of PPARγ, a key negative regulator of Th17 differentiation [36]. In CD4-PPARγ^{KO} mice, Th17 differentiation is strongly increased, while Th1, Th2, or T$_{reg}$ cell differentiation is not affected [36]. We subjected CD4-PPARγ^{KO} mice and CD4-PPARγ^{WT} controls to social defeat and analyzed social as well as anxiety-related behavior. Socially defeated CD4-PPARγ^{KO} mice showed an equivalently reduced interaction ratio in the social interaction test compared to CD4-PPARγ^{WT} controls (Figure 6A, main effect of stress: $p = 0.005$). In the open-field test, CD4-PPARγ^{KO} mice also showed comparably reduced center entries and time spent in the center when compared to CD4-PPARγ^{WT} controls after ten days of social defeat (Figure 6B,C; Center entries: main effect of stress: $p < 0.001$; Center time: main effect of stress: $p < 0.001$). These data indicate similar anxiety-related behavior in both genotypes that were equally affected by stress exposure. The distance traveled in the open field test was not affected by social defeat or genotype (Figure 6D). These findings suggest that PPARγ−mediated changes in T cell differentiation and function do not modulate social and anxiety-like behavior, neither under control conditions nor after ten days of social defeat. We also analyzed these behaviors subdividing the defeated groups into susceptible and resilient mice. Again, no effects of genotype or interaction effects of genotype and stress exposure could be detected (Figure S1) suggesting that an enhanced Th17 differentiation status induced by CD4-specific deficiency of PPARγ is not sufficient to alter emotional behavior or stress vulnerability.

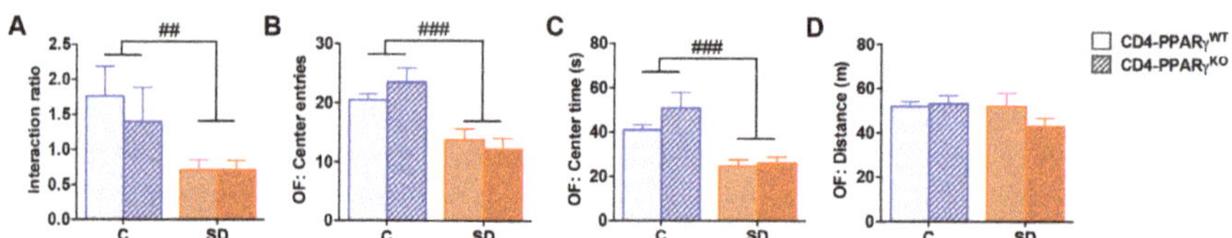

Figure 6. Behavioral data of CD4-specific PPARγ knockout (CD4-PPARγ^{KO}) mice and Cre-negative floxed controls (CD4-PPARγ^{WT}) after 10 days of social defeat. (**A**) The interaction ratio of the social interaction test. (**B**) The number of center entries, (**C**) the time spent in the center, and (**D**) the distance traveled in the open-field test. Data represent mean + SEM. CD4-PPARγ^{WT} Control (C): $n = 6$, CD4-PPARγ^{WT} social defeat (SD): $n = 11$, CD4-PPARγ^{KO} Control: $n = 6$, CD4-PPARγ^{KO} SD: $n = 8$, ##: main effect of stress, $p < 0.01$, ###: $p < 0.001$.

3. Discussion

Stress is known to evoke long-term alterations of the adaptive immune response [1–3]. At the same time, it is well established that individuals exhibit considerable variability in behavioral responses to stressors, and even genetically identical inbred mouse strains show an individual variability in the sensitivity to social stressors [48]. How specific adaptive immune alterations are correlated with individual stress susceptibility and resilience is poorly characterized. A better understanding of the underlying mechanisms is of major importance due to the enormous health burden of stress-related affective disorders.

In this study, we characterized alterations of the adaptive immune response associated with stress susceptibility and resilience in a mouse model of social defeat. Our data show reduced T cell percentages and altered expression levels of T cell differentiation and effector genes in the spleen of mice exposed to social defeat irrespective of a stress susceptible or resilient behavioral phenotype. We further observed greatly increased numbers of splenic IL-17 producing CD4$^+$ and CD8$^+$ T cells in susceptible animals compared to controls while regulatory T cells were reduced after social defeat.

A decrease in T cell numbers or percentages has been reported before in various animal models of acute and chronic stress [49–52]. Stress exposure results in an enhanced release of glucocorticoids and catecholamines that may induce apoptosis in peripheral T cells [53,54]. Enhanced T cell apoptosis following stressful events has further been attributed to lack of tryptophan, an essential factor in T cell proliferation [55]. In addition, inputs from adrenergic nerves have been shown to affect T cell trafficking, since stimulation of β2-adrenergic receptors on T cells reduces their egress from lymph nodes [56]. Furthermore, T cells from MDD patients showed lower expression of the chemokine receptors CXCR3 and CCR6 that modulate T cell differentiation and trafficking [57]. Of note, it has been suggested that T cells exert stress protective effects based on observations that T cell deficient BALB/c nude mice were more vulnerable to brief exposures to foot shocks than T cell competent mice on the same genetic background [58]. Our findings of reduced percentages but not numbers of splenic T cells point toward an altered cellular composition in the spleen after social defeat. It has frequently been shown that stress results in an increase in the numbers of innate immune cells including natural killer cells, neutrophils, and monocytes [49,59]. In particular, our earlier findings demonstrating that mice after chronic social defeat show higher numbers of splenic myeloid cells [43] may explain the here observed higher cellularity of the spleen.

In the present study, IL-17-producing CD4$^+$ T cells, classified before as pathogenic Th17 cells in inflammatory responses and neuroinflammation, were markedly enhanced in susceptible mice in response to social defeat stress. Accordingly, levels of the gene encoding IL-27, a cytokine mediating suppressive effects on the Th17 lineage, were lower in the spleen of these animals [47,60]. These findings are in line with previous studies demonstrating that the vulnerability for the development of learned helplessness was dependent on increased Th17 responses [30]. It has also been shown that the cytokine IL-6, which is required for the induction of Th17 differentiation [24,61], is indicative of stress susceptibility [42]. However, in CD4-specific PPARγ deficient animals exhibiting enhanced Th17 differentiation [36], social defeat stress had the same effect on the behavioral level when compared to PPARγ-competent controls. It is important to note that the underlying mechanism for increased Th17 differentiation induced by CD4-specific deficiency of PPARγ might differ from the mechanism responsible for the stress-induced Th17 shift in our model. Further research will focus on defining potential regulators involved in Th17 differentiation after social defeat, e.g., the signal transducer and activator of transcription 3 (STAT3) and the transcription factors IFN regulatory factor 4 (IRF4), c-Rel and RelA/p65 required for Th17 differentiation and responses. In addition, analyses of the Th1-specific T-box transcription factor T-bet and the Th17 specific RAR-related orphan receptor gamma t (RORγt) will provide better insights into the differentiation of T helper cells in this model.

Our findings suggest that CD4$^+$ cell-specific deletion of PPARγ may not be sufficient to promote stress resilience in this model. In contrast, an anxiolytic effect of neuronal deletion of PPARγ on the emotional response to acute stress has been described before [62]. It is, therefore, likely that PPARγ operates on various cellular levels in modulating anxiety- and depressive-like behavior. It is still unclear, however, whether alterations in the here-studied T cell subsets are the cause or consequence of stress susceptibility and whether IL-17 producing T cells have differential implications in stress susceptible and resilient mice.

T_{reg} cells are predominantly viewed as mediators for immune tolerance and suppression [24]. We observed reduced T_{reg} numbers in both, susceptible and to a minor extent also in resilient socially defeated mice. In analogy, reduced T_{reg} numbers associated with systemic T cell activation have been found during subordinate colony housing [63], suggesting that a reduction in these cells might contribute to an overall pro-inflammatory state in these animals. In the same model, T_{reg} cells are necessary to induce stress resilience by immunization with *Mycobacterium vaccae* [64] suggesting a functional role of these cells in mediating stress vulnerability and resilience. However, also controversial findings on T_{reg} cells in murine stress models have been reported and may be explained by the diversity of models and the different time points studied. For example, T_{reg} cell proportions have been found increased in mice due to chronic unpredictable mild stress [31] and enhanced frequencies of peripheral T_{reg} cells and an elevated suppressive function of these cells have further been found after chronic immobilization of mice [65]. In a model of learned helplessness induced by mild inescapable foot shocks, no difference was found in percentages of T_{reg} cells between controls and mice exhibiting learned helplessness [30]. Future studies will therefore have to focus on the impact of T_{reg} cells in this model to better understand the role of these cells in the regulation of emotional behavior.

In the spleen of resilient mice, we observed an "intermediate" immune pattern characterized by lower numbers and percentages of splenic IL-17 secreting T cells when compared to susceptible mice but higher values in these categories than controls, albeit those values did not reach significance. Thus, our analysis of the adaptive immune status after social defeat did not reveal overt differences in susceptible and resilient animals in contrast to our earlier findings regarding the innate immune system. Herein we demonstrated specific alterations in susceptible mice among those an enhanced maturation of dendritic cells in the spleen, and increased brain immigration of $CCR2^+$ $Ly6C^{hi}$ monocytes representing an inflammatory phenotype [43].

In conclusion, our study provides evidence that specific alterations of the adaptive immune responses, which are involved in maintaining brain function, plasticity and behavior, are induced by social defeat stress. Future studies in this model may close the knowledge gap concerning the link between adaptive immune responses and stress vulnerability by analyzing the impact of T_{reg} cells and TGF-β on pro-inflammatory responses, kynurenine metabolism, and microglial activation. In addition, longitudinal immune studies in rodents exposed to chronic stress and humans during clinical course of MDD are necessary to yield a better understanding of the pathophysiology of affective disorders.

4. Materials and Methods

4.1. Mice and Housing Conditions

Five-week-old, male C57BL/6J mice were purchased at Charles River (Sulzfeld, Germany). After a habituation period of two weeks, the social defeat experiments started. CD-1 mice from our in-house breeding facility were used as resident animals for the social defeat paradigm and social interaction partners. These mice were older than 3 months and most of them had mating experience. Their level of aggressive behavior was tested before chosen for the experiment (latency to attack intruder should be less than 30 s). All animals were housed at 22 ± 2 °C and humidity of 55 ± 10% under a 12 h:12 h light-dark cycle, with lights on at 6 am. Food and water were available ad libitum. This study was performed in accordance with the regulations covering animal experimentation in Germany and the EU (European Communities Council Directive 2010/63/EU). The project was approved by the local authority and the Animal Welfare Officer of the University of Münster (84-02.04.2013.A320, 31 October 2013). All efforts were made in order to minimize animal suffering and reduce the number of animals used.

4.2. Social Defeat Paradigm

The social defeat paradigm was performed as described before [43]. Briefly, experimental mice were inserted into the cage of an aggressive, older and heavier CD-1 mouse for 10 min per day. After 10 min direct physical contact, animals were separated by a perforated Plexiglas wall and kept on opposite sides of the same cage for 24 h to maintain visual and olfactory contact. This procedure was repeated daily with another CD-1 mouse. After the final confrontation on day 10, experimental mice were housed singly in Makrolon type II cages. Control mice were housed in the same type of cage as experimental mice. The degree of agonistic interactions was observed by an experienced observer who terminated the sessions and separated the animals immediately in case that escalated fighting occurred before 10 min passed [43].

4.3. Social Interaction Test

One day after the last social defeat session, the social interaction test was conducted as described before [43,66]. Briefly, it comprised two trials of 150 s each, one with an empty enclosure, the second with an unfamiliar CD-1 mouse therein. The time spent in the interaction zone, defined as the area surrounding the exploration enclosure 8 cm to each side was recorded in both trials by ANY-maze tracking software (Stoelting, Dublin, Ireland). An interaction ratio was calculated as time spent in the interaction zone during the second trial with mouse divided by the time spent in the zone during the first trial with the empty enclosure. When the interaction ratio was less than 0.5, animals were defined as susceptible, otherwise as resilient [43].

4.4. Open Field Test

The open field test was conducted as described before [67]. Briefly, mice were introduced into one corner of an 80 cm × 80 cm wooden box with 40 cm high walls and allowed to freely explore the box for 10 min. The distance traveled, the number of entries into the 40 cm × 40 cm center area and the time spent therein were automatically recorded by ANY-maze tracking software (Stoelting, Dublin, Ireland).

4.5. Gene Expression Analysis

The day after the social interaction test, spleens of control ($n = 6$), susceptible ($n = 6$) and resilient ($n = 4$) mice were dissected and immediately snap-frozen in liquid nitrogen. RNA was extracted from half spleens using the RNeasy Midi kit (Qiagen, Hilden, Germany). RNA was reverse transcribed to generate cDNA using the RT^2 HT First Strand kit (Qiagen). RT2 Profiler PCR Arrays for mouse cytokines and chemokines were run for control ($n = 3$), susceptible ($n = 3$) and resilient mice ($n = 2$) according to the manufacturer's instructions on an ABI 7900 HT PCR system (Life Technologies, Darmstadt, Germany). Based on the results of this array, candidate genes were chosen and Taqman assays were run in triplicates as described before [68]. The following candidate genes were assessed using inventoried assays (Life Technologies, Darmstadt, Germany): *Csf2* (Mm01290062_m1), *Il12a* (Mm00434165_m1), *Ifng* (Mm01168134_m1), *Il27* (Mm00461162_m1), *Il17f* (Mm00521423_m1), *Tgfb2* (Mm00436955_m1). To calculate ΔCt levels for each of the candidate genes, average Cts of housekeeping genes *Hsp90ab1* (Mm00833431_g1) and *Gapdh* (Mm99999915_g1) were used. Fold changes were calculated as $2^{-\Delta\Delta Ct}$ using the non-defeated control group as reference.

4.6. Flow Cytometry

Spleens of control ($n = 15$), susceptible ($n = 16$) and resilient ($n = 11$) mice were homogenized after transcardial perfusion and a single cell suspension was received as described before [43]. Due to limitations in the number of animals that could be dissected on a single day, the group of animals was divided into three cohorts as shown in Table S3.

The following antibodies (purchased at Biolegend, San Diego, CA, USA) were used for fluorescent staining of splenocytes: FITC or PerCP-Cy5.5-conjugated anti-mouse TCR β-chain (clone H57-597),

APC-Cy7-conjugated anti-mouse CD4 (clone RM4-5), APC or PE-Cy7-conjugated anti-mouse CD8a (clone 53-6.7), BV510-conjugated anti-mouse IFN-γ (clone XMG1.2), PE-conjugated anti-mouse IL-17A (clone eBio17B7), PE-Cy7-conjugated anti-mouse CD25 (clone PC-61) and PE-conjugated anti-mouse FoxP3 (clone FJK-16s).

Intracellular staining was performed according to the manufacturer's instructions using the Fixation/Permeabilization kit (BD Cytofix/Cytoperm), intranuclear staining using the FOXP3 Fix/Perm buffer set (Biolegend, London, UK). For ex vivo stimulation of lymphocytes, 5×10^6 splenocytes were incubated with PMA (10 ng/mL) and ionomycin (500 ng/mL) plus Monensin and Brefeldin A (Biolegend, London, UK) for 10 h overnight. Samples were acquired on a FACSCanto II (BD Biosciences, East Rutherford, NJ) flow cytometer and analyzed by FlowJo v10. The gating strategy comprised life gates (SSC-A vs- FSC-A) to exclude debris and dead cells. Subsequently, doublets were gated out by comparing sideward and forward scatter height and width (SSC-H vs SSC-W and FSC-H vs. FSC-W). TCR^+ cells were considered as T cells, $CD4^+$ and $CD8^+$ T cells were determined on TCR^+ pregates. $IFN-\gamma^+$, $IL\text{-}17^+$ cells were determined on $CD4^+$ or $CD8^+$ T cell pregates, respectively. Accordingly, $CD25^+$ and $FoxP3^+$ Th cells were assessed on $CD4^+$ T cell pregates.

4.7. Statistics

Data obtained in independent cohorts were combined and analyzed by analysis of covariance (ANCOVA) with stress phenotype as fixed factor and cohort as covariate. In case of significant effects of the stress phenotype, Bonferroni post hoc tests were calculated. The null-hypothesis was rejected for $p < 0.05$. All analyzes were calculated with SPSS 24 (IBM).

Supplementary Materials: Supplementary materials can be found at http://www.mdpi.com/1422-0067/20/14/3512/s1.

Author Contributions: Conceptualization, O.A., L.K., B.T.B., V.A., S.S., and J.A.; methodology, O.A., C.R., L.K., S.S., and J.A.; formal analysis, O.A., C.R., J.A.; investigation, O.A., C.R., J.A.; resources, O.A., P.Z., L.K., B.T.B., V.A., S.S., and J.A.; writing—original draft preparation, O.A., S.S., and J.A.; writing—review and editing, O.A., P.Z., L.K., B.T.B., V.A., S.S., and J.A.; supervision, O.A., S.S., and J.A.; funding acquisition, O.A., S.S., and J.A.

Funding: This work was supported by the fund "Innovative Medical Research" of the University of Münster Medical School (grant number: IMF AM211515 to O.A.), by the Cells in Motion–Cluster of Excellence of the German Research Society (grant number: EXC 1003 FF-2014-01 to J.A.), by the German Research Society (grant number: DFG SCHE692/3-1 to S.S. and DFG FOR2107 AL1145/5-2 to J.A.), and by the Strategic Research Fund of the Heinrich Heine University Düsseldorf (SFF-F2012/79-5-Scheu to S.S).

Acknowledgments: The authors thank Arezoo Fattahi-Mehr and Christiane Schettler for exceptional technical support.

Conflicts of Interest: The authors declare no conflict of interest.

Abbreviations

CD	Cluster of differentiation
IFN	Interferone
IL	Interleukin
MDD	Major depressive disorder
PPAR-γ	Peroxisome proliferator-activated receptor γ
TGF-β	Transforming growth factor-β
Th cell	T helper cell
T_{reg} cell	T regulatory cell

References

1. Steptoe, A.; Hamer, M.; Chida, Y. The effects of acute psychological stress on circulating inflammatory factors in humans: A review and meta-analysis. *Brain Behav. Immun.* **2007**, *21*, 901–912. [CrossRef] [PubMed]
2. Coelho, R.; Viola, T.W.; Walss-Bass, C.; Brietzke, E.; Grassi-Oliveira, R. Childhood maltreatment and inflammatory markers: A systematic review. *Acta Psychiatr. Scand* **2014**, *129*, 180–192. [CrossRef] [PubMed]

3. Baumeister, D.; Akhtar, R.; Ciufolini, S.; Pariante, C.M.; Mondelli, V. Childhood trauma and adulthood inflammation: A meta-analysis of peripheral C-reactive protein, interleukin-6 and tumour necrosis factor-alpha. *Mol. Psychiatry* **2016**, *21*, 642–649. [CrossRef] [PubMed]

4. Kessler, R.C.; Berglund, P.; Demler, O.; Jin, R.; Merikangas, K.R.; Walters, E.E. Lifetime prevalence and age-of-onset distributions of DSM-IV disorders in the National Comorbidity Survey Replication. *Arch. Gen. Psychiatry* **2005**, *62*, 593–602. [CrossRef] [PubMed]

5. Krishnan, V.; Nestler, E.J. The molecular neurobiology of depression. *Nature* **2008**, *455*, 894–902. [CrossRef] [PubMed]

6. Dowlati, Y.; Herrmann, N.; Swardfager, W.; Liu, H.; Sham, L.; Reim, E.K.; Lanctot, K.L. A meta-analysis of cytokines in major depression. *Biol. Psychiatry* **2010**, *67*, 446–457. [CrossRef] [PubMed]

7. Eyre, H.A.; Stuart, M.J.; Baune, B.T. A phase-specific neuroimmune model of clinical depression. *Prog. Neuropsychopharmacol Biol. Psychiatry* **2014**, *54*, 265–274. [CrossRef] [PubMed]

8. Kohler, C.A.; Freitas, T.H.; Maes, M.; de Andrade, N.Q.; Liu, C.S.; Fernandes, B.S.; Stubbs, B.; Solmi, M.; Veronese, N.; Herrmann, N.; et al. Peripheral cytokine and chemokine alterations in depression: A meta-analysis of 82 studies. *Acta Psychiatr Scand.* **2017**, *135*, 373–387. [CrossRef] [PubMed]

9. Gibney, S.M.; Drexhage, H.A. Evidence for a dysregulated immune system in the etiology of psychiatric disorders. *J. Neuroimmune Pharmacol.* **2013**, *8*, 900–920. [CrossRef] [PubMed]

10. Müller, N. Immunology of major depression. *Neuroimmunomodulation* **2014**, *21*, 123–130. [CrossRef]

11. Toben, C.; Baune, B.T. An Act of Balance Between Adaptive and Maladaptive Immunity in Depression: A Role for T Lymphocytes. *J. Neuroimmune Pharmacol.* **2015**, *10*, 595–609. [CrossRef] [PubMed]

12. Beurel, E.; Lowell, J.A. Th17 cells in depression. *Brain Behav. Immun.* **2018**, *69*, 28–34. [CrossRef] [PubMed]

13. Miller, A.H. Depression and immunity: A role for T cells? *Brain Behav. Immun.* **2010**, *24*, 1–8. [CrossRef] [PubMed]

14. Ziv, Y.; Ron, N.; Butovsky, O.; Landa, G.; Sudai, E.; Greenberg, N.; Cohen, H.; Kipnis, J.; Schwartz, M. Immune cells contribute to the maintenance of neurogenesis and spatial learning abilities in adulthood. *Nat. Neurosci.* **2006**, *9*, 268–275. [CrossRef] [PubMed]

15. Ziv, Y.; Schwartz, M. Orchestrating brain-cell renewal: The role of immune cells in adult neurogenesis in health and disease. *Trends Mol. Med.* **2008**, *14*, 471–478. [CrossRef] [PubMed]

16. Niebling, J.; Rünker, A.E.; Schallenberg, S.; Kretschmer, K.; Kempermann, G. Myelin-specific T helper 17 cells promote adult hippocampal neurogenesis through indirect mechanisms. *F1000Resarch* **2014**, *3*, 169. [CrossRef]

17. Zhang, L.; Zhang, J.; You, Z. Switching of the Microglial Activation Phenotype Is a Possible Treatment for Depression Disorder. *Front. Cell Neurosci.* **2018**, *12*, 306. [CrossRef] [PubMed]

18. Delpech, J.C.; Madore, C.; Nadjar, A.; Joffre, C.; Wohleb, E.S.; Laye, S. Microglia in neuronal plasticity: Influence of stress. *Neuropharmacology* **2015**, *96*, 19–28. [CrossRef]

19. Lee, M.M.; Reif, A.; Schmitt, A.G. Major depression: A role for hippocampal neurogenesis? *Curr. Top. Behav. Neurosci.* **2013**, *14*, 153–179.

20. Irwin, M.R.; Miller, A.H. Depressive disorders and immunity: 20 years of progress and discovery. *Brain Behav. Immun.* **2007**, *21*, 374–383. [CrossRef]

21. Zorrilla, E.P.; Luborsky, L.; McKay, J.R.; Rosenthal, R.; Houldin, A.; Tax, A.; McCorkle, R.; Seligman, D.A.; Schmidt, K. The relationship of depression and stressors to immunological assays: A meta-analytic review. *Brain Behav. Immun.* **2001**, *15*, 199–226. [CrossRef] [PubMed]

22. Slyepchenko, A.; Maes, M.; Kohler, C.A.; Anderson, G.; Quevedo, J.; Alves, G.S.; Berk, M.; Fernandes, B.S.; Carvalho, A.F. T helper 17 cells may drive neuroprogression in major depressive disorder: Proposal of an integrative model. *Neurosci. Biobehav. Rev.* **2016**, *64*, 83–100. [CrossRef] [PubMed]

23. Waisman, A.; Hauptmann, J.; Regen, T. The role of IL-17 in CNS diseases. *Acta Neuropathol.* **2015**, *129*, 625–637. [CrossRef] [PubMed]

24. Hall, B.M. T Cells: Soldiers and Spies–The Surveillance and Control of Effector T Cells by Regulatory T Cells. *Clin J. Am. Soc. Nephrol.* **2015**, *10*, 2050–2064. [CrossRef] [PubMed]

25. Grosse, L.; Hoogenboezem, T.; Ambree, O.; Bellingrath, S.; Jorgens, S.; de Wit, H.J.; Wijkhuijs, A.M.; Arolt, V.; Drexhage, H.A. Deficiencies of the T and natural killer cell system in major depressive disorder: T regulatory cell defects are associated with inflammatory monocyte activation. *Brain Behav. Immun.* **2016**, *54*, 38–44. [CrossRef] [PubMed]

26. Chen, Y.; Jiang, T.; Chen, P.; Ouyang, J.; Xu, G.; Zeng, Z.; Sun, Y. Emerging tendency towards autoimmune process in major depressive patients: A novel insight from Th17 cells. *Psychiatry Res.* **2011**, *188*, 224–230. [CrossRef] [PubMed]

27. Li, Y.; Xiao, B.; Qiu, W.; Yang, L.; Hu, B.; Tian, X.; Yang, H. Altered expression of CD4(+)CD25(+) regulatory T cells and its 5-HT(1a) receptor in patients with major depression disorder. *J. Affect. Disord.* **2010**, *124*, 68–75. [CrossRef]

28. Grosse, L.; Carvalho, L.A.; Birkenhager, T.K.; Hoogendijk, W.J.; Kushner, S.A.; Drexhage, H.A.; Bergink, V. Circulating cytotoxic T cells and natural killer cells as potential predictors for antidepressant response in melancholic depression. Restoration of T regulatory cell populations after antidepressant therapy. *Psychopharmacology (Berl)* **2016**, *233*, 1679–1688. [CrossRef]

29. Dhabhar, F.S. Effects of stress on immune function: The good, the bad, and the beautiful. *Immunol. Res.* **2014**, *58*, 193–210. [CrossRef]

30. Beurel, E.; Harrington, L.E.; Jope, R.S. Inflammatory T helper 17 cells promote depression-like behavior in mice. *Biol. Psychiatry* **2013**, *73*, 622–630. [CrossRef]

31. Hong, M.; Zheng, J.; Ding, Z.Y.; Chen, J.H.; Yu, L.; Niu, Y.; Hua, Y.Q.; Wang, L.L. Imbalance between Th17 and Treg cells may play an important role in the development of chronic unpredictable mild stress-induced depression in mice. *Neuroimmunomodulation* **2013**, *20*, 39–50. [CrossRef] [PubMed]

32. Santarelli, L.; Saxe, M.; Gross, C.; Surget, A.; Battaglia, F.; Dulawa, S.; Weisstaub, N.; Lee, J.; Duman, R.; Arancio, O.; et al. Requirement of hippocampal neurogenesis for the behavioral effects of antidepressants. *Science* **2003**, *301*, 805–809. [CrossRef] [PubMed]

33. Snyder, J.S.; Soumier, A.; Brewer, M.; Pickel, J.; Cameron, H.A. Adult hippocampal neurogenesis buffers stress responses and depressive behaviour. *Nature* **2011**, *476*, 458–461. [CrossRef] [PubMed]

34. Poletti, S.; de Wit, H.; Mazza, E.; Wijkhuijs, A.J.; Locatelli, C.; Aggio, V.; Colombo, C.; Benedetti, F.; Drexhage, H.A. Th17 cells correlate positively to the structural and functional integrity of the brain in bipolar depression and healthy controls. *Brain Behav. Immun.* **2016**, *61*, 317–325. [CrossRef] [PubMed]

35. Hucke, S.; Flossdorf, J.; Grutzke, B.; Dunay, I.R.; Frenzel, K.; Jungverdorben, J.; Linnartz, B.; Mack, M.; Peitz, M.; Brustle, O.; et al. Licensing of myeloid cells promotes central nervous system autoimmunity and is controlled by peroxisome proliferator-activated receptor gamma. *Brain* **2012**, *135*, 1586–1605. [CrossRef]

36. Klotz, L.; Burgdorf, S.; Dani, I.; Saijo, K.; Flossdorf, J.; Hucke, S.; Alferink, J.; Nowak, N.; Beyer, M.; Mayer, G.; et al. The nuclear receptor PPAR gamma selectively inhibits Th17 differentiation in a T cell-intrinsic fashion and suppresses CNS autoimmunity. *J. Exp. Med.* **2009**, *206*, 2079–2089. [CrossRef] [PubMed]

37. Eissa Ahmed, A.A.; Al-Rasheed, N.M.; Al-Rasheed, N.M. Antidepressant-like effects of rosiglitazone, a PPARgamma agonist, in the rat forced swim and mouse tail suspension tests. *Behav. Pharmacol.* **2009**, *20*, 635–642. [CrossRef]

38. Sadaghiani, M.S.; Javadi-Paydar, M.; Gharedaghi, M.H.; Fard, Y.Y.; Dehpour, A.R. Antidepressant-like effect of pioglitazone in the forced swimming test in mice: The role of PPAR-gamma receptor and nitric oxide pathway. *Behav. Brain Res.* **2011**, *224*, 336–343. [CrossRef]

39. Kurhe, Y.; Mahesh, R. Pioglitazone, a PPARgamma agonist rescues depression associated with obesity using chronic unpredictable mild stress model in experimental mice. *Neurobiol. Stress* **2016**, *3*, 114–121. [CrossRef]

40. Colle, R.; de Larminat, D.; Rotenberg, S.; Hozer, F.; Hardy, P.; Verstuyft, C.; Feve, B.; Corruble, E. Pioglitazone could induce remission in major depression: A meta-analysis. *Neuropsychiatr. Dis. Treat.* **2017**, *13*, 9–16. [CrossRef]

41. Colle, R.; de Larminat, D.; Rotenberg, S.; Hozer, F.; Hardy, P.; Verstuyft, C.; Feve, B.; Corruble, E. PPAR-gamma Agonists for the Treatment of Major Depression: A Review. *Pharmacopsychiatry* **2017**, *50*, 49–55. [PubMed]

42. Hodes, G.E.; Pfau, M.L.; Leboeuf, M.; Golden, S.A.; Christoffel, D.J.; Bregman, D.; Rebusi, N.; Heshmati, M.; Aleyasin, H.; Warren, B.L.; et al. Individual differences in the peripheral immune system promote resilience versus susceptibility to social stress. *Proc. Natl. Acad. Sci. USA* **2014**. [CrossRef]

43. Ambree, O.; Ruland, C.; Scheu, S.; Arolt, V.; Alferink, J. Alterations of the Innate Immune System in Susceptibility and Resilience After Social Defeat Stress. *Front. Behav. Neurosci.* **2018**, *12*, 141. [CrossRef] [PubMed]

44. Berton, O.; McClung, C.A.; Dileone, R.J.; Krishnan, V.; Renthal, W.; Russo, S.J.; Graham, D.; Tsankova, N.M.; Bolanos, C.A.; Rios, M.; et al. Essential role of BDNF in the mesolimbic dopamine pathway in social defeat stress. *Science* **2006**, *311*, 864–868. [CrossRef] [PubMed]

45. Poppensieker, K.; Otte, D.M.; Schurmann, B.; Limmer, A.; Dresing, P.; Drews, E.; Schumak, B.; Klotz, L.; Raasch, J.; Mildner, A.; et al. CC chemokine receptor 4 is required for experimental autoimmune encephalomyelitis by regulating GM-CSF and IL-23 production in dendritic cells. *Proc. Natl. Acad. Sci. USA* **2012**, *109*, 3897–3902. [CrossRef]

46. Becher, B.; Tugues, S.; Greter, M. GM-CSF: From Growth Factor to Central Mediator of Tissue Inflammation. *Immunity* **2016**, *45*, 963–973. [CrossRef] [PubMed]

47. Hall, A.O.; Silver, J.S.; Hunter, C.A. The immunobiology of IL-27. *Adv. Immunol* **2012**, *115*, 1–44.

48. Ebner, K.; Singewald, N. Individual differences in stress susceptibility and stress inhibitory mechanisms. *Curr. Opin Behav. Sci.* **2017**, *14*, 54–64. [CrossRef]

49. Engler, H.; Dawils, L.; Hoves, S.; Kurth, S.; Stevenson, J.R.; Schauenstein, K.; Stefanski, V. Effects of social stress on blood leukocyte distribution: The role of alpha- and beta-adrenergic mechanisms. *J. Neuroimmunol.* **2004**, *156*, 153–162. [CrossRef]

50. Stefanski, V.; Solomon, G.F.; Kling, A.S.; Thomas, J.; Plaeger, S. Impact of social confrontation on rat CD4 T cells bearing different CD45R isoforms. *Brain Behav. Immun.* **1996**, *10*, 364–379. [CrossRef]

51. Frick, L.R.; Arcos, M.L.; Rapanelli, M.; Zappia, M.P.; Brocco, M.; Mongini, C.; Genaro, A.M.; Cremaschi, G.A. Chronic restraint stress impairs T-cell immunity and promotes tumor progression in mice. *Stress* **2009**, *12*, 134–143. [CrossRef] [PubMed]

52. Frick, L.R.; Rapanelli, M.; Cremaschi, G.A.; Genaro, A.M. Fluoxetine directly counteracts the adverse effects of chronic stress on T cell immunity by compensatory and specific mechanisms. *Brain Behav. Immun.* **2009**, *23*, 36–40. [CrossRef] [PubMed]

53. Herold, M.J.; McPherson, K.G.; Reichardt, H.M. Glucocorticoids in T cell apoptosis and function. *Cell Mol. Life Sci.* **2006**, *63*, 60–72. [CrossRef] [PubMed]

54. Del Rey, A.; Kabiersch, A.; Petzoldt, S.; Besedovsky, H.O. Sympathetic abnormalities during autoimmune processes: Potential relevance of noradrenaline-induced apoptosis. *Ann. N Y Acad Sci* **2003**, *992*, 158–167. [PubMed]

55. Lee, G.K.; Park, H.J.; Macleod, M.; Chandler, P.; Munn, D.H.; Mellor, A.L. Tryptophan deprivation sensitizes activated T cells to apoptosis prior to cell division. *Immunology* **2002**, *107*, 452–460. [CrossRef] [PubMed]

56. Nakai, A.; Hayano, Y.; Furuta, F.; Noda, M.; Suzuki, K. Control of lymphocyte egress from lymph nodes through beta2-adrenergic receptors. *J. Ex.p Med.* **2014**, *211*, 2583–2598. [CrossRef]

57. Patas, K.; Willing, A.; Demiralay, C.; Engler, J.B.; Lupu, A.; Ramien, C.; Schafer, T.; Gach, C.; Stumm, L.; Chan, K.; et al. T Cell Phenotype and T Cell Receptor Repertoire in Patients with Major Depressive Disorder. *Front. Immunol* **2018**, *9*, 291. [CrossRef]

58. Han, A.; Yeo, H.; Park, M.J.; Kim, S.H.; Choi, H.J.; Hong, C.W.; Kwon, M.S. IL-4/10 prevents stress vulnerability following imipramine discontinuation. *J. Neuroinflammation* **2015**, *12*, 197. [CrossRef]

59. Engler, H.; Bailey, M.T.; Engler, A.; Sheridan, J.F. Effects of repeated social stress on leukocyte distribution in bone marrow, peripheral blood and spleen. *J. Neuroimmunol* **2004**, *148*, 106–115. [CrossRef]

60. Bi, Y.; Liu, G.; Yang, R. Reciprocal modulation between TH17 and other helper T cell lineages. *J. Cell Physiol.* **2011**, *226*, 8–13. [CrossRef]

61. Heink, S.; Yogev, N.; Garbers, C.; Herwerth, M.; Aly, L.; Gasperi, C.; Husterer, V.; Croxford, A.L.; Moller-Hackbarth, K.; Bartsch, H.S.; et al. Trans-presentation of IL-6 by dendritic cells is required for the priming of pathogenic TH17 cells. *Nat. Immunol* **2017**, *18*, 74–85. [CrossRef] [PubMed]

62. Domi, E.; Uhrig, S.; Soverchia, L.; Spanagel, R.; Hansson, A.C.; Barbier, E.; Heilig, M.; Ciccocioppo, R.; Ubaldi, M. Genetic Deletion of Neuronal PPARgamma Enhances the Emotional Response to Acute Stress and Exacerbates Anxiety: An Effect Reversed by Rescue of Amygdala PPARgamma Function. *J. Neurosci.* **2016**, *36*, 12611–12623. [CrossRef] [PubMed]

63. Schmidt, D.; Reber, S.O.; Botteron, C.; Barth, T.; Peterlik, D.; Uschold, N.; Mannel, D.N.; Lechner, A. Chronic psychosocial stress promotes systemic immune activation and the development of inflammatory Th cell responses. *Brain Behav. Immun.* **2010**, *24*, 1097–1104. [CrossRef] [PubMed]

64. Reber, S.O.; Siebler, P.H.; Donner, N.C.; Morton, J.T.; Smith, D.G.; Kopelman, J.M.; Lowe, K.R.; Wheeler, K.J.; Fox, J.H.; Hassell, J.E., Jr.; et al. Immunization with a heat-killed preparation of the environmental bacterium Mycobacterium vaccae promotes stress resilience in mice. *Proc. Natl. Acad. Sci. USA* **2016**, *113*. [CrossRef] [PubMed]

65. Kim, H.R.; Moon, S.; Lee, H.K.; Kang, J.L.; Oh, S.; Seoh, J.Y. Immune dysregulation in chronic stress: A quantitative and functional assessment of regulatory T cells. *Neuroimmunomodulation* **2012**, *19*, 187–194. [CrossRef] [PubMed]

66. Ambree, O.; Klassen, I.; Forster, I.; Arolt, V.; Scheu, S.; Alferink, J. Reduced locomotor activity and exploratory behavior in CC chemokine receptor 4 deficient mice. *Behav Brain Res.* **2016**, *314*, 87–95. [CrossRef]

67. Sakalem, M.E.; Seidenbecher, T.; Zhang, M.; Saffari, R.; Kravchenko, M.; Wordemann, S.; Diederich, K.; Schwamborn, J.C.; Zhang, W.; Ambree, O. Environmental enrichment and physical exercise revert behavioral and electrophysiological impairments caused by reduced adult neurogenesis. *Hippocampus* **2017**, *27*, 36–51. [CrossRef]

68. Buschert, J.; Sakalem, M.E.; Saffari, R.; Hohoff, C.; Rothermundt, M.; Arolt, V.; Zhang, W.; Ambree, O. Prenatal immune activation in mice blocks the effects of environmental enrichment on exploratory behavior and microglia density. *Prog Neuropsychopharmacol Biol Psychiatry* **2016**, *67*, 10–20. [CrossRef]

International Journal of
Molecular Sciences

Article

Effect of Acute Stress on the Expression of BDNF, trkB, and PSA-NCAM in the Hippocampus of the Roman Rats: A Genetic Model of Vulnerability/Resistance to Stress-Induced Depression

Maria Pina Serra [1], Laura Poddighe [1], Marianna Boi [1], Francesco Sanna [2],
Maria Antonietta Piludu [2], Fabrizio Sanna [3], Maria G. Corda [2], Osvaldo Giorgi [2]
and Marina Quartu [1,*]

[1] Department of Biomedical Sciences, Section of Cytomorphology, University of Cagliari,
   Cittadella Universitaria di Monserrato, 09042 Monserrato (CA), Italy; mpserra@unica.it (M.P.S.);
   laura.poddighe@gmail.com (L.P.); marianna.boi@unica.it (M.B.)
[2] Department of Life and Environmental Sciences, Section of Pharmaceutical, Pharmacological and
   Nutraceutical Sciences, University of Cagliari, 09042 Monserrato (CA), Italy; francesco.sanna@unica.it (F.S.);
   maripiludu@tiscali.it (M.A.P.); mgcorda@unica.it (M.G.C.); giorgi@unica.it (O.G.)
[3] Department of Biomedical Sciences, Section of Neurosciences and Clinical Pharmacology,
   University of Cagliari, Cittadella Universitaria di Monserrato, 09042 Monserrato (CA), Italy;
   fabrizio.sanna@unica.it
* Correspondence: quartu@unica.it; Tel.: +39-070-675-4084

Received: 10 October 2018; Accepted: 21 November 2018; Published: 24 November 2018

Abstract: The Roman High-Avoidance (RHA) and the Roman Low-Avoidance (RLA) rats, represent two psychogenetically-selected lines that are, respectively, resistant and prone to displaying depression-like behavior, induced by stressors. In the view of the key role played by the neurotrophic factors and neuronal plasticity, in the pathophysiology of depression, we aimed at assessing the effects of acute stress, i.e., forced swimming (FS), on the expression of brain-derived neurotrophic factor (BDNF), its trkB receptor, and the Polysialilated-Neural Cell Adhesion Molecule (PSA-NCAM), in the dorsal (dHC) and ventral (vHC) hippocampus of the RHA and the RLA rats, by means of western blot and immunohistochemical assays. A 15 min session of FS elicited different changes in the expression of BDNF in the dHC and the vHC. In RLA rats, an increment in the CA2 and CA3 subfields of the dHC, and a decrease in the CA1 and CA3 subfields and the dentate gyrus (DG) of the vHC, was observed. On the other hand, in the RHA rats, no significant changes in the BDNF levels was seen in the dHC and there was a decrease in the CA1, CA3, and DG of the vHC. Line-related changes were also observed in the expression of trkB and PSA-NCAM. The results are consistent with the hypothesis that the differences in the BDNF/trkB signaling and neuroplastic mechanisms are involved in the susceptibility of RLA rats and resistance of RHA rats to stress-induced depression.

Keywords: forced swimming; Roman rat lines; depression; stress; hippocampus; BDNF; trkB; PSA-NCAM; western blot; immunohistochemistry

---

1. Introduction

It is well-established that stressors may elicit different behavioral and neurochemical adaptive responses in each individual [1–5], depending on the genetically-determined pre-existing differences in temperament, cognition, and autonomic physiology [6]. It is, therefore, most likely that the interactions between the genes and the stressors play a crucial role in the individual responsiveness to adverse life-events and the vulnerability to stress-induced depression [7].

Several animal models have been designed to investigate the impact of the interactions between genetic and environmental factors on the neural substrates of depression. One of these models, the outbred Roman High-Avoidance (RHA) and the Roman Low-Avoidance (RLA) rats, were selected for rapid (RHA) vs. extremely poor (RLA) acquisition of active avoidance, in a shuttle-box [8–10]. It has been shown that emotional reactivity is the most prominent behavioral difference between the two lines, with the RLA rats being more fearful/anxious than their RHA counterparts. Thus, during avoidance training, the RLA rats display hypomotility and freezing, whereas the RHA rats exhibit an active coping behavior that leads to the rapid acquisition of the avoidance response [11]. Consistently, the RLA rats are more emotional/fearful than the RHA rats in different anxiety-related tasks and display a passive coping strategy, when exposed to aversive situations [4,12–14].

Moreover, the Roman lines exhibit divergent neuroendocrine responses to stressors, with the RLA rats showing a higher activation of the hypothalamus-pituitary-adrenal (HPA) axis than the RHA rats, as reflected by a larger increase in the corticotropin and corticosterone secretion, following exposure to mild stress [4,15,16]. Notably, a combination of the dexamethasone suppression test (DST), with a corticotropin releasing hormone (CRH) challenge, has shown that the RLA rats are more responsive to a CRH administration than the RHA rats [17].

The behavioral and neuroendocrine responses of the RLA rats, to drug treatments and environmental challenges (particularly, the combined DST/CRH test), resemble some of the key symptoms of depression [18], suggesting that this rat line may be more susceptible to develop depression-like behaviour, in the face of stressors [17]. Conversely, the RHA phenotype displays proactive coping, high impulsivity/sensation seeking, low HPA axis reactivity, and resilience to stress-induced depression [4,19–24]. Accordingly, in the forced swim test (FST), a paradigm used to assess antidepressant activity in rodents [25,26], the RLA rats display a depression-like behavior characterized by long-lasting immobility and very little escape-directed behaviors, whereas the RHA rats predominantly show active behaviors, such as swimming and climbing, but minimal freezing. Notably, the subacute and chronic treatment, with antidepressant drugs, normalizes the depression-like behavior of the RLA rats in the FST, but does not affect the behavior of the hypoemotional RHA rats [23,24]. Hence, the RLA and the RHA rats may be considered as a genetic model to investigate the neural circuits and molecular mechanisms underlying vulnerability and resistance to stress-induced depression, respectively.

Despite significant advances over the last decades, the causes of depression and the molecular basis of treatments are still poorly understood. Various hypotheses have been proposed to account for the overall pathophysiological state or particular symptoms of depression, based on the dysfunction of monoamine neurotransmission [27], the HPA axis [28], or the neuroimmune processes [29]. Another—the neurotrophic hypothesis—posits that depression may be caused by a dysfunction of the mechanisms underlying the plasticity of the neuronal networks [30,31], and that the susceptibility to depression, elicited by stress, results from the abnormal expression of genes that encode the trophic factors in neurons, which are modulated by monoaminergic inputs [32,33]. Another tenet of this hypothesis is that the hippocampal expression of specific growth factors, such as the brain-derived neurotrophic factor (BDNF), is negatively modulated by stressors and positively modulated by chronic antidepressant treatments. BDNF is a member of the neurotrophin family [34] that supports neuronal viability, during development and in adulthood [35], upon a high-affinity binding to the trkB receptor, a member of the trk family of the tyrosine kinase receptors [36,37]. BDNF, trkB mRNA, and protein immunoreactivity have a widespread distribution in rats' [38–41] and humans' [42–49] central nervous system. Under baseline conditions, BDNF and trkB are densely-expressed in the hippocampal formation [38–40,42,48,50,51], wherein they are implicated in depression-related development of maladaptive behavior and plasticity [52–54].

BDNF/trkB signaling promotes monoaminergic and glutamatergic neurotransmission in brain regions involved in the regulation of mood and emotion [53–61]. It is, therefore, not surprising that the expression of BDNF in the hippocampus, is decreased, upon exposure to stressors and increased by treatment with antidepressants [62]. Accordingly, the BDNF Met polymorphism, which results in a loss of function, is associated with a reduced volume of the hippocampus in depressed patients [63]. It is noteworthy, however, that the way in which BDNF is involved in the pathogenesis of depression has not yet been precisely established [63]. Thus, the local infusion of BDNF in the hippocampus mimics the behavioral effects of antidepressants [64], whereas the intra-VTA infusions of BDNF produce a depression-like effect [65].

A large body of experimental evidence indicates that neurons in certain areas of the adult brain can modify their connections through modulation of dendritic arbors and spine/synapse numbers, in response to experience, with several effects on cognition, emotional regulation, self-regulatory behaviors, and neuroendocrine and autonomic functions [66]. Many of these structural changes are mediated by the proteins involved in cell adhesion, such as the neural cell adhesion molecule (NCAM). This protein is able to incorporate long chains of polysialic acid (PSA), which confers upon NCAM its anti-adhesive properties. The presence of PSA on the extracellular domain of the NCAM, has been related to plastic events, such as neuronal migration, neurite extension/retraction, and synaptogenesis, under normal circumstances or after different physiological, behavioral, or pharmacological stimulations. In addition, Polysialilated-Neural Cell Adhesion Molecule (PSA-NCAM) may generate binding sites for soluble extracellular BDNF to concentrate it nearby its receptor, and promote clustering and aggregation of the trkB receptor molecules, thereby, facilitating the BDNF signaling (see Reference [67], for review). Importantly, the expression of PSA-NCAM in the hippocampus, is down-regulated upon contextual fear conditioning [68]. Conversely, chronic antidepressant treatment positively modulates the expression of NCAM in the hippocampus [69,70].

The intrinsic organization of the hippocampus is highly conserved, but its afferent and efferent projections are markedly different along the septo-temporal axis. Functionally, the dorsal hippocampus (dHC) is preferentially involved in the processing of sensory signals into memories; in contrast, the ventral hippocampus (vHC) has distinct afferent/efferent connections, including pathways to the amygdala, which may enhance the emotional salience of memories [71]. In this context, it has been shown that acute stress induces different effects on the protein levels in the dHC versus the vHC [72,73]. Moreover, we have recently reported that, under basal conditions, the densitometric analysis of the immunostained brain slices shows that, in the dHC of the RLA rats, the BDNF-like immunoreactivity (LI) is lower in the Ammon's horn, whereas, trkB-LI is lower in the dentate gyrus (DG), as compared to their RHA counterparts [74]. As for the vHC, the BDNF-LI in naïve animals is lower in the CA3 and DG of RLA versus the RHA rats, while no differences across the lines are observed for trkB [74]. Hence, on the basis of the above findings, the present study was undertaken to investigate the impact of acute forced swimming (FS) on the expression of BDNF, its high affinity trkB receptor, and PSA-NCAM in the dHC and vHC of the RHA and RLA rats.

2. Results

2.1. Behavioral Measurements During Forced Swimming

In line with our previous studies [23,24], the RHA and RLA rats exhibited markedly different behavioral performances, when exposed for the first time to a 15 min session of FS (Figure 1). Thus, the RLA rats displayed significantly longer immobility than the RHA rats ($p < 0.001$), while the RHA rats spent more time climbing ($p < 0.001$) and diving ($p < 0.05$), than their RLA counterparts. No significant differences between the lines were found in the other behavioral parameters.

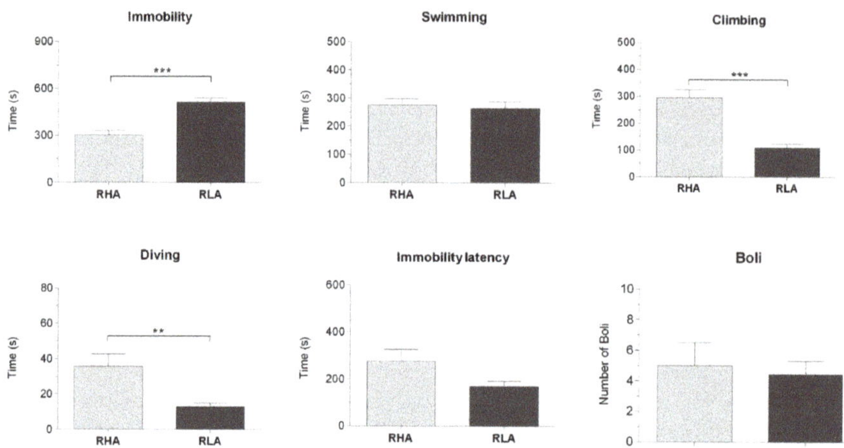

Figure 1. Behavioral performance of the Roman High-Avoidance (RHA) and the Roman Low-Avoidance (RLA) rats, during the 15 min forced swimming session. The columns and bars represent the mean ± SEM (N = 7 rats in each experimental group). ** $p < 0.01$, *** $p < 0.001$ (Student's t test for independent samples).

2.2. Western Blot Assays

2.2.1. The BDNF Protein Levels

The anti-BDNF antibody recognized a protein band with a relative molecular weight (mw) of about 13 kDa (Figures 2A and 3A), in agreement with the reported mw of the monomeric form of the protein [75]. Assessment of the densitometric values of BDNF, in the tissue homogenates from the dHC, by a two-way ANOVA (between groups factors—rat line and treatment [i.e., FS]), revealed a significant interaction line x FS but no significant effects of line and FS (Table 1). Consistent with our previous study [75], the relative levels of the BNDF protein, in the basal conditions, were lower in the RLA vs. the RHA, but did not reach statistical significance (Figure 2B). Additional pair-wise contrasts showed that, after FS, the relative level of the BDNF-LI of the RLA rats was 175% higher than the basal (control) value, whereas, no significant changes were observed in the RHA rats. In the vHC, a two-way ANOVA revealed a significant effect of the FS but not of line or the interaction line × FS (Table 1). Post hoc contrasts showed that after FS, the relative level of the BDNF-LI of the RLA rats was 88% lower than the respective control, while in the RHA rats, the basal level of the BDNF-LI remained unchanged (Figure 3B). Additional post hoc contrasts showed that, after FS, the relative level of BDNF-LI was 79% lower in the RLA than the RHA rats.

Table 1. F values and significance levels of two-way ANOVAs performed on western blot data, shown in Figures 2 and 3.

Brain Area	Marker	Line		FS		Line × FS		
		F	p	F	p	F	p	d.f.
Dorsal Hippocampus	BDNF	0.03923	n.s.	1.428	n.s.	8.524	0.0068	1.28
	trkB	0.3064	n.s.	3.393	n.s.	6.086	0.002	1.28
	PSA-NCAM	24.55	<0.0001	3.505	n.s.	10.23	0.003	1.28
Ventral Hippocampus	BDNF	2.481	n.s.	14.41	0.0007	3.869	n.s.	1.28
	trkB	3.514	n.s.	0.032	n.s.	0.018	n.s.	1.28
	PSA-NCAM	4.815	0.0367	3.010	n.s.	7.334	0.0114	1.28

n.s.—not significant; *d.f.*—degrees of freedom.

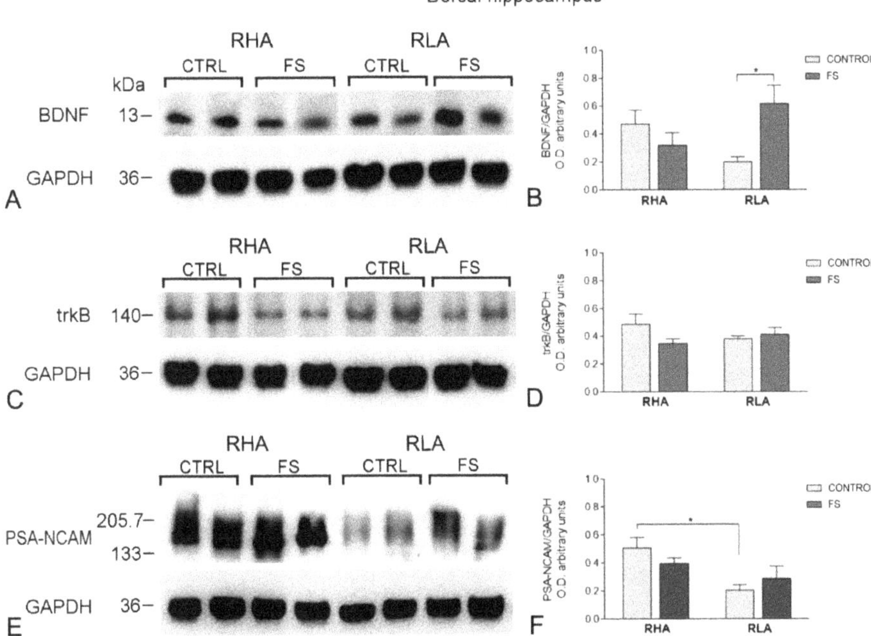

Figure 2. Western blot analysis of the brain-derived neurotrophic factor (BDNF) (**A**,**B**), trkB (**C**,**D**), and the polysialilated-neural cell adhesion molecule (PSA-NCAM) (**E**,**F**) in the dorsal hippocampus of the RHA and the RLA rats, under the baseline conditions (CONTROL), and after forced swimming (FS). (**A**,**C**,**E**): BDNF-immunostained blots (**A**), trkB-immunostained blots (**C**), and PSA-NCAM-immunostained blots (E), showing representative samples from two rats; (**B**,**D**,**F**): Densitometric analysis of the BDNF/GAPDH (B), trkB/GAPDH (D), and the PSA-NCAM/GAPDH band gray optical density (O.D.) ratios (F). Columns and bars denote the mean ± S.E.M. of eight rats in each experimental group. *: $p < 0.05$. (Tukey's post hoc test for, multiple comparisons).

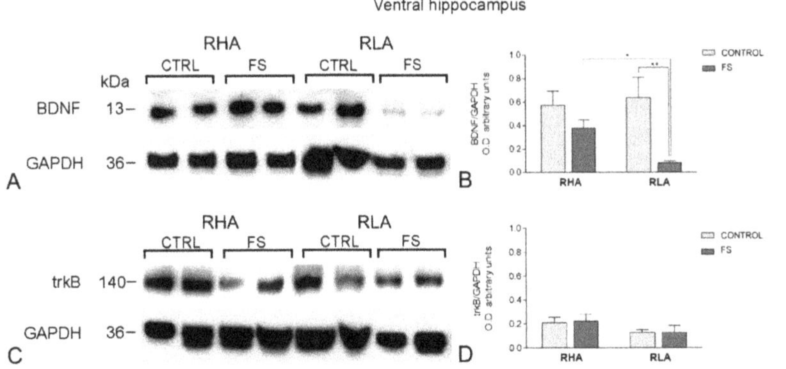

Figure 3. *Cont.*

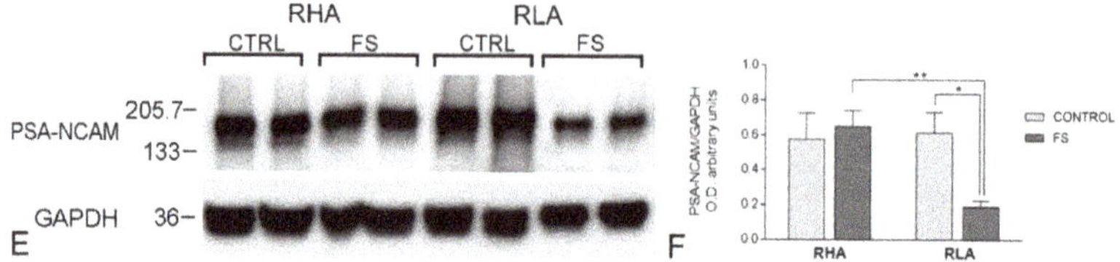

Figure 3. Western blot analysis of the BDNF (**A,B**), trkB (**C,D**), and the PSA-NCAM (**E,F**), in the ventral hippocampus of the RHA and the RLA rats, under the baseline conditions (CONTROL), and after forced swimming (FS). (**A,C,E**): BDNF-immunostained blots (**A**), trkB-immunostained blots (**C**), and PSA-NCAM-immunostained blots (**E**) showing representative samples from two rats; (**B,D,F**): Densitometric analysis of the BDNF/GAPDH (**B**), trkB/GAPDH (**D**), and the PSA-NCAM/GAPDH band gray optical density (O.D.) ratios (**F**). The columns and bars denote the mean ± S.E.M. of eight rats, in each experimental group. *: $p < 0.05$; **: $p < 0.01$ (Tukey's post hoc test or Sidak's correction, for multiple comparisons).

2.2.2. The trkB Protein Levels

The antibody against the full-length form of the trkB labeled a protein band with a relative mw = ~140 kDa (Figures 2C and 3C), consistent with the reported mw of the receptor protein [76]. A two-way ANOVA of the densitometric values of the trkB in the dHC, revealed a significant interaction line x FS, but no significant effects of line or FS (Table 1); moreover, pair-wise post hoc contrasts showed that, after FS, the relative level of the trkB-LI of the RHA rats was tendentially lower than the respective control (−37%) (Figure 2D), but did not reach statistical significance, whereas, no significant changes were observed in the relative basal level of the trkB-LI of RLA rats (Figure 2). On the other hand, in the vHC, the two-way ANOVA showed no significant effects of line, FS, or their interactions (Table 1; Figure 3D).

2.2.3. The PSA-NCAM Protein Levels

The anti-PSA-NCAM antibody labeled a single broad band (Figures 2E and 3E), corresponding to the expected mw [49,77,78]. In the dHC, a two-way ANOVA of the densitometric values of the PSA-NCAM, revealed a significant effect of the line and the interaction of line × FS (Table 1) and post hoc contrasts indicated that the relative levels of the PSA-NCAM protein in the basal conditions were significantly lower (−71%) in the RLA vs. the RHA rats, while the basal level of the PSA-NCAM-LI remained unchanged, upon FS, in both lines (Figure 2F).

In the vHC, the two-way ANOVA (Table 1) revealed an effect of line and a line × FS interaction; pair-wise contrasts indicated that, upon FS, the PSA-NCAM-LI was 69% lower than the respective control value in the RLA rats, whereas, no significant changes in the basal PSA-NCAM-LI were observed, upon FS, in the RHA rats (Figure 3F). Additional post hoc contrasts showed that, after FS, the RLA rats displayed a relative level of PSA-NCAM-LI, which was 71% lower in the RLA vs. the RHA rats.

2.3. Immunohistochemistry

The immunoreactivities for the BDNF (Figures S1 and S2), the trkB (Figures S3 and S4), and the PSA-NCAM (Figures S5 and S6) were unevenly distributed within the hippocampal formation. Immunostained structures were represented by labeled cell bodies, neuronal proximal processes, and nerve fibers distributed within the Ammon's horn and the dentate gyrus. BDNF-LI, trkB-LI, and PSA-NCAM-LI were also observed in the nerve fibers, in the alveus and the fimbria.

2.3.1. BDNF-Like Immunoreactivity

The bulk of BDNF-like immunoreactive nerve fiber networks occurred in the Ammon's horn (Figures S1 and S2) where the immunostained structures had mostly the aspect of filamentous elements running in between the neuronal perikarya of the pyramidal layer and in the molecular layers of CA1 (Figures S1A–D and S2A–D), CA2 (Figure S1E–H), and CA3 sectors (Figures S1I–L and S2E–H). The BDNF-positive cell bodies were also observed in the pyramidal, molecular, and oriens layers (Figure S1I–K). Under the baseline conditions, the BDNF-like immunoreactive elements appeared to be denser in the RHA than in RLA rats. In the dentate gyrus, the BDNF-like immunoreactive nerve fibers appeared as loose meshes and punctate elements distributed in the molecular layer, with increasing density from its outer-third to an inner-narrow band—bordering the granule cell layer (Figure 4M–P)—and in the hilus (Figures S1M–P and S2I–L). The BDNF-labeled neuronal cell bodies were observed within the granular layer, at the interface between the granule cell layer and the polymorphic layer, and in the hilus (Figures S1M–P and S2I–L).

The densitometric analysis in the CA sectors of the hippocampus proper and in the dentate gyrus (Figures 4 and 5) revealed significant differences in the BDNF-LI between the Roman lines, between the baseline and FS conditions, and between the dHC and vHC. Thus, as shown in Table 2, in the dHC, the two-way ANOVA revealed a line effect in the CA2 and CA3 sectors, a FS effect in the CA2 and CA3 sectors, and a significant line × FS interaction in the CA3 sector. Moreover, pair-wise contrasts showed that in the CA3 sector, the basal BDNF-LI was significantly lower (−31%) in the RLA vs. the RHA rats. After FS, the BDNF-LI of the RLA rats was significantly higher (+42% and +43%) than the respective basal values in the CA2 and CA3 sectors, respectively (Figure 4).

In the vHC, two-way ANOVAs revealed a significant FS effect in the CA1, CA3, and DG, as well as a line x FS interaction in the DG (Table 2). In addition, post hoc contrasts showed that upon FS, the BDNF-LI decreased by 61% and 62% in CA1, 66% and 51% in CA3, and 66% and 45% in the DG of the RHA and the RLA rats, respectively (Figure 5).

Table 2. F values and significance levels of two-way ANOVAs performed on data obtained from the densitometric analysis of tissue section distribution of the BDNF- like immunoreactivity (LI), the trkB-LI, and the PSA-NCAM-LI, shown in Figures S1–S6.

Brain Area	Marker	Line		FS		Line × FS		
		F	*p*	F	*p*	F	*p*	*d.f.*
Dorsal Hippocampus								
CA1	BDNF	3.784	n.s.	2.088	n.s.	1.126	n.s.	1.44
	trkB	1.703	n.s.	0.002	n.s.	2.289	n.s.	1.44
	PSA-NCAM	4.391	0.0419	2.9	n.s.	0.4548	n.s.	1.44
CA2	BDNF	4.643	0.0367	8.959	0.0045	0.738	n.s.	1.44
	trkB	0.269	n.s.	1.759	n.s.	1.248	n.s.	1.44
	PSA-NCAM	2.214	n.s.	13.17	0.0007	0.01217	n.s.	1.44
CA3	BDNF	7.824	0.0076	5.807	0.0202	9.001	0.0044	1.44
	trkB	0.067	n.s.	1.304	n.s.	1.025	n.s.	1.44
	PSA-NCAM	23.09	0.0001	6.969	0.0114	1.230	n.s.	1.44
DG	BDNF	0.080	n.s.	1.837	n.s.	1.290	n.s.	1.44
	trkB	34.75	<0.0001	39.80	<0.0001	15.99	0.0002	1.44
	PSA-NCAM	13.83	0.0006	1.471	n.s.	0.1820	n.s.	1.44
Ventral Hippocampus								
CA1	BDNF	0.089	n.s.	47.34	<0.0001	0.562	n.s.	1.44
	trkB	0.0048	n.s.	27.47	<0.0001	0.0027	n.s.	1.44
	PSA-NCAM	2.473	n.s.	9.359	0.0038	0.7689	n.s.	1.44
CA3	BDNF	2.955	n.s.	39.99	<0.0001	3.274	n.s.	1.44
	trkB	0.3182	n.s.	4.087	0.0493	0.3437	n.s.	1.44
	PSA-NCAM	6.608	0.0136	17.32	0.0001	0.3567	n.s.	1.44
DG	BDNF	1.984	n.s.	40.85	<0.0001	4.728	0.0351	1.44
	trkB	0.3049	n.s.	64.58	<0.0001	0.04395	n.s.	1.44
	PSA-NCAM	1.198	n.s.	24.36	<0.0001	1.753	n.s.	1.44

n.s.—not significant; *d.f.*—degrees of freedom.

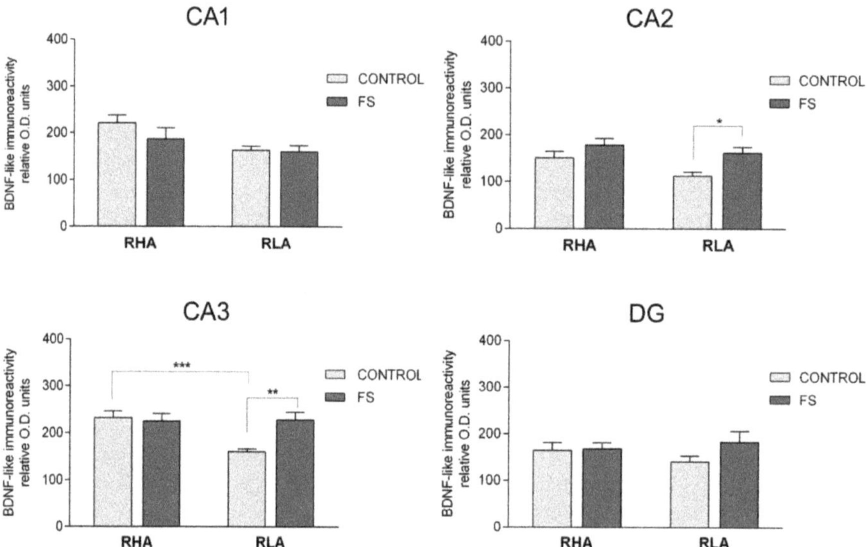

Figure 4. Densitometric analysis of the BDNF-like immunoreactivity in the CA1–CA3 sectors of the Ammon's horn, and in the dentate gyrus (DG) of the dorsal hippocampus in the baseline conditions (CONTROL) and after forced swimming (FS). Columns and bars denote the mean ± S.E.M. of six rats, in each experimental group. Two different sections were analyzed for each rat. *: $p < 0.05$; **: $p < 0.01$; ***: $p < 0.001$ (Tukey's post hoc test or Sidak's correction for multiple comparisons).

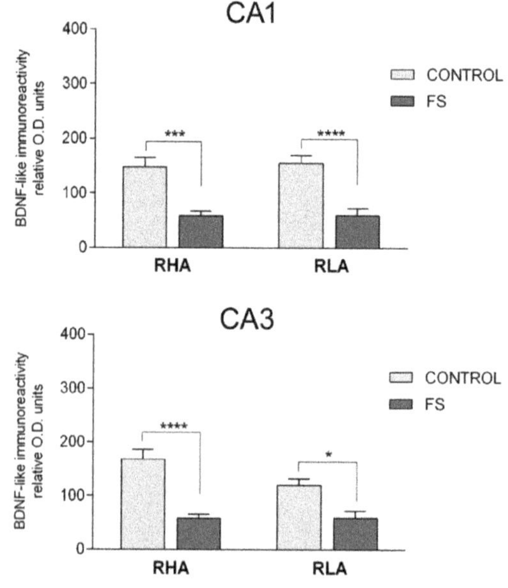

Figure 5. *Cont.*

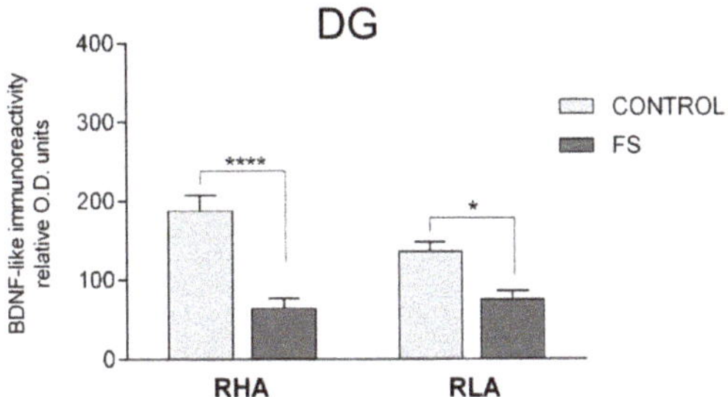

Figure 5. Densitometric analysis of the BDNF-like immunoreactivity in the CA1 and CA3 sectors of the Ammon's horn, and in the dentate gyrus (DG) of the ventral hippocampus in the baseline conditions (CONTROL) and after forced swimming (FS). Columns and bars denote the mean ± S.E.M. of six rats, in each experimental group. Two different sections were analyzed for each rat. *: $p < 0.05$; ***: $p = 0.0001$; ****: $p < 0.0001$ (Tukey's post hoc test or Sidak's correction for multiple comparisons).

2.3.2. The trkB-Like Immunoreactivity

The TrkB-LI also labeled extensive nerve fiber systems, mostly appearing as filaments, short hollow tubules, and coarse punctate elements (Figures S3 and S4). In the Ammon's horn, the occasional trkB-immunolabeled neuronal cell bodies or their proximal processes were observed (Figure S3A–L). In the DG, the trkB-immunolabeling was localized to the filaments and punctate structures distributed in between the granule cell bodies, deep in the molecular layer, and with lesser density, in the hilus (Figure S3M–P). The trkB-positive neuronal perikarya were observed in the hilus (Figures S3M–P and S4I–L). Overall, under the baseline conditions, trkB-LI appeared to be lower in the RLA vs. the RHA rats, in the DG of the dHC (Figure 3M–P), while upon FS, a marked decrease of immunoreactivity vs. the respective controls was observed in the vHC of both Roman lines (Figure S4I–L).

The densitometric analysis in the CA sectors of the hippocampus proper and the DG (Figures 6 and 7) revealed significant differences in the trkB-LI between the Roman lines, between the control and stressed rats, and between the dHC and vHC. In the dHC, the two-way ANOVA revealed the effects of line, FS, and line × FS interaction in the DG (Table 2), and pair-wise contrasts showed that the basal trkB-LI was significantly lower (-35%) in the DG of the RLA vs. the RHA rats. Moreover, after FS, the trkB-LI in the DG of the RHA rats was significantly lower (-36%) than the control value (Figure 6). In the vHC, the two-way ANOVA revealed a significant FS effect in the CA1 sector and the DG (Table 2). Moreover, post hoc contrasts showed that, upon FS, the trkB-LI was significantly lower than the respective control value in the CA1 and the DG, of both lines (RHA -60% and RLA -57% in the CA1; RHA -71% and RLA -74% in the DG) (Figure 7).

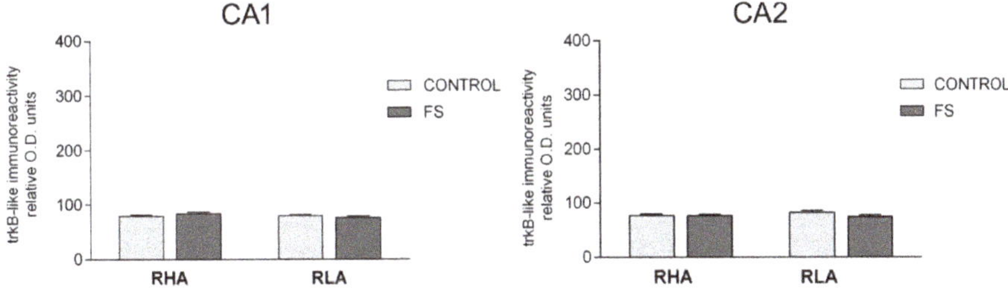

Figure 6. *Cont.*

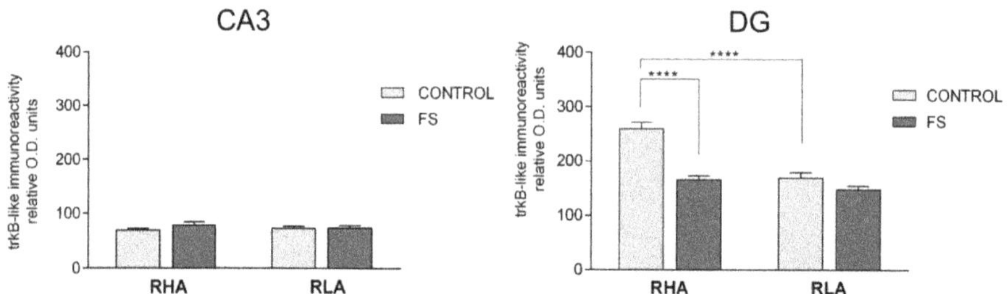

Figure 6. Densitometric analysis of the trkB-like immunoreactivity in the CA1–CA3 sectors of the Ammon's horn and in the dentate gyrus (DG) of the dorsal hippocampus, in the baseline conditions (CONTROL) and after forced swimming (FS). Columns and bars denote the mean ± S.E.M. of six rats, in each experimental group. Two different sections were analyzed for each rat. ****: $p < 0.0001$ (Tukey's post hoc test or Sidak's correction for multiple comparisons).

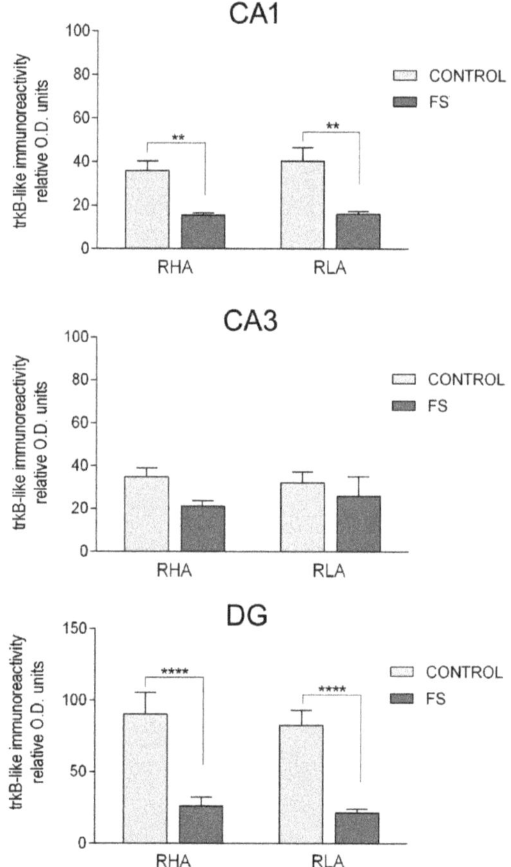

Figure 7. Densitometric analysis of the trkB-like immunoreactivity in the CA1 and CA3 sectors of the Ammon's horn and in the dentate gyrus (DG) of the ventral hippocampus, in the baseline conditions (CONTROL) and after forced swimming (FS). Columns and bars denote the mean ± S.E.M. of six rats, in each experimental group. Two different sections were analyzed for each rat. **: $p < 0.01$; ****: $p < 0.0001$ (Tukey's post hoc test or Sidak's correction for multiple comparisons).

2.3.3. The PSA-NCAM-Like Immunoreactivity

PSA-NCAM-LI was distributed throughout the hippocampus with a prevalent aspect of a diffusely-spread labeling in the neuropil of both the Ammon's horn and the DG, over which stood out a number of neuronal cell bodies and nerve fibers (Figures S5 and S6). In the Ammon's horn, the PSA-NCAM-LI was represented mainly by tiny dust-like elements producing a diffuse labeling, distributed throughout (Figures S5A–L and S6A–D). Rare neuronal perikarya, showing an intense cytoplasmic labeling were observed in the pyramidal (Figure S5B,D,J), the molecular (Figure S5F–J), and the oriens (Figure S5B,F,J) layers. Positive filamentous elements, with a course resembling that of mossy fibers, run parallel to the pyramidal layer of the CA3 sector (Figure S5I–L). In the DG, labeling was mostly localized to the neuronal perikarya (Figures S5M–P and S6E–H) in the infragranular layer of the dHC, where they often showed a peripheral staining, suggestive of membrane-labeling (Figure S5M–P), and to some multipolar neurons in the hilus of both the dHC (Figure S5M–P) and the vHC (Figure S6E–H). A fine network of nerve fibers was observed around the non-immunoreactive granular cells (Figures S5M–P and S6E–H), and a light dust-like immunostaining was present in the neuropil of the outer part of the molecular layer (Figures S5M–P and S6E–H).

Densitometric analysis of the CA sectors of the hippocampus proper and the DG (Figures 8 and 9; Table 2) revealed differences in the PSA-NCAM-LI, between the Roman lines, between the baseline and FS conditions, and between the dHC and vHC. As shown in Table 2, in the dHC, the ANOVAs revealed a line effect in the CA1 and CA3 sectors and in the DG, as well as an effect of FS in the CA2 and CA3 sectors. Moreover, post hoc comparisons indicated that, upon FS, the PSA-NCAM-LI of the RHA rats was 68% higher than the respective controls, in the CA2 and CA3 sectors (Figure 8). Furthermore, in the CA3 sector and the DG, the PSA-NCAM-LI, upon FS, was 52% and 29% lower in the RLA vs. the RHA rats, respectively (Figure 8). In the vHC, two-way ANOVAs revealed an effect of line in the CA3 sector, and of FS in the CA3 sector and the DG (Table 2). Pair-wise contrasts showed that upon FS, the PSA-NCAM-LI was 33% and 50% lower than the corresponding control values in the CA1 and CA3 sector, 30% lower in the DG of the RLA rats, and 39% and 63% lower in the CA3 and DG of the RHA rats, respectively (Figure 9). Moreover, as a general trend, the PSA-NCAM-LI was higher (without reaching statistical significance) in the RLA vs. the RHA rats, in the CA3 sector (+78%) of the control rats, and in the CA1 (+62%), CA3 (+47%), and DG (+81%) of the stressed rats.

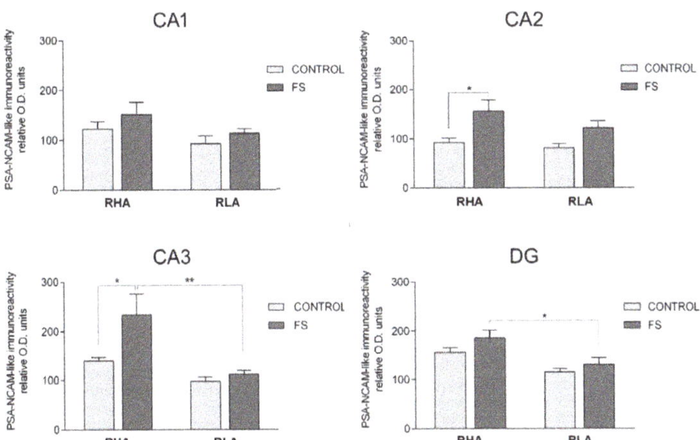

Figure 8. Densitometric analysis of the PSA-NCAM-like immunoreactivity in the CA1–CA3 sectors of the Ammon's horn and in the dentate gyrus (DG) of the dorsal hippocampus, in baseline conditions (CONTROL) and after forced swimming (FS). Columns and bars denote the mean ± S.E.M. of six rats, in each experimental group. Two different sections were analyzed for each rat. *: $p < 0.05$; **: $p < 0.01$ (Tukey's post hoc test or Sidak's correction for multiple comparisons).

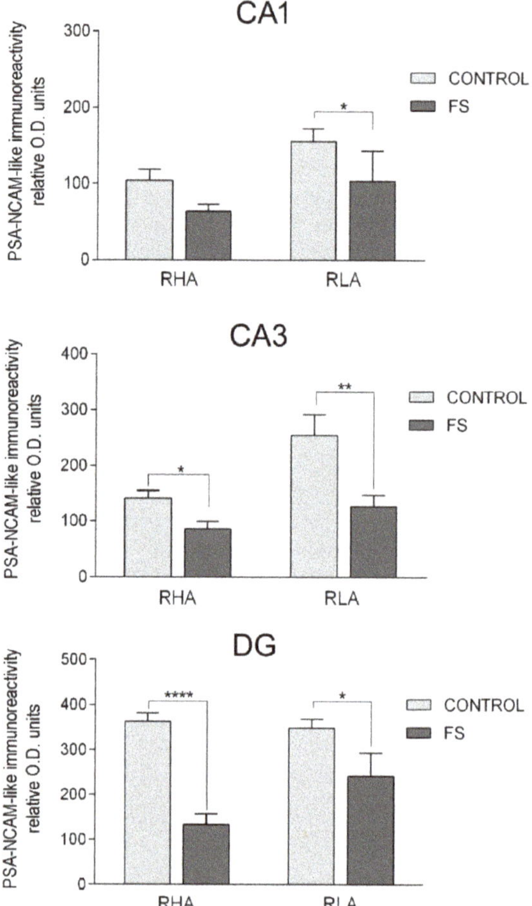

Figure 9. Densitometric analysis of the PSA-NCAM-like immunoreactivity in the CA1 and CA3 sectors of the Ammon's horn and in the dentate gyrus (DG) of the ventral hippocampus, in baseline conditions (CONTROL) and after forced swimming (FS). Columns and bars denote the mean ± S.E.M. of six rats, in each experimental group. Two different sections were analyzed for each rat. *: $p < 0.05$; ** $p < 0.01$; ****: $p < 0.0001$ (Tukey's post hoc test or Sidak's correction for multiple comparisons).

3. Discussion

The RLA and RHA rats represent two divergent phenotypes, respectively, prone and resistant to display depression-like behavior, in the face of aversive environmental conditions like FS-induced acute stress. The present results confirmed and extended that of our previous studies [23,24] showing that, during FS, the RLA rats exhibited longer lasting immobility and fewer climbing and diving counts, when compared to their RHA counterparts, which exhibit a proactive coping style. Since the reactive coping behavior exhibited by the RLA rats, during the FS session, is normalized by chronic treatment with antidepressant drugs [23,24], we decided to characterize the neural substrates and mechanisms, such as the BDNF/trkB signaling, underlying the vulnerability to stress-induced behaviors in the RLA rats, as well as the molecular adaptations mediating the resistance to such changes in the RHA rats. Accordingly, we have recently shown that in the basal conditions the protein levels of the BDNF and trkB, in the hippocampus of the RLA rats, are lower than those of their RHA counterparts [74], consistent with the susceptibility of the RLA line to stress-induced depression.

Extending our previous work [74], we show here that an acute 15 min session of forced swimming elicits line-dependent changes in the expression of the BDNF, the trkB, and the PSA-NCAM, a protein that modulates neuroplastic processes [70,79,80] and influences the BDNF/trkB signaling (see [67] and references, therein) Most importantly, FS induces different modifications in the levels of BDNF, trkB, and PSA-NCAM, in the dHC, versus the vHC.

3.1. Effect of Acute Stress on the BDNF, trkB, and PSA-NCAM Protein Levels in the Dorsal and Ventral Hippocampus

Interestingly, the densitometric analysis of the WBs of tissue homogenates showed that, in the RLA rats, FS elicited opposite changes on the BDNF levels in the hippocampal subregions examined—an increment in the dHC versus a decrease in the vHC. This finding supports the view that stress can modulate hippocampal plasticity, in opposite directions, along the longitudinal septotemporal axis [81]. Accordingly, it has been shown that adult rats that had experienced juvenile stress, expressed an impaired long term potentiation (LTP) in the dHC, while LTP was enhanced in the vHC; in addition, juvenile stress induced a reduction in the sensitivity to the β-adrenergic receptor agonist isoproterenol, in the dHC of adult rats, whereas, in the vHC the sensitivity to isoproterenol was increased [82].

Our results are consistent with ample evidence suggesting that a dynamic and rapid regulation of the BDNF expression and signaling, is implicated in the effect of acute stress on hippocampal structure and connectivity [62,83–87]. In particular, an increase in the BDNF protein levels in the dHC of the RLA rats, upon FS, is in agreement with the increment in the BDNF mRNA or protein levels caused by different types of acute stress [88–90] and may be considered to be an adaptive neuronal plasticity response to FS. On the other hand, in the vHC of the RLA rats, the BDNF protein levels were decreased upon FS, and this effect was associated with a reduction in the levels of PSA-NCAM. Conversely, no significant alterations in the levels of the BDNF, trkB, and PSA-NCAM were observed, upon FS, in the RHA rats, suggesting that acute stress may hinder plastic events, such as neuronal migration, neurite extension/retraction, and synaptogenesis in the vHC of the RLA rats, but not of their stress-resistant RHA counterparts. It is noteworthy, however, that the densitometric analysis of the immunostained slices from the vHC of the RHA rats revealed that, upon FS, the BDNF- and PSA-NCAM-LI decreased in the CA1, CA3, and DG, and the trkB-LI decreased in the CA1 and DG (see below).

3.2. Effect of Acute Stress on the Regional and Subregional Immunohistochemical Distribution of BDNF, trkB, and PSA-NCAM in the Dorsal and Ventral Hippocampus

The densitometric analysis of the immunostained brain slices revealed several differences between the subregions of the dHC and vHC of the RHA versus the RLA rats, either under the basal conditions or upon FS. Thus, in agreement with earlier observations, in the dHC of the control RLA rats, the BDNF-LI was lower in the CA3 sector of the Ammon's horn, while the trkB-LI was lower in the DG, when compared with their RHA counterparts. On the other hand, no significant differences were observed in the levels of the BDNF-LI and trkB-LI between the control RHA and the RLA rats, in the vHC. Notably, the distribution pattern of the PSA-NCAM-LI in the dHC and vHC of the controls, paralleled that of the BDNF-LI and trkB-LI, without marked differences across the two lines, suggesting that a similar capability of undergoing neuroplastic changes in the face of a stressful condition persists until adulthood, in both the RHA and the RLA rats.

In the dHC, FS elicited markedly different changes in the BDNF-LI, the trkB-LI, and the PSA-NCAM-LI, across the lines. In fact, in the RLA rats, the BDNF-LI was significantly higher than the respective control values in the CA2 and CA3 sectors, whereas trkB-LI and PSA-NCAM-LI remained unchanged in all sectors of the Ammon's horn and in the DG, consistent with the results of the densitometric analyses of the WBs. On the other hand, in the RHA rats, no changes were observed in the BDNF-LI but the trkB-LI was decreased in the DG, and the PSA-NCAM-LI was increased in the CA2 and CA3. In contrast, uniform changes were elicited by FS in the different subregions of the vHC.

Thus, upon FS, a decrease in the BDNF-LI, the trkB-LI and the PSA-NCAM-LI, was observed in the Ammon's horn and the DG, in both Roman lines.

The CA3 subfield is dynamically subjected to a process of continuous adjustment of its connectivity, due to persistent invasion of new mossy fiber projections, along with formation of new synaptic contacts, and contextual growth and retraction of the pyramidal dendritic arborizations [80,89]. Given the changes in the BDNF-LI in the CA3 of the dHC and vHC, and its co-occurrence with the PSA-NCAM-LI, it may be hypothesized that these two proteins may interact to modulate FS-induced plastic events in the RLA rats. Accordingly, the modifications in the BDNF-immunostaining, in the pyramidal and molecular layers, upon FS, and the presence of the PSA-NCAM-immunostained fibers in the stratum lucidum, suggests that both proteins may be localized on the continuously growing mossy fibers. Indeed, studies on cell and organotypic cultures demonstrated the existence of synergistic effects between the PSA and the BDNF [67,70,91], supporting the view of a possible interplay between them. Thus, it has been proposed that, due to the chemical characteristics of these molecules, PSA would facilitate the BDNF-trkB interaction by inducing an increase in the soluble BDNF protein concentration in the proximity of the PSA-NCAM-positive cells [67,91], or by acting on the trkB receptor, either increasing its signaling efficacy or mediating cis interactions at the cell surface, thereby, causing a reorganization of the signaling complexes [67,92]. This hypothesis is supported by the finding that the BDNF and the PSA-NCAM are co-localized in the hilar neurons of the human hippocampal formation [51].

We previously proposed that the lower protein levels of the BDNF in the CA3 subfield of the dHC and the vHC of the control RLA rats versus their RHA counterparts, could be due to a slower synthesis rate of neurotrophin, in that region [74]. BDNF is both locally-produced and anterogradely-transported, along the mossy fibers, in the CA3 sector. Hence, according to the neurotrophic hypothesis of depression, the presumably slower production of the BDNF protein in the RLA rats may, in turn, lead to a deficit in the synaptic release and a reduced target-derived support to promote the synaptic contacts with the mossy fibers. In fact, besides the potential autocrine/paracrine effects within the granule cell population, the BDNF potently regulates the synaptic plasticity of mossy fibers. Thus, in mouse hippocampal slices, BDNF stimulates the sprouting of mossy fibers, expands their innervation of the CA3 stratum oriens when infused in vivo, and regulates the extension of their infrapyramidal and suprapyramidal projections [93].

In the present study, we have shown that the 15 min FS seems to interfere with the baseline "neurotrophic" setting, eliciting different changes in the dHC versus the vHC, and between the two lines. In fact, the levels of BDNF-LI increased in the CA2 and CA3 subfields of the dHC of the RLA rats while in the vHC it decreased, markedly, in the CA1 and CA3 subfields of both lines. Consistently, a transient small reduction of BDNF in the CA3 subfield has been observed after acute immobilization stress (2 h) [94]. As for the local production of BDNF, the possibility of concurrent mechanisms of the BDNF-mediated trophic support is corroborated by studies on BDNF targeting on hippocampal CA3 dendrites [95–97]. Thus, the endogenous BDNF secreted during neuronal activity may contribute to local mechanisms of trophic support that direct the accumulation of BDNF/trkB mRNAs, towards specific subcellular compartments of the CA3 principal neurons [95], by means of its anterograde transport, along the mossy fibers [98–100]. Further studies are needed to assess the co-localization of the BDNF and its trkB receptor, and to characterize the hippocampal neural circuitry involved in the trophic activity of the BDNF/trkB signaling in the Roman rats.

The concurrent marked decrease in the expression of BDNF, its receptor trkB, and the PSA-NCAM, in the vHC, suggests that acute stress exerts a strong disruptive effect on the capability of vHC neurons to engage in neuroplastic processes. To our knowledge this is the first study on the basal expression and FS-induced regulation of the PSA-NCAM, in a genetic model of susceptibility/resistance to stress-induced depression. Further experimental evidence, in terms of different stress modalities and duration (i.e., acute or chronic) [68], is warranted, to understand the role of the observed changes in the PSA-NCAM-LI on the hippocampal structural plasticity of Roman rats. Of note in this context, the marked decrease in the PSA-NCAM-LI in the vHC, induced by FS, is consistent with a previous study showing that the expression of PSA-NCAM is significantly reduced in the synaptosomal fraction of the vHC, 30 min after water-maze training [101]. Furthermore, in rats submitted to contextual fear conditioning, the levels of PSA-NCAM in the hippocampus, are significantly reduced 24 h after training [68]. Considered together, these findings suggest that the PSA-NCAM plays a role in the effects of stressors involved in both 'emotional learning' (i.e., contextual fear conditioning) and spatial learning/memory (i.e., water-maze training), corresponding to the different functional involvement of the vHC and dHC, in learning experiences.

3.3. Acute Stress-Induced Expression Changes of the BDNF, the trkB, and the PSA-NCAM in the Dentate Gyrus

BDNF-LI and trkB-LI occur in the DG of both Roman lines, where they label nerve fibers and terminals in the molecular layer and the neuronal cells, at the interface between the granule cell and the polymorphic layers, and in the hilus [74]. Here we show that the DG is also enriched with PSA-NCAM-LI, whose labeling is localized to similar neuronal cells within the subgranular layer and the hilus. The occurrence of the BDNF-containing granule cells is consistent with a local production of BDNF mRNA, in this hippocampal region [42]. However, the small number of labeled granule cells, in our preparations, does not allow us to evaluate possible quantitative differences in their occurrence between the RLA and the RHA rats.

After the acute forced swimming, a significant decrease in trkB-LI, without changes in BDNF-LI and PSA-NCAM-LI, occurred in the DG of the dHC of the RHA rats, whereas, in the RLA rats, no significant changes in the BDNF-LI, the trkB-LI, and the PSA-NCAM-LI were observed in the DG of the dHC. However, in the vHC, FS induced a marked decrease in the BDNF-LI, the trkB-LI, and the PSA-NCAM-LI, in the DG of both Roman lines. It has been proposed that the neural substrate of the therapeutic efficacy of antidepressants consists in the integration of the newly-generated neurons in the subgranular zone of the DG to the neural circuitry of the hippocampus [70,96,102]. In this process, synaptic connections of mature granule cells are established between their dendritic trees, extending in the molecular layer and axon terminals of extrinsic projections from the entorhinal cortex. Furthermore, granule cells send their axonal projections (which terminate in the characteristic giant boutons) to the pyramidal neurons of the CA3 region [80,103–105].

PSA-NCAM is recognized as a key marker of most developmental stages, during the adult hippocampal neurogenesis [106]. Applied anatomical and genetic studies further demonstrate that newly born neurons contribute mainly to the highly plastic infrapyramidal mossy fiber projections, and that both mossy fiber plasticity and adult neurogenesis are co-regulated by extrinsic stimuli, such as environmental enrichment and seizure activity [106,107]. Importantly, the size of the infrapyramidal mossy fiber projection has been shown to correlate positively with performance, in a variety of behavioral tasks; on the other hand, the suprapyramidal mossy fibers represent the majority of the connecting fibers and are relatively more stable than the extremely plastic infrapyramidal fibers [105,107]. Although there is a gap between studies of the DG circuitry and studies of the DG-dependent behavior, it is well known that the subgranular zone receives synaptic monoaminergic input from the ventral tegmental area and the raphe nuclei, cholinergic projections from the septum, γ-aminobutyric acid (GABA)ergic connections from local interneurons, and commissural/associational inputs [104,105]. In this context, we show that, in the DG of the dHC, the BDNF-LI and the trkB-LI nerve fibers are mostly detectable in the inner third of the molecular layer, where the axons originating from the hilar mossy cells play a commissural/associational role [103,105]. Further studies are required to establish whether the commissural fibers and ventro-dorsal projections contribute to the differences between the RHA and the RLA rats, in terms of BDNF/trkB signaling, and the PSA-NCAM-LI in the DG. Interestingly, the DG of the rat hippocampus shows increased BDNF levels, after chronic antidepressant treatment [55,64], while the selective loss of BDNF in the DG, but not in the CA1 sector, is essential for the effectiveness of antidepressants. This appears to be due to the supporting effect of BDNF on the survival and differentiation of newborn granule cells [108]. Furthermore, since both stress and antidepressant treatment have been shown to produce rapid regionally specific patterns of chromatin remodeling in the hippocampus [109,110], alternative mechanisms implying the epigenetic control of the BDNF transcription may also play a role in the mechanism of action of antidepressant drugs [111,112]. Accordingly, we have recently shown that FS induces distinctive patterns of the phosphorylated form of histone H3, in the neurons of the prefrontal cortex and the DG of the dHC, of the RHA versus the RLA rats [113]. Notably, the phosphorylation of histone H3, in turn, activates the expression of immediate early genes, such as *c-fos* and *Egr-1*, thereby, contributing to the consolidation of memories for adaptive responses, such as increased immobility in the FS [114–116].

4. Materials and Methods

4.1. Animals

Outbred male Roman rats (N = 28 for each line) obtained from the colony established in 1998, at the University of Cagliari, Italy [117], were used throughout and were four months old (weight = 400–450 g), at the beginning of the experiments.

Rats were housed in groups of four, per cage, and maintained under temperature- and humidity-controlled environmental conditions (23 °C $\pm$ 1 °C and 60% $\pm$ 10%, respectively) and with a 12 h light–dark cycle (lights on at 8:00 a.m.). Standard laboratory food and water were available ad libitum. To avoid stressful stimuli resulting from manipulation, the maintenance activities in the animal house were carried out by a single attendant and bedding in the home cages was not changed on the two days preceding the test. All procedures were performed according to the guidelines and protocols of the European Union (Directive 2010/63/EU) and the Italian legislation (D.L. 04/04/2014, n. 26), and were approved by the Ethical Committee for Animal Care and Use of the University of Cagliari (authorization No. 684/2015 PR, 15/09/2015). Every possible effort was made to minimize animal pain and discomfort and to reduce the number of experimental subjects.

4.2. FS and Behavioral Measurements

The RHA and RLA rats were randomly assigned to the control or FS groups and were processed in parallel, according to a schedule that was counterbalanced for animal line and treatment. All animals (N = 28 for each line) were naive at the beginning of the experiments and were used only once. Rats in the FS groups (N = 14 for each line) were singly moved from the animal house to a sound-attenuated, dimly-illuminated test room, whereas, the controls (N = 14 for each line) were kept in their home cages in the animal house, until sacrifice. All testing was performed between 10:00 a.m. and 6:00 p.m. and consisted of a 15 min session of acute forced swimming, according to the experimental conditions previously described [113]. Briefly, rats were placed individually in plastic cylinders (58 cm tall × 32 cm diameter) which were filled with water at 24–25 °C to a 40-cm depth, to ensure that they were unable to touch the bottom of the cylinder with their tails or hind paws. At the end of the 15 min swimming sessions, rats were removed from the cylinders, gently dried with paper towels, placed in a heated cage for 15 min, and singly-transferred to an adjacent room where they were sacrificed. The water in the cylinders was replaced before starting the next test session. All the behaviors were quantified by a single well-trained observer who was blind to rat line. A time-sampling technique was used to record the predominant behavior in each 15 s period of the FS session. The following behaviors were recorded: (1) Immobility—floating passively in the water without struggling and doing only those movements necessary to keep the head above water. (2) Immobility latency—the time from the beginning of the test until the first immobility episode. (3) Swimming—showing moderate active motions all around, in the cylinder, more than necessary to simply keep the head above water. (4) Climbing—making active vigorous movements with forepaws in and out of the water, usually directed against the walls. (5) Diving—swimming under water looking for a way out of the cylinder. (6) Boli—number of fecal boli excreted. The behaviors were recorded only in a representative sample of animals that were subsequently used for the Western Blot (four RHA and four RLA) or immunohistochemical assays (three RHA and three RLA).

In the present report it was not examined whether the RLA rats were more susceptible than the unselected reference rats (i.e., the external controls) to exhibiting FS-induced depression-like behavior. We believe that this issue cannot be addressed by simply comparing the behavior during the FS session of the RHA and the RLA rats bred in our laboratory, to that of rats bred in an animal farm, under different pre- and post-natal housing conditions. This is because such environmental differences can significantly alter rodent anxiety- and depression-related behavior [118–120], thereby, affecting the outcome of behavioral experiments. Ideally, unselected Wistar rats, bred together with the RLA and the RHA rats, in the same animal care facility should be used to avoid confounding environmental differences but such a stock of rats was not available in our colony.

4.3. Sampling

Forty five minutes after the end of the FS session, the animals used for the WBs were killed by decapitation whereas the animals used for the immunohistochemical assays were deeply anesthetized with chloral hydrate (500 mg/kg, i.p., 2 mL/kg) and transcardially-perfused with ice-cold PBS (Phosphate Buffered Saline: 137 mM NaCl, 2.7 mM KCl, 10 mM Na_2HPO_4, 2 mM KH_2PO_4, pH 7.3) and 4% paraformaldehyde (PFA).

Immediately after sacrifice, the brains were rapidly removed from the skull and processed for either WB or immunohistochemistry. For WB, the brains were cooled in dry ice for 15 s, placed in a brain matrix, and cut in 2 mm thick coronal slices, using the stereotaxic coordinates of the rat brain atlas of Paxinos and Watson [121] as a reference. The AP coordinates (from Bregma) were approximately −3.30 mm and −6.04 mm, for the dorsal and vHC, respectively. Bilateral punches (diameter 2.5 mm) of the dHC and vHC were taken, as described by Palkovits [122] (Figure 10). For each rat, the tissue punches from both hemispheres were pooled, rapidly frozen at −80 °C, and homogenized in distilled water containing 2% sodium dodecylsulfate (SDS) (300 μL/100 mg of tissue) and a cocktail of protease inhibitors (cOmpleteTM, Mini Protease Inhibitor Cocktail Tablets, Cat# 11697498001, Roche, Basel, Switzerland). For immunohistochemistry, brains were post-fixed by immersion in a freshly prepared 4% phosphate-buffered PFA, pH 7.3, for 4–6 h at 4 °C, and then rinsed until they sank in 0.1 M phosphate buffer (PB), pH 7.3, containing 20% sucrose.

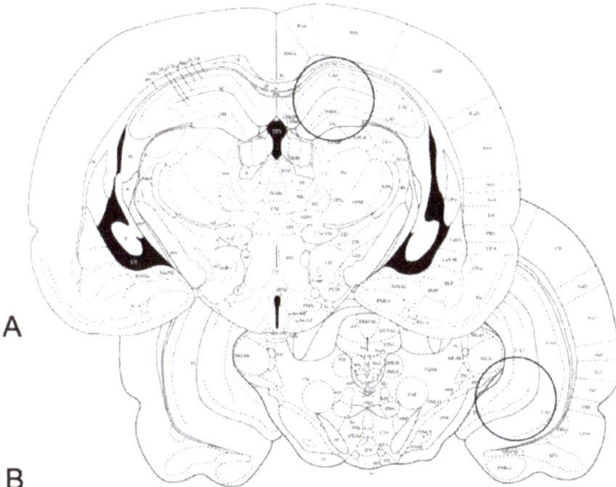

Figure 10. Schematic representation of two rat brain coronal sections (Figures 33 and 44, modified from Paxinos & Watson [121]). The circles denote the areas of the dorsal (**A**) and ventral (**B**) hippocampus, taken for western blot analysis, by means of a 2.5 mm punch. Stereotaxic coordinates (from Bregma): (**A**) −3.30 mm, (**B**) −6.04 mm.

4.4. Western Blot

Total protein concentrations were determined as described by Lowry et al. [123], using bovine serum albumin as a standard. Proteins from each tissue homogenate (40 μg), diluted 3:1 in 4× loading buffer (NuPAGE LDS Sample Buffer 4×, Cat# NP0008, Novex by Life Technologies, Carlsbad, CA, USA), were heated to 95 °C for 7 min, and separated by sodium dodecyl sulfate (SDS)-polyacrilamide gel electrophoresis (SDS-PAGE), using precast polyacrylamide gradient gel (NuPAGE 4–12% Bis-Tris Gel Midi, Cat# NP0321, Novex by Life Technologies, Carlsbad, CA, USA), in the XCell4 Sure LockTM Midi-Cell chamber (Life Technologies). Internal mw standards (Precision Plus Protein Western C Standards, Cat# 161-0376, Bio-Rad, Hercules, CA, USA) were run in parallel. Blots were blocked by immersion in 20 mM Tris base and 137 mM sodium chloride (TBS), containing 0.1% Tween 20 (TBS-T) and 5% milk powder, for 60 min, at room temperature. The primary antibodies were rabbit polyclonal antibodies against BDNF (Cat# N-20 sc-546, RRID:AB_630940, Santa Cruz Biotechnology, Dallas, TX, USA) and trkB (Cat# (794) sc-12, RRID:AB_632557, Santa Cruz Biotechnology), both diluted 1:1000, and a mouse monoclonal antibody against PSA-NCAM (Cat# MAB5324, RRID:AB_95211, Merck Millipore, Darmstadt, Germany), diluted 1:1000, in TBS containing 5% milk powder and 0.02% sodium azide. Incubations with primary antiserum were carried out for two nights at 4 °C. After rinsing in TBS/T, blots were incubated at room temperature, for 60 min, with peroxidase-conjugated goat anti-rabbit serum (Cat#9169, RRID:AB_258434, Sigma Aldrich, St. Louis, MO, USA), diluted 1:10,000, and anti-mouse serum (AP124P, RRID:AB_90456, Millipore, Darmstadt, Germany), diluted 1:5000 in TBS/T. Controls for equal-loading of the wells were obtained by immunostaining the membranes, as above, using a mouse monoclonal antibody against glyceraldehyde-3-phosphate dehydrogenase (GAPDH) (MAB374, RRID:AB_2107445, EMD Millipore, Darmstadt, Germany), diluted 1:1000, as the primary antiserum, and a peroxidase-conjugated goat anti-mouse serum (AP124P, RRID:AB_90456, Millipore, Darmstadt, Germany), diluted 1:5000, as the secondary antiserum. In order to control for non-specific staining, blots were stripped and incubated with the relevant secondary antiserum. In order to check for antibody specificity and cross-reactivity, the anti-BDNF antibody was challenged with 200 ng of rhBDNF (Cat# B-257, Alomone Labs, Jerusalem, Israel) [74], while the anti-PSA-NCAM antibody was preabsorbed with 500 ng of the alfa-2-8-linked sialic polymer colominic acid (Cat# sc-239576, Santa Cruz Biotechnology, USA). After rinsing in TBS/T, protein bands were developed using the Western Lightning Plus ECL (Cat# 103001EA, PerkinElmer, Waltham, MA, USA), according to the protocol provided by the manufacturer, and visualized using the ImageQuant LAS-4000 (GE Healthcare, Little Chalfont, UK). Approximate molecular weight (mw) and relative optical density (O.D.) of the labeled protein bands were evaluated by a blinded examiner. The ratio of the intensity of the BDNF-positive, trkB-positive, and PSA-NCAM-positive bands, to the intensity of the GAPDH-positive ones was used to compare the relative expression levels of these proteins in the RHA and the RLA lines. The O.D. was quantified by the Image Studio Lite Software (RRID:SCR_014211, Li-Cor, http://www.licor.com/bio/products/software/image_studio_lite/).

4.5. Immunohistochemistry

Coronal brain sections from the RLA and RHA rats were examined in pairs, on the same slide. Semiconsecutive cryostat sections (14 μm thick) were collected on chrome alum-gelatin coated slides and processed by the avidin–biotin–peroxidase complex (ABC) immunohistochemical technique. The endogenous peroxidase activity was blocked with 0.1% phenylhydrazine (Cat# 101326606, Sigma Aldrich, St. Louis, MO, USA) in phosphate-buffered saline (PBS), containing 0.2% Triton X-100 (PBS/T), followed by incubation with 20% of normal goat serum (Cat# S-1000, Vector, Burlingame, CA, USA). The same antibodies were used for WB, i.e. rabbit polyclonal antibodies against BDNF and trkB (Santa Cruz Biotechnology, Santa Cruz, CA, USA), both diluted 1:500, and mouse monoclonal antibody against PSA-NCAM (Millipore, Darmstadt, Germany), diluted 1:400, were used as primary antibody. Biotin-conjugated goat anti-rabbit (BA-1000, RRID:AB_2313606, Vector, Burlingame, CA, USA), and anti-mouse sera (BA-9200, RRID:AB_2336171, Vector, Burlingame, CA, USA), diluted 1:400, were used as secondary antiserum. The reaction product was revealed with the ABC (Cat#G011-61, BioSpa Div. Milan, Italy), diluted 1:250, followed by incubation with a solution of 0.1 M PB, pH 7.3, containing 0.05% 3,3′-diaminobenzidine (Sigma Aldrich, St. Louis, MO, USA), 0.04% nickel ammonium sulfate and 0.01% hydrogen peroxide. All antisera and the ABC were diluted in PBS/T. Incubation with primary antibodies was carried out overnight at 4 °C. Incubations with secondary antiserum and ABC lasted 60 min and 40 min, respectively, and were performed at the room temperature. Negative control preparations were obtained by incubating tissue sections in parallel with either PBS/T, alone, or in one of the following four ways—(i) with the relevant primary antiserum pre-absorbed with an excess of the corresponding peptide antigen (Cat# sc-546P and sc-12 P, for the BDNF and the trkB, respectively, Santa Cruz Biotechnology, Santa Cruz, CA, USA); (ii) with colominic acid (as described above); (iii) by omitting the primary antibody; or (iv) by substituting it with normal goat serum. Slides were observed with an Olympus BX61 microscope and digital images were captured with a Leica DFC450C camera.

4.6. Image Densitometry

For the quantitative evaluation of the BDNF, the trkB, and the PSA-NCAM immunohistochemical labeling, representative 10× magnification microscopic fields, were taken from twelve coronal sections of six animals, for each condition. The sections corresponded, approximately, to the AP coordinates used to obtain the tissue samples used for the WB assays, and were blindly analyzed with ImageJ (http://rsb.info.nih.gov/ij/; RRID:SCR_003070) to calculate the density of immunoreactivity per μm^2. Mean gray values from the unstained areas were subtracted from the gray values of the immunostained regions, to exclude the background staining.

4.7. Statistical Analyses

Behavioral measurements were statistically evaluated using the Student's *t* test for independent samples. WB and immunohistochemical data were statistically evaluated, using the two-way ANOVA (see Tables 1 and 2). Before performing both Student's *t* tests and the ANOVAs, data sets of each experimental condition were inspected for normal distribution of data and homogeneity of variances, with the Shapiro-Wilk's test and the Bartlett's test, respectively. Among the behavioral measurements, the diving data set showed statistically significant unequal variances and, therefore, were analysed with the Welch's *t* test. Data sets that did not show homogeneity of variances, were log-transformed and then analysed by two-way ANOVA, as previously described [124]. When two-way ANOVAs revealed statistically significant interactions, the sources of significance were ascertained by pair-wise post hoc contrasts with the HSD Tukey's test. In all the other cases, pair-wise comparisons were performed by using two-tailed *t* tests with Sidak's corrected alpha values. Statistical analyses were all carried out with the PRISM, GraphPad 6 Software (San Diego, CA, USA), with the significance level set at $p < 0.05$.

5. Conclusions

The present results confirmed our previous finding that in the basal conditions, the protein levels of BDNF and trkB, in the hippocampus of RLA rats, are lower than those of their RHA counterparts, consistent with the susceptibility of the RLA line to stress-induced depression. Moreover, exposure to FS elicits line-dependent changes in the expression of BDNF, trkB, and PSA-NCAM, a protein that plays a prominent role in different forms of neural plasticity, and influences the BDNF/trkB signaling.

Alterations in the cognitive processes, as well as psychiatric disorders, can be precipitated when the hippocampal functions are acutely disrupted by acute stressors. The cellular and synaptic modular organization of the hippocampus remains constant, along its septo-temporal axis; however, this brain region can be functionally subdivided into a dorsal (dHC) and a ventral (vHC) compartment, inasmuch as the dHC plays a key role in the spatial navigation and memory storage, whereas, the vHC is involved in the expression of emotion-related behaviors. Notably, stressors can elicit opposite plastic adaptations in the hippocampal compartments; for instance stressors impair the long-term potentiation in the dHC but enhances it in the vHC. Likewise, the densitometric analysis of WBs and immunohistochemical assays showed that, in the RLA rats, FS elicited opposite changes on the BDNF levels in the hippocampal subregions examined—an increment in the CA2 and CA3 subfields of the Ammon's horn of the dHC, versus a decrease in the CA1 and CA3 subfields, and the DG of the vHC. In contrast, in the RHA rats, FS failed to elicit significant changes in the dHC but decreased the BDNF levels in CA1, CA3, and DG of the vHC.

A large body of preclinical and clinical evidence indicates that depression may be caused by alterations of the mechanisms underlying the plasticity of neuronal networks, and that the vulnerability to stress-induced depression is due to the dysfunctional expression of genes encoding neurotrophic factors, like BDNF. In addition, exposure to stress and antidepressant treatments modulates the expression of specific growth factors that support neuronal viability, during development and in adulthood. Several experimental findings support this hypothesis: (i) The expression of neurotrophic factors is decreased in the hippocampus, in animal models of depression and in depressed patients. (ii) In animal models, clinically-effective antidepressant drugs are able to normalize behaviors that are reminiscent of symptoms of depression. (iii) Chronic treatments with antidepressant drugs increase the expression of neurotrophic factors in the hippocampus. In agreement with the above findings and with the results of the immunostaining assays, the RLA rats displayed a depression-like behavior characterized by immobility and freezing, when exposed to aversive conditions, while their RHA counterparts exhibited a proactive coping style, characterized by active behaviors aimed at gaining control over the stressor. Moreover, subacute and chronic treatment with antidepressant drugs normalizes the depression-like behavior of RLA rats, in the FS, but does not affect the behavior of the RHA rats in this task.

A widely-held view, regarding mood disorders, is that the individual responsiveness to environmental challenges plays an important role in the vulnerability to depression. Thus, a reduced capability to cope with an acute and severe stressful event or with mild but persistent aversive challenges, is considered to be critical in determining the vulnerability to stress-induced depression and post-traumatic stress disorder [3,7,125]. Hence, to further characterize the impact of the interaction between the genotype and the environmental factors, on the pathophysiology of depression, it would be interesting to evaluate the behavioral and neurochemical consequences of the long-term exposure of the RHA and the RLA rats, to mild stressors, using the Chronic Mild Stress (CMS) paradigm [126].

Moreover, epidemiologic studies indicate that clinical depression is more frequent in women than men; thus, it has been recently reported that the aggregate prevalence of depression in the community, from thirty countries, between 1994 and 2014, was 14.4% for women and 11.5% for men [127]. Therefore, another important issue to be addressed is the evaluation of the effect of acute and chronic stress on the neurotrophic factor signaling and neural plasticity in the hippocampus of the female RHA and RLA rats.

In closing, the present results add experimental support to the view that RLA and RHA rats provide a useful genetic model to investigate the neural substrates of the susceptibility and resistance to stress-induced depression, respectively, as well as the molecular mechanisms involved in the effects of antidepressant treatments. Furthermore, the results underscore the differences in the impact of stress on the neuroplastic adaptive responses of the dHC and vHC.

Supplementary Materials: Supplementary materials can be found at http://www.mdpi.com/1422-0067/19/12/3745/s1.

Author Contributions: Conceptualization, M.Q., M.G.C. and O.G.; Data curation, M.Q., M.P.S., L.P., M.B. and F.S.; Formal analysis, M.Q., M.P.S. and M.B.; Funding acquisition, M.Q., M.G.C. and O.G.; Investigation, M.Q., M.P.S., L.P., M.B. and F.S.; Methodology, M.P.S., L.P., M.B., F.S., M.A.P. and F.S.; Project administration, M.Q., M.G.C. and O.G.; Supervision, M.Q.; Validation, M.P.S., M.G.C. and O.G.; Visualization, M.Q.; Writing—original draft, M.Q., M.G.C. and O.G.; Writing—review & editing, M.Q., M.P.S., F.S., M.G.C. and O.G.

Funding: This research was funded by a grant to O.G. and M.Q. from ARS (Autonomous Region of Sardinia, L.R. 7/2007, Project Code No. CRP-59842). L.P. was the recipient of a fellowship co-funded by Fondazione Banco di

Sardegna and Progetti di Ricerca di Interesse Dipartimentale (PRID 2015). F.S. is the recipient of a PhD fellowship funded by the Italian Ministry of University and Research (MIUR).

Conflicts of Interest: The authors declare no conflict of interest.

Abbreviations

ABC	Avidin–Biotin–peroxidase Complex
trkB	tyrosine receptor kinase B
GAPDH	Glyceraldehyde-3-Phosphate Dehydrogenase
O.D.	Relative optical density
BDNF	Brain Derived Neurotrophic Factor
HPA	Hypothalamus–Pituitary–Adrenal
DG	Dentate Gyrus
PBS	Phosphate Buffered Saline
RHA	Roman High-Avoidance
RLA	Roman Low Avoidance
SDS-PAGE	Sodium dodecyl sulphate-polyacrylamide gel electrophoresis
TBS-T	Tris base, Sodium chloride, Tween 2
VTA	Ventral Tegmental Area
WB	Western Blot
FS	Forced Swimming
dHC	dorsal Hippocampus
vHC	ventral Hippocampus
PFA	Paraformaldehyde

References

1. Aan het Rot, M.; Mathew, S.J.; Charney, D.S. Neurobiological mechanisms in major depressive disorder. *CMAJ* **2009**, *180*, 305–313. [CrossRef] [PubMed]

2. Anisman, H.; Matheson, K. Stress, depression, and anhedonia: Caveats concerning animal models. *Neurosci. Biobehav. Rev.* **2005**, *29*, 525–546. [CrossRef] [PubMed]

3. Hoge, C.W.; Clark, J.C.; Castro, C.A. Commentary: Women in combat and the risk of post-traumatic stress disorder and depression. *Int. J. Epidemiol.* **2007**, *36*, 327–329. [CrossRef] [PubMed]

4. Steimer, T.; Driscoll, P. Divergent stress responses and coping styles in psychogenetically selected Roman high-(RHA) and low-(RLA) avoidance rats: Behavioural, neuroendocrine and developmental aspects. *Stress* **2003**, *6*, 87–100. [CrossRef] [PubMed]

5. Úbeda-Contreras, J.; Marín-Blasco, I.; Nadal, R.; Armario, A. Brain c-fos expression patterns induced by emotional stressors differing in nature and intensity. *Brain Struct. Funct.* **2018**, *223*, 2213–2227. [CrossRef] [PubMed]

6. Caspi, A.; Moffitt, T.E. Gene-environment interactions in psychiatry: Joining forces with neuroscience. *Nat. Rev. Neurosci.* **2006**, *7*, 583–590. [CrossRef] [PubMed]

7. Charney, D.S.; Manji, H.K. Life stress, genes, and depression: Multiple pathways lead to increased risk and new opportunities for intervention. *Sci. STKE* **2004**. [CrossRef] [PubMed]

8. Broadhurst, P.L.; Bignami, G. Correlative effects of psychogenetic selection: A study of the Roman high and low avoidance strains of rats. *Behav. Res. Ther.* **1965**, *3*, 273–280. [CrossRef]

9. Driscoll, P.; Bättig, K. Behavioral, emotional and neurochemical profiles of rats selected for extreme differences in active, two-way avoidance performance. In *Genetics of the Brain*; Lieblich, I., Ed.; Elsevier Biomedical Press: Amsterdam, NL, USA, 1982; pp. 95–123, ISBN 10:0444804382.

10. Giorgi, O.; Piras, G.; Corda, M.G. The psychogenetically selected Roman high-and low-avoidance rat lines: A model to study the individual vulnerability to drug addiction. *Neurosci. Biobehav. Rev.* **2007**, *31*, 148–163. [CrossRef] [PubMed]

11. Escorihuela, R.M.; Tobeña, A.; Driscoll, P.; Fernández-Teruel, A. Effects of training, early handling, and perinatal flumazenil on shuttle box acquisition in Roman low-avoidance rats: Toward overcoming a genetic deficit. *Neurosci. Biobehav. Rev.* **1995**, *19*, 353–367. [CrossRef]

12. Escorihuela, R.M.; Fernández-Teruel, A.; Gil, L.; Aguilar, R.; Tobeña, A.; Driscoll, P. Inbred Roman high-and low-avoidance rats: Differences in anxiety, novelty-seeking, and shuttlebox behaviours. *Physiol. Behav.* **1999**, *67*, 19–26. [CrossRef]

13. Fernández-Teruel, A.; Escorihuela, R.M.; Gray, J.A.; Aguilar, R.; Gil, L.; Giménez-Llort, L.; Tobeña, A.; Bhomra, A.; Nicod, A.; Mott, R.; et al. A quantitative trait locus influencing anxiety in the laboratory rat. *Genome Res.* **2002**, *12*, 618–626. [CrossRef] [PubMed]

14. Ferré, P.; Fernández-Teruel, A.; Escorihuela, R.M.; Driscoll, P.; Corda, M.G.; Giorgi, O.; Tobeña, A. Behavior of the Roman/Verh high- and low-avoidance rat lines in anxiety tests: Relationship with defecation and self-grooming. *Physiol. Behav.* **1995**, *58*, 1209–1213. [CrossRef]

15. Carrasco, J.; Márquez, C.; Nadal, R.; Tobeña, A.; Fernández-Teruel, A.; Armario, A. Characterization of central and peripheral components of the hypothalamus-pituitary-adrenal axis in the inbred Roman rat strains. *Psychoneuroendocrinology* **2008**, *33*, 437–445. [CrossRef] [PubMed]

16. Gentsch, C.; Lichtsteiner, M.; Feer, H. Genetic and environmental influences on reactive and spontaneous locomotor activities in rats. *Experientia* **1991**, *47*, 998–1008. [CrossRef] [PubMed]

17. Steimer, T.; Python, A.; Schulz, P.E.; Aubry, J.M. Plasma corticosterone, dexamethasone (DEX) suppression and DEX/CRH tests in a rat model of genetic vulnerability to depression. *Psychoneuroendocrinology* **2007**, *32*, 575–579. [CrossRef] [PubMed]

18. Holsboer, F. Stress, hypercortisolism and corticosteroid receptors in depression: Implications for therapy. *J. Affect Disord.* **2001**, *62*, 77–91. [CrossRef]

19. Fernández-Teruel, A.; Driscoll, P.; Gil, L.; Aguilar, R.; Tobeña, A.; Escorihuela, R.M. Enduring effects of environmental enrichment on novelty seeking, saccharin and ethanol intake in two rat lines (RHA/Verh and RLA/Verh) differing in incentive seeking behavior. *Pharmacol. Biochem. Behav.* **2002**, *73*, 225–231. [CrossRef]

20. Giorgi, O.; Lecca, D.; Piras, G.; Driscoll, P.; Corda, M.G. Dissociation between mesocortical dopamine release and fear related behaviors in two psychogenetically selected lines of rats that differ in coping strategies to aversive conditions. *Eur. J. Neurosci.* **2003**, *17*, 2716–2726. [CrossRef] [PubMed]

21. Moreno, M.; Cardona, D.; Gómez, M.J.; Sánchez-Santed, F.; Tobeña, A.; Fernández-Teruel, A.; Campa, L.; Suñol, C.; Escarabajal, M.D.; Torres, C.; et al. Impulsivity characterization in the Roman high-and low-avoidance rat strains: Behavioral and neurochemical differences. *Neuropsychopharmacology* **2010**, *35*, 1198–1208. [CrossRef] [PubMed]

22. Siegel, J. Augmenting and reducing of visual evoked potentials in high- and low-sensation seeking humans, cats, and rats. *Behav. Genet.* **1997**, *27*, 557–563. [CrossRef] [PubMed]

23. Piras, G.; Giorgi, O.; Corda, M.G. Effects of antidepressants on the performance in the forced swim test of two psychogenetically selected lines of rats that differ in coping strategies to aversive conditions. *Psychopharmacology* **2010**, *211*, 403–414. [CrossRef] [PubMed]

24. Piras, G.; Piludu, M.A.; Giorgi, O.; Corda, M.G. Effects of chronic antidepressant treatments in a putative genetic model of vulnerability (Roman low-avoidance rats) and resistance (Roman high-avoidance rats) to stress-induced depression. *Psychopharmacology* **2014**, *231*, 43–53. [CrossRef] [PubMed]

25. Porsolt, R.D.; Le Pichon, M.; Jalfre, M. Depression: A new animal model sensitive to antidepressant treatments. *Nature* **1977**, *266*, 730–732. [CrossRef] [PubMed]

26. Detke, M.J.; Johnson, J.; Lucki, I. Acute and chronic antidepressant drug treatment in the rat forced swimming test model of depression. *Exp. Clin. Psychopharmacol.* **1997**, *5*, 107–112. [CrossRef] [PubMed]

27. Prins, J.; Olivier, B.; Korte, S.M. Triple reuptake inhibitors for treating subtypes of major depressive disorder: The monoamine hypothesis revisited. *Exp. Op. Investig. Drugs* **2011**, *20*, 1107–1130. [CrossRef] [PubMed]

28. Anacker, C.; Zunszain, P.A.; Carvalho, L.A.; Pariante, C.M. The glucocorticoid receptor: Pivot of depression and of antidepressant treatment? *Psychoneuroendocrinology* **2011**, *36*, 415–425. [CrossRef] [PubMed]

29. Miller, A.H. Depression and immunity: A role for T cells? *Brain Behav. Immun.* **2010**, *24*, 1–8. [CrossRef] [PubMed]

30. Castrén, E. Is mood chemistry? *Nat. Rev. Neurosci.* **2005**, *6*, 241–246. [CrossRef] [PubMed]

31. Duman, R.S.; Malberg, J.; Thome, J. Neural plasticity to stress and antidepressant treatment. *Biol. Psychiatry* **1999**, *46*, 1181–1191. [CrossRef]

32. Stahl, S.M. Blue genes and the mechanism of action of antidepressants. *J. Clin. Psychiatry* **2000**, *61*, 164–165. [CrossRef] [PubMed]

33. Nestler, E.J.; Gould, E.; Manji, H.; Buncan, M.; Duman, R.S.; Greshenfeld, H.K.; Hen, R.; Koester, S.; Lederhendler, I.; Meaney, M.; et al. Preclinical models: Status of basic research in depression. *Biol. Psychiatry* **2002**, *52*, 503–528. [CrossRef]

34. Barde, Y.A.; Edgar, D.; Thoenen, H. Purification of a new neurotrophic factor from mammalian brain. *EMBO J.* **1982**, *1*, 549–553. [CrossRef] [PubMed]

35. Ibáñez, C.F. Neurotrophic factors: From structure-function studies to designing effective therapeutics. *Trends Biotechnol.* **1995**, *13*, 217–227. [CrossRef]

36. Binder, D.K.; Scharfman, H.E. Brain-derived neurotrophic factor. *Growth Factors* **2004**, *22*, 123–131. [CrossRef] [PubMed]

37. Reichardt, L.F. Neurotrophin-regulated signalling pathways. *Philos. Trans. R. Soc. Lond. B Biol. Sci.* **2006**, *361*, 1545–1564. [CrossRef] [PubMed]

38. Conner, J.M.; Lauterborn, J.C.; Yan, Q.; Gall, C.M.; Varon, S. Distribution of brain-derived neurotrophic factor (BDNF) protein and mRNA in the normal adult rat CNS: Evidence for anterograde axonal transport. *J. Neurosci.* **1997**, *17*, 2295–2313. [CrossRef] [PubMed]

39. Yan, Q.; Radeke, M.J.; Matheson, C.R.; Talvenheimo, J.; Welcher, A.A.; Feinstein, S.C. Immunocytochemical localization of trkB in the central nervous system of the rat. *J. Comp. Neurol.* **1997**, *378*, 135–157. [CrossRef]

40. Yan, Q.; Rosenfeld, R.D.; Matheson, C.R.; Hawkins, N.; Lopez, O.T.; Bennett, L.; Welcher, A.A. Expression of brain-derived neurotrophic factor protein in the adult rat central nervous system. *Neuroscience* **1997**, *78*, 431–448. [CrossRef]

41. Drake, C.T.; Milner, T.A.; Patterson, S.L. Ultrastructural localization of full-length trkB immunoreactivity in rat hippocampus suggests multiple roles in modulating activity-dependent synaptic plasticity. *J. Neurosci.* **1999**, *19*, 8009–8026. [CrossRef] [PubMed]

42. Phillips, H.S.; Hains, J.M.; Armanini, M.; Laramee, G.R.; Johnson, S.A.; Winslow, J.W. BDNF mRNA is decreased in the hippocampus of individuals with Alzheimer's disease. *Neuron* **1991**, *7*, 695–702. [CrossRef]

43. Connor, B.; Young, D.; Yan, Q.; Faull, R.L.; Synek, B.; Dragunow, M. Brain-derived neurotrophic factor is reduced in Alzheimer's disease. *Mol. Brain Res.* **1997**, *49*, 71–81. [CrossRef]

44. Benisty, S.; Boissiere, F.; Faucheux, B.; Agid, Y.; Hirsch, E.C. trkB messenger RNA expression in normal human brain and in the substantia nigra of parkinsonian patients: An in situ hybridization study. *Neuroscience* **1997**, *86*, 813–826. [CrossRef]

45. Quartu, M.; Setzu, M.D.; Del Fiacco, M. trk-like immunoreactivity in the human trigeminal ganglion and subnucleus caudalis. *Neuroreport* **1996**, *10*, 1013–1019. [CrossRef]

46. Quartu, M.; Lai, M.L.; Del Fiacco, M. Neurotrophin-like immunoreactivity in the human hippocampal formation. *Brain Res. Bull.* **1999**, *48*, 375–382. [CrossRef]

47. Quartu, M.; Serra, M.P.; Manca, A.; Follesa, P.; Ambu, R.; Del Fiacco, M. High affinity neurotrophin receptors in the human pre-term newborn, infant, and adult cerebellum. *Int. J. Dev. Neurosci.* **2003**, *21*, 309–320. [CrossRef]

48. Quartu, M.; Serra, M.P.; Manca, A.; Follesa, P.; Lai, M.L.; Del Fiacco, M. Neurotrophin-like immunoreactivity in the human pre-term newborn, infant, and adult cerebellum. *Int. J. Dev. Neurosci.* **2003**, *21*, 23–33. [CrossRef]

49. Quartu, M.; Serra, M.P.; Boi, M.; Melis, T.; Ambu, R.; Del Fiacco, M. Brain-derived neurotrophic factor (BDNF) and polysialylated-neural cell adhesion molecule (PSA-NCAM): Codistribution in the human brainstem precerebellar nuclei from prenatal to adult age. *Brain Res.* **2010**, *1363*, 49–62. [CrossRef] [PubMed]

50. Webster, M.J.; Herman, M.M.; Kleinman, J.E.; Weickert, C.S. BDNF and trkB mRNA expression in the hippocampus and temporal cortex during the human lifespan. *Gene Expr. Patterns* **2006**, *6*, 941–951. [CrossRef] [PubMed]

51. Quartu, M.; Serra, M.P.; Boi, M.; Demontis, R.; Melis, T.; Poddighe, L.; Del Fiacco, M. Polysialylated-neural cell adhesion molecule (PSA-NCAM) in the human nervous system at prenatal, postnatal and adult ages. In *Recent Advances in Adhesion Research, Series Human Anatomy and Physiology-Materials Science and Technologies*; McFarland, A., Akins, M., Eds.; Nova Science Publishers: Hauppauge, NY, USA, 2013; pp. 27–54, ISBN 978-1-62417-447-6.

52. Vaidya, V.A.; Siuciak, J.A.; Du, F.; Duman, R.S. Hippocampal mossy fiber sprouting induced by chronic electroconvulsive seizures. *Neuroscience* **1999**, *89*, 157–166. [CrossRef]

53. Malberg, J.E.; Eisch, A.J.; Nestler, E.J.; Duman, R.S. Chronic antidepressant treatment increases neurogenesis in adult rat hippocampus. *J. Neurosci.* **2000**, *20*, 9104–9110. [CrossRef] [PubMed]

54. Dias, B.G.; Banerjee, S.B.; Duman, R.S.; Vaidya, V.A. Differential regulation of brain derived neurotrophic factor transcripts by antidepressant treatments in the adult rat brain. *Neuropharmacology* **2003**, *45*, 553–563. [CrossRef]

55. Nibuya, M.; Morinobu, S.; Duman, R.S. Regulation of BDNF and trkB mRNA in rat brain by chronic electroconvulsive seizure and antidepressant drug treatments. *J. Neurosci.* **1995**, *15*, 7539–7547. [CrossRef] [PubMed]

56. Duman, R.S.; Heninger, G.R.; Nestler, E.J. A molecular and cellular theory of depression. *Arch. Gen. Psychiatry* **1997**, *54*, 597–606. [CrossRef] [PubMed]

57. Vaidya, V.A.; Duman, R.S. Depression-emerging insights from neurobiology. *Br. Med. Bull.* **2001**, *57*, 61–79. [CrossRef] [PubMed]

58. Duman, R.S. Role of neurotrophic factors in the etiology and treatment of mood disorders. *Neuromol. Med.* **2004**, *5*, 11–25. [CrossRef]

59. Nestler, E.J.; Carlezon, W.A., Jr. The mesolimbic dopamine reward circuit in depression. *Biol. Psychiatry* **2006**, *59*, 1151–1159. [CrossRef] [PubMed]

60. Burke, T.F.; Advani, T.; Adachi, M.; Monteggia, L.M.; Hensler, J.G. Sensitivity of hippocampal 5-HT1A receptors to mild stress in BDNF-deficient mice. *Int. J. Neuropsychopharmacol.* **2013**, *16*, 631–645. [CrossRef] [PubMed]

61. Duman, R.S.; Voleti, B. Signaling pathways underlying the pathophysiology and treatment of depression: Novel mechanisms for rapid-acting agents. *Trends Neurosci.* **2012**, *35*, 47–56. [CrossRef] [PubMed]

62. Autry, A.E.; Monteggia, L.M. Brain-derived neurotrophic factor and neuropsychiatric disorders. *Pharmacol. Rev.* **2012**, *64*, 239–245. [CrossRef] [PubMed]

63. Castrén, E.; Rantamaki, T. The role of BDNF and its receptors in depression and antidepressant drug action: Reactivation of developmental plasticity. *Dev. Neurobiol.* **2010**, *70*, 289–297. [CrossRef] [PubMed]

64. Shirayama, Y.; Chen, A.C.; Nakagawa, S.; Russell, D.S.; Duman, R.S. Brain-derived neurotrophic factor produces antidepressant effects in behavioral models of depression. *J. Neurosci.* **2002**, *22*, 223251–223261. [CrossRef]

65. Eisch, A.J.; Bolaños, C.A.; de Wit, J.; Simonak, R.D.; Pudiak, C.M.; Barrot, M.; Verhaagen, J.; Nestler, E.J. Brain-derived neurotrophic factor in the ventral midbrain-nucleus accumbens pathway: A role in depression. *Biol. Psychiatry* **2003**, *54*, 994–1005. [CrossRef] [PubMed]

66. McEwen, B.S.; Morrison, J.H. The brain on stress: Vulnerability and plasticity of the prefrontal cortex over the life course. *Neuron* **2013**, *79*, 16–29. [CrossRef] [PubMed]

67. Bonfanti, L. PSA-NCAM in mammalian structural plasticity and neurogenesis. *Prog. Neurobiol.* **2006**, *80*, 129–164. [CrossRef] [PubMed]

68. Merino, J.J.; Cordero, M.I.; Sandi, C. Regulation of hippocampal cell adhesion molecules NCAM and L1 by contextual fear conditioning is dependent upon time and stressor intensity. *Eur. J. Neurosci.* **2000**, *12*, 3283–3290. [CrossRef] [PubMed]

69. Bessa, J.M.; Ferreira, D.; Melo, I.; Marques, F.; Cerqueira, J.J.; Palha, J.A.; Almeida, O.F.; Sousa, N. The mood-improving actions of antidepressants do not depend on neurogenesis but are associated with neuronal remodeling. *Mol. Psychiatry* **2009**, *14*, 764–773. [CrossRef] [PubMed]

70. Wainwright, S.R.; Galea, L.A. The neural plasticity theory of depression: Assessing the roles of adult neurogenesis and PSA-NCAM within the hippocampus. *Neural. Plast.* **2013**, *2013*, 805497. [CrossRef] [PubMed]

71. Tanti, A.; Belzung, C. Neurogenesis along the septo-temporal axis of the hippocampus: Are depression and the action of antidepressants region-specific? *Neuroscience* **2013**, *252*, 234–252. [CrossRef] [PubMed]

72. Maras, P.M.; Molet, J.; Chen, Y.; Rice, C.; Ji, S.G.; Solodkin, A.; Baram, T.Z. Preferential loss of dorsal-hippocampus synapses underlies memory impairments provoked by short, multimodal stress. *Mol. Psychiatry* **2014**, *19*, 811–822. [CrossRef] [PubMed]

73. Floriou-Servou, A.; von Ziegler, L.; Stalder, L.; Sturman, O.; Privitera, M.; Rassi, A.; Cremonesi, A.; Thöny, B.; Bohacek, J. Distinct Proteomic, Transcriptomic, and Epigenetic Stress Responses in Dorsal and Ventral Hippocampus. *Biol. Psychiatry* **2018**, *84*, 531–541. [CrossRef] [PubMed]

74. Serra, M.P.; Poddighe, L.; Boi, M.; Sanna, F.; Piludu, M.A.; Corda, M.G.; Giorgi, O.; Quartu, M. Expression of BDNF and trkB in the hippocampus of a rat genetic model of vulnerability (Roman low-avoidance) and resistance (Roman high-avoidance) to stress-induced depression. *Brain Behav.* **2017**, *7*, e00861. [CrossRef] [PubMed]

75. Rosenthal, A.; Goeddel, D.V.; Nguyen, T.; Martin, E.; Burton, L.E.; Shih, A.; Laramee, G.R.; Wurm, F.; Mason, A.; Nikolics, K.; et al. Primary structure and biological activity of human brain-derived neurotrophic factor. *Endocrinology* **1991**, *129*, 1289–1294. [CrossRef] [PubMed]

76. Klein, R.; Parada, L.F.; Coulier, F.; Barbacid, M. trkB, a novel tyrosine protein kinase receptor expressed during mouse neural development. *EMBO J.* **1989**, *8*, 3701–3709. [CrossRef] [PubMed]

77. Dubois, C.; Figarella-Branger, D.; Pastoret, C.; Rampini, C.; Karpati, G.; Rougon, G. Expression of NCAM and its polysialylated isoforms during mdx mouse muscle regeneration and in vitro myogenesis. *Neuromuscul. Disord.* **1994**, *4*, 171–182. [CrossRef]

78. Quartu, M.; Serra, M.P.; Boi, M.; Ibba, V.; Melis, T.; Del Fiacco, M. Polysialylated-neural cell adhesion molecule (PSA-NCAM) in the human trigeminal ganglion and brainstem at prenatal and adult ages. *BMC Neurosci.* **2008**, *9*, 108. [CrossRef] [PubMed]

79. Seki, T.; Arai, Y. Distribution and possible roles of the highly polysialylated neural cell adhesion molecule (NCAM-H) in the developing and adult central nervous system. *Neurosci. Res.* **1993**, *17*, 265–290. [CrossRef]

80. Seki, T.; Rutishauser, U. Removal of polysialic acid-neural cell adhesion molecule induces aberrant mossy fiber innervation and ectopic synaptogenesis in the hippocampus. *J. Neurosci.* **1998**, *18*, 3757–3766. [CrossRef] [PubMed]

81. Maggio, N.; Segal, M. Striking variations in corticosteroid modulation of long-term potentiation along the septotemporal axis of the hippocampus. *J. Neurosci.* **2007**, *27*, 5757–5765. [CrossRef] [PubMed]

82. Grigoryan, G.; Ardi, Z.; Albrecht, A.; Richter-Levin, G.; Segal, M. Juvenile stress alters LTP in ventral hippocampal slices: Involvement of noradrenergic mechanisms. *Behav. Brain Res.* **2015**, *278*, 559–562. [CrossRef] [PubMed]

83. Murakami, S.; Imbe, H.; Morikawa, Y.; Kubo, C.; Senba, E. Chronic stress, as well as acute stress, reduces BDNF mRNA expression in the rat hippocampus but less robustly. *Neurosci. Res.* **2005**, *53*, 129–139. [CrossRef] [PubMed]

84. Nair, A.; Vadodaria, K.C.; Banerjee, S.B.; Benekareddy, M.; Dias, B.G.; Duman, R.S.; Vaidya, V.A. Stressor-specific regulation of distinct brain-derived neurotrophic factor transcripts and cyclic AMP response element-binding protein expression in the postnatal and adult rat hippocampus. *Neuropsychopharmacology* **2007**, *32*, 1504–1519. [CrossRef] [PubMed]

85. Pittenger, C.; Duman, R.S. Stress, depression, and neuroplasticity: A convergence of mechanisms. *Neuropsychopharmacology* **2008**, *33*, 88–109. [CrossRef] [PubMed]

86. Kozisek, M.E.; Middlemas, D.; Bylund, D.B. The differential regulation of BDNF and TrkB levels in juvenile rats after four days of escitalopram and desipramine treatment. *Neuropharmacology* **2008**, *54*, 251–257. [CrossRef] [PubMed]

87. Molteni, R.; Calabrese, F.; Cattaneo, A.; Mancini, M.; Gennarelli, M.; Racagni, G.; Riva, M.A. Acute stress responsiveness of the neurotrophin BDNF in the rat hippocampus is modulated by chronic treatment with the antidepressant duloxetine. *Neuropsychopharmacology* **2009**, *34*, 1523–1532. [CrossRef] [PubMed]

88. Shi, S.; Shao, S.; Yuan, B.; Pan, F.; Li, Z. Acute Stress and Chronic Stress Change Brain-Derived Neurotrophic Factor (BDNF) and Tyrosine Kinase-Coupled Receptor (TrkB) Expression in Both Young and Aged Rat Hippocampus. *Yonsei Med. J.* **2010**, *51*, 661–671. [CrossRef] [PubMed]

89. Uysal, N.; Sisman, A.R.; Dayi, A.; Ozbal, S.; Cetin, F.; Baykara, B.; Aksu, I.; Tas, A.; Cavus, S.A.; Gonenc-Arda, S.; Buyuk, E. Acute footshock-stress increases spatial learning-memory and correlates to increased hippocampal BDNF and VEGF and cell numbers in adolescent male and female rats. *Neurosci. Lett.* **2012**, *514*, 141–146. [CrossRef] [PubMed]

90. Marmigère, F.; Givalois, L.; Rage, F.; Arancibia, S.; Tapia-Arancibia, L. Rapid induction of BDNF expression in the hippocampus during immobilization stress challenge in adult rats. *Hippocampus* **2003**, *13*, 646–655. [CrossRef] [PubMed]

91. Muller, D.; Djebbara-Hannas, Z.; Jourdain, P.; Vutskits, L.; Durbec, P.; Rougon, G.; Kiss, J.Z.; Durbec, P.; Rougon, G.; Kiss, J.Z. Brain-derived neurotrophic factor restores long-term potentiation in polysialic acid-neural cell adhesion molecule-deficient hippocampus. *Proc. Natl. Acad. Sci. USA* **2000**, *97*, 4315–4320. [CrossRef] [PubMed]

92. Durbec, P.; Cremer, H. Revisiting the function of PSA-NCAM in the nervous system. *Mol. Neurobiol.* **2001**, *24*, 53–64. [CrossRef] [PubMed]

93. Isgor, C.; Pare, C.; McDole, B.; Coombs, P.; Guthrie, K. Expansion of the dentate mossy fiber-CA3 projection in the brain-derived neurotrophic factor-enriched mouse hippocampus. *Neuroscience* **2015**, *288*, 10–23. [CrossRef] [PubMed]

94. Lakshminarasimhan, H.; Chattarji, S. Stress Leads to Contrasting Effects on the Levels of Brain Derived Neurotrophic Factor in the Hippocampus and Amygdala. *PLoS ONE* **2012**, *7*, e30481. [CrossRef] [PubMed]

95. Righi, M.; Tongiorgi, E.; Cattaneo, A. Brain-Derived Neurotrophic Factor (BDNF) Induces Dendritic Targeting of BDNF and Tyrosine KinaseB mRNAs in Hippocampal Neurons through a Phosphatidylinositol-3 Kinase-Dependent Pathway. *J. Neurosci.* **2000**, *20*, 3165–3174. [CrossRef] [PubMed]

96. Baj, G.; D'Alessandro, V.; Musazzi, L.; Mallei, A.; Sartori, C.R.; Sciancalepore, M.; Tardito, D.; Langone, F.; Popoli, M.; Tongiorgi, E. Physical exercise and antidepressants enhance BDNF targeting in hippocampal CA3 dendrites: Further evidence of a spatial code for BDNF splice variants. *Neuropsychopharmacology* **2012**, *37*, 1600–1611. [CrossRef] [PubMed]

97. Baj, G.; Pinhero, V.; Vaghi, V.; Tongiorgi, E. Signaling pathways controlling activity-dependent local translation of BDNF and their localization in dendritic arbors. *J. Cell. Sci.* **2016**, *129*, 2852–2864. [CrossRef] [PubMed]

98. Altar, C.A.; Cai, N.; Bliven, T.; Juhasz, M.; Conner, J.M.; Acheson, A.L.; Lindsay, R.M.; Wiegand, S.J. Anterograde transport of brain-derived neurotrophic factor and its role in the brain. *Nature* **1997**, *389*, 856–860. [CrossRef] [PubMed]

99. Altar, C.A.; DiStefano, P.S. Neurotrophin trafficking by anterograde transport. *Trends Neurosci.* **1998**, *21*, 431–437. [CrossRef]

100. Dieni, S.; Matsumoto, T.; Dekkers, M.; Rauskolb, S.; Ionescu, M.S.; Deogracias, R.; Gundelfinger, E.D.; Kojima, M.; Nestel, S.; Frotscher, M.; et al. BDNF and its pro-peptide are stored in presynaptic dense core vesicles in brain neurons. *J. Cell. Biol.* **2012**, *196*, 775–788. [CrossRef] [PubMed]

101. Conboy, L.; Tanrikut, C.; Zoladz, P.R.; Campbell, A.M.; Park, C.R.; Gabriel, C.; Mocaer, E.; Sandi, C.; Diamond, D.M. The antidepressant agomelatine blocks the adverse effects of stress on memory and enables spatial learning to rapidly increase neural cell adhesion molecule (NCAM) expression in the hippocampus of rats. *Int. J. Neuropsychopharmacol.* **2009**, *12*, 329–341. [CrossRef] [PubMed]

102. Duman, R.S.; Monteggia, L.M. A neurotrophic model for stress-related mood disorders. *Biol. Psychiatry* **2006**, *59*, 1116–1127. [CrossRef] [PubMed]

103. Amaral, D.G.; Scharfman, H.E.; Lavenex, P. The dentate gyrus: Fundamental neuroanatomical organization (dentate gyrus for dummies). *Prog. Brain Res.* **2007**, *163*, 3–22. [CrossRef]

104. Kempermann, G.; Song, H.; Gage, F.H. Neurogenesis in the adult hippocampus. *Cold Spring Harb. Perspect. Biol.* **2015**, *7*, a018812. [CrossRef] [PubMed]

105. Scharfman, H.E. The enigmatic mossy cell of the dentate gyrus. *Nat. Rev. Neurosci.* **2016**, *17*, 562–575. [CrossRef] [PubMed]

106. Römer, B.; Krebs, J.; Overall, R.W.; Fabel, K.; Babu, H.; Overstreet-Wadiche, L.; Brandt, M.D.; Williams, R.W.; Jessberger, S.; Kempermann, G. Adult hippocampal neurogenesis and plasticity in the infrapyramidal bundle of the mossy fiber projection: I. Co-regulation by activity. *Front. Neurosci.* **2011**, *5*, 107. [CrossRef] [PubMed]

107. Krebs, J.; Römer, B.; Overall, R.W.; Fabel, K.; Babu, H.; Brandt, M.D.; Williams, R.W.; Jessberger, S.; Kempermann, G. Adult Hippocampal Neurogenesis and Plasticity in the Infrapyramidal Bundle of the Mossy Fiber Projection: II. Genetic Covariation and Identification of Nos1 as Linking Candidate Gene. *Front. Neurosci.* **2011**, *5*, 106. [CrossRef] [PubMed]

108. Adachi, M.; Barrot, M.; Autry, A.E.; Theobald, D.; Monteggia, L.M. Selective loss of brain-derived neurotrophic factor in the dentate gyrus attenuates antidepressant efficacy. *Biol. Psychiatry* **2008**, *63*, 642–649. [CrossRef] [PubMed]

109. Hunter, R.G.; McCarthy, K.J.; Milne, T.A.; Pfaff, D.W.; McEwen, B.S. Regulation of hippocampal H3 histone methylation by acute and chronic stress. *Proc. Natl. Acad. Sci. USA* **2009**, *106*, 20912–20917. [CrossRef] [PubMed]

110. Hunter, R.G.; Murakami, G.; Dewell, S.; Seligsohn, M.; Baker, M.E.R.; Datson, N.A.; Pfaff, D.W.; McEwen, B.S. Stress induced hippocampal transposon silencing. *Proc. Natl. Acad. Sci. USA* **2012**, *109*, 17657–17662. [CrossRef] [PubMed]

111. Ninan, P.T.; Shelton, R.C.; Bao, W.; Guico-Pabia, C.J. BDNF, interleukin-6, and salivary cortisol levels in depressed patients treated with desvenlafaxine. *Prog. Neuropsychopharmacol. Biol. Psychiatry* **2014**, *48*, 86–91. [CrossRef] [PubMed]

112. Hing, B.; Sathyaputri, L.; Potash, J.B. A comprehensive review of genetic and epigenetic mechanisms that regulate BDNF expression and function with relevance to major depressive disorder. *Am. J. Med. Genet. B Neuropsychiatr Genet.* **2018**, *177*, 143–167. [CrossRef] [PubMed]

113. Morello, N.; Plicato, O.; Piludu, M.A.; Poddighe, L.; Serra, M.P.; Quartu, M.; Corda, M.G.; Giorgi, O.; Giustetto, M. Effects of forced swimming stress on ERK and histone H3 phosphorylation in limbic areas of Roman high- and low-avoidance rats. *PLoS ONE* **2017**, *12*, e0170093. [CrossRef] [PubMed]

114. Chandramohan, Y.; Droste, S.K.; Arthur, J.S.; Reul, J.M. The forced swimming-induced behavioural immobility response involves histone H3 phospho-acetylation and c-Fos induction in dentate gyrus granule neurons via activation of the N-methyl-D-aspartate/extracellular signal-regulated kinase/mitogen- and stress-activated kinase signalling pathway. *Eur. J. Neurosci.* **2008**, *27*, 2701–2713. [CrossRef] [PubMed]

115. Bilang-Bleuel, A.; Ulbricht, S.; Chandramohan, Y.; De Carli, S.; Droste, S.K.; Reul, J.M. Psychological stress increases histone H3 phosphorylation in adult dentate gyrus granule neurons: Involvement in a glucocorticoid receptor-dependent behavioural response. *Eur. J. Neurosci.* **2005**, *22*, 1691–1700. [CrossRef] [PubMed]

116. Saunderson, E.A.; Spiers, H.; Mifsud, K.R.; Gutierrez-Mecinas, M.; Trollope, A.F.; Shaikh, A.; Mill, J.; Reul, J.M.H.M. Stress-induced gene expression and behavior are controlled by DNA methylation and methyl donor availability in the dentate gyrus. *Proc. Natl. Acad. Sci. USA* **2016**, *113*, 4830–4835. [CrossRef] [PubMed]

117. Giorgi, O.; Piras, G.; Lecca, D.; Corda, M.G. Differential activation of dopamine release in the nucleus accumbens core and shell after acute or repeated amphetamine injections: A comparative study in the Roman high-and low-avoidance rat lines. *Neuroscience* **2005**, *135*, 987–998. [CrossRef] [PubMed]

118. Crabbe, J.C.; Wahlsten, D.; Dudek, B.C. Genetics of mouse behavior: Interactions with laboratory environment. *Science* **1999**, *284*, 1670–1672. [CrossRef] [PubMed]

119. Driscoll, P.; Fernández-Teruel, A.; Corda, M.G.; Giorgi, O.; Steimer, T. Some guidelines for defining personality differences in rats. In *Handbook of Behavior Genetic*; Kim, Y.-K., Ed.; Springer: New York, NY, USA, 2009; pp. 281–300.

120. Escorihuela, R.M.; Tobeña, A.; Fernández-Teruel, A. Environmental enrichment and postnatal handling prevent spatial learning deficits in aged hypoemotional (Roman high-avoidance) and hyperemotional (Roman low-avoidance) rats. *Learn. Memory* **1995**, *2*, 40–48. [CrossRef]

121. Paxinos, G.; Watson, C. *The Rat Brain in Stereotaxic Coordinates*, 4th ed.; Academic Press: San Diego, CA, USA, 1998; p. 237, ISBN 10: 0125476191.

122. Palkovits, M. Punch sampling biopsy technique. *Methods Enzymol.* **1983**, *103*, 368–376. [CrossRef] [PubMed]

123. Lowry, O.H.; Rosebrough, N.J.; Farr, A.L.; Randall, R.J. Protein measurements with the Folin phenol reagent. *J. Biol. Chem.* **1951**, *193*, 265–275. [PubMed]

124. Sanna, F.; Poddighe, L.; Serra, M.P.; Boi, M.; Bratzu, J.; Sanna, F.; Corda, M.G.; Giorgi, O.; Melis, M.R.; Argiolas, A.; et al. c-Fos, ΔFosB, BDNF, trkB and Arc expression in the limbic system of male Roman High and Low Avoidance rats that show differences in sexual behaviour: Effect of sexual activity. *Neuroscience* **2018**. [CrossRef] [PubMed]

125. Kendler, K.S.; Karkowski, L.M.; Prescott, C.A. Causal relationship between stressful life events and the onset of major depression. *Am. J. Psychiatry* **1999**, *156*, 837–841. [CrossRef] [PubMed]

126. Vitale, G.; Ruggieri, V.; Filaferro, M.; Frigeri, C.; Alboni, S.; Tascedda, F.; Brunello, N.; Guerrini, R.; Cifani, C.; Massi, M. Chronic treatment with the selective NOP receptor antagonist [Nphe 1, Arg 14, Lys 15]N/OFQ-NH 2 (UFP-101) reverses the behavioural and biochemical effects of unpredictable chronic mild stress in rats. *Psychopharmacology* **2009**, *207*, 173–189. [CrossRef] [PubMed]

127. Lim, G.Y.; Tam, W.W.; Lu, Y.; Ho, C.S.; Zhang, M.W.; Ho, R.C. Prevalence of Depression in the Community from 30 Countries between 1994 and 2014. *Sci. Rep.* **2018**, *12*, 2861. [CrossRef] [PubMed]

Review

DNA Methylation as a Biomarker of Treatment Response Variability in Serious Mental Illnesses: A Systematic Review Focused on Bipolar Disorder, Schizophrenia, and Major Depressive Disorder

Charanraj Goud Alladi [1,2], Bruno Etain [2,3,4], Frank Bellivier [2,3,4] and Cynthia Marie-Claire [2,*]

[1] Department of Pharmacology, Jawaharlal Institute of Postgraduate Medical Education and Research, Puducherry 605006, India; charan.raj002@gmail.com
[2] INSERM U1144 Variabilité de réponse aux psychotropes, Université Paris Descartes, Sorbonne Paris Cité, 75006 Paris, France; Bruno.Etain@inserm.fr (B.E.); frank.bellivier@inserm.fr (F.B.)
[3] AP-HP, GH Saint-Louis—Lariboisière—F. Widal, Pôle de Psychiatrie et de Médecine Addictologique, 75475 Paris CEDEX 10, France
[4] Fondation Fondamental, 94000 Créteil, France
* Correspondence: cynthia.marie-claire@parisdescartes.fr; Tel.: +33-1-53-73-99-90

Received: 5 September 2018; Accepted: 29 September 2018; Published: 4 October 2018

Abstract: So far, genetic studies of treatment response in schizophrenia, bipolar disorder, and major depression have returned results with limited clinical utility. A gene × environment interplay has been proposed as a factor influencing not only pathophysiology but also the treatment response. Therefore, epigenetics has emerged as a major field of research to study the treatment of these three disorders. Among the epigenetic marks that can modify gene expression, DNA methylation is the best studied. We performed a systematic search (PubMed) following Preferred Reporting Items for Systematic Reviews and Meta-Analyses (PRISMA guidelines for preclinical and clinical studies focused on genome-wide and gene-specific DNA methylation in the context of schizophrenia, bipolar disorders, and major depressive disorder. Out of the 112 studies initially identified, we selected 31 studies among them, with an emphasis on responses to the gold standard treatments in each disorder. Modulations of DNA methylation levels at specific CpG sites have been documented for all classes of treatments (antipsychotics, mood stabilizers, and antidepressants). The heterogeneity of the models and methodologies used complicate the interpretation of results. Although few studies in each disorder have assessed the potential of DNA methylation as biomarkers of treatment response, data support this hypothesis for antipsychotics, mood stabilizers and antidepressants.

Keywords: schizophrenia; bipolar disorder; major depressive disorder; DNA methylation; response variability

---

1. Introduction

Schizophrenia, bipolar disorder, and major depressive disorder are severe mental illnesses (SMI) defined by classifications such as Diagnostic and Statistical Manual of Mental Disorders (DSM-5) and International Classification of Diseases (ICD10), of sufficient duration to meet diagnostic criteria and resulting in significant functional impairment [1]. Schizophrenia (SCZ), bipolar disorder (BD), and major depressive disorder (MDD) are associated with poor health outcomes, global disability, and public health burden [2,3]. While SCZ affects 1% of the population [4], 2% of the population is affected by BD worldwide [5]. Depression is the second leading cause of global health problem with 6.6–21% of the population in high-income countries and 6.5–18.4% of the population in low–middle-income countries prone to succumb this disorder [6,7]. Briefly, SCZ, BD, and MDD are characterized by acute episodes including psychotic, depressive, and/or manic features that are superimposed, in some patients, with a chronic course that mainly includes negative symptoms in SCZ and potential cognitive decline in these three SMI. Even though these disorders are defined as separated entities, their clinical boundaries remain unclear as they share common symptomatic and functional impairments. For example, abnormalities in neurocognitive functioning are associated with BD and SCZ [8]. Psychotic symptoms, such as delusions or hallucinations predominantly associated with SCZ, are also frequently experienced by patients during severe mood episodes belonging to BD or MDD [9]. These observed phenotypic similarities might be underpinned by shared brain alterations such as impairments in white and grey matter in BD and MDD [10]. SMI reduce patients' life expectancy by 10 to 20 years [11], emphasizing the need for a better stratification of patients and identification of patients that are more likely to respond to a given treatment.

1.1. Pharmacogenetics Studies

The etiology of SCZ, BD and MDD disorders remains largely unknown, and numerous gene-specific or genome wide association studies have been conducted to investigate the possible genetic inheritance to these disorders [12,13]. However, conflicting results have been obtained, and many of the results could not be replicated, due to the complexity of the disease phenotypes, the implication of environmental factors in interaction with vulnerability genes and the polygenic nature of these disorders with more than 600 genes possibly involved in SCZ [12,13], more than 800 genes in BD [14,15] and more than 102 genes identified in MDD [16]. Some of these genes have been reported to be shared among these SMI in a meta-analysis [17]. Hence, pharmacogenetic studies have been used to identify genetic variants that are associated with treatment response in BD, SCZ and MDD. Several genetic variants have been identified, either in candidate gene approaches or in genome-wide association analyses (for review see [18–20]). This is consistent with the complexity of the response phenotypes and the polygenic nature of these SMI and raises the question of the transferability to clinical practice. Furthermore, transcriptomic modulations observed retrospectively or after the initiation of treatment in responders and non- (or poor) responders have been identified in BD, SCZ and MDD in animal and in vitro models, but also in patients [21–25]. Emerging evidences suggest that epigenetic marks could represent relevant biomarkers to be used as predictors of treatment response in several pathologies [26–28].

1.2. Epigenetic Mechanisms

Epigenetics is the study of mechanisms that control gene expression, irrespective of changes in the DNA sequence. Epigenetic mechanisms such as DNA methylation and histone modifications regulate chromatin and thereby the access of transcription factors. Another mechanism involves miRNAs (microRNAs) that target via seed sequences of single, or multiple, mRNAs leading to their down-regulation. All these mechanisms contribute to gene expression alterations, and they have been extensively reviewed [29,30]. These mechanisms can regulate the expression of several genes at the same time and for some of them (DNA methylation and miRNA) undergo transgenerational transmission [31]. Therefore, they may account for the polygenic nature, the partial heredity, and the differential gene expression modulations observed. DNA methylation is one of the most studied epigenetic mechanisms; it consists of the addition of methyl groups from *S*-adenylyl methionine (SAM) to the fifth carbon position of the cytosine residue in DNA by a family of DNA methyl transferase (DNMT) enzymes [29]. DNA methylation regulates gene expression through gene activation or gene silencing in the nervous system [32] and as a result, it may play a role not only in neurogenesis, but also in brain maturation and functioning [33–35].

1.3. Variability of Response to Treatments

In SMI, the treatment of acute phases and the long-term prevention of recurrence strategies mainly rely on four major classes of medications: atypical antipsychotics, mood stabilizers (anticonvulsants and lithium carbonate), and antidepressants. Long-term therapy with atypical antipsychotic drugs is the first line choice of treatment in SCZ. However, due to better tolerabilities and safety profiles, they may also be used in BD and MDD, for bipolar and unipolar depressions, but also for manic episodes [36,37]. Antipsychotics act mainly by binding to dopamine (DRD2) and serotonin (HTR2) receptors [38]. However, 30–40% of the patients with SCZ do not respond to the treatment in the acute phase and experience severe adverse effects [39]. Along with mood stabilizers such as lithium, antiepileptic drugs, valproate, lamotrigine, and carbamazepine, but also in combination with atypical antipsychotics quetiapine, olanzapine, and aripiprazole are used in BD management [40]. However, only 30% of the patients respond well in the chronic phase, and 70% show various degrees of treatment response in BD [41–44]. Tricyclic antidepressants, selective serotonin reuptake inhibitors (SSRIs), serotonin, and norepinephrine reuptake inhibitors (SNRIs) and mono amino oxidase inhibitors are the most commonly used medication in the treatment of depression. However, nearly 60% of the patients do not respond to the antidepressant therapy, and 30% do not respond at all [45,46].

In these three SMI, the lack of predictive biomarkers for treatment response sometimes results in a substantial proportion of patients experiencing potential adverse effects with ineffective therapies. Increasing data show that these drugs not only modulate DNA methylation in animal and cellular models, but also that the observed modulations can be associated with the response phenotypes in patients. Here, we will review the available data suggesting a role of DNA methylation in response to the treatment in three major psychiatric disorders: MDD, SCZ and BD.

2. DNA Methylation Patterns in SCZ, BD and MDD

Multiple environmental factors have been shown to influence the pathogenesis of psychiatric disorders. Available preclinical and in-vitro studies indicate that altered epigenetic mechanisms such as DNA methylation, histone modifications, and miRNA regulation are associated with altered gene expression in major psychiatric disorders. Indeed, alterations of DNA methylation of genes important for the physiopathological aspects of SCZ, BD, and MDD, such as dopaminergic, serotoninergic, and Brain-Derived Neurotrophic Factor (BDNF) pathways have been reported [47–50]. In line with the shared genetic and environmental risks in SCZ and BD, hypomethylation of *FAM63B*, and an intergenic region on chromosome 16 have been proposed as common epigenetic risk factors in these two pathologies [51,52]. However, it is not possible to discriminate in these studies between the effects of the vulnerability to the disorders, the course of the disease, and those induced by the treatments. We concluded from the overall picture that the candidate gene, and the genome-wide approaches of DNA methylation patterns in SCZ, BD and MDD is of low contribution for the investigation of treatment response variability in these disorders.

Therefore, in this review, we will focus on the reported effects of treatments on DNA methylation and their potential influence on the therapeutic response. Therapeutic strategies in SCZ, MDD and BD are generally based on similar classes of molecules (antidepressant, antipsychotics, and mood stabilizers) that are used in distinct dose and temporal combinations. Indeed, the lack of predictive markers for response lead to lengthy trial and error processes, and delayed optimal care of patients. For example, in SCZ, the current guidelines on the delay before evaluating the non-response to an antipsychotic and the switch for another differ substantially, but they are estimated at around two to three months [53–55]. In MDD, antidepressant acute response can be characterized after a mean delay of eight weeks; however, a recent meta-analysis found that their effects are stable over six months, as compared to the placebo [56]. In the case of BD, the characterization of lithium response can take up to two years to be characterized [57]. Interestingly though, several recent studies reported that these drugs can modulate epigenetic mechanisms at several levels of regulation. For instance, histone deacetylase 1 (*HDAC1*) was shown to be directly inhibited by valproic acid (VPA), while its expression can be downregulated by lithium [58,59]. Both mechanisms result in a decreased HDAC activity in cells. Similarly, the glutamatergic agonist LY379268-induced demethylation effects in *Reln*, *BDNF*, and *Gad67* genes may underlie antipsychotic effects in a mouse model [60]. These preclinical and clinical evidences indicate that pre-existing or psychotropic drug-induced epigenetic mechanisms may play a novel role in therapeutic response [61,62]. This review will focus on the most commonly studied epigenetic modification: DNA methylation, as a biomarker of treatment response in SCZ, BD and MDD.

3. Treatment-Induced DNA Methylation Modifications in Animal Models

Results obtained from several preclinical investigations examining the changes of DNA methylation in response to antipsychotics, suggest that this epigenetic mark may play a role in the therapeutic response to these drugs (Table 1). Melka and colleagues showed that olanzapine could induce DNA methylation changes in dopamine receptors (*DRD1*, *DRD2*, and *DRD5*) and cadherin gene families, and that these modifications could be associated with an enhancement of the response (reduced stress-induced locomotor activity) in a rat model [63]. Another study, showed that olanzapine-induced DNA methylation changes of cadherin/pro-cadherin genes can impact the antipsychotic response (reduced stress-induced locomotor activity) in rats [64]. Another atypical antipsychotic, quetiapine, has been found to modulate DNA methylation at the promoter of *SLC64A* in SK-N-SH cells [65]. Furthermore, treatment with the typical neuroleptic haloperidol decreased the methyl cytosine (mC) content specifically in the brain tissue of female rats, and increased mC content specifically in the liver of male rats [66]. These latter results highlight the possible discrepancies between brain and peripheral DNA methylation modulations by drugs, as well as differences between males and females. Other drugs have also been shown to modulate DNA methylation in rodent model of SCZ (Table 1). Hence, VPA alone or in combination with atypical antipsychotic drugs induced demethylation effects of the *Reln* and *GAD67* promoters selectively in mouse brains [67]. Moreover, in line with the hypermethylation of *Reln* observed in the brain of patients with SCZ, imidazenil or VPA can reverse the hypermethylation of the *Reln* gene promoter in a mouse model of SCZ [68]. The discrepancies observed in blood vs. brain or female vs. male as well as the brain region specificities of DNA methylation observed in the rodent model will represent key issues to be considered for a transfer to bedside treatments.

As previously detailed, this has been tested in mice, and the observed demethylation of *GAD67* and *Reln* genes may also influence the therapeutic response in BD [67] (Table 1). In addition to modulations of histone acetylation, VPA also decreases DNA methylation at the *Reln* promoter in vitro [69–71]. Since the *BDNF* gene has been associated with the pathophysiology and symptoms of BD, its promoters have been the best studied. Incubation with lithium for 48 h was found to significantly decrease DNA methylation at the *BDNF* promoter IV in cultured rat hippocampal neurons; in addition, simultaneous increase of *BDNF* mRNA levels was also reported [72,73]. Fourteen days of treatment of mice with VPA induced a specific decrease of DNA methylation at the distal CpG island of the *Cdkn* p21 (cyclin dependent kinase inhibitor) promoter in the hippocampus, which could explain the observed increase in mRNA levels of this gene [74]. Similarly, the observed increase of *Glt1* (Glutamate transporter) transcript induced by VPA in rat primary astrocyte cell cultures could be attributed to several epigenetic changes, including a decrease of DNA methylation at the promoter of this gene [75]. Modulation of DNA methylation at imprinted loci have also been reported in stem cells after incubation with lithium at concentrations greater that the therapeutic range [76]. At therapeutic concentrations, lithium, carbamazepine, and VPA were found to modulate a large number of genes in SK-N-SH neuronal cells [77]. Most of the genes affected by carbamazepine and VPA were common, while lithium influences DNA methylation, not only for these genes, but also additional specific genes [77].

Antidepressants' effects on DNA methylation have also been studied (Table 1). Administration of the SSRI escitalopram was associated with a decrease in DNA methylation at the *S100a10* (S100 Calcium Binding Protein A10) gene promoter region, and an increase of its transcripts in the prefrontal cortex in rat [78].

Table 1. Animal and cellular studies.

Model/Tissue	Method	Main Findings	Reference
Leukocytes/brain/liver tissues of rat	High-performance liquid chromatography	• Decreased mC in the brain of haloperidol-treated female rats Increased mC in liver DNA in haloperidol-treated male rats	[66]
Frontal cortex from a mouse model receiving L-methionine	Bisulfite conversion + PCR + sequencing	• VPA corrects *Reln* promoter hypermethylation induced in this SCZ-like model	[68]
Brian/liver tissues of rats	MeDIP	• 19 days of treatment with olanzapine increases DMA methylation at several dopaminergic genes	[63]
Mouse prefrontal cortex and striatum	MeDIP ChIP followed by qPCR	• VPA, clozapine, sulpiride, VPA + clozapine, and VPA + sulpiride treatment induce DNA demethylation of *GAD67* and *Reln* genes	[67]
Human neuroblastoma cell lines SK-N-SH	Infinium HumanMethylation27 BeadChip + bisulfite sequencing	• Quetiapine decreases DNA methylation of the CpG3 island of *SLC6A4*	[65]
Brian/liver tissues of rats	MeDIP	• Olanzapine alters DNA methylation at several cadherin/procadherin promoter region	[64]
Cultured rat hippocampal neurons	Methylation-specific PCR	• 48 h exposition to 1 and 2 mM lithium: 0.6-fold decrease DNA methylation at the promoter IV of BDNF • 2-fold increase of mRNA	[73]
Mouse hippocampal cells	PCR methylation-sensitive restriction site analysis	• 30% decrease of DNA methylation at the distal CpG island of the Cdkn p21 gene	[74]
Rat primary astrocytes	luminometric methylation analysis (LUMA)	• Decrease of DNA methylation at the Glt-1 promoter	[75]
Mouse embryonic and neural stem cells	Bisulfite sequencing	• 10 and 20 mM of lithium induce a 66% decrease of DNA methylation at the *Igf2/H19* differentially methylated domain (DMD) in embryonic stem cells • 5 mM of lithium induce a 33% decrease of DNA methylation at the *Igf2/H19* DMD in neural stem cells	[76]
Human neuroblastoma cell lines SK-N-SH	Infinium HumanMethylation27 BeadChip	• Hypermethylation of 345 genes (lithium), 64 genes (VPA), and 64 genes (carbamazepine) • Hypomethylation of 138 genes (lithium), 36 genes (VPA), and 14 genes (carbamazepine)	[77]

Polymerase Chain reaction (PCR), quantitative PCR (qPCR), Methylated DNA Immunoprecipitation (MeDIP), Chromatin ImmunoPrecipitation (ChIP).

It is important to note that there are only a few studies published on cell lines or in animal models of depression. Several rodent models are available, and they could help in understanding the central epigenetic effects of antidepressants. However, the growing body of evidence in rodents and cell lines show that antipsychotics, antidepressants, and mood stabilizers influence DNA methylation at promoters of genes involved in their targeted pathways, drug metabolism, but also in various other cellular pathways. The role of these induced changes in treatment response have been assessed in cross-sectional and longitudinal human studies.

4. DNA Methylation Modifications and Responses to Treatment in Human Studies

4.1. Cross-Sectional Studies

- Schizophrenia: There is evidence that antipsychotics induce the alteration of the methylation status of several genes, including those coding for proteins and pathways targeted by these drugs. Very few human studies have investigated the influence of epigenetic mechanisms on the response to antipsychotics, but the DNA methylation status of candidate genes have been found to be differentially modulated by treatment in patients with SCZ according to their response phenotypes (Table 2). In a cohort of 177 SCZ patients, Melas and colleagues reported that patients treated with haloperidol (n = 16) displayed a significantly higher level of global DNA methylation in blood [49], as compared to other antipsychotic drugs. Interestingly though, correlations between the DNA methylation levels and the response status of SCZ patients have been found in several studies. However, a tendency to reverse hypermethylation at the *DTNBP1* (Dysbindin) gene promoter in post-mortem brain samples of schizophrenic patients was found with antipsychotic treatments [79].

- Bipolar disorder: The decreased DNA methylation of the *DTNBP1* gene promoter with antipsychotic drug treatment found in the post-mortem brain samples of patients with SCZ was not seen in post-mortem brains from patients with BD, probably due to small number of patients using classic antipsychotics [79] (Table 2). A lithium-induced decrease of global DNA methylation was found in lymphoblast cell lines derived from 14 lithium-responder BD patients, as compared to 16 healthy controls [80]. Recently, Houtepen and colleagues investigated the effects of antipsychotics (olanzapine and quetiapine) and mood stabilizers (lithium, VPA, carbamazepine) on genome-wide DNA methylation in blood samples from 172 patients with BD. After adjustment for drugs effects on blood cell types, composition-only VPA and quetiapine modified the DNA methylation status significantly [81]. In a study of global DNA methylation in the leukocytes of BD patients, no differences were found compared to the healthy control. However a significantly lower DNA methylation level was observed in patients on lithium monotherapy, compared to controls or BD patients treated with a combination of lithium + VPA [82]. In this study, the DNA methylation level could not be correlated with the lithium response as assessed with the Alda scale. However, since global DNA methylation studies provide an imprecise picture of the effect of a given drug, gene-specific effects have been investigated. A decrease of DNA methylation at the promoter I of *BDNF* was observed in the Peripheral Blood Mononuclear Cells (PBMC) of BD patients with antidepressant therapy, compared to no antidepressant therapy [50]. Similarly, patients treated with lithium or VPA displayed a decrease of DNA methylation levels, as compared to other medications [50,83]. A trend, yet not significant, for a decreased DNA methylation level at the *BDNF* promoter I was found in another study with a slightly different design [84].

- Major depressive disorder: A large study explored the DNA methylation of the *BDNF* gene in patients with MDD (n = 207), BD (n = 59), and controls (n = 278). They reported an increased methylation of *BDNF* gene in patients with MDD compared to those with BD and controls. Somewhat surprisingly, they also found that the increased methylation of *BDNF* is associated with antidepressant therapy, but not with the clinical features of MDD [85]. Although very informative, these studies did not discriminate between the antidepressant classes. The only available study on therapeutic response to the SSRI, paroxetine, reported an association with the methylation level of the *PPFIA4* (Protein Tyrosine Phosphatase, Receptor Type, F Polypeptide, Interacting Protein, Alpha 4) and *HS3ST1* (heparin sulfate-glucosamine 3-sulfotransferase 1) genes in MDD responders, compared to the worst responders ($n = 10$ per group) [86].

These data indicate that DNA methylation can be influenced by antipsychotics, mood stabilizers, and antidepressants in SCZ, BD and MDD patients. Moreover, these modulations of DNA methylation can be associated with a clinical response. However, these data obtained in retrospective studies do not allow for differentiation between pre-existing differences in DNA methylation (influenced by heredity and/or environmental factors) and those induced by the treatments. Longitudinal designs are therefore required in order to understand the role of DNA methylation in treatment responses, and their potential use as predictive or diagnostic biomarkers in these disorders.

Table 2. Cross-sectional studies.

Model/Tissue	Method	Main Findings	Reference
Leukocytes of SCZ patients ($n = 177$) and controls ($n = 171$)	luminometric methylation analysis (LUMA)	• Increased global methylation in patients treated with haloperidol compared to other treatments	[49]
Saliva samples of SCZ ($n = 30$), first-degree relatives of SCZ ($n = 15$), controls ($n = 30$)/postmortem brain samples of patients with SCZ ($n = 35$) and BD ($n = 35$)	Quantitative methylation specific PCR	• Increased DNA methylation of DTNBP1 promoter in the saliva of patients with SCZ compared to controls • Inverse correlation between DTNBP1 methylation and expression in post-mortem brains of SCZ patients • Trend to reduced DNA methylation of DTNBP1 by antipsychotics treatment	[79]
Postmortem brain samples of patients with BD ($n = 35$) and controls ($n = 35$)	Quantitative methylation specific PCR	• No significant difference of the *DTNBP1* promoter region associated with antipsychotic treatment in patients with BD	[79]
Transformed lymphoblast cell lines from: lithium responders BD ($n = 14$), affected relatives ($n = 14$), unaffected relatives ($n = 16$), Healthy controls ($n = 16$)	ELISA	• Decreased DNA methylation in cell lines of BD patients, affected and unaffected relatives, compared to healthy controls • Lithium-induced decrease in global DNA methylation in BD patients (lithium responders) compared to controls	[80]
Whole blood of patients with BD ($n = 172$), Human	Infinium Human-Methylation450 BeadChip	• Quetiapine, VPA showed significant DNA methylation alterations patients with BD	[81]
Peripheral blood from BD patients on Li monotherapy ($n = 29$), Lithium + VPA ($n = 11$), Lithium + antipsychotics ($n = 21$), healthy controls ($n = 26$)	ELISA	• Hypomethylation of DNA in BD patients treated with lithium monotherapy vs. lithium + VPA or healthy controls • No significant relation between DNA methylation and lithium response	[82]
PBMC from BDI ($n = 45$), BDII ($n = 49$), and control subjects ($n = 52$)	Methylation-specific qPCR	• Decrease of DNA methylation at *BDNF* promoter I in patients with antidepressant therapy vs. controls • Decrease of DNA methylation at *BDNF* promoter I in patients with lithium therapy vs. other medications • Decrease of DNA methylation at *BDNF* promoter I in patients with VPA therapy vs. other medications	[50]
PBMC from BDI ($n = 45$), BDII ($n = 45$)	methylation specific qPCR	• Decrease of DNA methylation at *PDYN* promoter I in patients with lithium or VPA therapy ($n = 25$) vs. other medications	[83]
PBMC from BDI ($n = 61$), BDII ($n = 50$), MDD ($n = 43$) patients	methylation specific qPCR	• Not significant trend for a decrease of DNA methylation in patients treated with lithium and VPA	[84]
PBMC from MDD ($n = 207$), BD ($n = 59$) and controls ($n = 278$)	Methylation-specific quantitative PCR	• *BDNF* gene exon I promoter methylation increased in MDD compared to BD and controls • Increased *BDNF* DNA methylation in MDD patients associated with antidepressant therapy	[85]
Peripheral leukocytes from MDD patients (10 best responders and 10 worst responders to paroxetine)	Infinium Human-Methylation450 BeadChip	• Methylation levels of the CpG sites in *PPFIA4* and *HS3ST1* gene can discriminate between best and worst responders	[86]

ELISA (Enzyme-Linked Immunosorbent Assay), Polymerase Chain reaction (PCR), quantitative PCR (qPCR), Bipolar disorder type 1 (BDI), Bipolar disorder type 2 (BDII).

4.2. Longitudinal Studies

- Schizophrenia: DNA methylation change at the 13th CpG site of *HTR1A* is associated with negative symptoms in patients with SCZ after 10 weeks of treatment with antipsychotic drugs ($n = 82$) [87] (Table 3). Likewise, clozapine-induced DNA methylation changes in the CREB-binding protein (*CREBBP*) gene are inversely correlated with the percentage of Positive and Negative Syndrome Scale (PANSS) changes in treatment-resistant SCZ patients ($n = 21$) [88]. A recent study in Chinese Han schizophrenic patients investigated not only genes that were involved in the dopaminergic and serotoninergic pathways, but also in the metabolism and transport of risperidone. They found no significant CpG sites in *HTR2A*, *ABCB1*, and *DRD2* gene promoters associated with responses, while differentially methylated CpG of the drug-metabolizing enzymes *CYP3A4* and *CYP2D6* genes promoter regions were associated with a response to risperidone [89]. Furthermore, a whole-genome study of DNA methylation modifications before and after treatment with antipsychotics found gender-specific differences in the methylation profiles of patients with SCZ. Significant differences were observed in the male patient group in complete remission [90]. In this study methylation levels of six genes (*APIS3*, *C16orf59*, *KCNK15*, *LOC146336*, *MGC16384* and *XRN2*) and nine genes (*C16orf70*, *CST3*, *DDRGK1*, *FA2H*, *FLJ30058*, *MFSD2B*, *RFX4*, *UBE2J1* and *ZNF311*) were respectively identified as good markers of treatment-induced effects, and good predictive markers of treatment response [90].

- Major depressive disorder: Several of the drugs used to treat an MDD target, a serotonin transporter, an association between its DNA methylation levels before treatment and impaired treatment response after 12 weeks of antidepressant therapy in patients with MDD ($n = 108$) have been reported [91] (Table 3). Another study comparing the methylation levels before and after six weeks of antidepressant therapy showed that an increased DNA methylation at the third CpG site of *SLC6A4* was associated with better therapeutic response in patients with MDD [92]. The results of Okada and colleagues were confirmed in a naturalistic study of MDD patients ($n = 94$) treated with escitalopram [93]. They found that higher methylation at the *SLC6A4* gene was associated with better treatment response after six weeks of treatment (Table 3). However, the response status to escitalopram was found not associated with the DNA methylation level of another gene *MAO-A* (mono amino oxidase A) in 61 MDD patients [94]. Recently, hypomethylation at two CpGs sites (*HTR1A* CpG 668 and *HTR1B* CpG 1401) was found to significantly differ in remitter and non-remitter Chinese Han patients with MDD (n = 85) with escitalopram treatment [95]. Very promising results have been obtained in MDD patients treated with escitalopram ($n = 80$) or with the tricyclic antidepressant nortriptyline ($n = 33$) [96]. In this study, higher DNA methylation level at the fourth CpG island of the interleukin-11 (*IL-11*) gene before treatment was associated with a better response to escitalopram, while hypomethylation at the same site was associated with a better nortriptyline response (Table 3). These results suggest that DNA methylation levels before treatment could be a predictor of the best suited antidepressant for an individual.

Few longitudinal studies have been published so far, and the results were obtained from peripheral samples. Nonetheless, they suggest that DNA methylation might be used as a predictive biomarker before the initiation of treatment or a monitoring biomarker of the efficacy of therapies in SCZ and MDD. Unfortunately, there is, for the moment, no available prospective study examining genome-wide DNA methylation modulation by specific mood stabilizers in BD patients. Further prospective studies are required to better understand how epigenetics could help physician in the prediction or the monitoring of treatment response in SCZ, BD and MDD.

Table 3. Longitudinal studies.

Model/Tissue	Method	Main Findings	Reference
Peripheral blood of patients with SCZ (n = 82)	Bisulfite conversion + PCR + pyrosequencing	• Decreased DNA methylation at CpG13 of *HTR1A* associated with poorer response to antipsychotics	[87]
Peripheral blood from SCZ patients (n = 21)	Infinium Human-Methylation450 BeadChip	• Clozapine-induced DNA methylation changes in the *CREBBP* gene were significantly correlated with clinical improvements	[88]
Peripheral blood from SCZ patients; good responders (n = 88), poor responders (n = 54)	Methylation-specific PCR + mass spectrometry	• Seven CpGs at *CYP3A4* and *CYP2D6* genes were differentially methylated in good vs. poor responders to risperidone therapy	[89]
Peripheral blood from SCZ patients n = 20 (12 M/8 F)	MeDIP ChIP	• Before treatment: nine genes with DMR in male SCZ patients in complete remission after treatment (vs. matched control subjects) • After treatment: six genes with DMR in male SCZ patients in complete remission after treatment (vs. matched control subjects) • Before treatment: one gene (M1R181C) with DMR in female SCZ patients in complete remission after treatment (vs. matched control subjects) • After treatment: one gene (BCOR) with DMR in female SCZ patients in complete remission after treatment (vs. matched control subjects)	[90]
Leukocytes from patients with MDD (n = 108)	Bisulfite conversion followed by PCR	• Increased of SLC6A4 DNA methylation level associated with impaired treatment response to antidepressants	[91]
Whole blood from patients with MDD (n = 50 before treatment and n = 40 after 6 weeks of treatment) and controls (n = 50)	Methylation-specific PCR + mass spectrometry	• Methylation level of the third CpG site of SLC6A4 gene association with better therapeutic response to antidepressant therapy in patients MD	[92]
Whole blood from patients with MDD (n = 94), (Human)	Bisulfite conversion + PCR	• Increased DNA methylation of *SLC6A4* gene associated with better treatment response to escitalopram	[93]
Whole blood from patients with MDD, (n = 61) (Human)	Bisulfite conversion + PCR	• No major influence of mono amino oxidase (MAO-A) gene methylation status on escitalopram response	[94]
Peripheral blood of patients with MDD (n = 85)	Bisulfite sequencing	• Significant association of 668 CpG sites of HTR1A and 1401 CpG sites of HTR1B gene methylation with treatment response to escitalopram	[95]
Peripheral blood samples of patients with MDD treated with escitalopram (n = 80) or nortriptyline (n = 33)	Bisulfite conversion + PCR	• Fourth CpG island hypomethylation of IL-11 gene associated with better response to nortriptyline • Hypermethylation of fourth CpG island of IL-11 gene associated with better response to escitalopram	[96]

Polymerase Chain reaction (PCR), quantitative PCR (qPCR), Methylated DNA Immunoprecipitation (MeDIP), Chromatin ImmunoPrecipitation (ChIP).

5. Materials and Methods

We conducted a literature search in the PubMed database until April 2018 using combinations of the following keywords: "DNA methylation", "antipsychotic response", "treatment response", "bipolar disorder", "major depressive disorder", "schizophrenia", "epigenetics", and "antidepressants". We followed the Preferred Reporting Items for Systematic Reviews and Meta-Analyses (PRISMA) guidelines [97]. Studies were included according to the following criteria: (a) being an original paper in a peer-reviewed journal; and (b) containing an epigenetic analysis of response treatment in BD, MDD and/or SCZ samples. Figure 1 summarizes the search strategy used for selecting the studies (identification, screening, eligibility, inclusion process) in the present review. Two blinded independent researchers (GAC and CMC) conducted a two-step literature search. Discrepancies were resolved by consultations with the other authors. The reference lists of the articles were also manually checked for additional relevant studies. We identified 112 articles on the basis of their titles; 67 abstracts focused on DNA methylation as a biomarker of treatment response in cellular or rodent models, or in BD, SCZ and/or MDD patients were selected. Exclusion criteria included (i) studies not written in English; (ii) review articles, book chapters, conference abstracts, and case studies; (iii) studies not related to response to treatment. The final selection consisted of 31 original articles specifically related to DNA methylation in treatment responses to BD, SCZ and MDD. This systematic review will first present an overview of the DNA methylation and its role in gene expression and treatment response. We will then summarize studies on DNA methylation patterns in SCZ, BD and MDD as results of potential interest for the investigation of treatment response variability. Then, we will present treatment-induced DNA methylation modifications in animal models and DNA methylation modifications in response to psychotropic treatments in human studies. Finally, the overall picture will be discussed.

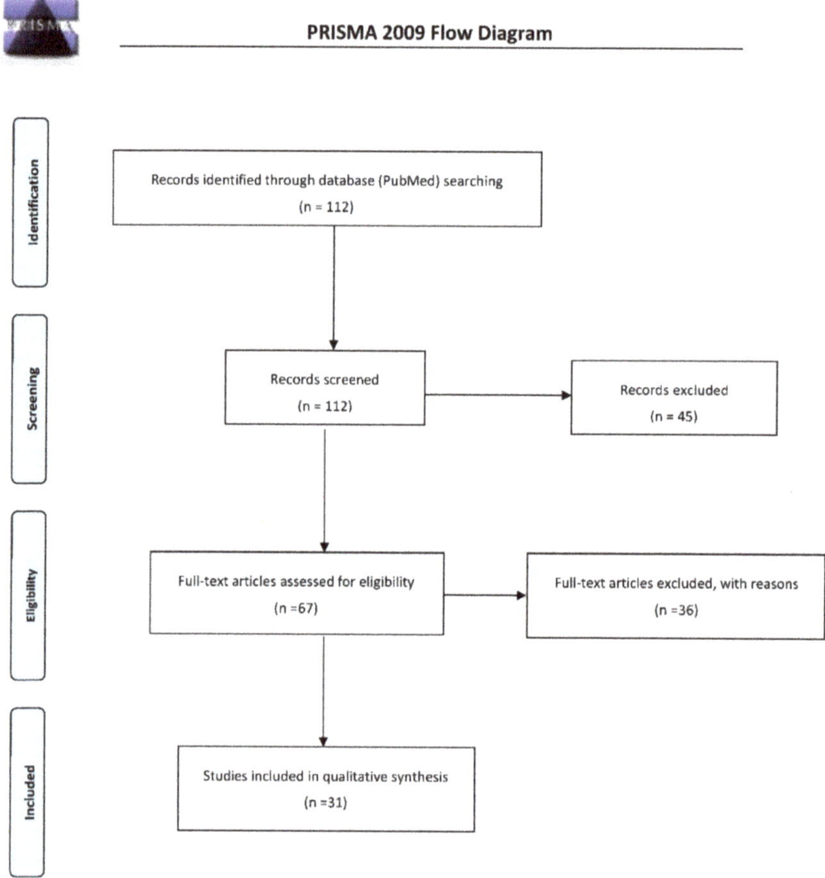

Figure 1. PRISMA (Preferred Reporting Items for Systematic Reviews and Meta-Analyses) flow-diagram of the screening strategy.

6. Conclusions

The characterization of biomarkers throughout the disease course is a key element, not only for understanding its pathophysiology, but also to monitor treatment responses. Case control studies have shown differential DNA methylation levels in patients, as compared to control subjects. Recent data in high-risk offspring from BD patients suggest that the observed differences in DNA methylation are directly related to the familial environment [98]. Likewise, the significant decreased methylation of the CpG site at the −1438A/G polymorphism site of the *HTR2A* (5-hydroxytryptamine receptor 2A) receptor observed in SCZ and BD patients' saliva, but also in their first degree relatives, suggests an effect of environmental factors independent of the onset and course of the disorders, as well as the treatments [99]. The course of the disease might also influence the DNA methylation levels, as suggested by the accelerated epigenetic aging observed in older BD patients as compared to control subjects, while no significant difference was observed in younger patients [100]. Despite this very complex landscape, available evidence supports the hypothesis of the DNA methylation role in therapeutic response to antipsychotics, mood stabilizers, and antidepressants in the treatment of SMI, such as SCZ, BD and MDD. Finding novel pathways by targeting the epigenome in the context of a treatment response may also help to understand the role of these pathways in diseases.

These pathways might also be targeted by drug repurposing strategies in order to identify already known medications that could alter the identified epigenetic marks [101]. Finally, this may also help clinicians to specifically prescribe psychoactive treatments targeting the identified biological pathways to move towards a more tailored and personalized strategy of medicine in psychiatry. However, findings from in vitro and in vivo models studies suggesting the DNA methylation as a therapeutic target may not be relevant to human subjects. Notably, gender specificities and appropriate sample plan designs are key issues for transferability to clinical practice. SMI often require a complex mixture of several co-medications to improve patients' conditions. However, the main limitations of this review is the fact that only a limited number of human studies are available for the moment, and they target the effects of one type of drug, or even one specific drug at a time, whereas patients often receive polypharmacy. Therefore, the combined effects of antidepressants, antipsychotics, and mood stabilizers are not taken into account. The reported epigenetic effects might not represent real life. Moreover, as a result of the cross-sectional design of most studies it cannot be concluded that the epigenetics marks observed in presence of a given medication are solely the effect of the drug, and not the sum of the effects of drugs and other environmental factors that are likely to modify DNA methylation. Very few studies have been reported in human subjects, and the use of several co-medications, retrospective design, and various response evaluation time points represent a serious limitation. Further studies with monotherapy treatment, prospective study designs considering the duration of the response assessment, co-medications, and addressing the environmental factors during the illness as well as treatment, will better explain the DNA methylation signatures. These studies will be useful in the implementation and the development of efficient and personalized therapeutic strategies.

Funding: C.G.A. was supported by Indo-French Centre for the Promotion of Advanced Research (IFCPAR/CEFIPRA) for promoting the collaboration (Grant No. IFC/4098/RCF2016/853) between Jawaharlal Institute of Postgraduate Medical Education and Research (JIPMER) and the University of Sorbonne universities, Paris, France. The group of F.B. is a member of the Bio-Psy LabEx under reference ANR-11-IDEX-0004-02.

Conflicts of Interest: The authors declare no conflict of interest.

Abbreviations

SMI	Severe mental illnesses
SCZ	Schizophrenia
BD	Bipolar disorder
MDD	Major depressive disorder

References

1. World Health Organization. *The 10th Revision of the International Classification of Diseases and Related Health Problems (ICD-10)*; WHO: Geneva, Switzerland, 1992.
2. Ferrari, A.J.; Saha, S.; McGrath, J.J.; Norman, R.; Baxter, A.J.; Vos, T.; Whiteford, H.A. Health states for schizophrenia and bipolar disorder within the Global Burden of Disease 2010 Study. *Popul. Health Metr.* **2012**, *10*. [CrossRef] [PubMed]
3. Reddy, M.S. Depression: The Disorder and the Burden. *Indian J. Psychol. Med.* **2010**, *32*, 1–2. [CrossRef] [PubMed]
4. McGrath, J.; Saha, S.; Chant, D.; Welham, J. Schizophrenia: A concise overview of incidence, prevalence, and mortality. *Epidemiol. Rev.* **2008**, *30*, 67–76. [CrossRef] [PubMed]
5. Merikangas, K.R.; Jin, R.; He, J.-P.; Kessler, R.C.; Lee, S.; Sampson, N.A.; Viana, M.C.; Andrade, L.H.; Hu, C.; Karam, E.G.; et al. Prevalence and Correlates of Bipolar Spectrum Disorder in the World Mental Health Survey Initiative. *Arch. Gen. Psychiatry* **2011**, *68*, 241–251. [CrossRef] [PubMed]
6. Murray, C.J.; Lopez, A.D. Global mortality, disability, and the contribution of risk factors: Global Burden of Disease Study. *Lancet* **1997**, *349*, 1436–1442. [CrossRef]
7. Kessler, R.C.; Bromet, E.J. The epidemiology of depression across cultures. *Annu. Rev. Public Health* **2013**, *34*, 119–138. [CrossRef] [PubMed]

8. Kuswanto, C.N.; Sum, M.Y.; Sim, K. Neurocognitive Functioning in Schizophrenia and Bipolar Disorder: Clarifying Concepts of Diagnostic Dichotomy vs. Continuum. *Front. Psychiatry* **2013**, *4*. [CrossRef] [PubMed]

9. Patel, K.R.; Cherian, J.; Gohil, K.; Atkinson, D. Schizophrenia: Overview and Treatment Options. *Pharm. Ther.* **2014**, *39*, 638–645.

10. Serafini, G.; Pompili, M.; Borgwardt, S.; Houenou, J.; Geoffroy, P.A.; Jardri, R.; Girardi, P.; Amore, M. Brain changes in early-onset bipolar and unipolar depressive disorders: A systematic review in children and adolescents. *Eur. Child Adolesc. Psychiatry* **2014**, *23*, 1023–1041. [CrossRef] [PubMed]

11. Chesney, E.; Goodwin, G.M.; Fazel, S. Risks of all-cause and suicide mortality in mental disorders: A meta-review. *World Psychiatry* **2014**, *13*, 153–160. [CrossRef] [PubMed]

12. Need, A.C.; Goldstein, D.B. Schizophrenia genetics comes of age. *Neuron* **2014**, *83*, 760–763. [CrossRef] [PubMed]

13. Harrison, P.J. Recent genetic findings in schizophrenia and their therapeutic relevance. *J. Psychopharmacol.* **2015**, *29*, 85–96. [CrossRef] [PubMed]

14. Piletz, J.E.; Zhang, X.; Ranade, R.; Liu, C. Database of genetic studies of bipolar disorder. *Psychiatr. Genet.* **2011**, *21*, 57–68. [CrossRef] [PubMed]

15. Serretti, A.; Mandelli, L. The genetics of bipolar disorder: Genome "hot regions," genes, new potential candidates and future directions. *Mol. Psychiatry* **2008**, *13*, 742–771. [CrossRef] [PubMed]

16. López-León, S.; Janssens, A.C.; González-Zuloeta Ladd, A.M.; Del-Favero, J.; Claes, S.J.; Oostra, B.A.; van Duijn, C.M. Meta-analyses of genetic studies on major depressive disorder. *Mol. Psychiatry* **2008**, *13*, 772–785. [CrossRef] [PubMed]

17. Gatt, J.M.; Burton, K.L.O.; Williams, L.M.; Schofield, P.R. Specific and common genes implicated across major mental disorders: A review of meta-analysis studies. *J. Psychiatr. Res.* **2015**, *60*, 1–13. [CrossRef] [PubMed]

18. Avramopoulos, D. Recent Advances in the Genetics of Schizophrenia. *Mol. Neuropsychiatry* **2018**, *4*, 35–51. [CrossRef] [PubMed]

19. Ikeda, M.; Saito, T.; Kondo, K.; Iwata, N. Genome-wide association studies of bipolar disorder: A systematic review of recent findings and their clinical implications. *Psychiatry Clin. Neurosci.* **2018**, *72*, 52–63. [CrossRef] [PubMed]

20. Shadrina, M.; Bondarenko, E.A.; Slominsky, P.A. Genetics Factors in Major Depression Disease. *Front. Psychiatry* **2018**, *9*. [CrossRef] [PubMed]

21. Lin, E.; Tsai, S.-J. Genome-wide microarray analysis of gene expression profiling in major depression and antidepressant therapy. *Prog. Neuro-Psychopharmacol. Biol. Psychiatry* **2016**, *64*, 334–340. [CrossRef] [PubMed]

22. Mamdani, F.; Martin, M.V.; Lencz, T.; Rollins, B.; Robinson, D.G.; Moon, E.A.; Malhotra, A.K.; Vawter, M.P. Coding and noncoding gene expression biomarkers in mood disorders and schizophrenia. *Dis. Markers* **2013**, *35*, 11–21. [CrossRef] [PubMed]

23. Molteni, R.; Calabrese, F.; Racagni, G.; Fumagalli, F.; Riva, M.A. Antipsychotic drug actions on gene modulation and signaling mechanisms. *Pharmacol. Ther.* **2009**, *124*, 74–85. [CrossRef] [PubMed]

24. Bellivier, F.; Marie-Claire, C. Lithium response variability: New avenues and hypotheses. In *The Science and Practice of Lithium Therapy*; Malhi, G.S., Masson, M., Bellivier, F., Eds.; Springer: Basel, Switzerland, 2017; pp. 157–178.

25. Fabbri, C.; Hosak, L.; Mössner, R.; Giegling, I.; Mandelli, L.; Bellivier, F.; Claes, S.; Collier, D.A.; Corrales, A.; Delisi, L.E.; et al. Consensus paper of the WFSBP Task Force on Genetics: Genetics, epigenetics and gene expression markers of major depressive disorder and antidepressant response. *World J. Biol. Psychiatry* **2017**, *18*, 5–28. [CrossRef] [PubMed]

26. Bock, C. Epigenetic biomarker development. *Epigenomics* **2009**, *1*, 99–110. [CrossRef] [PubMed]

27. Chan, T.A.; Baylin, S.B. Epigenetic biomarkers. *Curr. Top. Microbiol. Immunol.* **2012**, *355*, 189–216. [CrossRef] [PubMed]

28. Heerboth, S.; Lapinska, K.; Snyder, N.; Leary, M.; Rollinson, S.; Sarkar, S. Use of epigenetic drugs in disease: An overview. *Genet. Epigenet.* **2014**, *6*, 9–19. [CrossRef] [PubMed]

29. Moore, L.D.; Le, T.; Fan, G. DNA methylation and its basic function. *Neuropsychopharmacology* **2013**, *38*, 23–38. [CrossRef] [PubMed]

30. Afonso-Grunz, F.; Müller, S. Principles of miRNA-mRNA interactions: Beyond sequence complementarity. *Cell. Mol. Life Sci.* **2015**, *72*, 3127–3141. [CrossRef] [PubMed]

31. Yeshurun, S.; Hannan, A.J. Transgenerational epigenetic influences of paternal environmental exposures on brain function and predisposition to psychiatric disorders. *Mol. Psychiatry* **2018**. [CrossRef] [PubMed]

32. Cholewa-Waclaw, J.; Bird, A.; von Schimmelmann, M.; Schaefer, A.; Yu, H.; Song, H.; Madabhushi, R.; Tsai, L.-H. The role of epigenetic mechanisms in the regulation of gene expression in the nervous system. *J. Neurosci.* **2016**, *36*, 11427–11434. [CrossRef] [PubMed]

33. Feng, J.; Chang, H.; Li, E.; Fan, G. Dynamic expression of de novo DNA methyltransferases DNMT3A and DNMT3B in the central nervous system. *J. Neurosci. Res.* **2005**, *79*, 734–746. [CrossRef] [PubMed]

34. Guo, J.U.; Ma, D.K.; Mo, H.; Ball, M.P.; Jang, M.-H.; Bonaguidi, M.A.; Balazer, J.A.; Eaves, H.L.; Xie, B.; Ford, E.; et al. Neuronal activity modifies the DNA methylation landscape in the adult brain. *Nat. Neurosci.* **2011**, *14*, 1345–1351. [CrossRef] [PubMed]

35. Martinowich, K.; Hattori, D.; Wu, H.; Fouse, S.; He, F.; Hu, Y.; Fan, G.; Sun, Y.E. DNA methylation-related chromatin remodeling in activity-dependent *BDNF* gene regulation. *Science* **2003**, *302*, 890–893. [CrossRef] [PubMed]

36. Poo, S.X.; Agius, M. Atypical anti-psychotics in adult bipolar disorder: Current evidence and updates in the NICE guidelines. *Psychiatr. Danub.* **2014**, *26*, 322–329. [PubMed]

37. Patkar, A.A.; Pae, C.-U. Atypical antipsychotic augmentation strategies in the context of guideline-based care for the treatment of major depressive disorder. *CNS Drugs* **2013**, *27*, 29–37. [CrossRef] [PubMed]

38. Richtand, N.M.; Welge, J.A.; Logue, A.D.; Keck, P.E.; Strakowski, S.M.; McNamara, R.K. Role of serotonin and dopamine receptor binding in antipsychotic efficacy. *Prog. Brain Res.* **2008**, *172*, 155–175. [CrossRef] [PubMed]

39. Simonsen, E.; Friis, S.; Opjordsmoen, S.; Mortensen, E.L.; Haahr, U.; Melle, I.; Joa, I.; Johannessen, J.O.; Larsen, T.K.; Røssberg, J.I.; et al. Early identification of non-remission in first-episode psychosis in a two-year outcome study. *Acta Psychiatr. Scand.* **2010**, *122*, 375–383. [CrossRef] [PubMed]

40. Goodwin, G.M.; Haddad, P.M.; Ferrier, I.N.; Aronson, J.K.; Barnes, T.; Cipriani, A.; Coghill, D.R.; Fazel, S.; Geddes, J.R.; Grunze, H.; et al. Evidence-based guidelines for treating bipolar disorder: Revised third edition recommendations from the British Association for Psychopharmacology. *J. Psychopharmacol.* **2016**, *30*, 495–553. [CrossRef] [PubMed]

41. Baldessarini, R.J.; Tondo, L. Does lithium treatment still work? Evidence of stable responses over three decades. *Arch. Gen. Psychiatry* **2000**, *57*, 187–190. [CrossRef] [PubMed]

42. Garnham, J.; Munro, A.; Slaney, C.; Macdougall, M.; Passmore, M.; Duffy, A.; O'Donovan, C.; Teehan, A.; Alda, M. Prophylactic treatment response in bipolar disorder: Results of a naturalistic observation study. *J. Affect. Disord.* **2007**, *104*, 185–190. [CrossRef] [PubMed]

43. Rybakowski, J.K.; Chlopocka-Wozniak, M.; Suwalska, A. The prophylactic effect of long-term lithium administration in bipolar patients entering treatment in the 1970s and 1980s. *Bipolar Disord.* **2001**, *3*, 63–67. [CrossRef] [PubMed]

44. Grandjean, E.M.; Aubry, J.-M. Lithium: Updated human knowledge using an evidence-based approach: Part III: Clinical safety. *CNS Drugs* **2009**, *23*, 397–418. [CrossRef] [PubMed]

45. Crisafulli, C.; Fabbri, C.; Porcelli, S.; Drago, A.; Spina, E.; De Ronchi, D.; Serretti, A. Pharmacogenetics of antidepressants. *Front. Pharmacol.* **2011**, *2*. [CrossRef] [PubMed]

46. Trivedi, M.H.; Fava, M.; Wisniewski, S.R.; Thase, M.E.; Quitkin, F.; Warden, D.; Ritz, L.; Nierenberg, A.A.; Lebowitz, B.D.; Biggs, M.M.; et al. Medication augmentation after the failure of SSRIs for depression. *N. Engl. J. Med.* **2006**, *354*, 1243–1252. [CrossRef] [PubMed]

47. Abdolmaleky, H.M.; Nohesara, S.; Ghadirivasfi, M.; Lambert, A.W.; Ahmadkhaniha, H.; Ozturk, S.; Wong, C.K.; Shafa, R.; Mostafavi, A.; Thiagalingam, S. DNA hypermethylation of serotonin transporter gene promoter in drug naïve patients with schizophrenia. *Schizophr. Res.* **2014**, *152*, 373–380. [CrossRef] [PubMed]

48. Schneider, I.; Kugel, H.; Redlich, R.; Grotegerd, D.; Bürger, C.; Bürkner, P.-C.; Opel, N.; Dohm, K.; Zaremba, D.; Meinert, S.; et al. Association of serotonin transporter gene *AluJb* methylation with major depression, amygdala responsiveness, 5-HTTLPR/rs25531 polymorphism, and stress. *Neuropsychopharmacology* **2018**, *43*, 1308–1316. [CrossRef] [PubMed]

49. Melas, P.A.; Rogdaki, M.; Ösby, U.; Schalling, M.; Lavebratt, C.; Ekström, T.J. Epigenetic aberrations in leukocytes of patients with schizophrenia: Association of global DNA methylation with antipsychotic drug treatment and disease onset. *FASEB J.* **2012**, *26*, 2712–2718. [CrossRef] [PubMed]

50. D'Addario, C.; Dell'Osso, B.; Palazzo, M.C.; Benatti, B.; Lietti, L.; Cattaneo, E.; Galimberti, D.; Fenoglio, C.; Cortini, F.; Scarpini, E.; et al. Selective DNA methylation of BDNF promoter in bipolar disorder: Differences among patients with BDI and BDII. *Neuropsychopharmacology* **2012**, *37*, 1647–1655. [CrossRef] [PubMed]

51. Starnawska, A.; Demontis, D.; McQuillin, A.; O'Brien, N.L.; Staunstrup, N.H.; Mors, O.; Nielsen, A.L.; Børglum, A.D.; Nyegaard, M. Hypomethylation of FAM63B in bipolar disorder patients. *Clin. Epigenetics* **2016**, *8*. [CrossRef] [PubMed]

52. Sugawara, H.; Murata, Y.; Ikegame, T.; Sawamura, R.; Shimanaga, S.; Takeoka, Y.; Saito, T.; Ikeda, M.; Yoshikawa, A.; Nishimura, F.; et al. DNA methylation analyses of the candidate genes identified by a methylome-wide association study revealed common epigenetic alterations in schizophrenia and bipolar disorder. *Psychiatry Clin. Neurosci.* **2018**, *72*, 245–254. [CrossRef] [PubMed]

53. Castle, D.J.; Galletly, C.A.; Dark, F.; Humberstone, V.; Morgan, V.A.; Killackey, E.; Kulkarni, J.; McGorry, P.; Nielssen, O.; Tran, N.T.; et al. The 2016 Royal Australian and New Zealand College of Psychiatrists guidelines for the management of schizophrenia and related disorders. *Med. J. Aust.* **2017**, *206*, 501–505. [CrossRef] [PubMed]

54. Falkai, P.; Wobrock, T.; Lieberman, J.; Glenthoj, B.; Gattaz, W.F.; Möller, H.-J. WFSBP Task Force on Treatment Guidelines for Schizophrenia. World Federation of Societies of Biological Psychiatry (WFSBP) guidelines for biological treatment of schizophrenia, part 2: Long-term treatment of schizophrenia. *World J. Biol. Psychiatry* **2006**, *7*, 5–40. [CrossRef] [PubMed]

55. Lehman, A.F.; Lieberman, J.A.; Dixon, L.B.; McGlashan, T.H.; Miller, A.L.; Perkins, D.O.; Kreyenbuhl, J. American Psychiatric Association; Steering Committee on Practice Guidelines. Practice guideline for the treatment of patients with schizophrenia, second edition. *Am. J. Psychiatry* **2004**, *161*, 1–56. [PubMed]

56. Henssler, J.; Kurschus, M.; Franklin, J.; Bschor, T.; Baethge, C. Long-term acute-phase treatment with antidepressants, 8 weeks and beyond. *J. Clin. Psychiatry* **2018**, *79*, 60–68. [CrossRef] [PubMed]

57. Malhi, G.S.; Gessler, D.; Outhred, T. The use of lithium for the treatment of bipolar disorder: Recommendations from clinical practice guidelines. *J. Affect. Disord.* **2017**, *217*, 266–280. [CrossRef] [PubMed]

58. Phiel, C.J.; Zhang, F.; Huang, E.Y.; Guenther, M.G.; Lazar, M.A.; Klein, P.S. Histone deacetylase is a direct target of valproic acid, a potent anticonvulsant, mood stabilizer, and teratogen. *J. Biol. Chem.* **2001**, *276*, 36734–36741. [CrossRef] [PubMed]

59. Wu, S.; Zheng, S.-D.; Huang, H.-L.; Yan, L.-C.; Yin, X.-F.; Xu, H.-N.; Zhang, K.-J.; Gui, J.-H.; Chu, L.; Liu, X.-Y. Lithium down-regulates histone deacetylase 1 (HDAC1) and induces degradation of mutant huntingtin. *J. Biol. Chem.* **2013**, *288*, 35500–35510. [CrossRef] [PubMed]

60. Matrisciano, F.; Dong, E.; Gavin, D.P.; Nicoletti, F.; Guidotti, A. Activation of group II metabotropic glutamate receptors promotes DNA demethylation in the mouse brain. *Mol. Pharmacol.* **2011**, *80*, 174–182. [CrossRef] [PubMed]

61. Labermaier, C.; Masana, M.; Müller, M.B. Biomarkers predicting antidepressant treatment response: How can we advance the field? *Dis. Markers* **2013**, *35*, 23–31. [CrossRef] [PubMed]

62. Fond, G.; d'Albis, M.-A.; Jamain, S.; Tamouza, R.; Arango, C.; Fleischhacker, W.W.; Glenthøj, B.; Leweke, M.; Lewis, S.; McGuire, P.; et al. The promise of biological markers for treatment response in first-episode psychosis: A systematic review. *Schizophr. Bull.* **2015**, *41*, 559–573. [CrossRef] [PubMed]

63. Melka, M.G.; Castellani, C.A.; Laufer, B.I.; Rajakumar, R.N.; O'Reilly, R.; Singh, S.M. Olanzapine induced DNA methylation changes support the dopamine hypothesis of psychosis. *J. Mol. Psychiatry* **2013**, *1*. [CrossRef] [PubMed]

64. Melka, M.G.; Castellani, C.A.; Rajakumar, N.; O'Reilly, R.; Singh, S.M. Olanzapine-induced methylation alters cadherin gene families and associated pathways implicated in psychosis. *BMC Neurosci.* **2014**, *15*, 112. [CrossRef] [PubMed]

65. Sugawara, H.; Bundo, M.; Asai, T.; Sunaga, F.; Ueda, J.; Ishigooka, J.; Kasai, K.; Kato, T.; Iwamoto, K. Effects of quetiapine on DNA methylation in neuroblastoma cells. *Prog. Neuro-Psychopharmacol. Biol. Psychiatry* **2015**, *56*, 117–121. [CrossRef] [PubMed]

66. Shimabukuro, M.; Jinno, Y.; Fuke, C.; Okazaki, Y. Haloperidol treatment induces tissue- and sex-specific changes in DNA methylation: A control study using rats. *Behav. Brain Funct.* **2006**, *2*, 37. [CrossRef] [PubMed]

67. Dong, E.; Nelson, M.; Grayson, D.R.; Costa, E.; Guidotti, A. Clozapine and sulpiride but not haloperidol or olanzapine activate brain DNA demethylation. *Proc. Natl. Acad. Sci. USA* **2008**, *105*, 13614–13619. [CrossRef] [PubMed]

68. Tremolizzo, L.; Doueiri, M.-S.; Dong, E.; Grayson, D.R.; Davis, J.; Pinna, G.; Tueting, P.; Rodriguez-Menendez, V.; Costa, E.; Guidotti, A. Valproate corrects the schizophrenia-like epigenetic behavioral modifications induced by methionine in mice. *Biol. Psychiatry* **2005**, *57*, 500–509. [CrossRef] [PubMed]

69. Carrard, A.; Salzmann, A.; Malafosse, A.; Karege, F. Increased DNA methylation status of the serotonin receptor 5HTR1A gene promoter in schizophrenia and bipolar disorder. *J. Affect. Disord.* **2011**, *132*, 450–453. [CrossRef] [PubMed]

70. Cruceanu, C.; Alda, M.; Grof, P.; Rouleau, G.A.; Turecki, G. Synapsin II is involved in the molecular pathway of lithium treatment in bipolar disorder. *PLoS ONE* **2012**, *7*, e32680. [CrossRef] [PubMed]

71. Mitchell, C.P.; Chen, Y.; Kundakovic, M.; Costa, E.; Grayson, D.R. Histone deacetylase inhibitors decrease reelin promoter methylation in vitro. *J. Neurochem.* **2005**, *93*, 483–492. [CrossRef] [PubMed]

72. Pandey, G.; Dwivedi, Y.; SridharaRao, J.; Ren, X.; Janicak, P.G.; Sharma, R. Protein kinase C and phospholipase C activity and expression of their specific isozymes is decreased and expression of MARCKs is increased in platelets of bipolar but not in unipolar patients. *Neuropsychopharmacology* **2002**, *26*, 216–228. [CrossRef]

73. Dwivedi, T.; Zhang, H. Lithium-induced neuroprotection is associated with epigenetic modification of specific *BDNF* gene promoter and altered expression of apoptotic-regulatory proteins. *Front. Neurosci.* **2014**, *8*, 457. [CrossRef] [PubMed]

74. Aizawa, S.; Yamamuro, Y. Valproate administration to mice increases hippocampal p21 expression by altering genomic DNA methylation. *Neuroreport* **2015**, *26*, 915–920. [CrossRef] [PubMed]

75. Perisic, T.; Zimmermann, N.; Kirmeier, T.; Asmus, M.; Tuorto, F.; Uhr, M.; Holsboer, F.; Rein, T.; Zschocke, J. Valproate and amitriptyline exert common and divergent influences on global and gene promoter-specific chromatin modifications in rat primary astrocytes. *Neuropsychopharmacology* **2010**, *35*, 792–805. [CrossRef] [PubMed]

76. Popkie, A.P.; Zeidner, L.C.; Albrecht, A.M.; D'Ippolito, A.; Eckardt, S.; Newsom, D.E.; Groden, J.; Doble, B.W.; Aronow, B.; McLaughlin, K.J.; et al. Phosphatidylinositol 3-kinase (PI3K) signaling via glycogen synthase kinase-3 (Gsk-3) regulates DNA methylation of imprinted loci. *J. Biol. Chem.* **2010**, *285*, 41337–41347. [CrossRef] [PubMed]

77. Asai, T.; Bundo, M.; Sugawara, H.; Sunaga, F.; Ueda, J.; Tanaka, G.; Ishigooka, J.; Kasai, K.; Kato, T.; Iwamoto, K. Effect of mood stabilizers on DNA methylation in human neuroblastoma cells. *Int. J. Neuropsychopharmacol.* **2013**, *16*, 2285–2294. [CrossRef] [PubMed]

78. Melas, P.A.; Rogdaki, M.; Lennartsson, A.; Björk, K.; Qi, H.; Witasp, A.; Werme, M.; Wegener, G.; Mathé, A.A.; Svenningsson, P.; et al. Antidepressant treatment is associated with epigenetic alterations in the promoter of P11 in a genetic model of depression. *Int. J. Neuropsychopharmacol.* **2012**, *15*, 669–679. [CrossRef] [PubMed]

79. Abdolmaleky, H.M.; Pajouhanfar, S.; Faghankhani, M.; Joghataei, M.T.; Mostafavi, A.; Thiagalingam, S. Antipsychotic drugs attenuate aberrant DNA methylation of DTNBP1 (dysbindin) promoter in saliva and post-mortem brain of patients with schizophrenia and Psychotic bipolar disorder. *Am. J. Med. Genet. B Neuropsychiatr. Genet.* **2015**, *168*, 687–696. [CrossRef] [PubMed]

80. Huzayyin, A.A.; Andreazza, A.C.; Turecki, G.; Cruceanu, C.; Rouleau, G.A.; Alda, M.; Young, L.T. Decreased global methylation in patients with bipolar disorder who respond to lithium. *Int. J. Neuropsychopharmacol.* **2014**, *17*, 561–569. [CrossRef] [PubMed]

81. Houtepen, L.C.; van Bergen, A.H.; Vinkers, C.H.; Boks, M.P.M. DNA methylation signatures of mood stabilizers and antipsychotics in bipolar disorder. *Epigenomics* **2016**, *8*, 197–208. [CrossRef] [PubMed]

82. Backlund, L.; Wei, Y.B.; Martinsson, L.; Melas, P.A.; Liu, J.J.; Mu, N.; Östenson, C.-G.; Ekström, T.J.; Schalling, M.; Lavebratt, C. Mood stabilizers and the influence on global leukocyte DNA methylation in bipolar disorder. *Mol. Neuropsychiatry* **2015**, *1*, 76–81. [CrossRef] [PubMed]

83. D'Addario, C.; Palazzo, M.C.; Benatti, B.; Grancini, B.; Pucci, M.; di Francesco, A.; Camuri, G.; Galimberti, D.; Fenoglio, C.; Scarpini, E.; et al. Regulation of gene transcription in bipolar disorders: Role of DNA methylation in the relationship between prodynorphin and brain derived neurotrophic factor. *Prog. Neuropsychopharmacol. Biol. Psychiatry* **2018**, *82*, 314–321. [CrossRef] [PubMed]

84. Dellosso, B.; Daddario, C.; Palazzo, M.C.; Benatti, B.; Camuri, G.; Galimberti, D.; Fenoglio, C.; Scarpini, E.; Di Francesco, A.; Maccarrone, M.; et al. Epigenetic modulation of *BDNF* gene: Differences in DNA methylation between unipolar and bipolar patients. *J. Affect. Disord.* **2014**, *166*, 330–333. [CrossRef] [PubMed]

85. Carlberg, L.; Scheibelreiter, J.; Hassler, M.R.; Schloegelhofer, M.; Schmoeger, M.; Ludwig, B.; Kasper, S.; Aschauer, H.; Egger, G.; Schosser, A. Brain-derived neurotrophic factor (BDNF)-epigenetic regulation in unipolar and bipolar affective disorder. *J. Affect. Disord.* **2014**, *168*, 399–406. [CrossRef] [PubMed]

86. Takeuchi, N.; Nonen, S.; Kato, M.; Wakeno, M.; Takekita, Y.; Kinoshita, T.; Kugawa, F. Therapeutic response to paroxetine in major depressive disorder predicted by DNA methylation. *Neuropsychobiology* **2017**, *75*, 81–88. [CrossRef] [PubMed]

87. Tang, H.; Dalton, C.F.; Srisawat, U.; Zhang, Z.J.; Reynolds, G.P. Methylation at a transcription factor-binding site on the 5-HT1A receptor gene correlates with negative symptom treatment response in first episode schizophrenia. *Int. J. Neuropsychopharmacol.* **2014**, *17*, 645–649. [CrossRef] [PubMed]

88. Kinoshita, M.; Numata, S.; Tajima, A.; Yamamori, H.; Yasuda, Y.; Fujimoto, M.; Watanabe, S.; Umehara, H.; Shimodera, S.; Nakazawa, T.; et al. Effect of clozapine on DNA methylation in peripheral leukocytes from patients with treatment-resistant schizophrenia. *Int. J. Mol. Sci.* **2017**, *18*, 632. [CrossRef] [PubMed]

89. Shi, Y.; Li, M.; Song, C.; Xu, Q.; Huo, R.; Shen, L.; Xing, Q.; Cui, D.; Li, W.; Zhao, J.; et al. Combined study of genetic and epigenetic biomarker risperidone treatment efficacy in Chinese Han schizophrenia patients. *Transl. Psychiatry* **2017**, *7*, e1170. [CrossRef] [PubMed]

90. Rukova, B.; Staneva, R.; Hadjidekova, S.; Stamenov, G.; Milanova, V.; Toncheva, D. Whole genome methylation analyses of schizophrenia patients before and after treatment. *Biotechnol. Biotechnol. Equip.* **2014**, *28*, 518–524. [CrossRef] [PubMed]

91. Kang, H.-J.; Kim, J.-M.; Stewart, R.; Kim, S.-Y.; Bae, K.-Y.; Kim, S.-W.; Shin, I.-S.; Shin, M.-G.; Yoon, J.-S. Association of SLC6A4 methylation with early adversity, characteristics and outcomes in depression. *Prog. Neuro-Psychopharmacol. Biol. Psychiatry* **2013**, *44*, 23–28. [CrossRef] [PubMed]

92. Okada, S.; Morinobu, S.; Fuchikami, M.; Segawa, M.; Yokomaku, K.; Kataoka, T.; Okamoto, Y.; Yamawaki, S.; Inoue, T.; Kusumi, I.; et al. The potential of *SLC6A4* gene methylation analysis for the diagnosis and treatment of major depression. *J. Psychiatr. Res.* **2014**, *53*, 47–53. [CrossRef] [PubMed]

93. Domschke, K.; Tidow, N.; Schwarte, K.; Deckert, J.; Lesch, K.-P.; Arolt, V.; Zwanzger, P.; Baune, B.T. Serotonin transporter gene hypomethylation predicts impaired antidepressant treatment response. *Int. J. Neuropsychopharmacol.* **2014**, *17*, 1167–1176. [CrossRef] [PubMed]

94. Domschke, K.; Tidow, N.; Schwarte, K.; Ziegler, C.; Lesch, K.-P.; Deckert, J.; Arolt, V.; Zwanzger, P.; Baune, B.T. Pharmacoepigenetics of depression: No major influence of MAO-A DNA methylation on treatment response. *J. Neural Transm.* **2015**, *122*, 99–108. [CrossRef] [PubMed]

95. Wang, P.; Lv, Q.; Mao, Y.; Zhang, C.; Bao, C.; Sun, H.; Chen, H.; Yi, Z.; Cai, W.; Fang, Y. HTR1A/1B DNA methylation may predict escitalopram treatment response in depressed Chinese Han patients. *J. Affect. Disord.* **2018**, *228*, 222–228. [CrossRef] [PubMed]

96. Powell, T.R.; Smith, R.G.; Hackinger, S.; Schalkwyk, L.C.; Uher, R.; McGuffin, P.; Mill, J.; Tansey, K.E. DNA methylation in interleukin-11 predicts clinical response to antidepressants in GENDEP. *Transl. Psychiatry* **2013**, *3*, e300. [CrossRef] [PubMed]

97. Liberati, A.; Altman, D.G.; Tetzlaff, J.; Mulrow, C.; Gøtzsche, P.C.; Ioannidis, J.P.A.; Clarke, M.; Devereaux, P.J.; Kleijnen, J.; Moher, D. The PRISMA statement for reporting systematic reviews and meta-analyses of studies that evaluate healthcare interventions: Explanation and elaboration. *BMJ* **2009**, *339*, b2700. [CrossRef] [PubMed]

98. Fries, G.R.; Quevedo, J.; Zeni, C.P.; Kazimi, I.F.; Zunta-Soares, G.; Spiker, D.E.; Bowden, C.L.; Walss-Bass, C.; Soares, J.C. Integrated transcriptome and methylome analysis in youth at high risk for bipolar disorder: A preliminary analysis. *Transl. Psychiatry* **2017**, *7*, e1059. [CrossRef] [PubMed]

99. Ghadirivasfi, M.; Nohesara, S.; Ahmadkhaniha, H.-R.; Eskandari, M.-R.; Mostafavi, S.; Thiagalingam, S.; Abdolmaleky, H.M. Hypomethylation of the serotonin receptor type-2A Gene (HTR2A) at T102C polymorphic site in DNA derived from the saliva of patients with schizophrenia and bipolar disorder. *Am. J. Med. Genet. Part B Neuropsychiatr. Genet.* **2011**, *156*, 536–545. [CrossRef] [PubMed]

100. Fries, G.R.; Bauer, I.E.; Scaini, G.; Wu, M.-J.; Kazimi, I.F.; Valvassori, S.S.; Zunta-Soares, G.; Walss-Bass, C.; Soares, J.C.; Quevedo, J. Accelerated epigenetic aging and mitochondrial DNA copy number in bipolar disorder. *Transl. Psychiatry* **2017**, *7*. [CrossRef] [PubMed]
101. Kidnapillai, S.; Bortolasci, C.C.; Udawela, M.; Panizzutti, B.; Spolding, B.; Connor, T.; Sanigorski, A.; Dean, O.M.; Crowley, T.; Jamain, S.; et al. The use of a gene expression signature and connectivity map to repurpose drugs for bipolar disorder. *World J. Biol. Psychiatry* **2018**, 1–9. [CrossRef] [PubMed]

International Journal of
Molecular Sciences

Article

Classical Risk Factors and Inflammatory Biomarkers: One of the Missing Biological Links between Cardiovascular Disease and Major Depressive Disorder

Thomas C. Baghai [1,2,*], Gabriella Varallo-Bedarida [3], Christoph Born [1], Sibylle Häfner [1], Cornelius Schüle [1], Daniela Eser [1], Peter Zill [1], André Manook [2], Johannes Weigl [2], Somayeh Jooyandeh [2], Caroline Nothdurfter [2], Clemens von Schacky [3], Brigitta Bondy [1] and Rainer Rupprecht [1,2,4]

[1] Department of Psychiatry and Psychotherapy, Ludwig-Maximilian-University of Munich, Nußbaumstraße 7, D-80336 Munich, Germany; Christoph.Born@vivantes.de (C.B.); sibylle.haefner@med.uni-heidelberg.de (S.H.); Cornelius.Schuele@med.uni-muenchen.de (C.S.); Daniela.Eser@med.uni-muenchen.de (D.E.); Peter.Zill@med.uni-muenchen.de (P.Z.); brigitta.bondy@med.uni-muenchen.de (B.B.); Rainer.Rupprecht@medbo.de (R.R.)

[2] Department of Psychiatry and Psychotherapy, University Regensburg, Universitätsstraße 84, D-93053 Regensburg, Germany; Andre.Manook@medbo.de (A.M.); Johannes.Weigl@medbo.de (J.W.); Somayeh.MohammadiJooyandeh@medbo.de (S.J.); Caroline.Nothdurfter@medbo.de (C.N.)

[3] Department of Internal Medicine—Preventive Cardiology, Ludwig-Maximilian-University of Munich, Ziemssenstraße 1, D-80336 Munich, Germany; Gabriella.Bedarida@pfizer.com (G.V.-B.); Clemens.vonSchacky@med.uni-muenchen.de (C.v.S.)

[4] Max-Planck Fellow at the Max-Planck-Institute for Psychiatry, Kraepelinstraße 2-10, D-80804 Munich, Germany

[*] Correspondence: Thomas.Baghai@medbo.de; Tel.: +49-941-941-1012; Fax: +49-941-941-1645

Received: 28 May 2018; Accepted: 8 June 2018; Published: 12 June 2018

Abstract: Background: Cardiovascular disorders (CVD) and major depressive disorder (MDD) are the most frequent diseases worldwide responsible for premature death and disability. Behavioral and immunological variables influence the pathophysiology of both disorders. We therefore determined frequency and severity of MDD in CVD and studied whether MDD without CVD or other somatic diseases influences classical and inflammatory biomarkers of cardiovascular risk. In addition, we investigated the influence of proinflammatory cytokines on antidepressant treatment outcome. Methods: In a case-control design, 310 adults (MDD patients without CVD, CVD patients, and cardiologically and psychiatrically healthy matched controls) were investigated. MDD patients were recruited after admission in a psychiatric university hospital. Primary outcome criteria were clinical depression ratings (HAM-D scale), vital signs, classical cardiovascular risk factors and inflammatory biomarkers which were compared between MDD patients and healthy controls. Results: We detected an enhanced cardiovascular risk in MDD. Untreated prehypertension and signs directing to a metabolic syndrome were detected in MDD. Significantly higher inflammatory biomarkers such as the high sensitivity C-reaktive protein (hsCRP) and proinflammatory acute phase cytokines interleukine-1β (IL-1β) and interleukine-6 (IL-6) underlined the higher cardiovascular risk in physically healthy MDD patients. Surprisingly, high inflammation markers before treatment were associated with better clinical outcome and faster remission. The rate of MDD in CVD patients was high. Conclusions: Patients suffering from MDD are at specific risk for CVD. Precise detection of cardiovascular risks in MDD beyond classical risk factors is warranted to allow effective prophylaxis and treatment of both conditions. Future studies of prophylactic interventions may help to provide a basis for prophylactic treatment of both MDD and CVD. In addition, the high risk for MDD in CVD patients was confirmed and underlines the requirement for clinical attention.

Keywords: cardiovascular disease; cell adhesion molecules; immunology; inflammation; nervous system

---

1. Introduction

Cardio- and cerebrovascular disorders are recognized worldwide as the most frequent causes for death and major depressive disorder (MDD) for disability [1]. In recent cross-sectional studies, repeatedly an association of current depressive symptoms and lifetime depressive episodes with cardiovascular disease (CVD), could be confirmed [2]. MDD represents a major risk factor for CVD and for myocardial infarction (MI) [3] independent of traditional risk factors [4], this seems to be especially true in geriatric depression [5]. The MDD subtype, severity of depression [6], and an activation of central stress regulatory systems [7] seem to be relevant in this context. Thereby, the influence of MDD on CVD is equivalent to somatic risk factors [8] and a bi-directional relationship between the cardiovascular system and altered mood states related to an inflammatory status has been suggested [9].

CVD is a chronic inflammatory disease [10] that can be monitored using inflammatory biomarkers such as C-reactive protein (CRP), adhesive cell-surface glycoproteins (sVCAM-1), the monocyte chemotactic protein-1 (MCP-1), the pro-inflammatory acute phase cytokines inteleukin-1 and -6 (IL-1, IL-6), cell adhesion molecules (leukocyte (L-), endothelial (E-), and platelet (P-)selectin), and the intracellular adhesion molecule-1 (sICAM-1) [11], which is acting together with VCAM-1 as monocyte and T cells receptors [11]. Inflammatory mediators and adhesion molecules are peripheral clinical markers for vascular wall inflammation [12] and represent risk factors associated strongly with atherosclerosis [13] and the risk for MI or stroke [14]. They offer prognostic relevance both on the short-run and in the long term in chronic inflammatory states [15].

High CRP seems to reflect also the activity of bipolar disorder: It was associated with high activity of the disorder in both conditions manic and depressed states [16]. Whereas high plasma cholesterol levels are a risk factor in CVD, the use as a clinical marker for affective disorders was a matter of debate and is still unclear [17]. The association of low cholesterol levels with the risk for suicide in psychiatric patients seems to be a more stable finding, but is also discussed controversially up to now [18].

MDD symptoms can be mimicked by excessive secretion of the pro-inflammatory macrophage cytokines [19] and are often accompanied by a hyperactivity of the hypothalamic-pituitary-adrenal (HPA) axis [20] which includes direct stimulatory influence of interleukins (IL-1, IL-6) on hypothalamic corticotropin-releasing factor (CRF) and pituitary corticotropin (ACTH) secretion [19]. Pro-inflammatory cytokines can influence glucocorticoid receptor resistance and serotonergic neurotransmission [21]. Therefore, a causal relationship between inflammation and MDD seems plausible [22,23].

Both population based [23–25] and prospective [26] studies implied that MDD belongs to the group of chronic inflammatory diseases, but also controversial results have been obtained in cross-sectional studies which assessed correlations of inflammatory markers and depressive symptoms assessed using self-rating scores [27].

Therefore, the presented comparative prospective case-control study in both diagnostic groups MDD and CVD which directly compares immunological and clinical cardiovascular risk variables may help to clarify these interdependencies. We prospectively investigated both MDD and CVD patients in comparison to psychiatrically and medically healthy controls. We studied frequency and severity of depression in CVD and compared clinical and immunological cardiovascular risk factors between CVD and MDD in an interdisciplinary study. Moreover, we studied to what extent MDD is related to cardiovascular risk factors and to inflammatory biomarkers in cardiovascularly healthy patients suffering from MDD in relation to healthy controls without CVD in order to identify biomarkers which may be suitable for estimation of the cardiovascular risk in MDD. In addition, we investigated the influence of inflammatory markers on antidepressant treatment outcome.

2. Results

2.1. Classical Cardiovascular Risk Markers

Comparison of mean values showed significant differences in a variety of risk markers indicating the elevated cardiovascular risk of MDD patients in comparison to healthy controls (Table 1). Blood pressure (BP) determinations according to the Joint National Committee on Prevention, Detection, Evaluation and Treatment of High Blood Pressure (JNC)-7 definitions [28] (Figure 1) revealed significantly more normal values in healthy controls, prehypertensive BP in MDD and stage 1 and 2 hypertension in CVD patients (χ^2 = 25.6, d.f. = 6, p < 0.001).

Table 1. Demographic, clinical, and medical characteristics of samples 1–3 including psychiatric ratings, vital signs, and selected classical cardiovascular risk factors. Vital signs as well as classical cardiovascular risk factors and behavioral variables are indicating an elevated cardiovascular risk in MDD.

Variable	Sample 1 * MDD	Sample 2 CVD	Sample 3 † Controls	Kruskal-Wallis- or χ^2-Test ‡ χ^2, d.f., p
n	100	106	104	
age (mean ± SD)	46.6 ± 14.8	66.9 ± 7.3	54.7 ± 14.4	100, 2, p < 0.001
sex (male/female)	37.0%/63.0%	81.1%/18.9%	45.2%/54.8%	46.3, 2, p < 0.001 ‡
clinical ratings (baseline)				
CGI-1	5.3 ± 0.5	1.3 ± 0.8	1.0 ± 0.0	205, 2, **p < 0.001**
HAM-D17 (mean ± SD)	22.0 ± 5.3	2.4 ± 4.4	0.7 ± 1.2	160, 2, **p < 0.001**
MADRS (mean ± SD)	31.8 ± 7.4	3.2 ± 6.0	0.7 ± 1.3	162, 2, **p < 0.001**
BDI (mean ± SD)	26.4 ± 9.1	7.0 ± 5.1	2.8 ± 3.2	130, 2, **p < 0.001**
Vital signs and classical cardiovascular risk factors				
blood pressure (systolic)	129.2 ± 13.5	138.6 ± 19.1	128.6 ± 19.1	18.7, 2, **p < 0.001**
blood pressure (diastolic)	79.7 ± 8.2	78.3 ± 10.9	76.8 ± 11.9	4.63, 2, n.s. §
heart rate (beats/minute)	85.7 ± 14.7	63.8 ± 10.7	68.9 ± 10.1	104, 2, **p < 0.001**
total cholesterol (mg/dL)	206.2 ± 58.1	n.d.	223.8 ± 48.1	4.15, 1, p = 0.042
LDL (mg/dL)	125.2 ± 42.8	n.d.	136.9 ± 45.6	2.30, 1, n.s.
triglycerides (mg/dL)	148.8 ± 100.3	n.d.	100.6 ± 51.6	10.5, 1, **p = 0.001**
HDL (mg/dL)	58.4 ± 17.1	n.d.	66.1 ± 19.8	5.67, 1, **p = 0.017**
fasting glucose (mg/dL)	95.1 ± 18.2	n.d.	86.9 ± 11.6	5.41, 1, **p = 0.02**
body weight (kg)	72.9 ± 13.6	82.8 ± 14.7	72.3 ± 11.8	32.9, 2, p < 0.001
body mass index (kg/m^2)	25.3 ± 4.0	28.0 ± 4.4	24.4 ± 2.9	35.8, 2, **p < 0.001**
waist circumference (cm)	96.9 ± 12.3	101.9 ± 12.9	90.1 ± 12.1	34.2, 2, **p < 0.001**
hip circumference (cm)	105.3 ± 10.7	105.8 ± 9.5	100.6 ± 8.4	12.5, 2, p = 0.002
waist-hip-ratio	0.92 ± 0.12	0.96 ± 0.07	0.90 ± 0.09	35.9, 2, p < 0.001
smoker/non-smoker (%)	42.9%/57.1%	9.0%/91.0%	14.6%/85.4%	34.1, 2, **p < 0.001** ‡
pack years (20 cigarettes/day * years)	18.8 ± 13.4	35.7 ± 30.7	18.8 ± 20.3	9.96, 2, p = 0.007
Framingham-index (total)	2.77 ± 5.74	9.69 ± 2.69	4.09 ± 5.86	99.3, 2, p < 0.001
Framingham-index, 10 years-risk (%)	4.94 ± 5.62	11.47 ± 7.11	5.85 ± 5.78	96.4, 2, p < 0.001

dropouts: n = 16; MDD = major depressive disorder; CVD = cardiovascular disorder; ANOVA = univariate analysis of variance; d.f. = degrees of freedom; SD = standard deviation; n.s. = not statistically significant; n.d. = not done (due to treatment with statins); CGI-1 = Clinical global impression scale, Item 1—severity of disease; HAM-D17 = Hamilton rating scale for depression, 17-item version; MADRS = depression rating scale; BDI = Beck depression inventory; LDL = low density lipoproteins; HDL = high density lipoproteins; * medical comorbidities were exclusion criteria for sample 1; † medical and psychiatric comorbidities were exclusion criteria for sample 3; bold = statistical significant differences between MDD and controls confirmed (Mann-Whitney-U test) after exact age and sex matching; ‡ χ^2-test in case of categorical variables (continuous variables were evaluated by univariate analysis of variance); § Man-Whitney-U test: p = 0.005.

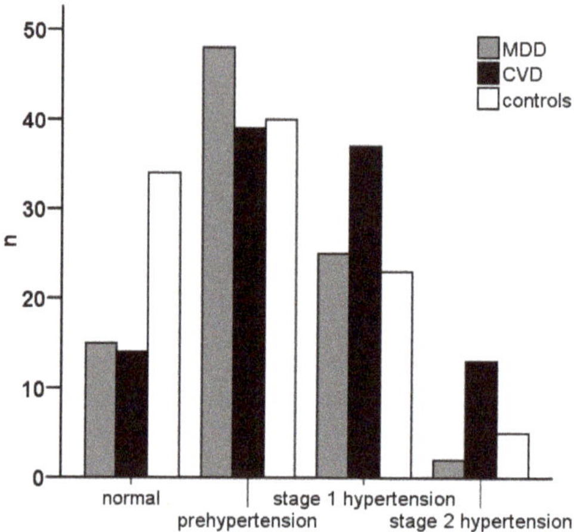

Figure 1. JNC 7 classification of blood pressure. Classification of blood pressure baseline values according "The Seventh Report of the Joint National Committee on Prevention, Detection, Evaluation, and Treatment of High Blood Pressure" (JNC-7) [28] showed significantly different distribution of hypertension categories in major depressive disorder (MDD) and cardiovascular disorder (CVD) patients as well as in healthy controls (n = number of subjects in either group).

Post-hoc comparison of depressed patients and controls confirmed the significantly elevated diastolic blood pressure (BP) (U = 3777, p = 0.034) and heart rates (U = 1574, p < 0.001). Metabolic risk factors included significantly lower total cholesterol (U = 2622, p = 0.042) in view of significantly higher triglycerides (U = 2221, p = 0.001) and fasting glucose values (U = 1917, p = 0.020) in combination with lower high density lipoproteins (HDL) (U = 1711, p = 0.017). Lifetime smoking habits did not differ, whereas more MDD patients are current smokers. After correction for age using exact matching all significant differences except for cholesterol (p = 0.24) were confirmed. Moreover, systolic BP was significantly higher in MDD patients at baseline (U = 2396, p = 0.040).

CVD patients in comparison to controls showed significantly higher systolic BP values (U = 3677, p < 0.001) and heart rates (U = 3622, p < 0.001). Increased body weight (U = 3034, p < 0.001), BMI (U = 2794, p < 0.001), waist (U = 2573, p < 0.001), hip (U = 3417, p < 0.001), and waist/hip ratio (U = 2505, p < 0.001) were indicators of the higher metabolic risk. We also registered a higher lifetime cigarette consumption with more pack years (U = 733, p = 0.003). The Framingham sum index (U = 2042, p < 0.001) and the 10 years risk (U = 2078, p < 0.001) were significantly higher in CVD patients. After exact matching for age and sex all statistically significant differences with the exception of systolic BP differences were confirmed.

2.2. Inflammation Biomarkers in MDD in Comparison to Healthy Controls

Among inflammatory biomarkers, the general inflammatory marker high sensitivity (hs)CRP, the pro-inflammatory cytokines IL-6 and IL-1β as well as the adhesion molecule sICAM-1 were significantly higher in depressed patients in comparison to controls (Table 2). To rule out possible effects of a divergent age and gender distribution we repeated all comparisons after exact matching for age and sex. All differences remained statistically significant, except for higher sICAM-1 values in MDD patients, which changed to a nonsignificant trend (T = −1.94, p = 0.055).

To evaluate putative improvements of cardiovascular risk factors, all inflammatory markers were determined also after remission of depression shortly before discharge of the hospital. Only IL-6 and P selectin declined significantly after treatment with antidepressants (Table 2).

Table 2. Inflammation biomarkers and risk factors.

Samples Variable	Sample 1 MDD Baseline	Discharge	*t*-Test * T, *p*	Sample 3 Controls Baseline	*t*-Test † T, *p*
inflammation marker hsCRP (mg/L)	3.07 ± 3.7	3.97 ± 4.4	0.23, n.s.	1.37 ± 1.2	**4.25, *p* < 0.001**
pro-inflammatory cytokines interleukin 1β (IL-1 β) (pg/mL) interleukin 6 (IL-6) (pg/mL)	1.08 ± 1.1 1.58 ± 1.5	1.34 ± 1.0 1.35 ± 1.7	0.88, n.s. 2.26, *p* = 0.027	0.54 ± 0.6 1.32 ± 1.3	**4.10, *p* < 0.001** **2.64, *p* = 0.009**
adhesion molecules P selectin (ng/mL) E selectin (ng/mL) MCP-1 (pg/mL) sICAM-1 (ng/mL) sVCAM-1 (ng/mL)	150.4 ± 102.4 54.6 ± 29.0 221.3 ± 149.7 535.4 ± 210.0 486.1 ± 182.0	114.4 ± 78.9 58.0 ± 32.0 301.8 ± 179.2 555.0 ± 209.5 533.2 ± 221.4	2.56, *p* = 0.013 −0.69, n.s. −2.92, *p* = 0.005 −0.92, n.s. −1.62, n.s.	184.2 ± 146.0 46.5 ± 27.2 256.5 ± 140.8 360.6 ± 107.7 552.3 ± 142.2	−1.47, n.s. 1.76, n.s. −1.54, n.s. **5.90, *p* < 0.001** **−4.41, *p* = 0.017**
costimulatory glycoprotein sCD40 (ng/mL)	10.4 ± 3.7	11.3 ± 3.7	−2.42, *p* = 0.019	10.1 ± 4.2	0.59, n.s.

Inflammation biomarkers in MDD patients. Comparison to healthy controls: hsCRP, IL-1β, and sICAM-1 were significantly elevated in comparison to healthy controls. Other markers showed only nonsignificant trends or were even lower in MDD. Comparison before and after treatment of depression (at baseline and before discharge of the hospital): Reduced IL-6 and P selectin after antidepressant treatment. * Student's *t*-test for dependent samples: sample 1 baseline vs. discharge; † Student's *t*-test for independent samples: baseline sample 1 vs. baseline sample 3; n.s. = not statistically significant; bold = statistical significant differences between MDD and controls confirmed after exact age and sex matching (*t*-test and Mann-Whitney-U test).

2.3. Cardiovascular Risk, Severity of Depression, and Time to Remission in MDD

There was correlative coherence between classical clinical and inflammatory cardiovascular risk factors and the severity of depressive symptoms: Baseline values showed positive correlations with HAM-D17 scores for heart rate (Spearman's ρ = 0.37, *p* < 0.001), triglycerides (ρ = 0.20, *p* = 0.001), and fasting glucose levels (ρ = 0.25, *p* < 0.001). Weak, but statistically significant negative correlations could be seen for HDL (ρ = −0.15, *p* = 0.02) and for the Framingham sum score (ρ = −0.15, *p* = 0.002) as well as for the 10-years risk score (ρ = −0.15, *p* = 0.003). The current smoking status (numbers of cigarettes) was correlated with hsCRP levels at the time of admission in MDD patients (ρ = 0.59, *p* = 0.045).

Baseline values of inflammation biomarkers revealed weak, but statistically significant positive correlations of HAM-D17 sum scores with hsCRP (ρ = 0.23, *p* < 0.001), sICAM-1 (ρ = 0.35, *p* < 0.001) and IL-1β (ρ = 0.26, *p* = 0.002). IL-6 showed a positive correlation only before discharge (ρ = 0.30, *p* = 0.025).

Surprisingly, a specific combination of psychopathology and inflammatory markers seems to affect the time to complete remission (HAM-D17 scores ≤ 7) from MDD: patients more likely to experience a more rapid relief from depressive symptoms suffered from lesser severe MDD (weak positive correlation of HAM-D17 at baseline with time to remission: ρ = 0.28, *p* = 0.045), were younger (ρ = −0.36, *p* < 0.001) and showed a specific inflammatory biomarker profile at baseline. This included higher LDL cholesterol (ρ = 0.39, *p* = 0.037) together with higher IL-1β- (ρ = −0.33, *p* = 0.030) and hsCRP-values (ρ = −0.32, *p* = 0.029) indicating a higher level of chronic inflammation. Also high hsCRP levels >2.0 mg/dL are considered as sign of subclinical inflammation and a biomarker of an enhanced cardiovascular risk [29,30]. Baseline hsCRP values were elevated in 38.5% of our MDD patients. As indicated in Figure 2 the time to reach remission was significantly shorter (mean ± SD high vs. low: 25.6 ± 17.5 vs. 40.0 ± 27.3 days) in patients with elevated baseline CRP (Kaplan-Meyer-analysis, Log Rank (Mantel-Cox) d.f. = 1, *p* = 0.025).

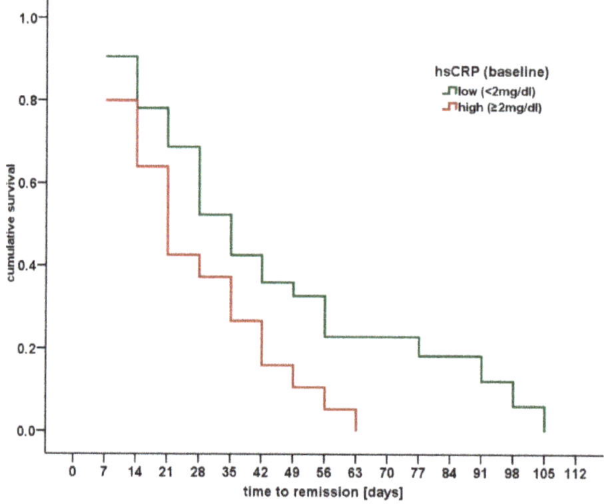

Figure 2. Kaplan-Meyer survival function of time to full remission of MDD for hsCRP status. Estimated likelihood of full remission (HAM-D17 score < 7) based on Mantel-Cox's regression analysis for hsCRP concentrations at baseline in patients suffering from MDD. Time to full remission was significantly shorter in patients with elevated baseline hsCRP concentrations.

3. Discussion

Patients with stable CVD suffered from additional MDD in 15.1% of the cases. Our rate of MDD in CVD patients is in line with other studies investigating the simultaneous presence of both diseases. In the Enhancing Recovery in Coronary Heart Disease (ENRICHD) study CVD patients one month after acute MI, 19.8% suffered from a depressive syndrome and 10.5% fulfilled the criteria for MDD [31]. Also Sørensen et al. found a rate of 10.0% of CVD patients fulfilling ICD-10 criteria for depressive disorders and 7.2% suffered from moderate to severe depression [32]. Our higher incidence rate may be due to the selection of CVD patients suffering from chronic somatic disease, because we recruited the patients not shortly after MI as done in other studies, but in a stable condition during their regular outpatient care in the department of preventive cardiology. An additional reason may be an improved detection of MDD in comparison to other studies due to highly trained psychiatrists on duty in our study. Due to the known elevated inflammation markers in CVD without clear connection to depressive symptoms [33] we omitted the measurement of hsCRP and cytokines in patients suffering from CVD.

Even if depression has been considered as an additional risk factor for CVD independent of classical known risks such as age, gender, vital signs, and blood lipid levels [4,34] the presence of MDD without CVD affects also classical risk factors. In particular, we found a higher number of patients with prehypertensive BP in MDD patients without antihypertensive medication. In spite of antihypertensive treatments, hypertensive states according to JNC-7 criteria [28] were more frequent in CVD. In line with our results, Yan et al. described a higher incidence of hypertension associated with higher depression rating scores in young adults [35].

We detected higher triglycerides and low HDL cholesterol levels in depressed patients in comparison to healthy controls. This profile indicates the considerably enhanced risk for CVD in MDD patients. Differences in fasting glucose levels and waist circumferences as well as a higher rate of smokers in MDD point towards an unhealthy life style (diet, smoking, and low activity) associated with depressive symptoms [36]. Our results suggest that the higher cardiovascular risk in depressive patients is mediated by four of the five component of the metabolic syndrome (triglycerides, fasting glucose, waist circumference, and HDL) rather than by other classical risk factors. Therefore, our findings contribute new data to the current ongoing open debate on the association between depression and metabolic syndrome [37].

Higher inflammatory biomarkers in MDD in comparison to healthy controls support the theory of chronic inflammation and an enhanced cardiovascular risk in MDD. hsCRP represents a stable plasma biomarker for a low-grade systemic inflammation [38]. Elevated IL-6, IL-1β, and sICAM-1 levels in our MDD patients without CVD support the hypothesis of chronic inflammation in MDD. IL-6 stimulates HPA-axis overdrive well known in MDD and CRP production which represents an important risk factor for CVD [39]. This is in line with other studies showing an elevation of inflammatory biomarkers in depression [40] even if underlying mechanisms still are not fully understood and we still don't know whether the inflammation drives depression or vice versa [41].

In contrast to other studies [26], we could not detect a consistent significant impact of successful antidepressant treatment on all inflammatory markers. Only IL-6 concentrations were significantly lower after treatment. Possible reasons are the divergent influence of antidepressant medication or of MDD itself on differential stages of inflammatory processes, which should be systematically studied in further investigations. We could not detect a reduction of IL-1β concentrations in the peripheral blood. Nevertheless we cannot rule out reduced concentrations after treatment in the brains of our patients which are suggested due to results of preclinical studies: rats exposed to chronic mild stress and treated with the antidepressant fluoxetine for up to 12 days demonstrated lower IL-1β in both plasma and brain [42]. Therefore, these effects should be investigated systematically in further studies.

However, our data indicate that patients remitted from an acute phase of MDD may still have a persistent elevated cardiovascular risk independently from the state of the depressive disorder, because high IL6 indicates an up to three-fold higher risk of sudden cardiac death and high CRP is associated with nonfatal coronary heart disease [43]. This implies a modification of the actually used and possibly insufficient prophylactic strategies preventing CVD in remitted MDD patients.

Moreover, we provide first evidence in our MDD sample that higher hsCRP levels, which indicate subclinical inflammation and higher risks for CVD, are associated with significantly shorter time to remission during antidepressant treatment. This further fosters the discussion about depression influencing immune dysregulation [44] or pathological alterations of the immune system causing depressive symptoms and MDD [26]. In contrast to our results, in the study of Lanquillon et al. in 24 MDD patients higher IL-6 levels predicted worse outcome during the treatment of MDD [45]. However, IL-6 concentrations were not elevated in comparison to healthy controls and elevated CRP was not useful for the prediction of treatment response.

In a recently published study, patients suffering primarily from hypertension and a metabolic syndrome showed also depressive symptoms, but no significant relationship between depression scores and cardiovascular or metabolic risk factors was detected [36]. In our study, smoking showed a positive correlation with hsCRP [46], but we could not confirm correlations with other inflammatory markers. However, correlations of inflammatory biomarkers with severity of depression confirmed a higher cardiovascular risk in more severely depressed patients.

Summarizing the results, our study revealed elevated inflammatory biomarkers in physically completely healthy MDD patients pointing to an enhanced risk for CVD related to the severity of depression. However, a positive impact of inflammation on outcome after antidepressant treatment could be observed. Thus, the immune system's inflammatory response may not solely facilitate the symptoms of depression [26] but, similar to infection or injury [44], may also represent a constructive consequence of the human body facilitating clinical remission in case of MDD.

Patients suffering from MDD are at specific risk for the development of CVD. For the assessment of this risk the observance of classical cardiovascular risk factors including the Framingham index appears not to be sufficient. Our study suggests that the enhanced cardiovascular risk is mediated additionally by non-traditional risk factors including inflammatory biomarkers and a metabolic syndrome. MDD facilitates an unhealthy life style and consequently favors the development of a metabolic syndrome. On the other hand, biomarkers indicating a chronic inflammation, such as hsCRP, interleukins, and others, suggest that MDD patients are a high risk group for CVD. As a consequence a close monitoring of smoking status, vital signs (blood pressure and heart rate), triglycerides, HDL cholesterol, hsCRP, and, if possible, pro-inflammatory cytokines (IL-1β) and adhesion molecules (sICAM-1) is recommended for cardiovascular risk assessment.

In patients suffering from stable CVD a relatively high rate of MDD could be detected. Due to the fact that MDD in CVD represents a risk factor for subsequent cardiovascular events sufficient treatment of a putative depressive disorder is warranted. Effective antidepressant treatment supports the relief of depression, whereas an improvement of the prognosis and the reduction of cardiovascular mortality could not be proven sufficiently in prospective studies.

For patients suffering from CVD in combination with MDD a cardiac rehabilitation program is recommended, because it was found to increase the levels of physical and mental quality of life and lower also levels of depression [47].

Limitations of our study were the limited case number in comparison to epidemiologic studies. Therefore we were not able to detect small differences in the investigated biomarkers between our three groups. Moreover, up until now, no follow-up investigations to detect longitudinal developments of risk factors and biomarkers and their clinical long-term consequences for the investigated MDD patient group have been performed. Such further investigations are warranted, because they may help to better classify the investigated risk factors and deduce better clinical recommendations.

To meet the requirement of an effective prophylaxis of both cardiovascular disorders in MDD and for diminishing the risk for MDD in CVD, more secondary and primary prevention studies are needed. In MDD early assessment of cardiovascular risk and an early induction of preventive arrangements may be useful. These steps may start with nutritional and life style educations, proceed to dietary arrangements and supplements, and, in case of increased cardiovascular risks, may even include cardio-protective medication. In CVD early diagnosis, treatment and prophylaxis of depression is desirable. This may include psychoeducation, psychosocial therapies, psychotherapeutic approaches, and suitable antidepressant medication.

Even if evidence from randomized controlled trials RCTs still is lacking, in most cases a combination of the mentioned treatments may be of use. From a clinical point of view only the awareness for the described risks, sufficient diagnostics and adequate treatment may help to go against the predicted worldwide rise of the impact of both disorders, CVD and MDD.

4. Material and Methods

4.1. Study Samples

A total of 333 subjects were recruited within the funding period of four years, 23 patients and controls were excluded due to somatic or psychiatric diagnoses detected after study inclusion or due to withdrawal of consent. Therefore, a total of 310 patients were investigated in the study. Table 1 shows clinical and demographic data for patients and controls.

One hyndred unrelated in-patients suffering from unipolar MDD were recruited. Patients were diagnosed by experienced and trained psychiatrists according to DSM-IV [48] using the Structured Clinical Interview for DSM-IV (SCID). Only patients over 18 years old with an at least moderately severe depressive episode were included. The main inclusion criteria were unipolar depression and a score in the Hamilton Rating Scale for Depression [49] (17-item version, HAM-D17) of at least 17. Exclusion criteria were all other psychiatric comorbidities including e.g., schizophrenia, addiction or mental retardation. Prior to inclusion in the study blood samples were obtained for routine laboratory screening, a medical history was taken and a physical examination was performed by a physician to exclude medical disorders. Clinically relevant medical illness and the concomitant use of antihypertensive medication as well as hormone replacement therapies, alcohol or drug abuse within the last 6 month prior to study inclusion or withdrawal signs led to exclusion from the study. After washout period of at least 3 days prior to the blood sampling for the study patients received non-standardized antidepressant treatments (predominantly Mirtazapine up to 45 mg/day and Venlafaxine up to 300 mg/day) according to clinical requirements, cognitive behavioral treatment and social support. Changes in the depressive state were monitored using the HAM-D17 which was the primary outcome variable, the Montgomery-Åsberg-depression rating scale (MADRS) and the Beck depression inventory (BDI).

One hundred and six outpatients suffering from CVD were recruited from the Department of Internal Medicine—Preventive Cardiology of the Ludwig-Maximilian-University of Munich (sample 2—CVD). Their psychiatric evaluation included also the SCID for DSM-IV diagnoses together with CGI, HAM-D17, MADRS and BDI evaluation. Criteria for inclusion were evidence of CVD documented by confirmed diagnosis of previous myocardial infarction as per hospital discharge summary, or history of CABG or PCI or evidence of ischemic heart disease based on stress electrocardiography confirmed by diagnostic imaging. Patients had to be in stable conditions defined as at least 3 months from an acute episode, intervention or hospitalization for CVD. Thyroid disease, diabetes, hypertension and other chronic conditions had to be well controlled on a stable medical regimen for a minimum of 3 months. Inflammatory biomarkers were not determined in sample 2 due to ongoing treatments with statins.

One hundred and four age- and sex-matched controls of Caucasian ethnicity were recruited at the LMU and screened for psychiatric (SCID) and for medical disorders. A complete medical and social history, a detailed review of systems and complete physical examination were obtained to assess absence of cardiac, cerebral or peripheral vascular disease. Subjects with diabetes were excluded as this is considered a cardiovascular-equivalent disorder. Only healthy individuals negative for both psychiatric and medical disorders entered the study.

4.2. Assessment of Vital Signs and Calculation of the Framingham-Index

Height and weight were measured and the body mass index (BMI, kg/m^2) was calculated. Hip and waist circumference were measured to calculate the hip/waist ratio. At least two blood pressure determinations were made after the patient or control subject had been sitting for at least 5 min with the arm at the heart level. Average values were used for further analysis. The calculation of the Framingham index including the risk factors age, gender, total cholesterol, HDL, systolic and diastolic blood pressure, and smoking status was performed as described elsewhere [50]. It was expressed as a total score and the 10-years risk for CVD incidence.

4.3. Biochemical Analyses

Inflammatory Risk Factors

All measurements were performed twice from single blinded personal not knowing sample affiliation or clinical details. hsCRP concentrations were determined with a commercially available ELISA (sensitivity 0.1–10 mg/L; DRG Diagnostics, Germany). High sensitivity IL-6 and IL-1β concentrations were measured using ELISA of R&D Systems, Minneapolis according the protocol delivered from the manufacturer. MCP-1, VCAM-1, ICAM-1, E-Selectin, P-Selectin, sCD40-L were determined using ELISAs obtained from IBL (Immuno Biological Lab), Minneapolis, MN, USA according to standard protocols.

4.4. Statistical Analysis

All analyses were performed using the Statistical Package for the Social Sciences (SPSS) for Windows (Releases 15-24, SPSS Inc., Chicago, IL, USA and IBM Deutschland GmbH, Ehningen, Germany). The One-Sample Kolmogorov-Smirnov Test was used to test about normal distribution of all variables. In case of a non-normal distribution, the corresponding variables were transformed with the log-transformation to reach a normal distribution before entering parametric testing. In case of persistent deviations from normal distribution non-parametric comparisons of mean values using the Mann-Whitney U test for comparison of means were performed. In case of normal distributions mean differences in demographic and clinical variables between patients and controls were compared using univariate analyses of variance (ANOVA procedure) or Student's t-tests in case of continuous variables. In case of categorical variables the frequencies were compared using χ^2-tests. In addition, we screened cardiovascular risk factors for correlations with predominantly nonparametric variables using Spearman's rho coefficient. To test for influence of hsCRP status on treatment outcome in MDD patients Kaplan-Meyer survival analysis including Cox-regression was applied.

To rule out significant age and gender effects all comparisons were done after exact matching.

4.5. Ethical Approval

The study was approved by the ethics committee of the medical faculty of the Ludwig-Maximilian-University Munich (Project No. 207/03, approval 29 August 2003). Written informed consent was obtained from all patients and control subjects. Patients' data were anonymized.

Author Contributions: T.C.B., G.V.-B., R.R., B.B., and C.v.S. were responsible for study concept and design, analysis and interpretation of data, drafting the manuscript and revising it for important intellectual content, and study supervision. A.M., J.W., S.J. and C.N. contributed a substantial manuscript revision, C.B., S.H., C.S., and D.E., acquired the data, T.C.B. and G.V.-B. obtained funding. T.C.B. and B.B. analyzed the data. T.C.B., G.V.-B., P.Z., R.R., B.B., and C.v.S. gave administrative and technical support. T.C.B. is the guarantor.

Acknowledgments: This project is supported by a grant from the *Deutsche Forschungsgemeinschaft* (DFG BA 2309/1-1, funding period 2004–2008). The authors would like to thank Sylvia de Jonge and Klaus Neuner for expert technical assistance.

Conflicts of Interest: The authors declare no conflict of interest.

References

1. Mathers, C.D.; Stein, C.; Ma Fat, D.; Rao, C.; Inoue, M.; Tomijima, N.; Bernard, C.; De Lopez, A.; Murray, C.J.L. *Global Burden of Disease 2000: Version 2 Methods and Results*; Global Programme on Evidence for Health Policy Discussion Paper; World Health Organization: Geneva, Switzerland, 2002.
2. Ivanovs, R.; Kivite, A.; Ziedonis, D.; Mintale, I.; Vrublevska, J.; Rancans, E. Association of depression and anxiety with cardiovascular co-morbidity in a primary care population in Latvia: A cross-sectional study. *BMC Public Health* **2018**, *18*, 328. [CrossRef] [PubMed]
3. Frasure-Smith, N.; Lesperance, F.; Talajic, M. Depression following myocardial infarction. Impact on 6-month survival. *JAMA* **1993**, *270*, 1819–1825. [CrossRef] [PubMed]
4. Barefoot, J.C.; Schroll, M. Symptoms of depression, acute myocardial infarction, and total mortality in a community sample. *Circulation* **1996**, *93*, 1976–1980. [CrossRef] [PubMed]
5. Zhang, Y.; Chen, Y.; Ma, L. Depression and cardiovascular disease in elderly: Current understanding. *J. Clin. Neurosci.* **2018**, *47*, 1–5. [CrossRef] [PubMed]
6. Penninx, B.W.; Beekman, A.T.; Honig, A.; Deeg, D.J.; Schoevers, R.A.; van Eijk, J.T.; van Tilburg, W. Depression and cardiac mortality: Results from a community-based longitudinal study. *Arch. Gen. Psychiatry* **2001**, *58*, 221–227. [CrossRef] [PubMed]
7. Lederbogen, F.; Deuschle, M.; Heuser, I. Depression—A cardiovascular risk factor. *Internist* **1999**, *40*, 1119–1121. [CrossRef] [PubMed]
8. Wulsin, L.R.; Singal, B.M. Do depressive symptoms increase the risk for the onset of coronary disease? A systematic quantitative review. *Psychosom. Med.* **2003**, *65*, 201–210. [CrossRef] [PubMed]
9. Grippo, A.J.; Johnson, A.K. Biological mechanisms in the relationship between depression and heart disease. *Neurosci. Biobehav. Rev.* **2002**, *26*, 941–962. [CrossRef]
10. Libby, P. Inflammation in atherosclerosis. *Nature* **2002**, *420*, 868–874. [CrossRef] [PubMed]
11. Ross, R. Atherosclerosis—An inflammatory disease. *N. Engl. J. Med.* **1999**, *340*, 115–126. [CrossRef] [PubMed]
12. Blake, G.J.; Ridker, P.M. Novel clinical markers of vascular wall inflammation. *Circ. Res.* **2001**, *89*, 763–771. [CrossRef] [PubMed]

13. Blann, A.D.; Lip, G.Y.; McCollum, C.N. Changes in von Willebrand factor and soluble ICAM, but not soluble VCAM, soluble E selectin or soluble thrombomodulin, reflect the natural history of the progression of atherosclerosis. *Atherosclerosis* **2002**, *165*, 389–391. [CrossRef]

14. Ridker, P.M. Inflammatory biomarkers, statins, and the risk of stroke: Cracking a clinical conundrum. *Circulation* **2002**, *105*, 2583–2585. [CrossRef] [PubMed]

15. Ridker, P.M. Clinical application of C-reactive protein for cardiovascular disease detection and prevention. *Circulation* **2003**, *107*, 363–369. [CrossRef] [PubMed]

16. De Berardis, D.; Conti, C.M.; Campanella, D.; Carano, A.; Scali, M.; Valchera, A.; Serroni, N.; Pizzorno, A.M.; D'Albenzio, A.; Fulcheri, M.; et al. Evaluation of C-reactive protein and total serum cholesterol in adult patients with bipolar disorder. *Int. J. Immunopathol. Pharmacol.* **2008**, *21*, 319–324. [CrossRef] [PubMed]

17. De Berardis, D.; Conti, C.M.; Serroni, N.; Moschetta, F.S.; Carano, A.; Salerno, R.M.; Cavuto, M.; Farina, B.; Alessandrini, M.; Janiri, L.; et al. The role of cholesterol levels in mood disorders and suicide. *J. Biol. Regul. Homeost. Agents* **2009**, *23*, 133–140. [PubMed]

18. De Berardis, D.; Marini, S.; Piersanti, M.; Cavuto, M.; Perna, G.; Valchera, A.; Mazza, M.; Fornaro, M.; Iasevoli, F.; Martinotti, G.; et al. The Relationships between Cholesterol and Suicide: An Update. *ISRN Psychiatry* **2012**. [CrossRef] [PubMed]

19. Smith, R.S. The macrophage theory of depression. *Med. Hypotheses* **1991**, *35*, 298–306. [CrossRef]

20. Holsboer, F. The corticosteroid receptor hypothesis of depression. *Neuropsychopharmacology* **2000**, *23*, 477–501. [CrossRef]

21. Leonard, B.E. The HPA and immune axes in stress: The involvement of the serotonergic system. *Eur. Psychiatry* **2005**, *20* (Suppl. 3), S302–S306. [CrossRef]

22. Dantzer, R.; O'Connor, J.C.; Freund, G.G.; Johnson, R.W.; Kelley, K.W. From inflammation to sickness and depression: When the immune system subjugates the brain. *Nat. Rev. Neurosci.* **2008**, *9*, 46–56. [CrossRef] [PubMed]

23. Leonard, B.E.; Myint, A. The psychoneuroimmunology of depression. *Hum. Psychopharmacol.* **2009**, *24*, 165–175. [PubMed]

24. Bremmer, M.A.; Beekman, A.T.; Deeg, D.J.; Penninx, B.W.; Dik, M.G.; Hack, C.E.; Hoogendijk, W.J. Inflammatory markers in late-life depression: Results from a population-based study. *J. Affect. Disord.* **2008**, *106*, 249–255. [CrossRef] [PubMed]

25. Leonard, B.E. Psychopathology of depression. *Drugs Today* **2007**, *43*, 705–716. [CrossRef] [PubMed]

26. Kim, Y.K.; Na, K.S.; Shin, K.H.; Jung, H.Y.; Choi, S.H.; Kim, J.B. Cytokine imbalance in the pathophysiology of major depressive disorder. *Prog. Neuropsychopharmacol. Biol. Psychiatry* **2007**, *31*, 1044–1053. [CrossRef] [PubMed]

27. Ovaskainen, Y.; Koponen, H.; Jokelainen, J; Keinanen-Kiukaanniemi, S.; Kumpusalo, E.; Vanhala, M. Depressive symptomatology is associated with decreased interleukin-1 beta and increased interleukin-1 receptor antagonist levels in males. *Psychiatry Res.* **2009**, *167*, 73–79. [CrossRef] [PubMed]

28. Chobanian, A.V.; Bakris, G.L.; Black, H.R.; Cushman, W.C.; Green, L.A.; Izzo, J.L., Jr.; Jones, D.W.; Materson, B.J.; Oparil, S.; Wright, J.T., Jr.; et al. The Seventh Report of the Joint National Committee on Prevention, Detection, Evaluation, and Treatment of High Blood Pressure: The JNC 7 report. *JAMA* **2003**, *289*, 2560–2572. [CrossRef] [PubMed]

29. Danner, M.; Kasl, S.V.; Abramson, J.L.; Vaccarino, V. Association between depression and elevated C.-reactive protein. *Psychosom. Med.* **2003**, *65*, 347–356. [CrossRef] [PubMed]

30. Ridker, P.M. High-sensitivity C-reactive protein, inflammation, and cardiovascular risk: From concept to clinical practice to clinical benefit. *Am. Heart J.* **2004**, *148*, S19–S26. [CrossRef] [PubMed]

31. The ENRICHD investigators. Enhancing recovery in coronary heart disease (ENRICHD): Baseline characteristics. *Am. J. Cardiol.* **2001**, *88*, 316–322.

32. Sorensen, C.; Brandes, A.; Hendricks, O.; Thrane, J.; Friis-Hasche, E.; Haghfelt, T.; Bech, P. Psychosocial predictors of depression in patients with acute coronary syndrome. *Acta Psychiatr. Scand.* **2005**, *111*, 116–124. [CrossRef] [PubMed]

33. Duivis, H.E.; de Jonge, P.; Penninx, B.W.; Na, B.Y.; Cohen, B.E.; Whooley, M.A. Depressive symptoms, health behaviors, and subsequent inflammation in patients with coronary heart disease: Prospective findings from the heart and soul study. *Am. J. Psychiatry* **2011**, *168*, 913–920. [CrossRef] [PubMed]

34. Kannel, W.B.; McGee, D.; Gordon, T. A general cardiovascular risk profile: The Framingham Study. *Am. J. Cardiol.* **1976**, *38*, 46–51. [CrossRef]

35. Yan, L.L.; Liu, K.; Matthews, K.A.; Daviglus, M.L.; Ferguson, T.F.; Kiefe, C.I. Psychosocial factors and risk of hypertension: The Coronary Artery Risk Development in Young Adults (CARDIA) study. *JAMA* **2003**, *290*, 2138–2148. [CrossRef] [PubMed]

36. Bonnet, F.; Irving, K.; Terra, J.L.; Nony, P.; Berthezene, F.; Moulin, P. Depressive symptoms are associated with unhealthy lifestyles in hypertensive patients with the metabolic syndrome. *J. Hypertens.* **2005**, *23*, 611–617. [CrossRef] [PubMed]

37. Richter, N.; Juckel, G.; Assion, H.J. Metabolic syndrome: A follow-up study of acute depressive inpatients. *Eur. Arch. Psychiatry Clin. Neurosci.* **2009**, *260*, 41–49. [CrossRef] [PubMed]

38. Tsimikas, S.; Willerson, J.T.; Ridker, P.M. C-reactive protein and other emerging blood biomarkers to optimize risk stratification of vulnerable patients. *J. Am. Coll. Cardiol.* **2006**, *47*, C19–C31. [CrossRef] [PubMed]

39. Papanicolaou, D.A.; Wilder, R.L.; Manolagas, S.C.; Chrousos, G.P. The pathophysiologic roles of interleukin-6 in human disease. *Ann. Intern. Med.* **1998**, *128*, 127–137. [CrossRef] [PubMed]

40. Lesperance, F.; Frasure-Smith, N.; Theroux, P.; Irwin, M. The association between major depression and levels of soluble intercellular adhesion molecule 1, interleukin-6, and C-reactive protein in patients with recent acute coronary syndromes. *Am. J. Psychiatry* **2004**, *161*, 271–277. [CrossRef] [PubMed]

41. Chrysohoou, C.; Kollia, N.; Tousoulis, D. The link between depression and atherosclerosis through the pathways of inflammation and endothelium dysfunction. *Maturitas* **2018**, *109*, 1–5. [CrossRef] [PubMed]

42. Lu, Y.; Ho, C.S.; Liu, X.; Chua, A.N.; Wang, W.; McIntyre, R.S.; Ho, R.C. Chronic administration of fluoxetine and pro-inflammatory cytokine change in a rat model of depression. *PLoS ONE* **2017**, *12*, e0186700. [CrossRef] [PubMed]

43. Empana, J.P.; Jouven, X.; Canoui-Poitrine, F.; Luc, G.; Tafflet, M.; Haas, B.; Arveiler, D.; Ferrieres, J.; Ruidavets, J.B.; Montaye, M.; et al. C-reactive protein, interleukin 6, fibrinogen and risk of sudden death in European middle-aged men: The PRIME study. *Arterioscler. Thromb. Vasc. Biol.* **2010**, *30*, 2047–2052. [CrossRef] [PubMed]

44. Kiecolt-Glaser, J.K.; Glaser, R. Depression and immune function: Central pathways to morbidity and mortality. *J. Psychosom. Res.* **2002**, *53*, 873–876. [CrossRef]

45. Lanquillon, S.; Krieg, J.C.; Bening-Abu-Shach, U.; Vedder, H. Cytokine production and treatment response in major depressive disorder. *Neuropsychopharmacology* **2000**, *22*, 370–379. [CrossRef]

46. Aronson, D.; Avizohar, O.; Levy, Y.; Bartha, P.; Jacob, G.; Markiewicz, W. Factor analysis of risk variables associated with low-grade inflammation. *Atherosclerosis* **2008**, *200*, 206–212. [CrossRef] [PubMed]

47. Choo, C.C.; Chew, P.K.H.; Lai, S.M.; Soo, S.C.; Ho, C.S.; Ho, R.C.; Wong, R.C. Effect of Cardiac Rehabilitation on Quality of Life, Depression and Anxiety in Asian Patients. *Int. J. Environ. Res. Public Health* **2018**, *15*. [CrossRef] [PubMed]

48. American Psychiatric Association. *Diagnostic and Statistical Manual of Mental Disorders*, 4rd ed.; American Psychiatric Association: Washington, DC, USA, 1994.

49. Hamilton, M. Development of a rating scale for primary depressive illness. *Br. J. Soc. Clin. Psychol.* **1967**, *6*, 278–296. [CrossRef] [PubMed]

50. Wilson, P.W.; D'Agostino, R.B.; Levy, D.; Belanger, A.M.; Silbershatz, H.; Kannel, W.B. Prediction of coronary heart disease using risk factor categories. *Circulation* **1998**, *97*, 1837–1847. [CrossRef] [PubMed]

 International Journal of
Molecular Sciences

Article

Peripheral Acid Sphingomyelinase Activity Is Associated with Biomarkers and Phenotypes of Alcohol Use and Dependence in Patients and Healthy Controls

Christiane Mühle [1,*], Christian Weinland [1], Erich Gulbins [2,3], Bernd Lenz [1,†] and Johannes Kornhuber [1,†]

[1] Department of Psychiatry and Psychotherapy, Friedrich-Alexander University Erlangen-Nürnberg (FAU), D-91054 Erlangen, Germany; christian.weinland@uk-erlangen.de (C.W.); bernd.lenz@uk-erlangen.de (B.L.); johannes.kornhuber@uk-erlangen.de (J.K.)

[2] Department of Molecular Biology, University of Duisburg-Essen, D-45259 Essen, Germany; erich.gulbins@uni-due.de

[3] Department of Surgery, University of Cincinnati, Cincinnati, OH 45267-0558, USA

* Correspondence: christiane.muehle@uk-erlangen.de or christiane.muehle@molbio-research.eu; Tel.: +49-9131-85-44738; Fax: +49-9131-85-36381

† These authors contributed equally to this work.

Received: 8 November 2018; Accepted: 4 December 2018; Published: 13 December 2018

Abstract: By catalyzing the hydrolysis of sphingomyelin into ceramide, acid sphingomyelinase (ASM) changes the local composition of the plasma membrane with effects on receptor-mediated signaling. Altered enzyme activities have been noted in common human diseases, including alcohol dependence. However, the underlying mechanisms remain largely unresolved. Blood samples were collected from early-abstinent alcohol-dependent in-patients ($n[\male]$ = 113, $n[\female]$ = 87) and matched healthy controls ($n[\male]$ = 133, $n[\female]$ = 107), and analyzed for routine blood parameters and serum ASM activity. We confirmed increased secretory ASM activities in alcohol-dependent patients compared to healthy control subjects, which decreased slightly during detoxification. ASM activity correlated positively with blood alcohol concentration, withdrawal severity, biomarkers of alcohol dependence (liver enzyme activities of gamma-glutamyl transferase, alanine aminotransferase, aspartate aminotransferase; homocysteine, carbohydrate-deficient transferrin; mean corpuscular volume, and creatine kinase). ASM activity correlated negatively with leukocyte and thrombocyte counts. ASM and gamma-glutamyl transferase were also associated in healthy subjects. Most effects were similar for males and females with different strengths. We describe previously unreported associations between ASM activity and markers of liver damage and myelosuppression. Further research should investigate whether this relationship is causal, or whether these parameters are part of a common pathway in order to gain insights into underlying mechanisms and develop clinical applications.

Keywords: acid sphingomyelinase; alcohol dependence; liver enzymes; sphingolipid metabolism; withdrawal

---

Part of this work has been presented as a poster at the 1st International Symposium "Neurodevelopment and CNS vulnerability" in Erlangen, Germany in September 2018.

1. Introduction

Yearly, nearly 2.5 million deaths worldwide are attributable to alcohol use, in addition to other consequences of alcohol-related diseases and injuries (World Health Organization: Global status report on alcohol and health). Various risk factors and mechanisms have been suggested to play a role in the development and maintenance of alcohol-use disorders [1]. Genetic components account for 50%–60% according to twin, adoption, and family studies [2]. Prenatal hyperandrogenization [3,4] may partially be responsible for the two-fold higher prevalence of alcohol dependence in males compared to females. Disturbances in sphingolipid metabolism have been identified in human studies and animal models of psychiatric disorders [5], including major and mild depression [6–8], which share a high comorbidity with alcohol dependence.

Sphingolipids play an increasingly recognized role in neuronal function in the brain, not only by serving as a membrane component to form a physical barrier. They also influence the local composition of the plasma membrane, the localization and activity of proteins and, thus receptor-mediated signaling, in addition to their own actions as ligands [9]. Current research specifically focuses on enzymes crucial for maintaining the balance between the pro-apoptotic ceramide and its anti-apoptotic metabolite sphingosine 1-phosphate (the so-called "rheostat") in the context of various physiological and pathophysiological conditions [10]. Sphingomyelinases and ceramidases are involved, for example, in ceramide-mediated signal transduction required for apoptosis, differentiation, and other cellular (including inflammatory) responses, in intracellular cholesterol trafficking and metabolism, as well as in lysosomal degradation of sphingomyelin and ceramide [11]. Two further pathways, de novo biosynthesis and the salvage pathway, lead to the generation of ceramide.

Acid sphingomyelinase (ASM, EC 3.1.4.12), encoded by the gene *SMPD1*, catalyzes the hydrolysis of the abundant membrane lipid sphingomyelin into ceramide and phosphorylcholine. Altered enzyme activities have been noted in a variety of common human diseases [12]. In alcohol dependence, levels of both the lysosomal [13] and secretory form (S-ASM) of the enzyme are increased, and decrease gradually during withdrawal treatment in male, as well as female, patients [14,15]. Consequently, plasma glycerophospholipid and sphingolipid species are also dysregulated in alcohol-dependent patients [16]. In ethanol-fed mice, tissue ASM activity is increased [17,18]. To our knowledge, there are no further published studies on the influence of the exposure of animals or cultured cells to ethanol on peripheral or culture supernatant S-ASM levels, except for one report. No alteration of serum S-ASM activity was detected in both transgenic mice overexpressing ASM and wildtype mice in a two-bottle free-choice drinking paradigm with a gender-balanced design [19]. Stimulation of the neutral sphingomyelinase by ethanol likewise contributes to alterations in the sphingomyelin/ceramide balance [20,21].

Hepatotoxicity is a major consequence of alcohol misuse. Of note, liver damage could also result from components of alcoholic beverages beyond their ethanol content. Both in experimental models of chronic ethanol-induced steatohepatitis, and patients with severe chronic alcohol-related liver disease, the immunoreactivity and ceramide content are increased [22]. Interestingly, an accumulation of ceramide and elevated levels of S-ASM have also been found in non-alcoholic fatty liver disease [23]. ASM knockout mice are resistant to alcohol-mediated fatty liver and cell death [24]. Inhibition of ASM by imipramine blocked the ethanol-induced ASM activation and ceramide generation, resulting in amelioration of hepatic steatosis in ethanol-fed mice [18]. Likewise, treatment of ethanol-fed rats with antioxidants, for example, *N*-acetylcysteine, reduced the severity of chronic alcohol-related steatohepatitis, possibly attributable to the observed decreased expression of inflammatory mediators, reduced acid sphingomyelinase activity, and lowered ceramide load [25]. The role of ASM likely involves sensitization of hepatocytes to the cytotoxic effects of TNFalpha [24] and regulation of autophagy [26].

The lack of reliable markers of alcohol consumption is a major obstacle to the diagnosis and treatment of alcohol dependence. Interviews and subjective questionnaires have their limitations, particularly because subjects are known to downplay the extent of their drinking behavior. Direct measurement of alcohol concentration in the breath, blood, or urine does not provide information more than a few hours beyond the most recent consumption of alcohol [27], or it requires special equipment with a high cost (ethyl glucuronide, ethyl sulfate, phosphatidylethanol) [28]. Currently available indirect biochemical markers, including carbohydrate-deficient transferrin (CDT, a form of the serum iron-carrying protein transferrin, with altered carbohydrate composition), mean corpuscular volume of erythrocytes (MCV), as well as the liver enzymes gamma-glutamyl transferase (GGT), alanine aminotransferase (ALT, also glutamic-pyruvic transaminase GPT), and aspartate aminotransferase (AST, also glutamic-oxaloacetic transaminase GOT), react to steady and significant alcohol intake over weeks or months, but suffer from relatively low sensitivity and specificity, and an uncertain time window of detection [28,29].

Furthermore, there is a need for reliable predictors of relapse after withdrawal treatment, which is a common problem in alcohol dependence, resulting in a rate of up to 85% in the absence of further support after the initial detoxification phase [30]. A number of known risk factors have been identified, albeit with limited accuracy and a high cost and time investment, which restricts their clinical applications for the identification of patients at risk of relapse and for individualized treatment [31].

In our large and sex-balanced cohort of alcohol-dependent patients and matched healthy controls, we aimed at characterizing the readily quantifiable activity of peripheral S-ASM with respect to the phenotype and known biomarkers of alcohol dependence, with particular emphasis on liver parameters. Moreover, we evaluated the diagnostic performance of this enzyme in discriminating between patients and controls, and predicting relapse as assessed as alcohol-related readmissions to the hospital.

2. Results

2.1. Elevated S-ASM Activity in Early-Abstinent Alcohol-Dependent Patients

We quantified the serum S-ASM activity in our cohort of 200 severely alcohol-dependent patients, and 240 control subjects matched for age and sex (Table 1). We replicated previous findings of significantly increased levels of S-ASM activity in early-abstinent alcohol-dependent patients [14,15] in this considerably larger cohort. At recruitment during early abstinence, male as well as female patients both presented with 1.5-fold higher serum activities, compared to healthy controls ($p < 0.001$, Figure 1). In both patients and controls, the enzyme activity was 11% higher in males than in females ($p = 0.049$ and $p = 0.100$, respectively). On average, levels decreased slightly during withdrawal treatment by 9.4% ($p = 0.007$) and 1.5% ($p = 0.308$) for male and female patients, respectively, to levels that were still significantly higher than those of control subjects ($p < 0.001$). However, a decrease was only observed in about half of the patients (63% of males and 52% of females) during the approximately 5-day interval.

Table 1. Demographic and laboratory data for male and female alcohol-dependent patients and healthy control subjects.

Parameter	Alcohol-Dependent Patients		Healthy Control Subjects		p Patients vs. Controls		p ♂ vs. ♀	
	♂	♀	♂	♀	♂	♀	Patients	Controls
n	113	87	133	107	-	-	-	-
Age (years)	48 (40–53)	48 (42–55)	48 (38–56)	49 (39–55)	0.794	0.772	0.308	0.762
BMI (kg/m^2) [b]	24.9 (22.1–27.8)	24.4 (22.1–29.3)	27.7 (24.9–29.5)	25.0 (21.7–28.6)	**<0.001**	0.961	0.963	**<0.001**
Active smokers (%) [b]	77.9	76.9	21.8	18.7	**<0.001**	**<0.001**	**<0.001**	**<0.001**
Age at onset of alcohol dependence (years) [c]	30 (24–39)	35 (28–42)	-	-	-	-	**0.015**	-
Previous withdrawal treatments (n) [c]	6 (2–12)	5 (2–11)	-	-	-	-	0.892	-
Alcohol concentration at admission (‰) [b]	1.7 (0.5–2.4)	1.2 (0.1–1.8)	-	-	-	-	**0.020**	-
CIWA-Ar total score [c]	18 (15–23)	16 (14–21)	-	-	-	-	0.163	-
CAGE score	-	-	0 (0–1)	0 (0–0)	-	-	-	**0.024**
AUDIT score [b]	-	-	4 (3–6)	3 (2–4)	-	-	-	**<0.001**
Days until first alcohol-related readmission	285 (57–730)	625 (90–730)	-	-	-	-	**0.047**	-
Number of alcohol-related readmissions	2 (0–4)	1 (0–3)	-	-	-	-	**0.021**	-
S-ASM activity during recruitment (fmol/h/μL serum)	224 (168–318)	202 (157–256)	150 (112–194)	134 (97–181)	**<0.001**	**<0.001**	**0.049**	0.100
S-ASM activity during follow-up (fmol/h/μL serum) [b]	214 (142–299)	183 (159–251)	-	-	-	-	0.175	-
GGT (U/L)	109 (50–275)	57 (34–230)	28 (21–41)	17 (14–25)	**<0.001**	**<0.001**	**0.022**	**<0.001**
ALT (U/L)	48 (28–84)	28 (20–50)	30 (23–39)	18 (14–24)	**<0.001**	**<0.001**	**<0.001**	**<0.001**
AST (U/L)	51 (36–91)	37 (28–67)	29 (25–34)	23 (20–26)	**<0.001**	**<0.001**	**0.002**	**<0.001**
CDT (nephelometry, %) [a]	2.8 (1.9–4.0)	1.9 (1.6–2.5)	1.5 (1.3–1.7)	1.5 (1.3–1.6)	**<0.001**	**<0.001**	**<0.001**	0.486
MCV (fl) [a]	93 (90–96)	95 (91–97)	88 (85–91)	88 (85–91)	**<0.001**	**<0.001**	0.166	0.734
Homocysteine (μmol)	15 (12–23)	15 (11–21)	12 (10–14)	10 (9–12)	**<0.001**	**<0.001**	0.161	**<0.001**
CK (U/L)	132 (81–220)	89 (72–141)	151 (112–213)	92 (75–110)	0.058	0.422	**0.006**	**<0.001**
Leukocytes (per nL) [a]	7.2 (5.5–8.4)	6.9 (5.2–8.6)	5.8 (5.0–7.2)	5.8 (4.7–6.7)	**<0.001**	**<0.001**	0.341	0.173
Thrombocytes (per nL) [a]	198 (145–254)	218 (181–266)	230 (196–257)	251 (214–287)	**<0.001**	**<0.001**	**0.021**	**<0.001**
Triglycerides (mg/dL)	165 (96–236)	135 (95–221)	135 (98–192)	117 (80–165)	0.244	**0.012**	0.479	**0.008**
Total cholesterol (mg/dL)	214 (185–252)	216 (180–255)	210 (189–239)	223 (191–252)	0.833	0.530	0.926	0.106
High-density lipoprotein cholesterol (mg/dL)	61 (51–80)	68 (54–85)	48 (43–57)	62 (51–74)	**<0.001**	**0.015**	0.123	**<0.001**
Low-density lipoprotein cholesterol (mg/dL) [a]	130 (96–160)	122 (98–157)	143 (125–171)	146 (122–166)	**<0.001**	**<0.001**	0.750	0.676

The table shows medians with interquartile range or percentages, and p-values from Mann–Whitney U tests and χ^2 tests comparing sex-specifically early-abstinent alcohol-dependent patients with healthy control subjects, and comparing group-specific male to female individuals. The subgroup at follow-up (94 male and 69 female patients willing and able to participate 5 days (3–6) after the initial recruitment) is representative for the total cohort of alcohol-dependent patients, since there are no significant differences in sociodemographic characteristics [32] or regarding S-ASM activity (all sex-specific $p < 0.1$). Data on alcohol-related readmissions were extracted from medical records for the 24-month period after study recruitment. For patients without readmission, this value was set to 730 days. Missing values: a $\leq$ 1%, b $\leq$ 10%, c $\leq$ 30%. $p < 0.05$ in bold. BMI: body mass index, CIWA-Ar score: German version of the Clinical Institute Withdrawal Assessment for Alcohol revised, CAGE: acronym for 4-item questionnaire indicating potential problems with alcohol abuse, AUDIT: Alcohol Use Disorders Identification Test, S-ASM: secretory acid sphingomyelinase, GGT: gamma-glutamyl transferase, ALT: alanine aminotransferase (glutamic-pyruvic transaminase, GPT), AST: aspartate aminotransferase (glutamic-oxaloacetic transaminase, GOT), CDT: carbohydrate-deficient transferrin, MCV: mean corpuscular volume, CK: creatine kinase.

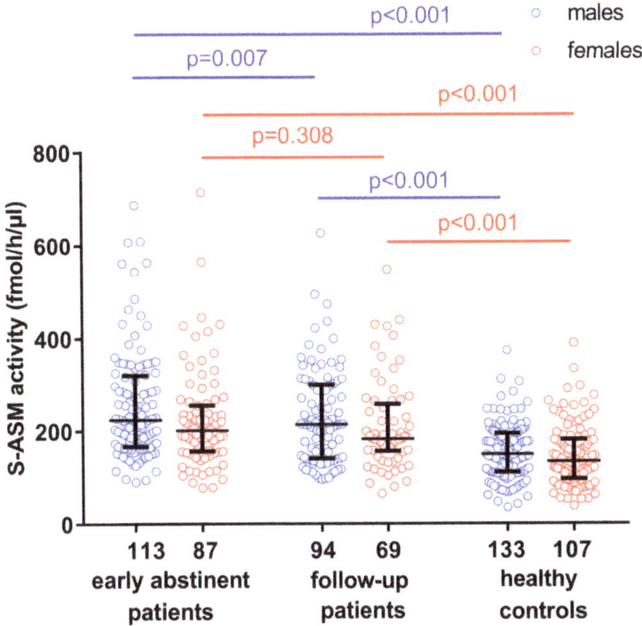

Figure 1. Sex-specific activity of the secretory acid sphingomyelinase (S-ASM) in alcohol-dependent male and female patients during early abstinence and follow-up, compared to healthy control subjects. Boxplots show individual data, and the median and interquartile range. The numbers of male and female individuals are provided below the x-axis. *p*-values were calculated using the Mann–Whitney *U* test except for the pair-wise comparison for patients where the Wilcoxon signed-rank test was applied.

2.2. S-ASM Activity Is Positively Associated with Alcohol Levels at Admission and Withdrawal Severity

Blood alcohol levels determined during recruitment varied from 0‰ to 3.7‰, and they correlated significantly with S-ASM activity (Rho = 0.315, $p = 8.3 \times 10^{-6}$, Figure 2a). This moderate effect was mainly driven by the male subgroup (Rho = 0.371, $p = 7.7 \times 10^{-5}$) and was much smaller in female patients (Rho = 0.214, $p = 0.049$). In previous experiments, we had ensured that, in our assay, the analysis of S-ASM activity was not confounded by the remaining alcohol concentrations in serum samples [14]. After subdividing the patients according to their predominantly consumed type of alcoholic beverage (beer $n = 83$, wine $n = 43$, hard liquor $n = 25$), no significant difference in alcohol concentrations at admission or in S-ASM activity was observed between these subgroups for all patients and for sex-specific analyses (all $p > 0.05$, Supplemental Figure S1).

To investigate the relationship between S-ASM activity and withdrawal severity, patients were asked at the follow-up visit to report their strongest withdrawal symptoms since study inclusion (CIWA-Ar scale). The cumulative score for ten sub-items (Supplemental Table S1) showed a significant correlation with S-ASM activity in the total cohort (Rho = 0.242, $p = 0.003$, Figure 2b). This correlation derived from the male patients (Rho = 0.194, $p = 0.079$), and, to a larger extent, from female patients (Rho = 0.267, $p = 0.034$). There was a particularly notable association between S-ASM and the sub-item scores for nausea/vomiting (Rho = 0.232, $p = 0.005$), and most strongly for tremor (Rho = 0.351, $p = 1.4 \times 10^{-5}$) in the total cohort with considerably stronger effects in the male patient subgroup (Figure 2c, Rho = 0.395, $p = 2.2 \times 10^{-4}$ in males vs. Rho = 0.250, $p = 0.048$ in females).

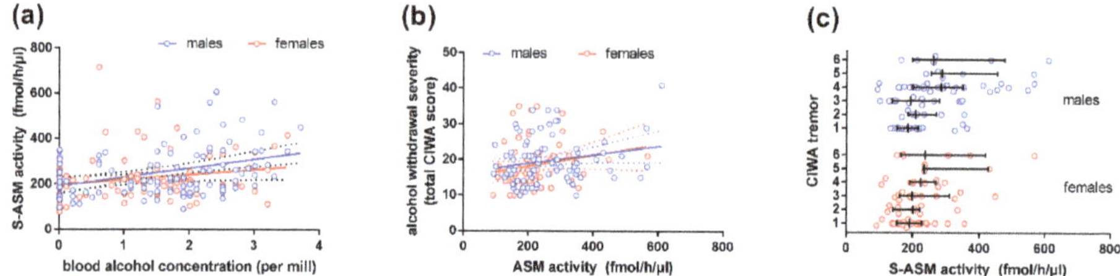

Figure 2. Correlation of serum S-ASM activity with blood alcohol levels and withdrawal in alcohol-dependent patients: (**a**) S-ASM activity is positively associated with blood alcohol levels of male and female patients at study inclusion; (**b**) S-ASM activity is positively associated with withdrawal severity in male and female patients assessed by the Clinical Institute Withdrawal Assessment for Alcohol revised scale (CIWA-Ar total score); (**c**) S-ASM activity is strongly positively associated with the CIWA-Ar sub-item tremor. S-ASM: secretory acid sphingomyelinase. Individual data with linear regression line and 95% confidence intervals (**a**,**b**) and boxplots with median and interquartile range (**c**).

2.3. S-ASM Activity Is Strongly Associated with Liver Enzymes in Alcohol-Dependent Patients

The liver is severely damaged by excessive consumption of alcoholic beverages as indicated by an increase in activities of the liver enzymes GGT, ALT, and AST. In line with the elevated S-ASM activities in patients, we have found strong and highly significant correlations of S-ASM with these liver enzyme activities (Rho > 0.37 and $p < 6 \times 10^{-8}$ for all three enzymes, Table 2, Figure 3). Interestingly, the strength of these associations was similar for both male and female patients.

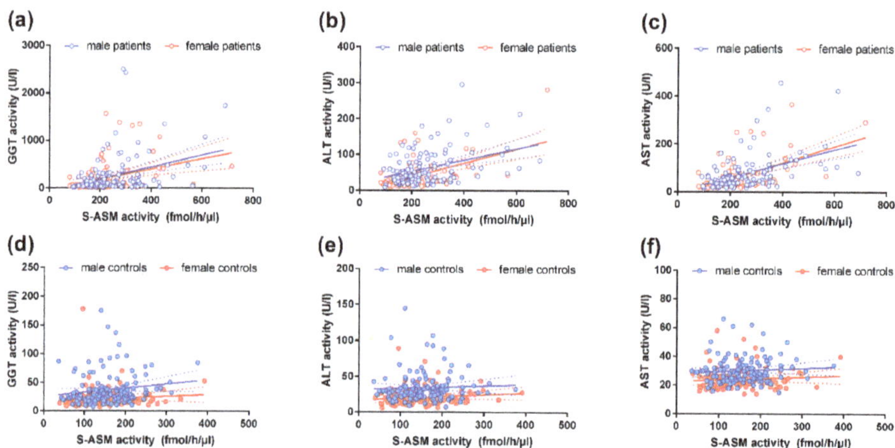

Figure 3. Correlation of serum S-ASM activity with liver enzymes in alcohol-dependent patients and healthy controls: (**a–c**) Associations for patients; (**d–f**) Associations for healthy controls. For control graphs, the strongest outliers deviating from the mean by more than two standard deviations were excluded from the graph (one for GGT, two each for ALT and AST each) to avoid distortion of the linear regression line and compression of the dataset. S-ASM: secretory acid sphingomyelinase, GGT: gamma-glutamyl transferase, ALT: alanine aminotransferase (glutamic-pyruvic transaminase, GPT), AST: aspartate aminotransferase (glutamic-oxaloacetic transaminase, GOT). Individual data with linear regression line and 95% confidence intervals.

2.4. S-ASM Activity Is Associated with GGT and ALT Activity in Healthy Controls

Despite the much lower variations in GGT and ALT liver enzyme activities in healthy controls compared to patients, both were also positively (albeit weaker) correlated to the S-ASM activity

(Table 2, Figure 3). The female subgroup contributed more to this effect. While alcohol blood levels or consumption were not assessed in the healthy control subjects and could be the mediators, S-ASM, however, was not associated with commonly utilized scales to detect alcohol misuse, using AUDIT (Rho = 0.085, p = 0.208) or CAGE scores (Rho = 0.038, p = 0.555). This was also true for sex-specific analyses. Of note, further exploratory analysis revealed an influence of age on S-ASM activity in this group of healthy individuals (Rho = 0.268, p = 2.6 $\times$ 10^{-5}) which was clearly more relevant for males (Rho = 0.320, p = 1.8 $\times$ 10^{-4}) than females (Rho = 0198, p = 0.041). Moreover, GGT activity was found to be age-dependent in this group (Rho = 0.286, p = 7.0 $\times$ 10^{-6}), with a definitely larger effect from males (Rho = 0.352, p = 3.3 $\times$ 10^{-5}) than females (Rho = 0.299, p = 0.002). In consideration of these two positive correlations, we cannot exclude that the observed association of S-ASM with GGT in healthy subjects is partially driven by an age effect. For ALT, a dependence on age was exclusively found in the female subgroup (Rho = 0.336, p = 4.0 $\times$ 10^{-4}).

When we subdivided the group of healthy control subjects into those with at least one binge-drinking episode ($\geq$5 standard drinks of ~13 g of alcohol per drink within 2 h) during the past 24 months (n = 36) versus those without a binge-drinking episode (n = 204), we unexpectedly observed slightly decreased S-ASM levels in binge-drinkers in comparison to the increased S-ASM levels in alcohol-dependent patients (Supplemental Table S2). However, considering the influence of age, these lower S-ASM activities might be due to the naturally lower age of the binge-drinking group, particularly the males.

Another unanticipated observation was that the association of S-ASM levels with liver enzymes strengthened in the healthy control sample after exclusion of binge-drinkers (GGT: Rho = 0.310, p = 6.3 $\times$ 10^{-6}; ALT: Rho = 0.193, p = 0.006 for n = 204, compared to Table 2) whereas there was no statistical trend observed for the small binge-drinking group (n = 36, p > 0.3 for both enzymes). This was also true for the sex-specific analysis of GGT.

Table 2. Total and sex-specific correlations of peripheral S-ASM activity with blood parameters altered in alcohol-dependent patients.

Parameter	Patients		Controls		Patients				Controls			
					♂		♀		♂		♀	
	Rho	p	Rho	p	Rho	p	Rho	p	Rho	p	Rho	p
GGT	0.373	5.6×10^{-8}	0.254	7.0×10^{-5}	0.327	4.4×10^{-4}	0.378	3.1×10^{-4}	0.217	**0.012**	0.307	**0.001**
ALT	0.373	5.4×10^{-8}	0.170	**0.008**	0.327	4.0×10^{-4}	0.368	4.6×10^{-4}	0.124	0.154	0.177	0.068
AST	0.499	5.4×10^{-14}	0.122	0.060	0.465	2.2×10^{-7}	0.524	1.9×10^{-7}	0.092	0.291	0.093	0.340
CDT	0.378	3.4×10^{-8}	−0.062	0.341	0.311	7.9×10^{-4}	0.400	1.2×10^{-4}	0.030	0.732	−0.162	0.095
MCV	0.172	**0.016**	0.139	**0.032**	0.154	0.104	0.248	**0.022**	0.170	**0.050**	0.085	0.388
Homocysteine	0.304	1.1×10^{-5}	0.142	**0.028**	0.369	5.7×10^{-5}	0.201	0.062	0.123	0.158	0.092	0.347
CK	0.218	**0.002**	0.008	0.902	0.167	0.076	0.228	**0.033**	0.015	0.868	−0.114	0.243
Leukocytes	−0.152	**0.033**	0.145	**0.025**	−0.248	**0.008**	−0.059	0.591	0.093	0.287	0.209	**0.032**
Thrombocytes	−0.292	3.0×10^{-5}	0.015	0.812	−0.229	**0.015**	−0.365	6.0×10^{-4}	0.030	0.732	0.070	0.475
Triglycerides	−0.214	**0.002**	0.126	0.052	−0.233	**0.013**	−0.188	0.082	−0.004	0.965	0.252	**0.009**
Cholesterol	0.055	0.435	0.027	0.679	0.097	0.308	0.022	0.838	0.034	0.698	0.058	0.551
HDL cholesterol	0.229	**0.001**	0.004	0.954	0.272	**0.004**	0.215	**0.045**	0.139	0.112	−0.086	0.379
LDL cholesterol	−0.050	0.485	0.017	0.797	−0.035	0.715	−0.072	0.509	−0.034	0.700	0.081	0.405

Rho and p-values from Spearman correlations. $p < 0.05$ in bold. S-ASM: secretory acid sphingomyelinase, GGT: gamma-glutamyl transferase, ALT: alanine aminotransferase (glutamic-pyruvic transaminase, GPT), AST: aspartate aminotransferase (glutamic-oxaloacetic transaminase, GOT), CDT: carbohydrate-deficient transferrin, MCV: mean corpuscular volume, CK: creatine kinase, HDL high-density lipoprotein, LDL low-density lipoprotein.

2.5. Comparison of S-ASM Activity with Additional Biomarkers of Alcohol Dependence

In addition to a clear elevation of the liver enzymes GGT, ALT, and AST, pathophysiological processes in alcohol-dependent patients lead to further alterations that can be detected in peripheral blood samples, including CDT, MCV (both often serving as biomarkers for alcohol dependence), and homocysteine. Of note, S-ASM activities were also associated with these parameters, but these relationships were weaker than those with liver enzymes, and they also showed different strengths for male and female subgroups (Table 2). For CDT, the association was strong and highly significant for the total group, as well as for males and females separately. For homocysteine, it was moderate and highly significant for the total group, but only for males separately. For MCV, the association was weak and only found for females.

We next analyzed the quality of S-ASM as a biomarker for alcohol dependence by comparing sensitivity and specificity with those of established parameters. The predictive power of S-ASM activity in our cohort, as judged from the area under the curve (AUC = 0.771), was in a similar range, but it was lower than that of classically used biomarkers CDT (0.868), GGT (0.853), and MCV (0.799) in our sample (Table 3). Due to its high correlation with these parameters, adding S-ASM to the biomarkers listed in Table 3, in a binary logistic regression model that included sex and age, did not clearly improve the prediction (correct classification of one more patient). This marker would, thus, probably not add clinically relevant information to the current practice to separate alcohol-dependent patients from non-alcohol-dependent subjects.

Table 3. ROC analysis for S-ASM, common alcohol biomarkers, and alcohol-dependent routine blood parameters.

	Total Cohort				♂				♀			
	AUC	Y	Sens	Spec	AUC	Y	Sens	Spec	AUC	Y	Sens	Spec
S-ASM (fmol/h/μL)	0.771	151	0.850	0.554	0.787	224	0.504	0.910	0.753	151	0.816	0.607
CDT (%)	0.868	1.72	0.760	0.849	0.890	1.77	0.796	0.841	0.848	1.71	0.690	0.888
GGT (U/L)	0.853	39.5	0.740	0.817	0.843	44.0	0.779	0.774	0.886	33.5	0.770	0.869
MCV (fl)	0.799	90.1	0.787	0.695	0.778	90.0	0.777	0.684	0.824	91.8	0.729	0.830
ALT (U/L)	0.687	47.5	0.410	0.904	0.687	47.5	0.522	0.857	0.725	22.5	0.621	0.738
AST (U/L)	0.806	33.5	0.700	0.821	0.816	35.5	0.752	0.812	0.837	26.5	0.782	0.776
Hcy (μmol)	0.758	13.2	0.650	0.750	0.742	13.0	0.717	0.669	0.783	12.2	0.667	0.785

ROC: receiver operating characteristic, AUC: area under the curve, Y: Youden cut-point, Sens: sensitivity, Spec: specificity, S-ASM: secretory acid sphingomyelinase, CDT: carbohydrate-deficient transferrin, GGT: gamma-glutamyl transferase, MCV: mean corpuscular volume, ALT: alanine aminotransferase (glutamic-pyruvic transaminase, GPT), AST: aspartate aminotransferase (glutamic-oxaloacetic transaminase, GOT).

When comparing males and females, the differentiation capacity of the analyzed parameters, as indicated by the AUC, was similar, with a maximum of 5% difference in AUC. S-ASM and CDT differentiated slightly better for males, and GGT, MCV, ALT, AST, and Hcy for females. This was despite considerable variation in the cut-point determined by the Youden index. For the S-ASM activity, the optimal cut-point was nearly 50% higher (224 fmol/h/μL) in males than in females (151 fmol/h/μL), which might reflect the weak sex difference in patients and trend difference in controls (Table 1). The Youden cut-point for males was also markedly higher than for females for the liver enzyme activities GGT, ALT, and AST, with a factor between 1.3 and 2.1 (Table 3) related to the sex differences for these enzymes in patients and controls (Table 1).

We included creatine kinase (CK) in our analysis, which is typically assayed as a marker of muscle damage, because alcohol has been found to lead to a rapid increase in plasma CK activity in rats [33]. Moreover, raised levels of CK have been detected in various psychiatric conditions, including alcohol dependence [34]. In addition, CK differentiated between alcohol dependence, alcohol withdrawal, and delirium tremens (with increasing levels in that order) [35]. We also found significantly lower CK activity in females compared to males for patients, as well as for healthy controls (Table 1), which was similarly observed in the animal model [33]. Contrary to the expectation of increased levels, CK was slightly, but not significantly, decreased in patients compared to healthy subjects (Table 1). Thus, CK would not be useful as a biomarker for alcohol dependence, based on our data. However, we observed a positive correlation between CK and S-ASM activity within the group of patients (Table 2), characterized by a stronger contribution from the female group.

2.6. S-ASM Activity Is Differentially Associated with Myelosuppression in Patients and Controls

As expected from the strong association of S-ASM activity with hepatotoxicity, as demonstrated by the correlation with liver enzymes, the enzyme activity was also related to myelosuppression in patients (Table 2). In the total cohort of patients (and, particularly, in the male subgroup), higher S-ASM levels correlated with lower leukocyte numbers, which are indicative of the toxic effect of alcohol on hematopoiesis. In contrast, higher S-ASM levels in healthy controls (especially in females) were associated with higher leukocyte numbers, which could reflect inflammatory processes that are known to be related to elevated S-ASM levels [12]. For thrombocytes, higher S-ASM levels were also associated with lower cell counts, and the strongest effects were found in female patients. Here, we did not observe an effect in controls.

2.7. S-ASM Activity Is Associated with Alterations in Triglycerides in HDL Cholesterol in Patients

Triglycerides and cholesterol subspecies might interact with the activity of a lipid metabolizing enzyme, such as S-ASM. Levels of triglycerides were slightly increased in female patients and unaltered in males, compared to controls, in contrast to a decrease found in Japanese [36] and in European Americans [37] for low-to-moderate alcohol consumption (Table 1). Triglyceride concentrations correlated negatively with S-ASM activity in our total patient group, as well as in male patients (Table 2), in accordance with previous observations in a smaller study [14].

Both male and female patients' samples contained significantly higher levels of high-density lipoprotein (HDL) cholesterol (Table 1), in line with published reports for the effect of alcohol intake [36–38]. The significant positive correlation between HDL cholesterol levels and S-ASM activity in patients (Table 2), in agreement with previous reports [14], could reflect a causal relationship or an independent influence of alcohol consumption on both parameters. Interestingly, while the concentration of low-density lipoprotein (LDL) cholesterol was also significantly altered in both male and female patients, in line with published data on reduced levels [36,37], unlike for HDL cholesterol, it was not associated with S-ASM activity (Tables 1 and 2), in accordance with previous results [14]. The analysis of HDL and LDL cholesterol subfractions could provide further insights into these apparently differential relationships with S-ASM activity.

In healthy control subjects, we detected only a relationship between S-ASM and triglycerides in females, with an opposite direction compared to patients, i.e., higher S-ASM activity was associated with higher triglyceride levels (Table 2).

2.8. S-ASM Activity Does Not Predict Alcohol-Related Readmission for Patients

Medical records of patients were checked for 24 months after recruitment, to assess alcohol-related readmissions. However, S-ASM activity did not differ between patients with at least one alcohol-related readmission (n = 122), and those without a readmission (n = 78; Mann–Whitney U test, U = 4451, p = 0.442). Moreover, the S-ASM activity did not predict the days until the first readmission (Rho = -0.042, p = 0.555), nor did it predict the number of readmissions (Rho = 0.025, p = 0.723). This was also true when male and female patients were analyzed separately. On the other hand, one of the strongest known predictors—the number of previous withdrawal treatments—also strongly predicted alcohol-related readmission in this cohort (Mann–Whitney U test, U = 1368, p = 2.0×10^{-5}), as well as days until first readmission (Rho = -0.345, $p < 1.8 \times 10^{-5}$) and the number of readmissions (Rho = 0.400, $p < 5.3 \times 10^{-7}$).

3. Discussion

We have confirmed both the previously described increased activity of S-ASM in alcohol-dependent patients, and the decrease in S-ASM activity during detoxification treatment in a large and sex-balanced cohort [14,15]. However, in a smaller previous study consisting predominantly of males, S-ASM activity in patients was 3-fold higher compared to healthy controls, and it declined in every single individual over 7–10 days of withdrawal treatment by 52% of the initial value, on average [14]. In another small mixed gender study without controls, S-ASM activity fell by 15%–20% for females to 22%–29% for males during 2–7 days of treatment. A possible explanation for the smaller effects, in this study, could be our time window for inclusion during early abstinence (i.e., 24 to 72 h after the last consumption of alcohol), which could already be too late to detect the high initial drop in ASM activity.

While the increase in S-ASM activity in alcohol-dependent patients has been replicated, its origin remains elusive. A wide variety of cells have been demonstrated to secrete substantial amounts of this Zn^{2+}-dependent enzyme, including human vascular endothelial cells, macrophages, and platelets [12], resulting in detectable levels not only in the blood, but also in cerebrospinal fluid [39]. In mice fed on an atherogenic diet containing saturated fats and cholesterol, an increased macrophage secretion seemed to be responsible for the elevated S-ASM activity [40]. The enzyme might be released into the bloodstream when cells are injured by ethanol, or other components of alcoholic beverages or as a response to systemic changes induced by these factors. ASM is a key regulator of ceramide-dependent signaling pathways, and it can be induced by cellular stress resulting from inflammation or infection [12]. Mechanisms of ASM activation by ethanol could involve post-translational, as well as transcriptional effects [41].

On the other hand, alcohol-dependent patients could carry risk factors for endogenously higher ASM levels. Genetically determined ASM activity is already known to influence the susceptibility for common human diseases, such as allergy [42]. However, because known single nucleotide polymorphisms or variations in the repeat number within the special signal peptide negatively affect ASM activity [12,43,44], it is rather unlikely that a frequent, but so far undetected, variant within the *SMPD1* gene would predispose carriers to developing alcohol dependence, and be the cause of higher S-ASM levels. On the other hand, processes associated with posttranslational modifications, and regulation that modulate ASM trafficking, maturation, or secretion [45], as well as those leading to degradation of the enzyme, could permanently or temporarily be altered in patients. Additionally, gene variants encoding proteins further upstream in the lipid synthesis pathway, such as in *SERINC2*, could alter ASM levels. This gene was identified as a top-ranked risk gene for alcohol dependence [46,47], and the encoded protein incorporates serine into membranes, facilitating the synthesis of phosphatidylserine and sphingolipids [48]. Moreover, *SMPD1* splicing [49] has been reported to influence ASM activity and be altered in major depression [50]. However, it has not yet been analyzed in alcohol dependence.

Remarkably, GGT values of a considerable proportion of healthy controls (17% of males, 7% of females) were above the normal reference range of the analyzing laboratory, with the upper limit of 60 U/L for males and 40 U/L for females. Moreover, there is evidence from data on blood pressure, pulse rate, relative body weight, and serum insulin, that call for an even lower upper limit of 10 U/L, compared to the 28 U/L limit at the time of the investigation [51]. Consumption of small amounts of alcoholic beverages as "social" drinking, that is not yet detected by the CAGE or AUDIT questionnaires, but is probably the main cause of elevated GGT activity which reflects a low level of liver damage, seems to also be associated with higher S-ASM levels in control subjects. However, the influence of chronic low-dose alcohol exposure appears to be different from the impact of acutely high levels, suggested by the different correlations between GGT and S-ASM after subdividing the healthy control sample according to the presence of binge-drinking episodes. These associations of S-ASM activity with liver enzyme activities warrant further investigation.

We detected a significant correlation between S-ASM and each patient's alcohol levels at admission. A greater effect was observed in the male subsample, but no relationship was found with the predominantly consumed type of alcoholic beverage. It is also noteworthy that there were different correlation strengths for S-ASM and various classical biomarkers of alcohol-dependence. While the association was very strong for liver enzymes and CDT, it was much weaker for the well-accepted indicator, MCV. Similarly, significant correlations between ASM and the indicators of hepatic injury (GGT, ALT, and AST) have also been described for patients with a hepatitis C virus infection where ASM showed a high discriminative power [23]. There could be a common link between ASM and hepatotoxicity that involves endoplasmic reticulum stress and cholesterol loading of mitochondria [52] that are highly abundant in hepatocytes, possibly via mechanisms of transcriptional regulation recently identified for mitochondrial defects in lysosomal storage disorders, like Niemann–Pick disease caused by a genetic defect in the ASM encoding gene (Yambire K.F. et al. preprint under revision).

Withdrawal symptoms, assessed during the follow-up visit using the CIWA-Ar scale, were significantly related to S-ASM levels, and this correlation was stronger in females than in males. On the other hand, the strong and highly significant positive relationship, between the sub-item tremor during detoxification and S-ASM, was most prominent in males. Interestingly, when beta-endorphin levels from the same cohort were analyzed with respect to withdrawal severity, the female subgroup contributed mostly to the correlation of the sub-item score for impaired concentration with higher initial beta-endorphin levels, as well as a stronger decline during withdrawal [53]. These sex-specific effects once more emphasize the importance of separate analyses for males and females in the field of alcohol addiction.

Although there is a need for very cautious interpretation due to the clearly different mechanisms at work, there is an intriguing association between tremor and ASM in a very different disorder—Parkinson's disease—with tremor being one of the primary early symptoms. Following the initial identification of the rare p.L302P mutation in *SMPD1* as a strong risk factor for Parkinson's disease in Ashkenazi Jews [54], two additional *SMPD1* founder mutations were identified in this population [55]. A pathogenic mechanism for Parkinson's disease has been hypothesized that involves alterations of the autophagy–lysosome pathway based on additional genetic factors encoding lysosomal enzymes [56].

Biomarkers for alcohol consumption, that are more reliable than self-reports and physiological assessments, are essential not only for diagnosis and treatment of alcohol-related disorders, but also for epidemiological studies of the health effects of alcohol itself, or of other exposure events with alcohol as a cofactor. No gold standard is yet available, and commonly applied biochemical markers are far from ideal with respect to their discriminatory power, as indicated by AUC values ranging from 0.21 to 0.67 [29]. While new approaches, including protein markers [29], DNA methylation patterns [57], aldehyde-induced DNA, and protein adducts [58], and even neuroimaging [59,60] enhance the detection of heavy drinkers, there is still room for improvement, for example, by utilizing novel biomarkers or by developing of a composite score for excessive alcohol use screening [61]. In our study population, S-ASM activity alone did not perform better than any of the analyzed markers (Table 2, AUC = 0.77 compared to AUC ≥ 0.80 for conventional biomarkers in our cohort), and it also did not appear to be beneficial when this information was added to a combined model. However, S-ASM elevation might respond to a different threshold of alcohol consumption and/or span a different time window than CDT, MCV, or GGT. So far, only the gradual decrease during the short period of withdrawal treatment of up to 10 days has been described, characterized by a reduction ranging from 6% to 52% of the initial value, in this and previous studies [14,15]. The rate of decline would be relevant in monitoring for relapse. The response time of S-ASM to different amounts and types of alcohol intake has not been investigated yet.

There was no indication that S-ASM levels are suitable to predict 24-month alcohol relapse, although the collected data appeared valid, as suggested by the observed and known strong predictive value of the number of previous withdrawal treatments. Further predictive factors have also been identified in this NOAH cohort. For example, a single nucleotide variant in *OPRM1*, which encodes the mu opioid receptor binding the endogenous ligand beta-endorphin, was associated with an increased risk of more and earlier alcohol-related hospital readmissions [53]. Moreover, a higher body mass index in male patients and higher craving scores [62], as well as clinical Cloninger and Lesch typology classifications [63] are suggested as easily accessible risk factors and promising tools. The failure of

S-ASM to serve as a useful predictor of alcohol relapse, however, does not imply that other components and enzymes of the sphingolipid pathway could not serve as biomarkers. Analysis of their activities or of the serum sphingolipid profile has, thus, some potential, given the relationships of S-ASM with the alcohol biomarkers observed in this study.

Although offering promising results, our study has some limitations beyond the limited sample size. The group of healthy controls was not abstinent and might not be representative because they were recruited from a largely academic environment. They also differ from patients in parameters such as BMI, smoking, and certainly nutrition, including supply with vitamins, which could have an additional influence on the observed effects. Many patients had to be excluded during the screening process and, thus, the generalization of the patient data might also be limited. The frequency of relapse is certainly underestimated, and could contribute to the observed lack of an effect of ASM because we relied on medical records for readmission to the two study centers. We, therefore, may have missed patients treated at other centers, or who did not seek out medical advice at all. Our data need to be interpreted with caution because they do not reflect causal relationships, but are instead associational. Future studies should investigate the potential causation of these findings. Some aspects warrant verification in cell culture or animal models. Due to the explorative nature of the study, we have not corrected the *p*-values for multiple testing. As such, some strong and nominally highly significant associations would survive strict corrections, but still require independent verification in larger cohorts of mixed sexes, if possible.

4. Materials and Methods

4.1. Cohort Characteristics

This investigation was part of the bicentric, cross-sectional, and prospective Neurobiology of Alcoholism (NOAH) study [32]. In 2013 and 2014, a sex-balanced cohort of 200 alcohol-dependent in-patients seeking withdrawal treatment was recruited at the Universitätsklinikum Erlangen Department of Psychiatry and Psychotherapy, and the Klinikum am Europakanal Clinic for Psychiatry, Addiction, Psychotherapy, and Psychosomatic Medicine in Erlangen, Germany. Each patient was diagnosed with an alcohol-use disorder according to the fifth edition of the Diagnostic and Statistical Manual of Mental Disorders (DSM-5, American Psychiatric Association, 2013) and alcohol dependence, according to the tenth revision of the International Classification of Diseases (ICD-10, World Health Organization, 1992). In addition, we recruited 240 healthy control subjects who underwent a multi-step screening procedure to exclude severe somatic and psychiatric morbidity (with the exclusion of nicotine dependence) (for details see [32]). The study was approved by the Ethics Committee of the Medical Faculty of the Friedrich-Alexander University Erlangen-Nürnberg (NOAH study ID 81_12 B, 19 April 2012). All participants provided written informed consent.

The study inclusion with the first blood draw took place during early abstinence (24 to 72 h of abstinence). Afterwards, 81.5% of the patients participated at a direct follow-up at median 5 days later (interquartile range (IQR) 3–6), which included a second blood draw. Whole blood, behavioral scores, and other parameters [32] were collected at the time of recruitment. The German version of the Clinical Institute Withdrawal Assessment for Alcohol revised (CIWA-Ar) scale was used to measure alcohol withdrawal severity in patients at the follow-up [64]. The patients' records were followed for 24 months post-inclusion, to investigate alcohol-related readmissions. In the healthy controls, potential problems of alcohol consumption were assessed by the 4-item CAGE questionnaire [65], and the 10-item Alcohol Use Disorders Identification Test (AUDIT) [66]. The study sample characteristics are provided in Table 1.

4.2. Blood Analysis

Blood samples were collected in the morning for all individuals to minimize circadian effects on hormone levels. Serum vials were centrifuged (10 min and 2000× *g* at room temperature), and serum was aliquoted and placed into storage at $-80\ ^{\circ}$C for later S-ASM activity assays. Glutamic oxaloacetic transaminase (GOT), glutamic-pyruvic transaminase (GPT), gamma-glutamyl transferase (GGT), and creatine kinase (CK) activities, as well as leukocyte and thrombocyte counts, triglycerides, total, HDL and LDL cholesterol, were quantified at the Central Laboratory of the Universitätsklinikum Erlangen, Germany (DIN EN ISO 15189 accredited), from separately collected serum and EDTA vials. Except for one patient who underwent a direct measurement, blood alcohol concentrations were estimated from breath alcohol content that was determined and documented upon admission to the hospital.

4.3. Determination of S-ASM Activity

The activity of S-ASM was quantified using the fluorescent substrate BODIPY-FL-C12-SM (*N*-(4,4-difluoro-5,7-dimethyl-4-bora-3a,4a-diaza-s-indacene-3-dodecanoyl)sphingosyl phosphocholine, D-7711, Thermo Fisher Scientific, Waltham, MA, USA), as described previously [67]. Briefly, the reaction was performed in 96-well polystyrene plates with 58 pmol sphingomyelin in a reaction buffer totaling 50 µL in volume, in the following composition: 200 mM sodium acetate buffer (pH 5.0), 500 mM NaCl, 0.2% IGEPAL®CA-630 (NP 40), and 500 µM $ZnCl_2$. The reaction was initiated by the addition of 6 µL of a 1:10 dilution of serum in physiological 154 mM NaCl solution. After incubation at 37 $^{\circ}$C for 24 h, reactions were stopped by freezing at $-20\ ^{\circ}$C, and stored until further processing. For direct chromatography, 1.5 µL of the reaction was spotted directly without further purification on silica gel 60 thin layer chromatography plates (ALUGRAM SIL G, 818232, Macherey-Nagel, Düren, Germany). Product and uncleaved substrate were separated using ethyl acetate with 1% (*v/v*) acetic acid as a solvent. Spot intensities were detected on a Typhoon Trio scanner, and quantified using the ImageQuant software (GE Healthcare Life Sciences, Buckinghamshire, UK). All enzyme activity assays were carried out with four replicate dilutions of each sample, and using the same lot of reagents and consumables and performed by a single operator.

4.4. Statistics

Data were analyzed using IBM SPSS Statistics Version 21 for Windows (SPSS Inc., Chicago, IL, USA) and GraphPad Prism 7.00 (GraphPad Software Inc., San Diego, CA, USA). Continuous data are presented as the median and IQR in tables as calculated by the custom tables function of SPSS. In the case of missing data points (percentage indicated in Table 1), study subjects were excluded from the specific analyses. Spearman correlations were employed to evaluate associations between two continuous variables. Differences between groups were tested using the Mann–Whitney U test because the values were not normally distributed according to the Kolmogorov–Smirnov test. Differences between alcohol-dependent patients' enzyme activities at different time points were tested using the Wilcoxon signed-rank test. p-Values less than 0.05 for two-sided tests were considered statistically significant. Receiver operating characteristic curve analysis was used to estimate S-ASM activity and alcohol biomarkers required to separate alcohol-dependent patients from healthy controls (including AUC, Youden cut-point, and related sensitivity and specificity). Female and male patients were analyzed separately because of the well-established sex differences in alcohol dependence [68,69] and the highly significant differences in many of the investigated parameters between males and females (Table 1).

5. Conclusions

We replicated our previous observation of increased S-ASM activity in male and female alcohol-dependent patients by examining a large cohort. We characterized previously unreported associations among S-ASM, alcohol levels, and alcohol withdrawal, as well as biomarkers of alcohol dependence. These associations were not only observed in patients but, to some extent, in healthy controls as well. While most effects were similarly present in male and female subgroups, some differences emphasized the necessity to sex-specificity when analyzing the data. Further research should investigate whether there is a causal relationship, or whether these parameters are part of a common pathway, in order to gain insights into the underlying mechanisms and to develop clinical applications.

Increased lysosomal and S-ASM activity in alcohol-dependent patients could lead to elevated ceramide concentrations in the brain. Ceramide is assumed to act as a negative regulator of neurogenesis, neuronal maturation, and survival [70]. Thus, a reduction by pharmacological means (most common antidepressants act as functional inhibitors of ASM [71,72]) to normalize levels could improve these pathologies. A less direct effect could be achieved by nutritional changes, such as supplementation with docosahexaenoic acid during detoxification, to potentially protect against dependence-related neuroinjury [73,74]. A more comprehensive molecular understanding of the alterations of sphingolipid metabolism in the context of alcohol dependence would provide a basis for the rational development of new drugs and treatments.

Supplementary Materials: The following are available online at http://www.mdpi.com/1422-0067/19/12/4028/s1, Table S1: Total and sex-specific correlations of peripheral S-ASM activity with sub-items of the Clinical Institute Withdrawal Assessment for Alcohol revised scale (CIWA-Ar) in alcohol-dependent patients, Table S2: Activity of serum acid sphingomyelinase (S-ASM) in control subjects subdivided according to the presence or absence of at least one binge-drinking episode within the past 24 months. Figure S1: Sex-specific activity of the secretory acid sphingomyelinase (S-ASM) in alcohol-dependent male and female patients subdivided according to their predominantly consumed type of alcoholic beverage during early abstinence. Boxplots show individual data, and the median and interquartile range. The numbers of male and female individuals is provided below the x-axis.

Author Contributions: Conceptualization, C.M., B.L. and J.K.; Data curation, B.L.; Formal analysis, C.M. and B.L.; Funding acquisition, E.G. and J.K.; Investigation, C.M.; Methodology, C.M.; Resources, C.W. and B.L.; Visualization, C.M.; Writing—original draft, C.M.; Writing—review & editing, C.M., C.W., E.G., B.L. and J.K.

Funding: This research was funded by grants of the Deutsche Forschungsgemeinschaft DFG grants GU 335/32-2 to E.G. and KO 947/15-2 to J.K. The work was also supported by intramural grants from the Universitätsklinikum of the Friedrich-Alexander University Erlangen-Nürnberg (FAU). C.M. is an associated fellow of the research training group 2162 "Neurodevelopment and Vulnerability of the Central Nervous System" of the DFG (GRK2162/1). In addition, we acknowledge support by the DFG and the FAU within the funding program Open Access Publishing.

Acknowledgments: We thank Andreas Ahnert, Ute Hamers, and Kristina Bayerlein for the support to recruit patients at the Klinik für Psychiatrie, Sucht, Psychotherapie und Psychosomatik of the Klinikum am Europakanal Erlangen. We gratefully appreciate the support of Juliane Behrens, Sarah Kubis, Katrin Mikolaiczik, Sarah Saigali, Marina Sibach, and Petya Tanovska in recruiting patients and healthy control subjects. We are thankful to Franziska Kress, Hedya Riesop and Marcel-René Muschler for excellent technical support.

Conflicts of Interest: The authors declare no conflict of interest. The funders had no role in the design of the study; in the collection, analyses, or interpretation of data; in the writing of the manuscript, or in the decision to publish the results.

Abbreviations

ALT	alanine aminotransferase (glutamic-pyruvic transaminase, GPT)
AST	aspartate aminotransferase (glutamic-oxaloacetic transaminase, GOT)
AUC	area under the curve
AUDIT	Alcohol Use Disorders Identification Test
BMI	body mass index
CAGE	acronym for 4-item questionnaire indicating potential problems with alcohol abuse
CDT	carbohydrate-deficient transferrin
CIWA-Ar score	Clinical Institute Withdrawal Assessment for Alcohol revised score
CK	creatine kinase
GGT	gamma-glutamyl transferase
Hcy	homocysteine
HDL	high-density lipoprotein
IQR	interquartile range
LDL	low-density lipoprotein
MCV	mean corpuscular volume
S-ASM	secretory acid sphingomyelinase
SMPD1	sphingomyelin phosphodiesterase 1 gene encoding ASM

References

1. Tabakoff, B.; Hoffman, P.L. The neurobiology of alcohol consumption and alcoholism: An integrative history. *Pharmacol. Biochem. Behav.* **2013**, *113*, 20–37. [CrossRef]
2. Foroud, T.; Edenberg, H.; Crabbe, J.C. Genetic Research—Who is at risk for alcoholism? *Alcohol Res. Health* **2010**, *33*, 64–75.
3. Lenz, B.; Bouna-Pyrrou, P.; Mühle, C.; Kornhuber, J. Low digit ratio (2D:4D) and late pubertal onset indicate prenatal hyperandrogenziation in alcohol binge drinking. *Prog. Neuropsychopharmacol. Biol. Psychiatry* **2018**, *86*, 370–378. [CrossRef]
4. Huber, S.E.; Zoicas, I.; Reichel, M.; Mühle, C.; Büttner, C.; Ekici, A.B.; Eulenburg, V.; Lenz, B.; Kornhuber, J.; Müller, C.P. Prenatal androgen receptor activation determines adult alcohol and water drinking in a sex-specific way. *Addict. Biol.* **2018**, *23*, 904–920. [CrossRef]
5. Mühle, C.; Reichel, M.; Gulbins, E.; Kornhuber, J. Sphingolipids in psychiatric disorders and pain syndromes. *Handb. Exp. Pharmacol.* **2013**, 431–456. [CrossRef]
6. Huston, J.P.; Kornhuber, J.; Mühle, C.; Japtok, L.; Komorowski, M.; Mattern, C.; Reichel, M.; Gulbins, E.; Kleuser, B.; Topic, B.; et al. A sphingolipid mechanism for behavioral extinction. *J. Neurochem.* **2016**, *137*, 589–603. [CrossRef]
7. Müller, C.P.; Reichel, M.; Mühle, C.; Rhein, C.; Gulbins, E.; Kornhuber, J. Brain membrane lipids in major depression and anxiety disorders. *Biochim. Biophys. Acta* **2015**, *1851*, 1052–1065. [CrossRef]
8. Kornhuber, J.; Reichel, M.; Tripal, P.; Groemer, T.W.; Henkel, A.W.; Mühle, C.; Gulbins, E. The role of ceramide in major depressive disorder. *Eur. Arch. Psychiatry Clin. Neurosci.* **2009**, *259* (Suppl. 2), 199–204. [CrossRef]
9. Van Meer, G.; Voelker, D.R.; Feigenson, G.W. Membrane lipids: Where they are and how they behave. *Nat. Rev. Mol. Cell Boil.* **2008**, *9*, 112–124. [CrossRef]
10. Hait, N.C.; Oskeritzian, C.A.; Paugh, S.W.; Milstien, S.; Spiegel, S. Sphingosine kinases, sphingosine 1-phosphate, apoptosis and diseases. *Biochim. Biophys. Acta* **2006**, *1758*, 2016–2026. [CrossRef]
11. Hannun, Y.A.; Obeid, L.M. Sphingolipids and their metabolism in physiology and disease. *Nat. Rev. Mol. Cell Boil.* **2018**, *19*, 175–191. [CrossRef]
12. Kornhuber, J.; Rhein, C.; Müller, C.P.; Mühle, C. Secretory sphingomyelinase in health and disease. *Biol. Chem.* **2015**, *396*, 707–736. [CrossRef]
13. Reichel, M.; Greiner, E.; Richter-Schmidinger, T.; Yedibela, O.; Tripal, P.; Jacobi, A.; Bleich, S.; Gulbins, E.; Kornhuber, J. Increased acid sphingomyelinase activity in peripheral blood cells of acutely intoxicated patients with alcohol dependence. *Alcohol. Clin. Exp. Res.* **2010**, *34*, 46–50. [CrossRef]
14. Reichel, M.; Beck, J.; Mühle, C.; Rotter, A.; Bleich, S.; Gulbins, E.; Kornhuber, J. Activity of secretory sphingomyelinase is increased in plasma of alcohol-dependent patients. *Alcohol. Clin. Exp. Res.* **2011**, *35*, 1852–1859. [CrossRef]
15. Mühle, C.; Amova, V.; Biermann, T.; Bayerlein, K.; Richter-Schmidinger, T.; Kraus, T.; Reichel, M.; Gulbins, E.; Kornhuber, J. Sex-dependent decrease of sphingomyelinase activity during alcohol withdrawal treatment. *Cell. Physiol. Biochem.* **2014**, *34*, 71–81. [CrossRef]
16. Reichel, M.; Hönig, S.; Liebisch, G.; Lüth, A.; Kleuser, B.; Gulbins, E.; Schmitz, G.; Kornhuber, J. Alterations of plasma glycerophospholipid and sphingolipid species in male alcohol-dependent patients. *Biochim. Biophys. Acta* **2015**, *1851*, 1501–1510. [CrossRef]
17. Deaciuc, I.V.; Nikolova-Karakashian, M.; Fortunato, F.; Lee, E.Y.; Hill, D.B.; McClain, C.J. Apoptosis and dysregulated ceramide metabolism in a murine model of alcohol-enhanced lipopolysaccharide hepatotoxicity. *Alcohol. Clin. Exp. Res.* **2000**, *24*, 1557–1565. [CrossRef]
18. Liangpunsakul, S.; Rahmini, Y.; Ross, R.A.; Zhao, Z.; Xu, Y.; Crabb, D.W. Imipramine blocks ethanol-induced ASMase activation, ceramide generation, and PP2A activation, and ameliorates hepatic steatosis in ethanol-fed mice. American journal of physiology. *Gastrointest. Liver Physiol.* **2012**, *302*, G515–G523. [CrossRef]
19. Müller, C.P.; Kalinichenko, L.S.; Tiesel, J.; Witt, M.; Stöckl, T.; Sprenger, E.; Fuchser, J.; Beckmann, J.; Praetner, M.; Huber, S.E.; et al. Paradoxical antidepressant effects of alcohol are related to acid sphingomyelinase and its control of sphingolipid homeostasis. *Acta Neuropathol.* **2017**, *133*, 463–483. [CrossRef]

20. Liu, J.J.; Wang, J.Y.; Hertervig, E.; Cheng, Y.; Nilsson, A.; Duan, R.D. Activation of neutral sphingomyelinase participates in ethanol-induced apoptosis in Hep G2 cells. *Alcohol Alcohol.* **2000**, *35*, 569–573. [CrossRef]
21. Pascual, M.; Valles, S.L.; Renau-Piqueras, J.; Guerri, C. Ceramide pathways modulate ethanol-induced cell death in astrocytes. *J. Neurochem.* **2003**, *87*, 1535–1545. [CrossRef]
22. Yang, L.; Jin, G.H.; Zhou, J.Y. The role of ceramide in the pathogenesis of alcoholic liver disease. *Alcohol Alcohol.* **2016**, *51*, 251–257. [CrossRef]
23. Grammatikos, G.; Mühle, C.; Ferreiros, N.; Schroeter, S.; Bogdanou, D.; Schwalm, S.; Hintereder, G.; Kornhuber, J.; Zeuzem, S.; Sarrazin, C.; et al. Serum acid sphingomyelinase is upregulated in chronic hepatitis C infection and non alcoholic fatty liver disease. *Biochim. Biophys. Acta* **2014**, *1841*, 1012–1020. [CrossRef]
24. Fernandez, A.; Colell, A.; Garcia-Ruiz, C.; Fernandez-Checa, J.C. Cholesterol and sphingolipids in alcohol-induced liver injury. *J. Gastroenterol. Hepatol.* **2008**, *23* (Suppl. 1), S9–S15. [CrossRef]
25. Setshedi, M.; Longato, L.; Petersen, D.R.; Ronis, M.; Chen, W.C.; Wands, J.R.; de la Monte, S.M. Limited therapeutic effect of N-acetylcysteine on hepatic insulin resistance in an experimental model of alcohol-induced steatohepatitis. *Alcohol. Clin. Exp. Res.* **2011**, *35*, 2139–2151. [CrossRef]
26. Garcia-Ruiz, C.; Mato, J.M.; Vance, D.; Kaplowitz, N.; Fernandez-Checa, J.C. Acid sphingomyelinase-ceramide system in steatohepatitis: A novel target regulating multiple pathways. *J. Hepatol.* **2015**, *62*, 219–233. [CrossRef]
27. Bearer, C.F.; Bailey, S.M.; Hoek, J.B. Advancing alcohol biomarkers research. *Alcohol. Clin. Exp. Res.* **2010**, *34*, 941–945. [CrossRef]
28. Gonzalo, P.; Radenne, S.; Gonzalo, S. Biomarkers of chronic alcohol misuse. *Curr. Biomark. Find.* **2014**, *4*, 9–22. [CrossRef]
29. Liangpunsakul, S.; Lai, X.; Ross, R.A.; Yu, Z.; Modlik, E.; Westerhold, C.; Heathers, L.; Paul, R.; O'Connor, S.; Crabb, D.W.; et al. Novel serum biomarkers for detection of excessive alcohol use. *Alcohol. Clin. Exp. Res.* **2015**, *39*, 556–565. [CrossRef]
30. Batra, A.; Muller, C.A.; Mann, K.; Heinz, A. Alcohol dependence and harmful use of alcohol. *Dtsch. Arzteblatt Int.* **2016**, *113*, 301–310. [CrossRef]
31. Heinz, A.; Deserno, L.; Zimmermann, U.S.; Smolka, M.N.; Beck, A.; Schlagenhauf, F. Targeted intervention: Computational approaches to elucidate and predict relapse in alcoholism. *NeuroImage* **2017**, *151*, 33–44. [CrossRef]
32. Lenz, B.; Mühle, C.; Braun, B.; Weinland, C.; Bouna-Pyrrou, P.; Behrens, J.; Kubis, S.; Mikolaiczik, K.; Muschler, M.R.; Saigali, S.; et al. Prenatal and adult androgen activities in alcohol dependence. *Acta Psychiatr. Scand.* **2017**, *136*, 96–107. [CrossRef]
33. Spargo, E. The acute effects of alcohol on plasma creatine kinase (CK) activity in the rat. *J. Neurol. Sci.* **1984**, *63*, 307–316. [CrossRef]
34. Gupta, A.; Gupta, C.; Khurana, S. Evaluation of total creatine kinase levels in a spectrum of neuro-psychiatric disorders in a tertiary neurosciences centre. *Int. J. Med. Public Health* **2015**, *5*, 362–366. [CrossRef]
35. Segal, M.; Avital, A.; Rusakov, A.; Sandbank, S.; Weizman, A. Serum creatine kinase activity differentiates alcohol syndromes of dependence, withdrawal and delirium tremens. *Eur. Neuropsychopharmacol.* **2009**, *19*, 92–96. [CrossRef]
36. Tabara, Y.; Arai, H.; Hirao, Y.; Takahashi, Y.; Setoh, K.; Kawaguchi, T.; Kosugi, S.; Ito, Y.; Nakayama, T.; Matsuda, F.; et al. The causal effects of alcohol on lipoprotein subfraction and triglyceride levels using a Mendelian randomization analysis: The Nagahama study. *Atherosclerosis* **2017**, *257*, 22–28. [CrossRef]
37. Vu, K.N.; Ballantyne, C.M.; Hoogeveen, R.C.; Nambi, V.; Volcik, K.A.; Boerwinkle, E.; Morrison, A.C. Causal role of alcohol consumption in an improved lipid profile: The Atherosclerosis Risk in Communities (ARIC) Study. *PLoS ONE* **2016**, *11*, e0148765. [CrossRef]
38. Sillanaukee, P.; Koivula, T.; Jokela, H.; Myllyharju, H.; Seppa, K. Relationship of alcohol consumption to changes in HDL-subfractions. *Eur. J. Clin. Investig.* **1993**, *23*, 486–491. [CrossRef]
39. Mühle, C.; Huttner, H.B.; Walter, S.; Reichel, M.; Canneva, F.; Lewczuk, P.; Gulbins, E.; Kornhuber, J. Characterization of acid sphingomyelinase activity in human cerebrospinal fluid. *PLoS ONE* **2013**, *8*, e62912. [CrossRef]
40. Deevska, G.M.; Sunkara, M.; Morris, A.J.; Nikolova-Karakashian, M.N. Characterization of secretory sphingomyelinase activity, lipoprotein sphingolipid content and LDL aggregation in ldlr−/− mice fed on a high-fat diet. *Biosci. Rep.* **2012**, *32*, 479–490. [CrossRef]

41. Nikolova-Karakashian, M. Alcoholic and non-alcoholic fatty liver disease: Focus on ceramide. *Adv. Biol. Regul.* **2018**, *70*, 40–50. [CrossRef]

42. Reichel, M.; Richter-Schmidinger, T.; Mühle, C.; Rhein, C.; Alexopoulos, P.; Schwab, S.G.; Gulbins, E.; Kornhuber, J. The common acid sphingomyelinase polymorphism p.G508R is associated with self-reported allergy. *Cell. Physiol. Biochem.* **2014**, *34*, 82–91. [CrossRef]

43. Rhein, C.; Mühle, C.; Kornhuber, J.; Reichel, M. Alleged detrimental mutations in the *SMPD1* gene in patients with Niemann-Pick disease. *Int. J. Mol. Sci.* **2015**, *16*, 13649–13652. [CrossRef]

44. Rhein, C.; Reichel, M.; Mühle, C.; Rotter, A.; Schwab, S.G.; Kornhuber, J. Secretion of acid sphingomyelinase is affected by its polymorphic signal peptide. *Cell. Physiol. Biochem.* **2014**, *34*, 1385–1401. [CrossRef]

45. Reagan, J.W., Jr.; Hubbert, M.L.; Shelness, G.S. Posttranslational regulation of acid sphingomyelinase in Niemann-Pick type C1 fibroblasts and free cholesterol-enriched Chinese hamster ovary cells. *J. Boil. Chem.* **2000**, *275*, 38104–38110. [CrossRef]

46. Zuo, L.; Wang, K.; Zhang, X.Y.; Krystal, J.H.; Li, C.S.; Zhang, F.; Zhang, H.; Luo, X. *NKAIN1-SERINC2* is a functional, replicable and genome-wide significant risk gene region specific for alcohol dependence in subjects of European descent. *Drug Alcohol Depend.* **2013**, *129*, 254–264. [CrossRef]

47. Zuo, L.J.; Wang, K.S.; Zhang, X.Y.; Li, C.S.R.; Zhang, F.Y.; Wang, X.P.; Chen, W.A.; Gao, G.M.; Zhang, H.P.; Krystal, J.H.; et al. Rare *SERINC2* variants are specific for alcohol dependence in individuals of European descent. *Pharmacogenet. Genom.* **2013**, *23*, 395–402. [CrossRef]

48. Inuzuka, M.; Hayakawa, M.; Ingi, T. Serinc, an activity-regulated protein family, incorporates serine into membrane lipid synthesis. *J. Boil. Chem.* **2005**, *280*, 35776–35783. [CrossRef]

49. Rhein, C.; Tripal, P.; Seebahn, A.; Konrad, A.; Kramer, M.; Nagel, C.; Kemper, J.; Bode, J.; Mühle, C.; Gulbins, E.; et al. Functional implications of novel human acid sphingomyelinase splice variants. *PLoS ONE* **2012**, *7*, e35467. [CrossRef]

50. Rhein, C.; Reichel, M.; Kramer, M.; Rotter, A.; Lenz, B.; Mühle, C.; Gulbins, E.; Kornhuber, J. Alternative splicing of *SMPD1* coding for acid sphingomyelinase in major depression. *J. Affect. Disord.* **2017**, *209*, 10–15. [CrossRef]

51. Kornhuber, J.; Kornhuber, H.H.; Backhaus, B.; Kornhuber, A.; Kaiserauer, C.; Wanner, W. The normal values of gamma-glutamyltransferase are falsely defined up to now: On the diagnosis of hypertension, obesity and diabetes with reference to "normal" consumption of alcohol. *Versicherungsmedizin* **1989**, *41*, 78–81.

52. Fernandez, A.; Matias, N.; Fucho, R.; Ribas, V.; Von Montfort, C.; Nuno, N.; Baulies, A.; Martinez, L.; Tarrats, N.; Mari, M.; et al. ASMase is required for chronic alcohol induced hepatic endoplasmic reticulum stress and mitochondrial cholesterol loading. *J. Hepatol.* **2013**, *59*, 805–813. [CrossRef]

53. Gegenhuber, B.; Weinland, C.; Kornhuber, J.; Mühle, C.; Lenz, B. *OPRM1* A118G and serum beta-endorphin interact with sex and digit ratio (2D:4D) to influence risk and course of alcohol dependence. *Eur. Neuropsychopharmacol.* **2018**. [CrossRef]

54. Gan-Or, Z.; Ozelius, L.J.; Bar-Shira, A.; Saunders-Pullman, R.; Mirelman, A.; Kornreich, R.; Gana-Weisz, M.; Raymond, D.; Rozenkrantz, L.; Deik, A.; et al. The p.L302P mutation in the lysosomal enzyme gene *SMPD1* is a risk factor for Parkinson disease. *Neurology* **2013**, *80*, 1606–1610. [CrossRef]

55. Dagan, E.; Schlesinger, I.; Ayoub, M.; Mory, A.; Nassar, M.; Kurolap, A.; Peretz-Aharon, J.; Gershoni-Baruch, R. The contribution of Niemann-Pick *SMPD1* mutations to Parkinson disease in Ashkenazi Jews. *Park. Relat. Disord.* **2015**, *21*, 1067–1071. [CrossRef]

56. Gan-Or, Z.; Dion, P.A.; Rouleau, G.A. Genetic perspective on the role of the autophagy-lysosome pathway in Parkinson disease. *Autophagy* **2015**, *11*, 1443–1457. [CrossRef]

57. Liu, C.; Marioni, R.E.; Hedman, A.K.; Pfeiffer, L.; Tsai, P.C.; Reynolds, L.M.; Just, A.C.; Duan, Q.; Boer, C.G.; Tanaka, T.; et al. A DNA methylation biomarker of alcohol consumption. *Mol. Psychiatry* **2018**, *23*, 422–433. [CrossRef]

58. Heymann, H.M.; Gardner, A.M.; Gross, E.R. Aldehyde-induced DNA and protein adducts as biomarker tools for alcohol use disorder. *Trends Mol. Med.* **2018**, *24*, 144–155. [CrossRef]

59. Logan, C.; Asadi, H.; Kok, H.K.; Looby, S.T.; Brennan, P.; O'Hare, A.; Thornton, J. Neuroimaging of chronic alcohol misuse. *J. Med. Imaging Radiat. Oncol.* **2017**, *61*, 435–440. [CrossRef]

60. Zou, Y.; Murray, D.E.; Durazzo, T.C.; Schmidt, T.P.; Murray, T.A.; Meyerhoff, D.J. White matter microstructural correlates of relapse in alcohol dependence. *Psychiatry Res. Neuroimaging* **2018**, *281*, 92–100. [CrossRef]

61. Tu, W.; Chu, C.; Li, S.; Liangpunsakul, S. Development and validation of a composite score for excessive alcohol use screening. *J. Investig. Med.* **2016**, *64*, 1006–1011. [CrossRef]
62. Weinland, C.; Mühle, C.; Kornhuber, J.; Lenz, B. Body mass index and craving predict 24-month hospital readmissions of alcohol-dependent in-patients following withdrawal. *Prog. Neuropsychopharmacol. Biol. Psychiatry* **2018**. [CrossRef]
63. Weinland, C.; Braun, B.; Mühle, C.; Kornhuber, J.; Lenz, B. Cloninger type 2 score and Lesch typology predict hospital readmission of female and male alcohol-dependent inpatients during a 24-month follow-up. *Alcohol. Clin. Exp. Res.* **2017**, *41*, 1760–1767. [CrossRef]
64. Stuppaeck, C.H.; Barnas, C.; Falk, M.; Guenther, V.; Hummer, M.; Oberbauer, H.; Pycha, R.; Whitworth, A.B.; Fleischhacker, W.W. Assessment of the alcohol withdrawal syndrome—Validity and reliability of the translated and modified Clinical Institute Withdrawal Assessment for Alcohol scale (CIWA-A). *Addiction* **1994**, *89*, 1287–1292. [CrossRef]
65. Rumpf, H.-J.; Hapke, U.; John, U. Deutsche Version des CAGE Fragebogens (CAGE-G). In *Elektronisches Handbuch zu Erhebungsinstrumenten im Suchtbereich (EHES)*; Version 3.00; Glöckner-Rist, A.F., Rist, H.K., Eds.; Zentrum für Umfragen, Methoden und Analysen: Mannheim, Germany, 2003.
66. Rumpf, H.-J.; Meyer, C.; Hapke, U.; John, U. Deutsche Version des Alcohol Use Disorders Identification Test (AUDIT-G-L). In *Elektronisches Handbuch zu Erhebungsinstrumenten im Suchtbereich (EHES)*; Version 3.00; Glöckner-Rist, A.F., Rist, H.K., Eds.; Zentrum für Umfragen, Methoden und Analysen: Mannheim, Germany, 2003.
67. Mühle, C.; Kornhuber, J. Assay to measure sphingomyelinase and ceramidase activities efficiently and safely. *J. Chromatogr. A* **2017**, *1481*, 137–144. [CrossRef]
68. Lenz, B.; Müller, C.P.; Stoessel, C.; Sperling, W.; Biermann, T.; Hillemacher, T.; Bleich, S.; Kornhuber, J. Sex hormone activity in alcohol addiction: Integrating organizational and activational effects. *Prog. Neurobiol.* **2012**, *96*, 136–163. [CrossRef]
69. Clayton, J.A.; Collins, F.S. Policy: NIH to balance sex in cell and animal studies. *Nature* **2014**, *509*, 282–283. [CrossRef]
70. Gulbins, E.; Palmada, M.; Reichel, M.; Lüth, A.; Böhmer, C.; Amato, D.; Müller, C.P.; Tischbirek, C.H.; Groemer, T.W.; Tabatabai, G.; et al. Acid sphingomyelinase-ceramide system mediates effects of antidepressant drugs. *Nat. Med.* **2013**, *19*, 934–938. [CrossRef]
71. Kornhuber, J.; Muehlbacher, M.; Trapp, S.; Pechmann, S.; Friedl, A.; Reichel, M.; Mühle, C.; Terfloth, L.; Groemer, T.W.; Spitzer, G.M.; et al. Identification of novel functional inhibitors of acid sphingomyelinase. *PLoS ONE* **2011**, *6*, e23852. [CrossRef]
72. Kornhuber, J.; Tripal, P.; Reichel, M.; Mühle, C.; Rhein, C.; Muehlbacher, M.; Groemer, T.W.; Gulbins, E. Functional Inhibitors of Acid Sphingomyelinase (FIASMAs): A novel pharmacological group of drugs with broad clinical applications. *Cell. Physiol. Biochem.* **2010**, *26*, 9–20. [CrossRef]
73. Babenko, N.A.; Semenova Ya, A. Sphingolipid turnover in the hippocampus and cognitive dysfunction in alcoholized rats: correction with the help of alimentary n-3 fatty acids. *Neurophysiology* **2010**, *42*, 206–212. [CrossRef]
74. Collins, M.A. Alcohol abuse and docosahexaenoic acid: Effects on cerebral circulation and neurosurvival. *Brain Circ.* **2015**, *1*, 63–68. [CrossRef]

International Journal of
Molecular Sciences

Communication

Alpha-Synuclein RNA Expression is Increased in Major Depression

Andrea Rotter, Bernd Lenz, Ruben Pitsch, Tanja Richter-Schmidinger, Johannes Kornhuber and Cosima Rhein *

Department of Psychiatry and Psychotherapy, Friedrich-Alexander-University Erlangen-Nürnberg (FAU), D-91054 Erlangen, Germany; andrea.rotter-neubert@Landratsamt-Roth.de (A.R.);
bernd.lenz@uk-erlangen.de (B.L.); rubenpitsch@gmail.com (R.P.);
tanja.richter-schmidinger@uk-erlangen.de (T.R.-S.); johannes.kornhuber@uk-erlangen.de (J.K.)
* Correspondence: cosima.rhein@uk-erlangen.de; Tel.: +49-9131-85-44604

Received: 3 April 2019; Accepted: 20 April 2019; Published: 25 April 2019

Abstract: Alpha-synuclein (SNCA) is a small membrane protein that plays an important role in neuro-psychiatric diseases. It is best known for its abnormal subcellular aggregation in Lewy bodies that serves as a hallmark of Parkinson's disease (PD). Due to the high comorbidity of PD with depression, we investigated the role of SNCA in patients suffering from major depressive disorder (MDD). *SNCA* mRNA expression levels were analyzed in peripheral blood cells of MDD patients and a healthy control group. *SNCA* mRNA expression was positively correlated with severity of depression as indicated by psychometric assessment. We found a significant increase in *SNCA* mRNA expression levels in severely depressed patients compared with controls. Thus, SNCA analysis could be a helpful target in the search for biomarkers of MDD.

Keywords: alpha-synuclein; SNCA; major depression; Hamilton Scale of Depression

---

1. Introduction

Alpha-synuclein (SNCA) is a small membrane protein (~14 kDa) consisting of 140 amino acids encoded on chromosome 4q21 [1–4]. It was shown that SNCA is localized close to synaptic vesicles and interacts with the cell membrane. A specific role in the regulation of dopamine transmission was suggested [5]. SNCA was further found to localize at neuronal growth cones, which indicates a role in neuronal plasticity [6–10]. Abnormal subcellular SNCA aggregation is a hallmark of neurodegenerative diseases (Parkinson's disease, dementia with Lewy bodies, and multiple system atrophy) that are recognized as alpha-synucleinopathies [11]. In addition, there is increasing evidence that SNCA could also be involved in the pathophysiology of major depressive disorder (MDD). MDD is a severe psychiatric disorder with a lifetime prevalence of approximately 10% that is characterized by depressed mood, a decline in motivation and the loss of feelings of pleasure and interest, resulting in increased suicide rates [12]. Due to the unclear pathogenesis of MDD it is suggested that environmental factors such as psychosocial stress and genetic characteristics trigger dysregulation of the cytokine system, the neurotransmitter systems, the hormonal systems and the circadian rhythm [13–16]. It has been shown that in all alpha-synucleinopathies, there is a 30–60% comorbidity with MDD [11]. An involvement of SNCA in psychiatric disorders was first detected in a study about eating disorders that correlated *SNCA* mRNA levels positively with the severity of depressive symptoms [17]. The connecting link between SNCA and MDD could be its modulating effect on monoamine transporters [18]. SNCA influences the expression and, thereby, the activity of dopamine, serotonin and norepinephrine transporters through direct binding and influence on trafficking, and helps to maintain the homeostasis of monoamine neurotransmitters in the brain [19–21]. SNCA was also associated with stress in a rat model of depression [22]. In addition, antidepressant therapy influences the SNCA system. Desipramine has been shown to modulate SNCA and the norepinephrine transporter in an animal model of depression [23]. Antidepressant therapy was found to influence *SNCA* mRNA expression in the hippocampus of rats [24]. Rats that were treated with paroxetine showed decreased protein expression of SNCA [25]. Several *SNCA* single nucleotide polymorphisms (SNPs) have been identified and seem to play an important role in the pathophysiology of psychiatric diseases. Alcohol craving, for example, was shown to be associated with an SNP in the *SNCA* gene [4] and higher SNCA protein levels in patients [26]. Moreover, significantly longer alleles of the repeat NACP-REP1 were detected in alcohol-dependent patients compared with healthy controls [27]. The NACP-REP1 length polymorphism was also found to correlate with depressive symptoms in healthy volunteers [15]. GWAS studies have identified *SNCA* as one of the top genes relevant to psychiatric disorders [28].

Therefore, we hypothesized a role for SNCA in MDD and investigated the link between *SNCA* mRNA expression in peripheral blood and depressive symptoms in depressed patients.

2. Results

2.1. SNCA mRNA Expression Correlates Positively with the Severity of Depressive Symptoms

The mRNA expression levels of *SNCA* were determined in peripheral blood cells of MDD patients. The severity of depressive symptoms in patients was assessed using two psychometric scales: the Hamilton depression rating scale (HAM-D-17) for clinician-administered rating and Beck's Depression Inventory—revised (BDI-II) for self-report rating. To investigate the relationship between *SNCA* mRNA expression and the severity of depressive symptoms in patients, we conducted a correlative analysis. Using Pearson correlation, we found a significant correlation between *SNCA* mRNA expression levels and BDI-II ($r = 0.281$, $p = 0.026$) and HAM-D-17 scores in the patients ($r = 0.273$, $p = 0.028$). Thus, the severity of depressive symptoms in MDD patients, as indicated by a higher psychometric score in the self-report rating as well as in the clinician-administered rating, was positively correlated with the measured *SNCA* mRNA expression in their blood cells.

2.2. SNCA mRNA Expression is Increased in Patients with Severe Depression

When analyzing the data of the MDD patients more closely, it turned out that the two patient subgroups, "ADT" (MDD patients recruited for the study "AntiDepressive Therapy" (ADT)) and "BLADe" (MDD patients participating in the study "Blood Lipid Alterations in Depression" (BLADe), Table 1) differed not only with regard to treatment, but also with regard to severity of symptoms. The patients of the ADT study, who were untreated, had an average HAM-D score of 17.7 ± 8.2 points, which characterizes this group as being moderately depressed. In contrast, the patients of the BLADe study, who were already treated at the beginning of the study, had an average HAM-D score of 21.4 ± 5.2 points, which falls into the category of severe depression. The difference in the severity of symptoms between both groups was statistically significant, as the ADT group exhibited significantly lower HAM-D scores than the BLADe group (*t*-test, df = 63, T = 2.2, $p = 0.031$; Table 2). There was no significant difference between females and males regarding the severity of depression (HAM-D score of 21.3 ± 6.6 and 18.0 ± 6.7, respectively; *t*-test, df = 63, T = −1.9, $p > 0.05$).

In a further analysis, we assessed the difference in *SNCA* mRNA expression between patients and controls. Due to the significant difference in age between the BLADe patient group and the healthy volunteers (Table 1), age was included as a covariate in all analyses. *SNCA* mRNA expression values were normally distributed. Compared with the control group, which had a mean normalized *SNCA* mRNA expression level of 17.4 ± 5.4 in their blood cells, MDD patients in the BLADe and ADT studies displayed increased *SNCA* mRNA expression levels (mean normalized expression of 31.9 ± 15.3 and 24.3 ± 13.8, respectively; analysis of variance (ANOVA) df = 2, F = 5.9, $p = 0.004$; Table 2). Pairwise comparison analysis revealed that the significant difference resulted from the comparison of the control group with the BLADe patient group ($p = 0.001$), but not with the ADT patient group ($p = 0.114$).

Moreover, the comparison of *SNCA* mRNA expression between both patient groups revealed that the BLADe subgroup showed significantly higher *SNCA* mRNA expression levels than the ADT subgroup ($p = 0.031$). Even though females had a higher level of *SNCA* mRNA expression compared with males (mean normalized expression of 31.0 ± 17.2 and 21.3 ± 8.1, respectively; *t*-test, df = 63, T = -3.4, $p = 0.001$), there was no interaction effect between the groups and sex. Therefore, *SNCA* mRNA expression differs between healthy controls and depressed patients and seems to increase with the severity of depressive symptoms.

Table 1. Demographic overview. Differences in sex distribution were calculated using the chi quadrat test. Differences regarding age were calculated using analysis of variance (ANOVA). SD, standard deviation. ADT, MDD patients recruited for the study "AntiDepressive Therapy"; BLADe, MDD patients participating in the study "Blood Lipid Alterations in Depression".

	BLADe	**ADT**	**Healthy Controls**	*p*-**Value**
N (male/female)	39 (15/24)	31 (15/16)	18 (13/5)	0.060
Age (years $\pm$ SD)	46.3 ± 14.2	39.7 ± 16.5	30.4 ± 8.8	0.001

Table 2. Values for *SNCA* mRNA expression and psychometric scores. *SNCA* mRNA expression and Hamilton depression rating scale (HAM-D) scores differ significantly between groups (ANOVA). HAM-D was not conducted in the control group.

	BLADe	**ADT**	**Healthy Controls**	*p*-**Value**
SNCA expression $\pm$ SD	31.9 ± 15.3	24.3 ± 13.8	17.4 ± 5.4	0.004
HAM-D scores $\pm$ SD	21.4 ± 5.2	17.9 ± 8.2	-	0.034

3. Discussion

Our study shows for the first time a significant increase in *SNCA* mRNA expression levels in severely depressed patients compared with healthy controls. This is in line with a parallel study in which increased SNCA protein levels were measured in the blood serum of depressed patients [29]. Our results showing a positive correlation between *SNCA* mRNA expression and BDI-II scores confirm a study by Frieling and colleagues in which this relationship was found in eating disorders [17]. Of note, other data have shown that patients who exhibited an early remission upon antidepressant treatment had increased *SNCA* mRNA expression levels at baseline compared with a non-responder group, but *SNCA* mRNA expression was not monitored during and after treatment [30]. The insights from clinical studies are derived from analyses of peripheral blood cells and do not address central mechanisms. In a murine study, the overexpression of SNCA in midbrain dopaminergic neurons resulted in depressive-like behavior [31]. The mediating effect of increased SNCA on the development of depression may be related to impaired adult neurogenesis in the hippocampus. In a mouse model overexpressing mutant A53T SNCA, adult neurogenesis in the dentate gyrus of the hippocampus was significantly impaired due to a reduction in proliferation of neural stem and precursor cells [32]. Another link could involve compromised neurotransmitter release associated with increased SNCA. In a stress model of depression in rats, several proteins were found to be differentially expressed and associated with deficits in synaptic vesicle release involving SNCA, synapsin I and the adaptor protein-3 complex, which were hypothesized to contribute to the pathomechanisms of psychiatric diseases [22].

Interestingly, increased mRNA expression of *SNCA* in patients were also detected in studies focusing on other neuro-psychiatric diseases: in neuronal disorders [3], in alcohol dependence [4] and cocaine dependence [33]. A common hallmark of these diseases is the impairment of cognition. The high comorbidity of PD with dementia and depression thus points to a common pathway. In 30–60% of PD patients, depressive symptoms occur and often precede motor symptoms [34]. The lack of studies investigating PD patients with depression makes further insights difficult. The treatment of depression in PD was investigated in two studies that found better outcomes for tricyclic antidepressants than for selective serotonin reuptake inhibitors (SSRIs) [35,36]. In murine studies, it was shown that treatment with fluoxetine did not influence *SNCA* mRNA expression levels [32], whereas paroxetine decreased *SNCA* mRNA levels [25]. It could be hypothesized that antidepressants in PD work via an influence on SNCA levels and that fluoxetine and paroxetine classified as SSRI are not optimized for this effect. Additionally, in a mouse model overexpressing SNCA, serotonergic projections in the hippocampus seem to be compromised, and the high protein levels of SNCA affected responsiveness to SSRIs [37]. These conflicting results indicate the need for human studies that monitor SNCA levels after antidepressant treatment.

One limitation of the present study may be that the number of patients in this study was relatively small, and we did not monitor treatment effects on *SNCA* mRNA expression. Moreover, the healthy volunteers differed from one of the patient groups regarding age.

In summary, we show a significant increase in *SNCA* mRNA expression levels in patients suffering from severe depression. Further studies with larger sample sizes and treatment monitoring are warranted to elucidate the clinical relevance of SNCA in MDD.

4. Materials and Methods

4.1. Ethics Statement

The collection of blood samples was approved by the Ethics Committee of the Friedrich-Alexander-University Erlangen-Nürnberg (FAU) (ID 4194, renewal of 3412, approval date: 20 April 2010) and conducted in concordance with the Declaration of Helsinki. Written informed consent was obtained from all participants.

4.2. Study Sample

All patients had an established diagnosis of MDD according to the International Statistical Classification of Diseases and Related Health Problems (ICD-10) and the Diagnostic and Statistical Manual of Mental Disorders (DSM-IV) criteria. After hospital admission, diagnosis was confirmed by conducting a diagnostic interview using the *Strukturiertes Klinisches Interview für DSM-IV* (SKID-I). Further, the following clinical scales were administered: the Hamilton depression rating scale (HAM-D-17) for clinician-administered rating and Beck's Depression Inventory—revised (BDI-II) for self-report rating. All participants were carefully screened to rule out the existence of inflammatory, cardiac, endocrine, renal and hepatic disease by means of a structured medical history, physical examination, routine laboratory testing, and electrocardiography. Patients were excluded if comorbidity of alcohol or drug dependence was detected. MDD patients participating in the BLADe (Blood Lipid Alterations in DEpression) study ($n = 39$) were already treated with a standard antidepressant therapy at admission. MDD patients recruited for the ADT (AntiDepressive Therapy) study ($n = 31$) had not been treated with antidepressants, and standard antidepressant therapy was initiated after taking blood samples. A group of 18 healthy subjects without a personal history of psychiatric and somatic disorders served as a control group [38]. The BLADe patient group differed significantly from the healthy volunteers in terms of age (ANOVA, df = 2, F = 7.8, $p = 0.001$; post hoc analysis revealed significant difference only between controls and the BLADe group, $p = 0.001$; Table 1).

4.3. RNA Isolation and cDNA Synthesis

For patients in the ADT study, blood of fasting patients was taken for RNA isolation in the morning to secure for stable experimental conditions. For RNA isolation, the PAXgene system was employed (PreAnalytiX GmbH, Hombrechikon, Switzerland). PAXgene tubes containing blood samples were incubated at room temperature for 2 h, stored at $-80\ ^{\circ}$C, and RNA was isolated according to manufacturer's instructions. For patients in the BLADe study and for control samples, total RNA was extracted from whole blood in EDTA using Qiacube and the accordant protocol (QIAGEN GmbH,

Hilden, Germany). RNA quality and quantity were analyzed using the Experion TM Automated Electrophoresis System and Nanodrop 1000 (PEQLAB, Erlangen, Germany). Reverse transcription was performed using the Bio-Rad Laboratories' iScript cDNA Synthesis Kit (Bio-Rad, Munich, Germany).

4.4. Quantitative PCR

The expression of *SNCA* was analyzed by quantitative PCR using the LightCycler System (LightCycler® SW 1.5, Roche Diagnostics GmbH, Mannheim, Germany) as previously described [39]. Briefly, *SNCA* expression was assessed using SYBR green technology (Bio-Rad, Munich, Germany), and the mean of beta-actin (*B-Actin*), beta-2-microglobulin (*B2M*) and ornithine decarboxylase 1 (*ODC1*) expression values, assessed using specific probes of the Roche Universal Probe Library (Roche Diagnostics GmbH, Mannheim, Germany), served as reference values (Table 3). Mean normalized expression was calculated using the "Abs Quant/2nd Derivative Max" analysis method provided by Roche (Mannheim, Germany).

Table 3. Sequences of oligonucleotides employed.

SNCA-F	5′-CTC CTT TTC CTT CTT CTT TCC T-3′
SNCA-R	5′-TGT TTG GTT TTC TCA GCA GC-3′
B-Actin-F	5′-GTC TTC CCC TCC ATC GTG-3′
B-Actin-R	5′-AGG TGT GGT GCC AGA TTT TC-3′
B-Actin-probe	5′ Cy5-GAG CAA GAG AGG CAT CCT CAC CCT GAA GTA-Eclipse 3′
ODC1-F	5′-CGC TTA CAC TGT TGC TGC TG-3′
ODC1-R	5′-CAT CCT GTT CCT CTA CTT CGG G-3′
ODC1-probe	5′ HEX-TCC AGA GGC CGA CGA TCT ACT ATG TGA TGT-BHQ1 3′
B2M-F	5′-CGC TAC TCTC TCT TTC TGG C-3′
B2M-R	5′-GTC AAC TTC AAT GTC GGA TGG AT-3′
B2M-probe	#42 of Roche Universal Probe Library

4.5. Statistical Analysis

Variables were tested for deviation from the normal distribution using the Kolmogorov-Smirnov test. Differences in sex distribution were calculated using the chi quadrat test. Correlative analyses were conducted using Pearson correlation coefficient. T-test and analysis of variance (ANOVA) were used to test for differences between the groups. A two-sided p-value ≤ 0.05 was considered indicative of statistical significance. The data were analyzed using SPSS TM for Windows 18.0 (SPSS Inc., Chicago, Ill., USA).

Author Contributions: Conceptualization, A.R., B.L., T.R.-S. and J.K.; Data curation, A.R., R.P. and T.R.-S.; Formal analysis, A.R., B.L., R.P., T.R.-S., J.K. and C.R.; Funding acquisition, J.K. and C.R.; Methodology, A.R., B.L., T.R.-S. and C.R.; Resources, A.R., B.L. and J.K.; Supervision, A.R. and C.R.; Validation, C.R.; Writing—original draft, A.R. and C.R.; Writing—review and editing, B.L. and J.K.

Funding: This research was funded by Forschungsstiftung Medizin at the University Hospital Erlangen, and the Scholarship Program 'Equality for Women in Research and Teaching' at the Friedrich-Alexander-University Erlangen-Nürnberg (FAU), to C.R.

Acknowledgments: We thank Alice Konrad for her excellent technical assistance.

Conflicts of Interest: The authors declare no conflict of interest. The funders had no role in the design of the study; in the collection, analyses, or interpretation of data; in the writing of the manuscript, or in the decision to publish the results.

References

1. Iwai, A.; Masliah, E.; Yoshimoto, M.; Ge, N.; Flanagan, L.; de Silva, H.A.; Kittel, A.; Saitoh, T. The precursor protein of non-Aβ component of Alzheimer's disease amyloid is a presynaptic protein of the central nervous system. *Neuron* **1995**, *14*, 467–475. [CrossRef]
2. Iwai, A.; Yoshimoto, M.; Masliah, E.; Saitoh, T. Non-Aβ component of Alzheimer's disease amyloid (NAC) is amyloidogenic. *Biochemistry* **1995**, *34*, 10139–10145. [CrossRef] [PubMed]
3. Bayer, T.A.; Jakala, P.; Hartmann, T.; Egensperger, R.; Buslei, R.; Falkai, P.; Beyreuther, K. Neural expression profile of alpha-synuclein in developing human cortex. *Neuroreport* **1999**, *10*, 2799–2803. [CrossRef] [PubMed]
4. Agrawal, A.; Wetherill, L.; Bucholz, K.K.; Kramer, J.; Kuperman, S.; Lynskey, M.T.; Nurnberger, J.I., Jr.; Schuckit, M.; Tischfield, J.A.; Edenberg, H.J.; et al. Genetic influences on craving for alcohol. *Addict. Behav.* **2013**, *38*, 1501–1508. [CrossRef]
5. Pfefferkorn, C.M.; Lee, J.C. Tryptophan probes at the α-synuclein and membrane interface. *J. Phys. Chem. B* **2010**, *114*, 4615–4622. [CrossRef] [PubMed]
6. Quilty, M.C.; Gai, W.-P.; Pountney, D.L.; West, A.K.; Vickers, J.C. Localization of α-, β-, and γ-synuclein during neuronal development and alterations associated with the neuronal response to axonal trauma. *Exp. Neurol.* **2003**, *182*, 195–207. [CrossRef]
7. Madine, J.; Doig, A.J.; Middleton, D.A. A study of the regional effects of α-synuclein on the organization and stability of phospholipid bilayers. *Biochemistry* **2006**, *45*, 5783–5792. [CrossRef]
8. Hsu, L.J.; Mallory, M.; Xia, Y.; Veinbergs, I.; Hashimoto, M.; Yoshimoto, M.; Thal, L.J.; Saitoh, T.; Masliah, E. Expression pattern of synucleins (non-β component of Alzheimer's disease amyloid precursor protein/α-synuclein) during murine brain development. *J. Neurochem.* **1998**, *71*, 338–344. [CrossRef]
9. George, J.M.; Jin, H.; Woods, W.S.; Clayton, D.F. Characterization of a novel protein regulated during the critical period for song learning in the zebra finch. *Neuron* **1995**, *15*, 361–372. [CrossRef]
10. Gureviciene, I.; Gurevicius, K.; Tanila, H. Aging and α-synuclein affect synaptic plasticity in the dentate gyrus. *J. Neural Transm.* **2009**, *116*, 13–22. [PubMed]
11. Stefanova, N.; Seppi, K.; Scherfler, C.; Puschban, Z.; Wenning, G.K. Depression in alpha-synucleinopathies: Prevalence, pathophysiology and treatment. *J. Neural Transm. Suppl.* **2000**, 335–343.
12. DeRubeis, R.J.; Siegle, G.J.; Hollon, S.D. Cognitive therapy versus medication for depression: Treatment outcomes and neural mechanisms. *Nat. Rev. Neurosci.* **2008**, *9*, 788–796. [CrossRef] [PubMed]
13. Howren, M.B.; Lamkin, D.M.; Suls, J. Associations of depression with C-reactive protein, IL-1, and IL-6: A meta-analysis. *Psychosom. Med.* **2009**, *71*, 171–186. [CrossRef]
14. Dowlati, Y.; Herrmann, N.; Swardfager, W.; Liu, H.; Sham, L.; Reim, E.K.; Lanctôt, K.L. A meta-analysis of cytokines in major depression. *Biol. Psychiatry* **2010**, *67*, 446–457. [CrossRef] [PubMed]
15. Lenz, B.; Sysk, C.; Thuerauf, N.; Clepce, M.; Reich, K.; Frieling, H.; Winterer, G.; Bleich, S.; Kornhuber, J. Erratum to: NACP-Rep1 relates to Beck Depression Inventory scores in healthy humans. *J. Mol. Neurosci.* **2013**, *50*, 376–377. [CrossRef]
16. Zhang, X.; Beaulieu, J.-M.; Sotnikova, T.D.; Gainetdinov, R.R.; Caron, M.G. Tryptophan hydroxylase-2 controls brain serotonin synthesis. *Science* **2004**, *305*, 217. [CrossRef]
17. Frieling, H.; Gozner, A.; Römer, K.D.; Wilhelm, J.; Hillemacher, T.; Kornhuber, J.; de Zwaan, M.; Jacoby, G.E.; Bleich, S. Alpha-synuclein mRNA levels correspond to beck depression inventory scores in females with eating disorders. *Neuropsychobiology* **2008**, *58*, 48–52. [CrossRef]
18. Oaks, A.W.; Sidhu, A. Synuclein modulation of monoamine transporters. *FEBS Lett.* **2011**, *585*, 1001–1006. [CrossRef]
19. Jeannotte, A.M.; Sidhu, A. Regulation of the norepinephrine transporter by α-synuclein-mediated interactions with microtubules. *Eur. J. Neurosci.* **2007**, *26*, 1509–1520. [CrossRef]
20. Wersinger, C.; Jeannotte, A.; Sidhu, A. Attenuation of the norepinephrine transporter activity and trafficking via interactions with α-synuclein. *Eur. J. Neurosci.* **2006**, *24*, 3141–3152. [CrossRef]
21. Wersinger, C.; Rusnak, M.; Sidhu, A. Modulation of the trafficking of the human serotonin transporter by human alpha-synuclein. *Eur. J. Neurosci.* **2006**, *24*, 55–64. [CrossRef]
22. Henningsen, K.; Palmfeldt, J.; Christiansen, S.; Baiges, I.; Bak, S.; Jensen, O.N.; Gregersen, N.; Wiborg, O. Candidate hippocampal biomarkers of susceptibility and resilience to stress in a rat model of depression. *Mol. Cell Proteom.* **2012**, *11*, M111 016428. [CrossRef]

23. Jeannotte, A.M.; McCarthy, J.G.; Redei, E.E.; Sidhu, A. Desipramine modulation of α-, γ-synuclein, and the norepinephrine transporter in an animal model of depression. *Neuropsychopharmacology* **2009**, *34*, 987–998. [CrossRef]

24. Lee, J.H.; Ko, E.; Kim, Y.E.; Min, J.Y.; Liu, J.; Kim, Y.; Shin, M.; Hong, M.; Bae, H. Gene expression profile analysis of genes in rat hippocampus from antidepressant treated rats using DNA microarray. *BMC Neurosci.* **2010**, *11*, 152. [CrossRef]

25. McHugh, P.C.; Rogers, G.R.; Glubb, D.M.; Joyce, P.R.; Kennedy, M.A. Proteomic analysis of rat hippocampus exposed to the antidepressant paroxetine. *J. Psychopharmacol* **2010**, *24*, 1243–1251. [CrossRef]

26. Bönsch, D.; Greifenberg, V.; Bayerlein, K.; Biermann, T.; Reulbach, U.; Hillemacher, T.; Kornhuber, J.; Bleich, S. α-Synuclein protein levels are increased in alcoholic patients and are linked to craving. *Alcohol Clin. Exp. Res.* **2005**, *29*, 763–765. [CrossRef]

27. Bönsch, D.; Lederer, T.; Reulbach, U.; Hothorn, T.; Kornhuber, J.; Bleich, S. Joint analysis of the NACP-REP1 marker within the alpha synuclein gene concludes association with alcohol dependence. *Hum. Mol. Genet.* **2005**, *14*, 967–971. [CrossRef]

28. Levey, D.F.; Le-Niculescu, H.; Frank, J.; Ayalew, M.; Jain, N.; Kirlin, B.; Learman, R.; Winiger, E.; Rodd, Z.; Shekhar, A.; et al. Genetic risk prediction and neurobiological understanding of alcoholism. *Transl. Psychiatry* **2014**, *4*, e391. [CrossRef]

29. Ishiguro, M.; Baba, H.; Maeshima, H.; Shimano, T.; Inoue, M.; Ichikawa, T.; Yasuda, S.; Shukuzawa, H.; Suzuki, T.; Arai, H. Increased serum levels of α-synuclein in patients with major depressive disorder. *Am. J. Geriatr. Psychiatry* **2019**, *27*, 280–286. [CrossRef]

30. Eyre, H.A.; Eskin, A.; Nelson, S.F.; Cyr, N.M. St.; Siddarth, P.; Baune, B.T.; Lavretsky, H. Genomic predictors of remission to antidepressant treatment in geriatric depression using genome-wide expression analyses: A pilot study. *Int. J. Geriatr. Psychiatry* **2016**, *31*, 510–517. [CrossRef]

31. Caudal, D.; Alvarsson, A.; Bjorklund, A.; Svenningsson, P. Depressive-like phenotype induced by AAV-mediated overexpression of human α-synuclein in midbrain dopaminergic neurons. *Exp. Neurol.* **2015**, *273*, 243–252. [CrossRef]

32. Kohl, Z.; Winner, B.; Ubhi, K.; Rockenstein, E.; Mante, M.; Münch, M.; Barlow, C.; Carter, T.; Masliah, E.; Winkler, J. Fluoxetine rescues impaired hippocampal neurogenesis in a transgenic A53T synuclein mouse model. *Eur. J. Neurosci.* **2012**, *35*, 10–19. [CrossRef]

33. Brenz Verca, M.S.; Bahi, A.; Boyer, F.; Wagner, G.C.; Dreyer, J.L. Distribution of α- and γ-synucleins in the adult rat brain and their modification by high-dose cocaine treatment. *Eur. J. Neurosci.* **2003**, *18*, 1923–1938. [CrossRef]

34. Gallagher, D.A.; Lees, A.J.; Schrag, A. What are the most important nonmotor symptoms in patients with Parkinson's disease and are we missing them? *Mov. Disord.* **2010**, *25*, 2493–2500. [CrossRef]

35. Devos, D.; Dujardin, K.; Poirot, I.; Moreau, C.; Cottencin, O.; Thomas, P.; Destée, A.; Bordet, R.; Defebvre, L. Comparison of desipramine and citalopram treatments for depression in Parkinson's disease: A double-blind, randomized, placebo-controlled study. *Mov. Disord.* **2008**, *23*, 850–857. [CrossRef]

36. Menza, M.; Dobkin, R.D.; Marin, H.; Mark, M.H.; Gara, M.; Buyske, S.; Bienfait, K.; Dicke, A. A controlled trial of antidepressants in patients with Parkinson disease and depression. *Neurology* **2009**, *72*, 886–892. [CrossRef]

37. Deusser, J.; Schmidt, S.; Ettle, B.; Plötz, S.; Huber, S.; Müller, C.P.; Masliah, E.; Winkler, J.; Kohl, Z. Serotonergic dysfunction in the A53T alpha-synuclein mouse model of Parkinson's disease. *J. Neurochem.* **2015**, *135*, 589–597. [CrossRef]

38. Rotter, A.; Asemann, R.; Decker, A.; Kornhuber, J.; Biermann, T. Orexin expression and promoter-methylation in peripheral blood of patients suffering from major depressive disorder. *J. Affect. Disord.* **2011**, *131*, 186–192. [CrossRef]

39. Lenz, B.; Klafki, H.W.; Hillemacher, T.; Frieling, H.; Clepce, M.; Gossler, A.; Thuerauf, N.; Winterer, G.; Kornhuber, J.; Bleich, S. ERK1/2 protein and mRNA levels in human blood are linked to smoking behavior. *Addict. Biol.* **2012**, *17*, 1026–1035. [CrossRef]

International Journal of
Molecular Sciences

Review

Depression and Sleep

Axel Steiger [1,*] and Marcel Pawlowski [1,2]

[1] Max Planck Institute of Psychiatry, Research Group Sleep Endocrinology, 80804 Munich, Germany;
    p@wlowski.de
[2] Centre of Mental Health, 85049 Ingolstadt, Germany
* Correspondence: steiger@psych.mpg.de; Tel.: +49-89-30622-236

Received: 30 November 2018; Accepted: 7 January 2019; Published: 31 January 2019

Abstract: Impaired sleep is both a risk factor and a symptom of depression. Objective sleep is assessed using the sleep electroencephalogram (EEG). Characteristic sleep-EEG changes in patients with depression include disinhibition of rapid eye movement (REM) sleep, changes of sleep continuity, and impaired non-REM sleep. Most antidepressants suppress REM sleep both in healthy volunteers and depressed patients. Various sleep-EEG variables may be suitable as biomarkers for diagnosis, prognosis, and prediction of therapy response in depression. In family studies of depression, enhanced REM density, a measure for frequency of rapid eye movements, is characteristic for an endophenotype. Cordance is an EEG measure distinctly correlated with regional brain perfusion. Prefrontal theta cordance, derived from REM sleep, appears to be a biomarker of antidepressant treatment response. Some predictive sleep-EEG markers of depression appear to be related to hypothalamo-pituitary-adrenocortical system activity.

Keywords: depression; sleep; sleep EEG; biomarkers; antidepressants; cordance

---

1. Introduction

Insomnia is a frequent symptom of depression. Conversely, it is a risk factor for the development of a depressive episode [1]. Objective sleep is assessed by polysomnography, also named sleep electroencephalogram (EEG). Sleep EEG appears to be suitable method of gaining biomarkers of depression, and these biomarkers may contribute to the nosology, prognosis, and prediction of therapy response in depression.

There are two reasons why psychiatrists became interested in sleep EEG in the 1970s. Rapid eye movement (REM) latency was suggested to indicate depression [2]. Furthermore, it was found that most antidepressants suppressed REM sleep [3]. Previously it was thought that REM latency may distinguish between certain subtypes of depression. In addition, it was hypothesized that REM suppression is the mechanism of action of antidepressants [4]. Since then, research has provided much more complex results.

Mammalian sleep consists of alternating periods of REM and non-REM sleep. Infants sleep in a polyphasic fashion. During human development, a mostly monophasic sleep-wake pattern emerges. The criteria by Rechtschaffen and Kales [5] differentiate between four stages of non-REM sleep, whereas the more recent classification by the American Academy of Sleep Medicine (AASM) shows three stages [6]. Stages 3 and 4 according to Rechtschaffen and Kales, or N3 according to AASM, are also termed slow-wave sleep (SWS). Shortly after going to bed, young normal subjects enter the lighter sleep stages N1 and N2 of non-REM sleep, followed by N3 (SWS). The major portion of SWS occurs during the first non-REM period. After a mean duration of 90 min of the first non-REM period, the first REM period occurs. The first REM period is relatively short; but its duration consecutively increases during the night. Accordingly, during the first half of the night, SWS preponderates. During the second half of the night, however, stage N2 and REM sleep dominate. Most subjects show four to five sleep cycles during the night, each consisting of one period of non-REM sleep and one period of REM sleep.

2. Sleep EEG in Patients with Depression

Most patients with depression suffer from impaired sleep, about 80 percent suffer from insomnia and 15–35 percent from hypersomnia [7,8]. Patients with depression show characteristic sleep-EEG changes [8–10] including:

(i) Impaired sleep continuity (prolonged sleep latency, increased intermittent awakenings, early morning awakenings).
(ii) Disinhibition of REM sleep: shortened REM latency, or sleep onset REM periods (SOREMs, REM latency 0–20 min), prolonged first REM period, enhanced REM density (measure of frequency of rapid eye movements) particularly during first REM period.
(iii) Changes in non-REM sleep (decreased stage N2 and SWS, in younger patients shift of SWS from the first to the second sleep cycle).

Patients with depression show reduced EEG delta power, also termed slow wave activity (SWA) throughout the night [11–14]. Sleep EEG is modulated by age and gender in healthy volunteers and in depressed patients as well. In the third decade of the life, SWS and SWA start to decrease. Menopause is the major turning point in sleep quality in women. In male subjects, however, sleep quality declines continuously during aging. In patients with depression, age and illness exert a synergistic effect on sleep EEG. The effect of aging on sleep EEG in patients with depression and normal control subjects were investigated in two studies [15,16]. These studies showed a clear effect of age on REM latency, whereas patients and healthy subjects did not differ until the middle of the fourth decade. On the other hand, REM density was enhanced in all investigated age groups in patients when compared with controls. SWS declined throughout the lifespan without differences between patients and controls.

In two longitudinal studies, no changes in sleep-EEG variables of depressed patients were found between acute depression and remission [17,18]. In one study, sleep stage 4 decreased after remission when compared to baseline [18]. Similar results were reported in depressed adolescents [19]. Increasing abnormality of REM sleep variables was observed during a long-term study on repetitive episodes of depression. SWS did not differ between episodes [20].

The view that shortened REM latency is a specific marker of depression [2] was challenged by other studies reporting similar changes in other psychiatric disorders including mania [21], schizophrenia [22], schizoaffective disorder [23], obsessive-compulsive disorder [24], panic disorder [25], eating disorders [26], and sexual impotence [27]. The finding that sleep-EEG changes persist in remitted patients [17,18] may explain that comorbidity with depression or a history of depression result in a shortened REM latency in these disorders. This view is supported by two studies by Lauer et al. [28,29]. These authors compared three groups of patients with major depression, anorexia nervosa, and bulimia with healthy subjects. The latter two groups of patients were never depressed. REM density was enhanced in the patients with depression [28]. In the other study, depressed patients, patients with panic disorder without a history of depression, and normal controls were compared. Differences were observed during the first sleep cycle. SWS was reduced and REM time and REM density were increased during this interval in the depressed patients. In patients with panic disorder this cycle was shortened. REM latency was shorter in both groups of patients than in healthy controls [29].

3. Sleep EEG in High-Risk Probands for Affective Disorders

In the Munich Vulnerability Study on affective disorders, a prospective high-risk design was applied. In order to identify premorbid vulnerability factors for affective disorder, high-risk probands were examined. They had a high genetic load for affective disorders due to a positive family history. Comparison of the high-risk probands with healthy subjects without a family history for this disease showed enhanced REM density and reduced time spent in non-REM sleep during the first sleep cycle [30]. This finding remained stable at the follow-up investigation four years later [31]. In a subgroup of the high-risk probands, the cholinergic REM sleep induction test was performed using the cholinomimetic RS86. At baseline, REM latency did not differ between high-risk probands and the controls. After RS86, REM latency was decreased in the high-risk probands [32]. This finding points to a threshold cholinergic dysfunction in the high-risk probands. The response pattern in the cholinergic REM sleep induction test predicted the onset of the first episode of depression [33]. Twenty subjects of the initial sample of 83 high-risk probands of this study developed an affective disorder during the follow-up period. In these subjects, the premorbid sleep EEG showed increased REM density during the total night and during the first REM period when compared to healthy volunteers [34]. These findings show that increased REM density meets all requirements for biological vulnerability markers of affective disorders. The authors recommend REM density as a possible endophenotype in family studies [34].

4. Sleep EEG and Risk Genes for Depression

P2RX7 is a susceptibility gene for affective disorders. It is located on chromosome 12 q24, which appears to be associated with major depression [35] and bipolar disorder [36]. *P2RX7* is found in immune, endothelial, and epithelial cells, and regulates various aspects of immune function, as expression and secretion of cytokines [37]. The single nucleotide polymorphism (SNP) rs2230912 in the *P2RX7* gene (base change 1405A>G) leading to substitution of glutamine (Gln, Q) by arginine (Arg, R), at codon 460 (Gln 460 Arg, Q 460 R), has been associated with mood disorders [38–40]. To clarify whether elevated risk for depression related to this SNP shows sleep-EEG changes, young healthy volunteers who were free of psychiatric disorders in their own and family history, were investigated in the sleep laboratory. Homozygous (A/A) subjects and heterozygous (A/G) carriers of the risk variant were compared. Significant differences in sleep-EEG were found between groups. In the heterozygous (A/G) subjects, prolonged sleep latency and shortened sleep period time was found; the number of entries from stage N2 into N1 and wakefulness was enhanced during the first sleep cycle; in the lower spindle range frequencies were elevated, particularly in parietal regions; peak frequencies of all sleep spindles were lower during non-REM sleep. In particular, elevated parietal variations during stage N2 beta frequencies were reported. These data show that healthy volunteers with a potential risk for affective disorders related to their *P2RX7* genotype differ in sleep EEG from subjects with lower risk [41].

Mice that harbor *P2RX7*-Gln 460 AG and the wild-type *P2RX7* showed, compared to homozygous *P2RX7* wildtype and *P2RX7*[hQ460R] mice an increase of entries to REM sleep during the light period, suggesting a stronger drive towards REM sleep and more fragmented sleep cycles. Furthermore, SWA was lower and the amount of deep non-REM sleep was only small in heterozygous mice. Taken together, heterogeneous mice show altered sleep architecture and reduced sleep quality compared to homozygous mice [41].

5. Effects of Antidepressants on Sleep EEG

Most antidepressants suppress REM sleep in patients and in healthy volunteers. REM suppression includes prolonged REM latency, reduced time spent in REM sleep, and decreased REM density. Withdrawal of REM suppressing antidepressants is followed by REM rebound. Decreased REM latency, increased REM time, and enhanced REM density are the components of REM rebound. All these variables exceed baseline values. Withdrawal of antidepressants after two weeks of treatment prompted a REM rebound that persisted after one week [42]. REM suppression occurs after tricyclics [43,44], tetracyclics [3], selective serotonin reuptake inhibitors (SSRIs) [45,46], selective noradrenaline reuptake inhibitors (NRI) [47], selective serotonin and noradrenaline reuptake inhibitors (SNRI) [48], reversible [49–51], and short acting reversible [52] monoamine oxidase inhibitors. Only some antidepressants do not suppress REM sleep including trimipramine [53], bupropion [54], the serotonin reuptake enhancer tianeptine [55], and the noradrenergic and specific serotonergic antidepressant (NaSSA) mirtazapine [56,57].

Various antidepressants differ in the potency to suppress REM sleep. Total REM suppression was found after clomipramine [58] and the irreversible monoamine oxidase inhibitors phenelzine and tranylcypromine [59]. Additionally, distinct REM suppression was observed in healthy volunteers following the combined SSRI and serotonin 5-HT1A receptor agonist vilazodone [60]. Also the dosage and plasma concentrations of the substances influence the amount of changes in REM sleep [58].

After selective REM sleep deprivation, but not after selective non-REM sleep deprivation for three weeks, antidepressant effects were observed [4]. This finding and the observation that most antidepressants suppress REM sleep resulted in the hypothesis that REM suppression is the mechanism of action of antidepressant drugs. This theory however was challenged by the lack of antidepressant effect of selective REM suppression for the first eleven days of treatment [61]. In addition, the fact that some antidepressants do not suppress REM sleep, like trimipramine, tianeptine, and mirtazapine, contradicts the hypothesis by Vogel et al. [4].

The comparison of the effects of the stereoisomeres of oxaprotiline, R(−)oxaprotiline, and S(+)oxaprotiline on sleep support the view that REM suppression is a distinct, but not absolute requirement for antidepressant effects of a substance. S(+)oxaprotiline suppressed REM sleep in patients with depression, whereas R(−)oxaprotiline did not share this effect. S(+)oxaprotiline had better antidepressant effects then R(−)oxaprotiline [62]. The effects of most antidepressants on REM sleep are similar. In contrast, substances differ in their effect on sleep continuity and on non-REM sleep. Whereas most tricyclics elevate SWS [3], clomipramine [58], and imipramine [53] diminished SWS. SSRIs do not modulate SWS, but impair sleep continuity and enhance intermittent wakefulness [63,64]. In addition, the NaRI reboxetine diminishes sleep efficiency and elevates intermittent wakefulness and stage 2 sleep [47]. After vilazodone, REM sleep in healthy volunteers was distinctly suppressed together with increases in SWS and SWA in the first and the last third of the night [60]. After the SSNRI duloxetine, stage 3 increased in depressed patients [48]. On the second day of mirtazapine treatment, patients with depression showed an increase in total sleep time and sleep efficiency and a decrease in time awake. These effects persisted after four weeks, when SWS, low delta, theta, and alpha activity increased [57]. After two days of treatment with amitriptyline, the increase seen in REM latency correlated with the clinical outcome after four weeks [65]. A single observation was reported for imipramine [53], but not after clomipramine [66].

6. Contribution of the HPA System to Sleep-EEG Abnormalities in Depression

It is well established that over-activity of the hypothalamo-pituitary-adrenocortical (HPA) system plays a key role in the pathophysiology of affective disorders [67]. In two longitudinal studies, nocturnal cortisol [18,68] and ACTH [68] concentrations were compared between acute depression and recovery. In comparison to healthy controls, cortisol and ACTH levels were elevated in patients with acute depression [68]. After treatment with electroconvulsive therapy or amitriptyline and remission, ACTH levels decreased [68]. Similarly, comparison of cortisol levels between acute depression and recovery in patients who were drug-free at both examinations, showed a decrease in cortisol levels [18]. These findings show that enhanced nocturnal HPA hormone secretion is a state marker of acute depression. Administration of the key hormone of the HPA system, corticotropin-releasing hormone (CRH), prompted more shallow sleep in rats [69], rabbits [70], and mice [71,72]. Similarly, after repetitive intravenous (iv) injections of CRH around sleep onset, SWS decreased and endocrine changes that are characteristic for depression (i.e., elevated cortisol levels, blunted growth hormone (GH) peak) were observed in young male volunteers [73]. Mouse mutants overexpressing CRH in the entire central nervous system or only in the forebrain showed increased REM sleep compared to wild-type mice [74].

In healthy women, the effects of pulsatile CRH injections on sleep EEG were more distinct than in healthy males, as intermittent wakefulness increased during the total night and the sleep efficiency index decreased. Furthermore, during the first third of the night, REM sleep and stage 2 sleep increased and sleep stage 4 was diminished. Cortisol levels were elevated throughout the night, whereas GH secretion remained unchanged [75].

Already in kindergarten children, associations were found between unfavorable sleep-EEG patterns, elevated HPA activity, and more difficult behavioral psychosocial dimensions [76]. In preschool children, sleep EEG was recorded and saliva samples were collected after awakening and before and after a psychological challenge for cortisol analysis. Children labeled as "poor" sleepers showed significantly increased morning cortisol values in comparison to "good" sleepers. Increased cortisol values after stress were significantly associated with an increased number of awakenings after sleep onset and an increased amount of sleep stages 1 and 2. Furthermore, psychological difficulties, such as impulsivity, over-anxiousness, and social inhibition, showed a significant association with low sleep efficiency.

In a clinical trial of the CRH receptor type 1 (CRHR1) antagonist R121919, a random subgroup of 10 patients had their sleep EEG assessed. Sleep-EEG recordings were performed at baseline, before treatment, after one week of active treatment, and at the end of the fourth week of treatment. SWS increased after week 1 and after week 4 compared to baseline. During the same period, the number of awakenings and REM density decreased. Separate evaluation of these changes for two different dose ranges showed no significant effects with the lower dose, whereas with the higher dose, REM density decreased, and SWS increased significantly between baseline and week 4. Positive associations were found between the Hamilton-Depression-Score and SWS at the end of active treatment. These results support the hypothesis that CRH is involved in the pathophysiology of sleep-EEG changes in depressed patients. In addition, these findings suggest that CRHR1 antagonism induces normalization of the sleep EEG in depressed patients [77].

Multiple sclerosis patients receiving subchronic administration of the synthetic glucocorticoid receptor agonist methylprednisolone showed similar sleep-EEG changes as in patients with depression. These changes included shortened REM latency, enhanced REM density, and shift of SWS and SWA from the first to the second non-REM period [78].

In male human subjects and in male animals, GH-releasing hormone (GHRH) exerts effects on sleep that are opposite to those of CRH. SWS increases after intracerebroventricular (icv) administration of GHRH in rats [69,79], after injection into the medial preoptic area of rats [80], and after iv administration to rats [81]. Similarly in young male healthy volunteers, in a protocol analogous to the study by Holsboer et al. [73] repetitive iv administration of GHRH increased SWS and GH and decreased cortisol [82]. In women, however, sleep was impaired after GHRH and cortisol and ACTH was enhanced [83,84], which is similar to the effects of CRH [73]. It is thought that at least in male patients a balance exists between GHRH and CRH in sleep regulation. GHRH appears to be active at the beginning of the night as mirrored by the high amounts of SWS and GH. During the second half of the night, CRH appears to preponderate and to induce more REM sleep and elevated cortisol. During depressive episodes (and during normal aging as well) the GHRH/CRH ratio is changed in favor of CRH due to CRH overactivity in affective disorders (or to declining GHRH activity during aging) (see Figure 1). A synergism of CRH and cortisol may contribute to REM sleep disinhibition.

7. Amyloid-β and Sleep

Aggregation and accumulation of amyloid-β (Aβ) contributes to the development of Alzheimer's disease [85]. Several recent studies address the interaction of Aβ and sleep. Using positron emission tomography Shokri-Kojori et al. (2018) showed significant increases in Aβ burden in the right hippocampus and thalamus after a night of sleep deprivation in healthy controls. These increases were associated with worsening of mood after sleep deprivation [86]. In rats, sleep deprivation impaired cognitive function and elevated Aβ levels [87]. The effect of sleep on overnight cerebrospinal fluid (CSF) Aβ kinetics was tested in healthy volunteers using intracerebroventricular (icv) lumbar catheters for serial sampling of CSF while subjects were sleep deprived, received sleep promoting sodium oxybate or slept normally. To measure Aβ kinetics all participants were infused with $^{13}C_6$-leucine. Sleep deprivation increased overnight Aβ38, Aβ40, and Aβ42 levels by 25–30% via increased overnight Aβ production relative to sleeping subjects. The authors concluded that disrupted sleep increases Alzheimer's disease risk by increased Aβ production [88]. In order to elucidate whether chronic sleep restriction potentiates the brain impact of Aβ oligomers (AβOs) studies in mice were performed. A single icv infusion of Aβ oligomers disturbed sleep pattern in mice. Conversely, chronically sleep restricted mice showed higher brain expression of pro-inflammatory mediators, reduced levels of pre- and post-synaptic marker proteins. Furthermore, this study exhibited increased susceptibility to the sub-toxic dose of AβOs on performance in a novel object recognition memory task. After sleep restriction, elevated brain tumor necrosis factor α (TNF-α) levels were found in response to AβOs. Neuronal impairment in sleep restricted AβOs infused mice was prevented by a TNF-α neutralizing monoclonal antibody. The authors discuss a dual relationship between sleep and Alzheimer's disease with disruption of sleep wake patterns by AβOs and increased brain vulnerability to AβOs after chronic sleep restriction [89]. In Alzheimer's disease model mice, chronic sleep fragmentation was induced by a running-wheel-based device that resulted in increased Aβ deposition in the mouse brain. The severity of Aβ deposition showed a significant positive correlation with the extent of sleep fragmentation [90]. Specific disruption of SWA in healthy adults without sleep disorders correlated with an increase in Aβ [91]. In patients with insomnia CSF Aβ levels were significantly higher than in healthy controls [92].

Interestingly, there is some overlap between the pathophysiology of depression, Alzheimer's disease and sleep. Human neuroblastoma cells produced more Aβ after treatment with CRH [93]. Morgese et al. (2017) discuss that chronic stress may represent common biological bases linking Alzheimer's dementia and depression [94]. The interaction of sleep and Aβ in patients with depression is an open topic on the research agenda.

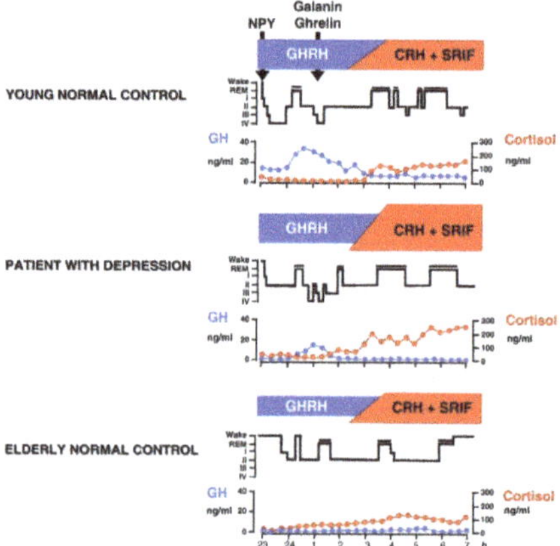

Figure 1. Patterns of normal and impaired peptidergic sleep regulation. Characteristic hypnograms and patterns of cortisol and GH secretion are shown in a young and in an elder healthy subject and in a patient with depression It is thought that GHRH is active during the first hours of sleep resulting in GH peak and the major portion of SWS during the night. During the second half of the night the influence of CRH preponderates which prompts increases of cortisol and REM sleep. Galanin and ghrelin may act as co-factors of GHRH. Somatostatin (SRIF) may impair sleep. The balance between GHRH and CRH changes during normal aging, when GHRH activity declines and during depressive episodes, when CRH activity is enhanced. Reprinted with permission from Springer, Nervenarzt, Schlafendokrinologie, Axel Steiger, 1995.

8. State and Vulnerability Markers Related to Antidepressant Therapy

In a clinical trial, the effects of the serotonin reuptake enhancer tianeptine and the SSRI paroxetine were compared. The effects of these substances on sleep EEG were investigated in a subgroup of these patients. Sleep EEG was recorded at days 7 and 42 after the start of treatment with either substance. In male treatment responders, a distinct decline in the higher sigma frequency range (14–16 Hz) during non-REM sleep was found independently of medication. In contrast, male and female non-responders did not show marked changes in this frequency range. This finding supports the view that gender should be taken into account when the biological effects of drugs are studied. After paroxetine, the amount of REM sleep was reduced and intermittent wakefulness was increased in comparison to tianeptine. In the total sample after one week of treatment, REM density was a predictor of treatment response. The change in REM density showed an inverse correlation to changes in the Hamilton Depression Score in the patients who received paroxetine, but not in those who received tianeptine [55].

Patients with depression who had participated in an earlier study with trimipramine were involved in an exploratory follow-up study. The retrospectively-assessed long-term course of depression in these patients was related with sleep-EEG variables during the acute episode. The lower the sleep continuity (total sleep time, sleep efficiency index, time spent awake, number of awakenings), the higher was the number of previous episodes of depression. This association disappeared at the end of drug treatment with a distinct association found between reduced SWS, particularly during the first third of the sleep period, elevated REM density (by trend), and the number of previous episodes. A clear association was observed between the prospective long-term course and sleep EEG, as increased REM density and decreased SWS at the end of treatment were associated with an elevated recurrence rate between the end of the trial and the follow-up study. These sleep-EEG variables showed an association with impaired HPA system, evident by abnormal results of the dexamethasone/CRH (DEX/CRH) test. Patients with an unfavorable long-term course of depression appear to show increasing aberrant sleep regulation. These changes seem predictive not only for treatment response during the acute episode, but also for recurrences in the long-term. These predictive sleep-EEG markers may relate with HPA system activity, since the more sleep-EEG markers were disturbed, the more the HPA system was impaired [95].

9. Cordance Derived from REM Sleep as a Predictor of Therapy Response

Cordance is a quantitative EEG measure that combines information from absolute and relative EEG spectral power. It correlates with regional brain activity. Theta frequency band of cordance shows positive correlation with cerebral blood perfusion [96]. Prefrontal theta cordance, derived from the awake EEG, correlates with antidepressant treatment outcome. After one week of drug treatment, prefrontal theta cordance decreased in several studies, irrespective of the investigated drugs [97–100]. It is thought that prefrontal theta cordance reflects activity of prefrontal cortex and anterior cingulate cortex (ACC) [101]. Both appear to be crucially involved in major depression [102]. During REM sleep, ACC activity is maximal. In contrast, the surrounding frontal cortex activity is minimal [103,104]. During REM sleep, ACC shows distinct oscillatory activity in the theta frequency band [105]. Therefore, prefrontal theta cordance is an ideal way to detect theta frequency band. Prefrontal theta cordance of depressed patients was measured during tonic REM sleep. In responders (of totally 20 depressive in-patients on various antidepressants), prefrontal theta cordance was significantly higher after the first week of antidepressant medication than in non-responders. This result was still significant after controlling for age, gender, and the number of previous episodes of depression. In addition, prefrontal cordance in all patients showed a significant positive correlation with the improvement of the Hamilton Depression Score between inclusion week and the first week of drug treatment [106].

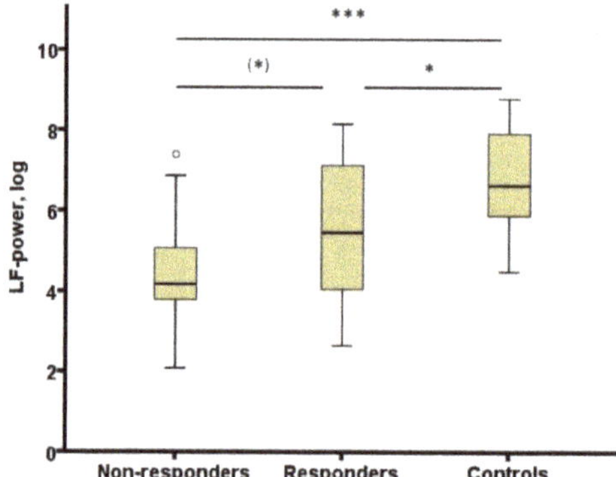

Figure 2. HRV (LF-power, log) in non-responders and responders at week 4 and controls. HRV: heart rate variability; LF-power, log: power in low frequency range (0.04–0.15 Hz) transformed with natural logarithm, (*) $p < 0.10$; * $p < 0.05$; *** $p < 0.001$. From [108] with permission from Elsevier.

10. Heart Rate Variability Derived from REM Sleep in Depressed Patients

The study by Adamczyk et al. [106] was extended to test whether heart rate variability (HRV) derived from REM sleep could represent a biomarker of antidepressant treatment response. A meta-analysis showed that major depression is associated with blunted HRV [107]. It was expected that an association of HRV and depression would be stronger in offline conditions like sleep. In patients with depression, HRV was derived from 3 min artefact free electrocardiogram sequence during REM sleep. In comparison to controls, HRV during REM sleep was decreased in depressed patients (responders as well as non-responders) during the fourth week of treatment in comparison to controls (see Figure 2). It showed a negative correlation with REM density in healthy subjects and in patients at week four. HRV derived from REM sleep appears to categorize healthy subjects and patients with depression [108].

11. Perspectives

This review presents sleep EEG as a promising tool for psychiatric research and clinical application in affective disorders.

The observation of subtle influence of the *P2RX7* genotype on sleep-EEG pattern should be extended to studies of the association of other risk genes of depression on sleep EEG in healthy and in depressed patients. This approach may support the efforts to establish a new nosology of depression related to neurobiology.

Cordance appears to help to differentiate early during treatment between responders and non-responders to antidepressant therapy. The next step will be to test the capacity of cordance to shorten the long way to recovery that many patients with depression suffer. This would be possible if the expected response to a certain antidepressant in a patient is tested using cordance after one week of treatment. If non-response is predicted, medication could be changed much earlier than in the traditional way of assessing response related to psychopathology after about four weeks.

Some antidepressants promote, and others impair sleep. However, it is not yet clear whether stability of remission is influenced by such differences in medication.

Author Contributions: M.P. and A.S. jointly wrote this review article.

Funding: Parts of the research from the authors' laboratory reviewed in § 6 was funded by the Deutsche Forschungsgemeinschaft grant number [Ste 486/1-1 to 5/4]".

Conflicts of Interest: The authors declare no conflict of interest.

References

1. Baglioni, C.; Battagliese, G.; Feige, B.; Spiegelhalder, K.; Nissen, C.; Voderholzer, U.; Lombardo, C.; Riemann, D. Insomnia as a predictor of depression: A meta-analytic evaluation of longitudinal epidemiological studies. *J. Affect. Disord.* **2011**, *135*, 10–19. [CrossRef] [PubMed]
2. Kupfer, D.; Foster, F.G. Interval between onset of sleep and rapid-eye-movement sleep as an indicator of depression. *Lancet* **1972**, *300*, 684–686. [CrossRef]
3. Chen, C.-N. Sleep, depression and antidepressants. *Br. J. Psychiatry* **1979**, *135*, 385–402. [CrossRef] [PubMed]
4. Vogel, G.W.; Thurmond, A.; Gibbons, P.; Sloan, K.; Boyd, M.; Walker, M. REM sleep reduction effects on depression syndromes. *Arch. Gen. Psychiatry* **1975**, *32*, 765–777. [CrossRef] [PubMed]
5. Kales, A.; Rechtschaffen, A. *A Manual of Standardized Terminology, Techniques and Scoring System for Sleep Stages of Human Subjects*; US Department of Health, Education and Welfare, Public Health Service, National Institutes of Health, National Institute of Neurological Diseases and Blindness, Neurological Information Network: Bethesda, MD, USA, 1968.
6. Iber, C.; Ancoli-Israel, S.; Chesson, A.L.; Quan, S.F. *The AASM Manual for the Scoring of Sleep and Associated Events: Rules, Terminology and Technical Specifications*, 1st ed.; American Academy of Sleep Medicine: Westchester, IL, USA, 2007.
7. Hawkins, D.R.; Taub, J.M.; Van de Castle, R.L. Extended sleep (hypersomnia) in young depressed patients. *Am. J. Psychiatry* **1985**, *142*, 905–910. [CrossRef] [PubMed]
8. Armitage, R. Sleep and circadian rhythms in mood disorders. *Acta Psychiatr. Scand.* **2007**, *115*, 104–115. [CrossRef]
9. Reynolds, C.F.; Kupfer, D.J. Sleep research in affective illness: State of the art circa 1987. *Sleep* **1987**, *10*, 199–215. [CrossRef]
10. Benca, R.M.; Okawa, M.; Uchiyama, M.; Ozaki, S.; Nakajima, T.; Shibui, K.; Obermeyer, W.H. Sleep and mood disorders. *Sleep Med. Rev.* **1997**, *1*, 45–56. [CrossRef]
11. Borbély, A.A.; Tobler, I.; Loepfe, M.; Kupfer, D.J.; Ulrich, R.F.; Grochocinski, V.; Doman, J.; Matthews, G. All-night spectral analysis of the sleep EEG in untreated depressives and normal controls. *Psychiatry Res.* **1984**, *12*, 27–33. [CrossRef]
12. Kupfer, D.J.; Ulrich, R.F.; Coble, P.A.; Jarrett, D.B.; Grochocinski, V.; Doman, J.; Matthews, G.; Borbély, A.A. Application of automated REM and slow wave sleep analysis: II. Testing the assumptions of the two-process model of sleep regulation in normal and depressed subjects. *Psychiatry Res.* **1984**, *13*, 335–343. [CrossRef]
13. Kupfer, D.J.; Reynolds, C.F.; Ulrich, R.F.; Grochocinski, V.J. Comparison of automated REM and slow-wave sleep analysis in young and middle-aged depressed subjects. *Biol. Psychiatry* **1986**, *21*, 189–200. [CrossRef]
14. Kupfer, D.J.; Reynolds, C.F., III; Grochocinski, V.J.; Ulrich, R.F.; McEachran, A. Aspects of short REM latency in affective states: A revisit. *Psychiatry Res.* **1986**, *17*, 49–59. [CrossRef]
15. Lauer, C.J.; Riemann, D.; Wiegand, M.; Berger, M. From early to late adulthood changes in EEG sleep of depressed patients and healthy volunteers. *Biol. Psychiatry* **1991**, *29*, 979–993. [CrossRef]
16. Riemann, D.; Lauer, C.; Hohagen, F.; Berger, M. Longterm evolution of sleep in depression. In *Sleep and Aging*; Masson Press: Milano, Italy, 1991; pp. 195–204.
17. Rush, A.J.; Erman, M.K.; Giles, D.E.; Schlesser, M.A.; Carpenter, G.; Vasavada, N.; Roffwarg, H.P. Polysomnographic findings in recently drug-free and clinically remitted depressed patients. *Arch. Gen. Psychiatry* **1986**, *43*, 878–884. [CrossRef] [PubMed]
18. Steiger, A.; von Bardeleben, U.; Herth, T.; Holsboer, F. Sleep EEG and nocturnal secretion of cortisol and growth hormone in male patients with endogenous depression before treatment and after recovery. *J. Affect. Disord.* **1989**, *16*, 189–195. [CrossRef]
19. Rao, U.; Poland, R.E. Electroencephalographic sleep and hypothalamic–pituitary–adrenal changes from episode to recovery in depressed adolescents. *J. Child Adolesc. Psychopharmacol.* **2008**, *18*, 607–613. [CrossRef]
20. Kupfer, D.J.; Ehlers, C.L.; Frank, E.; Grochocinski, V.J.; McEachran, A.B. EEG sleep profiles and recurrent depression. *Biol. Psychiatry* **1991**, *30*, 641–655. [CrossRef]
21. Hudson, J.I.; Lipinski, J.F.; Frankenburg, F.R.; Grochocinski, V.J.; Kupfer, D.J. Electroencephalographic sleep in mania. *Arch. Gen. Psychiatry* **1988**, *45*, 267–273. [CrossRef]
22. Zarcone, V.P.; Benson, K.L.; Berger, P.A. Abnormal rapid eye movement latencies in schizophrenia. *Arch. Gen. Psychiatry* **1987**, *44*, 45–48. [CrossRef]

23. Reich, L.; Weiss, B.L.; Coble, P.; McPartland, R.; Kupfer, D.J. Sleep disturbance in schizophrenia: A revisit. *Arch. Gen. Psychiatry* **1975**, *32*, 51–55. [CrossRef]

24. Insel, T.R.; Gillin, J.C.; Moore, A.; Mendelson, W.B.; Loewenstein, R.J.; Murphy, D.L. The sleep of patients with obsessive-compulsive disorder. *Arch. Gen. Psychiatry* **1982**, *39*, 1372–1377. [CrossRef] [PubMed]

25. Uhde, T.W.; Roy-Byrne, P.; Gillin, J.C.; Mendelson, W.B.; Boulenger, J.-P.; Vittone, B.J.; Post, R.M. The sleep of patients with panic disorder: A preliminary report. *Psychiatry Res.* **1984**, *12*, 251–259. [CrossRef]

26. Katz, J.L.; Kuperberg, A.; Pollack, C.P.; Walsh, B.T.; Zumoff, B.; Weiner, H. Is there a relationship between eating disorder and affective disorder? New evidence from sleep recordings. *Am. J. Psychiatry* **1984**. [CrossRef]

27. Schmidt, H.S.; Nofzinger, E.A. Short REM latency in impotence without depression. *Biol. Psychiatry* **1988**, *24*, 25–32. [CrossRef]

28. Lauer, C.J.; Krieg, J.-C.; Riemann, D.; Zulley, J.; Berger, M. A polysomnographic study in young psychiatric inpatients: Major depression, anorexia nervosa, bulimia nervosa. *J. Affect. Disord.* **1990**, *18*, 235–245. [CrossRef]

29. Lauer, C.J.; Krieg, J.-C.; Garcia-Borreguero, D.; Özdaglar, A.; Holsboer, F. Panic disorder and major depression: A comparative electroencephalographic sleep study. *Psychiatry Res.* **1992**, *44*, 41–54. [CrossRef]

30. Lauer, C.J.; Schreiber, W.; Holsboer, F.; Krieg, J.-C. In quest of identifying vulnerability markers for psychiatric disorders by all-night polysomnography. *Arch. Gen. Psychiatry* **1995**, *52*, 145–153. [CrossRef]

31. Modell, S.; Ising, M.; Holsboer, F.; Lauer, C.J. The Munich Vulnerability Study on Affective Disorders: Stability of polysomnographic findings over time. *Biol. Psychiatry* **2002**, *52*, 430–437. [CrossRef]

32. Schreiber, W.; Lauer, C.J.; Krumrey, K.; Holsboer, F.; Krieg, J.-C. Cholinergic REM sleep induction test in subjects at high risk for psychiatric disorders. *Biol. Psychiatry* **1992**, *32*, 79–90. [CrossRef]

33. Lauer, C.J.; Modell, S.; Schreiber, W.; Krieg, J.-C.; Holsboer, F. Prediction of the development of a first major depressive episode with a rapid eye movement sleep induction test using the cholinergic agonist RS 86. *J. Clin. Psychopharmacol.* **2004**, *24*, 356–357. [CrossRef] [PubMed]

34. Modell, S.; Ising, M.; Holsboer, F.; Lauer, C.J. The Munich vulnerability study on affective disorders: Premorbid polysomnographic profile of affected high-risk probands. *Biol. Psychiatry* **2005**, *58*, 694–699. [CrossRef] [PubMed]

35. Abkevich, V.; Camp, N.J.; Hensel, C.H.; Neff, C.D.; Russell, D.L.; Hughes, D.C.; Plenk, A.M.; Lowry, M.R.; Richards, R.L.; Carter, C. Predisposition locus for major depression at chromosome 12q22-12q23.2. *Am. J. Hum. Genet.* **2003**, *73*, 1271–1281. [CrossRef] [PubMed]

36. Degn, B.; Lundorf, M.; Wang, A.; Vang, M.; Mors, O.; Kruse, T.; Ewald, H. Further evidence for a bipolar risk gene on chromosome 12q24 suggested by investigation of haplotype sharing and allelic association in patients from the Faroe Islands. *Mol. Psychiatry* **2001**, *6*, 450–455. [CrossRef] [PubMed]

37. Wiley, J.; Sluyter, R.; Gu, B.; Stokes, L.; Fuller, S. The human P2X7 receptor and its role in innate immunity. *Tissue Antigens* **2011**, *78*, 321–332. [CrossRef] [PubMed]

38. Barden, N.; Harvey, M.; Gagné, B.; Shink, E.; Tremblay, M.; Raymond, C.; Labbé, M.; Villeneuve, A.; Rochette, D.; Bordeleau, L.; et al. Analysis of single nucleotide polymorphisms in genes in the chromosome 12Q24. 31 region points to P2RX7 as a susceptibility gene to bipolar affective disorder. *Am. J. Med. Genet. Part B Neuropsychiatr. Genet.* **2006**, *141*, 374–382. [CrossRef] [PubMed]

39. Lucae, S.; Salyakina, D.; Barden, N.; Harvey, M.; Gagné, B.; Labbé, M.; Binder, E.B.; Uhr, M.; Paez-Pereda, M.; Sillaber, I.; et al. *P2RX7*, a gene coding for a purinergic ligand-gated ion channel, is associated with major depressive disorder. *Hum. Mol. Genet.* **2006**, *15*, 2438–2445. [CrossRef] [PubMed]

40. Soronen, P.; Mantere, O.; Melartin, T.; Suominen, K.; Vuorilehto, M.; Rytsälä, H.; Arvilommi, P.; Holma, I.; Holma, M.; Jylhä, P.; et al. *P2RX7* gene is associated consistently with mood disorders and predicts clinical outcome in three clinical cohorts. *Am. J. Med. Genet. Part B Neuropsychiatr. Genet.* **2011**, *156*, 435–447. [CrossRef]

41. Metzger, M.W.; Walser, S.M.; Dedic, N.; Aprile-Garcia, F.; Jakubcakova, V.; Adamczyk, M.; Webb, K.J.; Uhr, M.; Refojo, D.; Schmidt, M.V. Heterozygosity for the mood disorder-associated variant Gln460Arg alters P2X7 receptor function and sleep quality. *J. Neurosci.* **2017**, *37*, 11688–11700. [CrossRef]

42. Steiger, A.; Von Bardeleben, U.; Guldner, J.; Lauer, C.; Rothe, B.; Holsboer, F. The sleep EEG and nocturnal hormonal secretion studies on changes during the course of depression and on effects of CNS-active drugs. *Prog. Neuro-Psychopharmacol. Biol. Psychiatry* **1993**, *17*, 125–137. [CrossRef]

43. Dunleavy, D.; Brezinova, V.; Oswald, I.; Maclean, A.; Tinker, M. Changes during weeks in effects of tricyclic drugs on the human sleeping brain. *Br. J. Psychiatry* **1972**, *120*, 663–672. [CrossRef]

44. Passouant, P.; Cadilhac, J.; Billiard, M.; Besset, A. La suppression du sommeil paradoxal par la clomipramine. *Thérapie* **1973**, *28*, 379–392. [PubMed]

45. Shipley, J.E.; Kupfer, D.J.; Dealy, R.S.; Griffin, S.J.; Coble, P.A.; McEachran, A.B.; Grochocinski, V.J. Differential effects of amitriptyline and of zimelidine on the sleep electroencephalogram of depressed patients. *Clin. Pharmacol. Ther.* **1984**, *36*, 251–259. [CrossRef] [PubMed]

46. Von Bardeleben, U.; Steiger, A.; Gerken, A.; Holsboer, F. Effects of fluoxetine upon pharmacoendocrine and sleep-EEG parameters in normal controls. *Int. Clin. Psychopharmacol.* **1989**, *4*, 1–5.

47. Künzel, H.; Murck, H.; Held, K.; Ziegenbein, M.; Steiger, A. Reboxetine induces similar sleep-EEG changes like SSRI's in patients with depression. *Pharmacopsychiatry* **2004**, *37*, 193–195. [CrossRef]

48. Kluge, M.; Schüssler, P.; Steiger, A. Duloxetine increases stage 3 sleep and suppresses rapid eye movement (REM) sleep in patients with major depression. *Eur. Neuropsychopharmacol.* **2007**, *17*, 527–531. [CrossRef] [PubMed]

49. Cramer, H.; Ohlmeier, D. Ein Fall von Tranylcypromin-und Trifluoperazin-(Jatrosom®)-Sucht: Psychopathologische, schlafphysiologische und biochemische Untersuchungen. *Archiv für Psychiatrie und Nervenkrankheiten* **1967**, *210*, 182–197. [CrossRef]

50. Wyatt, R.J.; Fram, D.H.; Kupfer, D.J.; Snyder, F. Total prolonged drug-induced REM sleep suppression in anxious-depressed patients. *Arch. Gen. Psychiatry* **1971**, *24*, 145–155. [CrossRef]

51. Landolt, H.-P.; Raimo, E.B.; Schnierow, B.J.; Kelsoe, J.R.; Rapaport, M.H.; Gillin, J.C. Sleep and sleep electroencephalogram in depressed patients treated with phenelzine. *Arch. Gen. Psychiatry* **2001**, *58*, 268–276. [CrossRef]

52. Steiger, A.; Benkert, O.; Holsboer, F. Effects of long-term treatment with the MAO-A inhibitor moclobemide on sleep EEG and nocturnal hormonal secretion in normal men. *Neuropsychobiology* **1994**, *30*, 101–105. [CrossRef]

53. Sonntag, A.; Rothe, B.; Guldner, J.; Yassouridis, A.; Holsboer, F.; Steiger, A. Trimipramine and imipramine exert different effects on the sleep EEG and on nocturnal hormone secretion during treatment of major depression. *Depression* **1996**, *4*, 1–13. [CrossRef]

54. Nofzinger, E.A.; Reynolds, C.F., III; Thase, M.E.; Frank, E. REM sleep enhancement by bupropion in depressed men. *Am. J. Psychiatry* **1995**, *152*, 274–276. [PubMed]

55. Murck, H.; Nickel, T.; Künzel, H.; Antonijevic, I.; Schill, J.; Zobel, A.; Steiger, A.; Sonntag, A.; Holsboer, F. State markers of depression in sleep EEG: Dependency on drug and gender in patients treated with tianeptine or paroxetine. *Neuropsychopharmacology* **2003**, *28*, 348–358. [CrossRef] [PubMed]

56. Ruigt, G.; Kemp, B.; Groenhout, C.; Kamphuisen, H. Effect of the antidepressant Org 3770 on human sleep. *Eur. J. clin. Pharmacol.* **1990**, *38*, 551–554. [CrossRef] [PubMed]

57. Schmid, D.A.; Wichniak, A.; Uhr, M.; Ising, M.; Brunner, H.; Held, K.; Weikel, J.C.; Sonntag, A.; Steiger, A. Changes of sleep architecture, spectral composition of sleep EEG, the nocturnal secretion of cortisol, ACTH, GH, prolactin, melatonin, ghrelin, and leptin, and the DEX-CRH test in depressed patients during treatment with mirtazapine. *Neuropsychopharmacology* **2006**, *31*, 832–844. [CrossRef] [PubMed]

58. Steiger, A. Effects of clomipramine on sleep EEG and nocturnal penile tumescence: A long-term study in a healthy man. *J. Clin. Psychopharmacol.* **1988**, *8*, 349–354. [CrossRef] [PubMed]

59. Akindele, M.; Evans, J.; Oswald, I. Mono-amine oxidase inhibitors, sleep and mood. *Electroencephalogr. Clin. Neurophysiol.* **1970**, *29*, 47–56. [CrossRef]

60. Murck, H.; Frieboes, R.; Antonijevic, I.; Steiger, A. Distinct temporal pattern of the effects of the combined serotonin-reuptake inhibitor and 5-HT 1A agonist EMD 68843 on the sleep EEG in healthy men. *Psychopharmacology* **2001**, *155*, 187–192. [CrossRef] [PubMed]

61. Grözinger, M.; Kögel, P.; Röschke, J. Effects of REM sleep awakenings and related wakening paradigms on the ultradian sleep cycle and the symptoms in depression. *J. Psychiatr. Res.* **2002**, *36*, 299–308. [CrossRef]

62. Steiger, A.; Gerken, A.; Benkert, O.; Holsboer, F. Differential effects of the enantiomers R (−) and S (+) oxaprotiline on major endogenous depression, the sleep EEG and neuroendocrine secretion: Studies on depressed patients and normal controls. *Eur. Neuropsychopharmacol.* **1993**, *3*, 117–126. [CrossRef]

63. Saletu, B.; Frey, R.; Krupka, M.; Anderer, P.; Grfulberger, J.; See, W.R. Sleep laboratory studies on the single-dose effects of serotonin reuptake inhibitors paroxetine and fluoxetine on human sleep and awakening qualities. *Sleep* **1991**, *14*, 439–447. [CrossRef]

64. Sharpley, A.; Williamson, D.; Attenburrow, M.; Pearson, G.; Sargent, P.; Cowen, P. The effects of paroxetine and nefazodone on sleep: A placebo controlled trial. *Psychopharmacology* **1996**, *126*, 50–54. [CrossRef] [PubMed]

65. Kupfer, D. REM latency: A psychobiologic marker for primary depressive disease. *Biol. Psychiatry* **1976**, *11*, 159–174. [PubMed]

66. Riemann, D.; Berger, M. The effects of total sleep deprivation and subsequent treatment with clomipramine on depressive symptoms and sleep electroencephalography in patients with a major depressive disorder. *Acta Psychiatr. Scand.* **1990**, *81*, 24–31. [CrossRef] [PubMed]

67. Holsboer, F.; Ising, M. Stress hormone regulation: Biological role and translation into therapy. *Annu. Rev. Psychol.* **2010**, *61*, 81–109. [CrossRef] [PubMed]

68. Linkowski, P.; Mendlewicz, J.; Kerkhofs, M.; Leclercq, R.; Golstein, J.; Brasseur, M.; Copinschi, G.; Cauter, E.V. 24-hour profiles of adrenocorticotropin, cortisol, and growth hormone in major depressive illness: Effect of antidepressant treatment. *J. Clin. Endocrinol. MeTable* **1987**, *65*, 141–152. [CrossRef] [PubMed]

69. Ehlers, C.L.; Reed, T.K.; Henriksen, S.J. Effects of corticotropin-releasing factor and growth hormone-releasing factor on sleep and activity in rats. *Neuroendocrinology* **1986**, *42*, 467–474. [CrossRef]

70. Opp, M.; Obal, F., Jr.; Krueger, J. Corticotropin-releasing factor attenuates interleukin 1-induced sleep and fever in rabbits. *Am. J. Physiol.-Regul. Integr. Comp. Physiol.* **1989**, *257*, R528–R535. [CrossRef] [PubMed]

71. Romanowski, C.; Fenzl, T.; Flachskamm, C.; Deussing, J.; Kimura, M. CRH-R1 is involved in effects of CRH on NREM, but not REM, sleep suppression. *Sleep Biol. Rhythms* **2007**, *5*, A53.

72. Sanford, L.; Yang, L.; Wellman, L.; Dong, E.; Tang, X. Mouse strain differences in the effects of corticotropin releasing hormone (CRH) on sleep and wakefulness. *Brain Res.* **2008**, *1190*, 94–104. [CrossRef]

73. Holsboer, F.; Von Bardeleben, U.; Steiger, A. Effects of intravenous corticotropin-releasing hormone upon sleep-related growth hormone surge and sleep EEG in man. *Neuroendocrinology* **1988**, *48*, 32–38. [CrossRef]

74. Kimura, M.; Müller-Preuss, P.; Lu, A.; Wiesner, E.; Flachskamm, C.; Wurst, W.; Holsboer, F.; Deussing, J. Conditional corticotropin-releasing hormone overexpression in the mouse forebrain enhances rapid eye movement sleep. *Mol. Psychiatry* **2010**, *15*, 154–165. [CrossRef] [PubMed]

75. Schüssler, P.; Kluge, M.; Gamringer, W.; Wetter, T.; Yassouridis, A.; Uhr, M.; Rupprecht, R.; Steiger, A. Corticotropin-releasing hormone induces depression-like changes of sleep electroencephalogram in healthy women. *Psychoneuroendocrinology* **2016**, *74*, 302–307. [CrossRef] [PubMed]

76. Hatzinger, M.; Brand, S.; Perren, S.; Stadelmann, S.; von Wyl, A.; von Klitzing, K.; Holsboer-Trachsler, E. Electroencephalographic sleep profiles and hypothalamic–pituitary–adrenocortical (HPA)-activity in kindergarten children: Early indication of poor sleep quality associated with increased cortisol secretion. *J. Psychiatr. Res.* **2008**, *42*, 532–543. [CrossRef] [PubMed]

77. Held, K.; Künzel, H.; Ising, M.; Schmid, D.; Zobel, A.; Murck, H.; Holsboer, F.; Steiger, A. Treatment with the CRH1-receptor-antagonist R121919 improves sleep-EEG in patients with depression. *J. Psychiatr. Res.* **2004**, *38*, 129–136. [CrossRef]

78. Antonijevic, I.A.; Steiger, A. Depression-like changes of the sleep-EEG during high dose corticosteroid treatment in patients with multiple sclerosis. *Psychoneuroendocrinology* **2003**, *28*, 780–795. [CrossRef]

79. Obal, F., Jr.; Alfoldi, P.; Cady, A.; Johannsen, L.; Sáry, G.; Krueger, J. Growth hormone-releasing factor enhances sleep in rats and rabbits. *Am. J. Physiol.-Regul. Integr. Comp. Physiol.* **1988**, *255*, R310–R316. [CrossRef] [PubMed]

80. Zhang, J.; Obál, F.; Zheng, T.; Fang, J.; Taishi, P.; Krueger, J.M. Intrapreoptic microinjection of GHRH or its antagonist alters sleep in rats. *J. Neurosci.* **1999**, *19*, 2187–2194. [CrossRef]

81. Obál, F., Jr.; Floyd, R.; Kapas, L.E.; Bodosi, B.; Krueger, J. Effects of systemic GHRH on sleep in intact and hypophysectomized rats. *Am. J. Physiol.-Endocrinol. MeTable* **1996**, *270*, E230–E237. [CrossRef]

82. Steiger, A.; Guldner, J.; Hemmeter, U.; Rothe, B.; Wiedemann, K.; Holsboer, F. Effects of growth hormone-releasing hormone and somatostatin on sleep EEG and nocturnal hormone secretion in male controls. *Neuroendocrinology* **1992**, *56*, 566–573. [CrossRef]

83. Antonijevic, I.A.; Murck, H.; Frieboes, R.-M.; Barthelmes, J.; Steiger, A. Sexually dimorphic effects of GHRH on sleep-endocrine activity in patients with depression and normal controls—p#art I: The sleep EEG. *Sleep Res. Online* **2000**, *3*, 5–13.

84. Antonijevic, I.A.; Murck, H.; Frieboes, R.-M.; Steiger, A. Sexually dimorphic effects of GHRH on sleep-endocrine activity in patients with depression and normal controls—p#art II: Hormone secretion. *Sleep Res. Online* **2000**, *3*, 15–21. [PubMed]

85. Mayeux, R.; Stern, Y. Epidemiology of Alzheimer disease. *Cold Spring Harb. Perspect. Med.* **2012**. [CrossRef] [PubMed]

86. Shokri-Kojori, E.; Wang, G.-J.; Wiers, C.E.; Demiral, S.B.; Guo, M.; Kim, S.W.; Lindgren, E.; Ramirez, V.; Zehra, A.; Freeman, C.; et al. β-Amyloid accumulation in the human brain after one night of sleep deprivation. *Proc. Natl. Acad. Sci. USA* **2018**, *115*, 4483–4488. [CrossRef] [PubMed]

87. Chen, L.; Huang, J.; Yang, L.; Zeng, X.-A.; Zhang, Y.; Wang, X.; Chen, M.; Li, X.; Zhang, Y.; Zhang, M. Sleep deprivation accelerates the progression of Alzheimer's disease by influencing Aβ-related metabolism. *Neurosci. Lett.* **2017**, *650*, 146–152. [CrossRef] [PubMed]

88. Lucey, B.P.; Hicks, T.J.; McLeland, J.S.; Toedebusch, C.D.; Boyd, J.; Elbert, D.L.; Patterson, B.W.; Baty, J.; Morris, J.C.; Ovod, V.; et al. Effect of sleep on overnight cerebrospinal fluid amyloid β kinetics. *Ann. Neurol.* **2018**, *83*, 197–204. [CrossRef] [PubMed]

89. Kincheski, G.C.; Valentim, I.S.; Clarke, J.R.; Cozachenco, D.; Castelo-Branco, M.T.; Ramos-Lobo, A.M.; Rumjanek, V.M.; Donato, J., Jr.; De Felice, F.G.; Ferreira, S.T. Chronic sleep restriction promotes brain inflammation and synapse loss, and potentiates memory impairment induced by amyloid-β oligomers in mice. *Brain. Behav. Immun.* **2017**, *64*, 140–151. [CrossRef] [PubMed]

90. Minakawa, E.N.; Miyazaki, K.; Maruo, K.; Yagihara, H.; Fujita, H.; Wada, K.; Nagai, Y. Chronic sleep fragmentation exacerbates amyloid β deposition in Alzheimer's disease model mice. *Neurosci. Lett.* **2017**, *653*, 362–369. [CrossRef]

91. Ju, Y.-E.S.; Ooms, S.J.; Sutphen, C.; Macauley, S.L.; Zangrilli, M.A.; Jerome, G.; Fagan, A.M.; Mignot, E.; Zempel, J.M.; Claassen, J.A. Slow wave sleep disruption increases cerebrospinal fluid amyloid-β levels. *Brain* **2017**, *140*, 2104–2111. [CrossRef]

92. Chen, D.-W.; Wang, J.; Zhang, L.-L.; Wang, Y.-J.; Gao, C.-Y. Cerebrospinal fluid amyloid-β levels are increased in patients with insomnia. *J. Alzheimers Dis.* **2018**, *61*, 645–651. [CrossRef]

93. Park, H.J.; Ran, Y.; Jung, J.I.; Holmes, O.; Price, A.R.; Smithson, L.; Ceballos-Diaz, C.; Han, C.; Wolfe, M.S.; Daaka, Y.; et al. The stress response neuropeptide CRF increases amyloid-β production by regulating γ-secretase activity. *EMBO J.* **2015**, *34*, 1674–1686. [CrossRef]

94. Morgese, M.G.; Schiavone, S.; Trabace, L. Emerging role of amyloid beta in stress response: Implication for depression and diabetes. *Eur. J. Pharmacol.* **2017**, *817*, 22–29. [CrossRef] [PubMed]

95. Hatzinger, M.; Hemmeter, U.M.; Brand, S.; Ising, M.; Holsboer-Trachsler, E. Electroencephalographic sleep profiles in treatment course and long-term outcome of major depression: Association with DEX/CRH-test response. *J. Psychiatr. Res.* **2004**, *38*, 453–465. [CrossRef] [PubMed]

96. Leuchter, A.F.; Uijtdehaage, S.H.; Cook, I.A.; O'Hara, R.; Mandelkern, M. Relationship between brain electrical activity and cortical perfusion in normal subjects. *Psychiatry Res. Neuroimaging* **1999**, *90*, 125–140. [CrossRef]

97. Bares, M.; Brunovsky, M.; Kopecek, M.; Stopkova, P.; Novak, T.; Kozeny, J.; Höschl, C. Changes in QEEG prefrontal cordance as a predictor of response to antidepressants in patients with treatment resistant depressive disorder: A pilot study. *J. Psychiatr. Res.* **2007**, *41*, 319–325. [CrossRef] [PubMed]

98. Bares, M.; Brunovsky, M.; Novak, T.; Kopecek, M.; Stopkova, P.; Sos, P.; Krajca, V.; Höschl, C. The change of prefrontal QEEG theta cordance as a predictor of response to bupropion treatment in patients who had failed to respond to previous antidepressant treatments. *Eur. Neuropsychopharmacol.* **2010**, *20*, 459–466. [CrossRef] [PubMed]

99. Cook, I.A.; Leuchter, A.F.; Morgan, M.; Witte, E.; Stubbeman, W.F.; Abrams, M.; Rosenberg, S.; Uijtdehaage, S.H. Early changes in prefrontal activity characterize clinical responders to antidepressants. *Neuropsychopharmacology* **2002**, *27*, 120–131. [CrossRef]

100. Cook, I.A.; Leuchter, A.F.; Morgan, M.L.; Stubbeman, W.; Siegman, B.; Abrams, M. Changes in prefrontal activity characterize clinical response in SSRI nonresponders: A pilot study. *J. Psychiatr. Res.* **2005**, *39*, 461–466. [CrossRef]

101. Asada, H.; Fukuda, Y.; Tsunoda, S.; Yamaguchi, M.; Tonoike, M. Frontal midline theta rhythms reflect alternative activation of prefrontal cortex and anterior cingulate cortex in humans. *Neurosci. Lett.* **1999**, *274*, 29–32. [CrossRef]

102. Drevets, W.C. Neuroimaging studies of mood disorders. *Biol. Psychiatry* **2000**, *48*, 813–829. [CrossRef]

103. Braun, A.R.; Balkin, T.; Wesenten, N.; Carson, R.; Varga, M.; Baldwin, P.; Selbie, S.; Belenky, G.; Herscovitch, P. Regional cerebral blood flow throughout the sleep-wake cycle. An H2(15)O PET study. *Brain* **1997**, *120*, 1173–1197. [CrossRef]

104. Hobson, J.A.; Pace-Schott, E.F. The cognitive neuroscience of sleep: Neuronal systems, consciousness and learning. *Nat. Rev. Neurosci.* **2002**, *3*, 679–693. [CrossRef] [PubMed]

105. Nishida, M.; Hirai, N.; Miwakeichi, F.; Maehara, T.; Kawai, K.; Shimizu, H.; Uchida, S. Theta oscillation in the human anterior cingulate cortex during all-night sleep: An electrocorticographic study. *Neurosci. Res.* **2004**, *50*, 331–341. [CrossRef] [PubMed]

106. Adamczyk, M.; Gazea, M.; Wollweber, B.; Holsboer, F.; Dresler, M.; Steiger, A.; Pawlowski, M. Cordance derived from REM sleep EEG as a biomarker for treatment response in depression—A naturalistic study after antidepressant medication. *J. Psychiatr. Res.* **2015**, *63*, 97–104. [CrossRef] [PubMed]

107. Kemp, A.H.; Quintana, D.S.; Gray, M.A.; Felmingham, K.L.; Brown, K.; Gatt, J.M. Impact of depression and antidepressant treatment on heart rate variability: A review and meta-analysis. *Biol. Psychiatry* **2010**, *67*, 1067–1074. [CrossRef] [PubMed]

108. Pawlowski, M.A.; Gazea, M.; Wollweber, B.; Dresler, M.; Holsboer, F.; Keck, M.E.; Steiger, A.; Adamczyk, M.; Mikoteit, T. Heart rate variability and cordance in rapid eye movement sleep as biomarkers of depression and treatment response. *J. Psychiatr. Res.* **2017**, *92*, 64–73. [CrossRef] [PubMed]

International Journal of
Molecular Sciences

Comparative Evaluation of Machine Learning Strategies for Analyzing Big Data in Psychiatry

Han Cao, Andreas Meyer-Lindenberg and Emanuel Schwarz *

Department of Psychiatry and Psychotherapy, Central Institute of Mental Health, Medical Faculty Mannheim, Heidelberg University, 68159 Mannheim, Germany; han.cao@zi-mannheim.de (H.C.); Andreas.Meyer-Lindenberg@zi-mannheim.de (A.M.-L.)
* Correspondence: emanuel.schwarz@zi-mannheim.de; Tel.: +49-621-1703-2368

Received: 9 August 2018; Accepted: 25 October 2018; Published: 29 October 2018

Abstract: The requirement of innovative big data analytics has become a critical success factor for research in biological psychiatry. Integrative analyses across distributed data resources are considered essential for untangling the biological complexity of mental illnesses. However, little is known about algorithm properties for such integrative machine learning. Here, we performed a comparative analysis of eight machine learning algorithms for identification of reproducible biological fingerprints across data sources, using five transcriptome-wide expression datasets of schizophrenia patients and controls as a use case. We found that multi-task learning (MTL) with network structure (MTL_NET) showed superior accuracy compared to other MTL formulations as well as single task learning, and tied performance with support vector machines (SVM). Compared to SVM, MTL_NET showed significant benefits regarding the variability of accuracy estimates, as well as its robustness to cross-dataset and sampling variability. These results support the utility of this algorithm as a flexible tool for integrative machine learning in psychiatry.

Keywords: multi-task learning; machine learning; biomarker discovery; psychiatry

---

1. Introduction

Biological research on psychiatric illnesses has highlighted the scale of investigations required to identify reproducible hallmarks of illness [1,2]. In schizophrenia, collaborative analysis of common genetic variants has exceeded 150,000 subjects [3], demonstrating the challenges tied to low-effect sizes of individual variants, large biological and clinical heterogeneity, and genetic complexity. Not surprisingly, these challenges are also found in other mental illnesses [4] and do not seem to be modality specific, as analysis of neuroimaging data, for example, faces similar problems [5,6].

The combined "mega-analysis" of data across cohorts and modalities has advantages compared to the more traditional meta-analysis [4,7], as it makes data amenable for a broader spectrum of computational analyses and allows consideration of confounders across studies. There is growing consensus that advanced computational strategies are required to extract biologically meaningful patterns from these data sources. Beyond functional analysis, a particular focus is on machine learning, which, in other areas, has shown substantial success in integrating weak signals into accurate classifiers [8]. In addition to potential clinical use of such classifiers, the discovery of robust biological patterns may uncover new insights into etiological processes. However, the increasing scale and complexity of big data in psychiatry requires careful evaluation of the most suitable computational strategies. A particularly intuitive and very timely problem is the optimal integration of multi-cohort data, where simple concatenation of datasets may give suboptimal results, and even more so when integration is performed across modalities.

The application of machine-learning techniques on biological problems in psychiatry has already yielded impressive results, including on the prediction of genetic risk, the identification of biomarker candidates, or the exploration of etiological mechanisms [9]. For example, the use of a Bayesian approach for the incorporation of linkage disequilibrium (LD) information during polygenic risk score determination led to a 5% improvement of accuracy in a large schizophrenia dataset [10]. In a study exploring the molecular basis of psychiatric comorbidity, an iterative LASSO approach was used for cross-tissue prediction and identified a schizophrenia expression signature that predicted a peripheral biomarker of T2D [11]. Beyond the analysis of individual data modalities, several machine-learning strategies have been developed for integrative multimodal analysis. For example, a study focusing on the IMAGEN cohort [12] applied an elastic net model to explore information patterns linked to binge drinking across multiple domains, including brain structure and function, personality traits, cognitive differences, candidate gene information, environmental factors, and life experiences. Similarly, another study [13] explored the inherent data sparsity of neuroimaging and psychiatric symptom data, and successfully stratified subjects using sparse canonical correlation analysis. The study found four dimensions of psychopathology with different patterns of connectivity. In the present study, we were particularly interested in multi-task learning (MTL), which aims to improve generalizability by simultaneously learning multiple tasks (such as case-control associations in different datasets) and these learning processes exchange information to achieve a globally optimal solution [14]. Historically, MTL was developed as an extension of neural networks [14], and has since been used across data-intensive research areas, including biomedical informatics [15–20], speech and natural language processing [21,22], image processing and computer vision [23,24], and web based applications [25,26]. In psychiatric research, MTL has been applied for integrating measures of cognitive functioning and structural neuroimaging [27], as well as for improved fMRI pattern recognition [28]. In other research fields, MTL approaches have been proposed to combine different sources of biological data, including the linking of MRI or expression with genetic data [29,30], as well as the integrative analysis of multi-cohort expression data [31].

In the present study, we used MTL to differentiate schizophrenia patients from controls across multiple transcriptome-wide expression datasets. We hypothesized that MTL is particularly suited for this task, since it allows the consideration of different cohorts as separate classification tasks. As MTL aims to identify predictive patterns that are shared across tasks, it should uncover expression patterns that are biologically reproducible across cohorts. This may result in better and biologically more relevant classifiers compared to those derived from conventional single task learning (STL), which may be unduly influenced by strong signals present in individual cohorts. To test this, we performed a comparative analysis of different MTL and STL approaches in five transcriptome-wide datasets of schizophrenia brain expression. A 'leave-dataset-out' procedure was applied to explore and compare the generalizability of the models, with specific focus on classification accuracy, and variability thereof, as well as model sensitivity to cross-dataset and sampling variability.

2. Results

2.1. Accuracy Comparison Between MTL and STL

Figure 1 shows a comparison of average classification accuracies when four out of five datasets were used for training and the remaining dataset for testing. The distributions of accuracies are shown for 10 repetitions of the classification procedure to assess the variability caused by parameter tuning via cross-validation. With an average accuracy of 0.73, MTL_NET outperformed all other methods, followed by SVM, which had a marginally inferior accuracy of 0.72. Moderate accuracies were observed for MTL_Trace (0.69), MTL_L21 (0.66) and RF (0.68). The sparse logistic regression performed worst (0.64). As an extension of MTL_NET and MTL_L21, respectively, MTL_SNET (0.71) and MTL_EN (0.66) achieved similar accuracies to their original algorithms. In the following analysis, we focused on the comparison of MTL_NET and SVM as representatives of MTL and STL, respectively.

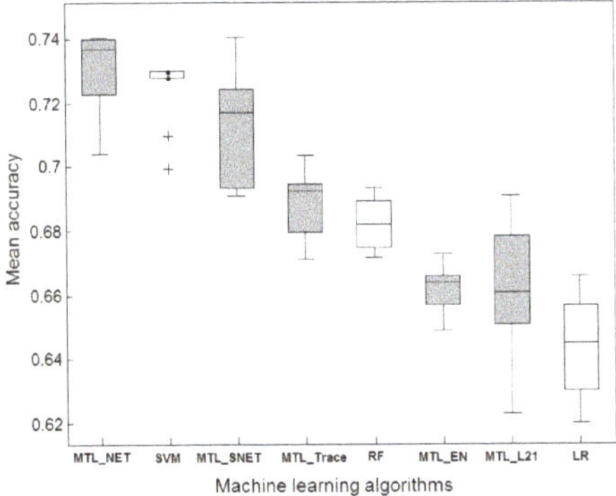

Figure 1. Predictive performance comparison between eight algorithms. The 'leave-dataset-out' procedure was used for comparison. Four out of five datasets were combined for training, and then the model was tested on the remaining dataset. The distribution of accuracy estimates indicated the variation of parameter selection across 10 repetitions. The boxplots in gray denote the multi-task learning algorithms.

In Figure 1, the standard error of accuracies for SVM (0.011) was slightly smaller than that for MTL_NET (0.012), indicating that SVM might be more robust regarding parameter selection. A possible reason was that SVM obtained higher statistical power by comparing cases and controls across datasets. In contrast, MTL_NET derived transcriptomic signatures using cases and controls within datasets, limiting the statistical power.

2.2. Dependency of Classification Performance on the Number of Training Datasets

We performed a side-by-side comparison of MTL_NET and SVM to explore the dependency of classification performance on the number of available training datasets. Figure 2a shows that increasing accuracy was observed for both MTL_NET and SVM with increasing numbers of training datasets. Notably, MTL_NET only outperformed SVM at $n_d = 4$ (four datasets used for training), suggesting that MTL required a higher dataset number to identify a reproducible biological pattern. However, we observed that the variation of accuracies for MTL_NET substantially decreased with increasing numbers of training datasets (Figure 2b), which was not the case for SVM. This suggested that MTL_NET was more conservative in that accuracy was not driven by highly successful prediction on an individual test set, but by improved predictability observed for all test sets.

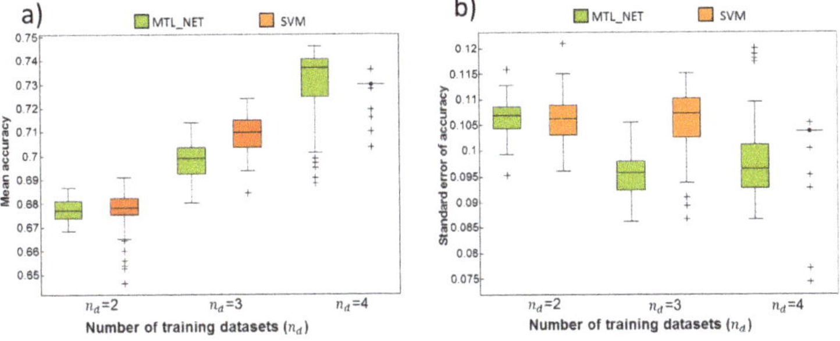

Figure 2. Distribution of classification accuracies and their standard errors across different numbers of training datasets. The Figure shows the mean (**a**) and standard error (**b**) of classification accuracies obtained for different numbers of training datasets (n_d). Performance was evaluated from the test datasets not used for training. The variation of the boxplot was due to the sampling variability during cross-validation.

2.3. Consistency and Stability of Trained Models

Figure 3a,b show that, in terms of vertical and horizontal consistency, MTL_NET outperformed SVM, independently of the number of training datasets. This indicated that similar discriminative patterns of genes were identified by MTL across training datasets, and implied strong robustness against cross-dataset variability. In particular, the superior performance of vertical consistency for MTL_NET showed that this algorithm was less sensitive to the small numbers of training datasets compared to SVM. Table 1 shows the mean consistency (both horizontal and vertical) across bootstrapping samples. Compared to SVM, MTL_NET achieved a higher mean consistency by approximately 1.6% for horizontal and 2.2% for vertical consistency. Notably, the success rate of consistency was 100%, independent of the number of training sets, showing that MTL_NET models consistently identified higher transcriptomic profile robustness across bootstrapping samples than SVM.

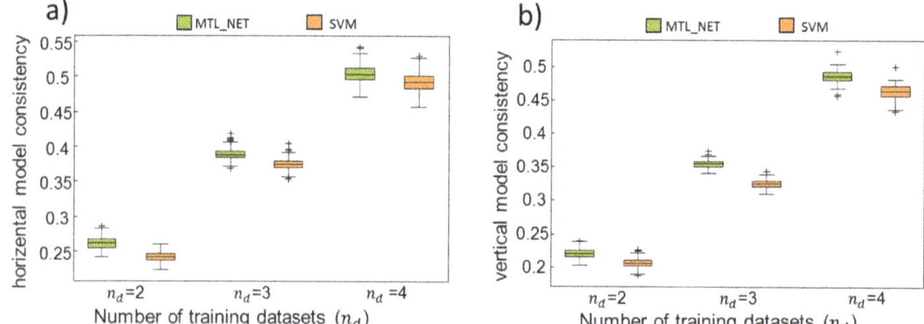

Figure 3. Horizontal and vertical model consistency. To analyze the consistency of a given machine-learning algorithm against the cross-dataset variability, we quantified the horizontal (**a**) and vertical (**b**) model consistency for different numbers (n_d) of training datasets. Specifically, horizontal consistency quantified the similarity between models trained using the same number of datasets, and vertical consistency quantified the pairwise similarity of models, where one was trained using all datasets and the other was trained using less datasets. Stratified 100-fold bootstrapping procedure was applied to quantify the variation of the consistency.

Table 1. Mean consistency, stability, and success rate across the number of training sets, n_d.

MTL_NET/SVM	$n_d = 2$	$n_d = 3$	$n_d = 4$	$n_d = 5$
Horizontal consistency	0.26/0.24	0.39/0.37	0.51/0.49	-
Vertical consistency	0.22/0.21	0.35/0.33	0.49/0.46	-
Stability	0.64/0.63	0.65/0.64	0.65/0.64	0.654/0.645
Success rate (horizontal consistency)	1	1	1	-
Success rate (vertical consistency)	1	1	1	-
Success rate (stability)	1	1	1	1

To further identify the robustness of models against sampling variability, we quantified the algorithms' stability. In Figure 4, across the number of training datasets, n_d, the increasing trend of stability demonstrated that both MTL_NET and SVM gained more robustness against sampling variability with an increasing number of subjects used for training. However, MTL_NET demonstrated higher stability than SVM independently of the number of training datasets (Figure 4). The mean stability across models also supported the result (Table 1). Moreover, the mean stability for MTL_NET was 1.2% higher than SVM (100% success rate of stability across all n_d, Table 2).

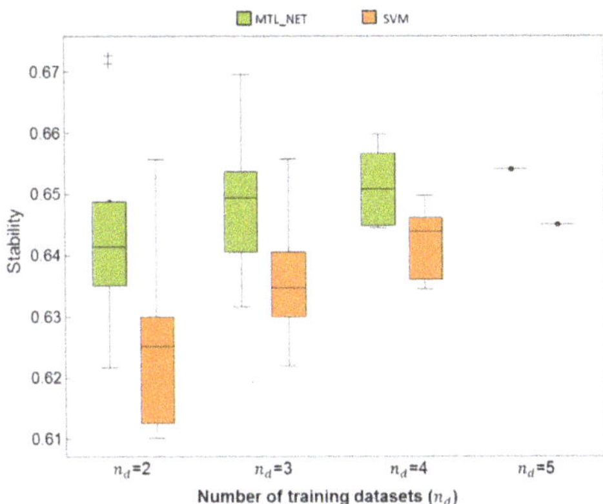

Figure 4. Stability comparison. The stability quantified the robustness of an algorithm against sampling variability. For each n_d, stability was computed as the pairwise similarity of models trained from two given bootstrap samples. The stability was then averaged across bootstrap samples. In the Figure, the distribution of the stability was due to the different combination of training datasets given, n_d.

We did not perform comparative functional analysis of markers identified by the two algorithms, since marker sets were quite similar. For example, using all five datasets for training, the average similarity over all bootstrapping samples was 98.75%, suggesting that similar functional implications would be derived for these algorithms.

3. Discussion

The present study provides a comparative evaluation of using MTL for integrative machine learning, compared to classical, single task learning in five transcriptome-wide datasets of schizophrenia brain expression. Overall, MTL showed similar accuracy, albeit with lower variability, compared to STL. Accuracy estimates varied by up to approximately 10% between algorithms, suggesting different sensitivities of algorithms to cross-dataset heterogeneity as well as sampling variability. Among all MTL formulations, MTL_NET was most predictive. This was likely due to the fact that it harmonized algorithms across tasks with respect to both predictor weight and sign of diagnosis association, resulting in biologically plausible predictive patterns. In contrast, MTL_L21 ignores the sign of association and MTL_Trace improves models' correlation in each subspace, but failed to modulate the cross-subspace correlation. Contrary to the usual assumption that simpler models show improved generalizability [32], a sparse version of MTL_NET (MTL_SNET) did not improve the prediction. This may be due to the fact that the sparse model was trained by constructing a solution tree among an unlimited number of optimal solution trees. Although these solution trees have similar performance on the training dataset, they may show differently predictive ability on a cross-modality test dataset because the "independent and identically distributed (i.i.d)" assumption may not hold. MTL_NET (as well as SVM) solves a strictly convex optimization problem, resulting in a uniform solution in the entire feature space, which may be equally effective when tested on independent test data.

The higher consistency and stability of MTL_NET implied that a set of similar differentially expressed genes were identified for multiple training datasets. In addition, these genes demonstrated higher predictability and robustness against study-specific effects, which is particularly important for data integration in multi-modal analyses, such as the integrative analysis of genetic and expression data [33] or the analysis of shared markers across multiple comorbid conditions [34–36].

Table 2. Overview of demographic details. Values are shown as mean $\pm$ sd.

	GSE12679	GSE35977	GSE17612	GSE21935	GSE21138
Reference	[37]	[38]	[39]	[40]	[41]
n SZ	11	50	22	19	29
n HC	11	50	22	19	29
age SZ	46.1 $\pm$ 5.9	42.4 $\pm$ 9.9	76 $\pm$ 12.9	77.6 $\pm$ 11.4	43.3 $\pm$ 17.3
age HC	41.7 $\pm$ 7.9	45.5 $\pm$ 9	68 $\pm$ 21.5	67.7 $\pm$ 22.2	44.7 $\pm$ 16.1
sex SZ (m/f)	7/4	37/13	16/6	11/8	23/6
sex HC (m/f)	8/3	35/15	11/11	10/9	24/5
PMI SZ	33 $\pm$ 6.7	31.8 $\pm$ 15.4	6.2 $\pm$ 4.1	5.5 $\pm$ 2.6	38.1 $\pm$ 10.8
PMI HC	24.2 $\pm$ 15.7	27.3 $\pm$ 11.8	10.1 $\pm$ 4.3	9.1 $\pm$ 4.3	40.5 $\pm$ 14
brain pH SZ	NA	6.4 $\pm$ 0.3	6.1 $\pm$ 0.2	6.1 $\pm$ 0.2	6.2 $\pm$ 0.2
brain pH HC	NA	6.5 $\pm$ 0.3	6.5 $\pm$ 0.3	6.5 $\pm$ 0.3	6.3 $\pm$ 0.2
Genechip	HGU	HuG	HGU	HGU	HGU
Brain Region	PFC	PC	APC	STC	PFC

HGU: HG-U133_Plus_2; HuG = HuGene-1_0-st; APC: Anterior prefrontal cortex; PFC: Prefrontal cortex; PC: Parietal cortex; STC: Superior temporal cortex; HC: Healthy control; SZ: Schizophrenia.

An interesting observation of the present study was that for MTL_NET, the variance of the classification accuracy substantially decreased with an increasing number of training datasets. This suggested that MTL_NET selected biological signatures with similar effect sizes across independent training datasets, further supporting the biological reproducibility of the identified patterns. In contrast, SVM did not show a decreasing accuracy variance with increasing numbers of training datasets. This indicates that despite the increasing classification accuracy, the identified signatures worked well only for some, but not other, test datasets. These results for these particular datasets highlight differences between single and multi-task learning regarding the variance of the test-set accuracy, which is a fundamentally important consideration for study design and interpretation of classifier reproducibility.

4. Materials and Methods

4.1. Datasets

In the present study, five transcriptome-wide expression datasets from schizophrenia post-mortem brains and controls were used for analysis. Details of the datasets are shown in Table 2. All datasets were downloaded from the GEO (Gene Expression Omnibus).

4.2. Preprocessing

Preprocessing was performed using the statistical software, R (https://cran.r-project.org/). First, raw expression data were read using the 'ReadAffy' function. Then RMA (Multi-Array Average [42]) was applied for background correction, quantile normalization, and $\log_2$-transformation. Subsequently, multiple probes associated to one gene symbol were averaged. This was followed by the selection of common genes across all datasets (17,061 genes). For each dataset, propensity score matching was used to obtain a sample with approximate 1:1 matching for diagnosis, sex, ph, age, and post-mortem interval (pmi). Next, all datasets were concatenated for quantile normalization and covariate correction. Specifically, the 'Combat' function from the R library *sva* [43] was applied to correct for covariates (sex, ph, age, age^2, pmi, and a dataset indicator). Finally, datasets were separated again for feature standardization (z-score) to remove bias from the expressed genes with large variance and for downstream machine learning analysis.

4.3. Machine Learning Approaches

For MTL, multiple across-task regularization strategies were tested, such as MTL with network structure (MTL_NET), sparse network structure (MTL_SNET), joint feature learning (MTL_L21), joint feature learning with elastic net (MTL_EN), and low-rank structure (MTL_Trace). As a comparison, we selected logistic regression with lasso (LR), linear support vector machines (SVM), and random forests (RF) as representatives of conventional STL methods. For all models (except for RF), stratified five-fold cross validation was used to select hyper-parameters. Methodological details of the respective methods are described below. All machine-learning analyses were performed using Matlab (R2016b).

4.3.1. Multi-Task Learning

For all MTL formulations, the logistic loss ($\mathcal{L}(\cdot)$) was used as the common loss function.

$$\mathcal{L}(W,C) = \frac{1}{n_i} \sum_{j=1}^{n_i} \log\left(1 + e^{(-Y_{i,j}(X_{i,j}W_i^T + C_i))}\right) \tag{1}$$

where X, Y, W, and C referred to the gene expression matrixes, diagnostic status, weight vectors, and constants of all tasks, respectively. In addition, i and j denoted the index of the dataset and subject respectively, i.e., n_i and W_i^T referred to the number of subject and weight vector of task i. This model aimed to estimate the effect size of each feature such that the likelihood (i.e., the rate of successful prediction in the training data) was maximized. During the prediction procedure, given the expression profile of a previously unseen individual, the model calculates the probability of belonging to the schizophrenia class (with subjects where the probability exceeded 0.5 being assigned to the patient group). Notably, while we focused on classification due to the categorical outcomes of the investigated datasets, the cross-task regularization strategies explored in the present study are not limited to classification, but can also be applied for regression. All MTL formulations were used as implemented in the Matlab library, Malsar [44], or based on custom Matlab implementations.

$$\min_{W,\,C} \sum_{i=1}^{t} \mathcal{L}(W,C) + \lambda \sum_{i=1}^{t} \left\| W_i - \frac{1}{t} \sum_{j=1}^{t} W_j \right\|_2^2 \tag{2}$$

We selected the mean-regularized multi-task learning method [45] as an algorithm for the MTL_NET framework. This algorithm assumes that a latent model exists underlying all tasks, which can be estimated as the mean model across tasks. Based on this assumption, the formulation attempts to identify the most discriminative pattern in the high-dimensional feature space, while limiting the dissimilarity between pairwise models. Dissimilarity is quantified with respect to the effect size of a given predictor and the sign of its association with diagnosis. We expected this combined dissimilarity measure to lead to biologically plausible predictive patterns that are characterized by consistent differences across tasks, both in terms of magnitude as well as directionality. Here, λ had a range of $10^{(-6:1:2)}$.

$$\min_{W,\,C} \sum_{i=1}^{t} \mathcal{L}(W,C) + \lambda \left(\alpha \sum_{i=1}^{t} \left\| W_i - \frac{1}{t} \sum_{j=1}^{t} W_j \right\|_2^2 + (1-\alpha)\|W\|_1 \right) \tag{3}$$

MTL_SNET was the sparse version of MTL_NET, and the sparsity was introduced by the l_1 norm (i.e., coefficients of predictors with low utility are set to 0). Here, λ controls the entire penalty and α distributes the penalty to full-sparse and non-sparse terms. λ had a range of $10^{(-6:1:2)}$ and α was chosen from the range [0:0.1:1].

$$\min_{W,C} \sum_{i=1}^{t} \mathcal{L}(W,C) + \lambda\|W\|_{2,1} \tag{4}$$

The formulation of MTL_L21 introduced the group sparse term, $\|W\|_{2,1} = \sum_{i=1}^{p} \|W_i\|_2$, which aimed to select or reject the same group of genes across datasets. λ controlled the level of sparsity with a range of $10^{(-6:0.1:0)}$.

$$\min_{W,\,C} \sum_{i=1}^{t} \mathcal{L}(W,C) + \lambda((1-\alpha)||W||_{2,1} + \alpha||W||_2^2) \tag{5}$$

The MTL_EN was formulated by adding the composite penalties, where $||W||_2^2$ is the squared Frobenius norm. Similar to elastic net in conventional STL, such regularization helped to stabilize the solution when multiple highly correlated genes existed in the high-dimensional space [46]. Here, λ had a range of $10^{(-6:0.1:0)}$ and α was chosen from the range [0:0.1:1].

$$\min_{W,\,C} \sum_{i=1}^{t} \mathcal{L}(W,C) + \lambda||W||_* \tag{6}$$

MTL_Trace encouraged a low-rank model, W, by penalizing the sum of its eigenvalues, $||W||_*$. λ had a range of $10^{(-6:0.1:1)}$. By compressing the subspace spanned by weight vectors, models were structured (i.e., clustered structure). Thus, the models that were clustered together demonstrated high pairwise correlation.

4.3.2. Conventional, Single-Task Machine Learning

LR_L1: We trained logistic regression with lasso using the package, "Glmnet". The lambda parameter was chosen among the set, $10^{(-10:0.5:1)}$.

SVM: Linear support vector machine was trained using the built-in Matlab function, 'fitcsvm', with the box constraints in the range of $10^{(-5:1:5)}$. We only used the linear kernel to facilitate determination of predictor importance.

RF: We used the Matlab built-in function, 'TreeBagger', to train a random forest model with 5000 trees. The predictor importance was calculated according to the average error decrement for all splits on a given predictor.

4.3.3. Assessment of Predictive Performance

To quantify predictive performance and capture stability of decision rules against cross-dataset and sampling variability, we used a leave-dataset-out procedure. Specifically, the set of five expression datasets was denoted as $D = \{d_1, d_2, \ldots, d_5\}$ and we calculated the power set, $\mathbb{P}(D)$, of D. Then for each subset, $d \in \mathbb{P}(D)$, we trained a given algorithm on d and tested the model on $D - d$. For example, for $d = \{d_1, d_2\}$, we trained using the combination of datasets, $\{d_1, d_2\}$, and then tested on $\{d_3, d_4, d_5\}$. For convenience, we organized these training procedures according to the size of d, noted as $n_d \in \{2, 3, \ldots, 5\}$. We thus obtained a series of models trained using all subsets of the five datasets (except for single dataset) and they are referred to using n_d.

The comparison of the predictive performance between methods was mainly based on $n_d = 4$, i.e., when all, but one, datasets were used for training. To understand how dataset-specific confounders affect the prediction, models were trained on a range of n_d from 2 to 4. Finally, to explore the convergence of genes' coefficients across different training datasets, we compared the models trained when $n_d = i$, $i \in \{2, 3 \ldots 5\}$.

During cross-validation (CV), as illustrated in Figure A1, subjects were randomly allocated to 5 folds, stratified for diagnosis and the dataset indicator. Subsequently, different strategies were specified for MTL and STL. For MTL, the $training_{cv}$ datasets were trained in parallel, and the models were tested on each $test_{cv}$ dataset by averaging the prediction scores. To determine the final accuracy of the current fold, the accuracies retrieved from all $test_{cv}$ datasets were averaged. For STL, the $training_{cv}$ datasets were combined to train a single algorithm that was then predicted on the combined $test_{cv}$ datasets. Similar to CV, in the training procedure, MTL trained on datasets in parallel, while combining the prediction scores for testing.

4.3.4. Consistency and Stability Analysis

To compare the consistency and stability of markers between algorithms, we used the correlation coefficient as the similarity measure of pairwise transcriptomic profiles (i.e., the coefficient vector for all genes) learnt by algorithms. A high similarity between profiles implied that models shared important predictors with respect to their weights and signs. Using this similarity measure, 'consistency' and 'stability' were defined, respectively. These measures were derived from 100-fold stratified bootstrapping of subjects from a set of datasets. In each bootstrapping sample, we tested across the number of training sets ($n_d = i$, $i \in \{2, 3, \ldots, 5\}$). For MTL, since the training procedure would output multiple coefficient vectors (i.e., training on three datasets would output three coefficient vectors), to compare the similarity between algorithms, the coefficient vectors were averaged.

Consistency: With 'consistency', we quantified the pairwise similarity of models trained using overlapping or non-overlapping (i.e., 2 training datasets) datasets. For this, we differentiated two types of consistencies: 'Horizontal' and 'vertical' consistency as illustrated in Figure A2a,b, respectively. Horizontal consistency quantified model robustness against cross-dataset variability. For this, we fixed the number of training datasets, (n_d), and determined the pairwise similarity between models. This was performed for all possible choices of n_d (see supplementary methods for details). Vertical consistency measured the sensitivity of models to the number of training datasets. For this, we varied n_d and quantified similarity between the model determined on all training datasets, ($n_d = 5$), and all models derived from lower training datasets numbers, ($n_d = i$, $i \in \{2, 3, 4\}$) (see supplementary methods for details). Low vertical consistency would, for example, be observed when models trained on two training datasets led to vastly different transcriptomic profiles compared to that using all five datasets for training.

Stability: To quantify the stability of an algorithm against the sampling variability, we observed the variation of transcriptomic profiles learnt from different bootstrapping samples as illustrated in Figure A3. Then the variation of all models given n_d was summarized as the stability (see supplementary methods for details).

Success rate: In addition to consistency and stability, to perform a side-by-side comparison of algorithms, we defined the success rate as the proportion of cases where one algorithm outperformed the other. For example, we quantified the success rate of consistency as the proportion of bootstrapping samples where the first algorithm demonstrated higher consistency than the second (see supplementary methods for details). The success rate of stability was quantified as the proportion of models, which were more stable for the first algorithm than that for the second (see supplementary methods for details).

5. Limitations and Future Work

This work evaluates the performance of MTL and STL for biomarker analysis across five transcriptomic schizophrenia expression datasets. Several quality control procedures were employed to remove unwanted variation in the investigated datasets and to improve the biological generalizability of the obtained results. Despite this, the presented results should be interpreted in the light of the specific datasets investigated. Since other data modalities, including neuroimaging or gene methylation, show similar cross-dataset heterogeneity and correlation structures across variables, the present results may not be limited to expression data, although this remains to be empirically demonstrated. Furthermore, future investigations should include systematic simulation studies to explore the performance of MTL and its robustness against factors typically affecting machine learning performance, including data dimensionality, predictor effect sizes, and biological as well as experimental variability across datasets.

6. Conclusions

The present study demonstrates the utility of MTL for integrative machine learning in high-dimensional datasets, compared to classical single-task learning. Mega-analyses that require integration of data across numerous datasets are becoming more frequent, but thus far, have rarely used machine learning approaches. The present study shows that MTL bears substantial promise for such applications. This particularly applies for scenarios where inter-dataset heterogeneity far outweighs the illness associated signal, a typical case for high-dimensional datasets in psychiatric research.

Author Contributions: Conceptualization, H.C. and E.S.; Formal analysis, H.C.; Funding acquisition, E.S.; Methodology, H.C. and E.S.; Supervision, A.M.-L. and E.S.; Writing–original draft, H.C., A.M.-L. and E.S.; Writing–review & editing, H.C., A.M.-L. and E.S.

Funding: This research was funded by the Deutsche Forschungsgemeinschaft (DFG), SCHW 1768/1-1.

Conflicts of Interest: A.M.-L. has received consultant fees from Blueprint Partnership, Boehringer Ingelheim, Daimler und Benz Stiftung, Elsevier, F. Hoffmann-La Roche, ICARE Schizophrenia, K.G. Jebsen Foundation, L.E.K. Consulting, Lundbeck International Foundation (LINF), R. Adamczak, Roche Pharma, Science Foundation, Synapsis Foundation—Alzheimer Research Switzerland, System Analytics, and has received lectures including travel fees from Boehringer Ingelheim, Fama Public Relations, Institut d'investigacions Biomèdiques August Pi i Sunyer (IDIBAPS), Janssen-Cilag, Klinikum Christophsbad, Göppingen, Lilly Deutschland, Luzerner Psychiatrie, LVR Klinikum Düsseldorf, LWL PsychiatrieVerbund Westfalen-Lippe, Otsuka Pharmaceuticals, Reunions i Ciencia S.L., Spanish Society of Psychiatry, Südwestrundfunk Fernsehen, Stern TV, and Vitos Klinikum Kurhessen. All other authors declare no potential conflicts of interest. The funders had no role in the design of the study; in the collection, analyses, or interpretation of data; in the writing of the manuscript, and in the decision to publish the results.

Abbreviations

MTL	Multi-task learning
STL	Single-task learning
RF	Random Forests
SVM	Support Vector Machine

Appendix A

Supplementary Methods
Consistency, stability and success rate
Notations:

- The model pairs trained using different (overlapping or non-overlapping) combinations of datasets were represented as M and $\tilde{M}$, respectively (i.e., M represented the model trained using the training set, $d = \{1, 2\}$; $\tilde{M}$ was trained using a different dataset combination, for example, $d = \{3, 4\}$ or $d = \{1, 2, \ldots, 5\}$)
- The notation of an algorithm: α, β (i.e., α = MTL_NET, β = SVM)
- The index of the bootstrapping sample: $b \in \{1, 2, \ldots 100\}$ and $\tilde{b} \in \{1, 2, \ldots 100\}$. For computational efficiency, bootstrapping was performed across all datasets, $d = \{1, 2, \ldots, 5\}$, and data subsets were selected from this sampling.

As an example, a model, M_b^α, could be trained based on bootstrap sample, $b = 3$, from which training sets, $d = \{1, 2\}$, were extracted, using the algorithm, α = SVM. The model trained on the same bootstrap sample based on a different combination of training sets and using the algorithm, α = SVM, would be denoted as $\tilde{M}_b^\alpha$.

Consistency

Given $n_d = i$, $i \in \{2, 3, 4\}$ and the algorithm, α, we calculated the expected similarity for each bootstrapping sample, b as:

$$C_b^{\alpha,\,n_d} = \mathbb{E}_{M,\,\tilde{M},\,M \neq \tilde{M}}\, Cor(M_b^\alpha, \tilde{M}_b^\alpha)$$

Then, the expected similarity list, $C^{\alpha,\,n_d} = \left[C_1^{\alpha,\,n_d}, C_2^{\alpha,\,n_d}, \ldots, C_{100}^{\alpha,\,n_d} \right]$, over b was the consistency list of algorithm, α, for a given n_d. Here, the expectation was calculated empirically by enumerating all pairs of models, M and $\tilde{M}$. By assigning different values to M and $\tilde{M}$, horizontal and vertical consistency were differentiated. For horizontal consistency, M and $\tilde{M}$ represented the pairwise models trained using the same number (n_d) of datasets. For vertical consistency, $\tilde{M}$ was trained using $n_d = 5$ datasets and M was trained using fewer datasets.

Stability

Given $n_d = i$, $i \in \{2, 3, 4\}$, and algorithm, α, we quantified the expected similarity between pairwise models (M_b^α and $M_{\tilde{b}}^\alpha$), which were trained using the same datasets (M), but different bootstrapping samples (b and $\tilde{b}$), as:

$$S_M^{\alpha,\,n_d} = \mathbb{E}_{b,\,\tilde{b},\,b \neq \tilde{b}} Cor(M_b^\alpha,\, M_{\tilde{b}}^\alpha)$$

Over all models, (M), $S^{\alpha,\,n_d} = [S_1^{\alpha,\,n_d},\, S_2^{\alpha,\,n_d}, \ldots, S_{\binom{5}{n_d}}^{\alpha,\,n_d}]$ was quantified as the stability list of algorithm, α, given n_d. The expectation was estimated empirically by enumerating all pairs of bootstrapping samples, b and $\tilde{b}$.

Success rate

The success rate compared algorithms, α and β, side-by-side, and was measured as the proportion of cases where algorithm, α, outperformed β.

For example, given the consistency list of algorithm, α and β, ($C^{\alpha,\,n_d}$ and $C^{\beta,\,n_d}$), we determined the proportion of bootstrapping samples where algorithm, α, demonstrated higher consistency than β, yielding the success rate of consistency:

$$SR_C^{n_d} = \mathbb{E}_b 1_{C_b^{\alpha,\,n_d} - C_b^{\beta,\,n_d} > 0}$$

Given the stability list of algorithm, α and β, ($S^{\alpha,\,n_d}$ and $S^{\beta,\,n_d}$), we determined the proportion of models, which demonstrated higher stability for algorithm, α, yielding the success rate of stability:

$$SR_S^{n_d} = \mathbb{E}_M 1_{S_M^{\alpha,\,n_d} - S_M^{\beta,\,n_d} > 0}$$

Appendix B

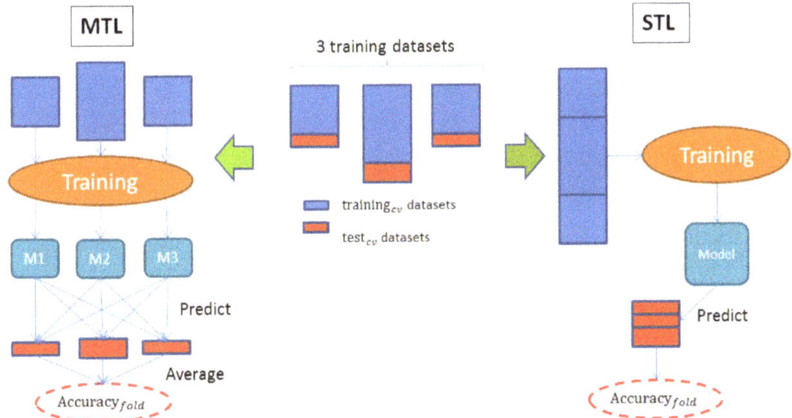

Figure A1. Procedure of five-fold-stratified-cross-validation for Single Task Learning (STL) and Multitask Learning (MTL) (showing one-fold as an example). Using $n_d = 3$ as an example, the specific procedure of the cross-validation procedure is shown. First, the subjects were randomly allocated to five folds, stratified for diagnosis per dataset. Subsequently, different strategies were specified for MTL and STL. For MTL, the training datasets were trained in parallel, and the three models (M1, M2, and M3) were tested on each test dataset by averaging the prediction score. The average across all accuracies was used as the final accuracy for the current fold. In contrast, for STL, the training datasets were combined to train a single algorithm that was then predicted on the combined test datasets.

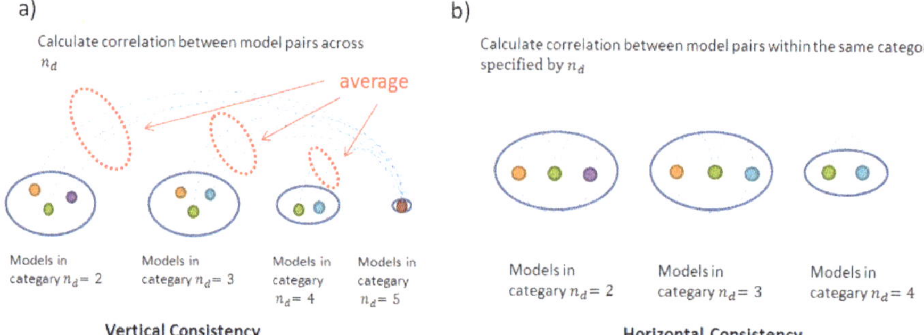

Figure A2. Illustration of model consistency calculation. Consistency quantified the robustness of an algorithm against the cross-dataset variability. To test this, we trained models using each subset of all five expression datasets and then categorized these models according to the number of training sets (n_d). Different models were rendered as colored circles, categorized by n_d. For vertical consistency, (**a**) the similarity was determined between the models learned on $n_d = 2$ to $n_d = 4$ and the model trained on $n_d = 5$. The resulting values were then averaged for a given category, n_d. For horizontal consistency, (**b**) the model similarity was calculated in each category, n_d, and then averaged.

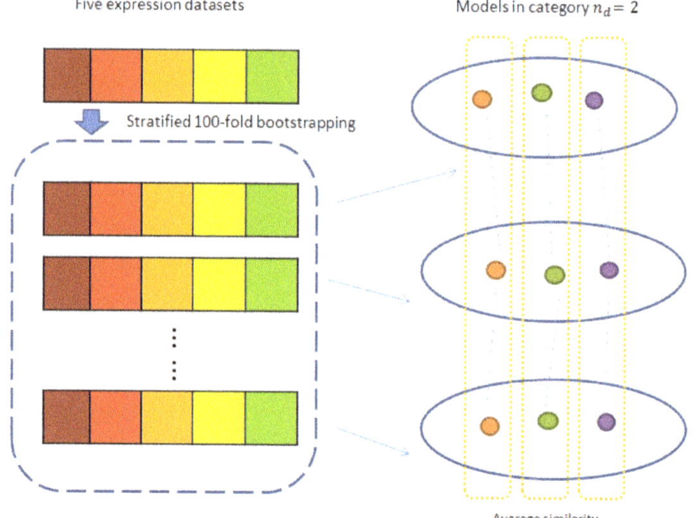

Figure A3. Illustration of model stability calculation. Stability quantified the robustness of an algorithm against sampling variability. This metric was computed by performing 100-fold-stratified-bootstrapping. In the left panel, five expression datasets are shown as colored boxes. Using $n_d = 2$ as an example, two out of five datasets were combined for training in each bootstrapping sample. Thus, a series of models were obtained as illustrated as the colored circles in the right panel. The stability was determined as the average pairwise similarity for each model, calculated across all pairs of bootstrapping samples.

References

1. Sullivan, P.F. The psychiatric GWAS consortium: Big science comes to psychiatry. *Neuron* **2010**, *68*, 182–186. [CrossRef] [PubMed]

2. Passos, I.C.; Mwangi, B.; Kapczinski, F. Big data analytics and machine learning: 2015 and beyond. *Lancet Psychiatry* **2016**, *3*, 13–15. [CrossRef]

3. Schizophrenia Working Group of the Psychiatric Genomics Consortium. Biological insights from 108 schizophrenia-associated genetic loci. *Nature* **2014**, *511*, 421–427. [CrossRef] [PubMed]

4. Major Depressive Disorder Working Group of the Psychiatric GWAS Consortium; Ripke, S.; Wray, N.R.; Lewis, C.M.; Hamilton, S.P.; Weissman, M.M.; Breen, G.; Byrne, E.M.; Blackwood, D.H.; Boomsma, D.I.; et al. A mega-analysis of genome-wide association studies for major depressive disorder. *Mol. Psychiatry* **2013**, *18*, 497–511. [PubMed]

5. Wolfers, T.; Buitelaar, J.K.; Beckmann, C.F.; Franke, B.; Marquand, A.F. From estimating activation locality to predicting disorder: A review of pattern recognition for neuroimaging-based psychiatric diagnostics. *Neurosci. Biobehav. Rev.* **2015**, *57*, 328–349. [CrossRef] [PubMed]

6. Franke, B.; Stein, J.L.; Ripke, S.; Anttila, V.; Hibar, D.P.; van Hulzen, K.J.E.; Arias-Vasquez, A.; Smoller, J.W.; Nichols, T.E.; Neale, M.C.; et al. Genetic influences on schizophrenia and subcortical brain volumes: Large-scale proof of concept. *Nat. Neurosci.* **2016**, *19*, 420–431. [CrossRef] [PubMed]

7. de Wit, S.J.; Alonso, P.; Schweren, L.; Mataix-Cols, D.; Lochner, C.; Menchón, J.M.; Stein, D.J.; Fouche, J.P.; Soriano-Mas, C.; Sato, J.R.; et al. Multicenter voxel-based morphometry mega-analysis of structural brain scans in obsessive-compulsive disorder. *Am. J. Psychiatry* **2014**, *171*, 340–349. [CrossRef] [PubMed]

8. Jordan, M.I.; Mitchell, T.M. Machine learning: Trends, perspectives, and prospects. *Science* **2015**, *349*, 255–260. [CrossRef] [PubMed]

9. Iniesta, R.; Stahl, D.; McGuffin, P. Machine learning, statistical learning and the future of biological research in psychiatry. *Psychol. Med.* **2016**, *46*, 2455–2465. [CrossRef] [PubMed]

10. Vilhjalmsson, B.J.; Yang, J.; Finucane, H.K.; Gusev, A.; Lindström, S.; Ripke, S.; Genovese, G.; Loh, P.R.; Bhatia, G.; Do, R.; et al. Modeling Linkage Disequilibrium Increases Accuracy of Polygenic Risk Scores. *Am. J. Hum. Genet.* **2015**, *97*, 576–592. [CrossRef] [PubMed]

11. Vos, T.; Flaxman, A.D.; Naghavi, M.; Lozano, R.; Michaud, C.; Ezzati, M.; Shibuya, K.; Salomon, J.A.; Abdalla, S.; et al. Years lived with disability (YLDs) for 1160 sequelae of 289 diseases and injuries 1990-2010: A systematic analysis for the Global Burden of Disease Study 2010. *Lancet* **2012**, *380*, 2163–2196. [CrossRef]

12. Whelan, R.; Watts, R.; Orr, C.A.; Althoff, R.R.; Artiges, E.; Banaschewski, T.; Barker, G.J.; Bokde, A.L.; Büchel, C.; Carvalho, F.M.; et al. Neuropsychosocial profiles of current and future adolescent alcohol misusers. *Nature* **2014**, *512*, 185–189. [CrossRef] [PubMed]

13. Xia, C.H.; Ma, Z.; Ciric, R.; Gu, S.; Betzel, R.F.; Kaczkurkin, A.N.; Calkins, M.E.; Cook, P.A.; García de la Garza, A.; Vandekar, S.N.; et al. Linked dimensions of psychopathology and connectivity in functional brain networks. *Nat. Commun.* **2018**, *9*, 3003. [CrossRef] [PubMed]

14. Caruana, R. Multitask Learning. In *Learning to Learn*; Springer: Boston, MA, USA, 1998; pp. 95–133.

15. Widmer, C.; Rätsch, G. Multitask Learning in Computational Biology. In Proceedings of the ICML Workshop on Unsupervised and Transfer Learning, PMLR, Bellevue, WA, USA, 2 July 2012; Volume 27, pp. 207–216.

16. Li, Y.; Wang, J.; Ye, J.P.; Reddy, C.K. A Multi-Task Learning Formulation for Survival Analysis. In Proceedings of the 22nd ACM SIGKDD International Conference on Knowledge Discovery and Data Mining, San Francisco, CA, USA, 13–17 August 2016.

17. Yuan, H.; Paskov, I.; Paskov, H.; González, J.A.; Leslie, S.C. Multitask learning improves prediction of cancer drug sensitivity. *Sci. Rep.* **2016**, *6*, 31619. [CrossRef] [PubMed]

18. Feriante, J. Massively Multitask Deep Learning for Drug Discovery. Master's Thesis, University of Wisconsin-Madison, Madison, WI, USA, 2015.

19. Xu, Q.; Pan, S.J.; Xue, H.H.; Yang, Q. Multitask Learning for Protein Subcellular Location Prediction. *IEEE/ACM Trans. Comput. Biol. Bioinform.* **2011**, *8*, 748–759. [PubMed]

20. Zhou, J.; Liu, J.; Narayan, V.A.; Ye, J.; Alzheimer's Disease Neuroimaging Initiative. Modeling disease progression via multi-task learning. *Neuroimage* **2013**, *78*, 233–248. [PubMed]

21. Collobert, R.; Weston, J. A unified architecture for natural language processing: Deep neural networks with multitask learning. In Proceedings of the 25th International Conference on Machine Learning, New York, NY, USA, 5–9 July 2008.

22. Wu, Z.; Valentini-Botinhao, C.; Watts, O.; King, S. Deep neural networks employing Multi-Task Learning and stacked bottleneck features for speech synthesis. In Proceedings of the 2015 IEEE International Conference on Acoustics, Speech and Signal Processing, Brisbane, Australia, 19–24 April 2015.

23. Wang, X.; Zhang, C.; Zhang, Z. Boosted multi-task learning for face verification with applications to web image and video search. In Proceedings of the 2009 IEEE International Conference on on Computer Vision and Pattern Recognition, Miami, FL, USA, 20–25 June 2009.

24. Zhang, Z.; Luo, P.; Loy, C.C.; Tang, X. Facial Landmark Detection by Deep Multi-task Learning. In Proceedings of the European Conference on Computer Vision, Zurich, Switzerland, 6–12 September 2014.

25. Chapelle, O.; Shivaswamy, P.; Vadrevu, P.; Weinberger, K.; Zhang, Y. Multi-task learning for boosting with application to web search ranking. In Proceedings of the 16th ACM SIGKDD International Conference on Knowledge Discovery and Data Mining, Washington, DC, USA, 25–28 July 2010.

26. Ahmed, A.; Aly, M.; Das, A.; Smola, J.A.; Anastasakos, T. Web-scale multi-task feature selection for behavioral targeting. In Proceedings of the 21st ACM International Conference on Information and Knowledge Management, Maui, HI, USA, 29 October–2 November 2012.

27. Marquand, A.F.; Brammer, M.; Williams, S.C.; Doyle, O.M. Bayesian multi-task learning for decoding multi-subject neuroimaging data. *Neuroimage* **2014**, *92*, 298–311. [CrossRef] [PubMed]

28. Jing, W.; Zhang, Z.L.; Yan, J.W.; Li, T.Y.; Rao, D.B.; Fang, S.F.; Kim, S.; Risacher, L.S.; Saykin, J.A.; Shen, L. Sparse Bayesian multi-task learning for predicting cognitive outcomes from neuroimaging measures in Alzheimer's disease. In Proceedings of the 2012 IEEE Conference on Computer Vision and Pattern Recognition, Providence, RI, USA, 16–21 June 2012.

29. Wang, H.; Nie, F.; Huang, H.; Kim, S.; Nho, K.; Risacher, S.L.; Saykin, A.J.; Shen, L.; Alzheimer's Disease Neuroimaging Initiative. Identifying quantitative trait loci via group-sparse multitask regression and feature selection: An imaging genetics study of the ADNI cohort. *Bioinformatics* **2012**, *28*, 229–237. [CrossRef] [PubMed]

30. Lin, D.; Zhang, J.; Li, J.; He, H.; Deng, H.W.; Wang, Y.P. Integrative analysis of multiple diverse omics datasets by sparse group multitask regression. *Front. Cell Dev. Biol.* **2014**, *2*, 62. [CrossRef] [PubMed]

31. Xu, Q.; Xue, H.; Yang, Q. Multi-platform gene-expression mining and marker gene analysis. *Int. J. Data Min. Bioinform.* **2011**, *5*, 485–503. [CrossRef] [PubMed]

32. O'Brien, C.M. Statistical Learning with Sparsity: The Lasso and Generalizations. *Int. Stat. Rev.* **2016**, *84*, 156–157. [CrossRef]

33. Gandal, M.J.; Haney, J.R.; Parikshak, N.N.; Leppa, V.; Ramaswami, G.; Hartl, C.; Schork, A.J.; Appadurai, V.; Buil, A.; Werge, T.M.; et al. Shared molecular neuropathology across major psychiatric disorders parallels polygenic overlap. *Science* **2018**, *359*, 693–697. [CrossRef] [PubMed]

34. Bulik-Sullivan, B.; Finucane, H.K.; Anttila, V.; Gusev, A.; Day, F.R.; Loh, P.R.; ReproGen Consortium; Psychiatric Genomics Consortium; Genetic Consortium for Anorexia Nervosa of the Wellcome Trust Case Control Consortium; Duncan, L.; et al. An atlas of genetic correlations across human diseases and traits. *Nat. Genet.* **2015**, *47*, 1236–1241. [CrossRef] [PubMed]

35. Cross-Disorder Group of the Psychiatric Genomics Consortium; Lee, S.H.; Ripke, S.; Neale, B.M.; Faraone, S.V.; Purcell, S.M.; Perlis, R.H.; Mowry, B.J.; Thapar, A.; Goddard, M.E.; et al. Genetic relationship between five psychiatric disorders estimated from genome-wide SNPs. *Nat. Genet.* **2013**, *45*, 984–994. [CrossRef] [PubMed]

36. International Schizophrenia Consortium; Purcell, S.M.; Wray, N.R.; Stone, J.L.; Visscher, P.M.; O'Donovan, M.C.; Sullivan, P.F.; Sklar, P. Common polygenic variation contributes to risk of schizophrenia and bipolar disorder. *Nature* **2009**, *460*, 748–752. [CrossRef] [PubMed]

37. Harris, L.W.; Wayland, M.; Lan, M.; Ryan, M.; Giger, T.; Lockstone, H.; Wuethrich, I.; Mimmack, M.; Wang, L.; Kotter, M.; et al. The cerebral microvasculature in schizophrenia: A laser capture microdissection study. *PLoS ONE* **2008**, *3*, e3964. [CrossRef] [PubMed]

38. Chen, C.; Cheng, L.; Grennan, K.; Pibiri, F.; Zhang, C.; Badner, J.A.; Members of the Bipolar Disorder Genome Study (BiGS) Consortium; Gershon, E.S.; Liu, C. Two gene co-expression modules differentiate psychotics and controls. *Mol. Psychiatry* **2013**, *18*, 1308–1314. [CrossRef] [PubMed]

39. Maycox, P.R.; Kelly, F.; Taylor, A.; Bates, S.; Reid, J.; Logendra, R.; Barnes, M.R.; Larminie, C.; Jones, N.; Lennon, M.; et al. Analysis of gene expression in two large schizophrenia cohorts identifies multiple changes associated with nerve terminal function. *Mol. Psychiatry* **2009**, *14*, 1083–1094. [CrossRef] [PubMed]

40. Barnes, M.R.; Huxley-Jones, J.; Maycox, P.R.; Lennon, M.; Thornber, A.; Kelly, F.; Bates, S.; Taylor, A.; Reid, J.; Jones, N.; et al. Transcription and pathway analysis of the superior temporal cortex and anterior prefrontal cortex in schizophrenia. *J. Neurosci. Res.* **2011**, *89*, 1218–1227. [CrossRef] [PubMed]

41. Narayan, S.; Tang, B.; Head, S.R.; Gilmartin, T.J.; Sutcliffe, J.G.; Dean, B.; Thomas, E.A. Molecular profiles of schizophrenia in the CNS at different stages of illness. *Brain Res.* **2008**, *1239*, 235–248. [CrossRef] [PubMed]

42. Irizarry, R.A.; Hobbs, B.; Collin, F.; Beazer-Barclay, Y.D.; Antonellis, K.J.; Scherf, U.; Speed, T.P. Exploration, normalization, and summaries of high density oligonucleotide array probe level data. *Biostatistics* **2003**, *4*, 249–264. [CrossRef] [PubMed]

43. Leek, J.T.; Johnson, W.E.; Parker, H.S.; Jaffe, A.E.; Storey, J.D. The sva package for removing batch effects and other unwanted variation in high-throughput experiments. *Bioinformatics* **2012**, *28*, 882–883. [CrossRef] [PubMed]

44. Zhou, J.; Chen, J.; Ye, J. *MALSAR: Multi-tAsk Learning via StructurAl Regularization*; Arizona State University: Tempe, AZ, USA, 2012.

45. Evgeniou, T.; Pontil, M. Regularized multi-task learning. In Proceedings of the Tenth ACM SIGKDD International Conference on Knowledge Discovery and Data Mining, Seattle, WA, USA, 22–25 August 2004.

46. Tibshirani, R.J. The lasso problem and uniqueness. *Electron. J. Statist.* **2013**, *7*, 1456–1490. [CrossRef]

International Journal of
Molecular Sciences

Review

Childhood-Onset Schizophrenia: Insights from Induced Pluripotent Stem Cells

Anke Hoffmann, Michael Ziller and Dietmar Spengler *

Department of Translational Research in Psychiatry, Max Planck Institute of Psychiatry, 80804 Munich, Germany;
hoffmann@psych.mpg.de (A.H.); michael_ziller@psych.mpg.de (M.Z.)
* Correspondence: spengler@psych.mpg.de; Tel.: +49-089-3062-2546

Received: 29 October 2018; Accepted: 27 November 2018; Published: 30 November 2018

Abstract: Childhood-onset schizophrenia (COS) is a rare psychiatric disorder characterized by earlier onset, more severe course, and poorer outcome relative to adult-onset schizophrenia (AOS). Even though, clinical, neuroimaging, and genetic studies support that COS is continuous to AOS. Early neurodevelopmental deviations in COS are thought to be significantly mediated through poorly understood genetic risk factors that may also predispose to long-term outcome. In this review, we discuss findings from induced pluripotent stem cells (iPSCs) that allow the generation of disease-relevant cell types from early brain development. Because iPSCs capture each donor's genotype, case/control studies can uncover molecular and cellular underpinnings of COS. Indeed, recent studies identified alterations in neural progenitor and neuronal cell function, comprising dendrites, synapses, electrical activity, glutamate signaling, and miRNA expression. Interestingly, transcriptional signatures of iPSC-derived cells from patients with COS showed concordance with postmortem brain samples from SCZ, indicating that changes in vitro may recapitulate changes from the diseased brain. Considering this progress, we discuss also current caveats from the field of iPSC-based disease modeling and how to proceed from basic studies to improved diagnosis and treatment of COS.

Keywords: childhood-onset schizophrenia (COS); induced pluripotent stem cell (iPSC); copy number variation (CNV); early neurodevelopment; neuronal differentiation; synapse; dendritic arborization; miRNAs

1. Introduction

Schizophrenia (SCZ) is a highly heritable, devastating mental disorder with a lifetime prevalence of ≈1% worldwide [1]. First episode psychosis typically manifest in early adulthood followed by recurrent episodes that frequently give way to a chronic course that confers substantial mortality and morbidity. As of yet, no cure is available and life expectancy of patients with SCZ is reduced by 15 to 30 years [2,3]. Around 4% of the patients experience early-onset schizophrenia (EOS) either during childhood prior to the 13th birthday (i.e., COS) or during adolescence up to the age of 17 years and carry a particular worse diagnosis [4].

Numerous hypothesis, observational, and experimental, have been put forward to explain the etiology and pathogenesis of SCZ with no consensus established so far [5]. Among these, the much-noticed neurodevelopmental hypothesis of SCZ posits that deviations in early brain development predispose to later vulnerability when critical processes of normal maturation call into operation damaged structures. Similar to other fields of early-onset disease, the study of patients with COS showed that early SCZ is characterized by increased symptom severity and a higher genetic load. This indicates a greater genetic salience for neurodevelopmental deviations and suggests that studying COS can also advance insight into disease-traits that develop more subtly in an adult-onset patient group [6]. Early-onset of disease reduces also the contribution of confounding environmental factors across life course and enables a less-clouded sight on the actual biology underpinning SCZ.

Due to the impossibility to isolate brain tissue from living patients, and the limitations of postmortem studies (scarcity of tissue availability, confounding effects from treatments, aging, and life history) patient-specific iPSCs offer a unique opportunity to study living human neuronal cells. iPSCs capture a donor's genotype including disease related genetic risk factors, known and unknown, and can be differentiated in virtually any cell type including early neural cells of potential relevance to COS. Differences in early cellular and molecular endophenotypes from case/control studies can inform on perturbations in neurodevelopmental pathways and on potential deviations in patients with COS. Patient-specific iPSC studies on COS are of particular interest given that early perturbations in vitro may couple more directly to early than later pathology and thus offer a better handle on cause–effect relationships.

Here, we will examine this hypothesis by considering the clinical picture and course of COS, and recent insights into the genetics of AOS and COS. Against this background, we discuss how iPSC-derived neuronal cells from early developmental stages differ between carriers of high-risk structural variations variants for COS or patients with COS vs. healthy donors. We further ask whether these molecular and cellular alterations do bridge to early brain development. Concluding, we address present caveats in patient-specific disease modeling and upcoming improvements from the field.

The literature selection process for this review was conducted in the databank PubMed via combinations of the search terms "schizophreni*", "childhood", "early-onset", "induced pluripotent stem cell*", "genetic*", and "psychosis" with date limits from 2007 (first report on iPSCs [7]) to September 2018. Additional searches included scrutiny of similar articles suggested by PubMed, of references from the identified publications, and of citatory publications identified by Google Scholar®.

2. The Neurodevelopmental Hypothesis of COS

COS is a rare disorder affecting 1 in 10,000–30,000 children [8]. Prior to the 20th century, bizarre behavior, social withdrawal, catatonia, and/or psychosis in children were regarded as undifferentiated conditions, labelled as "hereditary insanity", "dementia praecox", or "developmental idiocy" [9]. Today's diagnostic criteria are the same as in AOS and concern multiple domains of behavior and cognition with a prominent role of psychotic symptoms. Hallucinations, delusions, and disorganized thinking [10,11] frequently concur with impairments in social communication, as well as in motor, volitional, and emotional abnormalities [12]. Longitudinal studies have corroborated that diagnostic stability is high in EOS at around 80–90% [13,14]. Outcome of patients over 40 years with EOS is consistently worse relative to AOS [14–16] with the worst clinical and psychosocial outcomes in COS [17]. EOS manifests greater neurodevelopmental deviance early in life, yet it is clinically and neurobiological continuous with AOS [12,18,19].

The discovery of first-generation antipsychotics in the 1950s, known as typical antipsychotics [20], has transformed the treatment of SCZ. Although the first atypical antipsychotic, clozapine, was discovered in the 1960s and introduced clinically in the 1970s, most second-generation drugs, known as atypical antipsychotics, have been developed more recently. Both generations of medication are thought to block receptors in the brain's dopamine pathways with atypicals acting on serotonin receptors additionally. Recent data suggest a greater efficacy of clozapine, relative to other antipsychotics, in COS than in AOS [21] and raise the perspective that COS could offer a unique opportunity to learn to what degree neurodevelopmental deviations in SCZ could respond to current pharmacotherapy.

At the macroscopic scale, early structural MRI (magnetic resonance imaging) studies by the National Institute of Mental Health (NIMH) suggested a pattern of reduced cerebral volumes and larger ventricles in COS consistent with findings from AOS [22]. Longitudinal follow-up studies further showed that typically developing children undergo a small decrease in cortical gray matter ($\approx$2%) in the frontal and parietal regions throughout adolescence (Figure 1). By contrast, children with a history of COS experience exaggerated gray matter losses (≈8%) in frontal, parietal, and temporal lobes [22]. These losses originated in the parietal lobes and spread anteriorly over time until they leveled off in early adulthood when SCZ typically manifests [23]. This pattern fits well the hypothesis of an exaggerated synaptic pruning during critical neurodevelopmental time windows in SCZ [23,24] and supports that COS evolves from a vulnerable brain (Figure 1). It is also worth mentioning that these changes were specific for COS and were not shared with other age- and gender-matched patients with psychotic symptoms diagnosed as multidimensional impaired [25]. In addition, children with COS display losses in global gray matter and cortical thickness in childhood that with age approach those detected in AOS.

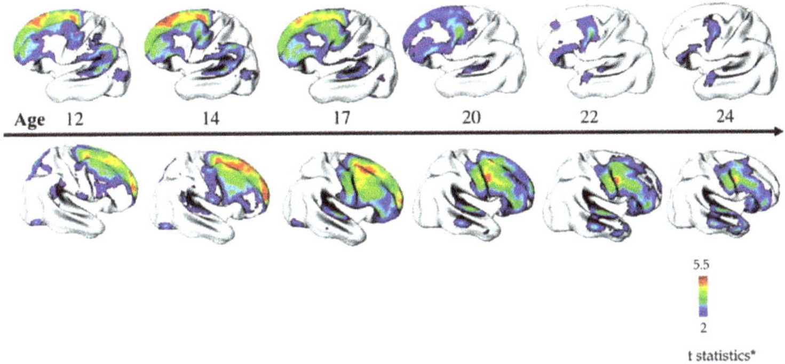

Figure 1. Progression of cortical gray matter loss in patients with childhood-onset schizophrenia (COS) (n = 70) relative to age-, sex-, and scan interval-matched healthy individuals (n = 72). Brain templates illustrate areas of significant thinning in patients with COS in a 'front-to-back' pattern from adolescence to young adulthood (age 12–24 years). Side bar shows t statistic with threshold to control for multiple comparisons. Figure 1 is reproduced from Gogtay [26] by permission of Oxford University Press, adapted by Greenstein et al. [27] by permission of John Wiley and Sons.

Interestingly, non-affected siblings of patients with COS show equally a pattern of decreased thickness in the frontal temporal and parietal lobes during childhood and adolescence that normalizes in early adulthood. This indicates that genetic risk factors underpinning COS interact in a complex manner with the environment leading to overt psychopathology or normalization of risk phenotypes [26,28]. Beyond structural changes, functional MRI studies on patients with COS suggest exaggerated long-range connectivity implicating greater global connectedness and efficiency. Concomitantly, short-range connectivity is impaired in patients with COS implicating disrupted modularity [29,30]. Similar structural deviations have been detected in neonates at high risk for SCZ re-enforcing that COS is contiguous to SCZ [31].

These neuroimaging studies raise the question how macroscopic findings can be explained microscopically. In the absence of neurodegenerative lesions and gliosis, histopathological studies have scrutinized the cytoarchitecture of the cerebral cortex for changes in the size, location, distribution, and packing density of neurons and their synaptic connections. Three putative alterations have caught particular attention: abnormal neuronal organization (dysplasia) in lamina II (pre-alpha cells) and lamina III of the entorhinal cortex [32], disarray of hippocampal neurons [33], and an altered distribution of neurons in the subcortical white matter [34]. These findings seemed to implicate impairments in neuronal migration and cytoarchitecture and were taken as strong evidence for the neurodevelopmental hypothesis of SCZ. Disappointingly, none of these findings has been firmly recapitulated so far. However, a bulk of histopathological studies collaborate the presence of smaller cortical and hippocampal pyramidal neurons, decreased cortical and hippocampal synaptic markers, and decreased dendritic spines as cardinal symptoms in AOS [35]. In light of our limited understanding of SCZ's neuropathology, future studies are needed to resolve the dynamic nature of the disorder.

A one-fits-all model is unlikely to reflect the complex nature of changes in development, adult plasticity, and aging. In fact, the diversity of postmortem cellular pathology may conform to accruing evidence for multi-factorial genetic heritability in SCZ.

3. The Genetic Architecture of AOS and COS

Heritability for AOS is about 60% and 80% in national family [36,37] and twin studies [38,39]. Similarly, twin studies on patients with COS indicate a heritability about 88% [40]. Additionally, family studies on patients with COS show an increased rate of schizophrenic spectrum disorders pointing to familial transmission [41].

Genome-wide association studies (GWAS) have identified many common genetic variants (mostly single nucleotide polymorphism, SNPs) of small effect size that explain between one-third and one-half of the genetic variance [42]. A seminal meta-analysis on 36,989 patients with AOS and 113,075 controls discovered 128 common variant associations encompassing 108 independent loci that met the criterion of genome-wide statistical significance (e.g., 5×10^{-8}) [43]. These loci covered multiple regions enriched in genes regulating glutamatergic, calcium, and G-protein coupled receptor signaling, neuronal ion channels, synaptic function and plasticity, and several neurodevelopmental regulators. A subsequent GWAS study has replicated 93 of these risk loci and identified additionally 52 new loci associated with AOS [44]. Most recently, a GWAS study for shared risk across major psychiatric disorders (including AOS) has highlighted fetal neurodevelopment as a key mediator of vulnerability: four genome-wide significant loci encompassed variants thought to regulate genes expressed in radial glia cells and interneurons in the developing cortex during midgestation [45].

Despite these advances, it is important to realize that risk-associated SNPs typically map to non-coding genomic regions equally represented by intergenic and intronic regions [46]. These SNPs are not necessarily the causal genetic variant underlying the association nor do they identify the causative gene(s). Future studies still have to identify those SNPs that encode a regulatory function and contribute causally to SCZ [47].

Over the last few years, an increasing number of copy number variations (CNVs) has been shown to increase the risk for SCZ. CNVs are typically caused by the presence of region-specific, repetitive DNA sequences, termed low copy repeats (LCRs). Recombination between adjacent and homologous LCRs via non-allelic homologous recombination (NAHR) results in deletions or duplications of the DNA stretches between the repeats. These CNVs tend to recur at the same chromosomal positions flanked by the LCRs, while other mechanism such as non-homologous end joining (NHEJ) can cause non-recurrent CNVs that contain different breakpoints. In any case, most of these de novo mutations, recurrent and non-recurrent, are likely to reduce fecundity and are therefore rarely transmitted [48].

With the advent of microarrays, it became feasible to interrogate the whole genomes of large case/control cohorts for the presence of CNVs that enhance the risk for SCZ. These risk CNVs comprise deletions at 1q21.1, 2p16.3 (contains only *Neurexin 1* with a role in neurotransmission and synaptic contact formation), 3q29, 15q11.2, 15q13.3, and 22q11.2, and duplications at 1q21.1, 7q11.23, 15q11.2-q13.1, 16p13.1, and proximal 16p11.2. A recent combined meta-analysis of 21,094 patients with SCZ and 20,227 controls has shown in a small fraction (1.4%) of the cases genome-wide significant association with CNVs and has confirmed the role of most previously implicated CNVs including 1q21.1, 2p16.3, 3q29, 7q11.2, 15q13.3, distal and proximal 16p11.2 and 22q11.2 (Table 1) [49]. Furthermore, the researchers identified another eight loci that showed suggestive evidence of association with SCZ (Table 1).

Table 1. Significant CNV loci in patients with AOS and COS.

Chr	Locus	Mechanism	CNV	Effect	OR (95% CI)	COS
1	1q21.1	NAHR	Loss + gain	Risk	3.8 (2.1–6.9)	
2	2p16.3 (*NRXN1*)	NHEJ	Loss	Risk	14.4 (4.2–46.9)	+
3	3q29	NAHR	Loss	Risk	Infinite	+
7	7p36.3	NAHR	Loss + gain	Risk	3.5 (1.3–9.0)	
7	7q11.21	NAHR	Loss + gain	Protective	0.66 (0.52–0.84)	
7	7q11.23	NAHR	Gain	Risk	16.1 (3.1–125.7)	
8	8q22.2	NHEJ	Loss	Risk	14.5 (1.7–122.1)	
9	9p24.3	NHEJ	Loss + gain	Risk	12.4 (1.6–98.1)	
13	13q12.11	NAHR	Gain	Protective	0.36 (0.19–0.67)	
15	15q11.2	NAHR	Loss	Risk	1.8 (1.2–2.6)	+
15	15q13.3	NAHR	Loss	Risk	15.6 (3.7–66.5)	+
16	16p11.2. proximal	NAHR	Gain	Risk	9.4 (4.2–20.9)	
16	16p11.2. distal	NAHR	Loss	Risk	20.6 (2.6–162.2)	+
22	22q11.21	NAHR	Loss	Risk	67.7 (9.3–492.8)	+
22	22q11.21	NAHR	Gain	Protective	0.15 (0.04–0.52)	
X	Xq28	NAHR	Gain	Protective	0.35 (0.18–0.68)	
X	Xq28. distal	NAHR	Gain	Risk	8.9 (2.0–39.9)	

Abbreviations are: AOS, adult-onset schizophrenia; Chr, chromosome; CI, confidence interval; COS, childhood-onset schizophrenia; CNV, copy number variation; NAHR, non-allelic homologous recombination; NHEJ, non-homologous end joining; OR, odds ratio; +, present. Adapted by Springer Nature (https://www.nature.com/nature/), Contribution of copy number variants to schizophrenia from a genome-wide study of 41,321 subjects, Christian R. Marshall, 2017 [49].

The aggregate CNV burden was enriched for genes controlling synaptic function (OR = 1.68, P = 2.8 × 10^{-11}) and neurobehavior (in mice). Carrying a CNV risk allele explains only 0.85% of the variance in SCZ liability relative to 3.4% by the 108 genome-wide significant loci [43]. However, risk CNVs show significantly greater effects on SCZ risk (Table 1) than common SNP variants (OR < 1.3). It is worth noting that these risk CNVs associate also with a distinct spectrum of disorders (autism spectrum disorder (ASD), developmental delay, and congenital malformation) indicating that deviations in early neurodevelopment are shared across these disorders [48].

An early study on CNVs on patients with AOS, COS, and ancestry-matched controls found that 15% of patient with AOS had novel structural variants compared with 5% of controls [50]. By contrast, 20% of patients with onset of SCZ before 18 years of age and 28% of patients with COS carried one or more rare structural variants. Structural variations in patients with SCZ were enriched in genes controlling brain development, especially those involving neuregulin and glutamate pathways. In support of this finding, the NIMH COS study showed that 10% of the patients with COS exhibited large chromosomal abnormalities at rates significant higher than those measured in the general population or in patients with AOS [51]. This finding has been collaborated in a follow-up study [6]: a total of 11.9% of patients with COS harbored at least one CNV and 26.7% had two. Among these, 4% showed a 2.5–3 Mb deletion mapping to 22q11.2, a rate higher than that reported for AOS (0.3–1%) or the general population (0.2%), and the highest rate reported for any clinical population to date. Patients with COS also carried additional genomic lesions at 8q11.2, 10q22.3, 16p11.2, and 17q21.3 that had been previously associated with intellectual disability or autism supporting the pleiotropic role of these CNVs in early brain development.

Beyond CNVs, polygenetic risk scores derived from selected common risk variants for SCZ predict effectively COS status: patients with COS had higher genetic risk scores for SCZ (and autism) than their siblings suggesting that patients with COS have more salient genetic risk than do patients with AOS [43].

Taken together, COS is a rare form of SCZ in which both common variants of small effect (SNP) and rare variants (CNV) of large effect conspire together. Common and rare risk variants are more frequent in patients with COS than in patients with AOS. At the same time, patients with COS share rare variants associated with ASD. In essence, patients with COS carry a particular high risk for SCZ that underpins earlier manifestation and a more severe course relative to AOS. Regarding the neurodevelopmental hypothesis of SCZ, patients with COS are therefore expected to manifest more salient neurodevelopmental deviations than patients with AOS. With this in mind, future studies are mandatory to link common and rare risk variants-to genes-to function in order to understand the biology underlying COS and to develop better treatments. To approach this daunting task, iPSC-based studies can provide an important tool to study the regulatory effects of genetic variants, candidate genes, and the overall effect of these variants and their interconnected networks [52], known and unknown, on cellular and molecular endophenotypes in disease-relevant human cells.

4. iPSCs Provide Unique Access to Early Neurodevelopment in AOS and COS

Human brain development starts with the differentiation of neuronal progenitor cells (NPCs) in the third gestational week and subsists through at least late adolescence (Figure 2). Neural tube formation, neural patterning, and NPC differentiation take place in embryonic and early fetal periods and proceed to neuron production, migration, and differentiation in later fetal and early postnatal periods.

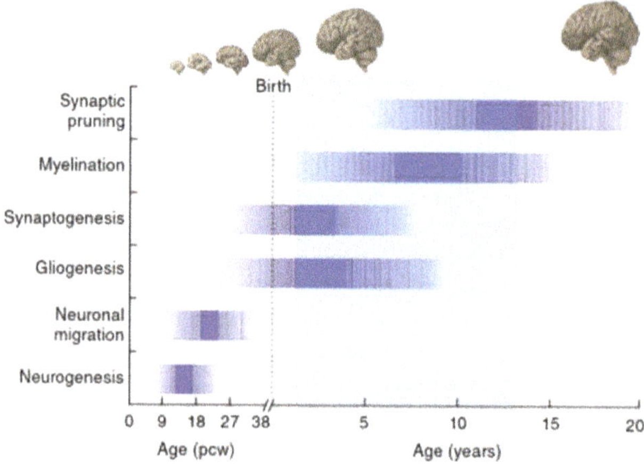

Figure 2. A timeline of human development during prenatal (in postconception weeks, pcw) and postnatal (in years) periods. The shaded horizontal bars represent the approximate timing of key neurobiological processes and developmental milestones. The light-blue overlay marks the period during which childhood-onset schizophrenia (COS) typically manifests. Gross anatomical features and the relative size of the brain at different stages are illustrated at the top. Adapted by Springer Nature (https://www.nature.com/nature/), Developmental timing and critical windows for the treatment of psychiatric disorders, Oscar Marín, 2016 [53].

Regressive and progressive neuronal processes, remodeling of synaptic contacts and circuitries, and myelination evolve postnatally and subsist beyond adolescence [54–56]. Cortical circuits are refined through pruning of excitatory synapses, proliferation of inhibitory circuits, and remodeling of pyramidal dendrites in early adulthood [57,58]. These modulatory processes serve to fine-tune excitatory–inhibitory cortical balance and appear perturbed in patients with SCZ.

Genetic studies on AOS and COS suggest combinatorial contributions of many variants across a host of loci, rather than one or a few penetrant single-gene mutations. These highly polygenic states cannot be engineered into animal models, as they demand replicating large portions, if not the entirety, of the human genome. Hence, human models are urgently needed to decode polygenic contribution to disease initiation and manifestation. Human iPSCs retain the unique genetic signature of the donor and provide insight into the relationship between the donor's genotype and an in vitro endophenotype. By now, human iPSCs are routinely generated from skin biopsies or peripheral blood mononuclear cells [59,60]. iPSCs can be differentiated in disease-relevant neurons and astroglia in order to re-enact altered trajectories of brain development in a diseased individual. In distinction to postmortem brain tissue, human iPSCs are not confounded by secondary disease processes, therapy, or life history. Therefore, iPSC studies are particular promising for the analysis of the effects of polygenic risk on programs underpinning cellular and molecular endophenotypes in the developing and early postnatal brain.

Encouragingly, comprehensive RNA expression profiling of human brain tissues from early embryonic to late adult postmortem stages has shown that neuronal cells produced from iPSCs closely recapitulate the progression from early embryogenesis to late fetal periods in vitro and yield neuronal cells of various stages of maturity [61–66]. Immature neurons and networks express molecules and processes that are not operative in the adult and follow a crucial developmental sequence that is instrumental in the formation of functional entities. While caution needs to be exercised to extrapolate from iPSC-derived cell stages to those in adolescents and adults, they provide unique access to explore molecular and cellular endophenotypes and cause–effect relationships in living disease-relevant cell types from early neurodevelopmental stages from patients with COS.

5. Tracing Early Neurodevelopment in Patients with COS

In 2011, Brennand and coworkers firstly reported the generation of iPSC-derived neuronal cells from patients with familial SCZ and detected significant reductions in neuronal connectivity, neurite outgrowth, and dendrite formation in forebrain neurons from patients relative to controls [67]. This influential work has prompted an increasing number of patient-specific iPSC studies on AOS [68], but also on bipolar disease [69]. Here, we consider iPSC-based case/control studies on carriers of high-risk structural variations associated with COS or on patients diagnosed COS vs. healthy controls. For clarity, experimental approaches, and key findings are summarized in a tabular format.

5.1. Role of the 22q11.2 Microdeletion as Risk Factor for AOS and COS

The 22q11.2 deletion syndrome (22q11.2.DS), also known as DiGeorge or velocardiofacial syndrome, has an incidence of 1 in 2000–4000 live births [70]. Typical microdeletions are either 3 Mb in size (covering about 60 known genes) or 1.5 Mb in size (covering about 35 known genes). Most of the genes inside these regions are expressed in the brain. The severity of the disorder is unrelated to the size of the deletions indicating that genes residing within the 1.5 Mb region are critical to the etiology of the syndrome. Frequent physical manifestations consist of craniofacial and cardiovascular anomalies and immunodeficiency among others symptoms. Patients with 22q11.2.DS also show cognitive and behavioral impairments and a high risk for ASD, neurodevelopmental delay, and SCZ [48]. In fact, the identification of rare and highly penetrant de novo structural variations at 22q11.2 in sporadic cases of SCZ provided the first evidence for the role of rare recurrent mutations in SCZ susceptibly [71]. This structural mutation is detected in up to 1% and 4% of AOS and COS cases, respectively [6] and up to one-third of all patients with 22q11.2.DS develop SCZ or schizoaffective disorder (SAD). Noteworthy, there are no major clinical differences in core psychopathology, treatment response, neurocognitive profile, and imaging anomalies between schizophrenic patients with 22q11.2.DS or an intact chromosome 22 [72]. In fact, many patients with 22q11.2.DS show no serious intellectual disability and congenital abnormalities can be so subtle that they appear undistinguishable from other patients with SCZ. Consistent with these findings, intellectual ability and length of the microdeletion do not appear to be major risk factors for SCZ associated with the 22q11.2 microdeletion.

In 2011, Pedrosa et al. [73] firstly reprogrammed fibroblasts (Table 2) from a patient with AOS carrying a 22q11.2 microdeletion, a high risk factor for COS, and two healthy controls. Two iPSC lines from patients with SCZ, one with adult-onset SCZ and one with COS (Table 3) were obtained additionally from Brennand et al. [67]. iPSC quality control (Table 2) consisted of immunocytochemistry (ICC), teratoma (Tera) and embryoid body formation (EB), and karyotype analysis (G-B, G-banding; FISH, fluorescence in-situ hybridization).

Table 2. iPSC generation and quality control.

Ref	Source	Factors	Method	*n*	Authentication	Karyotype	Pluripotency
[73]	Fibroblast	OKSM	RV	-	-	G-B, F	ICC, EB
[74]	As in [73]	OKSM	RV	-	-	G-B, F	ICC, EB
[75]	Fibroblast	OKSM	RV	-	-	CGH	ICC, Tera, EB
[76]	Fibroblast	OKSML	Epi	add	-	G-B, F	ICC, EB
[77]	As in [76]	OKSML	Epi	add	-	G-B, F, micro	ICC, EB
[78]	As in [75]	OKSM	RV	add	-	CGH, Taq	ICC, Tera, EB
[79]	Fibroblast	OKSML	Epi or Sen		-	G-B, F	ICC, Tera
[80]	Fibroblast	OKSM	Sen	add	-	G-B, CGH	ICC
[81]	hESC (H1)	na	na	-	-	na	na
[82]	Fibroblast	OKSML	Epi	na	CytoChip SNP	CGH, SNP	ICC
[83]	Fibroblast	OKSM	Sen	2–3	PsychChip SNP	G-B	FACS, PCR
[84]	As in [83]	OKSM	Sen	2–3	PsychChip SNP	G-B	FACS, PCR
[85]	As in [83]	OKSM	Sen	2–3	Verif-BamID [86]	G-B	FACS, PCR

Abbreviations are: add, additional iPSC clones for some donors; EB, embryoid body formation combined with ICC and/or qPCR; Epi, episomal plasmid; CGH, comparative genomic hybridization microarray; G-B, chromosomal G-banding; FACS, fluorescence activated cell sorting; F, fluorescence in-situ hybridization; ICC, immunocytochemistry; micro, microarray; *n*, numbers of independent clones per donor; na, non-applicable; OKSM, reprogramming factors OCT4, KLF4, SOX2, MYC; OKSML, reprogramming factors plus Lin28 and p53 shRNA; PCR, quantitative reversed transcribed polymerase chain reaction; Sen, Sendai virus; Ref, reference; RV, retroviral transduction; SNP, single nucleotide polymorphism; Taq, Taqman copy number assay; Tera, teratoma formation.

Table 3. Study design, cellular model, and neuronal cell types.

Ref	Case/Control	Deletion	Model	Major Cell Type
[73]	AOS ($n = 1$) AOS ($n = 1$) COS ($n = 1$) Ctr ($n = 2$)	22q11.2 - - -	iPSC	Forebrain glutamatergic neurons
[74]	AOS ($n = 1$) [73] Ctr ($n = 1$) [73]	22q11.2 -	iPSC	Early post-mitotic neurons
[75]	AOS ($n = 2$) Ctr ($n = 2$)	22q11.2 -	iPSC	Mixed early neuronal and glial cell types
[76]	AOS ($n = 1$) SAD ($n = 3$) COS ($n = 2$) Ctr ($n = 6$)	22q11.2 22q11.2 - -	iPSC	Mixed early glutamatergic and GABAergic neurons
[77]	As in [76] + COS ($n = 2$) + Ctr ($n = 1$)	As in [76] - -	iPSC	As in [76]
[78]	As in [75] + Ctr ($n = 1$)	As in [75] -	iPSC	As in [75]
[79]	COS ($n = 3$) Ctr ($n = 5$)	15q11.2 -	iPSC	Rosette-derived cortical NPCs
[80]	SAD ($n = 1$) Mother ($n = 1$) Ctr ($n = 1$)	15q11.2 15q11.2 –	iPSC	Rosette derived neurons
[81]	Isogenic hESCs	Mutated heterogeneous *NRXN1* alleles	hESC	Induced glutamatergic neurons, mixed forebrain neurons
[82]	ASD ($n = 1$) NSD ($n = 1$) NSD ($n = 1$) Autism ($n = 1$) Autism ($n = 1$) Autism ($n = 1$) Ctr ($n = 4$)	16p11.2 dup, de novo 16p11.2 dup, de novo 16p11.2 dup, inherited 16p11.2 del, de novo 16p11.2 del, unknown 16p11.2 del, inherited -	iPSC	NPCs, dorsal forebrain neurons, up to 14 weeks maturated

Table 3. *Cont.*

Ref	Case/Control	Deletion	Model	Major Cell Type
[83]	COS-1 ($n = 1$)	1p33	iPSC	NPCs
	COS-2 ($n = 1$)	2p16.3 del (*NRXN1*)		
	COS-3 ($n = 1$)	3p25.3		
	COS-4 ($n = 2$)	16p11.2		
	COS-5 ($n = 1$)	22q11.2		
	COS-6 ($n = 4$)	-		
	Ctr ($n = 10$)	-		
[84]	COS-1 to 4 ($n = 5$)	As in [83]	iPSC	NPCs, mixed glutamatergic and GABAergic forebrain neurons, Ngn2-induced excitatory neurons
	COS-6 ($n = 4$)	As in [83]		
	Ctr ($n = 8$)	As in [83]		
[85]	COS-1 to 5 ($n = 6$)	As in [83]		NPCs, mixed glutamatergic and GABAergic forebrain neurons
	COS-7 ($n = 1$)	18q22.1		
	COS-8 ($n = 1$)	8q12.3, 22q11		
	COS-9 ($n = 1$)	15q11.2, 2p25.3		
	COS-6 ($n = 4$)	As in [83]		
	COS-10 ($n = 3$)	-		
	Ctr ($n = 10$)	As in [83]		
	Ctr ($n = 2$)	-		

Abbreviations are: AOS, adult-onset schizophrenia; COS, childhood-onset schizophrenia; Ctr, control; n, number of case/control samples; NPC, neural progenitor cells; Ngn2, Neurogenin 2; NSD, non-spectrum disorder; Ref, reference; SAD, schizoaffective disorder; +, plus refers to new case/control samples in addition to those from the indicated reference, - refers to normal karyotype.

Neural induction involved embryoid body (EB) and neural rosette formation (an in vitro equivalent to the neural tube) (Table 4). Subsequently, NPCs were manually dissected and differentiated in mixed cultures of forebrain glutamatergic neurons that were able to fire action potentials after two months in culture. Expression profiling across undifferentiated and differentiated case/control iPSCs showed no gross differences except for the pluripotency markers OCT4 and NANOG. These markers declined more slowly during glutamatergic differentiation of the iPSCs derived from the patient with AOS and a 22q11.2 microdeletion relative to the other samples.

Table 4. Major differentiation methods.

Ref	Neural Induction	Patterning/Neural Progenitor Cells→Neural Cells
[73]	EB-/rosette formation	N2, WNT3A→N2, B27, BDNF, GDNF, IGF1, WNT3, cAMP
[74]	SB431542 + Dorsomorphin	N2, B27, bFGF→N2, B27, BDNF, GDNF
[75]	EB-formation + Noggin	FGF2, Shh or Wnt3a or BMP4→FGF2, EGF
[76]	EB-formation + Dorsomorphin	FGF2→N2, BDNF, GDNF, IGF1, WNT3, cAMP
[77]	As in [76]	As in [76]
[78]	As in [75]	As in [75]
[79]	SB431542 + CHIR99204	N2, B27, Dorsomorphin, RA
[80]	Rosette formation, N2, bFGF	N2, BDNF
[81]	Ngn2-mediated iN	N2, B27, BDNF, NT3→mouse glia, Ara-C
[82]	EB-/rosette formation	StemCell Induction medium™→as above
[83]	SB431542 + LDN-193189	N2, mTeSR™→BDNF, cAMP, AA→BrainPhys™
[84]	SB431542 + LDN-193189	N2, B27-RA, FGF2
	SB431542 + LDN-193189	N2, B27-RA, FGF2→B27-RA, BDNF, GDNF, cAMP, Ara-C, astrocytes, N2 B27-RA
[85]	Ngn2-mediated iN	BDNF, GDNF, cAMP, Ara-C, astrocytes
	SB431542 + LDN-193189	N2, B27-RA, FGF2→B27-RA, BDNF, GDNF, cAMP, Ara-C, astrocytes, N2

Abbreviations are: AA, ascorbic acid; Ara-C, arabinoside C; B27, B27 supplement; BDNF, brain derived neurotrophic factor; BMP, bone morphogenetic protein; cAMP, cyclic adenosine monophosphate; FGF2, fibroblast growth factor 2; EB, embryoid body; GDNF, glial cell derived neurotrophic factor; N2, N2 supplement; Ngn2, neurogenin 2; RA, retinoic acid; SHH, sonic hedgehog; Wnt, wingless.

Analysis of homogenized cultures from iPSC-based case/control studies can disguise the detection of disease-relevant signals due to the high heterogeneity of cell types, broadly varying maturation states, and of differences in differentiation capacity (see also Sections 5.5 and 6). In a follow-up study, Belinsky et al. [74] sought to address this concern by combining patch recording with single-cell PCR (polymerase chain reaction) for expression profiling of a selected panel of genes from neurodevelopment, GABAergic and glutamatergic signaling among others. Neurons derived from a patient with adult-onset SCZ carrying a 22q11.2 microdeletion and one control (Table 3) showed similar active and passive electrical activities across the entire time course of neuronal differentiation. At the same time, electrical activities were poorly synchronized due to varying maturation states. However, once patient-derived neurons developed electrical activities, the expression of genes relevant for GABAergic, glutamatergic, and dopaminergic specification appeared subtly deregulated relative to the control.

The development of complex behaviors and higher cognitive functions in human involves the development of highly specialized cell types and circuitries that may be impaired in psychiatric disorders such as COS/AOS. Accruing evidence suggests that genomic DNA in the brain contains characteristic somatic genetic variations relative to non-brain tissues [87]. These variations comprise mutations, chromosomal aneuploidy, or microdeletions, and the dynamics of non-long terminal repeat (LTR) retrotransposons. All of these variations contribute potentially to the production of functionally diversified brain cells. Among the known retrotransposons, only long interspersed nucleotide element-1 (L1) possesses autonomous retrotransposition activity that is required for the insertion of new L1 copies. L1 shows retrotransposition activity in rat hippocampal NPCs [88], human embryonic stem cells, and human fetal and adult brain [89]. Furthermore, increased L1 retrotransposition was detected in a mouse model of Rett syndrome and in Rett patients, suggesting a role in neurodevelopmental disorders [90].

Considering these findings, Bundo et al. [75] investigated L1 activity in postmortem prefrontal cortex from patients with AOS and iPSC-derived neurons from two patient with AOS carrying 22q11.2

microdeletion, a known high risk factor for COS. Whole-genome sequencing showed that brain-specific L1 insertion in patients with AOS localized preferentially to genes involved in synapse formation and function, cell adhesion, and cytoskeleton among other processes relevant to SCZ [75]. L1 copy number was unrelated to confounding factors (e.g., age, age of onset, and duration of illness) and emerged from early neurodevelopmental stages, at least in the prefrontal cortex (Table 5).

Table 5. Major methods and findings on COS associated CNVs and on COS.

Ref	Major Methods	Major Findings on COS and Associated CNVs
[73]	Microarray, WCPC	Delayed decline of pluripotency markers in AOS with 22q11.2
[74]	WCPC, single cell Ca^{2+} imaging and PCR	Dysregulation of genes relevant to GABAergic, glutamatergic, and dopaminergic in electrical active neurons
[75]	Whole genome sequencing, postmortem brain	Increased L1 retrotransposition in postmortem brain from patients with AOS and iPSC derived neurons from AOS patients with 22q11.2 deletion
[76]	MicroRNA profiling	32 miRNAs are upregulated in neurons with 22q11.2 microdeletion, miRNA deregulation is broadly shared across AOS, SAD, and COS
[77]	Paired-end mRNA sequencing	Perturbed neuronal MAPK signaling, differentially expressed genes from the 22q11.2 microdeletion act during critical periods of development
[78]	miRNA and mRNA arrays	Reduced neurosphere size, neural differentiation, neurite outgrowth, cellular migration, and expression of miR-17/92 cluster and miR-106a/b that inhibit p38a (MAPK14) expression, p38 inhibitors improve diminished neurogenic-to-gliogenic ratio
[79]	ICC/IHC, complementation and knock-down experiments	Defects in adherens junctions and apical polarity. Displacement of radial glia cells leads to cortical malformation during mouse development
[80]	ICC, IB	Lower expression of CYFIP1 and PSD-95, altered dendritic morphology
[81]	Gene editing, iNeurons, electrophysiology	Reduced spontaneous mEPSC frequency, but not amplitude, and decrease in evoked EPSC amplitude. Unaltered electrical properties of human neurons, synapse numbers, and dendritic arborization
[82]	Histomorphology, electro-physiology	16p del- and 16p dup-derived NPCs show opposing differences in soma size and arborization, reduced excitability in 16p del-derived neurons, increased potassium current density in 16p dup-derived neurons, lower density of excitatory synapses in 16p del- and 16p dup-derived neurons associates with increased amplitude of mEPSCs
[83]	digital miRNA profiling	Downregulation of miR-9, a regulator of neurogenesis and of radial migration
[84]	IB, IHC, IP, knock-down	Increased $STEP_{61}$ protein expression in forebrain neurons impairs NMDAR signaling
[85]	mRNA sequencing	Transcriptional signatures of NPCs and neurons show concordance with postmortem case/control brain samples from SCZ, BP, and ASD after adjusting for cell type composition

Abbreviations are: AOS, Adult Onset Schizophrenia; ASD, Autism Spectrum Disorder; BP, Bipolar Disorder; COS, Childhood Onset SCZ; IB, immunoblot; ICC, immunocytochemistry; IHC, immunohistochemistry; IP, immunoprecipitation; mEPSC, miniature excitatory postsynaptic current; NPC, neuronal progenitor cell; WCPC, whole cell patch clamp.

Interestingly, L1 insertion was also increased in iPSC-derived neurons containing the 22q11.2 microdeletion relative to controls supporting the role of this variation as risk factor for SCZ. Remember that iPSC-derived cells match early embryonic to early postnatal stages and thus provide in vitro evidence for a role of L1 retrotransposition during early neurodevelopment. In support of this hypothesis, immune activation by poly-I:C treatment of rat dams (a translational model for the generation of schizophrenia-like symptoms in the offspring) led to an increase of L1 copy number in the brain. Hence, an increase in L1 insertion in response to environmental or genetic risk factors may increase the vulnerability for SCZ by impairing synaptic and related functions in neurons, rather than representing a primary cause of the disease.

Beyond mRNAs, the developing human brain expresses also high levels of microRNAs (miRNAs) that regulate neural lineage and cell fate decisions, differentiation, and neuronal maturation [91,92]. miRNAs are noncoding RNAs of ~70 nucleotides in size (pri-miRNAs) that are cleaved by a nuclear protein complex encompassing DGCR8 and DROSHA into precursor RNAs (pre-miRNAs) [93]. Latter are further cleaved by DICER to yield single stranded ~22 nucleotide mature miRNAs that are incorporated into the RNA induced silencing complex (RISC). Subsequently, miRNAs target through a 6- to 8-base pair complementary 'seed region' one or more mRNAs with each miRNA potentially downregulating up to hundreds of downstream targets [93]. Changes in miRNA expression profiles have been detected in SCZ, autism, and major depressive disorder (MDD) [94]. For example, miRNA-137 (miR-137) maps to a risk locus of SCZ [95,96] and seems to downregulate disease related genes like *TCF4* (transcription factor 4) or *CACNA1C* (calcium channel, voltage-dependent, L-type, alpha-1C subunit) [97,98].

Noteworthy, *DGCR8* (DiGeorge syndrome critical region gene 8) resides inside the 22q11.2 microdeletion. Reduced expression of *DGCR8* slows the conversion of a subset of pri-miRNAs to pre-miRNAs and results in a dampened production of a particular subset of mature miRNAs [99]. Additionally, the 22q11.2 region harbors *MIR-185* that targets other candidate genes relevant to SCZ, to hippocampal dendritic spine density, and to synapse function [100].

Given these premises, Zhao et al. [76] sought to analyze the miRNA profiles in living neurons generated from patients with (i) SAD or AOS carrying the 22q11.2 microdeletion, a high risk factor for COS, or (ii) with COS carrying an intact chromosome 22 (Table 3). MiRNA sequencing of day 14 neurons (Table 5) from six controls (with multiple clones for two controls) and from six patients with SAD, AOS, or COS, detected 45 differentially expressed miRNAs (13 lower in SCZ; 32 higher). Among these miRNAs, six were significantly downregulated in neurons carrying the 22q11.2 microdeletion, including four miRNAs that map to the 22q11.2 microdeletion (miR-1306-3p, miR-1286, miR-1306-5p, and miR-185-5p), and two that did not (miR-3175 and miR-3158-3p). This result suggests that some miRNAs are downregulated independently of *DGCR8* possibly by one or more of the transcriptional and chromatin regulators that map to this chromosomal region. In support of this finding, 32 differentially expressed miRNAs were upregulated in the 22q11.2 microdeletion samples, rather than downregulated. Functional pathway analysis of the differentially expressed miRNAs showed enrichment for genes relevant to neurological and psychiatric disorders and neurodevelopment. For example, miR-34c, a member of the miR-34 family, is predicted to target *CNTNAP1, CNTNAP2, GABRA3, RELN, FOXP2, NRXN2,* and ANK3, while mi-R34a plays a role in neural stem cell (NSC) differentiation. Moreover, many of the differentially expressed miRNAs in iPSC-derived neurons carrying a 22q11.2 microdeletion were shared with clinical/autopsy samples drawn from the general population of AOS and ASD indicating that the underpinning molecular genetic networks are shared. Hence, deregulation of miRNA pathways extends well beyond the effects specific to *DGCR8* and applies broadly to patients with SAD or AOS carrying the 22q11.2 microdeletion and to patients with COS.

Given that each miRNA potentially downregulates up to hundreds of downstream targets [93], the researchers [76] further interrogated the mRNA expression profiles from iPSC-derived neurons (Table 3). Gene pathway and network analysis of differentially expressed genes (DEGs, $n = 42$) indicated a disruption of MAPK signaling in iPSC-derived neurons from patients with SCZ carrying the 22q11.2 microdeletion that may lead to perturbed neuronal proliferation and differentiation.

Beyond individual genes, weighted correlation network analysis (WGCNA) permits the detection of perturbed interactions between functionally interconnected genes that may represent only in part significant expression changes. In this study [76], WGCNA revealed, however, only subtle changes in 2 out of 15 gene modules identified. Accordingly, global wiring of functionally interconnected genes was unaffected in iPSC-derived neurons from patients. To uncover genes co-expressed with the DEGs, the researchers conducted a correlation analysis on different regions and developmental stages from human brain (i.e., BrainSpan database). Interestingly, DEGs were highly connected only during two developmental stages. The embryonic and the adolescent brain. Moreover, function enrichment analysis of the co-expression networks in the embryonic and adolescence brains showed that the embryonic cortex was enriched in genes critical to cell cycle, differentiation, and growth, while the adolescent cortex was enriched in genes critical to synaptic transmission and catabolism. In sum, these results support that a subset of the 22q11.2 microdeletion DEGs fulfill distinct functions during sensitive time-windows of brain development that become perturbed by haploinsufficiency.

The functional consequence of 22q11.2 haploinsufficiency for early neuronal and glial development has been assessed more recently by Toyoshima et al. [78] in neurosphere assays. These contain free-floating clusters of NSCs and provide a method to investigate iPSC-derived (Table 3) neural precursor cells in vitro. The researchers detected significant reductions in neurosphere size, neural differentiation efficiency, neurite outgrowth, and cellular migration in patient-derived cells. Although both patient- and control-derived neurospheres could be efficiently differentiated into neurons and astrocytes, the fraction of astrocytes among patient-derived differentiated cells was increased at the expense of neurons.

At the molecular scale, miRNA profiling [78] showed reduced expression of miRNAs belonging to the miR-17/92 cluster and miR-106a/b in patient-derived neurospheres. These miRNAs are predicted to target *MAPK14* transcripts encoding p38α, a member of the mitogen activated protein kinase family and regulator of neurogenic-to-gliogenic transition competence. Well-fitting this prediction, p38α was upregulated in patient-derived cells. Pharmacological inhibition of p38 in patient-derived neurospheres partially reinstated neurogenic competence. Moreover, mRNA expression profiling showed that DEGs between case/control neurospheres were enriched for genes relevant to cell differentiation, neuronal development, and microRNA processing. Specifically, upregulated genes in case neurospheres were significantly enriched for MAPK-mediated processes, neurotransmission, and signaling pathways. Collectively, these results indicate a 'reduced neurogenic' and 'elevated gliogenic' competence during early neurodevelopmental stages of patients with SCZ associated with a 22q11.2 microdeletion.

Taken together, different lines of evidence provide insight into the role of the 22q11.2 microdeletion as an early risk factor for AOS and COS: iPSC-derived neuronal cells show a delayed glutamatergic differentiation [73] and exhibit subtle deregulation of genes relevant for GABAergic, glutamatergic, and dopaminergic specification once they acquire electrical activities [74]. The effects of the 22q11.2 microdeletion appear to be mediated through different molecular mechanisms: an increase in the frequency of L1 insertion during early neurodevelopment through as yet unknown mechanisms may impair synaptic function and predispose for later disease [75]. Secondly, deregulation of miRNA pathways through *DGCR8* dependent and independent pathways control genes important to SCZ and neurodevelopment [76]. Such deregulation may disrupt MAPK signaling in iPSC-derived neurons from patients with SCZ and 22q11.2 microdeletion and lead to perturbed neuronal proliferation, differentiation, and increased gliogenic competence during early development [78]. Finally, deregulation of distinct coexpression gene networks at embryonic (cell cycle, differentiation, and growth) and adolescent (synaptic transmission and catabolism) stages.

5.2. Role of the 15q11.2 Microdeletion as Risk Factor for AOS and COS

The proximal long arm of chromosome 15 (15q11.2-q13) contains several CNVs that can increase the risk for common, severe neuropsychiatric disorders [101]. The CNVs arise from mis-paired low copy number repeats at three breakpoints denoted BP1, BP2, and BP3. The 15q11.2 BP1-BP2 microdeletion (Burnside-Butler syndrome) encloses four protein-encoding genes (*TUBGCP5*, *CFYIP1*, *NIPA1*, and *NIPA2*) and has a reported de novo frequency between 5–22%. On the other hand, about 35% and 51% of the carriers have inherited the microdeletion from an apparently affected or unaffected parent, respectively [102]. Genes inside the BP1-BP2 region are biallelically expressed, whereas the clinically related Prader-Willi/Angelman syndrome, defined by the distal breakpoint BP3 and the proximally located breakpoints BP1 or BP2, involves the deletion of a large genomically imprinted region between BP2-BP3.

Patients with the 15q11.2 BP1-BP2 microdeletion carry an increased risk for intellectual disability (ID), ASD, AOS, COS, and seizure disorders and manifest mild dysmorphic features and neurocognitive delay [102]. Discrete disabilities in learning, reading skills, and a marginally reduced intelligence quotient have been found among clinically affected, but also among normal individuals, with the 15q11.2 BP1-BP2 microdeletion [103]. Furthermore, this genetic variation affects brain structure in a pattern consistent with that observed during first-episode psychosis in SCZ [103].

The 15q11.2 BP1-BP2 microdeletion has a prevalence ranging from 0.57–1.27% as inferred from high resolution microarray analysis [102]. However, not all individuals with the deletion are clinically affected since this region harbors genetic material showing incomplete penetrance or low penetrance of pathogenicity along with variable expressivity. *NIPA1* (non-imprinted in Prader-Willi/Angelman syndrome 1 gene) is the best understood gene within this region and associates with autosomal dominant hereditary spastic paraplegia. It is highly expressed in neuronal tissues and serves the transport of Mg^{2+}. Likewise, NIPA2 (non-imprinted in Prader–Willi/Angelman syndrome 2 gene) regulates renal Mg^{2+} transport. The *TUBGCP5* (tubulin gamma complex associated protein 5) gene is required for microtubule nucleation at the centrosome and is thought to contribute to neurobehavioral disorders such as ADHD (attention deficit hyperactivity disorder) [104]. Finally, cytoplasmatic FMR1-interacting protein (CYFIP1), a binding partner of fragile X mental retardation protein (FMRP), is a leading candidate inside the BP1-BP2 domain. CYFIP1 has been found to interact with Rac1 (a RHO GTPase involved in modulation of the cytoskeleton, neuronal polarization, axonal growth, and differentiation), FMRP, and EIF4E (eukaryotic translation initiation factor 4E). In mice, complex formation between cyfip1, FMRP, and cap protein eiF4E serves to regulate activity-dependent protein translation in mature neurons [105]. Biochemical studies further suggest that CYFIP1 regulates the WAVE complex that controls Arp2/3-medited actin polymerization and membrane protrusion formation in non-neuronal cells.

15q11.2 microdeletion is one of the most frequent CNVs associated with an increased risk for AOS and COS (Table 1) [49]. To understand why 15q11.2 CNVs are prominent risk factors for SCZ, Yoon et al. [79] established iPSC lines from three individuals with COS carrying the microdeletion, and from five healthy individuals without the microdeletion (Table 3).

Immunostaining of iPSC-derived neural rosettes (an in vitro pendant of the neural tube) from COS cases displayed perturbed apical–basal polarity and disrupted adherens junctions relative to controls. The actin cytoskeleton acts as a cytoplasmatic anchor for cadherin/catenin proteins at adherens junctions and its proper organization is important for maintaining adherens junctions and polarity of NPCs. Consistent with CYFIP1's role as a regulator of the actin-modulating WAVE complex, biochemical analysis showed a specific defect of WAVE complex stabilization in NPCs carrying the 15q11.2 microdeletion. Gain-and-loss of function experiments for CYFIP1 in NPCs carrying the microdeletion and from control NPCs further supported this finding. In agreement with the in vitro experiments, cyfip1 was also necessary to sustain adherens junctions and apical polarity of NSCs in the developing mouse cortex as demonstrated by in vivo knockdown experiments. Moreover, deficits in cyfip1 led to false placement and pattern of mitosis of radial glial progenitor cells (RGCs) in the developing mouse cortex. This phenotype subsisted in intermediate progenitor cells (IPCs), the direct progeny of RGCs, as well as in glutamatergic projection neurons, resulting in cortical layer malformation.

Beyond NSC/NPC, the function of CYFP1 is known to extend to mature neurons. Recent reports showed that cyfip1 is enriched at mouse neuronal synapses and plays an important role in dendritic arborization as evidenced by gain-and-loss of function studies [106,107]. Given that human postmortem studies support a role for dendritic spine structure abnormalities in the pathogenesis of ID, ASD, and SCZ [108], haploinsufficiency of *CYFIP1* could present a mechanism whereby the 15q11.2 deletion confers risk for neuropsychiatric disorders. To address this topic, Das et al. [80] created iPSCs from a mother and her offspring, both carrying the 15q11.2 deletion, and a control with an intact chromosome 15. The offspring, but not the mother, additionally manifested SAD.

Neural rosettes derived from quality controlled iPSCs (Table 2) were dissected, expanded as neurospheres, subsequently kept as monolayers, and finally differentiated into neurons (Table 4). The expression of all four genes inside the deleted region as well as of PSD95, a key marker of synapses, was reduced during different stages of neuronal development in the mother and offspring when compared to the unrelated control. Moreover, at 10 weeks of differentiation qualitative analysis of iPSC-derived neurons provided tentative evidence that dendritic morphology was altered in 15q11.2-deletion carriers relative to control. In support of this view, Dimitrion et al. [109] observed in a follow up study that in low-density neuronal cultures the density of dendritic filopodia was strongly increased in neurons with the microdeletion (i.e., the maternally-derived iPSC line) relative to the control.

Collectively, these studies show that human iPSCs can serve as an entry point to investigate a common CNV risk factor for AOS, COS, and other neuropsychiatric disorders. Results from multiple levels of analysis also allowed prioritization of genes within the CNV and highlighted a role of CYFIP1 as contributing factor to biological processes implicated in the neurodevelopmental origins of these disorders. Specifically, CYFIP1 regulates apical–basal polarity and adherens junctions of NSCs, proper positioning of NSCs and their derivatives along neurodevelopmental trajectories, and dendritic arborization of mature neurons; all of these processes are key to AOS and COS.

5.3. Role of the 2p16.3 Microdeletion as Risk Factor for AOS and COS

AOS and COS has been associated with non-recurrent CNVs (Table 1) including those disrupting the *NRXN1* gene at 2p16.3. These deletions cluster in delineated regions and represent with variable size and unique breakpoints. The presence of short stretches of microhomology and additional base pair insertions at the breakpoint site [110] suggests that error-prone repair mechanisms referred to as NHEJ bridge, modify, and fuse free DNA ends at sites of double-stranded chromosomal breaks. In contradiction to NAHR, NHEJ does not depend on specific genomic architectural features such as LCR.

NRXN1 encodes neurexin-1 [111], an evolutionary conserved presynaptic cell-adhesion molecule. Humans contain three neurexin genes (*NRXN1*, *NRXN2*, and *NRXN3*) each of which harbors separate promoters for longer α- and shorter β-neurexins. These isoforms bind to postsynaptic cell-adhesion molecules such as neuroligins and LRRTMs that are also associated with ASD or SCZ.

Most *NRXN1* mutations represent heterozygous CNVs that delete only *NRXN1* due to the large size of the gene, while missense and truncation mutations are less frequent [110]. While *NRXN1* mutations are rare (≈0.18% of patients with SCZ [112]), they represent the most frequent-single gene mutation in AOS and COS. *NRXN1* polymorphisms have been also implicated in differential responses to antipsychotic medication in SCZ further strengthening the link between SCZ and *NRXN1* [113]. Individuals with 2p16.3 microdeletion can manifest developmental delay, especially in speech, abnormal behaviors, and mild dysmorphic features with epilepsy [114]. However, presence of *NRXN1* deletions in healthy parents and siblings indicates reduced penetrance and/or variable expressivity.

The variable clinical presentations and the observation that homozygous *Nrxn1α* mutations cause only a minor phenotype in mice [115], raise the question of whether heterozygous *NRXN1* mutations alone directly impair synaptic function. To address this question under conditions that control precisely for genetic background, Pak et al. [81] established isogenic human embryonic stem cell (ESC) lines carrying different heterozygous conditional *NRXN1* mutations and analyzed subsequently their effects on neuronal phenotypes and activities.

Loss-of-function mutations were generated by homologous recombination and consisted either of a conditional exon deletion that caused a frameshift and disrupted both neurexin-1α and -1β or a conditional truncation of neurexin-1α and -1β that introduced a stop codon and resulted in rapidly degraded protein. Both heterozygous conditional *NRXN1* mutations did not alter the electrical properties of human neurons, their synapse numbers, or dendritic arborization. Yet, they produced a severe and selective decrease in presynaptic neurotransmitter release concomitant with a reduction in spontaneous mEPSC (miniature excitatory postsynaptic current) frequency, but not amplitude, and a parallel decrease in evoked EPSC amplitude. Interestingly, the decrease in EPSC amplitude was rapidly relieved during a stimulus train indicating that this phenotype did not involve a general decline of the release probability, but exhibited a specific decrease in release probability for only the first stimulus. Moreover, key features of the *NRXN1* heterozygous mutant phenotype were detected in two different types of ESC-derived human cells: induced neurons (iN) consisting of a homogenous population of excitatory forebrain neurons, and a more heterogeneous population of neurons obtained from an NPC intermediate (Table 4). This observation strengthens the notion that heterozygous loss of *NRXN1* causes a selective impairment in synaptic transmission. A plausible explanation for this phenotype is impairment in presynaptic Ca^{2+} influx during an action potential that would only initially impair release in a high-frequency stimulus train due to the accumulation of residual Ca^{2+} later in the train.

Collectively, these results suggest that heterozygous *NRXN1* mutations may predispose to AOS, COS, and other neuropsychiatric disorders by impairing a highly specific synapse function.

5.4. Role of the 16p11.2 Microdeletion as Risk Factor for AOS and COS

The 16p11.2 CNV covers an $\approx$600 kb locus encompassing 29 annotated genes [116]. Carriers with either the deletion (16p-del) or the duplication (16p-dup) of this region manifest psychiatric disorders such as ASD, AOS, and COS (Table 1). Common developmental, cognitive, and behavioral symptoms are also equally shared by both genotypes. By contrast, they associate with opposing physical symptoms: individuals with 16p-del have normal birth weight, but develop a drastic increase in body mass index (BMI) by age 7 such that $\approx$75% of adult carriers are obese. Contrariwise, individuals with 16p-dup represent with below-normal weight at birth and an eightfold enhanced risk of underweight in adulthood. Additionally, carriers differ in head sizes: $\approx$17% of the individuals with 16p-del are macrocephalic, while $\approx$10% of the individuals with 16p-dup are microcephalic [116]. Neuroimaging studies on carriers suggest significant effects on gray matter volume, especially increase in the cortical surface area in individuals with 16p-del. On the other hand, a reciprocal decrease has been detected in individuals with 16p-dup [117]. Noteworthy, Lin et al. [118] predicted by dynamic protein interaction analysis profound changes in the 16p11.2 protein interaction networks throughout different stages of brain development and/or in different brain regions. Hereby, the late mid-fetal period of cortical development was most critical for establishing the connectivity of 16p11.2 proteins with their co-expressed partners.

To uncover cellular phenotypes due to 16p11.2 CNVs, Desphande et al. [82] generated iPSCs from donors with a diagnosis of ASD with gain (dup) or loss (del) of 16p11.2 CNV (Table 3). Quality control (Table 2) confirmed that 16p11.2 CNV carrier-derived iPSCs were comparable to control iPSCs regarding pluripotency, NPC proliferation, self-renewal, and the formation of forebrain neurons. By contrast, at three and six weeks post differentiation, 16p del-derived neurons showed neuronal hypertrophy with increases in soma size, total dendrite length and arborization, whereas 16p dup-derived neurons showed the opposite phenotype relative to controls, especially in excitatory neurons.

Functionally, 16p del-derived neurons exhibited reduced excitability with greatly reduced voltage responses and membrane resistance relative to 16p dup-derived neurons, which behaved undistinguishably to controls. On the other hand, 16p dup-derived neurons—but neither 16p del-derived neurons nor controls—showed an increased potassium current density at positive voltages indicating that they may compensate for their reduced somatic size by increasing the outward potassium current to stabilize intrinsic excitability. Finally, both 16p del- and dup-derived neurons revealed a lower density of excitatory synapses compared with controls that associated with a significant increase in the amplitude, but unaltered kinetic or frequency, of mEPSCs.

Collectively, reciprocal cellular phenotypes in 16p-dup/del iPSC-derived neurons may contribute to opposing brain size difference. In this respect a gene inside 16p11.2, namely *KCTD13*, encoding a nuclear protein that stimulates DNA polymerase activity at replication foci, has been shown to cause via proliferation dose-dependent macrocephaly in zebrafish [119]. Furthermore, KCTD13 plays a crucial role in the regulation of the KCTD13-Cul3-RhoA pathway in layer 4 of the inner cortical plate that controls brain size and connectivity [118]. At the same time, similar reductions in synapse density in either 16p11.2 genotype may contribute to the similarities in human clinical outcome and represent a major risk factor for the development of SCZ—the 'disease of the synapse' [120].

5.5. COS with or without CNVs

Despite the rare incidence of COS, recent studies [83–85] have taken a step forward toward the collection of larger sample sizes from patients with COS carrying risk CNVs and from those without known CNVs (i.e., idiopathic COS). In 2016, Topol et al. [83] reported the first COS/control study comprising each ten individuals: Patients with COS carried different CNVs (1p33, 2p16.3 del, 3p25.3, 16p11.2, and 22q11.2) or showed no detectable anomalies (*n* = 4) (Table 3). These patients were recruited from the longitudinal NIHM study (see Section 2) and showed across development reduced cortical thickness relative to controls. With increasing age developmental trajectory normalized in parietal regions but remained divergent in frontal and temporal regions, a pattern of loss similar to AOS [27].

Quality controlled COS/control iPSCs (Table 2) were differentiated via dual-SMAD inhibition into NPCs (Table 4), expanded, and harvested for miRNA profiling (Table 5). As noted before (Section 5.1), miRNAs play a pivotal role in the developing human brain [91,92] and altered miRNA expression profiles have been consistently detected in psychiatric disorders [94]. In parallel, the researchers conducted miRNA expression profiling also on previous AOS/control samples [61]. Among 800 miRNAs detected by digital expression profiling (Nanostring), miR-9, a regulator of neurogenesis in NSCs [121], was the most abundant and the most downregulated miRNA in NPCs from patients with either AOS or COS. Thereby, lower miR-9 levels in patient-derived NPCs relative to those from controls were largely driven by a subset of cases, which is not unexpected given the heterogeneity of a complex disorder like SCZ. Functionally, miR-9 enhanced radial migration as evidenced by gain-and-loss of function experiments in iPSC-derived NPCs.

Analysis of the AOS/control cohort, for which mRNA expression profiles from both iPSC-derived NPCs and neuronal cells were already available [61], suggested that known miR-9 target genes were significantly enriched (*n* = 84) among DEGs (56% upregulated, 44% downregulated). In this context it is interesting to note that previous SCZ GWAS gene-set enrichment analysis [43] has detected an enrichment on predicted miR-9 targets among SCZ-associated genes [122]. Together, these findings indicate that genetic variants in both miR-9 and its targets confer increased risk of SCZ.

Moving beyond miRNA profiling, the same case/control cohorts were also investigated for the expression of the brain-specific tyrosine phosphatase STEP (striatal-enriched protein tyrosine phosphatase) [84]. This membrane associated kinase is an important regulator of synaptic function: it counteracts synaptic strengthening by enhancing N-methyl-D-aspartate glutamate receptor (NMDAR) internalization through phosphorylation of the GluN2B subunit and inactivation of the extracellular signal-regulated kinase 1/2 and Fyn. Previous studies suggested that $STEP_{61}$ is higher expressed in postmortem anterior cingulate cortex and dorsolateral prefrontal cortex of patients with AOS, as well as in mice treated with the psychomimetic phencyclidine or the NMDAR antagonist MK-801 [123].

In a separate approach, the researchers [84] had originally found enhanced expression of $STEP_{61}$ in the cortices of $Nrg1^{+/-}$ (Neuregulin 1) and brain-specific ErbB2/4 knockout mice. Nrg1 signaling is a critical mediator of synaptic function and plasticity in glutamatergic signaling [124]. Therefore, $Nrg1^{+/-}$ knockout mice are deemed a valued translation model of SCZ.

In support of these findings, $STEP_{61}$ protein expression was also increased relative to controls in mixed or merely pure glutamatergic forebrain cultures generated from AOS- or COS-derived iPSCs. Similar to miR-9 expression, differences in $STEP_{61}$ protein expression were driven by a subset of patients with AOS (three out of four) or patients with COS (four out of nine) collaborating previous evidence for genetic heterogeneity in either cohort [83]. Notably, knock-down or pharmacological inhibition of STEP prevented the loss of NMDARs in iPSC-derived neurons from patients with AOS or mice brain and normalized behavior in $Nrg1^{+/-}$ mice.

Collectively, findings from transgenic mice models and patient-specific iPSCs support perturbed glutamate signaling in AOS and COS, and thus attest to the glutamate hypothesis of SCZ [125].

In a most recent study [85], an enlarged collection of COS/control iPSCs (Tables 2 and 3) was differentiated into NPCs and forebrain neurons to carry out mRNA sequencing (RNA-seq). A rigorous bioinformatic strategy was applied to adjust for technical variation and batch effects, spurious samples and samples that showed aberrant X-inactivation or contamination. Despite these precautions, the researcher observed large heterogeneity in cell type composition (CTC) between NPCs and neurons, even from the same individual. This indicates that differences in differentiation capacity led to unique neural compositions in each sample. Computational deconvolution analysis of CTC helped sharpening the distinction between NPCs and neurons; however, substantial heterogeneity remained, partly due to neural crest and mesenchymal contaminants. In fact, variation due to cell type heterogeneity surpassed variation due to donor effects and represented an important source of intra-donor expression variation that could hamper the analysis of inter-donor variation (i.e., case/control differences).

Owing these limitations, differential expression analysis of NPCs and neurons generated from COS- and control-derived iPSCs identified only few genes: one gene (*ENSG00000230847*; Occludin pseudogene) with FDR < 10% and one gene (*FZD6*, Frizzled Class Receptor 6) with FDR < 30% were both shared by NPCs and neurons. An additional three genes (*GTF2H2B*, General Transcription Factor IIH Subunit 2; *ELTD1*, EGF, latrophilin and seven transmembrane domain containing 1; *ENSG00000236725*, pseudogene RP11-154P18.1) with a FDR < 30% were specific to NPCs. Conversely, another three genes (*QPCT*, Glutaminyl-peptide cyclotransferase; *CBX2*, Chromobox homolog 2, drosophila Polycomb class; *INTS4P1*, integrator complex subunit 4 pseudogene 1) with a FDR < 30% were specific to neurons. Although plausible candidates in COS pathology such as FZD6, QPCT, and CBX2 were differentially expressed, no coherent set of biological pathways could be identified.

Moving beyond iPSCs, Hoffman et al. [85] therefore analyzed the concordance between gene expression in iPSCs-derived cells from patients with COS and differential expression results from post-mortem brain case/control studies from five psychiatric diseases: Alcoholism, MDD, BP, SCZ, and ASD. High concordance was observed for SCZ (higher in neurons vs. NPCs), BP, and ASD. By contrast, concordance was low for alcoholism and MDD. This outcome supports the specificity of gene expression data from iPSC-derived cells from patients with COS and agrees with current insight on cross-disorder genetic liability of psychiatric disorders [95,126].

Collectively, iPSC-based case/control studies on patients with COS have provided further insight into potential neurodevelopmental deviations. Reduced miR-9 expression in a subset of samples points to impaired radial migration of NPCs during early steps of development. In support of this view, increased STEP$_{61}$ protein expression in NPCs from patients with COS suggests perturbed glutamate signaling. Neuronal migration in the cortex is controlled by the paracrine action of the classical neurotransmitters glutamate and GABA (γ-aminobutyric acid) [127]. Glutamate controls radial migration of pyramidal neurons by acting primarily on NMDA receptors and regulates tangential migration of inhibitory interneurons by activating non-NMDA and NMDA receptors. In general, intra-donor and inter-donor differences in differentiation capacity of iPSCs can obscure detection of disease-relevant signals in case/control studies. However, subtle though statistically significant concordance between both NPCs and neurons generated from iPSCs derived from patients with COS and two recent SCZ post-mortem cohorts supports that in vitro findings can recapitulate processes from the diseased brain, at least in part.

6. Future Perspectives and Challenges

The possibility to generate patient-specific iPSCs has provided unique opportunities for the investigation of living disease-relevant cells from patients with COS and associated genetic risk factors. iPSC-based studies on CNVs associated with AOS and COS has helped to advance our insight in the biological underpinnings of these variations: CNVs enhance L1 retrotransposition to synaptic genes during early neurodevelopment [75] and perturb miRNA expression [76,78] thus contributing to impaired mitogenic signaling [77,78]. Furthermore, CNVs disrupt the formation of adherens junctions and apical polarity in early NPCs, especially RGCs, with long term effects on cortical organization [79]. The effects of CNVs on NPCs and neuronal cells are manifold; they reach from more subtle alterations in the expression of synaptic markers and dendritic morphology [80] to overt differences in NPC soma size, arborization, and excitatory synapses [82]. These morphological changes concur with distinct and selected changes in electrical activity such as reduced frequency [81] or increased amplitude of mEPCs [82]. In addition, more recent case/control studies on patients with COS have highlighted perturbed miRNA expression potentially affecting neurogenesis, radial glia migration [83], and glutamate signaling [84]. Importantly, transcriptional signatures of iPSC-derived NPCs and neurons from patients with COS show concordance with postmortem case/control samples from SCZ, but also with genetically related BP and ASD, and indicate that changes observed in vitro may reflect changes from the diseased brain. Unsurprisingly, there is no one-fits-all cellular or molecular phenotype emerging from these studies. While this is likely owed genetic heterogeneity in COS, it also raises questions as to the different differentiation protocols applied in current iPSC studies, especially, as to cellular heterogeneity that may obscure detection and reproducibility of disease-specific signals within and across COS/control studies. Therefore, we discuss next current caveats and further steps to be taken to improve the generation and design of patient-specific iPSC studies for COS and beyond.

6.1. High Resolution Karyotypes

Random mutations can arise along the reprogramming process and/or during in vitro culture at any time. Nowadays, non-integrating, so-called 'foot-print free', reprogramming techniques (i.e., Sendai virus, episomal, and mRNA transfection) (Table 2) are the method of choice to guard against random integration into the host genome. However, these techniques are not perfect: SNP array systems with an average genomic resolution of 43 KB (as opposed to 5 MB by traditional G-banding) showed the highest aneuploidy for retroviral (13.5%) and episomal (11.5%) derived iPSCs [128]. In-between aneuploidy was detected for lentiviral (4.5%) and Sendai virus (4.6%) derived iPSCs, and lowest aneuploidy for RNA (2.3%) derived iPSCs. Furthermore, whole exome sequencing suggests that clonal fibroblasts and iPSCs derived from the same fibroblast carry a similar number of mutations [129]. Accordingly, more than 90% of the mutations preexist randomly in small subsets of the parental unselected fibroblast population. Common genetic variations underpin molecular heterogeneity in iPSCs [130–135] and any genetic variation arising during reprogramming or in vitro culture can have potentially the same effect. Only recently, studies on COS (Table 2) have sought for donor-matched digital (e.g., SNP-based) karyotype maps to assess chromosomal anomalies, including copy number alterations [133], more precisely.

Digital karyotyping, but also mRNA-sequencing [85], can inform additionally on familial relationships and the proper assignment of iPSC lines and should be implemented in future iPSC studies on a routine basis.

6.2. Cellular Heterogeneity

Randomly distributed differences in genotype, expression profiles, and epigenetic state of individual iPSC lines [136] are known to influence the (neural) differentiation capacity of human embryonic stem cells and iPSCs from healthy donors [137–140]. Predictably, such variations will confound our ability to identify those related to disease status in a case/control design. Recent iPSC studies have therefore aimed to clarify to what degree variance across donors explains expression variation: Carcamo-Orrive [132] observed that ≈50% of genome wide expression variability in undifferentiated iPSCs (317 iPSCs from 101 healthy individuals) is explained by genetic variation across individuals. They also identified Polycomb targets to contribute significantly to the non-genetic variability seen within and across individuals [141]. By means of genome-wide profiling, Kilpinen et al. [133] determined that 5–46% of the variation (variation median ≈6) in different iPSC phenotypes (711 iPSCs from 301 healthy individuals), including differentiation capacity and cellular morphology, arise from differences between individuals. Relatedly, Schwartzentruber et al. [142]

observed that sample-to-sample ($n = 123$) variability in gene expression in iPSC derived sensory neurons from healthy donors clearly surpassed the one from in vivo dorsal root ganglia. Thereby, levels of variation for donor and reprogramming (23.2% in aggregate) were close to those from neuron differentiation batch (24.7%) reflecting varying mixtures of cell types across differentiation. Lastly, in a genetically heterogeneous and small cohort of patients with COS, Hoffman et al. [85] measured a smaller donor effect (2.2%) in iPSC-derived neurons that were obtained either by directed differentiation (dual-SMAD inhibition), known to give rise to various neuronal cell types, or by induced differentiation (Ngn2 overexpression), leading to mostly excitatory forebrain neurons (Table 4).

Recent advancements can help improving analysis of cellular heterogeneity in future case/control studies: induced neurons (iN) [143–146], generated by lentivirus-mediated overexpression of selected neuronal transcription factors, offer the benefit of less heterogeneous cell populations that may allow to detect more subtle albeit highly significant effects (e.g., [81]). Although iNs are more homogeneous with respect to cell type, they continue to display variable maturity as detected by single cell sequencing [143]. As an alternative to FACS-sorting, reporter gene assays can serve to select highly differentiated neurons with increased functionality for electrophysiology or transcriptional profiling [143]. Right now, large scale analysis of case/control samples by single cell sequencing appears still cost-prohibitive to most laboratories. In this situation, dissecting transcriptomic signatures of neuronal differentiation and maturation by improved computational skills (i.e., cellular deconvolution) may offer a more feasible alternative [62].

Implementation of these measures can help to substantially reduce or resolve cellular heterogeneity for improved detection of disease-specific signals. However, such improvements do not necessarily help to distinguish truly disease associated changes in (endo-) phenotypes from random line and culture artifacts. Testing of multiple cell clones per donor and of different differentiation protocols for the generation of the same or different cell types is strongly recommended, once preliminary results are obtained in case/control studies. Along the same line, postmortem analysis of case/control brain samples, despite known inherent limitations, is an important approach to collaborate iPSC-based findings [85]. Ideally, postmortem brain samples are not processed as bulk tissue, but as single cells, particularly for transcriptomics, to avoid anew pitfalls from cellular heterogeneity [147]. While still a matter of ongoing debate, these strategic guidelines can enhance the quality of iPSC studies on COS we should look for in the future.

6.3. Polygenic Disorders and the Environment

The presence of rare, highly penetrant genetic variants that associate with distinct cellular and molecular defects is a hallmark of Mendelian disorders. On the other hand, the basis of polygenic disorders such as COS is still less understood with numerous (non-) coding variants of small effect size converging jointly with rare variants of large effect size on highly complex phenotypes of varying expressivity. Despite this challenge, present iPSC studies on CNVs associated with AOS and COS have provided valuable information on cellular and molecular phenotypes (Table 5) of potential relevance to early neurodevelopment. Yet, given the small number of donors, for both cases and controls [148], we have to ask to what degree these observations can be generalized or specify only a subset of patients. In fact, iPSC-studies on patients with COS suggested considerable heterogeneity between phenotypes in vitro such miRNA expression [83] and glutamate signaling [84].

Although genetically-informed selection for patients with SCZ is thought to benefit detection of disease relevant signals in heterogeneous cell samples, or even to reduce cellular heterogeneity during differentiation, the size of cohorts needed to reach this goal is still a matter of uncertainty [148]. Schwartzentruber et al. [142] have provided provisional insight on this issue: they identified thousands of quantitative trait loci regulating gene expression, chromatin accessibility, and RNA splicing during neuronal differentiation in a large iPSC-derived cell sample ($n = 123$). In light of this finding, iPSCs from 20–80 donors appear sufficient to detect the effect of common regulatory variants of moderate to large effect sizes. Remember, that effect sizes of certain CNVs associated with SCZ are among the strongest known so far for this disorder suggesting that cohort sizes needed for genetically-stratified patients with COS are probably in the lower range of this estimate.

SCZ is a highly heritable (see Section 3); however, environmental risk factors are likewise important to SCZ pathogenesis. Insight into the mechanisms mediating the interaction of risk genes with environmental risk factors remains an important endeavor to attain a more comprehensive picture of this disease. Since the effects of environmental factors on specific disease-relevant cell-types cannot assessed in living patients, iPSC-based studies may provide a tractable model for this purpose. According to the neurodevelopmental hypothesis of SCZ, early deviations may stay latent until called into operation through maturational processes. In analogy, many SCZ-associated processes may be hidden in simple monolayer iPSC-derived NPC/neuron cultures and may be only detected through activity-dependent processes arising from neuronal-activity or transcriptional activation in response to stimuli mimicking environmental exposures. For example, exposure of iPSC-derived neurons to Δ9 tetrahydrocannabinol (THC, a major compound of cannabis), either acutely or chronically, dampened the neuronal transcription response following depolarization and was associated with significant synaptic, mitochondrial, and glutamate signaling alterations [149]. While the final verdict about the causal nature of the cannabis–psychosis association is still out [150], we may attain nevertheless a better understanding of its potential cellular and molecular underpinnings from iPSC-based studies.

6.4. Organoids—From Structure to Function?

As of yet, current differentiation protocols applied to iPSC-based case/control studies on COS do not recreate the three-dimensional organization of the human brain and well-known structure-function relationships [151]. Remember, MRI studies on patients with COS implicated greater global connectedness concomitant to impaired short-range connectivity and disrupted modularity [29,30]; a pattern barely portrayed in 2D-culture.

Recent advances on the generation of region-specific brain organoids is anticipated to address this challenge, at least in part [152,153]. In this approach, pluripotent cells are used to reproduce in vitro key aspects of human brain development and function within three-dimensional structures termed 'brain organoids'. As the name suggests, 'brain organoid' is not the same as a 'brain', but represents a reductionist cellular system that recapitulates some aspects of the cellular composition and activity of the brain, and that in its generation follows at least some of the steps of early human embryonic brain development. Although today's brain organoids can give rise to active neurons and functional circuits, they do not match the anatomical organization or connectivity of the living brain [152,153]. At the same time, organoid-to-organoid variability in architecture and cell-type composition imposes as yet a severe hurdle on case/control studies. In a nutshell, brain organoids are presently barley suited as first-line screening tool in iPSC-based case/control studies, but may allow deepening insight into findings from well-defined monolayer cultures in a model closer to neurodevelopment in vivo. As an alternative approach to organoids, transplantation of iPSC-derived neuronal cells into embryonic or adult mice may help to recapitulate the physiology of SCZ more closely than 2D culture, and ideally highlight associated behavioral phenotypes. In support of this view, iPSC-derived cortical neurons from patients with Down syndrome showed increased synaptic stability and reduced oscillation relative to controls when transplanted in the adult mouse cortex [154].

All in all, COS remains a major challenge with earlier onset, more severe course, and poorer outcome relative to AOS. The need for an improved understanding of the cellular and molecular underpinnings of COS pathology persists despite recent progress on genetics, neuroimaging, and therapy. The transformative discovery of iPSCs [7] has paved the way for new translational strategies to trace early neurodevelopment deviations of COS in vitro. iPSC-based studies on patients with COS do not recreate the complex cellular and spatio-temporal phenotypes from the perinatal and adult brain, nor do they mimic early or late clinical symptoms of patients with COS in a dish. However, they create new opportunities to deliver actionable knowledge, i.e., genetic findings whose biological implications can be used to improve diagnosis, to develop rationale therapies, and craft mechanistic approaches to primary prevention. For example, iPSC-based disease modeling has led to drug repurposing [155] in amyotrophic lateral sclerosis (ALS): Hyperexcitability of iPSC-derived motor neurons from patients with ALS could be reversed by retigabine resulting in better survival of ALS motor neurons. Previously approved by the Federal Drug Administration FDA for the treatment of epilepsy, retigabine is now in clinical trial in ALS, encouraging the effort to use iPSC-derived models for development of new therapies, including drug screening, drug repurposing, and tailored treatments [156] for patients with COS. Beyond present progress, generation of iPSC-derived living neurons from patients with COS will not only transform our mindscape of this disease, but can also help to improve the lives of patients and their families.

Author Contributions: All authors contributed to the writing of the manuscript.

Funding: Michael Ziller is supported by BMBF grant 01ZX1504.

Acknowledgments: We are grateful to our colleagues for thoughtful discussions. Figure 1 is reproduced from Gogtay ([26] by permission of Oxford University Press, license 4445310299419), adapted by Greenstein et al. ([27] by permission of John Wiley and Sons, license 4445320517171). Figure 2 is adapted by Oscar Marín [53] by permission from Springer Nature (https://www.nature.com/nature/, license 4418131132006). Table 1 is adapted by Christian R. Marshall [49] by permission from Springer Nature (https://www.nature.com/nature/, license 4431820839821).

Conflicts of Interest: The authors declare no conflict of interest.

References

1. World Health Organization Mental Disorders Fact Sheet Schizophrenia 2016. Available online: http://www.who.int/mediacentre/factsheets/fs397/en/ (accessed on 18 September 2018).
2. Olfson, M.; Gerhard, T.; Huang, C.; Crystal, S.; Stroup, T.S. Premature Mortality among Adults with Schizophrenia in the United States. *JAMA Psychiatry* **2015**, *72*, 1172–1181. [CrossRef]
3. Thornicroft, G. Physical health disparities and mental illness: The scandal of premature mortality. *Br. J. Psychiatry* **2011**, *199*, 441–442. [CrossRef] [PubMed]
4. McKenna, K.; Gordon, C.T.; Lenane, M.; Kaysen, D.; Fahey, K.; Rapoport, J.L. Looking for childhood-onset schizophrenia: The first 71 cases screened. *J. Am. Acad. Child Adolesc. Psychiatry* **1994**, *33*, 636–644. [CrossRef] [PubMed]
5. Weinberger, D.R.; Harrison, P.J. (Eds.) *Schizophrenia*, 3rd ed.; Wiley-Blackwell: Chichester, UK; Hoboken, NJ, USA, 2011; ISBN 978-1-4051-7697-2.
6. Ahn, K.; Gotay, N.; Andersen, T.M.; Anvari, A.A.; Gochman, P.; Lee, Y.; Sanders, S.; Guha, S.; Darvasi, A.; Glessner, J.T.; et al. High rate of disease-related copy number variations in childhood onset schizophrenia. *Mol. Psychiatry* **2014**, *19*, 568–572. [CrossRef]
7. Takahashi, K.; Tanabe, K.; Ohnuki, M.; Narita, M.; Ichisaka, T.; Tomoda, K.; Yamanaka, S. Induction of pluripotent stem cells from adult human fibroblasts by defined factors. *Cell* **2007**, *131*, 861–872. [CrossRef] [PubMed]
8. Clemmensen, L.; Vernal, D.L.; Steinhausen, H.-C. A systematic review of the long-term outcome of early onset schizophrenia. *BMC Psychiatry* **2012**, *12*, 150. [CrossRef] [PubMed]
9. Shorter, E.; Wachtel, L.E. Childhood catatonia, autism and psychosis past and present: Is there an "iron triangle"? *Acta Psychiatr. Scand.* **2013**, *128*, 21–33. [CrossRef]
10. Mueser, K.T.; Jeste, D.V. *Clinical Handbook of Schizophrenia*; Guilford Press: New York, NJ, USA, 2008; ISBN 978-1-60623-045-9.
11. NHS Choices. Schizophrenia—Symptoms. 2018. Available online: https://www.nhs.uk/conditions/schizophrenia/symptoms/ (accessed on 18 September 2018).
12. Kinros, J.; Reichenberg, A.; Frangou, S. The neurodevelopmental theory of schizophrenia: Evidence from studies of early onset cases. *Isr. J. Psychiatry Relat. Sci.* **2010**, *47*, 110–117.
13. Hollis, C. Adult outcomes of child- and adolescent-onset schizophrenia: Diagnostic stability and predictive validity. *Am. J. Psychiatry* **2000**, *157*, 1652–1659. [CrossRef]

14. Remschmidt, H.; Martin, M.; Fleischhaker, C.; Theisen, F.M.; Hennighausen, K.; Gutenbrunner, C.; Schulz, E. Forty-two-years later: The outcome of childhood-onset schizophrenia. *J. Neural Transm.* **2007**, *114*, 505–512. [CrossRef]

15. Eggers, C.; Bunk, D. The long-term course of childhood-onset schizophrenia: A 42-year followup. *Schizophr. Bull.* **1997**, *23*, 105–117. [CrossRef] [PubMed]

16. Jacobsen, L.K.; Giedd, J.N.; Castellanos, F.X.; Vaituzis, A.C.; Hamburger, S.D.; Kumra, S.; Lenane, M.C.; Rapoport, J.L. Progressive reduction of temporal lobe structures in childhood-onset schizophrenia. *Am. J. Psychiatry* **1998**, *155*, 678–685. [CrossRef] [PubMed]

17. Rabinowitz, J.; Levine, S.Z.; Häfner, H. A population based elaboration of the role of age of onset on the course of schizophrenia. *Schizophr. Res.* **2006**, *88*, 96–101. [CrossRef] [PubMed]

18. Gornick, M.C.; Addington, A.M.; Sporn, A.; Gogtay, N.; Greenstein, D.; Lenane, M.; Gochman, P.; Ordonez, A.; Balkissoon, R.; Vakkalanka, R.; et al. Dysbindin (DTNBP1, 6p22.3) is associated with childhood-onset psychosis and endophenotypes measured by the Premorbid Adjustment Scale (PAS). *J. Autism Dev. Disord.* **2005**, *35*, 831–838. [CrossRef] [PubMed]

19. Vourdas, A.; Pipe, R.; Corrigall, R.; Frangou, S. Increased developmental deviance and premorbid dysfunction in early onset schizophrenia. *Schizophr. Res.* **2003**, *62*, 13–22. [CrossRef]

20. Hippius, H. The history of clozapine. *Psychopharmacology* **1989**, *99*, S3–S5. [CrossRef] [PubMed]

21. Kasoff, L.I.; Ahn, K.; Gochman, P.; Broadnax, D.D.; Rapoport, J.L. Strong Treatment Response and High Maintenance Rates of Clozapine in Childhood-Onset Schizophrenia. *J. Child Adolesc. Psychopharmacol.* **2016**, *26*, 428–435. [CrossRef] [PubMed]

22. Frazier, J.A.; Giedd, J.N.; Hamburger, S.D.; Albus, K.E.; Kaysen, D.; Vaituzis, A.C.; Rajapakse, J.C.; Lenane, M.C.; McKenna, K.; Jacobsen, L.K.; et al. Brain anatomic magnetic resonance imaging in childhood-onset schizophrenia. *Arch. Gen. Psychiatry* **1996**, *53*, 617–624. [CrossRef] [PubMed]

23. Sporn, A.L.; Greenstein, D.K.; Gogtay, N.; Jeffries, N.O.; Lenane, M.; Gochman, P.; Clasen, L.S.; Blumenthal, J.; Giedd, J.N.; Rapoport, J.L. Progressive brain volume loss during adolescence in childhood-onset schizophrenia. *Am. J. Psychiatry* **2003**, *160*, 2181–2189. [CrossRef] [PubMed]

24. Giedd, J.N.; Jeffries, N.O.; Blumenthal, J.; Castellanos, F.X.; Vaituzis, A.C.; Fernandez, T.; Hamburger, S.D.; Liu, H.; Nelson, J.; Bedwell, J.; et al. Childhood-onset schizophrenia: Progressive brain changes during adolescence. *Biol. Psychiatry* **1999**, *46*, 892–898. [CrossRef]

25. Gogtay, N.; Sporn, A.; Clasen, L.S.; Nugent, T.F.; Greenstein, D.; Nicolson, R.; Giedd, J.N.; Lenane, M.; Gochman, P.; Evans, A.; et al. Comparison of progressive cortical gray matter loss in childhood-onset schizophrenia with that in childhood-onset atypical psychoses. *Arch. Gen. Psychiatry* **2004**, *61*, 17–22. [CrossRef] [PubMed]

26. Gogtay, N.; Greenstein, D.; Lenane, M.; Clasen, L.; Sharp, W.; Gochman, P.; Butler, P.; Evans, A.; Rapoport, J. Cortical brain development in nonpsychotic siblings of patients with childhood-onset schizophrenia. *Arch. Gen. Psychiatry* **2007**, *64*, 772–780. [CrossRef] [PubMed]

27. Greenstein, D.; Lerch, J.; Shaw, P.; Clasen, L.; Giedd, J.; Gochman, P.; Rapoport, J.; Gogtay, N. Childhood onset schizophrenia: Cortical brain abnormalities as young adults. *J. Child Psychol. Psychiatry* **2006**, *47*, 1003–1012. [CrossRef] [PubMed]

28. Mattai, A.A.; Weisinger, B.; Greenstein, D.; Stidd, R.; Clasen, L.; Miller, R.; Tossell, J.W.; Rapoport, J.L.; Gogtay, N. Normalization of cortical gray matter deficits in nonpsychotic siblings of patients with childhood-onset schizophrenia. *J. Am. Acad. Child Adolesc. Psychiatry* **2011**, *50*, 697–704. [CrossRef] [PubMed]

29. Alexander-Bloch, A.F.; Gogtay, N.; Meunier, D.; Birn, R.; Clasen, L.; Lalonde, F.; Lenroot, R.; Giedd, J.; Bullmore, E.T. Disrupted modularity and local connectivity of brain functional networks in childhood-onset schizophrenia. *Front. Syst. Neurosci.* **2010**, *4*, 147. [CrossRef] [PubMed]

30. Alexander-Bloch, A.F.; Vértes, P.E.; Stidd, R.; Lalonde, F.; Clasen, L.; Rapoport, J.; Giedd, J.; Bullmore, E.T.; Gogtay, N. The anatomical distance of functional connections predicts brain network topology in health and schizophrenia. *Cereb. Cortex* **2013**, *23*, 127–138. [CrossRef] [PubMed]

31. Shi, F.; Yap, P.-T.; Gao, W.; Lin, W.; Gilmore, J.H.; Shen, D. Altered structural connectivity in neonates at genetic risk for schizophrenia: A combined study using morphological and white matter networks. *Neuroimage* **2012**, *62*, 1622–1633. [CrossRef]

32. Jakob, H.; Beckmann, H. Prenatal developmental disturbances in the limbic allocortex in schizophrenics. *J. Neural Transm.* **1986**, *65*, 303–326. [CrossRef]

33. Kovelman, J.A.; Scheibel, A.B. A neurohistological correlate of schizophrenia. *Biol. Psychiatry* **1984**, *19*, 1601–1621.

34. Akbarian, S.; Bunney, W.E.; Potkin, S.G.; Wigal, S.B.; Hagman, J.O.; Sandman, C.A.; Jones, E.G. Altered distribution of nicotinamide-adenine dinucleotide phosphate-diaphorase cells in frontal lobe of schizophrenics implies disturbances of cortical development. *Arch. Gen. Psychiatry* **1993**, *50*, 169–177. [CrossRef]

35. Bakhshi, K.; Chance, S.A. The neuropathology of schizophrenia: A selective review of past studies and emerging themes in brain structure and cytoarchitecture. *Neuroscience* **2015**, *303*, 82–102. [CrossRef] [PubMed]

36. Lichtenstein, P.; Yip, B.H.; Björk, C.; Pawitan, Y.; Cannon, T.D.; Sullivan, P.F.; Hultman, C.M. Common genetic determinants of schizophrenia and bipolar disorder in Swedish families: A population-based study. *Lancet* **2009**, *373*, 234–239. [CrossRef]

37. Wray, N.R.; Gottesman, I.I. Using summary data from the danish national registers to estimate heritabilities for schizophrenia, bipolar disorder, and major depressive disorder. *Front. Genet.* **2012**, *3*, 118. [CrossRef] [PubMed]

38. Sullivan, P.F.; Kendler, K.S.; Neale, M.C. Schizophrenia as a complex trait: Evidence from a meta-analysis of twin studies. *Arch. Gen. Psychiatry* **2003**, *60*, 1187–1192. [CrossRef] [PubMed]

39. Sullivan, P.F.; Daly, M.J.; O'Donovan, M. Genetic architectures of psychiatric disorders: The emerging picture and its implications. *Nat. Rev. Genet.* **2012**, *13*, 537–551. [CrossRef] [PubMed]

40. Kallmann, F.J.; Roth, B. Genetic aspects of preadolescent schizophrenia. *Am. J. Psychiatry* **1956**, *112*, 599–606. [CrossRef]

41. Asarnow, R.F.; Forsyth, J.K. Genetics of childhood-onset schizophrenia. *Child Adolesc. Psychiatr. Clin. N. Am.* **2013**, *22*, 675–687. [CrossRef]

42. Purcell, S.M.; Moran, J.L.; Fromer, M.; Ruderfer, D.; Solovieff, N.; Roussos, P.; O'Dushlaine, C.; Chambert, K.; Bergen, S.E.; Kähler, A.; et al. A polygenic burden of rare disruptive mutations in schizophrenia. *Nature* **2014**, *506*, 185–190. [CrossRef]

43. Schizophrenia Working Group of the Psychiatric Genomics Consortium. Biological insights from 108 schizophrenia-associated genetic loci. *Nature* **2014**, *511*, 421–427. [CrossRef]

44. Pardiñas, A.F.; Holmans, P.; Pocklington, A.J.; Escott-Price, V.; Ripke, S.; Carrera, N.; Legge, S.E.; Bishop, S.; Cameron, D.; Hamshere, M.L.; et al. Common schizophrenia alleles are enriched in mutation-intolerant genes and in regions under strong background selection. *Nat. Genet.* **2018**, *50*, 381–389. [CrossRef]

45. Schork, A.J.; Won, H.; Appadurai, V.; Nudel, R.; Gandal, M.; Delaneau, O.; Hougaard, D.; Baekved-Hansen, M.; Bybjerg-Grauholm, J.; Pedersen, M.G. A genome-wide association study for shared risk across major psychiatric disorders in a nation-wide birth cohort implicates fetal neurodevelopment as a key mediator. *bioRxiv* **2017**. [CrossRef]

46. Freedman, M.L.; Monteiro, A.N.A.; Gayther, S.A.; Coetzee, G.A.; Risch, A.; Plass, C.; Casey, G.; De Biasi, M.; Carlson, C.; Duggan, D.; et al. Principles for the post-GWAS functional characterization of cancer risk loci. *Nat. Genet.* **2011**, *43*, 513–518. [CrossRef] [PubMed]

47. Forrest, M.P.; Zhang, H.; Moy, W.; McGowan, H.; Leites, C.; Dionisio, L.E.; Xu, Z.; Shi, J.; Sanders, A.R.; Greenleaf, W.J.; et al. Open Chromatin Profiling in hiPSC-Derived Neurons Prioritizes Functional Noncoding Psychiatric Risk Variants and Highlights Neurodevelopmental Loci. *Cell Stem Cell* **2017**, *21*, 305–318.e8. [CrossRef]

48. Kirov, G. CNVs in neuropsychiatric disorders. *Hum. Mol. Genet.* **2015**, *24*, R45–R49. [CrossRef] [PubMed]

49. Marshall, C.R.; Howrigan, D.P.; Merico, D.; Thiruvahindrapuram, B.; Wu, W.; Greer, D.S.; Antaki, D.; Shetty, A.; Holmans, P.A.; Pinto, D.; et al. Contribution of copy number variants to schizophrenia from a genome-wide study of 41,321 subjects. *Nat. Genet.* **2017**, *49*, 27–35. [CrossRef] [PubMed]

50. Walsh, T.; McClellan, J.M.; McCarthy, S.E.; Addington, A.M.; Pierce, S.B.; Cooper, G.M.; Nord, A.S.; Kusenda, M.; Malhotra, D.; Bhandari, A.; et al. Rare structural variants disrupt multiple genes in neurodevelopmental pathways in schizophrenia. *Science* **2008**, *320*, 539–543. [CrossRef]

51. Addington, A.M.; Rapoport, J.L. The genetics of childhood-onset schizophrenia: When madness strikes the prepubescent. *Curr. Psychiatry Rep.* **2009**, *11*, 156–161. [CrossRef]

52. Boyle, E.A.; Li, Y.I.; Pritchard, J.K. An Expanded View of Complex Traits: From Polygenic to Omnigenic. *Cell* **2017**, *169*, 1177–1186. [CrossRef]

53. Marín, O. Developmental timing and critical windows for the treatment of psychiatric disorders. *Nat. Med.* **2016**, *22*, 1229–1238. [CrossRef]

54. Kandel, E.R.; Schwartz, J.H.; Jessel, T.M.; Siegelbaum, S.A.; Hudspeth, A.J.; Mack, S. (Eds.) *Principles of Neural Science*, 15th ed.; McGraw-Hill Medical: New York, NY, USA; Lisbon, Portugal; London, UK, 2013; ISBN 978-0-07-139011-8.

55. Molyneaux, B.J.; Arlotta, P.; Menezes, J.R.L.; Macklis, J.D. Neuronal subtype specification in the cerebral cortex. *Nat. Rev. Neurosci.* **2007**, *8*, 427–437. [CrossRef]

56. Taverna, E.; Götz, M.; Huttner, W.B. The cell biology of neurogenesis: Toward an understanding of the development and evolution of the neocortex. *Annu. Rev. Cell Dev. Biol.* **2014**, *30*, 465–502. [CrossRef] [PubMed]

57. Lewis, D.A.; González-Burgos, G. Neuroplasticity of neocortical circuits in schizophrenia. *Neuropsychopharmacology* **2008**, *33*, 141–165. [CrossRef] [PubMed]

58. Rakic, P.; Bourgeois, J.P.; Eckenhoff, M.F.; Zecevic, N.; Goldman-Rakic, P.S. Concurrent overproduction of synapses in diverse regions of the primate cerebral cortex. *Science* **1986**, *232*, 232–235. [CrossRef] [PubMed]

59. Turksen, K.; Nagy, A. (Eds.) *Induced Pluripotent Stem (iPS) Cells: Methods and Protocols*; Methods in Molecular Biology; Humana Press: New York, NY, USA, 2016; ISBN 978-1-4939-3055-5.

60. Verma, P.J.; Sumer, H. (Eds.) *Cell Reprogramming: Methods and Protocols*; Methods in Molecular Biology; Humana Press: New York, NY, USA, 2015; ISBN 978-1-4939-2847-7.

61. Brennand, K.; Savas, J.N.; Kim, Y.; Tran, N.; Simone, A.; Hashimoto-Torii, K.; Beaumont, K.G.; Kim, H.J.; Topol, A.; Ladran, I.; et al. Phenotypic differences in hiPSC NPCs derived from patients with schizophrenia. *Mol. Psychiatry* **2015**, *20*, 361–368. [CrossRef] [PubMed]

62. Burke, E.E.; Chenoweth, J.G.; Shin, J.H.; Collado-Torres, L.; Kim, S.K.; Micali, N.; Wang, Y.; Straub, R.E.; Hoeppner, D.J.; Chen, H.-Y. Dissecting transcriptomic signatures of neuronal differentiation and maturation using iPSCs. *bioRxiv* **2018**. [CrossRef]

63. Mariani, J.; Simonini, M.V.; Palejev, D.; Tomasini, L.; Coppola, G.; Szekely, A.M.; Horvath, T.L.; Vaccarino, F.M. Modeling human cortical development in vitro using induced pluripotent stem cells. *Proc. Natl. Acad. Sci. USA* **2012**, *109*, 12770–12775. [CrossRef] [PubMed]

64. Nicholas, C.R.; Chen, J.; Tang, Y.; Southwell, D.G.; Chalmers, N.; Vogt, D.; Arnold, C.M.; Chen, Y.-J.J.; Stanley, E.G.; Elefanty, A.G.; et al. Functional maturation of hPSC-derived forebrain interneurons requires an extended timeline and mimics human neural development. *Cell Stem Cell* **2013**, *12*, 573–586. [CrossRef]

65. Paşca, A.M.; Sloan, S.A.; Clarke, L.E.; Tian, Y.; Makinson, C.D.; Huber, N.; Kim, C.H.; Park, J.-Y.; O'Rourke, N.A.; Nguyen, K.D.; et al. Functional cortical neurons and astrocytes from human pluripotent stem cells in 3D culture. *Nat. Methods* **2015**, *12*, 671–678. [CrossRef] [PubMed]

66. Stein, J.L.; de la Torre-Ubieta, L.; Tian, Y.; Parikshak, N.N.; Hernández, I.A.; Marchetto, M.C.; Baker, D.K.; Lu, D.; Hinman, C.R.; Lowe, J.K.; et al. A quantitative framework to evaluate modeling of cortical development by neural stem cells. *Neuron* **2014**, *83*, 69–86. [CrossRef] [PubMed]

67. Brennand, K.J.; Simone, A.; Jou, J.; Gelboin-Burkhart, C.; Tran, N.; Sangar, S.; Li, Y.; Mu, Y.; Chen, G.; Yu, D.; et al. Modelling schizophrenia using human induced pluripotent stem cells. *Nature* **2011**, *473*, 221–225. [CrossRef]

68. Ahmad, R.; Sportelli, V.; Ziller, M.; Spengler, D.; Hoffmann, A. Tracing Early Neurodevelopment in Schizophrenia with Induced Pluripotent Stem Cells. *Cells* **2018**, *7*, 140. [CrossRef] [PubMed]

69. Hoffmann, A.; Sportelli, V.; Ziller, M.; Spengler, D. From the Psychiatrist's Couch to Induced Pluripotent Stem Cells: Bipolar Disease in a Dish. *Int. J. Mol. Sci.* **2018**, *19*, 770. [CrossRef] [PubMed]

70. Karayiorgou, M.; Simon, T.J.; Gogos, J.A. 22q11.2 microdeletions: Linking DNA structural variation to brain dysfunction and schizophrenia. *Nat. Rev. Neurosci.* **2010**, *11*, 402–416. [CrossRef] [PubMed]

71. Karayiorgou, M.; Morris, M.A.; Morrow, B.; Shprintzen, R.J.; Goldberg, R.; Borrow, J.; Gos, A.; Nestadt, G.; Wolyniec, P.S.; Lasseter, V.K. Schizophrenia susceptibility associated with interstitial deletions of chromosome 22q11. *Proc. Natl. Acad. Sci. USA* **1995**, *92*, 7612–7616. [CrossRef]

72. Bassett, A.S.; Chow, E.W.C. Schizophrenia and 22q11.2 deletion syndrome. *Curr. Psychiatry Rep.* **2008**, *10*, 148–157. [CrossRef]

73. Pedrosa, E.; Sandler, V.; Shah, A.; Carroll, R.; Chang, C.; Rockowitz, S.; Guo, X.; Zheng, D.; Lachman, H.M. Development of patient-specific neurons in schizophrenia using induced pluripotent stem cells. *J. Neurogenet.* **2011**, *25*, 88–103. [CrossRef]

74. Belinsky, G.S.; Rich, M.T.; Sirois, C.L.; Short, S.M.; Pedrosa, E.; Lachman, H.M.; Antic, S.D. Patch-clamp recordings and calcium imaging followed by single-cell PCR reveal the developmental profile of 13 genes in iPSC-derived human neurons. *Stem Cell Res.* **2014**, *12*, 101–118. [CrossRef]

75. Bundo, M.; Toyoshima, M.; Okada, Y.; Akamatsu, W.; Ueda, J.; Nemoto-Miyauchi, T.; Sunaga, F.; Toritsuka, M.; Ikawa, D.; Kakita, A.; et al. Increased l1 retrotransposition in the neuronal genome in schizophrenia. *Neuron* **2014**, *81*, 306–313. [CrossRef]

76. Zhao, D.; Lin, M.; Chen, J.; Pedrosa, E.; Hrabovsky, A.; Fourcade, H.M.; Zheng, D.; Lachman, H.M. MicroRNA Profiling of Neurons Generated Using Induced Pluripotent Stem Cells Derived from Patients with Schizophrenia and Schizoaffective Disorder, and 22q11.2 Del. *PLoS ONE* **2015**, *10*, e0132387. [CrossRef]

77. Lin, M.; Pedrosa, E.; Hrabovsky, A.; Chen, J.; Puliafito, B.R.; Gilbert, S.R.; Zheng, D.; Lachman, H.M. Integrative transcriptome network analysis of iPSC-derived neurons from schizophrenia and schizoaffective disorder patients with 22q11. 2 deletion. *BMC Syst. Biol.* **2016**, *10*, 105. [CrossRef]

78. Toyoshima, M.; Akamatsu, W.; Okada, Y.; Ohnishi, T.; Balan, S.; Hisano, Y.; Iwayama, Y.; Toyota, T.; Matsumoto, T.; Itasaka, N.; et al. Analysis of induced pluripotent stem cells carrying 22q11.2 deletion. *Transl. Psychiatry* **2016**, *6*, e934. [CrossRef] [PubMed]

79. Yoon, K.-J.; Nguyen, H.N.; Ursini, G.; Zhang, F.; Kim, N.-S.; Wen, Z.; Makri, G.; Nauen, D.; Shin, J.H.; Park, Y.; et al. Modeling a genetic risk for schizophrenia in iPSCs and mice reveals neural stem cell deficits associated with adherens junctions and polarity. *Cell Stem Cell* **2014**, *15*, 79–91. [CrossRef] [PubMed]

80. Das, D.K.; Tapias, V.; Chowdari, K.V.; Francis, L.; Zhi, Y.; Ghosh, A.; Surti, U.; Tischfield, J.; Sheldon, M.; Moore, J.C. Genetic and morphological features of human iPSC-derived neurons with chromosome 15q11. 2 (BP1-BP2) deletions. *Mol. Neuropsychiatry* **2015**, *1*, 116–123. [CrossRef] [PubMed]

81. Pak, C.; Danko, T.; Zhang, Y.; Aoto, J.; Anderson, G.; Maxeiner, S.; Yi, F.; Wernig, M.; Südhof, T.C. Human Neuropsychiatric Disease Modeling using Conditional Deletion Reveals Synaptic Transmission Defects Caused by Heterozygous Mutations in NRXN1. *Cell Stem Cell* **2015**, *17*, 316–328. [CrossRef] [PubMed]

82. Deshpande, A.; Yadav, S.; Dao, D.Q.; Wu, Z.-Y.; Hokanson, K.C.; Cahill, M.K.; Wiita, A.P.; Jan, Y.-N.; Ullian, E.M.; Weiss, L.A. Cellular Phenotypes in Human iPSC-Derived Neurons from a Genetic Model of Autism Spectrum Disorder. *Cell Rep.* **2017**, *21*, 2678–2687. [CrossRef] [PubMed]

83. Topol, A.; Zhu, S.; Hartley, B.J.; English, J.; Hauberg, M.E.; Tran, N.; Rittenhouse, C.A.; Simone, A.; Ruderfer, D.M.; Johnson, J.; et al. Dysregulation of miRNA-9 in a Subset of Schizophrenia Patient-Derived Neural Progenitor Cells. *Cell Rep.* **2016**, *15*, 1024–1036. [CrossRef] [PubMed]

84. Xu, J.; Hartley, B.J.; Kurup, P.; Phillips, A.; Topol, A.; Xu, M.; Ononenyi, C.; Foscue, E.; Ho, S.-M.; Baguley, T.D.; et al. Inhibition of STEP61 ameliorates deficits in mouse and hiPSC-based schizophrenia models. *Mol. Psychiatry* **2018**, *23*, 271–281. [CrossRef]

85. Hoffman, G.E.; Hartley, B.J.; Flaherty, E.; Ladran, I.; Gochman, P.; Ruderfer, D.M.; Stahl, E.A.; Rapoport, J.; Sklar, P.; Brennand, K.J. Transcriptional signatures of schizophrenia in hiPSC-derived NPCs and neurons are concordant with post-mortem adult brains. *Nat. Commun.* **2017**, *8*, 2225. [CrossRef]

86. Jun, G.; Flickinger, M.; Hetrick, K.N.; Romm, J.M.; Doheny, K.F.; Abecasis, G.R.; Boehnke, M.; Kang, H.M. Detecting and estimating contamination of human DNA samples in sequencing and array-based genotype data. *Am. J. Hum. Genet.* **2012**, *91*, 839–848. [CrossRef]

87. Muotri, A.R.; Gage, F.H. Generation of neuronal variability and complexity. *Nature* **2006**, *441*, 1087–1093. [CrossRef]

88. Muotri, A.R.; Chu, V.T.; Marchetto, M.C.N.; Deng, W.; Moran, J.V.; Gage, F.H. Somatic mosaicism in neuronal precursor cells mediated by L1 retrotransposition. *Nature* **2005**, *435*, 903–910. [CrossRef]

89. Coufal, N.G.; Garcia-Perez, J.L.; Peng, G.E.; Yeo, G.W.; Mu, Y.; Lovci, M.T.; Morell, M.; O'Shea, K.S.; Moran, J.V.; Gage, F.H. L1 retrotransposition in human neural progenitor cells. *Nature* **2009**, *460*, 1127–1131. [CrossRef] [PubMed]

90. Muotri, A.R.; Marchetto, M.C.N.; Coufal, N.G.; Oefner, R.; Yeo, G.; Nakashima, K.; Gage, F.H. L1 retrotransposition in neurons is modulated by MeCP2. *Nature* **2010**, *468*, 443–446. [CrossRef] [PubMed]

91. Rajman, M.; Schratt, G. MicroRNAs in neural development: From master regulators to fine-tuners. *Development* **2017**, *144*, 2310–2322. [CrossRef] [PubMed]

92. Sun, A.X.; Crabtree, G.R.; Yoo, A.S. MicroRNAs: Regulators of neuronal fate. *Curr. Opin. Cell Biol.* **2013**, *25*, 215–221. [CrossRef] [PubMed]

93. Bartel, D.P. MicroRNAs: Target recognition and regulatory functions. *Cell* **2009**, *136*, 215–233. [CrossRef]

94. O'Connor, R.M.; Gururajan, A.; Dinan, T.G.; Kenny, P.J.; Cryan, J.F. All Roads Lead to the miRNome: miRNAs Have a Central Role in the Molecular Pathophysiology of Psychiatric Disorders. *Trends Pharmacol. Sci.* **2016**, *37*, 1029–1044. [CrossRef]

95. Cross-Disorder Group of the Psychiatric Genomics Consortium Identification of risk loci with shared effects on five major psychiatric disorders: A genome-wide analysis. *Lancet* **2013**, *381*, 1371–1379. [CrossRef]

96. Ripke, S.; O'Dushlaine, C.; Chambert, K.; Moran, J.L.; Kähler, A.K.; Akterin, S.; Bergen, S.E.; Collins, A.L.; Crowley, J.J.; Fromer, M.; et al. Genome-wide association analysis identifies 13 new risk loci for schizophrenia. *Nat. Genet.* **2013**, *45*, 1150–1159. [CrossRef]

97. Kwon, E.; Wang, W.; Tsai, L.-H. Validation of schizophrenia-associated genes CSMD1, C10orf26, CACNA1C and TCF4 as miR-137 targets. *Mol. Psychiatry* **2013**, *18*, 11–12. [CrossRef]

98. Lett, T.A.; Chakravarty, M.M.; Chakavarty, M.M.; Felsky, D.; Brandl, E.J.; Tiwari, A.K.; Gonçalves, V.F.; Rajji, T.K.; Daskalakis, Z.J.; Meltzer, H.Y.; et al. The genome-wide supported microRNA-137 variant predicts phenotypic heterogeneity within schizophrenia. *Mol. Psychiatry* **2013**, *18*, 443–450. [CrossRef] [PubMed]

99. Petri, R.; Malmevik, J.; Fasching, L.; Åkerblom, M.; Jakobsson, J. miRNAs in brain development. *Exp. Cell Res.* **2014**, *321*, 84–89. [CrossRef]

100. Forstner, A.J.; Degenhardt, F.; Schratt, G.; Nöthen, M.M. MicroRNAs as the cause of schizophrenia in 22q11.2 deletion carriers, and possible implications for idiopathic disease: A mini-review. *Front. Mol. Neurosci.* **2013**, *6*, 47. [CrossRef] [PubMed]

101. Takumi, T.; Tamada, K. CNV biology in neurodevelopmental disorders. *Curr. Opin. Neurobiol.* **2018**, *48*, 183–192. [CrossRef]

102. Cox, D.M.; Butler, M.G. The 15q11.2 BP1-BP2 microdeletion syndrome: A review. *Int. J. Mol. Sci.* **2015**, *16*, 4068–4082. [CrossRef] [PubMed]

103. Stefansson, H.; Meyer-Lindenberg, A.; Steinberg, S.; Magnusdottir, B.; Morgen, K.; Arnarsdottir, S.; Bjornsdottir, G.; Walters, G.B.; Jonsdottir, G.A.; Doyle, O.M.; et al. CNVs conferring risk of autism or schizophrenia affect cognition in controls. *Nature* **2014**, *505*, 361–366. [CrossRef] [PubMed]

104. De Wolf, V.; Brison, N.; Devriendt, K.; Peeters, H. Genetic counseling for susceptibility loci and neurodevelopmental disorders: The del15q11.2 as an example. *Am. J. Med. Genet. A* **2013**, *161A*, 2846–2854. [CrossRef]

105. Abekhoukh, S.; Bardoni, B. CYFIP family proteins between autism and intellectual disability: Links with Fragile X syndrome. *Front. Cell. Neurosci.* **2014**, *8*, 81. [CrossRef]

106. Oguro-Ando, A.; Rosensweig, C.; Herman, E.; Nishimura, Y.; Werling, D.; Bill, B.R.; Berg, J.M.; Gao, F.; Coppola, G.; Abrahams, B.S.; et al. Increased CYFIP1 dosage alters cellular and dendritic morphology and dysregulates mTOR. *Mol. Psychiatry* **2015**, *20*, 1069–1078. [CrossRef]

107. Pathania, M.; Davenport, E.C.; Muir, J.; Sheehan, D.F.; López-Doménech, G.; Kittler, J.T. The autism and schizophrenia associated gene CYFIP1 is critical for the maintenance of dendritic complexity and the stabilization of mature spines. *Transl. Psychiatry* **2014**, *4*, e374. [CrossRef]

108. Forrest, M.P.; Parnell, E.; Penzes, P. Dendritic structural plasticity and neuropsychiatric disease. *Nat. Rev. Neurosci.* **2018**, *19*, 215–234. [CrossRef] [PubMed]

109. Dimitrion, P.; Zhi, Y.; Clayton, D.; Apodaca, G.L.; Wilcox, M.R.; Johnson, J.W.; Nimgaonkar, V.; D'Aiuto, L. Low-Density Neuronal Cultures from Human Induced Pluripotent Stem Cells. *Mol. Neuropsychiatry* **2017**, *3*, 28–36. [CrossRef] [PubMed]

110. Hoeffding, L.K.E.; Hansen, T.; Ingason, A.; Doung, L.; Thygesen, J.H.; Møller, R.S.; Tommerup, N.; Kirov, G.; Rujescu, D.; Larsen, L.A. Sequence analysis of 17 NRXN1 deletions. *Am. J. Med. Genet. B* **2014**, *165*, 52–61. [CrossRef] [PubMed]

111. Südhof, T.C. Synaptic Neurexin Complexes: A Molecular Code for the Logic of Neural Circuits. *Cell* **2017**, *171*, 745–769. [CrossRef] [PubMed]

112. Rees, E.; Walters, J.T.R.; Georgieva, L.; Isles, A.R.; Chambert, K.D.; Richards, A.L.; Mahoney-Davies, G.; Legge, S.E.; Moran, J.L.; McCarroll, S.A.; et al. Analysis of copy number variations at 15 schizophrenia-associated loci. *Br. J. Psychiatry* **2014**, *204*, 108–114. [CrossRef] [PubMed]

113. Jenkins, A.; Apud, J.A.; Zhang, F.; Decot, H.; Weinberger, D.R.; Law, A.J. Identification of candidate single-nucleotide polymorphisms in NRXN1 related to antipsychotic treatment response in patients with schizophrenia. *Neuropsychopharmacology* **2014**, *39*, 2170–2178. [CrossRef] [PubMed]

114. Dabell, M.P.; Rosenfeld, J.A.; Bader, P.; Escobar, L.F.; El-Khechen, D.; Vallee, S.E.; Dinulos, M.B.P.; Curry, C.; Fisher, J.; Tervo, R.; et al. Investigation of NRXN1 deletions: Clinical and molecular characterization. *Am. J. Med. Genet. A* **2013**, *161A*, 717–731. [CrossRef]

115. Etherton, M.R.; Blaiss, C.A.; Powell, C.M.; Südhof, T.C. Mouse neurexin-1α deletion causes correlated electrophysiological and behavioral changes consistent with cognitive impairments. *Proc. Natl. Acad. Sci. USA* **2009**, *106*, 17998. [CrossRef]

116. Deshpande, A.; Weiss, L.A. Recurrent reciprocal copy number variants: Roles and rules in neurodevelopmental disorders. *Dev. Neurobiol.* **2018**, *78*, 519–530. [CrossRef]

117. Qureshi, A.Y.; Mueller, S.; Snyder, A.Z.; Mukherjee, P.; Berman, J.I.; Roberts, T.P.L.; Nagarajan, S.S.; Spiro, J.E.; Chung, W.K.; Sherr, E.H.; et al. Opposing brain differences in 16p11.2 deletion and duplication carriers. *J. Neurosci.* **2014**, *34*, 11199–11211. [CrossRef]

118. Lin, G.N.; Corominas, R.; Lemmens, I.; Yang, X.; Tavernier, J.; Hill, D.E.; Vidal, M.; Sebat, J.; Iakoucheva, L.M. Spatiotemporal 16p11.2 protein network implicates cortical late mid-fetal brain development and KCTD13-Cul3-RhoA pathway in psychiatric diseases. *Neuron* **2015**, *85*, 742–754. [CrossRef] [PubMed]

119. Golzio, C.; Willer, J.; Talkowski, M.E.; Oh, E.C.; Taniguchi, Y.; Jacquemont, S.; Reymond, A.; Sun, M.; Sawa, A.; Gusella, J.F.; et al. KCTD13 is a major driver of mirrored neuroanatomical phenotypes of the 16p11.2 copy number variant. *Nature* **2012**, *485*, 363–367. [CrossRef] [PubMed]

120. Harrison, P.J.; Weinberger, D.R. Schizophrenia genes, gene expression, and neuropathology: On the matter of their convergence. *Mol. Psychiatry* **2005**, *10*, 40–68. [CrossRef] [PubMed]

121. Zhao, C.; Sun, G.; Li, S.; Shi, Y. A feedback regulatory loop involving microRNA-9 and nuclear receptor TLX in neural stem cell fate determination. *Nat. Struct. Mol. Biol.* **2009**, *16*, 365–371. [CrossRef] [PubMed]

122. Hauberg, M.E.; Roussos, P.; Grove, J.; Børglum, A.D.; Mattheisen, M. Schizophrenia Working Group of the Psychiatric Genomics Consortium Analyzing the Role of MicroRNAs in Schizophrenia in the Context of Common Genetic Risk Variants. *JAMA Psychiatry* **2016**, *73*, 369–377. [CrossRef] [PubMed]

123. Carty, N.C.; Xu, J.; Kurup, P.; Brouillette, J.; Goebel-Goody, S.M.; Austin, D.R.; Yuan, P.; Chen, G.; Correa, P.R.; Haroutunian, V.; et al. The tyrosine phosphatase STEP: Implications in schizophrenia and the molecular mechanism underlying antipsychotic medications. *Transl. Psychiatry* **2012**, *2*, e137. [CrossRef] [PubMed]

124. Mei, L.; Xiong, W.-C. Neuregulin 1 in neural development, synaptic plasticity and schizophrenia. *Nat. Rev. Neurosci.* **2008**, *9*, 437–452. [CrossRef]

125. Coyle, J.T. NMDA receptor and schizophrenia: A brief history. *Schizophr. Bull.* **2012**, *38*, 920–926. [CrossRef]

126. Cross-Disorder Group of the Psychiatric Genomics Consortium; Lee, S.H.; Ripke, S.; Neale, B.M.; Faraone, S.V.; Purcell, S.M.; Perlis, R.H.; Mowry, B.J.; Thapar, A.; Goddard, M.E.; et al. Genetic relationship between five psychiatric disorders estimated from genome-wide SNPs. *Nat. Genet.* **2013**, *45*, 984–994. [CrossRef]

127. Luhmann, H.J.; Fukuda, A.; Kilb, W. Control of cortical neuronal migration by glutamate and GABA. *Front. Cell. Neurosci.* **2015**, *9*, 4. [CrossRef]

128. Schlaeger, T.M.; Daheron, L.; Brickler, T.R.; Entwisle, S.; Chan, K.; Cianci, A.; DeVine, A.; Ettenger, A.; Fitzgerald, K.; Godfrey, M.; et al. A comparison of non-integrating reprogramming methods. *Nat. Biotechnol.* **2015**, *33*, 58–63. [CrossRef] [PubMed]

129. Kwon, E.M.; Connelly, J.P.; Hansen, N.F.; Donovan, F.X.; Winkler, T.; Davis, B.W.; Alkadi, H.; Chandrasekharappa, S.C.; Dunbar, C.E.; Mullikin, J.C.; et al. iPSCs and fibroblast subclones from the same fibroblast population contain comparable levels of sequence variations. *Proc. Natl. Acad. Sci. USA* **2017**, *114*, 1964–1969. [CrossRef] [PubMed]

130. Banovich, N.E.; Li, Y.I.; Raj, A.; Ward, M.C.; Greenside, P.; Calderon, D.; Tung, P.Y.; Burnett, J.E.; Myrthil, M.; Thomas, S.M.; et al. Impact of regulatory variation across human iPSCs and differentiated cells. *Genome Res.* **2018**, *28*, 122–131. [CrossRef] [PubMed]

131. Burrows, C.K.; Banovich, N.E.; Pavlovic, B.J.; Patterson, K.; Gallego Romero, I.; Pritchard, J.K.; Gilad, Y. Genetic Variation, Not Cell Type of Origin, Underlies the Majority of Identifiable Regulatory Differences in iPSCs. *PLoS Genet.* **2016**, *12*, e1005793. [CrossRef] [PubMed]

132. Carcamo-Orive, I.; Hoffman, G.E.; Cundiff, P.; Beckmann, N.D.; D'Souza, S.L.; Knowles, J.W.; Patel, A.; Papatsenko, D.; Abbasi, F.; Reaven, G.M.; et al. Analysis of Transcriptional Variability in a Large Human iPSC Library Reveals Genetic and Non-genetic Determinants of Heterogeneity. *Cell Stem Cell* **2017**, *20*, 518–532.e9. [CrossRef] [PubMed]

133. Kilpinen, H.; Goncalves, A.; Leha, A.; Afzal, V.; Alasoo, K.; Ashford, S.; Bala, S.; Bensaddek, D.; Casale, F.P.; Culley, O.J.; et al. Common genetic variation drives molecular heterogeneity in human iPSCs. *Nature* **2017**, *546*, 370–375. [CrossRef]

134. Rouhani, F.; Kumasaka, N.; de Brito, M.C.; Bradley, A.; Vallier, L.; Gaffney, D. Genetic background drives transcriptional variation in human induced pluripotent stem cells. *PLoS Genet.* **2014**, *10*, e1004432. [CrossRef]

135. Ziller, M.J.; Gu, H.; Müller, F.; Donaghey, J.; Tsai, L.T.-Y.; Kohlbacher, O.; De Jager, P.L.; Rosen, E.D.; Bennett, D.A.; Bernstein, B.E.; et al. Charting a dynamic DNA methylation landscape of the human genome. *Nature* **2013**, *500*, 477–481. [CrossRef]

136. Kyttälä, A.; Moraghebi, R.; Valensisi, C.; Kettunen, J.; Andrus, C.; Pasumarthy, K.K.; Nakanishi, M.; Nishimura, K.; Ohtaka, M.; Weltner, J.; et al. Genetic Variability Overrides the Impact of Parental Cell Type and Determines iPSC Differentiation Potential. *Stem Cell Rep.* **2016**, *6*, 200–212. [CrossRef]

137. Bock, C.; Kiskinis, E.; Verstappen, G.; Gu, H.; Boulting, G.; Smith, Z.D.; Ziller, M.; Croft, G.F.; Amoroso, M.W.; Oakley, D.H.; et al. Reference Maps of human ES and iPS cell variation enable high-throughput characterization of pluripotent cell lines. *Cell* **2011**, *144*, 439–452. [CrossRef]

138. Hu, B.-Y.; Weick, J.P.; Yu, J.; Ma, L.-X.; Zhang, X.-Q.; Thomson, J.A.; Zhang, S.-C. Neural differentiation of human induced pluripotent stem cells follows developmental principles but with variable potency. *Proc. Natl. Acad. Sci. USA* **2010**, *107*, 4335–4340. [CrossRef] [PubMed]

139. Osafune, K.; Caron, L.; Borowiak, M.; Martinez, R.J.; Fitz-Gerald, C.S.; Sato, Y.; Cowan, C.A.; Chien, K.R.; Melton, D.A. Marked differences in differentiation propensity among human embryonic stem cell lines. *Nat. Biotechnol.* **2008**, *26*, 313–315. [CrossRef] [PubMed]

140. Wu, H.; Xu, J.; Pang, Z.P.; Ge, W.; Kim, K.J.; Blanchi, B.; Chen, C.; Südhof, T.C.; Sun, Y.E. Integrative genomic and functional analyses reveal neuronal subtype differentiation bias in human embryonic stem cell lines. *Proc. Natl. Acad. Sci. USA* **2007**, *104*, 13821–13826. [CrossRef] [PubMed]

141. Hoffmann, A.; Sportelli, V.; Ziller, M.; Spengler, D. Switch-Like Roles for Polycomb Proteins from Neurodevelopment to Neurodegeneration. *Epigenomes* **2017**, *1*, 21. [CrossRef]

142. Schwartzentruber, J.; Foskolou, S.; Kilpinen, H.; Rodrigues, J.; Alasoo, K.; Knights, A.J.; Patel, M.; Goncalves, A.; Ferreira, R.; Benn, C.L.; et al. Molecular and functional variation in iPSC-derived sensory neurons. *Nat. Genet.* **2018**, *50*, 54–61. [CrossRef] [PubMed]

143. Nehme, R.; Zuccaro, E.; Ghosh, S.D.; Li, C.; Sherwood, J.L.; Pietilainen, O.; Barrett, L.E.; Limone, F.; Worringer, K.A.; Kommineni, S.; et al. Combining NGN2 Programming with Developmental Patterning Generates Human Excitatory Neurons with NMDAR-Mediated Synaptic Transmission. *Cell Rep.* **2018**, *23*, 2509–2523. [CrossRef] [PubMed]

144. Sagal, J.; Zhan, X.; Xu, J.; Tilghman, J.; Karuppagounder, S.S.; Chen, L.; Dawson, V.L.; Dawson, T.M.; Laterra, J.; Ying, M. Proneural transcription factor Atoh1 drives highly efficient differentiation of human pluripotent stem cells into dopaminergic neurons. *Stem Cells Transl. Med.* **2014**, *3*, 888–898. [CrossRef] [PubMed]

145. Yang, N.; Chanda, S.; Marro, S.; Ng, Y.-H.; Janas, J.A.; Haag, D.; Ang, C.E.; Tang, Y.; Flores, Q.; Mall, M.; et al. Generation of pure GABAergic neurons by transcription factor programming. *Nat. Methods* **2017**, *14*, 621–628. [CrossRef] [PubMed]

146. Zhang, Y.; Pak, C.; Han, Y.; Ahlenius, H.; Zhang, Z.; Chanda, S.; Marro, S.; Patzke, C.; Acuna, C.; Covy, J.; et al. Rapid single-step induction of functional neurons from human pluripotent stem cells. *Neuron* **2013**, *78*, 785–798. [CrossRef] [PubMed]

147. Birnbaum, K.D. Power in Numbers: Single-Cell RNA-Seq Strategies to Dissect Complex Tissues. *Annu. Rev. Genet.* **2018**. [CrossRef] [PubMed]

148. Hoekstra, S.D.; Stringer, S.; Heine, V.M.; Posthuma, D. Genetically-Informed Patient Selection for iPSC Studies of Complex Diseases May Aid in Reducing Cellular Heterogeneity. *Front. Cell. Neurosci.* **2017**, *11*, 164. [CrossRef] [PubMed]

149. Guennewig, B.; Bitar, M.; Obiorah, I.; Hanks, J.; O'Brien, E.A.; Kaczorowski, D.C.; Hurd, Y.L.; Roussos, P.; Brennand, K.J.; Barry, G. THC exposure of human iPSC neurons impacts genes associated with neuropsychiatric disorders. *Transl. Psychiatry* **2018**, *8*, 89. [CrossRef] [PubMed]

150. Ksir, C.; Hart, C.L. Cannabis and Psychosis: A Critical Overview of the Relationship. *Curr. Psychiatry Rep.* **2016**, *18*, 12. [CrossRef] [PubMed]

151. Honey, C.J.; Kötter, R.; Breakspear, M.; Sporns, O. Network structure of cerebral cortex shapes functional connectivity on multiple time scales. *Proc. Natl. Acad. Sci. USA* **2007**, *104*, 10240–10245. [CrossRef] [PubMed]

152. Arlotta, P. Organoids required! A new path to understanding human brain development and disease. *Nat. Methods* **2018**, *15*, 27. [CrossRef] [PubMed]

153. Brown, J.; Quadrato, G.; Arlotta, P. Studying the Brain in a Dish: 3D Cell Culture Models of Human Brain Development and Disease. *Curr. Top. Dev. Biol.* **2018**, *129*, 99–122. [CrossRef] [PubMed]

154. Real, R.; Peter, M.; Trabalza, A.; Khan, S.; Smith, M.A.; Dopp, J.; Barnes, S.J.; Momoh, A.; Strano, A.; Volpi, E.; et al. In vivo modeling of human neuron dynamics and Down syndrome. *Science* **2018**. [CrossRef]

155. Wainger, B.J.; Kiskinis, E.; Mellin, C.; Wiskow, O.; Han, S.S.W.; Sandoe, J.; Perez, N.P.; Williams, L.A.; Lee, S.; Boulting, G.; et al. Intrinsic membrane hyperexcitability of amyotrophic lateral sclerosis patient-derived motor neurons. *Cell Rep.* **2014**, *7*, 1–11. [CrossRef]

156. Gandal, M.J.; Leppa, V.; Won, H.; Parikshak, N.N.; Geschwind, D.H. The road to precision psychiatry: Translating genetics into disease mechanisms. *Nat. Neurosci.* **2016**, *19*, 1397–1407. [CrossRef]

International Journal of
Molecular Sciences

Article

FKBP5 Gene Expression Predicts Antidepressant Treatment Outcome in Depression

Marcus Ising [1,*], Giuseppina Maccarrone [1], Tanja Brückl [1], Sandra Scheuer [1], Johannes Hennings [2], Florian Holsboer [1,3], Christoph W. Turck [1], Manfred Uhr [1] and Susanne Lucae [1]

[1] Max Planck Institute of Psychiatry, 80804 Munich, Germany; maccarrone@psych.mpg.de (G.M.); brueckl@psych.mpg.de (T.B.); scheuer.sandra@gmail.com (S.S.); florian.holsboer@hmnc.de (F.H.); turck@psych.mpg.de (C.W.T.); uhr@psych.mpg.de (M.U.); lucae@psych.mpg.de (S.L.)
[2] kbo-Isar-Amper Clinical Center Munich East, 85540 Munich, Germany; johannes.hennings@kbo.de
[3] HMNC Brain Health GmbH, 80807 Munich, Germany
* Correspondence: ising@psych.mpg.de; Tel.: +49-89-30622-430

Received: 9 December 2018; Accepted: 20 January 2019; Published: 23 January 2019

Abstract: Adverse experiences and chronic stress are well-known risk factors for the development of major depression, and an impaired stress response regulation is frequently observed in acute depression. Impaired glucocorticoid receptor (GR) signalling plays an important role in these alterations, and a restoration of GR signalling appears to be a prerequisite of successful antidepressant treatment. Variants in genes of the stress response regulation contribute to the vulnerability to depression in traumatized subjects. Consistent findings point to an important role of *FKBP5*, the gene expressing FK506-binding protein 51 (FKBP51), which is a strong inhibitor of the GR, and thus, an important regulator of the stress response. We investigated the role of *FKBP5* and FKB51 expression with respect to stress response regulation and antidepressant treatment outcome in depressed patients. This study included 297 inpatients, who participated in the Munich Antidepressant Response Signature (MARS) project and were treated for acute depression. In this open-label study, patients received antidepressant treatment according to the attending doctor's choice. In addition to the *FKBP5* genotype, changes in blood FKBP51 expression during antidepressant treatment were analyzed using RT-PCR and ZeptoMARK[TM] reverse phase protein microarray (RPPM). Stress response regulation was evaluated in a subgroup of patients using the combined dexamethasone (dex)/corticotropin releasing hormone (CRH) test. As expected, increased FKBP51 expression was associated with an impaired stress response regulation at baseline and after six weeks was accompanied by an elevated cortisol response to the combined dex/CRH test. Further, we demonstrated an active involvement of FKBP51 in antidepressant treatment outcome. While patients responding to antidepressant treatment had a pronounced reduction of *FKBP5* gene and FKBP51 protein expression, increasing expression levels were observed in nonresponders. This effect was moderated by the genotype of the *FKBP5* single nucleotide polymorphism (SNP) rs1360780, with carriers of the minor allele showing the most pronounced association. Our findings demonstrate that *FKBP5* and, specifically, its expression product FKBP51 are important modulators of antidepressant treatment outcome, pointing to a new, promising target for future antidepressant drug development.

Keywords: depression; antidepressant treatment; HPA axis; gene expression; *FKBP5*; FKBP51

---

1. Introduction

Depression is a very serious and highly prevalent mental disorder. Epidemiological studies suggest an average annual prevalence rate of 5–6% across different cultures, which increases to 10–15% over a lifetime [1]. Depression is also a highly recurrent disorder with more than half of first-episode patients experiencing a second episode or more [2]. It is a highly disabling disorder, ranking third among all causes of time spent living in disability, with only low-back pain and headache disorders causing longer periods of disability [3]. As depression is a multifactorial disorder, genetic and environmental factors also contribute substantially to depression risk and outcome development [4]. Indeed, twin and family studies suggest a 35–40% contribution of genetic factors to disease liability, while the remaining risk variance is best explained by individual environmental events and biographic circumstances [4,5]. Specifically, early adverse life experience has been frequently identified as an important environmental risk factor for adult depression [6,7], and severe and/or long-lasting stressors can trigger new disease episodes in vulnerable individuals [6,8].

In fact, acute depression is frequently accompanied by a disturbed stress response regulation, which is indicated, for instance, by elevated endocrine responses to pharmacological challenges of the hypothalamus–pituitary–adrenocortical (HPA) axis, the major stress regulation system [9]. Putatively, the most sensitive challenge test of the HPA axis is the combined dexamethasone (dex)/corticotropin releasing hormone (CRH) test, which evaluates plasma cortisol responses to stimulation with 100 µg CRH under the suppressive effects of 1.5 mg dex [9,10]. This test sensitively detects impaired HPA axis regulation in acute depression, which improves during successful antidepressant treatment. Restored HPA axis regulation after successful treatment as indicated by a normalized cortisol response to the combined dex/CRH test is associated with sustained remission [11,12], while the recurrence of an impaired HPA axis regulation predicts increased relapse risk in remitted patients [13,14]. The glucocorticoid receptor (GR) complex plays a critical role in HPA axis regulation [9,15] as impaired GR signalling results in an attenuated negative-feedback inhibition of the HPA axis, finally leading to chronically elevated glucocorticoid levels.

The GR function is modulated by chaperone proteins forming a molecular complex that is required for proper ligand binding and receptor activation, as well as transcriptional regulation of the GR target genes [16,17]. The heat-shock protein HSP90 and its cochaperones play a key role in determining the sensitivity of the GR. While HSP90 is essential for GR steroid binding, the cochaperone FK506-binding protein 51 (FKBP51), coded by the *FKBP5* gene, exhibits inhibitory effects by reducing the binding affinity of the GR [18,19]. Genetic variations in the *FKBP5* gene were shown to be associated with the regulation of the HPA axis, with increased depression recurrence and rapid antidepressant treatment response [20,21]. The same genetic variations increased the risk for adult depression [22,23] and for post-traumatic stress disorder [22,24] in individuals reporting early exposure to an adverse environment. These findings suggest the involvement of *FKBP5* gene variants in depression risk and antidepressant treatment outcome; however, the role of *FKBP5* gene expression is yet to be elucidated. Cattaneo and colleagues [25] reported a 11% reduction in leukocyte *FKBP5* RNA expression in patients with major depression (N = 74), who responded to eight weeks of antidepressant treatment (citalopram or nortriptyline), while no change was observed in treatment non-responders. These findings were independent of the type of antidepressant used in this study—the selective serotonin reuptake inhibitor (SSRI), citalopram, or the noradrenergic tricyclic antidepressant (TCA), nortriptyline. A recent study did not find changes in leukocyte *FKBP5* RNA expression in female patients with major depression (N = 30), who were treated for eight weeks with the SSRI, sertraline, or with the selective serotonin noradrenalin reuptake inhibitor (SNRI), venlafaxine [26].

Both studies investigating *FKBP5* gene expression on antidepressant treatment outcome were conducted with relatively small patient samples and did not consider *FKBP5* genotypes previously identified as associated with antidepressant treatment outcome [27]. Therefore, we intended to investigate the effects of change in *FKBP5* gene expression on antidepressant treatment outcome in a large sample of depressed inpatients participating in the Munich Antidepressant Response Signature (MARS) study. In addition, we evaluated the association between *FKBP5* gene expression and impaired HPA axis regulation assessed with the combined dex/CRH test. Finally, we analyzed the moderating effects of rs1360780, a single nucleotide polymorphism (SNP) located in intron 2 of the *FKBP5* gene, for which the most consistent findings on depression risk and antidepressant treatment outcome have been reported [21]. Given the presumed role of restored GR signalling in successful antidepressant treatment in combination with the inhibitory function of FKBP51 on GR sensitivity, we hypothesized (1) that increasing *FKBP5* gene expression is associated with more pronounced dysregulation of the HPA axis and (2) that successful antidepressant treatment is accompanied by a reduced *FKBP5* expression in peripheral blood cells. We further postulated (3) that this effect should be moderated by the rs1360780 genotype previously identified as relevant for antidepressant treatment outcome in depression.

2. Results

This analysis included 297 participants of from the Munich Antidepressant Response Signature (MARS) project. MARS is a naturalistic open-label longitudinal treatment study with inpatients suffering from a depressive episode. Patients with a moderate to severe depressive episode were recruited from the hospital of the Max Planck Institute of Psychiatry and collaborating hospitals of Southern Bavaria and Switzerland. Antidepressant treatment outcome was monitored weekly for at least six weeks using the 21-items version of the Hamilton Depression Rating Scale (HAMD-21). Treatment was selected according to the attending doctor's choice and optimized according to symptom profile, plasma medication levels, and side effects. Mean HAMD-21 depression severity on admission to the hospital (baseline) was 26.1 (SD = 6.0), which decreased to 11.4 (SD = 7.9) after six weeks. Of the total participants, 173 patients (58%) responded to antidepressant treatment as indicated by a reduction of the HAMD-21 score of at least 50%, while the remaining 124 patients were classified as treatment non-responders (ΔHAMD-21 < 50%).

Table 1 presents the demographic and baseline characteristics of treatment responders and nonresponders. No significant group differences were observed ($p > 0.08$).

Table 1. Demographic and baseline characteristics of responders and non-responders to six weeks of antidepressant treatment.

Characteristics	Responders $N = 173$	Nonresponders $N = 124$	p
Female sex (%)	77 (44.5%)	63 (50.8%)	0.284
Mean age (SD)	48.8 (14.0)	47.0 (13.4)	0.270
Diagnosis [1] (%)			0.121
F31	22 (12.7%)	7 (5.6%)	
F32	36 (20.8%)	30 (24.2 %)	
F33	115 (66.5%)	87 (70.2%)	
HAMD-21 [2] (SD)	26.5 (6.63)	25.5 (5.07)	0.156
FKBP5 RNA [2] (SD)	1.83 (0.88)	1.75 (0.65)	0.404
FKBP51 [2,3] (SD)	0.035 (0.07)	0.031 (0.06)	0.080

[1] Diagnosis according to ICD10, [2] at baseline, [3] available only in a subgroup of $N = 39$ patients: 23 responders, 16 non-responders.

Blood samples were collected at baseline and after six weeks. RNA and DNA were extracted from whole-blood samples to evaluate the relative change (percentage from baseline) in *FKBP5* gene expression after six weeks of antidepressant treatment and to analyze the moderating effects of the intronic *FKBP5* variant rs1360780, respectively. In a subgroup of 39 patients (23 responders, 16 non-responders), the relative change in FKBP51 protein levels in blood mononuclear cells was additionally analyzed. FKBP51 protein levels tended to be higher in responders without reaching statistical significance ($p = 0.080$). In a further subgroup of 93 patients (56 responders, 37 non-responders), HPA axis regulation was evaluated at baseline and after six weeks using the combined dex/CRH test. No significant differences in demographic and clinical characteristics between responders and non-responders were found for both subgroups ($p > 0.175$, $p > 0.272$, respectively).

Given the inhibitory effects of FKBP51 on GR sensitivity, we assumed a positive correlation between *FKBP5* RNA expression and cortisol response to the combined dex/CRH test indicating an impaired stress response regulation with increasing *FKBP5* expression. Indeed, we observed significant correlations between *FKBP5* RNA levels and the overall cortisol response to the combined dex/CRH test at baseline (r = 0.214, *p* = 0.044) and after six weeks (r = 0.225, *p* = 0.032); albeit, the size of the effects was small. At the descriptive level, responders to six weeks of antidepressant treatment showed a 52% reduction SD= 145 of the cortisol response to the second combined dex/CRH, while non-responders reached a reduction of only 32% (SD = 115). Given the large variance within the group, this difference did not reach statistical significance ($F_{1,89}$ = 0.53, *p* = 0.468).

We observed reduced *FKBP5* gene expression after six weeks in antidepressant treatment responders, while nonresponders presented with increased expression levels ($F_{1,281}$ = 5.71, *p* = 0.018, f = 0.14). In all patients, RNA expression change was significant with a small to medium effect size. The subgroup analysis on change in FKBP51 protein levels revealed the same outcome pattern and was significant ($F_{1,35}$ = 4.64, *p* = 0.038, f = 0.36) approaching the border of a large effect (f = 0.40 [28]) (see Figure 1).

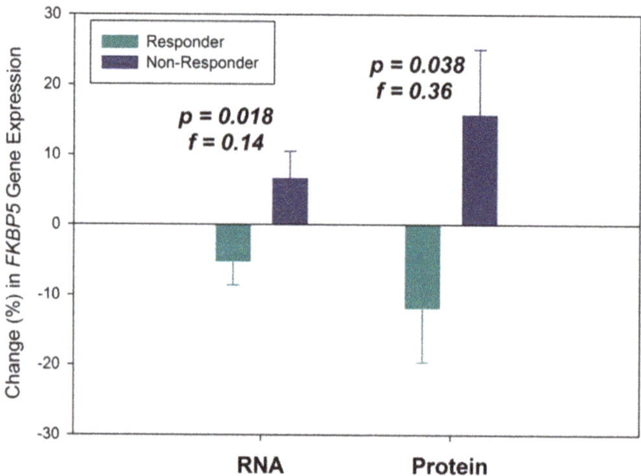

Figure 1. Change (% from baseline) in *FKBP5* gene expression for RNA and protein after six weeks of antidepressant treatment in responders and non-responders. Means ± standard errors of the mean (SEM) are presented.

It was found that while 47.8% of the patients carried the minor T allele of the intronic *FKBP5* variant rs1360780, 52.2% were noncarriers. In agreement with our expectations, patients carrying the T allele tended to show better treatment response with the strongest effect observed after four weeks (odds ratio (OR) = 1.56), which, however, failed to reach statistical significance (*p* = 0.066). This borderline effect was not observed after six weeks (OR = 1.31, *p* = 0.275). To investigate the moderating effect of this genotype, we reanalyzed the response data separately for T allele carriers and noncarriers. The results are presented in Figure 2. Patients of both genotype groups showed the expected effect pattern of reduced *FKBP5* RNA expression in treatment responders. However, the effect was statistically significant only in patients carrying the minor T allele ($F_{1,122}$ = 5.74, *p* = 0.018), presenting with a distinctly larger effect size (f = 0.22) than in the overall analysis (f = 0.14). In noncarriers, the effect pattern was less pronounced (f = 0.13), without reaching statistical significance ($F_{1,134}$ = 2.38, *p* = 0.125).

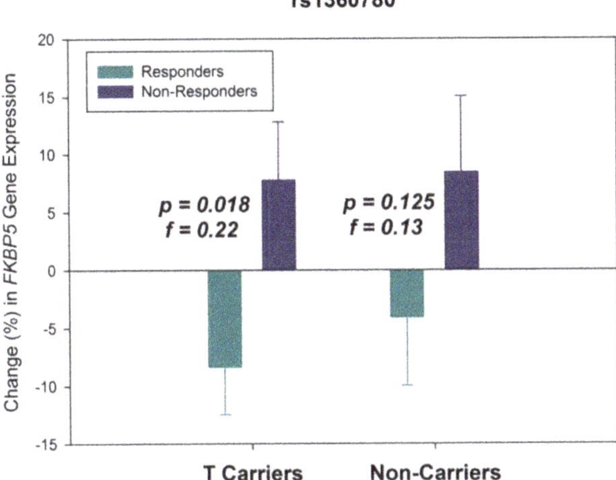

Figure 2. Change (% from baseline) in *FKBP5* gene expression at the RNA level after six weeks of antidepressant treatment in responders and non-responders for patients carrying the minor T allele of rs1360780 and for noncarriers. Means ± SEM are presented.

3. Discussion

While previous findings quite consistently suggested the involvement of *FKBP5* gene variants in depression risk and antidepressant treatment outcome, the role of *FKBP5* gene expression was less clear. Two previous clinical studies [25,26] investigating the association between *FKBP5* RNA expression and antidepressant treatment outcome showed conflicting results. These studies focused on RNA expression and did not analyze FKBP51 protein levels or the effects of *FKBP5* expression on HPA axis regulation. The genotype of the patients was also not considered. In addition, both studies were conducted with small sample sizes ($N = 74/N = 30$), and were restricted to specified treatments, which might reduce the generalizability of the findings. To address these limitations, we analyzed *FKBP5* gene expression changes in a large sample of depressed inpatients participating in the MARS study. The open character of the MARS study, which assured optimal treatment for all study participants in combination with liberal inclusion and exclusion criteria, resulted in a high participation rate with only about 15% of invited patients excluded or having refused participation. So it can be assumed that the study sample provided a good representation of patients hospitalized for depression treatment. In addition, we investigated the association between *FKBP5* gene expression and impaired HPA axis regulation in the combined dex/CRH test. Finally, we evaluated the moderating effects of rs1360780, a SNP located in intron 2 of the *FKBP5* gene, for which the most consistent findings on depression risk and antidepressant treatment outcome have been reported [21].

First, we could show that *FKBP5* expression correlated with the overall cortisol response to the combined dex/CRH test, although the effect size of the associations was rather small. This is in agreement with previous findings, suggesting that peripheral *FKBP5* RNA expression is induced by elevated glucocorticoid levels in patients with affective and anxiety disorders [20], which could also be confirmed in a human lymphoblastoid cell line model [29]. Contrary to previous findings, we did not find a reduced cortisol response to the combined dex/CRH test associated with response to antidepressant treatment. This could be related to the design of our study. In this study we investigated concomitant changes between HPA axis regulation, *FKBP5* gene expression, and antidepressant treatment outcome, while previous studies on the role of the combined dex/CRH test documented that a normalized HPA axis regulation precedes antidepressant treatment outcome by several weeks [12] and predicts future medium-term disease development [14]. Second, response to six weeks of antidepressant treatment was associated with a reduced *FKBP5* expression. The effect size was small to medium for the change in RNA expression, but bordered on a large effect for the change in protein levels. These findings confirm the results of a previous study reporting reduced peripheral *FKBP5* RNA expression in patients responding to eight weeks of citalopram or nortriptyline treatment [25]. A more recent study did not find changes in *FKBP5* RNA expression in patients treated for eight weeks with sertraline or venlafaxine [26]. However, this study was performed in a very small sample (N = 30) and restricted to female patients potentially limiting its power for documenting the general effects of *FKBP5* RNA expression on treatment outcome. In this regard, it is interesting to note that we observed the more substantial effect size for FKBP51 protein levels, which almost bordered on a large effect. This finding supports the biological relevance of the FKBP51 protein as the more proximal marker. Third, after stratifying patients into carriers and noncarriers of the minor allele of rs1360780, the *FKBP5* variant with, thus far, the most consistent findings in depression, the effects were distinctly more pronounced in carriers, while noncarriers showed a similar trend, albeit statistically not significant. It is interesting to note that patients carrying the minor allele also tended to show better treatment response after four weeks of treatment. This borderline effect, however, disappeared after six weeks. It is assumed that the rs1360780 risk variant, despite being intronic, might be associated with an increased glucocorticoid-related induction of *FKBP5* RNA expression [30], presumably, as a result of environmentally triggered changes in DNA demethylation of cytosine-phosphate-guanine (CpG) dinucleotide rich regions in intron 7 of the *FKBP5* gene [31]. Such allele-dependent epigenetic differences in the *FKBP5* gene could have contributed to the observed differences in *FKBP5* expression changes in rs1360780 risk-variant carriers and noncarriers.

The observed association between reduced *FKBP5* expression and antidepressant treatment response might be indirectly explained by an improved stress response regulation due to diminished inhibitory influences on GR sensitivity leading to a restored HPA axis regulation. However, improved HPA axis regulation evaluated with the combined dex/CRH test was not associated with concomitant antidepressant treatment response in our study suggesting that other, more direct pathways may be involved in the observed association between *FKBP5* gene expression and antidepressant treatment outcome. Indeed, preclinical studies using animal models, cell lines, and human specimens suggested the involvement of several pathways relevant for antidepressant action that are modulated by FKBP51. These pathways include the glycogen synthase kinase-3 (GSK-3) beta pathway [32], autophagy [33], and the modulation of enzymes of the epigenetic machinery [34]. While these findings suggest the potential of a direct causal link between altered FKBP51 and antidepressant action, the exact mechanisms are yet to be elucidated [35]. Nevertheless, FKBP51 inhibitors are currently in development as potential new antidepressant drugs, with early promising findings indicating the anxiolytic and potentially antidepressant effects of such compounds [36,37].

The present study has several strengths, but also some limitations. The strengths of the study are the sample size, which is larger than in previous studies, as well as the additional integration of protein data, genetic data, and a test on HPA axis regulation. There are also several limitations to be mentioned. First, depressed patients with different diagnoses (major depression, recurrent depression, or bipolar depression) were included in this study. However, the distribution of the diagnostic categories did not differ significantly between responders and non-responders (see Table 1). In addition, we repeated all analyses by including the type of diagnosis as additional covariates. The additional diagnostic covariates did not show any significant effects, and all findings could be replicated with similar effect sizes (see Supplementary Table S1). Second, due to the open study characteristic, patients received a variety of antidepressant drugs. Antidepressant treatment was carefully selected considering the current symptom profile of the individual patient, previous treatment history, and the plasma medication levels, with the aim of achieving the best possible antidepressant treatment. While the heterogeneity of the treatment might be a statistical limitation, it is also a strength with respect to the generalizability of the observed findings. In addition, previous studies using the combined dex/CRH test reported homogenous effects of effective antidepressant treatments on HPA axis regulation [11], which presumably can also be expected for FKBP51. Furthermore, we compared the applied classes of antidepressant drug treatment (selective serotonin reuptake inhibitors, tricyclic antidepressants, selective serotonin noradrenalin reuptake inhibitors, noradrenergic and specific serotonergic antidepressants, and other antidepressants) between responders and non-responders at baseline and at six weeks and did not find significant differences (see Supplementary Table S2). Third, *FKBP5* gene expression was analyzed using peripheral blood samples, which might not sufficiently reflect *FKBP5* expression in the pituitary or the brain. While variations in the gene expression pattern between different specimens or tissues cannot be ruled out, there is evidence that glucocorticoid exposure regulates FKBP51 expression via changes in DNA methylation in blood cells and in the brain in a very similar manner [38]. This suggests that peripheral *FKBP5* expression could be a proxy for expression changes in the brain.

4. Materials and Methods

4.1. Sample Description and Study Protocol

We recruited 297 patients suffering from a moderate to a severe depressive episode, who participated in the MARS project, for this study. MARS is a naturalistic open-label longitudinal study conducted at the hospital of the Max Planck Institute of Psychiatry and collaborating hospitals in Southern Bavaria and Switzerland to identify predictors for antidepressant drug response and to identify subgroups of depressed patients with common pathology benefitting from personalized treatment [39]. Patients were included during the first week after admission, and antidepressant treatment outcome was evaluated on a weekly basis with the HAMD-21. Only patients with a baseline HAMD-21 score of 14 or higher were included, setting the lower threshold at a moderate depression severity [40,41], while most patients were suffering from severe depression as indicated by an average HAMD-21 baseline score of 26.1 (SD = 6.0). Further exclusion criteria were depressive symptoms secondary to other medical or neurological disorders; presence of manic, hypomanic, or mixed affective symptoms; alcohol dependence or illicit drug abuse; and somatic treatments potentially affecting depression symptoms or HPA axis regulation (e.g., steroid medication). Treatment was selected according to the attending doctor's choice and optimized according to symptom profile, plasma medication levels, and side effects. Diagnosis according to the World Health Organization International Classification of Diseases, 10th revision (ICD10) [42], was obtained from trained psychiatrists at the end of the hospitalization considering patients reports, reports from relatives, and disease development.

At study inclusion (baseline) and after six weeks, morning fasting blood samples were collected for DNA extraction (S-Monovette, Sarstedt AG & Co., KG, Nümbrecht, Germany) and RNA extraction (PAXgene tubes; QUIAGEN GmbH, Hilden, Germany). In a subgroup of patients (N = 39), additional morning fasting serum samples (S-Monovette, Sarstedt AG & Co., KG, Nümbrecht, Germany) were collected at both time points—baseline and after six weeks—with mononuclear cells immediately isolated (ACCUSPIN System Histopaque-1077; Sigma-Aldrich, Merck KGaA, Darmstadt, Germany) for protein analysis. In another subgroup of 93 patients, a combined dex/CRH test was performed at baseline and after six weeks (the day after blood sample collection). All patients provided oral and written consent prior to study inclusion after all study details were explained. Ethical approval was provided by the permanent ethics committee of the Medical Faculty at the Ludwig–Maximilian University Munich, Germany (approval code: 318/00, 21/03/2001).

4.2. Laboratory Analysis

DNA was extracted using the Gentra PureGene extraction kit (QUIAGEN GmbH, Hilden, Germany) and genotyping was performed as part of a series of larger genotyping projects using Illumina Beadchip technology (Illumina Inc., San Diego, CA, USA). Rs1360780 was selected as the representative variant for the *FKBP5* risk genotype (minor allele frequency: 0.28; test for deviation from Hardy–Weinberg equilibrium: p = 0.667). RNA was isolated and purified with RNeasy kits (QUIAGEN GmbH, Hilden, Germany); expression analysis was performed with real-time polymerase chain reaction (RT-PCR) method using a TaqMan gene expression assay (Applied Biosystems Deutschland GmbH, Darmstadt, Germany). *FKBP5* RNA quantification was calculated against the activity of four housekeeping genes (*beta-glucuronidase, hypoxanthine–guanine phosphoribosyltransferase 1, phospholipase A2,* and *TATA-box binding protein*) using the delta cycle threshold (CT) method [43] for each housekeeping gene. Resulting deltas were then averaged across the four housekeeping genes showing excellent concordance for both time points (average intraclass correlation: $r = 0.937$). A ZeptoMARK^{TM} reverse phase protein microarray (RPPM) platform (Zeptosens AG, Witteswil, Switzerland) was used for FKBP51 protein profiling following the manufacturer's standard protocol [44] with array readout and quantification performed using the analysis software ZeptoVIEW 3.0 (Zeptosen AG).

The combined dex/CRH test was conducted as previously described [14]. Briefly, 1.5 mg dex was administered orally at 11 p.m. the evening before CRH stimulation. Blood samples were drawn the next day at 3:00, 3:30, 3:45, 4:00, and 4:15 p.m. while the subjects remained supine throughout the test. Within 30 s, after the collection of the first sample, 100 µg human CRH was injected. The cortisol response to the dex/CRH test was assessed by the total area under the curve (AUC) using the trapezoid rule across plasma cortisol concentrations of all sampling points. Plasma cortisol was determined by radioimmunoassay (ICN Biomedicals, Carson, CA, USA; detection limit 0.3 ng/mL).

Plasma medication levels were analyzed using liquid chromatography followed by mass spectrometry at the clinical laboratory of the Max Planck Institute of Psychiatry. Depending on the type of medication, metabolites of the active compound were also assessed to obtain a complete picture of the relevant drug concentrations. This information was available to the attending doctor to assist in finding the optimal drug dosage for the individual patient.

4.3. Statistical Analyses

Patients were classified as treatment responders and non-responders depending on the observed percent change in the HAMD-21 total score between baseline and six weeks with score improvements of 50% or more defined as response. Changes in *FKBP5* RNA expression and FKBP51 protein levels were also calculated as percent changes from baseline. RNA and protein values deviated from a normal distribution (Kolmogorov–Smirnov goodness-of-fit test, $p < 0.01$). To reduce skewness, scores were log-transformed to base 10 for the statistical analysis. Differences between responders and non-responders were evaluated by means of chi-square (categorical data) and t tests (continuous data) for demographic and clinical variables. Change in HAMD-21 scores over time, as well as baseline and change scores for the cortisol response to the dex/CRH test for *FKBP5* RNA expression and for FKBP51 protein levels were evaluated with analyses of covariance, controlling for the effects of sex and age as potential confounding variables. Cohen f scores [28] were calculated as effect-size measures. Associations between *FKBP5* RNA expression and the cortisol response to the combined dex/CRH test were expressed with partial correlation coefficients, which were controlled for the effects of sex and age. The level of statistical significance was set to $p = 0.05$. Means, standard deviations, or standard errors of uncorrected/untransformed values are reported in the tables and figures. All analyses were performed with PASW Statistics 18 (IBM, Armonk, NY, USA).

5. Conclusions

We were able to demonstrate that successful antidepressant treatment outcome in depressed patients is accompanied by a reduction in *FKBP5* gene and FKBP51 protein expression, particularly in those patients, who are carrying the risk allele of the *FKBP5* variant rs1360780. These findings further suggest an important role for *FKBP5* and FKBP51 in antidepressant treatment outcome and point to a new, promising target for future antidepressant drug development. However, further studies are warranted to fully understand the mechanism behind the observed effects.

Supplementary Materials: Supplementary materials can be found at http://www.mdpi.com/1422-0067/20/3/485/s1.

Author Contributions: Conceptualization: F.H., C.W.T., M.U., S.L., and M.I.; methodology: G.M., C.W.T., M.U., S.L., and M.I.; formal analysis: G.M., M.U., and M.I.; investigation: T.B., S.S., J.H., and S.L.; writing—original draft preparation: M.I.; writing—review and editing: M.I., G.M., T.B., S.S., J.H., F.H., C.W.T., M.U., and S.L.; supervision: F.H.; project administration: M.I. and S.L.; funding acquisition: F.H., C.W.T., and M.I.

Funding: This research was funded by the German Federal Ministry of Education and Research (BMBF), project No. 01ES0811 (Molecular Diagnostics).

Acknowledgments: The authors wish to cordially thank all collaborators and the clinical study and laboratory teams for their excellent work and support.

Conflicts of Interest: The authors F.H. and M.U. are coinventors of the patent "FKBP51: a novel target for antidepressant therapy" (WO2005054500). All other authors declare no conflict of interest.

Abbreviations

AUC	Area under the curve
CRH	Corticotropin releasing hormone
CpG	Cytosine-phosphate-guanine
Dex	Dexamethasone
DNA	Deoxyribonucleic acid
GR	Glucocorticoid receptor
FKBP5	FK506-binding protein 5 gene
FKBP51	FK506-binding protein 51
HAMD-21	Hamilton Depression Rating Scale, 21-items version
HPA	Hypothalamus–pituitary–adrenocortical
HSP90	Heat-shock protein 90
ICD10	International Classification of Diseases, 10th revision
MARS	Munich Antidepressant Response Signature
OR	Odds ratio
RNA	Ribonucleic acid
RPPM	Reverse phase protein microarray
RT-PCR	Real-time polymerase-chain-reaction
SD	Standard deviation
SEM	Standard error of the mean
SNP	Single nucleotide polymorphism
SNRI	Selective serotonin noradrenalin reuptake inhibitor
SSRI	Selective serotonin reuptake inhibitor
TCA	Tricyclic antidepressant

References

1. Kessler, R.C.; Bromet, E.J. The epidemiology of depression across cultures. *Annu. Rev. Public Health* **2013**, *34*, 119–138. [CrossRef] [PubMed]

2. Burcusa, S.L.; Iacono, W.G. Risk for recurrence in depression. *Clin. Psychol. Rev.* **2007**, *27*, 959–985. [CrossRef] [PubMed]

3. GBD 2017 Disease and Injury Incidence and Prevalence Collaborators. Global, regional, and national incidence, prevalence, and years lived with disability for 354 diseases and injuries for 195 countries and territories, 1990–2017: A systematic analysis for the Global Burden of Disease Study 2017. *Lancet* **2018**, *392*, 1789–1858. [CrossRef]

4. Kendler, K.S.; Prescott, C.A.; Myers, J.; Neale, M.C. The structure of genetic and environmental risk factors for common psychiatric and substance use disorders in men and women. *Arch. Gen. Psychiatry* **2003**, *60*, 929–937. [CrossRef] [PubMed]

5. Burmeister, M.; McInnis, M.G.; Zollner, S. Psychiatric genetics: Progress amid controversy. *Nat. Rev. Genet.* **2008**, *9*, 527–540. [CrossRef] [PubMed]

6. Hammen, C. Stress and depression. *Annu. Rev. Clin. Psychol.* **2005**, *1*, 293–319. [CrossRef] [PubMed]

7. Heim, C.; Newport, D.J.; Mletzko, T.; Miller, A.H.; Nemeroff, C.B. The link between childhood trauma and depression: Insights from HPA axis studies in humans. *Psychoneuroendocrinology* **2008**, *33*, 693–710. [PubMed]

8. Kendler, K.S.; Karkowski, L.M.; Prescott, C.A. Causal relationship between stressful life events and the onset of major depression. *Am. J. Psychiatry* **1999**, *156*, 837–841. [CrossRef] [PubMed]

9. Holsboer, F. The corticosteroid receptor hypothesis of depression. *Neuropsychopharmacology* **2000**, *23*, 477–501. [CrossRef]

10. Heuser, I.J.; Yassouridis, A.; Holsboer, F. The combined dexamethasone/CRH test: A refined laboratory test for psychiatric disorders. *J. Psychiatr. Res.* **1994**, *28*, 341–356. [CrossRef]

11. Ising, M.; Künzel, H.E.; Binder, E.B.; Nickel, T.; Modell, S.; Holsboer, F. The combined dexamethasone/CRH test as a potential surrogate marker in depression. *Prog. Neuropsychopharmacol. Biol. Psychiatry* **2005**, *29*, 1085–1093. [CrossRef] [PubMed]

12. Ising, M.; Horstmann, S.; Kloiber, S.; Lucae, S.; Binder, E.B.; Kern, N.; Kunzel, H.E.; Pfennig, A.; Uhr, M.; Holsboer, F. Combined dexamethasone/corticotropin releasing hormone test predicts treatment response in major depression-a potential biomarker? *Biol. Psychiatry* **2007**, *62*, 47–54. [CrossRef] [PubMed]

13. Appelhof, B.C.; Huyser, J.; Verweij, M.; Brouwer, J.P.; van Dyck, R.; Fliers, E.; Hoogendijk, W.J.G.; Tijssen, J.G.P.; Wiersinga, W.M.; Schene, A.H. Glucocorticoids and Relapse of Major Depression (Dexamethasone/Corticotropin-Releasing Hormone Test in Relation to Relapse of Major Depression). *Biol. Psychiatry* **2006**, *59*, 696–701. [CrossRef] [PubMed]

14. Zobel, A.W.; Nickel, T.; Sonntag, A.; Uhr, M.; Holsboer, F.; Ising, M. Cortisol response in the combined dexamethasone/CRH test as predictor of relapse in patients with remitted depression: A prospective study. *J. Psychiatr. Res.* **2001**, *35*, 83–94. [CrossRef]

15. De Kloet, E.R.; Joels, M.; Holsboer, F. Stress and the brain: From adaptation to disease. *Nat. Rev. Neurosci.* **2005**, *6*, 463–475. [CrossRef] [PubMed]

16. Criado-Marrero, M.; Rein, T.; Binder, E.B.; Porter, J.T.; Koren, J., III.; Blair, L.J. Hsp90 and FKBP51: Complex regulators of psychiatric diseases. *Philos. Trans. R. Soc. Lond B Biol. Sci.* **2018**, *373*, 20160532. [CrossRef] [PubMed]

17. Pratt, W.B.; Morishima, Y.; Murphy, M.; Harrell, M. Chaperoning of glucocorticoid receptors. In *Handbook of Experimental Pharmacology*; Starke, K., Gaestel, M., Eds.; Springer: Heidelberg, Germany, 2006; Volume 172, pp. 111–138.

18. Denny, W.B.; Valentine, D.L.; Reynolds, P.D.; Smith, D.F.; Scammell, J.G. Squirrel monkey immunophilin FKBP51 is a potent inhibitor of glucocorticoid receptor binding. *Endocrinology* **2000**, *141*, 4107–4113. [CrossRef]

19. Galigniana, N.M.; Ballmer, L.T.; Toneatto, J.; Erlejman, A.G.; Lagadari, M.; Galigniana, M.D. Regulation of the glucocorticoid response to stress-related disorders by the Hsp90-binding immunophilin FKBP51. *J. Neurochem.* **2012**, *122*, 4–18. [CrossRef]

20. Binder, E.B. The role of FKBP5, a co-chaperone of the glucocorticoid receptor in the pathogenesis and therapy of affective and anxiety disorders. *Psychoneuroendocrinology* **2009**, *34* (Suppl. 1), S186–S195. [CrossRef]

21. Fabbri, C.; Hosak, L.; Mossner, R.; Giegling, I.; Mandelli, L.; Bellivier, F.; Claes, S.; Collier, D.A.; Corrales, A.; Delisi, L.E.; et al. Consensus paper of the WFSBP Task Force on Genetics: Genetics, epigenetics and gene expression markers of major depressive disorder and antidepressant response. *World J. Biol. Psychiatry* **2017**, *18*, 5–28. [CrossRef]

22. Wang, Q.; Shelton, R.C.; Dwivedi, Y. Interaction between early-life stress and FKBP5 gene variants in major depressive disorder and post-traumatic stress disorder: A systematic review and meta-analysis. *J. Affect. Disord.* **2018**, *225*, 422–428. [CrossRef] [PubMed]

23. Zimmermann, P.; Bruckl, T.; Nocon, A.; Pfister, H.; Binder, E.B.; Uhr, M.; Lieb, R.; Moffitt, T.E.; Caspi, A.; Holsboer, F.; et al. Interaction of FKBP5 gene variants and adverse life events in predicting depression onset: Results from a 10-year prospective community study. *Am. J. Psychiatry* **2011**, *168*, 1107–1116. [CrossRef] [PubMed]

24. Binder, E.B.; Bradley, R.G.; Liu, W.; Epstein, M.P.; Deveau, T.C.; Mercer, K.B.; Tang, Y.; Gillespie, C.F.; Heim, C.M.; Nemeroff, C.B.; et al. Association of FKBP5 polymorphisms and childhood abuse with risk of posttraumatic stress disorder symptoms in adults. *JAMA* **2008**, *299*, 1291–1305. [CrossRef] [PubMed]

25. Cattaneo, A.; Gennarelli, M.; Uher, R.; Breen, G.; Farmer, A.; Aitchison, K.J.; Craig, I.W.; Anacker, C.; Zunsztain, P.A.; McGuffin, P.; et al. Candidate genes expression profile associated with antidepressants response in the GENDEP study: Differentiating between baseline 'predictors' and longitudinal 'targets'. *Neuropsychopharmacology* **2013**, *38*, 377–385. [CrossRef] [PubMed]

26. Banach, E.; Szczepankiewicz, A.; Leszczynska-Rodziewicz, A.; Pawlak, J.; Dmitrzak-Weglarz, M.; Zaremba, D.; Twarowska-Hauser, J. Venlafaxine and sertraline does not affect the expression of genes regulating stress response in female MDD patients. *Psychiatr. Pol.* **2017**, *51*, 1029–1038. [CrossRef] [PubMed]

27. Niitsu, T.; Fabbri, C.; Bentini, F.; Serretti, A. Pharmacogenetics in major depression: A comprehensive meta-analysis. *Prog. Neuropsychopharmacol. Biol. Psychiatry* **2013**, *45*, 183–194. [CrossRef] [PubMed]

28. Cohen, J. *Statistical Power Analysis for the Behavioral Sciences*, 2nd ed.; Lawrence Erlbaum Associates Inc.: Hillsdale, NJ, USA, 1988.

29. Vermeer, H.; Hendriks-Stegeman, B.I.; van der, B.B.; van Buul-Offers, S.C.; Jansen, M. Glucocorticoid-induced increase in lymphocytic FKBP51 messenger ribonucleic acid expression: A potential marker for glucocorticoid sensitivity, potency, and bioavailability. *J. Clin. Endocrinol. Metab* **2003**, *88*, 277–284. [CrossRef]

30. Binder, E.B.; Salyakina, D.; Lichtner, P.; Wochnik, G.; Ising, M.; Pütz, B.; Papiol, S.; Seaman, S.; Lucae, S.; Kohli, M.; et al. Polymorphisms in FKBP5 are associated with increased recurrence of depressive episodes and rapid response to antidepressant treatment. *Nat. Genet.* **2004**, *36*, 1319–1325. [CrossRef]

31. Klengel, T.; Mehta, D.; Anacker, C.; Rex-Haffner, M.; Pruessner, J.C.; Pariante, C.M.; Pace, T.W.; Mercer, K.B.; Mayberg, H.S.; Bradley, B.; et al. Allele-specific FKBP5 DNA demethylation mediates gene-childhood trauma interactions. *Nat. Neurosci.* **2013**, *16*, 33–41. [CrossRef]

32. Gassen, N.C.; Hartmann, J.; Zannas, A.S.; Kretzschmar, A.; Zschocke, J.; Maccarrone, G.; Hafner, K.; Zellner, A.; Kollmannsberger, L.K.; Wagner, K.V.; et al. FKBP51 inhibits GSK3beta and augments the effects of distinct psychotropic medications. *Mol. Psychiatry* **2016**, *21*, 277–289. [CrossRef]

33. Gassen, N.C.; Hartmann, J.; Zschocke, J.; Stepan, J.; Hafner, K.; Zellner, A.; Kirmeier, T.; Kollmannsberger, L.; Wagner, K.V.; Dedic, N.; et al. Association of FKBP51 with priming of autophagy pathways and mediation of antidepressant treatment response: Evidence in cells, mice, and humans. *PLoS. Med.* **2014**, *11*, e1001755. [CrossRef] [PubMed]

34. Gassen, N.C.; Fries, G.R.; Zannas, A.S.; Hartmann, J.; Zschocke, J.; Hafner, K.; Carrillo-Roa, T.; Steinbacher, J.; Preissinger, S.N.; Hoeijmakers, L.; et al. Chaperoning epigenetics: FKBP51 decreases the activity of DNMT1 and mediates epigenetic effects of the antidepressant paroxetine. *Sci. Signal.* **2015**, *8*, ra119. [CrossRef] [PubMed]

35. Rein, T. FK506 binding protein 51 integrates pathways of adaptation: FKBP51 shapes the reactivity to environmental change. *Bioessays* **2016**, *38*, 894–902. [CrossRef] [PubMed]

36. Hartmann, J.; Wagner, K.V.; Gaali, S.; Kirschner, A.; Kozany, C.; Ruhter, G.; Dedic, N.; Hausl, A.S.; Hoeijmakers, L.; Westerholz, S.; et al. Pharmacological Inhibition of the Psychiatric Risk Factor FKBP51 Has Anxiolytic Properties. *J. Neurosci.* **2015**, *35*, 9007–9016. [CrossRef] [PubMed]

37. Sabbagh, J.J.; Cordova, R.A.; Zheng, D.; Criado-Marrero, M.; Lemus, A.; Li, P.; Baker, J.D.; Nordhues, B.A.; Darling, A.L.; Martinez-Licha, C.; et al. Targeting the FKBP51/GR/Hsp90 Complex to Identify Functionally Relevant Treatments for Depression and PTSD. *ACS Chem. Biol.* **2018**, *13*, 2288–2299. [CrossRef] [PubMed]

38. Ewald, E.R.; Wand, G.S.; Seifuddin, F.; Yang, X.; Tamashiro, K.L.; Potash, J.B.; Zandi, P.; Lee, R.S. Alterations in DNA methylation of Fkbp5 as a determinant of blood-brain correlation of glucocorticoid exposure. *Psychoneuroendocrinology* **2014**, *44*, 112–122. [CrossRef] [PubMed]

39. Hennings, J.M.; Owashi, T.; Binder, E.B.; Horstmann, S.; Menke, A.; Kloiber, S.; Dose, T.; Wollweber, B.; Spieler, D.; Messer, T.; et al. Clinical characteristics and treatment outcome in a representative sample of depressed inpatients—Findings from the Munich Antidepressant Response Signature (MARS) project. *J. Psychiatr. Res.* **2009**, *43*, 215–229. [CrossRef]

40. Endicott, J.; Cohen, J.; Nee, J.; Fleiss, J.; Sarantakos, S. Hamilton Depression Rating Scale. Extracted from Regular and Change Versions of the Schedule for Affective Disorders and Schizophrenia. *Arch. Gen. Psychiatry* **1981**, *38*, 98–103. [CrossRef]

41. Kearns, N.P.; Cruickshank, C.A.; McGuigan, K.J.; Riley, S.A.; Shaw, S.P.; Snaith, R.P. A comparison of depression rating scales. *Br. J. Psychiatry* **1982**, *141*, 45–49. [CrossRef]

42. World Health Organization. *ICD-10: International Statistical Classification of Diseases and Related Health Problems, Tenth Revision*, 2nd ed.; World Health Organization: Geneva, Switzerland, 2004.

43. Schmittgen, T.D.; Livak, K.J. Analyzing real-time PCR data by the comparative C(T) method. *Nat. Protoc.* **2008**, *3*, 1101–1108. [CrossRef]
44. Pawlak, M.; Schick, E.; Bopp, M.A.; Schneider, M.J.; Oroszlan, P.; Ehrat, M. Zeptosens' protein microarrays: A novel high performance microarray platform for low abundance protein analysis. *Proteomics* **2002**, *2*, 383–393. [CrossRef]

International Journal of
Molecular Sciences

Review

Multi-Target Approach for Drug Discovery against Schizophrenia

Magda Kondej [1], Piotr Stępnicki [1] and Agnieszka A. Kaczor [1,2,*]

[1] Department of Synthesis and Chemical Technology of Pharmaceutical Substances, Faculty of Pharmacy with Division of Medical Analytics, Medical University of Lublin, 4A Chodźki St., Lublin PL-20093, Poland; magda.kondej@onet.pl (M.K.); piotr.stepnicki93@gmail.com (P.S.)

[2] School of Pharmacy, University of Eastern Finland, Yliopistonranta 1, P.O. Box 1627, Kuopio FI-70211, Finland

* Correspondence: agnieszka.kaczor@umlub.pl; Tel.: +48-81-448-7273

Received: 3 September 2018; Accepted: 6 October 2018; Published: 10 October 2018

Abstract: Polypharmacology is nowadays considered an increasingly crucial aspect in discovering new drugs as a number of original single-target drugs have been performing far behind expectations during the last ten years. In this scenario, multi-target drugs are a promising approach against polygenic diseases with complex pathomechanisms such as schizophrenia. Indeed, second generation or atypical antipsychotics target a number of aminergic G protein-coupled receptors (GPCRs) simultaneously. Novel strategies in drug design and discovery against schizophrenia focus on targets beyond the dopaminergic hypothesis of the disease and even beyond the monoamine GPCRs. In particular these approaches concern proteins involved in glutamatergic and cholinergic neurotransmission, challenging the concept of antipsychotic activity without dopamine D_2 receptor involvement. Potentially interesting compounds include ligands interacting with glycine modulatory binding pocket on *N*-methyl-D-aspartate (NMDA) receptors, positive allosteric modulators of α-Amino-3-hydroxy-5-methyl-4-isoxazolepropionic acid (AMPA) receptors, positive allosteric modulators of metabotropic glutamatergic receptors, agonists and positive allosteric modulators of α7 nicotinic receptors, as well as muscarinic receptor agonists. In this review we discuss classical and novel drug targets for schizophrenia, cover benefits and limitations of current strategies to design multi-target drugs and show examples of multi-target ligands as antipsychotics, including marketed drugs, substances in clinical trials, and other investigational compounds.

Keywords: antipsychotics; drug design; multi-target drugs; polypharmacology; schizophrenia

1. Introduction

Schizophrenia is a severe mental illness, affecting up to 1% of the population, with major public health implications. The causes of schizophrenia might be genetic or environmental or both but the complex pathomechanism of this disease is not sufficiently understood. The clinical picture of schizophrenia involves three groups of symptoms, i.e., positive, such as hallucinations, delusions and other thought disorders, negative, including social withdrawal, apathy and anhedonia, and cognitive deficits like memory and learning impairments or attention deficiencies [1]. It is generally agreed that the symptoms of schizophrenia result from disturbances in neurotransmission involving a significant number of receptors and enzymes, mainly within the dopaminergic, glutamatergic, serotoninergic, and adrenergic systems. In this regard, the dopaminergic hypothesis is still the main concept of the disease and all marketed antipsychotics target dopamine D_2 receptor. The dopaminergic hypothesis of schizophrenia evolved from the simple idea of excessive dopamine through the hypothesis combining prefrontal hypodopaminergia and striatal hyperdopaminergia and then to the current aberrant salience hypothesis [2]. However, novel findings in the field of neuroscience link schizophrenia with factors beyond the dopaminergic hypothesis and emphasize in particular the role of glutamatergic system in the development of the disease [3].

In order to treat efficiently complex neuropsychiatric diseases such as schizophrenia it is necessary to go beyond the "magic bullet" concept. This approach in drug discovery was based on the assumption that single-target drugs are safer as they have fewer side effects due to their selectivity. It turned out, however, that this is only true for single-gene diseases and the number of original single-target drugs were performing far behind expectations in the last ten years. Thus, "one-drug-one-target" paradigm has been gradually replaced by the concept of multi-target drugs (MTDs), sometimes termed "magic shotgun". From the historical perspective, MTDs, in contrast to clean single-target drugs, were sometimes referred to as dirty or promiscuous drugs. In the case of diseases with complex pathomechanisms, such as neuropsychiatric diseases or cancer, single-targets medications have been demonstrated to a great extent a failure. Most potent antipsychotics, in particular second generation or atypical antipsychotics, target simultaneously a number of aminergic G protein-coupled receptors (GPCRs). Clozapine, which is used to treat drug-resistant schizophrenia, has nanomolar affinity to several aminergic GPCRs.

In this scenario drug design and discovery today has moved from the molecular and cellular level to the systems-biology-oriented level [4] to reflect subtle events occurring on the biological networks which lead to the disease [5]. Network pharmacology involves important aspects such as connectivity, redundancy and pleiotropy of biological networks [6] which clearly shows that most drug interact with more than one target. MTDs have a number of advantages over single-target drugs, including improved efficacy due to synergistic or additive effects, better distribution in the target tissue, accelerated therapeutic efficacy in terms of clinical onset and achievement of full effect, predictable pharmacokinetic profile and fewer drug-drug interactions, lower risk of toxicity, improved patient compliance and tolerance and lower risk of target-based drug resistance due to modulation of a few targets [7]. However, it is not easy to design potent MTDs and problems arise starting from a proper target selection through affinity balancing to avoiding affinity to related off-targets.

In this review we present classical and novel drug targets for the treatment of schizophrenia, discuss benefits and limitations of MTDs and their design, as well as present multi-target antipsychotics including marketed compounds, compounds in clinical studies, and other investigational compounds. The literature search for this review was mainly based on searching PubMed database with the search terms: schizophrenia, schizophrenia drug targets, antipsychotics, multi-target antipsychotics, multi-target ligands, multi-target drugs with the focus on the references from the last five years, in particular regarding novel investigational compounds.

2. Drug Targets for the Treatment of Schizophrenia

2.1. Dopamine and Serotonin Receptors

Most of currently available antipsychotic drugs (excluding third generation drugs) act by blocking dopamine receptors in central nervous system, as seen in Table 1. This is the classical way to treat schizophrenia. The original dopamine hypothesis of schizophrenia was proposed by Carlsson (awarded a Nobel Prize in 2000) on the basis of indirect pharmacological evidence in humans and experimental animals. In humans, amphetamine causes the release of dopamine in the brain and can produce a behavioral syndrome that resembles an acute schizophrenic episode. Hallucinations are also a side effect of levodopa and dopamine agonists used in Parkinson's disease. In animals, dopamine release causes a specific pattern of stereotyped behavior that is reminiscent of the repetitive behaviors sometimes observed in patients suffering from schizophrenia. Potent D_2 receptor agonists, such as bromocriptine, lead to similar effects in animals, and these drugs, like amphetamine, aggravate the symptoms of schizophrenic patients. Moreover, dopamine antagonists and drugs blocking neuronal dopamine storage (e.g., reserpine) are effective in controlling the positive symptoms of schizophrenia and in preventing amphetamine-induced behavioral changes [8].

It is now thought that positive symptoms are the result of overactivity in the mesolimbic dopaminergic pathway (the neuronal projection from the ventral tegmental area (VTA) to the nucleus accumbens, amygdala and hippocampus) activating D_2 receptors, whereas negative symptoms may result from a lowered activity in the mesocortical dopaminergic pathway (the projection from the VTA to areas of the prefrontal cortex) where D_1 receptors predominate. Other dopaminergic pathways in the central nervous system (i.e., nigrostriatal and tuberoinfundibular) seem to function normally in schizophrenia. Thus, in terms of treatment it would be desirable to inhibit dopaminergic transmission in the limbic system but enhance this transmission in the area of prefrontal cortex [9].

Table 1. Potential clinical benefits and side effects related to the mechanisms of action of antipsychotics [10–12].

Mechanism of Action	Clinical Efficacy	Possible Side Effects
D_2 antagonism	↓Positive symptoms	Extrapyramidal symptoms (EPS) ↓Negative symptoms ↑Cognitive symptoms ↑Drowsiness
D_2 partial agonism	↓Positive symptoms ↓Negative symptoms ↓Cognitive symptoms	Little or no EPS Behavioral activation
D_3 antagonism		↑Endocrine dysfunction ↑Weight gain ↑Sexual dysfunction
5-HT$_{2A}$ antagonism	↓Negative symptoms	↓EPS ↓Hyperprolactinemia
5-HT$_{1A}$ partial agonism	↓Negative symptoms ↓Cognitive symptoms ↓Anxiety symptoms ↓Depressive symptoms	↓EPS ↓Hyperprolactinemia
5-HT$_{2C}$ antagonism		↑Weight gain ↑Appetite
M_1 antagonism	↓EPS	↑Anticholinergic symptoms, e.g., dry mouth, constipation, tachycardia ↑Drowsiness ↑Cognitive impairment
M_1 agonism	↓Psychotic symptoms ↓Cognitive symptoms	
M_3 antagonism		↑Type 2 diabetes mellitus ↑Hyperglycemic hyperosmolar syndrome ↑Diabetic ketoacidosis
H_1 antagonism		↑Weight gain ↑Drowsiness ↑Hypotension
α_1-antagonism		↑Dizziness ↑Drowsiness ↑Tachycardia ↓Blood pressure ↑Orthostatic hypotension
α_2-antagonism	↓Depressive symptoms	↑Anxiety ↑Tachycardia ↑Tremor ↑Dilated pupils ↑Sweating
β-antagonism		↑Orthostatic hypotension ↑Sedation ↑Sexual dysfunction
Glutamate modulation	↓Positive symptoms ↓Negative symptoms ↓Cognitive symptoms ↓Illness progression	

Legend: ↓ Decreasing ↑ Increasing.

Besides antagonism to the dopamine D_2 receptor, majority of antipsychotic drugs, especially those classified as second generation antipsychotics also block a wide range of other receptors, such as other dopamine receptors (D_1, D_3 or D_4), serotonin (especially 5-HT$_{2A}$ and 5-HT$_{2C}$), histamine (especially H_1) and α_1-adrenergic. Interaction of antipsychotics with those receptors is associated mainly with occurrence of side effects, such as sedation and drowsiness (H_1 receptors), weight gain (H_1 and 5-HT$_{2C}$), sexual dysfunction (5-HT$_2$), or orthostatic hypotension (α_1-adrenergic receptors). On the other hand, there are also hypotheses that antagonism to serotonin 5-HT$_{2A}$ receptor may have beneficial effects when it comes to occurrence of extrapyramidal side effects, as well as to reducing negative and cognitive symptoms of schizophrenia. Basis of schizophrenia is still poorly understood and there are several hypotheses, which involve different neurotransmitters and receptors and try to explain their role in the pathogenesis of the disorder [12].

The serotonin hypothesis of schizophrenia is based on the studies of interactions between the hallucinogenic drug, LSD, and serotonin. Observations of the antipsychotic effects of drugs which are serotonin and dopamine antagonists (e.g., risperidone, clozapine) have resulted in the increased interest in serotonin receptors as a possible target for drugs used in the treatment of schizophrenia.

There are evidences that the efficacy and tolerability of the atypical antipsychotic drugs, such as clozapine, olanzapine, quetiapine, risperidone, and ziprasidone in the treatment of schizophrenia may result, in part, from their interaction with various serotonin receptors, in particular 5-HT$_{2A}$ and 5-HT$_{1A}$ receptors, what is the reason of growing interest in the role, which serotonin plays in the mechanism of action of antipsychotics. The antagonism to 5-HT$_{2A}$ receptors, which is relatively potent, is connected with weaker antagonistic properties to dopamine D_2 receptors and is the only common pharmacologic feature of atypical antipsychotic drugs. The subtypes of serotonin 5-HT receptors, that are involved in the pharmacological action of second generation antipsychotics, such as clozapine, or that may potentially serve as targets for better tolerated and more effective new antipsychotic agents, include: 5-HT$_{1A}$, 5-HT$_{2A}$, 5-HT$_{2C}$, 5-HT$_3$, 5-HT$_6$, and 5-HT$_7$ receptors [13].

The distribution of serotonin 5-HT$_{2A}$ receptor in the central nervous system is wide, but the highest concentrations occur in the cortex. 5-HT$_{2A}$ as well as 5-HT$_{1A}$ receptors are located on the neurons that play significant role in schizophrenia. Those are cortical and hippocampal pyramidal glutamatergic neurons and γ-aminobutyric acid (GABA) interneurons. Serotonin 5-HT$_{2A}$ receptors localized on GABAergic interneurons stimulate the release of γ-aminobutyric acid and in that way play an important role in the regulation of the neuronal inhibition. 5-HT$_{2A}$ receptors are distributed also in the substantia nigra and ventral tegmentum from which arise the nigrostriatal and mesocorticolimbic dopaminergic neurons. 5-HT$_{2A}$ receptors modulate the activity of dopaminergic neurons. Antipsychotics that act by blocking serotonin 5-HT$_{2A}$ receptor (e.g., clozapine, risperidone) lead to the increased release of dopamine in the striatum by decreasing the inhibitory effect of serotonin, what manifests clinically in reducing extrapyramidal effects. It is also suggested that combined effects of antagonism at dopamine D_2 and serotonin 5-HT$_{2A}$ receptors in the mesolimbic circuit counteract the excessive dopamine transmission, which leads to occurrence of positive symptoms of schizophrenia. Moreover, improvement of the negative symptoms is associated with antagonism at 5-HT$_{2A}$ receptor, due to enhanced release of both dopamine and glutamate in the mesocortical pathway [9,13].

The behavioral evidence of interactions between serotonin 5-HT$_{2A}$ receptor and dopamine rests on the effect of 5-HT$_{2A}$ receptor antagonists on locomotor activity stimulated by amphetamine. Namely, giving low doses of amphetamine to rodents results in producing in them locomotor hyperactivity, which is mediated by the release of dopamine from the dopaminergic neurons in the mesolimbic circuit. This amphetamine stimulated hyperactivity is observed to be inhibited by first and second generation antipsychotic drugs and is thought to be an effect of antagonism to dopamine D_2 receptor, which all of those drugs share as a mechanism of action. However, some observations proved that compounds, such as amperozide, which are antagonists selective to serotonin 5-HT$_{2A}$ receptor and do not exhibit any affinity for dopamine D_2 receptor, also lead to lowering of hyperactivity in mice stimulated by administration of amphetamine [14]. These results support the concept that compounds that are antagonists to 5-HT$_{2A}$ receptor may improve behavioral states associated with excessive activity of dopaminergic neurons and may serve as effective antipsychotic medications.

Typical antipsychotic drugs, beside blocking dopamine D$_2$ receptors in the mesolimbic circuit, act also antagonistic to D$_2$ receptors localized in the nigrostriatal pathway, what is thought to result in occurrence of extrapyramidal side effects. Low doses of amphetamine administered to rodents lead to producing exploratory locomotor activity, whereas high doses of amphetamine causes the occurrence of repetitive, stereotyped behaviors, which are similar to those produced by the direct agonist of dopamine D$_2$ receptor, apomorphine. Those stereotyped behaviors are inhibited by first generation antipsychotics, what suggests that their antagonist properties are the cause of producing extrapyramidal side effects. Contrarily, amperozide and other antagonists of the serotonin 5-HT$_{2A}$ receptor do not reduce repetitive behaviors induced by apomorphine or high doses of amphetamine. These findings suggest that antipsychotic drugs which are antagonists to 5-HT$_{2A}$ receptor do not cause extrapyramidal side effects, in contrast to first generation drugs, which are devoid of activity to serotonin receptors.

The majority of clinical studies of serotonin 5-HT$_{2A}$ receptor antagonists have been carried out using ritanserin, the compound that exhibits antagonist properties to both 5-HT$_{2A}$ and 5-HT$_{2C}$ receptors. Its effectiveness has been studied in monotherapy, as well as an adjunct to existing treatment with antipsychotics. The studies have led to conclusions that ritanserin improves in particular negative symptoms of schizophrenia, which were poorly ameliorated in case of treatment with typical antipsychotic drugs [15].

To sum up, due to ability of antagonists of serotonin 5-HT$_{2A}$ receptor to interfere with elevated activity of dopamine, the antagonism of this receptor is believed to contribute to improvement of both positive and negative symptoms of schizophrenia and to causing less extrapyramidal side effects than older antipsychotics [16].

The 5-HT$_{1A}$ receptor is the subtype of serotonin receptors that is probably the best characterized in terms of functioning. It plays a significant role in modulating the activity of monoaminergic, inter alia dopaminergic, neurons. The functioning of 5-HT$_{1A}$ receptor may be described as antagonistic to the serotonin 5-HT$_{2A}$ receptor, when it comes to both presynaptic and postsynaptic its localization. Activation of serotonin 5-HT$_{1A}$ inhibitory autoreceptors located in the cells of raphe nucleus leads to inhibition of those neurons. In contrast, 5-HT$_{2A}$ receptors while activated in general cause the activation of serotonergic neurons by several mechanisms, which include a direct or indirect inhibition of GABAergic inhibitory interneurons, and a direct mechanism of excitation of other neurons, inter alia glutamatergic neurons. Both postsynaptical 5-HT$_{1A}$ and 5-HT$_{2A}$ receptors are located in the cortex on the pyramidal neurons. Activation of this 5-HT$_{1A}$ receptor results in neuronal inhibition through activation of potassium current, what leads to hyperpolarization. Contrary, 5-HT$_{2A}$ receptor while activated, facilitates neuronal output in the mechanism of activation of phospholipase C. Serotonin 5-HT$_{1A}$ receptors are suggested to be localized also presynaptically on GABA neurons terminals and pre- or postsynaptically on the GABAergic interneurons in the dentate gyrus in the hippocampus. Basing on the opposition between those two serotonin receptors, it is thought that agents acting as 5-HT$_{1A}$ receptor agonists are able to modulate dopaminergic transmission in the central nervous system in a similar way to antagonists to serotonin 5-HT$_{2A}$ receptor. Agonists to 5-HT$_{1A}$ receptor may both induce the dopamine release in the prefrontal cortex and potentiate the inhibiting effect on dopamine release of dopamine D$_2$ receptor antagonists [17].

In the brains of patients suffering from chronic schizophrenia, the density of serotonin 5-HT$_{1A}$ receptors is increased, what suggests a close correlation between pathogenesis of the disease and serotonin 5-HT$_{1A}$ receptors. These receptors are now considered as preferable target to treat schizophrenia, since there are evidences that stimulation of serotonin 5-HT$_{1A}$ receptors may contribute to decreasing of extrapyramidal side effects induced by antipsychotics [18] and ameliorating affective disorders such as depression or anxiety [19]. Moreover, blockade of 5-HT$_{1A}$ receptors may result in improvement of cognitive symptoms of schizophrenia [20].

It has been proved in different studies that agents which are selective agonists of serotonin 5-HT$_{1A}$ receptor, such as tandospirone or buspirone, reduced extrapyramidal side effects (e.g., bradykinesia, catalepsy) induced by antipsychotics from first generation [21]. Agonists of 5-HT$_{1A}$ receptor are thought to reduce extrapyramidal side effects induced by neuroleptics in the way of stimulating serotonin 5-HT$_{1A}$ receptors localized postsynaptically, since the inactivation of serotonergic neurons by p-chlorophenylalanine had no impact on the actions of 5-HT$_{1A}$ receptor agonists, when it comes to alleviating extrapyramidal side effects [22].

Reducing cognitive symptoms of schizophrenia is another significant role of serotonin 5-HT$_{1A}$ receptors. Cognitive dysfunction belongs to those symptoms of schizophrenia, whose treating with currently available drugs is still not very effective. Some of recently carried clinical studies have proved that the partial agonist properties of tandospirone regarding 5-HT$_{1A}$ receptor relevantly improved the deficits in cognition in schizophrenic patients. Studies carried on animals also showed that 5-HT$_{1A}$ receptor antagonists improved the cognitive deficits induced by antagonists to mACh receptor, such as scopolamine, or antagonists of *N*-methyl-D-aspartate (NMDA) receptor [23]. Although further studies are required, there are findings which suggest that serotonin 5-HT$_{1A}$ receptor antagonists may contribute to managing schizophrenia on account of ameliorating cognitive impairments [24].

Many compounds that bind to serotonin 5-HT$_{2A}$ receptors also exhibit an affinity to the structurally related serotonin 5-HT$_{2C}$ receptor. There are evidences that support the idea of an antipsychotic potential for antagonists of 5-HT$_{2C}$ receptor. One of them concerns meta-chlorophenylpiperazine (mCPP), which act as an agonist of serotonin 5-HT$_{2C}$ receptor [25]. The main action of mCPP in humans may be described as a selective activation of serotonin 5-HT$_{2C}$ receptors [26]. mCPP causes the worsening of positive symptoms in schizophrenic patients but pretreatment with mesulergine, which is an antagonist to 5-HT$_2$ receptor, results in decreased level of psychotic episodes, induced by the drug [27]. It is suspected that 5-HT$_{2C}$ receptor antagonists inhibit dopaminergic activity in mesolimbic and nigrostriatal pathways and thus contribute to reducing symptoms of schizophrenia and alleviating extrapyramidal side effects. Nonetheless, the role of this subtype of serotonin receptor in the pathogenesis of schizophrenia is still poorly understood and requires further studies [28].

Although dopamine and serotonin receptors are classical drug targets for the treatment of schizophrenia, novel drugs acting through these receptors can be developed based on novel signaling mechanisms typical for the family of GPCRs. These include allosteric modulators [29], biased ligands [30], compounds acting on receptor dimers, oligomers and mosaics [31–34] and last but not least intentionally promiscuous multi-target ligands [35].

2.2. Adrenergic and Histaminergic Receptors

Noradrenaline has a key role in the pathomechanism of schizophrenia although the specific role of α adrenergic receptors has been not well elucidated yet [36]. It has been hypothesized that interactions of atypical antipsychotics with α-adrenergic receptors contributes to their atypicality [37]. It was shown that antagonism at α_1 adrenergic receptors is beneficial to treat positive symptoms, in particular in acute schizophrenia while antagonism at α_2 adrenergic receptor, characteristic for clozapine and to some extent risperidone might be important to relieve negative symptoms and cognitive impairments [37]. Blockade of α adrenergic receptors may have a stabilizing effect on the dopaminergic neurotransmission in schizophrenia. In contrast, it was also reported that activation of α_{2A} adrenergic receptors in prefrontal cortex may improve cognitive functions [38]. Moreover, adjunctive α_2 adrenergic receptors antagonism increases the antipsychotic activity of risperidone and promotes cortical dopaminergic and glutamatergic, NMDA receptor-mediated neurotransmission [39]. It was also shown that blockade of α_{2C} adrenergic receptors alone or in combination with dopamine D$_2$ receptor blockade could be also beneficial in schizophrenia [38].

The histamine H$_1$ receptor is a classical off-target for antipsychotics as its blockade causes sedation and may be involved in weight gain. Although weight gain and metabolic disorders can also be attributed to blockade of adrenergic or cholinergic receptors, antagonism of histamine H_1 receptors is described as a key reason for second generation antipsychotics-induced obesity [40]. In contrast, the histamine H_3 receptor is an emerging target for novel antipsychotics [41] as selective antagonists or inverse agonists of this histamine receptor subtype are efficient in treatment cognitive deficiencies in schizophrenia [42].

2.3. Muscarinic and Nicotinic Receptors

Muscarinic receptors have a pivotal role in modulating synaptic plasticity in the prefrontal cortex and stimulation of these receptors results in long-term depression at the hippocampo-prefrontal cortex synapse [43]. A growing body of evidence indicates central role of disturbances in cholinergic neurotransmission in schizophrenia [44]. Postmortem studies indicate a reduced number of cholinergic interneurons in the ventral striatum in schizophrenia patients [45]. Furthermore, neuroimaging studies indicated that muscarinic receptors availability was significantly less in schizophrenia patients and positive symptoms of schizophrenia are negatively correlated with muscarinic receptors availability [46]. It should be emphasized that muscarinic receptor antagonists worsen cognitive and negative symptoms in schizophrenia patients and xanomeline, a muscarinic receptor agonist, ameliorates all symptoms in schizophrenia patients and corresponding animal models [43]. Based on these and other findings muscarinic hypothesis of schizophrenia has been suggested [47].

Involvement of nicotinic cholinergic receptors in the pathomechanism of schizophrenia can explain why schizophrenia patients are often heavy smokers [48,49]. It is assumed that smoking relieves particularly negative symptoms of schizophrenia. More and more evidence indicates that activation of α_7 nicotinic receptors [50] by agonists or positive allosteric modulators can be a promising strategy for the treatment of schizophrenia [51,52].

2.4. Metabotropic and Ionotropic Glutamatergic Receptors

Glutamate is one of the main excitatory neurotransmitters in the mammalian central nervous system [53]. Glutamatergic pathways linking to the cortex, the limbic system, and the thalamus regions are crucial in schizophrenia [54,55]. Abnormalities in the glutamatergic neurotransmission may influence synaptic plasticity and cortical microcircuitry, particularly NMDA receptor functioning [56]. NMDA receptors are ligand-gated ion channels, and are pivotal for excitatory neurotransmission, excitotoxicity and plasticity [57,58].

Glutamatergic hypothesis of schizophrenia is based on the observation that antagonists of N-methyl-D-aspartate (NMDA) receptors, such as phencyclidine or ketamine produce schizophrenia-like positive, negative, and cognitive symptoms in animal models and healthy individuals [59,60]. Glutamatergic hypothesis of schizophrenia is mainly a concept of hypofunction of NMDA receptors in this disease, however other ionotropic glutamate receptors (α-amino-3-hydroxy-5-methyl-4-isoazolepropionic acid, AMPA and kainate receptors) as well as metabotropic glutamate receptors are also involved.

In therapeutic trials compounds which promote NMDA receptor signaling were found relieve certain symptoms in patients with schizophrenia [61]. Moreover, in postmortem studies abnormalities in glutamatergic receptor density and subunit composition in the prefrontal cortex, thalamus, and temporal lobe were reported [62–64], and these are brain parts with altered stimulation during cognitive actions performed by schizophrenia patients [65]. NMDA receptor hypofunction may result in morphological and structural brain changes leading to the onset of psychosis [66,67]. It was suggested that levels of glutamate decrease with age in healthy people, but it was not found if they are influenced in case of chronic schizophrenia [68].

Antipsychotics may interfere with glutamatergic neurotransmission by influencing the release of glutamate, by modulation glutamatergic receptors, or by changing the density or subunit composition of glutamatergic receptors [55]. It was shown that antipsychotics blocking dopamine D_2 receptor increase the phosphorylation of the NR1 subunit of the NMDA receptor, thus promote its activation and consequent gene expression [69]. In this regard dopamine–glutamate interactions occur intraneuronally and intrasynaptically. There are also findings that certain second generation antipsychotics act on NMDA receptors in a distinct way than the first generation antipsychotics [70].

Abnormalities in glutamatergic neurotransmission constitute a possible drug target for schizophrenia, in particular for the treatment of cognitive impairment and negative symptoms [54,55]. Reports about hypoactivity of NMDA receptors in schizophrenia led to clinical trials with ligands stimulating this receptor [55]. Classical NMDA receptor agonists are not considered here due to excitotoxicity and neuron damage resulting from excessive NMDA receptor stimulation. In this regard, the glycine modulatory binding pocket on the NMDA receptor might be an attractive drug target [71]. Next, positive allosteric modulators of AMPA receptors [72,73] as well as orthosteric ligands and modulators of metabotropic glutamatergic receptors [74], in particular ligands acting on mGluR2/3 receptors [75] might be considered promising potential medications against schizophrenia in agreement with the glutamatergic hypothesis of this disease.

2.5. Other Drug Targets in Schizophrenia

There are also potential drug targets for the treatment of schizophrenia beyond transmembrane receptors. Most important enzymes with implications in schizophrenia include the serine/threonine kinase glycogen synthase kinase-3 (GSK-3) involved in cognitive-related processes such as neurogenesis, synaptic plasticity and neural cell survival [76], cyclic nucleotide (cNT) phosphodiestereases (PDEs)-intracellular enzymes which governs the activity of key second messenger signaling pathways in the brain [77] and acetylcholinesterase for treatment of cognitive impairments [78].

3. Multi-Target Compounds: Strategies of Design, Benefits, and Limitations

As has already been mentioned, during last twenty years most efforts in drug design and discovery followed the paradigm "one disease, one gene, one molecular target, one drug". However, novel findings in the field of systems biology and discoveries of molecular complexity of illnesses considerably moved current drug discovery efforts towards multi-target drugs [79,80]. Such compounds are able to exert numerous pharmacological actions and have emerged as magic shotguns in the treatment of multifactorial diseases in contrast to classical magic bullet approach [81].

3.1. Design of Multi-Target Compounds

Classical approaches to design multi-target ligands involve three different ways of combination of two pharmacophores, leading to a cleavable conjugate where two pharmacophores are connected by a linker (a modern form of combination therapy), a compound with overlapping pharmacophores or a highly integrated multi-target drug, as seen in Figure 1 [5]. Multi-target drugs, in particular those obtained by pharmacophore integration strategy are referred to as "master key compounds" [82,83]. Thus, MTDs are designed broadly as hybrid or conjugated drugs or as chimeric drugs from two or more pharmacophores/drugs having specific pharmacological activities [84].

Morphy and Rankovic [85] described two approaches for designing multi-target drugs: knowledge-based strategies and screening strategies. Knowledge-based techniques are based on available biological data from old drugs or other bioactive compounds, from either literature or proprietary company sources. Other methods include the screening of either diverse or focused compound libraries. Classical diversity based screening is the high-throughput screening (HTS) of large and differentiated compound collections versus one protein, and hits found are then triaged on the basis of activity at the other protein. In focused screening, compounds known to have robust activity at one protein are screened for activity at the other one. Even if only moderate activity is found for the second protein, it can supply a useful baseline for increasing that activity by incorporating structural elements from more potent selective ligands for this target [85].

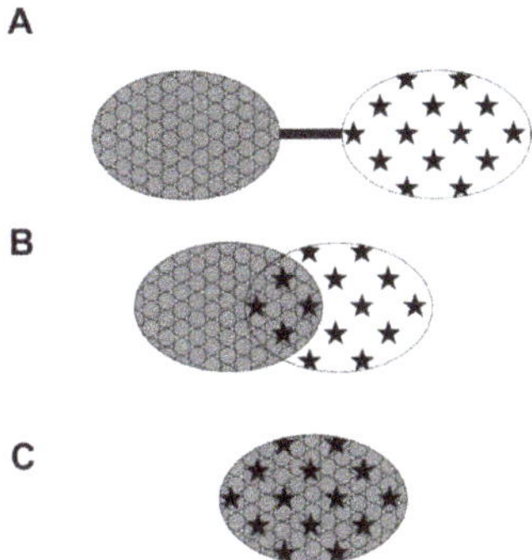

Figure 1. Strategies to design multi-target ligands. (**A**) linking of pharmacophores; (**B**) overlapping of pharmacophores; (**C**) integration of pharmacophores.

Modern in silico approaches can be also used to design multi-target ligands and can be classified into ligand-based and structure-based strategies [4]. Ligand-based target fishing strategies rely either on similarity-based screening or machine learning methods [4]. Moreover, ligand-based pharmacophores can be used. The advantage of this approach is independence from available structural information on the protein. These methods involve 2D or 3D similarity searches. Polyphramacological profiling of the compounds may also be based on three-dimensional structure-activity relationship (3D-QSAR) techniques [7]. Structure-based methods involve molecular docking (e.g., docking-based virtual screening [86] and inverse docking) or structure-based pharmacophores. The advantage of structure-based approaches in comparison to ligand-based approaches is that they do not rely on available activity data [4].

The main principle in designing multi-target compounds is the achievement of superior therapeutic efficacy and safety by targeting multiple players in pathogenic cascade simultaneously [4].

3.2. Advantages and Disadvantages of Multi-Target Ligands

Multifunctional ligands are particularly interesting as their molecules have common parts responsible for activity, and their structure is formed as a result use of pharmacophore fragments. Receiving such hybrid compounds allows not only to improve their activity, but also to positively affect pharmacokinetic parameters, similar to those shown by drugs used in therapy [81].

The main advantages of multi-target-drugs compared to single-target drugs and combination therapy include: (i) reflecting the complex pathomechanism of the disease and better therapeutic efficacy and (ii) better therapeutic safety avoidance of different bioavailabilities, pharmacokinetics, and metabolism of a combination regimen and avoidance of drug–drug interactions [87]. Multi-target mode of action is beneficial to combat drug resistance and development of tolerance and can be also a base of drug repurposing. The disadvantage of MTDs is the difficulty in designing compounds with balanced activity to multiple targets, sometimes resulting in a need to compromise activity at some targets. Moreover, compounds obtained in particular by pharmacophore linkage are often not drug-like due to high molecular mass.

4. Multi-Target Compounds to Treat Schizophrenia

4.1. Marketed Drugs—Second and Third Generation Antipsychotics

The second generation antipsychotics, which are nowadays the treatment of choice in cases of schizophrenia and also bipolar disorder, are essentially multi-target compounds. It should be emphasized, however, that many first generation antipsychotics have a complex pharmacological profile, including haloperidol, fluphenazine and even chlorpromazine, as seen in Table 2 [88].

Table 2. Relative neurotransmitter receptor affinities for first, second and third generation antipsychotics and involved side effects.

Drugs Generation	Examples	Receptors								Potential Side Effect
		D_1	D_2	D_3	D_4	5-HT$_{2A}$	α_1	H_1	M_1	
First	Chloropromazine	+	++	+++	+	+++	++	++	++	extrapyramidal symptoms such as dyskinesia, dystonias, akathisia, unwanted movements, muscle breakdown, tremors, rigidity and elevated prolactin
	Haloperidol	+	+++	+	+	0	+	0	0	
	Benperidol	0	+++	++		++	+	0	0	
	Fluspirilene	+	+++	+++		+	0	0	0	
	Thioridazine	+	++	++	+	++	++	+	+	
Second	Clozapine	++	+	++	+++	++	+		+++	hypotension, tachycardia, agranulocytosis
	Olanzapine	++	+++	+	++	+++	++	++	++	sedation, weight gain
	Risperidone	+	++	+	+	+++	++	+++	0	orthostatic hypotension, insomnia, restlessness, anxiety, headaches, agitation, extrapyramidal symptoms (EPS), rhinitis, sedation, fatigue, ocular disturbances, dizziness, palpitations, weight gain, diminished sexual desire, erectile and ejaculatory dysfunction, orthostatic dysregulation, reflex tachycardia, gastrointestinal complaints, nausea, rash, galactorrhea and amenorrhea
	Quetiapine	+	++	+++	++	+++	+++	+++	0	drowsiness, dizziness, headache, withdrawal symptoms, increased triglycerides, increased total cholesterol
	Ziprasidone	++	+++	+++	+++	+++	++	+	0	parkinsonism, headaches, rhinitis, orthostatic hypotension, tachykardia
Third	Aripiprazole	0	+++	++	+	+++	+	+	0	hyperglycaemia, headache, extrapyramidal symptoms

+++ = high, ++ = moderate, + = low, 0 = minimal to none.

Clozapine (**1**), Figure 2, is a classic example of a "dirty" drug which can be still considered a "gold standard" atypical antipsychotic due to absence of extrapyramidal syndrome (EPS), superiority in treatment of drug resistant schizophrenia and reducing suicidality [88]. Clozapine exerts severe side effects, in particular potentially life-threatening agranulocytosis, but also weight gain, diabetes, and seizures [89]. Both the effectiveness and side effects of clozapine result from its complex pharmacological profile, involving high affinity to many serotonin, dopamine, muscarinic, adrenergic, and other aminergic receptors, as seen in Figure 3 [90].

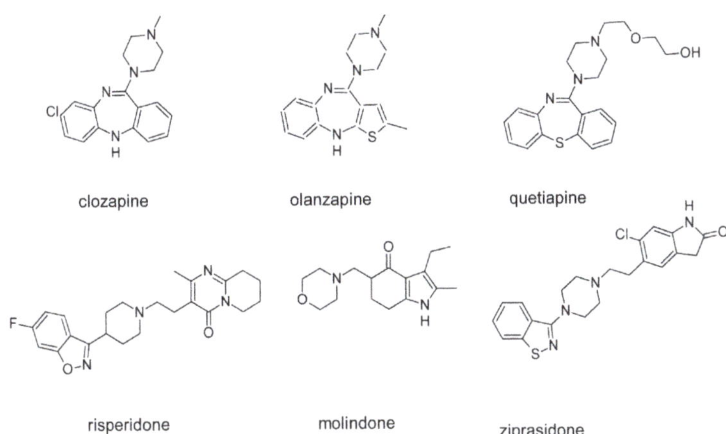

Figure 2. Examples of marketed multi-target second generation antipsychotics [1].

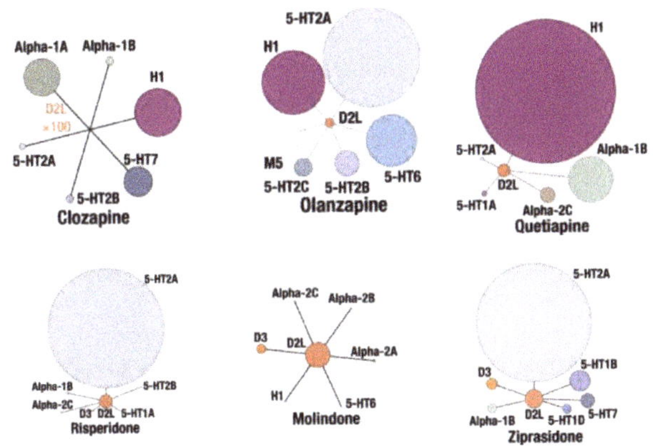

Figure 3. Pharmacological profiles of commonly used second generation antipsychotics, elaborated on the basis of [90] with modifications.

Some problems with side-effects of clozapine were solved with the introduction of another second generation antipsychotic, olanzapine (**2**), Figure 2. Olanzapine does not cause agranulocytosis but still has metabolic side effects leading to possible weight gain [91] which can be associated with histamine H_1 receptor signaling [92] and/or the $-759C/T$ and $-697G/C$ polymorphisms of the 5-HT_{2C} receptor gene [93]. Importantly, the side-effect profile of olanzapine can be considered beneficial, with a low incidence of EPS and little increase in prolactin during acute-phase trials [94]. Multi-receptor binding profile of olanzapine [95] involves a nanomolar affinity for dopaminergic, serotonergic, α_1 adrenergic, and muscarinic receptors, as seen in Figure 3. Olanzapine is also used to treat bipolar disorder.

Similarly, quetiapine (**3**), Figure 2, belongs to atypical antipsychotics, which, besides schizophrenia, are applied to treat bipolar disorder and major depressive disorder. Quetiapine is dopamine D_1, dopamine D_2 and serotonin 5-HT$_2$ receptor ligand, as seen in Figure 3. Antagonism to α_1 adrenergic and histamine H_1 receptor results in side effects like sedation and orthostatic hypotension. Moreover, there are reports about quetiapine misuse and abuse which can be linked with its high affinity for the H_1 receptor, as antihistamines agents cause rewarding action, compare Figure 3 [96].

Risperidone (**4**), Figure 2 was marketed as the first "non-clozapine" atypical antipsychotic and it is also used to treat the acute manic phase of bipolar disorder. Risperidone is a benzisoxazole derivative with nanomolar affinity for serotonin (5-HT$_{2A}$ and 5-HT$_7$) and dopamine D_2 receptors (its affinity for D_3 and D_4 receptors is three times lower), Figure 3 with a 5-HT$_{2A}$/D$_2$ affinity ratio of about 20 [11]. It also has a strong affinity for adrenergic (α_1 and α_2) receptors, and some affinity for histamine (H_1) receptors [11]. Pharmacological effect of risperidone is mainly a consequence of antagonism at D_2 and 5-HT$_{2A}$ receptors, as seen in Figure 3. Its multi-receptor profile resembles this of olanzapine, however risperidone causes sedation less frequent and orthostatic hypotension more often than olanzapine. There are also reports that this drug can increase the level of prolactin and cause arrhythmia.

Molindone (**5**), seen in Figure 2 is a dihydroindolone neuroleptic with dopamine D_2, D_3 and D_5 receptor antagonist activity and affects mainly dopaminergic neurotransmission in the CNS as seen in Figure 3. It is the second generation antipsychotic with atypical pharmacological profile. Its side effects rarely involve sedation and autonomic side effects but more often extrapyramidal side effects (more frequently than other new antipsychotics, although still less frequently than classical drugs). The application of molindone, in contrast to other atypical antipsychotics, does not usually lead to weight gain. Some patients with poor tolerance or response to other drugs can benefit from the treatment with molindone [97].

An example of modern second generation multi-target drug is ziprasidone (**6**), as shown in Figure 2. This antipsychotic is an optimized hybrid of dopamine receptor ligand (D_2 receptor agonist) and a lipophilic serotonin receptor ligand in which the D_2 agonist activity is transformed to D_2 receptor antagonist activity. It also exhibits desirable D_2/5-HT$_2$ ratio of 11 comparable to clozapine, as seen in Figure 3, and has lesser propensity of orthostatic hypotension. Moreover, ziprasidone has been reported not to cause significant weight gain and even to enable some weight loss in obese patients [98].

Some new second generation antipsychotics involve iloperidone (**7**), asenapine (**8**) and lurasidone (**9**), shown in Figure 4, however they have not gained popularity in clinical practice yet. Their pharmacological profiles are presented in Figure 5 [90]. From those three drugs lurasidone seems to be most important. Lurasidone has high antagonist activity at serotonin 5-HT$_{2A}$ and 5-HT$_7$ receptors and weaker antagonism at dopamine D_2 receptor [99]. It has also partial agonist activity at serotonin 5-HT$_{1A}$ receptor, considerable affinity to adrenergic α_{2A} and weaker affinity to muscarinic receptors [99]. Lurasidone is used for treatment of schizophrenia acute bipolar depression. It has low probability of side effects typical for second generation antipsychotics, but higher risk of akathisia in comparison to other atypicals [99].

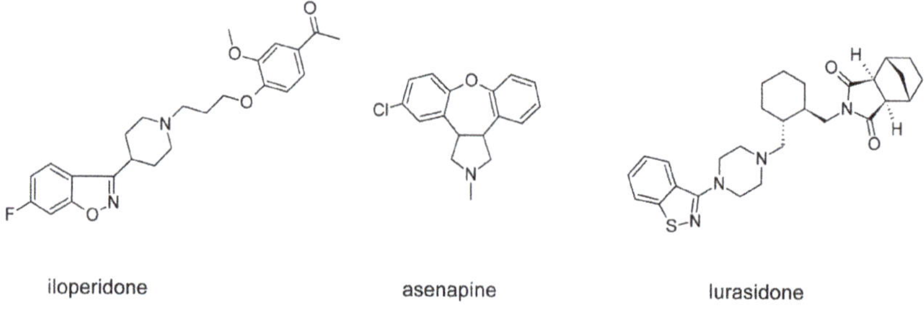

Figure 4. New second generation antipsychotics.

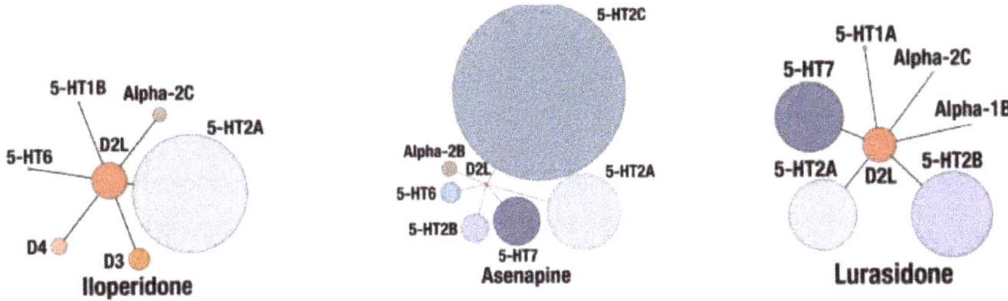

Figure 5. Pharmacological profiles of some new second generation antipsychotics, elaborated on the basis of [90] with modifications.

Third generation antipsychotics include aripiprazole (**10**), brexpiprazole (**11**) and cariprazine (**12**), as seen in Figure 6. The mechanism of action of these drugs is still mainly linked to the dopaminergic neurotransmission, shown in Figure 7, however, not to dopamine receptor antagonism but to partial or biased agonism (functional selectivity) [100,101]. Due to partial agonism properties aripiprazole is termed as "dopamine stabilizer" [102–104]. Aripiprazole was one of the first functionally selective D_2 receptor ligands identified that may stabilize the dopaminergic signaling through D_2 receptor. Although aripiprazole was first described as a partial D_2 receptor agonist, it was later demonstrated that aripiprazole could behave as a full agonist, a partial agonist, or an antagonist at D_2 receptor depending upon the signaling readout and cell type interrogated [105]. Aripiprazole is a partial agonist for inhibition of cyclic adenosine monophosphate (cAMP) accumulation through the D_2 receptor (i.e., $G\alpha$ signaling) [106–108]. In contrast, it has also been reported that aripiprazole is an antagonist in GTPγS binding assays with the D_2 receptor [107,109]. It was also revealed that aripiprazole failed to activate outward potassium currents following activation of the D_2 receptor in MES-23.5 cells, indirectly suggesting that it was inactive or possibly an antagonist for $G\beta\gamma$ signaling through the D_2 receptor [107]. Aripiprazole was also reported to be either an antagonist [110] or a partial agonist [111] for β-arrestin-2 recruitment.

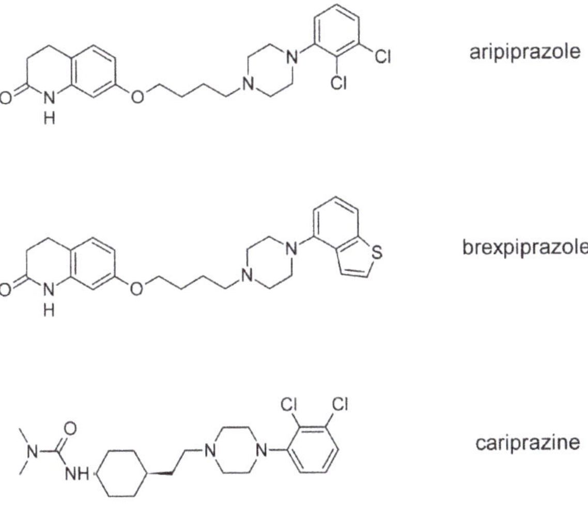

Figure 6. The third generation antipsychotics [1].

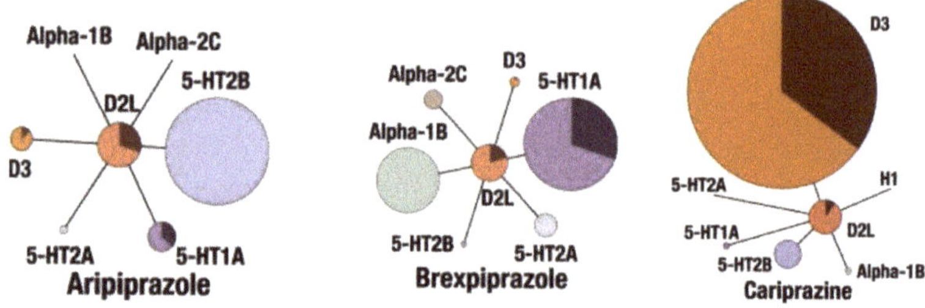

Figure 7. Pharmacological profiles of the third generation antipsychotics, elaborated on the basis of [90] with modifications.

Aripiprazole is also a partial agonist of 5-HT$_{1A}$ and 5-HT$_{2A}$ receptors (much weaker in the latter case) which results in functional antagonism at these receptors, as seen in Figure 7 [90]. In contrast to classical atypical drugs, aripiprazole has higher affinity for dopamine D$_2$ receptor than for serotonin 5-HT$_{2A}$ receptor. Clinical application of aripiprazole includes also bipolar disorder, major depression, obsessive-compulsive disorder, and autism. Aripiprazole is characterized by efficacy similar to that of both typical and atypical antipsychotic drugs (except olanzapine and amisulpride) [112]. Aripiprazole resulted in considerably lower weight gain and lower changes in glucose and cholesterol levels in comparison to clozapine, risperidone, and olanzapine [112]. Moreover, aripiprazole led to weaker EPS, less use of antiparkinsonian drugs, and akathisia, in comparison to typical antipsychotic drugs and risperidone [112]. Furthermore, aripiprazole is characterized by better tolerability compared to other antipsychotics [113]. Adverse effects of aripiprazole may include agitation, insomnia, anxiety, headache, constipation or nausea [103].

Brexpiprazole was approved by FDA in 2015 and is a partial agonist of dopamine D$_2$, D$_3$ and serotonin 5-HT$_{1A}$ receptors, as well as antagonist of 5-HT$_{2A}$, 5-HT$_{2B}$ and 5-HT$_7$ receptors, as seen in Figure 7 [114]. Its pharmacological properties are close to those of aripiprazole. In comparison to aripiprazole, brexpiprazole is more potent at 5-HT$_{1A}$ receptors and has less intrinsic activity at D$_2$ receptors [115]. Brexpiprazole is applied for treatment of schizophrenia and as an adjunct in major depressive disorder. The adverse effects of this drug invole akathisisa, weight gain, infections of upper respiratory tract, somnolence, headache, and nasopharyngitis.

Approval of both cariprazine and brexpiprazole was in 2015. Cariprazine is a new antipsychotic displaying unique pharmacodynamic and pharmacokinetic properties [116]. As aripiprazole and brexpiprazole, cariprazine is the dopamine D$_2$, D$_3$ and serotonin 5-HT$_{1A}$ receptors partial agonist, as seen in Figure 7. However, its affinity for dopamine D$_3$ receptor is approximately ten times higher than for D$_2$ receptors. It is metabolized to two equipotent metabolites, desmethyl cariprazine and didesmethyl cariprazine, of which didesmethyl cariprazine has a half-life of 1 to 3 weeks [116]. Available reports indicate that cariprazine is efficient in management of cognitive and negative symptoms of schizophrenia. It also seems to have antimanic properties and it has a potential to treat bipolar depression [117]. However, currently it is not possible to evaluate antipsychotic potential of cariprazine in comparison to other antipsychotics. Cariprazine may be associated with adverse effects such as sedation, akathisia, weight gain, nausea, constipation, anxiety, dizziness [117].

The problem with the third generation antipsychotics is that they deteriorate the patient's condition in some patients suffering from schizophrenia. Thus, multi-target second generation antipsychotics are nowadays a gold standard in the schizophrenia treatment, although some patients respond better to the first generation treatment.

4.2. Other Multi-Target Compounds for the Treatment of Schizophrenia

Although recently implemented antipsychotics (e.g., cariprazine and brexpiprazole) are the third generation drugs, attempts are still made to design new multi-target ligands, which can be developed into second generation antipsychotics or better third generation drugs. These efforts will be presented in this chapter.

4.2.1. Modifications of Marketed Drugs

In recent years, a number of research groups studied halogenated arylpiperazines as a privileged scaffold active in CNS resulting in antipsychotics such as aripiprazole, trazodone and cariprazine [118]. The multimodal receptor profile of aripiprazole (5-HT$_{1A}$, 5-HT$_{2A}$, 5-HT$_7$, D$_2$ and D$_3$ receptors), as well as its functional profile as a partial agonist of D$_2$ and 5-HT$_{1A}$ receptors and antagonist of 5-HT$_{2A}$ and 5-HT$_7$ sites, makes it a good starting point to design compounds with antipsychotic, antidepressant, and anxiolytic activity [119]. Expanding the concept of mixed serotonin/dopamine receptor agonists as novel antipsychotics, Butini et al. designed a series of aripiprazole analogs that combined high affinity for 5-HT$_{1A}$ and 5-HT$_{2A}$ receptors, low affinity for D$_2$ receptors and high affinity for D$_3$ receptors. The structures of the compounds were based mainly on the 2,3-dichlorophenylpiperazine core structure, which was functionalized with isoquinoline-amide and quinolone- and isoquinoline-ether moieties, e.g., compound (**13**), compared in Figure 8. The study revealed that the optimal serotonin/dopamine receptor affinity balance was characterized by compounds with isoquinoline or benzofurane rings as heteroatomic systems [120]. As a continuation of their studies they developed a series bishetero(homo)arylpiperazines as novel and potent multifunctional ligands characterized by high affinity to D$_3$, 5-HT$_{1A}$ and low occupancy at D$_2$ and 5-HT$_{2C}$ receptors [121].

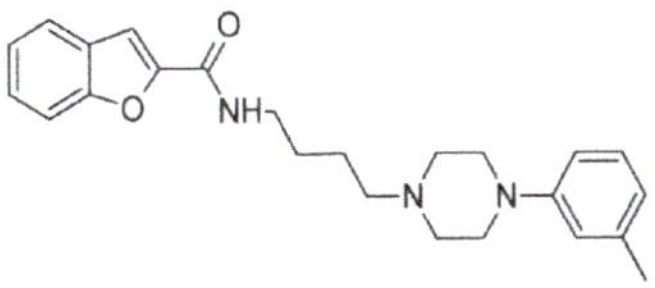

13

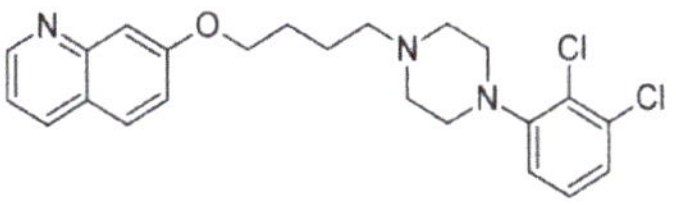

14

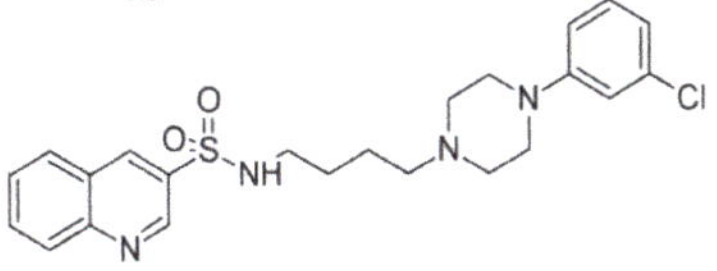

15

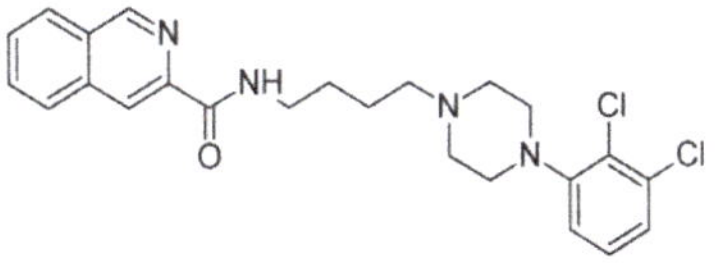

16

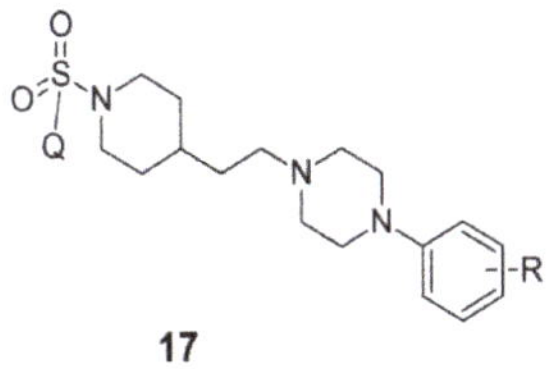

17

Figure 8. Novel potential multi-target antipsychotics derived from aripiprazole structure. Q: quinolone or isoquinoline.

In 2013 Zajdel et al. developed a series of new quinoline- and isoquinoline-sulfonamide analogs of aripiprazole to explore the effect of the replacement of the ether/amide moiety with sulfonamide, as well as the localization of a sulfonamide group in the azine moiety,

(**14–16**), see in Figure 8. In this study, two specific compounds displayed 5-HT$_{1A}$ agonistic, D$_2$ partial agonistic and 5-HT$_{2A}$/5-HT$_7$ antagonistic activity, thus resulting in significant antidepressant activity in mice models of depression [119]. Furthermore, the 4-isoquinolinyl analog (*N*-(4-(4-(2,3-dichlorophenyl)piperazin-1-yl)butyl)isoquinoline-4-sulfonamide) not only exhibited a similar receptor binding and functional profile but also displayed significant antipsychotic activity in MK-801-induced hyperlocomotor activity in mice [119]. These results supported the study previously conducted by Zajdel and coworkers in 2012, which reported on quinoline- and isoquinoline-sulfonamide derivatives of long-chain arylpiperazines with 3- or 4-chloro-phenylpiperazine moieties as potential antidepressant, anxiolytic and antipsychotic agents [122].

Partyka et al. inspired by previous findings on a group of *N*-alkylated azinesulfonamides, synthesized a series of 15 azinesulfonamides of phenylpiperazine derivatives, based on 4-(4-{2-[4-(4-chlorophenyl)-piperazin-1-yl]-ethyl}-piperidine-1-sulfonyl)-isoquinoline with semi-rigid alkylene spacer (**17**), as seen in Figure 8, and evaluated them as multimodal dopamine/serotonin receptor ligands. The study allowed to identify compound 5-(({4-(2-[4-(2,3-dichlorophenyl) piperazin-1-yl]ethyl)piperidin-1-yl}sulfonyl)quinolone which behaved as mixed D$_2$/5-HT$_{1A}$/5-HT$_7$ receptor antagonist. Preliminary pharmacological in vivo evaluation showed that compound was active in MK-801-evoked hyperactivity test in mice, and produced antidepressant-like activity in a mouse model of depression. Further studies in the area of CNS agents with multiple mode of action might confirmed its broad-based efficacy in the treatment of comorbid symptoms of schizophrenia/depression/anxiety [123].

In 2007 the atypical antipsychotic bifeprunox [1-(2-oxo-benzoxazolin-7-yl)-4-(3-biphenyl) methylpiperazine], with dual D$_2$ and 5-HT$_{1A}$ partial agonist activity, was filed for regulatory approval with the Food and Drug Administration (FDA), however the application was rejected owing to the weakness of evidence submitted and the death of a patient involved in the clinical trials. Nevertheless, through various molecular modification studies, it was established that the phenylpiperazine moiety is responsible for its antiserotonergic and antidopaminergic activity of this compound [120]. Based on these findings and the anti-inflammatory, nitric oxide synthase inhibitory activity, antidiabetic and antifungicidal activity of biphenyl compounds, a hybrid structure comprising a biphenyl and arylpiperazine moiety with an acetyl linker was designed [124]. In this study Bhosale et al. focused on combining the beneficial effects of the biphenyl moiety of bifeprunox with the methylpiperazine moiety of the aripiprazole. The newly designed hybrid antipsychotic scaffold (**18**) is presented in Figure 9.

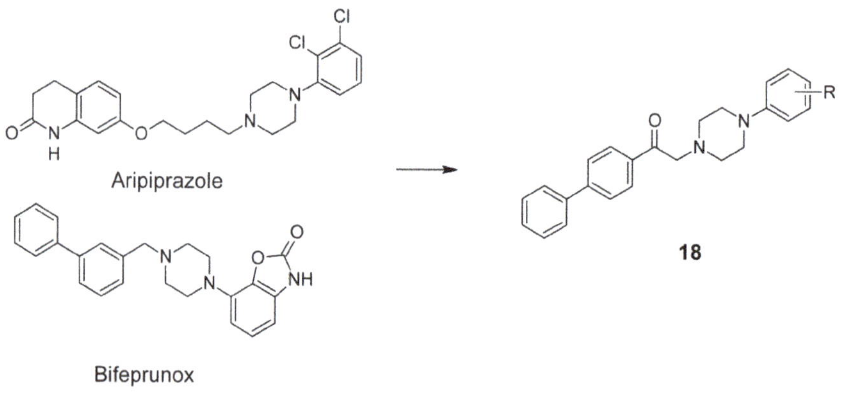

Figure 9. Design of multi-target hybrid compound based on aripiprazole and bifeprunox scaffold.

4.2.2. Other Multi-Target Compounds with Potential Application for the Treatment of Schizophrenia

It has been reported that the adjunctive usage of a neuroleptic together with selective serotonin reuptake inhibitor (SSRI), e.g., fluvoxamine, fluoxetine or citalopram is beneficial for the treatment of negative symptoms of schizophrenia without increasing EPS [125]. In this regard van Hes et al. elaborated SLV310, seen in Figure 10, (**19**), as a novel, potential antipsychotic displaying the interesting combination of potent dopamine D_2 receptor antagonism and serotonin reuptake receptor inhibition in one molecule which can be useful in treatment a broad range of symptoms in schizophrenia [126]. Subsequently the same research group obtained a series of compounds displaying D_2 receptor antagonism as well as SSRI properties by connecting the aryl piperazine of a neuroleptic with the indole moiety of a SSRI through alkyl chain in order to obtain promising antipsychotic agents, seen in Figure 10, (**20**). Optimization of length of the alkyl linker chain, substitution pattern of the indole moiety and bicyclic heteroaryl part has led to the maximally potent compound. Further, the molecular modelling studies have shown that the bifunctional activity of compound can be explained by its ability to adopt two different conformations fitting either D_2 receptor or SR pharmacophore without the disadvantages of potential pharmacokinetic interactions [127].

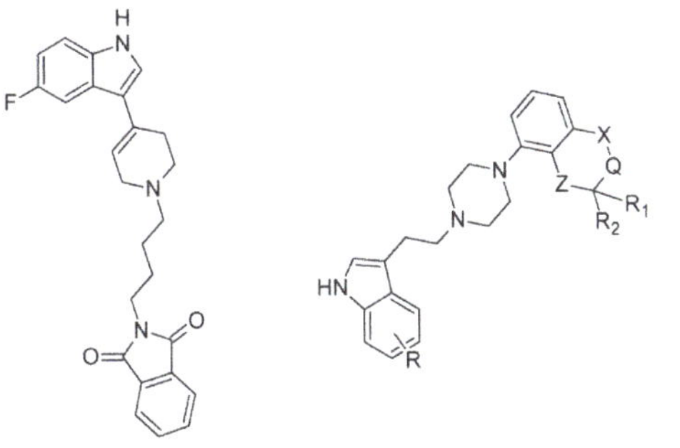

Figure 10. Dopamine D_2 receptor antagonists with selective serotonin reuptake inhibitor (SSRI) activity.

Li et al. reported synthesis and structure-activity relationships of a series of tetracyclic butyrophenones that display high affinities to serotonin 5-HT$_{2A}$ and dopamine D_2 receptors [128]. In particular, ITI-007 (4-((6bR,10aS)-3-methyl-2,3,6b,9,10,10a-hexahydro-1H,7H-pyrido[3′,4′:4,5]pyrrolo quinoxalin-8-yl)-1-(4-fluorophenyl)-butan-1-one 4-methylbenzenesulfonate), seen in Figure 11, (**21**), was found to be a potent 5-HT$_{2A}$ receptor antagonist, postsynaptic D_2 receptor antagonist and inhibitor of serotonin transporter [128].

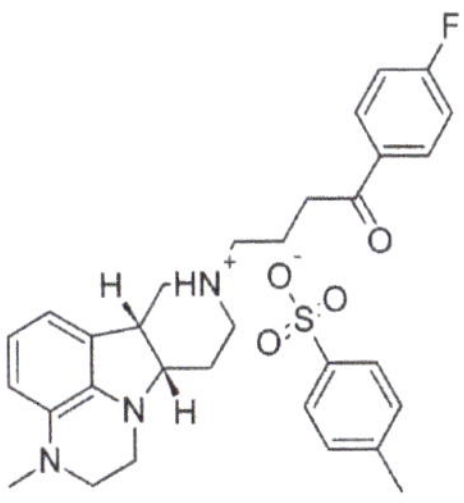

Figure 11. ITI-007, a potent 5-HT$_{2A}$ receptor antagonist, postsynaptic D_2 receptor antagonist and inhibitor of serotonin transporter.

In the latest study, Zajdel et al. [129] designed, synthesized and characterized a new series of azinesulfonamides of alicyclic amine derivatives with arylpiperazine/piperidine scaffold. Structure-activity studies of this compound series disclosed that the (isoquinolin-4-ylsulfonyl)-(S)-pyrrolidinyl fragment and the 1,2-benzothiazol-3-yl- and benzothiophen-4-yl-piperazine fragments were beneficial for affinity to 5-HT$_{1A}$, 5-HT$_{2A}$, 5-HT$_6$, 5-HT$_7$, D$_2$ and D$_3$ receptors. Furthermore, binding of these compounds with 5-HT$_6$ receptor depended on the stereochemistry of the alicyclic amine. Within this compound series, (S)-4-((2-(2-(4-(benzo[b]thiophen-4-yl)piperazin-1-yl)ethyl) pyrrolidin-1-yl) sulfonyl) isoquinoline, seen in Figure 12, (**22**), was identified as a potential novel antipsychotic. This compound is also characterized by blockade to SERT. Because it reverses PCP-induced hyperactivity and avoidance behavior in the CAR test, (**22**) it can be used to treat positive symptoms of schizophrenia. Next, its ability to reverse the social interaction deficit in a ketamine model and memory impairment in phencyclidine (PCP)- and ketamine-disrupted conditions reveals that that drug can improve the negative symptoms and has procognitive activity. Importantly, this compound did not have cardiac toxicity and tendency of inducing catalepsy [129].

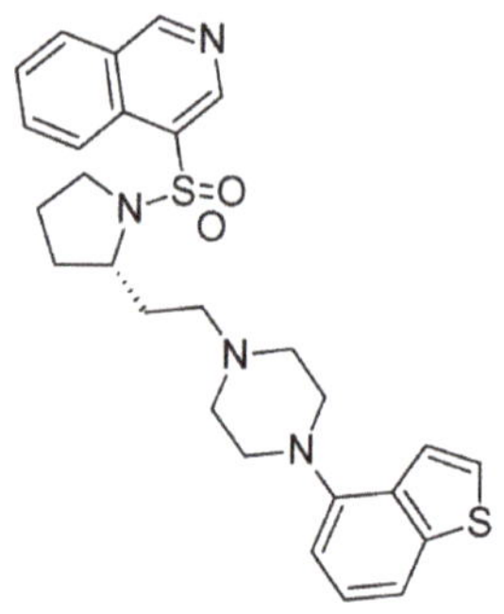

22

Figure 12. Multi-target ligand of aminergic G protein-coupled receptors (GPCRs) with SERT inhibitory properties.

In order to obtain novel antipsychotics Menegatti et al. designed and synthesized a series of N-phenylpiperazine derivatives [130]. A few compounds, i.e., 1-[1-(4-chlorophenyl)-1H-pyrazol-4-ylmethyl]-4-phenyl-piperazine (LASSBio-579, **23**, Figure 13), 1-phenyl-4-(1-phenyl-1H-[1,2,3]triazol-4-ylmethyl)-piperazine (LASSBio-580) and 1-[1-(4-chlorophenyl)-1H-[1,2,3]triazol-4-ylmethyl]-4-phenyl-piperazine (LASSBio-581) were selected based on potential antipsychotic activity. It was found that LASSBio-579 is the most promising of the three compounds, thanks to its affinity to both dopamine and serotonin receptors, in particular agonist activity at 5-HT$_{1A}$ receptor [131]. Thus, this multi-target compound was active in animal models of psychosis and reversed the catalepsy induced by WAY 100,635, Furthermore, co-administration of sub-effective doses of LASSBio-579 with sub-effective doses of clozapine or haloperidol prevented apomorphine-induced climbing without induction of catalepsy [131].

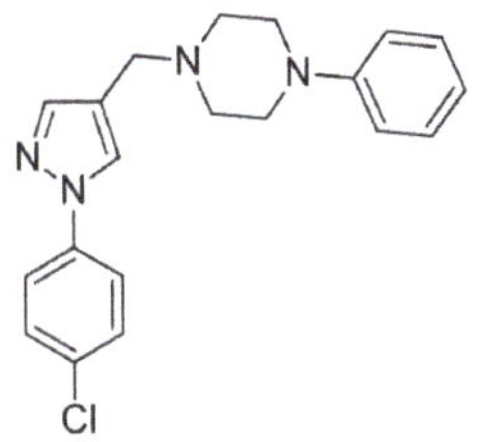

LASSBio-579 **23**

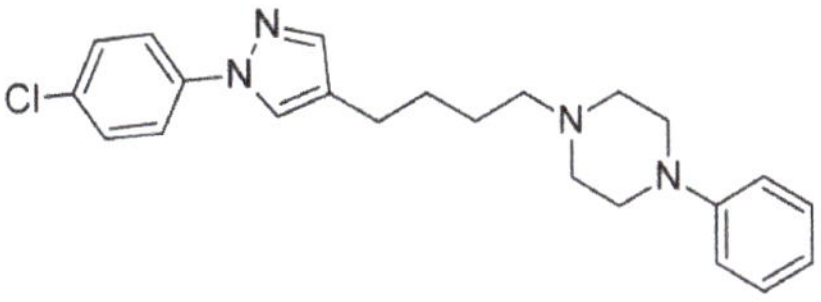

LASSBio-1635 **24**

Figure 13. Potential multi-target (dopamine and serotonin receptor ligands) antipsychotics.

In 2013, another team synthesized and made a pharmacological evaluation of the antipsychotic homologues of the lead compound LASSBio-579. The applied homologation approach turned out to be appropriate for increasing the affinity of these compounds to the 5-HT$_{2A}$ receptors, with no significant changes in the affinity for the D$_2$, D$_4$ and 5-HT$_{1A}$ receptors. In this context, (1-(4-(1-(4-chlorophenyl)-1*H*-pyrazol-4-yl) butyl)-4-phenylpiperazine) (LASSBio-1635, **24**), Figure 13 was the most promising derivative with a ten-fold higher affinity for the 5-HT$_{2A}$ receptor than its parent compound. Moreover, LASSBio-1635 displayed beneficial antagonistic efficacy at the 5-HT$_{2A}$ receptors. Next, LASSBio-1635 has also a 4-fold higher affinity for α_2 adrenergic receptors in comparison to LASSBio-579 and the favorable antagonistic efficacy. This multi-target ligand fully prevented the apomorphine-induced climbing in mice and prevented the ketamine-induced hyperlocomotion at doses with no effect on the mice locomotor activity [132].

In order to search for potential multi-target antipsychotics, Kaczor et al. [86] performed structure-based virtual screening using a D$_2$ receptor homology model in complex with olanzapine or chlorprothixene. As a result of a screen they selected 21 compounds, which were subjected to experimental validation. From 21 compounds tested, they found ten D$_2$ ligands (47.6% success rate, among them D$_2$ receptor antagonists as expected) possessing additional affinity to other receptors tested, in particular to 5-HT$_{1A}$ (partial agonists) and 5-HT$_{2A}$ receptors (antagonists). The affinity of the compounds ranged from 58 nM to about 24 µM. Similarity and fragmental analysis indicated a significant structural novelty of the identified compounds. The best compound (D2AAK1, **25**) has affinity of 58 nM to D$_2$ receptor and nanomolar or low micromolar affinity to D$_1$, D$_3$, 5-HT$_{1A}$ and 5-HT$_{2A}$ receptors. D2AAK1 is an antagonist at D$_2$ receptor and 5-HT$_{2A}$ receptor and a partial agonist at 5-HT$_{1A}$ receptor which is favorable for antipsychotic activity [131]. They found one D$_2$ receptor antagonist (D2AAK2, **26**) that did not have a protonatable nitrogen atom which is a key structural element of the classical D$_2$ pharmacophore model necessary to interact with the conserved Asp(3.32). This compound exhibited over 20-fold binding selectivity for the D$_2$ receptor compared to the D$_3$ receptor. The four best compounds (D2AAK1–D2AAK4, **25–28**, Figure 14) were subjected to in vivo evaluation. In particular compound D2AAK1 decreased amphetamine-induced hyperactivity (when compared to the amphetamine-treated group), measured as spontaneous locomotor activity in mice. In addition, in a passive avoidance test this compound improved memory consolidation after acute treatment in mice. Elevated plus maze tests indicated that D2AAK1 compound induced anxiogenic activity 30 min after acute treatment and anxiolytic activity 60 min after administration [133].

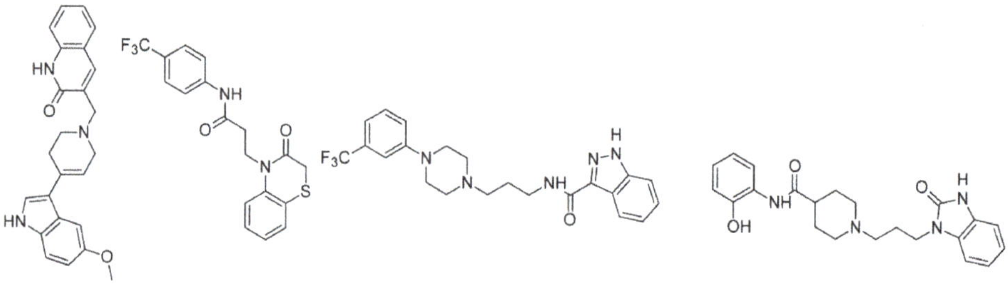

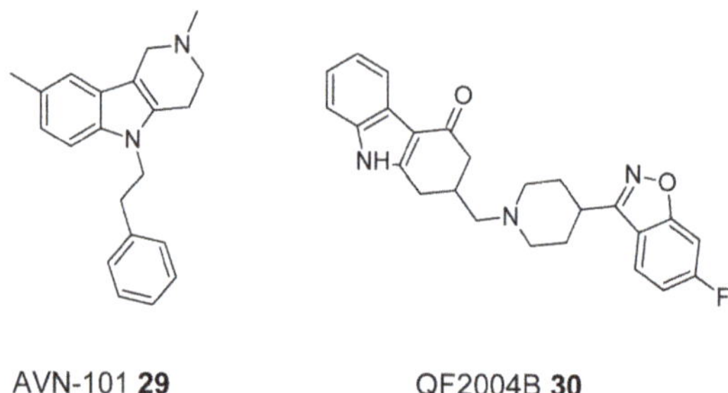

D2AAK1 **25**   D2AAK2 **26**   D2AAK3 **27**   D2AAK4 **28**

Figure 14. Multi-target compounds obtained in structure-based virtual screening.

AVN-101 (**29**, Figure 12) is another multi-target drug candidate that has an advantageous target fingerprint of activities with prevalent affinity to serotonin receptors, mainly 5-HT$_7$, 5-HT$_6$, 5-HT$_{2A}$, and 5-HT$_{2C}$, as well as to adrenergic α_{2B}, α_{2A}, and α_{2C} and histamine H$_1$ and H$_2$ receptors. The AVN-101 exhibits positive effects in the animal models of both impaired and innate cognition. It also exhibited significant anxiolytic and anti-depressant capabilities [134].

2-[4-(6-fluorobenzisoxazol-3-yl)piperidinyl]methyl-1,2,3,4-tetrahydro-carbazol-4-one (QF2004B), a conformationally constrained butyrophenone analog (**30**, Figure 15) has a multi-receptor profile with affinities similar to those of clozapine for serotonin (5-HT$_{2A}$, 5-HT$_{1A}$, and 5-HT$_{2C}$), dopamine (D$_1$, D$_2$, D$_3$ and D$_4$), alpha-adrenergic (α_1, α_2), muscarinic (M$_1$, M$_2$) and histamine H$_1$ receptors. In addition, QF2004B mirrored the antipsychotic activity and atypical profile of clozapine in a broad battery of in vivo tests including locomotor activity, apomorphine-induced stereotypies, catalepsy, apomorphine- and DOI (2,5-dimethoxy-4-iodoamphetamine)-induced prepulse inhibition (PPI) tests. These results point to QF2004B as a new lead compound with a relevant multi-receptor interaction profile for the discovery and development of new antipsychotics [135].

AVN-101 **29**   QF2004B **30**

Figure 15. Multi-target ligands of aminergic GPCRs as potential antipsychotics.

Searching for potential multi-target antipsychotics, Huang et al. [136] obtained a series of compounds bearing benzoxazole-piperidine (piperazine) scaffold with considerable dopamine D$_2$ and serotonin 5-HT$_{1A}$ and 5-HT$_{2A}$ receptor binding affinities. The best compound (**31**, Figure 16) had high affinity to D$_2$, 5-HT$_{1A}$ and 5-HT$_{2A}$ receptors, but low affinities for off-targets (the 5-HT$_{2C}$ and histamine H$_1$ receptors and human ether-a-go-go-related gene (hERG) channels). This compound diminished apomorphine-induced climbing and DOI-induced head twitching without observable catalepsy, even at the highest dose tested making it a promising candidate for multi-target antipsychotic treatment.

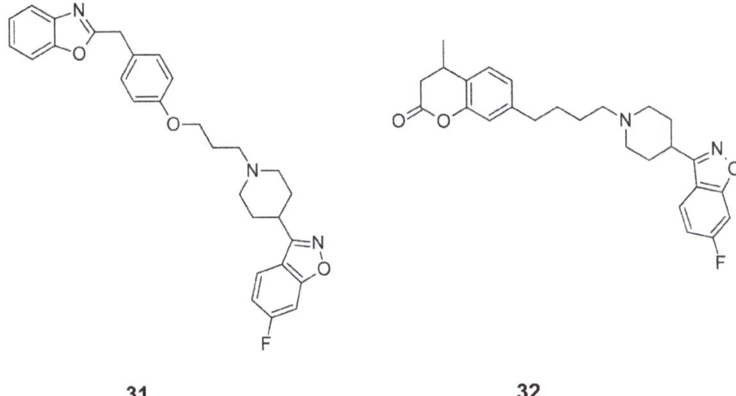

31   32

Figure 16. Multi-target ligands of aminergic GPCRs as potential antipsychotics with low affinity to off-targets.

Chen et al. [137] obtained potential antipsychotic coumarin derivatives, having potent dopamine D_2, D_3, and serotonin 5-HT$_{1A}$ and 5-HT$_{2A}$ receptor affinities. The best compound, seen in 32, Figure 16, also possesses low affinity for 5-HT$_{2C}$ and H$_1$ receptors and hERG channels. In behavioral studies this compound inhibited apomorphine-induced climbing behavior, MK-801-induced hyperactivity, and the conditioned avoidance response without observable catalepsy. Further, fewer preclinical side effects were observed for (**32**) in comparison to risperidone in assays that measured prolactin secretion and weight gain.

Another group synthesized a series of benzisothiazolylpiperazine derivatives combining potent dopamine D_2 and D_3, and serotonin 5-HT$_{1A}$ and 5-HT$_{2A}$ receptor affinities [138]. The best compound, as seen in (**33**), Figure 17, had significant affinity for D_2, D_3, 5-HT$_{1A}$, and 5-HT$_{2A}$ receptors, accompanied by a 20-fold selectivity for the D_3 versus D_2 subtype, and a low affinity for muscarinic M$_1$ and for hERG channels. In animal studies this compound blocked the locomotor-stimulating effects of phencyclidine, inhibited conditioned avoidance response, and improved the cognitive impairment in the novel object recognition tests in rats [138].

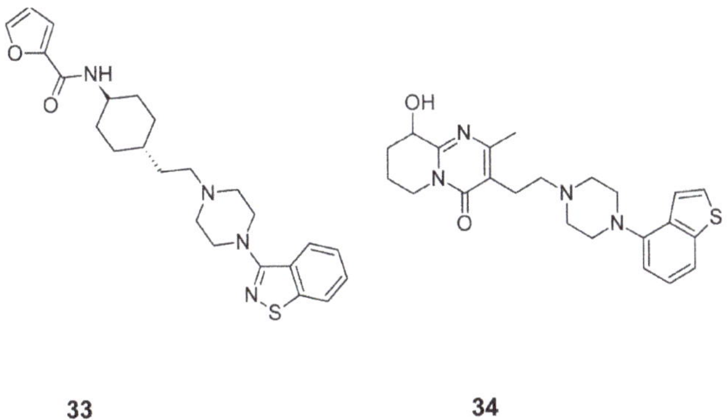

33   34

Figure 17. Potential multi-target antipsychotics with low probability of adverse effects.

In a recent study Xiamuxi et al. [139] reported a series of tetrahydropyridopyrimidinone derivatives, possessing potent dopamine D_2, serotonin 5-HT$_{1A}$ and 5-HT$_{2A}$ receptors affinities. The most promising compound, seen in (**34**), Figure 17, displayed high affinity to D_2, 5-HT$_{1A}$, and

5-HT$_{2A}$ receptors, with low affinity to α_{1A}, 5-HT$_{2C}$, H$_1$ receptors and hERG channels. In animal models, this compound diminished phencyclidine-induced hyperactivity with a high threshold for catalepsy induction.

In another new study Yang et al. [140] designed a series of benzamides, with potent dopamine D$_2$, serotonin 5-HT$_{1A}$ and 5-HT$_{2A}$ receptor affinity. Two best compounds, seen in (**35**) and (**36**), Figure 18, were not only potent D$_2$, 5-HT$_{1A}$, and 5-HT$_{2A}$ receptor ligands, but they were weak binders of 5-HT$_{2C}$, H$_1$ receptors and hERG channels. In behavioral studies these compounds decreased phencyclidine-induced hyperactivity with a high threshold for catalepsy induction.

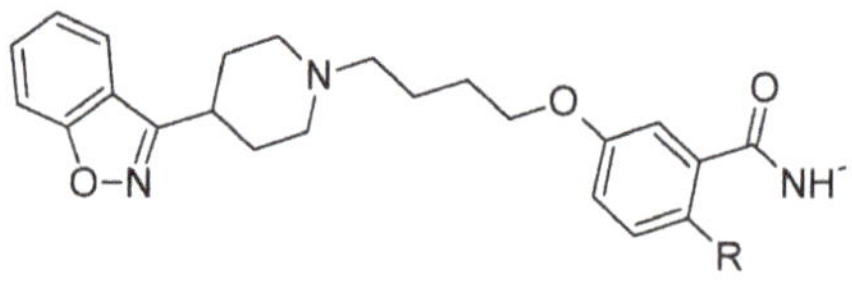

35: R=H
36: R=F

Figure 18. Benzamides as potential multi-target antipsychotics with low probability of side effects.

5. Conclusions and Perspectives

The growing pace of life promotes mental disorders. Pharmacotherapy for schizophrenia is nowadays very effective, in particular regarding treating positive symptoms of the disease, but at the same time there is a tremendous, unmet clinical need for the therapy of negative and cognitive symptoms, as well as for the management of drug resistant schizophrenia. Over the last half century, there has been only limited progress in the innovating mechanisms of action and the developing novel therapeutic agents for the treatment of schizophrenia. However, the breadth of potential goals and tested compounds clearly shows interest and importance in the pursuit of innovative drug development. A multi-target approach to drug design and discovery is now a hot topic in medicinal chemistry, in particular for the treatment of complex diseases such as schizophrenia. It should be emphasized that regarding management of schizophrenia, nothing more effective than multi-target treatment has been proposed. Involvement of nicotinic and glutamatergic targets in modern multi-target drugs can be beneficial for the treatment of negative symptoms and cognitive impairment. Another potential strategy is exploration novel signaling mechanisms concerning in particular GPCRs, such as allosteric modulation, biased signaling (functional selectivity), and receptor oligomerization. However, this approach will also be more promising when it involves multiple targets. In summary, as current multi-target antipsychotics are mainly orthosteric ligands of aminergic GPCRs with SSRI or SERT inhibitory activity in some cases, there is a huge unexplored area to include other receptors and enzymes as drug targets and to explore the wealth of signaling mechanism beyond the ternary complex model of GPCRs.

Author Contributions: M.K. wrote Section 4; P.S. wrote Section 2; A.A.K. wrote Sections 1, 3 and 5.

Funding: The work was performed under OPUS grant from National Science Center (NCN, Poland), grant number 2017/27/B/NZ7/01767. This article is based upon work from COST Action CA15135, supported by COST.

Conflicts of Interest: The authors declare no conflict of interest

Abbreviations

3D QSAR	Three-dimensional structure-activity relationship
AMPA	α-Amino-3-hydroxy-5-methyl-4-isoxazolepropionic acid
cAMP	Cyclic adenosine monophosphate
CAR	Conditioned avoidance response
CNS	Central nervous system
cNT PDEs	Cyclic nucleotide phosphodiestereases
EPS	Extrapyramidal symptoms
FDA	Food and Drug Administration
GABA	γ-Aminobutyric acid
GPCRs	G protein-coupled receptors
GSK-3	Glycogen synthase kinase-3
GTP	Guanosine-5′-triphosphate
LSD	Lysergic acid diethylamide
MTDs	Multi-target drugs
NMDA	*N*-methyl-D-aspartate
PCP	Phencyclidine
SSRI	Selective serotonin reuptake inhibitor
VTA	Ventral tegmental area

References

1. Stępnicki, P.; Kondej, M.; Kaczor, A.A. Current Concepts and Treatments of Schizophrenia. *Molecules* **2018**, *23*. [CrossRef] [PubMed]
2. Lau, C.-I.; Wang, H.-C.; Hsu, J.-L.; Liu, M.-E. Does the dopamine hypothesis explain schizophrenia? *Rev. Neurosci.* **2013**, *24*, 389–400. [CrossRef] [PubMed]
3. Hu, W.; MacDonald, M.L.; Elswick, D.E.; Sweet, R.A. The glutamate hypothesis of schizophrenia: Evidence from human brain tissue studies. *Ann. N. Y. Acad. Sci.* **2015**, *1338*, 38–57. [CrossRef] [PubMed]
4. Lavecchia, A.; Cerchia, C. In silico methods to address polypharmacology: Current status, applications and future perspectives. *Drug Discov. Today* **2016**, *21*, 288–298. [CrossRef] [PubMed]
5. Korcsmáros, T.; Szalay, M.S.; Böde, C.; Kovács, I.A.; Csermely, P. How to design multi-target drugs. *Expert Opin. Drug Discov.* **2007**, *2*, 799–808. [CrossRef] [PubMed]
6. Hopkins, A.L. Network pharmacology: The next paradigm in drug discovery. *Nat. Chem. Biol.* **2008**, *4*, 682–690. [CrossRef] [PubMed]
7. Nikolic, K.; Mavridis, L.; Djikic, T.; Vucicevic, J.; Agbaba, D.; Yelekci, K.; Mitchell, J.B.O. Drug Design for CNS Diseases: Polypharmacological Profiling of Compounds Using Cheminformatic, 3D-QSAR and Virtual Screening Methodologies. *Front. Neurosci.* **2016**, *10*, 265. [CrossRef] [PubMed]
8. Bakhshi, K.; Chance, S.A. The neuropathology of schizophrenia: A selective review of past studies and emerging themes in brain structure and cytoarchitecture. *Neuroscience* **2015**, *303*, 82–102. [CrossRef] [PubMed]
9. Rang, H.P.; Ritter, J.M.; Flower, R.J.; Henderson, G. *Rang and Dale's Pharmacology*, 8th ed.; Elsevier: Amsterdam, The Netherlands, 2016.
10. Miyamoto, S.; Duncan, G.E.; Marx, C.E.; Lieberman, J.A. Treatments for schizophrenia: A critical review of pharmacology and mechanisms of action of antipsychotic drugs. *Mol. Psychiatry* **2005**, *10*, 79–104. [CrossRef] [PubMed]
11. Mauri, M.C.; Paletta, S.; Maffini, M.; Colasanti, A.; Dragogna, F.; di Pace, C.; Altamura, A.C. Clinical pharmacology of atypical antipsychotics: An update. *EXCLI J.* **2014**, *13*, 1163–1191. [PubMed]
12. Orsolini, L.; Tomasetti, C.; Valchera, A.; Vecchiotti, R.; Matarazzo, I.; Vellante, F.; Iasevoli, F.; Buonaguro, E.F.; Fornaro, M.; Fiengo, A.L.C.; et al. An update of safety of clinically used atypical antipsychotics. *Expert Opin. Drug Saf.* **2016**, *15*, 1329–1347. [CrossRef] [PubMed]
13. Meltzer, H.Y.; Nash, J.F. Effects of antipsychotic drugs on serotonin receptors. *Pharmacol. Rev.* **1991**, *43*, 587–604. [PubMed]

14. Sorensen, S.M.; Kehne, J.H.; Fadayel, G.M.; Humphreys, T.M.; Ketteler, H.J.; Sullivan, C.K.; Taylor, V.L.; Schmidt, C.J. Characterization of the 5-HT$_2$ receptor antagonist MDL 100907 as a putative atypical antipsychotic: Behavioral, electrophysiological and neurochemical studies. *J. Pharmacol. Exp. Ther.* **1993**, *266*, 684–691. [PubMed]

15. Miller, C.H.; Fleischhacker, W.W.; Ehrmann, H.; Kane, J.M. Treatment of neuroleptic induced akathisia with the 5-HT$_2$ antagonist ritanserin. *Psychopharmacol. Bull.* **1990**, *26*, 373–376. [PubMed]

16. Schmidt, C.J.; Sorensen, S.M.; Kehne, J.H.; Carr, A.A.; Palfreyman, M.G. The role of 5-HT$_{2A}$ receptors in antipsychotic activity. *Life Sci.* **1995**, *56*, 2209–2222. [CrossRef]

17. Meltzer, H.Y.; Li, Z.; Kaneda, Y.; Ichikawa, J. Serotonin receptors: Their key role in drugs to treat schizophrenia. *Prog. Neuropsychopharmacol. Biol. Psychiatry* **2003**, *27*, 1159–1172. [CrossRef] [PubMed]

18. Millan, M.J. Improving the treatment of schizophrenia: Focus on serotonin 5-HT$_{1A}$ receptors. *J. Pharmacol. Exp. Ther.* **2000**, *295*, 853–861. [PubMed]

19. Akimova, E.; Lanzenberger, R.; Kasper, S. The serotonin-1A receptor in anxiety disorders. *Biol. Psychiatry* **2009**, *66*, 627–635. [CrossRef] [PubMed]

20. Ogren, S.O.; Eriksson, T.M.; Elvander-Tottie, E.; D'Addario, C.; Ekström, J.C.; Svenningsson, P.; Meister, B.; Kehr, J.; Stiedl, O. The role of 5-HT$_{1A}$ receptors in learning and memory. *Behav. Brain Res.* **2008**, *195*, 54–77. [CrossRef] [PubMed]

21. Prinssen, E.P.; Colpaert, F.C.; Koek, W. 5-HT$_{1A}$ receptor activation and anti-cataleptic effects: High-efficacy agonists maximally inhibit haloperidol-induced catalepsy. *Eur. J. Pharmacol.* **2002**, *453*, 217–221. [CrossRef]

22. Mignon, L.; Wolf, W.A. Postsynaptic 5-HT$_{1A}$ receptors mediate an increase in locomotor activity in the monoamine-depleted rat. *Psychopharmacology* **2002**, *163*, 85–94. [CrossRef] [PubMed]

23. Madjid, N.; Tottie, E.E.; Lüttgen, M.; Meister, B.; Sandin, J.; Kuzmin, A.; Stiedl, O.; Ogren, S.O. 5-Hydroxytryptamine 1A receptor blockade facilitates aversive learning in mice: Interactions with cholinergic and glutamatergic mechanisms. *J. Pharmacol. Exp. Ther.* **2006**, *316*, 581–591. [CrossRef] [PubMed]

24. Ohno, Y. Therapeutic role of 5-HT$_{1A}$ receptors in the treatment of schizophrenia and Parkinson's disease. *CNS Neurosci. Ther.* **2011**, *17*, 58–65. [CrossRef] [PubMed]

25. Kennett, G.A.; Curzon, G. Evidence that mCPP may have behavioural effects mediated by central 5-HT$_{1C}$ receptors. *Br. J. Pharmacol.* **1988**, *94*, 137–147. [CrossRef] [PubMed]

26. Kalkman, H.O. Hypersensitivity to meta-chlorophenylpiperazine (mCPP) in migraine and drug withdrawal. *Int. J. Clin. Pharmacol. Res.* **1997**, *17*, 75–77. [PubMed]

27. Krystal, J.H.; Seibyl, J.P.; Price, L.H.; Woods, S.W.; Heninger, G.R.; Aghajanian, G.K.; Charney, D.S. m-Chlorophenylpiperazine effects in neuroleptic-free schizophrenic patients. Evidence implicating serotonergic systems in the positive symptoms of schizophrenia. *Arch. Gen. Psychiatry* **1993**, *50*, 624–635. [CrossRef] [PubMed]

28. Wood, M.D.; Heidbreder, C.; Reavill, C.; Ashby, C.R., Jr.; Middlemiss, D.N. 5-HT$_{2C}$ receptor antagonists: Potential in schizophrenia. *Drug Dev. Res.* **2001**, *54*, 88–94.

29. Foster, D.J.; Conn, P.J. Allosteric Modulation of GPCRs: New Insights and Potential Utility for Treatment of Schizophrenia and Other CNS Disorders. *Neuron* **2017**, *94*, 431–446. [CrossRef] [PubMed]

30. Urs, N.M.; Gee, S.M.; Pack, T.F.; McCorvy, J.D.; Evron, T.; Snyder, J.C.; Yang, X.; Rodriguiz, R.M.; Borrelli, E.; Wetsel, W.C.; et al. Distinct cortical and striatal actions of a β-arrestin-biased dopamine D$_2$ receptor ligand reveal unique antipsychotic-like properties. *Proc. Natl. Acad. Sci. USA* **2016**, *113*, E8178–E8186. [CrossRef] [PubMed]

31. Kaczor, A.A.; Selent, J. Oligomerization of G protein-coupled receptors: Biochemical and biophysical methods. *Curr. Med. Chem.* **2011**, *18*, 4606–4634. [CrossRef] [PubMed]

32. Selent, J.; Kaczor, A.A. Oligomerization of G protein-coupled receptors: Computational methods. *Curr. Med. Chem.* **2011**, *18*, 4588–4605. [CrossRef] [PubMed]

33. Kaczor, A.A.; Jörg, M.; Capuano, B. The dopamine D$_2$ receptor dimer and its interaction with homobivalent antagonists: Homology modeling, docking and molecular dynamics. *J. Mol. Model.* **2016**, *22*, 203. [CrossRef] [PubMed]

34. Moreno, J.L.; Miranda-Azpiazu, P.; García-Bea, A.; Younkin, J.; Cui, M.; Kozlenkov, A.; Ben-Ezra, A.; Voloudakis, G.; Fakira, A.K.; Baki, L.; et al. Allosteric signaling through an mGlu2 and 5-HT$_{2A}$ heteromeric receptor complex and its potential contribution to schizophrenia. *Sci. Signal.* **2016**, *9*, ra5. [CrossRef] [PubMed]

35. Wong, E.H.F.; Tarazi, F.I.; Shahid, M. The effectiveness of multi-target agents in schizophrenia and mood disorders: Relevance of receptor signature to clinical action. *Pharmacol. Ther.* **2010**, *126*, 173–185. [CrossRef] [PubMed]

36. Maletic, V.; Eramo, A.; Gwin, K.; Offord, S.J.; Duffy, R.A. The Role of Norepinephrine and Its α-Adrenergic Receptors in the Pathophysiology and Treatment of Major Depressive Disorder and Schizophrenia: A. Systematic Review. *Front. Psychiatry* **2017**, *8*, 42. [CrossRef] [PubMed]

37. Svensson, T.H. α-Adrenoceptor modulation hypothesis of antipsychotic atypicality. *Prog. Neuropsychopharmacol. Biol. Psychiatry* **2003**, *27*, 1145–1158. [CrossRef] [PubMed]

38. Brosda, J.; Jantschak, F.; Pertz, H.H. α2-Adrenoceptors are targets for antipsychotic drugs. *Psychopharmacology* **2014**, *231*, 801–812. [CrossRef] [PubMed]

39. Marcus, M.M.; Wiker, C.; Frånberg, O.; Konradsson-Geuken, A.; Langlois, X.; Jardemark, K.; Svensson, T.H. Adjunctive α2-adrenoceptor blockade enhances the antipsychotic-like effect of risperidone and facilitates cortical dopaminergic and glutamatergic, NMDA receptor-mediated transmission. *Int. J. Neuropsychopharmacol.* **2010**, *13*, 891–903. [CrossRef] [PubMed]

40. He, M.; Deng, C.; Huang, X.-F. The role of hypothalamic H1 receptor antagonism in antipsychotic-induced weight gain. *CNS Drugs* **2013**, *27*, 423–434. [CrossRef] [PubMed]

41. Ellenbroek, B.A.; Ghiabi, B. Do Histamine receptor 3 antagonists have a place in the therapy for schizophrenia? *Curr. Pharm. Des.* **2015**, *21*, 3760–3770. [CrossRef] [PubMed]

42. Sadek, B.; Saad, A.; Sadeq, A.; Jalal, F.; Stark, H. Histamine H3 receptor as a potential target for cognitive symptoms in neuropsychiatric diseases. *Behav. Brain Res.* **2016**, *312*, 415–430. [CrossRef] [PubMed]

43. Ghoshal, A.; Rook, J.M.; Dickerson, J.W.; Roop, G.N.; Morrison, R.D.; Jalan-Sakrikar, N.; Lamsal, A.; Noetzel, M.J.; Poslusney, M.S.; Wood, M.R.; et al. Potentiation of M1 Muscarinic Receptor Reverses Plasticity Deficits and Negative and Cognitive Symptoms in a Schizophrenia Mouse Model. *Neuropsychopharmacol. Off. Publ. Am. Coll. Neuropsychopharmacol.* **2016**, *41*, 598–610. [CrossRef] [PubMed]

44. Dean, B.; Scarr, E. Possible involvement of muscarinic receptors in psychiatric disorders: A focus on schizophrenia and mood disorders. *Curr. Mol. Med.* **2015**, *15*, 253–264. [CrossRef] [PubMed]

45. Holt, D.J.; Bachus, S.E.; Hyde, T.M.; Wittie, M.; Herman, M.M.; Vangel, M.; Saper, C.B.; Kleinman, J.E. Reduced density of cholinergic interneurons in the ventral striatum in schizophrenia: An in situ hybridization study. *Biol. Psychiatry* **2005**, *58*, 408–416. [CrossRef] [PubMed]

46. Raedler, T.J.; Knable, M.B.; Jones, D.W.; Urbina, R.A.; Gorey, J.G.; Lee, K.S.; Egan, M.F.; Coppola, R.; Weinberger, D.R. In vivo determination of muscarinic acetylcholine receptor availability in schizophrenia. *Am. J. Psychiatry* **2003**, *160*, 118–127. [CrossRef] [PubMed]

47. Raedler, T.J.; Bymaster, F.P.; Tandon, R.; Copolov, D.; Dean, B. Towards a muscarinic hypothesis of schizophrenia. *Mol. Psychiatry* **2007**, *12*, 232–246. [CrossRef] [PubMed]

48. Freeman, T.P.; Stone, J.M.; Orgaz, B.; Noronha, L.A.; Minchin, S.L.; Curran, H.V. Tobacco smoking in schizophrenia: Investigating the role of incentive salience. *Psychol. Med.* **2014**, *44*, 2189–2197. [CrossRef] [PubMed]

49. Parikh, V.; Kutlu, M.G.; Gould, T.J. nAChR dysfunction as a common substrate for schizophrenia and comorbid nicotine addiction: Current trends and perspectives. *Schizophr. Res.* **2016**, *171*, 1–15. [CrossRef] [PubMed]

50. Adams, C.E.; Stevens, K.E. Evidence for a role of nicotinic acetylcholine receptors in schizophrenia. *Front. Biosci. J. Virtual Libr.* **2007**, *12*, 4755–4772. [CrossRef]

51. Li, Y.; Sun, L.; Yang, T.; Jiao, W.; Tang, J.; Huang, X.; Huang, Z.; Meng, Y.; Luo, L.; Wang, X.; et al. Design and Synthesis of Novel Positive Allosteric Modulators of α7 Nicotinic Acetylcholine Receptors with the Ability To Rescue Auditory Gating Deficit in Mice. *J. Med. Chem.* **2018**. [CrossRef] [PubMed]

52. Neves, G.A.; Grace, A.A. α7 Nicotinic receptor-modulating agents reverse the hyperdopaminergic tone in the MAM model of schizophrenia. *Neuropsychopharmacol. Off. Publ. Am. Coll. Neuropsychopharmacol.* **2018**, *43*, 1712–1720. [CrossRef] [PubMed]

53. Moghaddam, B.; Javitt, D. From Revolution to Evolution: The Glutamate Hypothesis of Schizophrenia and its Implication for Treatment. *Neuropsychopharmacology* **2012**, *37*, 4–15. [CrossRef] [PubMed]

54. Yang, A.C.; Tsai, S.-J. New Targets for Schizophrenia Treatment beyond the Dopamine Hypothesis. *Int. J. Mol. Sci.* **2017**, *18*. [CrossRef] [PubMed]

55. Goff, D.C.; Coyle, J.T. The emerging role of glutamate in the pathophysiology and treatment of schizophrenia. *Am. J. Psychiatry* **2001**, *158*, 1367–1377. [CrossRef] [PubMed]

56. Harrison, P.J.; Weinberger, D.R. Schizophrenia genes, gene expression, and neuropathology: On the matter of their convergence. *Mol. Psychiatry* **2005**, *10*, 40–68. [CrossRef] [PubMed]

57. Cull-Candy, S.; Brickley, S.; Farrant, M. NMDA receptor subunits: Diversity, development and disease. *Curr. Opin. Neurobiol.* **2001**, *11*, 327–335. [CrossRef]

58. Paoletti, P.; Neyton, J. NMDA receptor subunits: Function and pharmacology. *Curr. Opin. Pharmacol.* **2007**, *7*, 39–47. [CrossRef] [PubMed]

59. Saleem, S.; Shaukat, F.; Gul, A.; Arooj, M.; Malik, A. Potential role of amino acids in pathogenesis of schizophrenia. *Int. J. Health Sci. (Quassim)* **2017**, *11*, 63–68.

60. Farber, N.B. The NMDA receptor hypofunction model of psychosis. *Ann. N. Y. Acad. Sci.* **2003**, *1003*, 119–130. [CrossRef] [PubMed]

61. Goff, D.C. Glutamate receptors in schizophrenia and antipsychotic drugs, in Neurotransmitter Receptors. In *Actions of Antipsychotic Medications*; Lidow, M.S., Ed.; CRC Press: New York, NY, USA, 2000; pp. 121–136.

62. Meador-Woodruff, J.H.; Healy, D.J. Glutamate receptor expression in schizophrenic brain. *Brain Res. Brain Res. Rev.* **2000**, *31*, 288–294. [CrossRef]

63. Gao, X.M.; Sakai, K.; Roberts, R.C.; Conley, R.R.; Dean, B.; Tamminga, C.A. Ionotropic glutamate receptors and expression of *N*-methyl-D-aspartate receptor subunits in subregions of human hippocampus: Effects of schizophrenia. *Am. J. Psychiatry* **2000**, *157*, 1141–1149. [CrossRef] [PubMed]

64. Ibrahim, H.M.; Hogg, A.J.; Healy, D.J.; Haroutunian, V.; Davis, K.L.; Meador-Woodruff, J.H. Ionotropic glutamate receptor binding and subunit mRNA expression in thalamic nuclei in schizophrenia. *Am. J. Psychiatry* **2000**, *157*, 1811–1823. [CrossRef] [PubMed]

65. Heckers, S.; Goff, D.; Schacter, D.L.; Savage, C.R.; Fischman, A.J.; Alpert, N.M.; Rauch, S.L. Functional imaging of memory retrieval in deficit vs nondeficit schizophrenia. *Arch. Gen. Psychiatry* **1999**, *56*, 1117–1123. [CrossRef] [PubMed]

66. Kondziella, D.; Brenner, E.; Eyjolfsson, E.M.; Sonnewald, U. How do glial–neuronal interactions fit into current neurotransmitter hypotheses of schizophrenia? *Neurochem. Int.* **2007**, *50*, 291–301. [CrossRef] [PubMed]

67. Stone, J.M.; Morrison, P.D.; Pilowsky, L.S. Glutamate and dopamine dysregulation in schizophrenia—A synthesis and selective review. *J. Psychopharmacol. Oxf. Engl.* **2007**, *21*, 440–452. [CrossRef] [PubMed]

68. Kaiser, L.G.; Schuff, N.; Cashdollar, N.; Weiner, M.W. Age-related glutamate and glutamine concentration changes in normal human brain: ^{1}H MR spectroscopy study at 4 T. *Neurobiol. Aging* **2005**, *26*, 665–672. [CrossRef] [PubMed]

69. Leveque, J.C.; Macías, W.; Rajadhyaksha, A.; Carlson, R.R.; Barczak, A.; Kang, S.; Li, X.M.; Coyle, J.T.; Huganir, R.L.; Heckers, S.; et al. Intracellular modulation of NMDA receptor function by antipsychotic drugs. *J. Neurosci. Off. J. Soc. Neurosci.* **2000**, *20*, 4011–4020. [CrossRef]

70. Arvanov, V.L.; Liang, X.; Schwartz, J.; Grossman, S.; Wang, R.Y. Clozapine and haloperidol modulate *N*-methyl-D-aspartate- and non-*N*-methyl-D-aspartate receptor-mediated neurotransmission in rat prefrontal cortical neurons in vitro. *J. Pharmacol. Exp. Ther.* **1997**, *283*, 226–234. [PubMed]

71. Hashimoto, K.; Malchow, B.; Falkai, P.; Schmitt, A. Glutamate modulators as potential therapeutic drugs in schizophrenia and affective disorders. *Eur. Arch. Psychiatry Clin. Neurosci.* **2013**, *263*, 367–377. [CrossRef] [PubMed]

72. Menniti, F.S.; Lindsley, C.W.; Conn, P.J.; Pandit, J.; Zagouras, P.; Volkmann, R.A. Allosteric modulators for the treatment of schizophrenia: Targeting glutamatergic networks. *Curr. Top. Med. Chem.* **2013**, *13*, 26–54. [CrossRef] [PubMed]

73. Ward, S.E.; Pennicott, L.E.; Beswick, P. AMPA receptor-positive allosteric modulators for the treatment of schizophrenia: An overview of recent patent applications. *Future Med. Chem.* **2015**, *7*, 473–491. [CrossRef] [PubMed]

74. Herman, E.J.; Bubser, M.; Conn, P.J.; Jones, C.K. Metabotropic glutamate receptors for new treatments in schizophrenia. *Handb. Exp. Pharmacol.* **2012**, 297–365. [CrossRef]

75. Li, M.-L.; Hu, X.-Q.; Li, F.; Gao, W.-J. Perspectives on the mGluR2/3 agonists as a therapeutic target for schizophrenia: Still promising or a dead end? *Prog. Neuropsychopharmacol. Biol. Psychiatry* **2015**, *60*, 66–76. [CrossRef] [PubMed]

76. O'Leary, O.; Nolan, Y. Glycogen synthase kinase-3 as a therapeutic target for cognitive dysfunction in neuropsychiatric disorders. *CNS Drugs* **2015**, *29*, 1–15. [CrossRef] [PubMed]

77. Snyder, G.L.; Vanover, K.E. PDE Inhibitors for the Treatment of Schizophrenia. *Adv. Neurobiol.* **2017**, *17*, 385–409. [CrossRef] [PubMed]

78. Singh, J.; Kour, K.; Jayaram, M.B. Acetylcholinesterase inhibitors for schizophrenia. *Cochrane Database Syst. Rev.* **2012**, *1*, CD007967. [CrossRef] [PubMed]

79. Medina-Franco, J.L.; Giulianotti, M.A.; Welmaker, G.S.; Houghten, R.A. Shifting from the single to the multitarget paradigm in drug discovery. *Drug Discov. Today* **2013**, *18*, 495–501. [CrossRef] [PubMed]

80. Zheng, H.; Fridkin, M.; Youdim, M. From Single Target to Multitarget/Network Therapeutics in Alzheimer's Therapy. *Pharmaceuticals* **2014**, *7*, 113–135. [CrossRef] [PubMed]

81. Bansal, Y.; Silakari, O. Multifunctional compounds: Smart molecules for multifactorial diseases. *Eur. J. Med. Chem.* **2014**, *76*, 31–42. [CrossRef] [PubMed]

82. Méndez-Lucio, O.; Naveja, J.J.; Vite-Caritino, H.; Prieto-Martínez, F.D.; Medina-Franco, J.L.; Méndez-Lucio, O.; Naveja, J.J.; Vite-Caritino, H.; Prieto-Martínez, F.D.; Medina-Franco, J.L. Review. One Drug for Multiple Targets: A Computational Perspective. *J. Mex. Chem. Soc.* **2016**, *60*, 168–181.

83. Talevi, A. Multi-target pharmacology: Possibilities and limitations of the "skeleton key approach" from a medicinal chemist perspective. *Front. Pharmacol.* **2015**, *6*. [CrossRef] [PubMed]

84. Wichur, T.; Malawska, B. Multifunctional ligands—A new approach in the search for drugs against multi-factorial diseases. *Postepy Hig. Med. Doswiadczalnej Online* **2015**, *69*, 1423–1434.

85. Morphy, R.; Rankovic, Z. Designed Multiple Ligands. An Emerging Drug Discovery Paradigm. *J. Med. Chem.* **2005**, *48*, 6523–6543. [CrossRef] [PubMed]

86. Kaczor, A.A.; Silva, A.G.; Loza, M.I.; Kolb, P.; Castro, M.; Poso, A. Structure-Based Virtual Screening for Dopamine D_2 Receptor Ligands as Potential Antipsychotics. *ChemMedChem* **2016**, *11*, 718–729. [CrossRef] [PubMed]

87. Bawa, P.; Pradeep, P.; Kumar, P.; Choonara, Y.E.; Modi, G.; Pillay, V. Multi-target therapeutics for neuropsychiatric and neurodegenerative disorders. *Drug Discov. Today* **2016**, *21*, 1886–1914. [CrossRef] [PubMed]

88. Roth, B.L.; Sheffler, D.J.; Kroeze, W.K. Magic shotguns versus magic bullets: Selectively non-selective drugs for mood disorders and schizophrenia. *Nat. Rev. Drug Discov.* **2004**, *3*, 353–359. [CrossRef] [PubMed]

89. Naheed, M.; Green, B. Focus on Clozapine. *Curr. Med. Res. Opin.* **2001**, *17*, 223–229. [CrossRef] [PubMed]

90. Saklad, S.R. Graphic representation of pharmacology: Development of an alternative model. *Ment. Health Clin.* **2018**, *7*, 201–206. [CrossRef] [PubMed]

91. Boyda, H.N.; Ramos-Miguel, A.; Procyshyn, R.M.; Töpfer, E.; Lant, N.; Choy, H.H.T.; Wong, R.; Li, L.; Pang, C.C.Y.; Honer, W.G.; et al. Routine exercise ameliorates the metabolic side-effects of treatment with the atypical antipsychotic drug olanzapine in rats. *Int. J. Neuropsychopharmacol.* **2014**, *17*, 77–90. [CrossRef] [PubMed]

92. He, M.; Zhang, Q.; Deng, C.; Wang, H.; Lian, J.; Huang, X.-F. Hypothalamic histamine H1 receptor-AMPK signaling time-dependently mediates olanzapine-induced hyperphagia and weight gain in female rats. *Psychoneuroendocrinology* **2014**, *42*, 153–164. [CrossRef] [PubMed]

93. Godlewska, B.R.; Olajossy-Hilkesberger, L.; Ciwoniuk, M.; Olajossy, M.; Marmurowska-Michałowska, H.; Limon, J.; Landowski, J. Olanzapine-induced weight gain is associated with the -759C/T and -697G/C polymorphisms of the *HTR2C* gene. *Pharmacogenomics J.* **2009**, *9*, 234–241. [CrossRef] [PubMed]

94. Stephenson, C.M.; Pilowsky, L.S. Psychopharmacology of olanzapine. A. review. *Br. J. Psychiatry Suppl.* **1999**, *38*, 52–58. [CrossRef]

95. Selent, J.; López, L.; Sanz, F.; Pastor, M. Multi-receptor binding profile of clozapine and olanzapine: A structural study based on the new β2 adrenergic receptor template. *ChemMedChem* **2008**, *3*, 1194–1198. [CrossRef] [PubMed]

96. Piróg-Balcerzak, A.; Habrat, B.; Mierzejewski, P. Misuse and abuse of quetiapine. *Psychiatr. Pol.* **2015**, *49*, 81–93. [CrossRef] [PubMed]

97. George, B.; Craig, S. *Brenner and Stevens' Pharmacology*, 5th ed.; Elsevier: Amsterdam, The Netherlands, 2017.

98. Hasnain, M.; Vieweg, W.V.R. Weight considerations in psychotropic drug prescribing and switching. *Postgrad. Med.* **2013**, *125*, 117–129. [CrossRef] [PubMed]

99. Fornaro, M.; de Berardis, D.; Perna, G.; Solmi, M.; Veronese, N.; Orsolini, L.; Buonaguro, E.F.; Iasevoli, F.; Köhler, C.A.; Carvalho, A.F.; et al. Lurasidone in the Treatment of Bipolar Depression: Systematic Review of Systematic Reviews. *BioMed Res. Int.* **2017**, *2017*, 3084859. [CrossRef] [PubMed]

100. Urs, N.M.; Peterson, S.M.; Caron, M.G. New Concepts in Dopamine D_2 Receptor Biased Signaling and Implications for Schizophrenia Therapy. *Biol. Psychiatry* **2017**, *81*, 78–85. [CrossRef] [PubMed]

101. Brust, T.F.; Hayes, M.P.; Roman, D.L.; Watts, V.J. New functional activity of aripiprazole revealed. Robust antagonism of D_2 dopamine receptor stimulated by Gβγ antagonism. *Biochem Pharmacol.* **2015**, *93*, 85–91. [CrossRef] [PubMed]

102. Tamminga, C.A. Partial dopamine agonists in the treatment of psychosis. *J. Neural Transm. Vienna Austria 1996* **2002**, *109*, 411–420. [CrossRef] [PubMed]

103. Mailman, R.B.; Murthy, V. Third generation antipsychotic drugs: Partial agonism or receptor functional selectivity? *Curr. Pharm. Des.* **2010**, *16*, 488–501. [CrossRef] [PubMed]

104. Lieberman, J.A. Dopamine partial agonists: A new class of antipsychotic. *CNS Drugs* **2004**, *18*, 251–267. [CrossRef] [PubMed]

105. Allen, J.A.; Yost, J.M.; Setola, V.; Chen, X.; Sassano, M.F.; Chen, M.; Peterson, S.; Yadav, P.N.; Huang, X.; Feng, B.; et al. Discovery of β-arrestin-biased dopamine D_2 ligands for probing signal transduction pathways essential for antipsychotic efficacy. *Proc. Natl. Acad. Sci. USA* **2011**, *108*, 18488–18493. [CrossRef] [PubMed]

106. Burris, K.D.; Molski, T.F.; Xu, C.; Ryan, E.; Tottori, K.; Kikuchi, T.; Yocca, F.D.; Molinoff, P.B. Aripiprazole, a novel antipsychotic, is a high-affinity partial agonist at human dopamine D_2 receptors. *J. Pharmacol. Exp. Ther.* **2002**, *302*, 381–389. [CrossRef] [PubMed]

107. Shapiro, D.A.; Renock, S.; Arrington, E.; Chiodo, L.A.; Liu, L.-X.; Sibley, D.R.; Roth, B.L.; Mailman, R. Aripiprazole, a novel atypical antipsychotic drug with a unique and robust pharmacology. *Neuropsychopharmacol. Off. Publ. Am. Coll. Neuropsychopharmacol.* **2003**, *28*, 1400–1411. [CrossRef] [PubMed]

108. Lawler, C.P.; Prioleau, C.; Lewis, M.M.; Mak, C.; Jiang, D.; Schetz, J.A.; Gonzalez, A.M.; Sibley, D.R.; Mailman, R.B. Interactions of the novel antipsychotic aripiprazole (OPC-14597) with dopamine and serotonin receptor subtypes. *Neuropsychopharmacol. Off. Publ. Am. Coll. Neuropsychopharmacol.* **1999**, *20*, 612–627. [CrossRef]

109. Fell, M.J.; Perry, K.W.; Falcone, J.F.; Johnson, B.G.; Barth, V.N.; Rash, K.S.; Lucaites, V.L.; Threlkeld, P.G.; Monn, J.A.; McKinzie, D.L.; et al. In vitro and in vivo evidence for a lack of interaction with dopamine D_2 receptors by the metabotropic glutamate 2/3 receptor agonists 1S,2S,5R,6S-2-aminobicyclo[3.1.0]hexane-2,6-bicaroxylate monohydrate (LY354740) and (−)-2-oxa-4-aminobicyclo[3.1.0] Hexane-4,6-dicarboxylic acid (LY379268). *J. Pharmacol. Exp. Ther.* **2009**, *331*, 1126–1136. [CrossRef] [PubMed]

110. Masri, B.; Salahpour, A.; Didriksen, M.; Ghisi, V.; Beaulieu, J.-M.; Gainetdinov, R.R.; Caron, M.G. Antagonism of dopamine D_2 receptor/β-arrestin 2 interaction is a common property of clinically effective antipsychotics. *Proc. Natl. Acad. Sci. USA* **2008**, *105*, 13656–13661. [CrossRef] [PubMed]

111. Klewe, I.V.; Nielsen, S.M.; Tarpø, L.; Urizar, E.; Dipace, C.; Javitch, J.A.; Gether, U.; Egebjerg, J.; Christensen, K.V. Recruitment of β-arrestin2 to the dopamine D_2 receptor: Insights into anti-psychotic and anti-parkinsonian drug receptor signaling. *Neuropharmacology* **2008**, *54*, 1215–1222. [CrossRef] [PubMed]

112. Ribeiro, E.L.A.; de Mendonça Lima, T.; Vieira, M.E.B.; Storpirtis, S.; Aguiar, P.M. Efficacy and safety of aripiprazole for the treatment of schizophrenia: An overview of systematic reviews. *Eur. J. Clin. Pharmacol.* **2018**. [CrossRef] [PubMed]

113. Leucht, S.; Cipriani, A.; Spineli, L.; Mavridis, D.; Orey, D.; Richter, F.; Samara, M.; Barbui, C.; Engel, R.R.; Geddes, J.R.; et al. Comparative efficacy and tolerability of 15 antipsychotic drugs in schizophrenia: A multiple-treatments meta-analysis. *Lancet Lond. Engl.* **2013**, *382*, 951–962. [CrossRef]

114. Diefenderfer, L.A.; Iuppa, C. Brexpiprazole: A review of a new treatment option for schizophrenia and major depressive disorder. *Ment. Health Clin.* **2017**, *7*, 207–212. [CrossRef] [PubMed]

115. McEvoy, J.; Citrome, L. Brexpiprazole for the Treatment of Schizophrenia: A Review of this Novel Serotonin-Dopamine Activity Modulator. *Clin. Schizophr. Relat. Psychoses* **2016**, *9*, 177–186. [CrossRef] [PubMed]

116. Campbell, R.H.; Diduch, M.; Gardner, K.N.; Thomas, C. Review of cariprazine in management of psychiatric illness. *Ment. Health Clin.* **2017**, *7*, 221–229. [CrossRef] [PubMed]

117. De Berardis, D.; Orsolini, L.; Iasevoli, F.; Prinzivalli, E.; de Bartolomeis, A.; Serroni, N.; Mazza, M.; Valchera, A.; Fornaro, M.; Vecchiotti, R.; et al. The Novel Antipsychotic Cariprazine (RGH-188): State-of-the-Art in the Treatment of Psychiatric Disorders. *Curr. Pharm. Des.* **2016**, *22*, 5144–5162. [CrossRef] [PubMed]

118. Marciniec, K.; Kurczab, R.; Książek, M.; Bębenek, E.; Chrobak, E.; Satała, G.; Bojarski, A.J.; Kusz, J.; Zajdel, P. Structural determinants influencing halogen bonding: A case study on azinesulfonamide analogs of aripiprazole as 5-HT$_{1A}$, 5-HT$_7$, and D$_2$ receptor ligands. *Chem. Cent. J.* **2018**, *12*, 55. [CrossRef] [PubMed]

119. Zajdel, P.; Marciniec, K.; Maślankiewicz, A.; Grychowska, K.; Satała, G.; Duszyńska, B.; Lenda, T.; Siwek, A.; Nowak, G.; Partyka, A.; et al. Antidepressant and antipsychotic activity of new quinoline- and isoquinoline-sulfonamide analogs of aripiprazole targeting serotonin 5-HT$_{1A}$/5-HT$_{2A}$/5-HT$_7$ and dopamine D$_2$/D$_3$ receptors. *Eur. J. Med. Chem.* **2013**, *60*, 42–50. [CrossRef] [PubMed]

120. Butini, S.; Gemma, S.; Campiani, G.; Franceschini, S.; Trotta, F.; Borriello, M.; Ceres, N.; Ros, S.; Coccone, S.S.; Bernetti, M.; et al. Discovery of a New Class of Potential Multifunctional Atypical Antipsychotic Agents Targeting Dopamine D$_3$ and Serotonin 5-HT$_{1A}$ and 5-HT$_{2A}$ Receptors: Design, Synthesis, and Effects on Behavior. *J. Med. Chem.* **2009**, *52*, 151–169. [CrossRef] [PubMed]

121. Butini, S.; Campiani, G.; Franceschini, S.; Trotta, F.; Kumar, V.; Guarino, E.; Borrelli, G.; Fiorini, I.; Novellino, E.; Fattorusso, C.; et al. Discovery of bishomo(hetero)arylpiperazines as novel multifunctional ligands targeting dopamine D$_3$ and serotonin 5-HT$_{1A}$ and 5-HT$_{2A}$ receptors. *J. Med. Chem.* **2010**, *53*, 4803–4807. [CrossRef] [PubMed]

122. Zajdel, P.; Marciniec, K.; Maślankiewicz, A.; Satała, G.; Duszyńska, B.; Bojarski, A.J.; Partyka, A.; Jastrzbska-Wisek, M.; Wróbel, D.; Wesołowska, A.; et al. Quinoline- and isoquinoline-sulfonamide derivatives of LCAP as potent CNS multi-receptor—5-HT$_{1A}$/5-HT$_{2A}$/5-HT$_7$ and D$_2$/D$_3$/D$_4$—Agents: The synthesis and pharmacological evaluation. *Bioorg. Med. Chem.* **2012**, *20*, 1545–1556. [CrossRef] [PubMed]

123. Partyka, A.; Kurczab, R.; Canale, V.; Satała, G.; Marciniec, K.; Pasierb, A.; Jastrzębska-Więsek, M.; Pawłowski, M.; Wesołowska, A.; Bojarski, A.J.; et al. The impact of the halogen bonding on D$_2$ and 5-HT$_{1A}$/5-HT$_7$ receptor activity of azinesulfonamides of 4-[(2-ethyl)piperidinyl-1-yl]phenylpiperazines with antipsychotic and antidepressant properties. *Bioorg. Med. Chem.* **2017**, *25*, 3638–3648. [CrossRef] [PubMed]

124. Bhosale, S.H.; Kanhed, A.M.; Dash, R.C.; Suryawanshi, M.R.; Mahadik, K.R. Design, synthesis, pharmacological evaluation and computational studies of 1-(biphenyl-4-yl)-2-[4-(substituted phenyl)-piperazin-1-yl]ethanones as potential antipsychotics. *Eur. J. Med. Chem.* **2014**, *74*, 358–365. [CrossRef] [PubMed]

125. Van Dijk, H.G.; Dapper, E.A.; Vinkers, C.H. SSRIs and depressive symptoms in schizophrenia: A. systematic review. *Tijdschr. Voor Psychiatr.* **2017**, *59*, 40–46.

126. Van Hes, R.; Smid, P.; Stroomer, C.N.J.; Tipker, K.; Tulp, M.T.M.; van der Heyden, J.A.M.; McCreary, A.C.; Hesselink, M.B.; Kruse, C.G. SLV310, a novel, potential antipsychotic, combining potent dopamine D$_2$ receptor antagonism with serotonin reuptake inhibition. *Bioorg. Med. Chem. Lett.* **2003**, *13*, 405–408. [CrossRef]

127. Smid, P.; Coolen, H.K.A.C.; Keizer, H.G.; van Hes, R.; de Mos, J.-P.; den Hartog, A.P.; Stork, B.; Plekkenpol, R.H.; Niemann, L.C.; Stroomer, C.N.J.; et al. Synthesis, structure-activity relationships, and biological properties of 1-heteroaryl-4-[ω-(1H-indol-3-yl)alkyl]piperazines, novel potential antipsychotics combining potent dopamine D$_2$ receptor antagonism with potent serotonin reuptake inhibition. *J. Med. Chem.* **2005**, *48*, 6855–6869. [CrossRef] [PubMed]

128. Li, P.; Zhang, Q.; Robichaud, A.J.; Lee, T.; Tomesch, J.; Yao, W.; Beard, J.D.; Snyder, G.L.; Zhu, H.; Peng, Y.; et al. Discovery of a Tetracyclic Quinoxaline Derivative as a Potent and Orally Active Multifunctional Drug Candidate for the Treatment of Neuropsychiatric and Neurological Disorders. *J. Med. Chem.* **2014**, *57*, 2670–2682. [CrossRef] [PubMed]

129. Zajdel, P.; Kos, T.; Marciniec, K.; Satała, G.; Canale, V.; Kamiński, K.; Hołuj, M.; Lenda, T.; Koralewski, R.; Bednarski, M.; et al. Novel multi-target azinesulfonamides of cyclic amine derivatives as potential antipsychotics with pro-social and pro-cognitive effects. *Eur. J. Med. Chem.* **2018**, *145*, 790–804. [CrossRef] [PubMed]

130. Menegatti, R.; Cunha, A.C.; Ferreira, V.F.; Perreira, E.F.; El-Nabawi, A.; Eldefrawi, A.T.; Albuquerque, E.X.; Neves, G.; Rates, S.M.; Fraga, C.A.; et al. Design, synthesis and pharmacological profile of novel dopamine D$_2$ receptor ligands. *Bioorg. Med. Chem.* **2003**, *11*, 4807–4813. [CrossRef]

131. Neves, G.; Antonio, C.B.; Betti, A.H.; Pranke, M.A.; Fraga, C.A.M.; Barreiro, E.J.; Noël, F.; Rates, S.M.K. New insights into pharmacological profile of LASSBio-579, a multi-target *N*-phenylpiperazine derivative active on animal models of schizophrenia. *Behav. Brain Res.* **2013**, *237*, 86–95. [CrossRef] [PubMed]

132. Pompeu, T.E.T.; Alves, F.R.S.; Figueiredo, C.D.M.; Antonio, C.B.; Herzfeldt, V.; Moura, B.C.; Rates, S.M.K.; Barreiro, E.J.; Fraga, C.A.M.; Noël, F. Synthesis and pharmacological evaluation of new N-phenylpiperazine derivatives designed as homologues of the antipsychotic lead compound LASSBio-579. *Eur. J. Med. Chem.* **2013**, *66*, 122–134. [CrossRef] [PubMed]

133. Kaczor, A.A.; Targowska-Duda, K.M.; Budzyńska, B.; Biała, G.; Silva, A.G.; Castro, M. In vitro, molecular modeling and behavioral studies of 3-{[4-(5-methoxy-1H-indol-3-yl)-1,2,3,6-tetrahydropyridin-1-yl]methyl}-1,2-dihydroquinolin-2-one (D2AAK1) as a potential antipsychotic. *Neurochem. Int.* **2016**, *96*, 84–99. [CrossRef] [PubMed]

134. Ivachtchenko, A.V.; Lavrovsky, Y.; Okun, I. AVN-101: A Multi-Target Drug Candidate for the Treatment of CNS Disorders. *J. Alzheimers Dis.* **2016**, *53*, 583–620. [CrossRef] [PubMed]

135. Brea, J.; Castro, M.; Loza, M.I.; Masaguer, C.F.; Raviña, E.; Dezi, C.; Pastor, M.; Sanz, F.; Cabrero-Castel, A.; Galán-Rodríguez, B.; et al. QF2004B, a potential antipsychotic butyrophenone derivative with similar pharmacological properties to clozapine. *Neuropharmacology* **2006**, *51*, 251–262. [CrossRef] [PubMed]

136. Huang, L.; Zhang, W.; Zhang, X.; Yin, L.; Chen, B.; Song, J. Synthesis and pharmacological evaluation of piperidine (piperazine)-substituted benzoxazole derivatives as multi-target antipsychotics. *Bioorg. Med. Chem. Lett.* **2015**, *25*, 5299–5305. [CrossRef] [PubMed]

137. Chen, Y.; Wang, S.; Xu, X.; Liu, X.; Yu, M.; Zhao, S.; Liu, S.; Qiu, Y.; Zhang, T.; Liu, B.-F.; et al. Synthesis and biological investigation of coumarin piperazine (piperidine) derivatives as potential multireceptor atypical antipsychotics. *J. Med. Chem.* **2013**, *56*, 4671–4690. [CrossRef] [PubMed]

138. Chen, X.-W.; Sun, Y.-Y.; Fu, L.; Li, J.-Q. Synthesis and pharmacological characterization of novel *N*-(trans-4-(2-(4-(benzo[d]isothiazol-3-yl)piperazin-1-yl)ethyl)cyclohexyl)amides as potential multireceptor atypical antipsychotics. *Eur. J. Med. Chem.* **2016**, *123*, 332–353. [CrossRef] [PubMed]

139. Xiamuxi, H.; Wang, Z.; Li, J.; Wang, Y.; Wu, C.; Yang, F.; Jiang, X.; Liu, Y.; Zhao, Q.; Chen, W.; et al. Synthesis and biological investigation of tetrahydropyridopyrimidinone derivatives as potential multireceptor atypical antipsychotics. *Bioorg. Med. Chem.* **2017**, *25*, 4904–4916. [CrossRef] [PubMed]

140. Yang, F.; Jiang, X.; Li, J.; Wang, Y.; Liu, Y.; Bi, M.; Wu, C.; Zhao, Q.; Chen, W.; Yin, J.; et al. Synthesis, structure-activity relationships, and biological evaluation of a series of benzamides as potential multireceptor antipsychotics. *Bioorg. Med. Chem. Lett.* **2016**, *26*, 3141–3147. [CrossRef] [PubMed]

International Journal of
Molecular Sciences

Article

The Association Between Affective Temperament Traits and Dopamine Genes in Obese Population

Natalia Lesiewska [1,*], Alina Borkowska [1], Roman Junik [2], Anna Kamińska [2], Joanna Pulkowska-Ulfig [1], Andrzej Tretyn [3] and Maciej Bieliński [1]

[1] Chair and Department of Clinical Neuropsychology, Nicolaus Copernicus University in Toruń, Collegium Medicum, Bydgoszcz 85-094, Poland; alab@cm.umk.pl (A.B.); joanna.pulkowska@gmail.com (J.P.-U.); bielinskim@gmail.com (M.B.)
[2] Department of Endocrinology and Diabetology, Nicolaus Copernicus University in Toruń, Collegium Medicum, Bydgoszcz 85-094, Poland; junik@cm.umk.pl (R.J.); amikam@wp.pl (A.K.)
[3] Department of Biotechnology, Nicolaus Copernicus University, Toruń 87-100, Poland; prat@umk.pl
* Correspondence: n.lesiewska@gmail.com; Tel.: +48-52-585-37-03

Received: 31 March 2019; Accepted: 10 April 2019; Published: 15 April 2019

Abstract: Studies indicate the heritable nature of affective temperament, which shows personality traits predisposing to the development of mental disorders. Dopaminergic gene polymorphisms such as *DRD4*, *COMT*Val158Met, and *DAT1* have been linked to affective disorders in obesity. Due to possible correlation between the aforementioned polymorphisms and the affective temperament, the aim of our research was to investigate this connection in an obese population. The study enrolled 245 obese patients (178 females; 67 males). The affective temperament was assessed using the Temperament Evaluation of Memphis, Pisa, Paris, and San Diego autoquestionnaire (TEMPS-A). Genetic polymorphisms of *DAT1*, *COMT*Val158Met and *DRD4* were collected from peripheral blood sample and determined using a polymerase chain reaction (PCR). Only in COMT polymorphisms, the cyclothymic and irritable dimensions were significantly associated with Met/Val carriers ($p = 0.04$; $p = 0.01$). Another interesting finding was the correlation between the affective temperament and age in men and women. We assume that dopamine transmission in heterozygotes of COMT may determine the role of the affective temperament in obese persons. Dopaminergic transmission modulated by COMT may be responsible for a greater temperament expression in obese individuals. To our knowledge, this is the first study describing the role of affective temperament in the obese population, but more research is needed in this regard.

Keywords: dopaminergic gene polymorphisms; affective temperament; obesity

---

1. Introduction

Previous research devoted to eating disorders, mainly related to anorexia and bulimia, indicated the possibility of specific personality traits related to both the predisposition to the disease and those affecting the course and clinical picture of the disease [1]. The psychological aspects of predisposition to obesity are mostly: disorders of the self-regulation mechanism, beliefs and expectations of the individual, personality traits, difficulties in coping with stress and experienced emotions [2]. Recent psychiatric studies suggest that there is a link between obesity and mood disorders. The association between obesity and depression occurred in childhood. Previous research indicated that the symptoms of eating disorders are common and that patients with bipolar disorder are more obese than the control group [3–5]. The results indicate that the symptoms of eating disorders are common and that patients with bipolar disorder are more obese than the control group [6,7]. Along with the broadening of the limits of diagnostic criteria for bipolar disorder (BD) over the last years, research has pointed to the high prevalence of less severe forms of BDs, in particular hypomania, among obese patients [8].

Dopamine might be a factor linking obesity with mood disorders, especially given that maladaptive changes in dopaminergic transmission have been observed in obesity and [9–11].

Yokum et al. (2015) tested the multilocus genetic composite risk score—a proxy for dopaminergic signaling—and future changes in BMI values. The results of their study revealed that *DRD4*, *COMT*Val158Met and *DAT1* polymorphisms, putatively associated with a greater DA signaling capacity, were linked to greater increases in the BMI; hence the future weight gain [12].

According to the regulatory theory, the temperament is the basic, relatively permanent character traits that manifests in the formal specifics of behavior. These features are already present in early childhood and are common to humans and animals. Being originally determined by innate physiological mechanisms, temperament may change under the influence of puberty, aging and certain environmental factors. In their work, Serafini et al. (2015) showed that unpleasant events, inter alia: sexual abuse, physical abuse, child maltreatment or domestic violence, were associated with greater depression and suicidality in adolescents. It is worth noting that the type of events, as well as the frequency and the timing of maltreatment, may influence the risk of psychiatric disorders, including suicidal behavior, due to the disruption in the brain development connected to cognitive, social or emotional functioning [13].

According to Arnold Buss and Robert Plomin (1984), temperament is a set of inherited personality traits that are revealed in early childhood. The temperament understood in this way is the basis for shaping and developing personality [14]. According to the assumptions of modern psychiatry, temperament is considered a personality aspect that takes into account the constant behavior of the individual, predicts mood changes and is strongly genetically conditioned [15–17].

An important researcher in the field of psychiatry, Emil Kraepelin, believed that a depressive temperament, and a manic, irritable and cyclothymic temperament is not only represented by affective predispositions, but also by subclinical variations of manic and depressive disorders. Akiskal et al. distinguished four types of affective temperament: depressive, manic, irritable and cyclothymic. In later studies, manic temperament was changed to hyperthymic temperament, and anxiety temperament was added [18–20]. The conceptualization of these five types of temperament has led to the creation of a TEMPS psychometric tool (Temperament Evaluation of Memphis, Pisa, Paris and San Diego). In studies utilizing this tool, obese patients showed significantly higher results in cyclothymic, irritable and anxious temperaments compared to the control group [21]. Assuming that the cyclothymic temperament is part of the mild spectrum of BD, these results are consistent with previous studies suggesting a higher incidence of bipolar symptoms in people with obesity [8].

The relationship between temperamental traits in Cloninger's concept (Temperament and Character Inventory—TCI) and gene polymorphism for the serotonergic and dopaminergic systems was also found. Research is still under way to determine the role of genes in the regulation and emergence of bipolarity and affective temperament [22]. So far, in obesity, this type of research is scarce. Our previous study showed a significant contribution of the SERT gene in the regulation of temperament in the obese population [23].

There are few studies in the literature describing the connection between polymorphisms of the dopaminergic system genes with personality traits, character or temperament. Thus, the aim of this project is to determine the possible role of dopaminergic pathways in the regulation of the affective temperament in the obese population. In order to accomplish our objects, we formulated the following hypotheses:

1. Individuals with higher BMI values will have greater scores in cyclothymic, anxious and irritable temperaments, which are associated with the predisposition to psychiatric comorbidities.

2. A lower dopaminergic transmission modulated by the following gene polymorphisms: *COMT*Val158Met, *DRD4* and *DAT1*, will be associated with a higher BMI and more pronounced cyclothymic, anxious and irritable dimensions.

2. Results

Basic demographic data and TEMPS-A dimensions in a group of women and men are shown in Table 1. There were significant differences only in terms of more depressed and irritable dimensions in the group of men.

Table 1. Age, body mass index (BMI) and results on the TEMPS-A scale in study participants. Data are presented as medians, and 25th and 75th quartiles.

	Female (*n* = 178)	Male (*n* = 67)	P	Cohen's d
Age	41 (36.0–47.0)	42 (34.0–48.5)	0.11	0.24
BMI	40.7 (36.3–47.0)	41.4 (35.2–48.5)	0.8	0.03
TEMPS_D	0.38 (0.28–0.52)	0.43 (0.24–0.43)	0.04	0.36
TEMPS_C	0.36 (0.24–0.52)	0.47 (0.23–0.62)	0.09	0.29
TEMPS_H	0.52 (0.35–0.62)	0.52 (0.38–0.64)	0.53	0.1
TEMPS_I	0.14 (0.05–0.28)	0.23 (0.09–0.33)	0.001	0.42
TEMPS_A	0.33 (0.24–0.55)	0.35 (0.17–0.51)	0.12	0.18

BMI, body mass index; TEMPS_D—depressive subscale of TEMPS-A; TEMPS_C—cyclothymic subscale of TEMPS-A; hyperthymic subscale of TEMPS-A; TEMPS_I—irritable subscale of TEMPS-A; TEMPS_A—anxious subscale of TEMPS-A. Significance of differences between sexes was determined by the Mann—Whitney U test. Size effect was measured by Cohen's d method.

Table 2 shows the analysis of associations between the temperamental dimensions (according to TEMPS-A) and both the age and BMI. Our results revealed, that in the group of women, a greater age significantly correlated with more expressed dimensions of depression and anxiety. Regarding the BMI, we observed its positive correlation with a greater expression of the hyperthymic temperament and a smaller cyclothymia. On the other hand, in the group of men, there was a negative correlation between age and cyclothymia. In this group, the dimensions of cyclothymia and irritability were significantly more pronounced, as the BMI values increased. A partial Kendall's regression in the group of women showed the significance of the relationship between the age and depressive temperament, the age and anxiety temperament as well as between the BMI and cyclothymic temperament. On the other hand, in the group of men, the significance was confirmed for the BMI and cyclothymic temperament.

When analyzing the correlations of the studied COMT gene polymorphisms in the subgroups of both sexes (Table 3), no significant relationships were found. Thus, we performed an ANOVA for the entire group, and then conducted a post hoc analysis only for significant results for the ANOVA, which revealed a significantly greater expression of cyclothymia in the heterozygote subgroup. Similarly, the irritability was more pronounced in the heterozygous group.

A multiple testing procedure was then performed to confirm the validity of the relevant results. After applying the Bonferroni correction, it was confirmed that the still results for the COMT gene alleles and TEMPS-A cyclothymic ($p = 0.01$) and irritable ($p = 0.01$) dimensions are considered to be significant.

According to Table 4, the analyses carried out for the *DAT1* polymorphism did not show any significant relationships of temperament dimensions according to TEMPS-A.

Table 2. R-Spearman correlations of the age and BMI result with the TEMPS scores in women and men. Partial Kendall regression for significant correlations.

	Female (n = 178)		Male (n = 67)	
	Age	BMI	Age	BMI
TEMPS_D	r = 0.21 $p = 0.004$ Par. Kendall's tau Tau = -2.74; $p = 0.006$	r = −0.14 $p = 0.06$	r = −0.05 $p = 0.68$	r = 0.06 $p = 0.63$
TEMPS_C	r = −0.06 $p = 0.42$	r = −0.16 $p = 0.03$ Par. Kendall's tau Tau = −0.15; $p = 0.002$	r = −0.26 $p = 0.03$ Par. Kendall's tau Tau = −0.01; $p = 0.44$	r = 0.33 $p = 0.006$ Par. Kendall's tau Tau = −0,24; $p = 0.003$
TEMPS_H	r = −0.09 $p = 0.23$	r = 0.16 $p = 0.03$ Par. Kendall's tau Tau = 0.01; $p = 0.37$	r = 0.09 $p = 0.46$	r = −0.09 $p = 0.47$
TEMPS_I	r = 0.004 $p = 0.95$	r = −0.07 $p = 0.35$	r = −0.12 $p = 0.33$	r = 0.31 $p = 0.01$ Par. Kendall's tau Tau = −0.13; $p = 0.05$
TEMPS_A	r = 0.16 $p = 0.03$ Par. Kendall's tau Tau = −0.09; $p = 0.03$	r = −0.11 $p = 0.14$	r = −0.11 $p = 0.37$	r = 0.07 $p = 0.57$

BMI, body mass index; TEMPS_D—depressive subscale of TEMPS-A; TEMPS_C—cyclothymic subscale of TEMPS-A; TEMPS_H—hyperthymic subscale of TEMPS-A; TEMPS_I—irritable subscale of TEMPS-A; TEMPS_A—anxious subscale of TEMPS-A. Par. Kendall's tau—partial Kendall's tau. Bold values indicate statistical significance.

Table 3. COMT polymorphisms and TEMPS results in study group.

	All Group (n = 245)			
	G/G (*n* = 64)	**G/A** (*n* = 120)	**A/A** (*n* = 61)	*p*
BMI	40.9 (36.7–44.3)	42,5 (36.5–49.0)	42,4 (37.0–48.1)	0.52
TEMPS_D	0.36 (0.28–0.42)	0.42 (0.28–0.52)	0.38 (0.28–0.43)	0.36
TEMPS_C	0.28 (0.16–0.47)	0.47 (0.24–0.64)	0.38 (0.23–0.52)	0.04 — Post-hoc G/G vs G/A *p* = 0.014 G/A vs A/A ns G/Avs AA na
TEMPS_H	0.57 (0.50–0.67)	0.47 (0.28–0.61)	0.57 (0.38–0.57)	0.07
TEMPS_I	0.09 (0.04–0.16)	0.26 (0.09–0.33)	0.09 (0.05–0.24)	0.01 — Post-hoc G/G vs G/A *p* = 0.01 G/A vs A/A ns G/Avs AA ns
TEMPS_A	0,32 (0.20–0.52)	0,35 (0.22–0.59)	0.32 (0.24–0.52)	0.52

BMI, body mass index; TEMPS_D—depressive subscale of TEMPS-A; TEMPS_C—cyclothymic subscale of TEMPS-A; TEMPS_H—hyperthymic subscale of TEMPS-A; TEMPS_I—irritable subscale of TEMPS-A; TEMPS_A—anxious subscale of TEMPS-A. Significance of differences between subgroups was determined by the Kruskal-Wallis ANOVA. Post-hoc analysis was conducted with Fisher's NIR test.

Table 4. DAT polymorphisms and TEMPS-A scale results in study group.

	All Group (n =245)			
	L/L (*n* = 117)	**L/S** (*n* = 103)	**S/S** (*n* = 25)	*p*
BMI	41.2 (36.2–48.9)	41.6 (35.8–48.5)	40.7 (39.9–46.8)	0.9
TEMPS_D	0.42 (0.28–0.52)	0.38 (0.28–0.47)	0.38 (0.28–0.47)	0.71
TEMPS_C	0.38 (0.24–0.62)	0.38 (0.23–0.57)	0.33 (0.29–0.48)	0.86
TEMPS_H	0.52 (0.36–0.61)	0.52 (0.38–0.62)	0.57 (0.38–0.62)	0.87
TEMPS_I	0.19 (0.07–0.33)	0.14 (0.05–0.28)	0.09 (0.04–0.29)	0.23
TEMPS_A	0.32 (0.21–0.52)	0.33 (0.24–0.52)	0.44 (0.28–0.59)	0.41

BMI, body mass index; TEMPS_D—depressive subscale of TEMPS-A; TEMPS_C—cyclothymic subscale of TEMPS-A; TEMPS_H—hyperthymic subscale of TEMPS-A; TEMPS_I—irritable subscale of TEMPS-A; TEMPS_A—anxious subscale of TEMPS-A. Significance of differences between subgroups was determined by the Kruskal-Wallis ANOVA.

Due to a small group of *DRD4* L/L carriers, we combined groups of individuals with L/L and L/S together, tagged them as L-carriers, and performed proper calculations. Nevertheless, as shown in Table 5, the obtained results regarding the analysis of the dependencies for DRD4 polymorphisms did not show any significant associations, in the examined group of obese subjects.

Table 5. DRD4 polymorphisms and TEMPS-A results in subgroups of women and men.

	All Group (n = 245)		
	L/L; L/S (n = 84)	S/S 114 (n = 161)	*p*
BMI	42.9 (38.5–49.0)	41.8 (37.2–47.1)	0.21
TEMPS_D	0.4 (0.28–0.47)	0.33 (0.28–0.47)	0.25
TEMPS_C	0.38 (0.24–0.61)	0.47 (0.23–0.57)	0.64
TEMPS_H	0.47 (0.35–0.59)	0.19 (0.05–0.28)	0.15
TEMPS_I	0.16 (0.05–0.33)	0.19 (0.20–0.55)	0.27
TEMPS_A	0.32 (0.24–0.47)	0.35 (0.20–0.54)	0.75

BMI, body mass index; TEMPS_D—depressive subscale of TEMPS-A; TEMPS_C—cyclothymic subscale of TEMPS-A; TEMPS_H—hyperthymic subscale of TEMPS-A; TEMPS_I—irritable subscale of TEMPS-A; TEMPS_A—anxious subscale of TEMPS-A. Significance of differences between subgroups was determined by the Kruskal-Wallis ANOVA.

After making calculations of one-dimensional analyses on TEMPS-A (Table 6), we confirmed the significant interaction effect for the gender and following temperaments: depressive, cyclothymic and irritable; for the BMI and anxious temperament; and for *COMT* Val158Met and both the cyclothymic and irritable temperament. However, we did not observe any significance for the age and other examined polymorphisms (Table 6).

Table 6. Analyses of unidimensional interaction effects for TEMPS-A temperaments subscales.

	TEMPS-D			TEMPS-C			TEMPS-H			TEMPS-I			TEMPS-A		
	SS	F	*p*	SS	F	*p*	SS	F	*p*	SS	F	*p*	SS	F	*p*
Gender	0.15	5.3	0.02	0.23	4.6	0.03	0.01	0.36	0.54	0.18	6.1	0.01	0.04	0.94	0.33
Age	2.08	1.38	0.06	2.07	0.68	0.94	2.07	1.01	0.46	1.89	1.19	0.20	3.03	1.17	0.22
BMI	0.11	0.78	0.65	0.35	4.7	0.11	0.06	0.20	0.96	0.37	4.9	0.10	0.24	17.1	0.01
DAT1	0.02	0.38	0.68	0.22	0.22	0.79	0.007	0.1	0.90	0.09	1.53	0.21	0.05	0.52	0.59
COMT	0.07	1.49	0.22	0.32	3.2	0.04	0.22	2.9	0.05	0.20	3.4	0.03	0.10	1.07	0.34
DRD4	0.02	0.49	0.61	0.04	0.39	0.67	0.07	1.08	0.34	0.07	1.36	0.35	0.01	0.11	0.89

One-dimensional analysis of significance (ANOVA) F-test based on SS.

The Wald statistic in the logistic regression model indicated the coefficient of gender to be a significant predictor of the TEMPS-D results, and the *COMT* polymorphism to be a significant predictor of the TEMPS-H and TEMPS-I results (Table 7). These test results for *COMT* in predicting TEMPS-C and TEMPS-D, and for *DAT1* in predicting TEMPS-I, remained in the trend.

Table 7. Logistic regression model coefficients on TEMPS-A temperaments subscales.

TEMPS-D							
	B	S.E.	Wald	df	p	95%C.I. Lower	95%C.I. Upper
Gender	0.164	0.057	8.05	1	0.004	0.278	0.05
Age	0.003	0.004	0.69	1	0.4	0.011	−0.004
BMI	0.00008	0.006	0.01	1	0.89	0.013	−0.11
DAT1	0.036	0.06	0.37	2	0.82	0.157	−0.085
COMT	−0.126	0.07	5.7	2	0.057	0.018	−0.272
DRD4	−0.100	0.196	0.27	2	0.86	0.284	−0.485

TEMPS-C							
	B	S.E.	Wald	df	p	95%C.I. Lower	95%C.I. Upper
Gender	−0.13	0.066	3.95	1	0.04	−0.001	−0.262
Age	−0.0007	0.005	0.01	1	0.9	0.01	−0.012
BMI	0.008	0.008	1.09	1	0.29	0.296	−0.007
DAT1	−0.045	0.08	0.3	2	0.85	0.12	−0.21
COMT	−0.177	0.111	2.99	2	0.055	0.04	−0.39
DRD4	0.137	0.168	0.83	2	0.65	0.366	0.117

TEMPS-H							
	B	S.E.	Wald	df	p	95%C.I. Lower	95%C.I. Upper
Gender	−0.07	0.05	2.33	1	0.12	0.021	−0.175
Age	0.0008	0.003	0.05	1	0.81	0.008	−0.006
BMI	0.002	0.006	0.2	1	0.64	0.014	−0.009
DAT1	−0.019	0.06	0.27	2	0.87	0.098	−0.137
COMT	0.12	0.06	6.05	2	0.04	0.241	−0.002
DRD4	−0.07	0.18	2.45	2	0.29	0.282	−0.239

TEMPS-I							
	B	S.E.	Wald	df	p	95%C.I. Lower	95%C.I. Upper
Gender	0.032	0.094	0.11	1	0.73	0.217	−0.152
Age	0.003	0.008	0.17	1	0.67	0.021	−0.013
BMI	0.005	0.014	0.13	1	0.71	0.033	−0.023
DAT1	0.299	0.143	5.44	2	0.065	0.58	0.019
COMT	−0.35	0.211	5.92	2	0.04	0.074	−0.756
DRD4	0.322	0.207	3.13	2	0.2	−0.084	0.12

TEMPS-A							
	B	S.E.	Wald	df	p	95%C.I. Lower	95%C.I. Upper
Gender	0.085	0.078	1.18	1	0.27	0.238	−0.068
Age	0.005	0.005	0.8	1	0.37	0.016	−0.005
BMI	−0.008	0.008	0.95	1	0.32	0.008	−0.026
DAT1	−0.08	0.086	1.05	2	0.59	0.349	0.088
COMT	−0.219	0.1	4.2	2	0.12	0.04	−0.009
DRD4	−0.11	0.257	0.21	2	0.89	0.386	−0.623

BMI, body mass index; TEMPS_D—depressive subscale of TEMPS-A; TEMPS_C—cyclothymic subscale of TEMPS-A; TEMPS_H—hyperthymic subscale of TEMPS-A; TEMPS_I—irritable subscale of TEMPS-A; TEMPS_A—anxious subscale of TEMPS-A.

3. Discussion

To date, many studies point to the connection between obesity and mood disorders, such as depression or BD [24–28]. Oniszczenko et al. (2015) suggest that personality traits expressed by temperament may constitute specific risk factors for the development of obesity. Those traits might determine behaviors which hinder weight loss or cause excess eating. Moreover, mentioned temperaments may also contribute to the proneness to mood disorders associated with obesity [29].

Therefore, research on a neurobiological basis of affective temperament could convey essential details of how dopaminergic gene polymorphisms add to the pathogenesis of mood disorders in the obese population; it may, in particular, explain that changes in dopamine transmission may be a causative and a common factor in the development of obesity, as well as of affective diseases [30–32].

In this study, we analyzed affective temperament dimensions in an obese population using the TEMPS-A autoquestionnaire. Subsequently, we scrutinized correlations of affective temperament and dopaminergic gene polymorphisms which are involved in obesity and mood disorders. Those genes are comprised of *COMT* Val158Met, *DAT1* and *DRD4*. To our knowledge, this is the first study analyzing the affective temperament in the context of dopaminergic genes in an obese population.

Tables 1 and 2 show significant differences of affective temperament dimensions in both sexes. According to Table 1, men scored higher than women for the depressed and irritable temperament. The logistic regression model (Table 7) shows significant results for gender and TEMPS-D, but not for the irritable dimension. In our previous study, evaluating the affective temperament in an obese Polish population in the context of the serotonin transporter gene polymorphism (5-HTTLPR), we also observed a higher expression of the irritable temperament in men [23]. Studies show significant differences between temperament dimensions in patients suffering from BDs in comparison to healthy ones. Individuals with BD show greater scores in depressive, cyclothymic, irritable and anxious dimensions [33]. It has been shown that, among bipolar patients, cyclothymic and irritable temperaments may be connected with impulsivity [34]. The French study of Bénard et al. (2017) exhibits a stronger association between impulsivity and obesity in men than in women, suggesting the role of gender in weight status and eating behaviors [35]. Such results are interesting in the context of the proneness to affective disorders in this population, with a differentiation between both sexes.

The literature also shows that females may be more susceptible to depression than men [36]. This may stem from many factors, including sociocultural, psychosocial, or behavioral factors. Considering the molecular basis which connects gender, depression and obesity, the difference in sex hormones may affect a response to stressors and modulate immune responses, resulting in higher inflammation, eventually leading to depressive disorders [37–39]. Sex hormones affect the immune system by exerting pro-, or anti-inflammatory effects. This includes stimulating the immune cell activation, or an increased expression of cytokines which participate in the immune responses. Great evidence points to the link between elevated pro-inflammatory cytokines and depression. The data indicate that the immune system may contribute to depression pathogenesis in different ways due to sex differences. During puberty, a crucial period for depression development, the estradiol levels increase. Also, the interplay between sex hormones and the immune system may be seen in peri- and post-partum depression, where the level of estrogen is also augmented [40]. Androgens take part in the suppression of immune responses, but it has been shown that a greater activation of the immune system in males with a reduced testosterone concentration may contribute to mood disorders [41]. Even though the literature shows mixed results in this field, Byrne et al. (2015) conclude that the female sex may be the factor influencing immune responses and depression [38,42]. More research focusing on differences of affective temperament in both sexes would bring interesting data regarding the genetic and molecular basis of morbidity for mood disorders in men and women.

Affective temperament is considered a stable construct associated with genetic transmission and could serve as a phenotype to detect genes responsible for a susceptibility to affective disorders [18,43,44]. Surprisingly, we have observed the correlations between temperament dimensions and age in both men and women. A positive correlation between a depressive and anxious temperament and age may ensue from changes of a person's experience during their lifetime. The study of Caserta et al. (2011) showed no connections between depression and the immune system in young girls, although in older girls higher depression measures were associated with increased NK cells cytotoxicity [42]. It has been shown that a positive demeanor, i.e., extraversion, agreeableness, or being optimistic, may affect the immune system, by for example lowering the IL-6 response to the stress factors [45,46]. On the other hand, pessimism contributed to augmented markers of inflammation, like IL-6 and the C reactive protein [47]. We assume, that similar associations might be responsible for our results regarding TEMPS-A, and that sex and age might constitute potential modifiers of affective temperament dimensions. Furthermore, more research should be conducted in relation to the association between anxiety and depressive disorders, in the context of hypothalamic-pituitary-adrenal (HPA) axis dysregulation [48,49].

Epigenetics is a novel field describing alterations in gene functioning without changes within the genome sequence. It provides potential mechanisms explaining the adverse effects of environmental factors on modulatory mechanisms of gene expression, which may exert long-term effects and be putatively heritable [50,51]. Recent studies connect epigenetic changes with numerous diseases including cancer, while laying emphasis on their crucial role in the pathopsychology, by explaining the association between depressive and anxiety disorders, and adverse life events, or the impact of stress in childhood [52–56]. Additionally, in some studies, it has been corroborated that epigenetic changes may exert dysregulations in the HPA axis, by affecting its regulatory genes, thus contributing to stress-related disorders. The upregulation of the cortocotropin-releasing hormone expression or altered transcription of the glucocorticoid receptor in the brain regions may stem from stress-induced epigenetic modifications, and thus be responsible for HPA-axis dysfunction [57,58].

Therefore, we assume that epigenetics might be a putative link connecting received TEMPS-A results and age. Due to the scarce literature regarding this topic, we encourage more research engaged in psychoneuroimmunology or the influence of environmental factors on the affective temperament. Epigenetics constitutes a challenging field which may convey essential data explaining discrepancies in affective temperament investigations.

In the current study, an increased BMI positively correlated with a greater expression of hyperthymic temperament in women and a greater cyclothymic and irritable dimension in men. We can refer to our findings from our previous study. Temperament results between morbid obese (BMI > 40) and obese individuals (BMI ≤ 40) showed that morbidly obese scored greater in hyperthymic and cyclothymic dimensions [23]. In the study of Amann et al. (2009), patients with morbid obesity displayed higher scores in cyclothymic, irritable and anxious dimensions, which is partially consistent with our results [21]. Considering that studies show associations between the cyclothymic, irritable and hyperthymic temperament, and BD, the abovementioned data imply a heightened risk of this disease with a weight gain in obese patients [59–62]. In this study, the cyclothymic temperament in women showed a negative correlation with the BMI and with the age in males, which is inconsistent with findings in the literature [63]. We presume that the heterogeneity of the results may stem from the lack of the control group. It is possible that, when comparing with non-obese individuals, the study group could exhibit a more expressed cyclothymic dimension of the affective temperament.

The association between *COMT* Val158Met and mood disorders has been pointed out in the literature [64–66]. However, many researchers still show some concerns about the exact mechanism by which dopamine transmission, determined by *COMT*, contributes to the origin of affective disorders [67]. Some authors propose that the polymorphisms may influence the HPA axis reactivity and thus, by causing a dysregulation of the inflammation processes, may be involved in the pathogenesis of mood disorders and obesity in a reciprocal manner [68–70]. The literature also shows an association between *COMT* polymorphisms and personality traits in patients suffering from BD [71–73].

Some publications exhibit connections between Met alleles and vulnerability to stress and anxiety, and thus depression [65,74]. However, Massat et al. (2011) showed that the Val allele was more common in individuals with an early stage of depression [75]. The study performed on larger population showed mixed results: The Met allele occurred less frequently among men with depression in comparison to the control group [76].

During the analysis of the connection between affective temperament and dopaminergic gene polymorphisms, we have only observed the association between *COMT* Val158Met polymorphisms. Considering the affective temperament, *COMT* heterozygotes showed significant results only in irritable and cyclothymic dimensions. Using a logistic regression model (Table 7), we also received significant results concerning the irritable temperament and *COMT* polymorphism. Both temperaments were overrepresented in patients with bipolar disorders [59]. The irritable temperament has been linked with anxiety and agitation and found more often in persons with bipolar disorder, in comparison to healthy controls or patients with a major depressive disorder [62,77].

Previous studies on the COMT relationship with the dimensions of the temperament in Cloninger's concept were focused mainly on the novelty seeking dimension. These studies gave different results, the majority of which focused on the polymorphism rs4680 [78–80]. Golimbet et al. (2007) provided evidence that the *COMT* Met allele (which contributes to the reduction of enzyme activity and ultimately leads to an increase in dopamine levels) was associated with a greater severity of temperamental trait novelty seeking in women [78]. The repetition of this result was done by Tsai and co-workers (2004) on young Chinese women [81]. However, the association of the rs4680 polymorphism of the *COMT* gene with the novelty seeking dimension of temperament has not been confirmed. Searching in other studies conducted on the Caucasian population and the Japanese population [79,80]. In a study conducted on the Chinese population on drug addicts, the *COMT* gene polymorphism was shown to be related to the temperamental characteristics of novelty seeking and the tendency to addiction [82]. A decreased pre-dopaminergic activity and low control, associated with specific *COMT* genotypes, may increase impulsivity, which is a component of novelty seeking. Research by Kang and co-workers (2010) on the dimensions of character showed that the Val158Met *COMT* polymorphism may be related to a susceptibility to boredom and the need for strong sensations in women [83].

The TEMPS-A validation study showed a positive correlation between both the cyclothymic and irritable temperament and the higher novelty seeking scores; hence, our findings are consistent with the results of the abovementioned studies [84], in particular in relation to the fact that Cloninger's novelty seeking, as well as Akiskal's cylothymic and irritable dimension, are involved in affective disorders [85]. In their work, Parneix et al. (2014) found that patients with irritability related to major depressive episodes were characterized with atypical features like weight gain and showed greater novelty seeking. The authors suggested that such findings are indicative of a greater vulnerability to BD [86]. In another study, impulsivity seen in the bipolar spectrum was also described in the context of obesity and food addiction [87]. Thus, the affective temperament seems to be related to a susceptibility to mood disorders in obese individuals, and its evaluation might provide useful information considering treatment approaches.

Unfortunately, due to the observational design of our study and the lack of a control group, it is difficult to explain the molecular basis of the interplay between the dopaminergic transmission modulated by *COMT* and the affective temperament. We speculate that obese individuals, in comparison to healthy persons, show a disturbed dopamine transmission, and that dopaminergic signaling in heterozygotes gives rise to more pronounced affective temperament dimensions. This may constitute the link between *COMT* polymorphisms and affective disorders in the obese population. Moreover, individual changes in the dopaminergic transmission might bias the obtained results and influence the temperament expression or exert differences in one's behavior [88,89]. We propose that future researches of affective temperament should utilize neuroimaging, along with neurogenetic studies, and compare the obtained results with a control group. This measure might elucidate what kind of dopaminergic transmission, determined by *COMT*, is responsible for the pathogenesis of mood disorders in the obese population.

In Tables 4 and 5, we did not observe any statistically significant associations between the affective temperament and polymorphisms of *DAT1* nor *DRD4*.

The literature shows mixed results about the connection between the abovementioned polymorphism and temperament analyzed with various scales. According to Cloninger's theory, the dimension of temperament novelty seeking is, according to this concept, related to the *DRD4* gene. Previous studies on the association of the VNTR polymorphism in the *DRD4* gene suggested association with the dimension of novelty seeking of temperament [90]. However, further studies did not detect a similar relationship, but showed a correlation of the polymorphism (-521 C/T) of the DRD4 gene with impulsivity and novelty seeking. Other researchers have found a connection between the VNTR polymorphism and two mood temperaments: cyclothymic and irritable; however, this study was performed on a healthy volunteer of the Asian population, and therefore it may be difficult to compare the results to our group [87].

Regarding the *DAT1* gene, some studies indicate that the VNTR 3'UTR polymorphism of the *DAT1* gene is associated with novelty seeking; however, other researchers have not obtained similar results [91–93]. The research also indicates the interaction of *DAT1* gene polymorphisms, *DRD4* and neuroticism [94]. The literature shows little findings describing affective temperament measured with TEMPS-Am, and *DRDR4* or *DAT1* polymorphisms, and more studies are needed in this field.

In Table 6, the effect interaction was observed for the anxious dimension and BMI. However, by using a logistic regression we have not obtained significant results for the BMI and any temperament dimension. In the study of Amann et al. (2009), obese patients scored significantly higher in the anxious dimension, as well as for the irritable and cyclothymic factors [21]. Therefore, we assume that persons characterized by an anxious temperament might be at greater risk of further weight gain. Even though we did not find any associations between this dimension and the dopaminergic genes, it could be that an anxious temperament is related to the serotonergic transmission. It could be, in particular, that it has been linked to moderate novelty seeking and greater harm avoidance—which is connected to this type of signaling [95]. Amann et al. (2009) displayed an association between the S allele of the 5HTTLPRI polymorphism in the serotonin transporter gene and greater scores in the following TEMPS dimensions: cyclothymic, irritable and anxious [21]. Gonda et al. (2006) also obtained similar results in the group of women, which indicates the relationship between an affective temperament and the serotonergic transmission [96]. Additionally, in our previous study regarding the *5HTTLPR* polymorphism, subjects homozygous to the S allele exhibited higher scores in anxious and depressive dimensions in comparison to L allele carriers. Such results indicate a stronger connection between the affective temperament measured by TEMPS-A and the serotonergic transmission, instead of dopaminergic signaling in the obese population [23].

In this study we analyzed only one neurotransmitter signaling. We must take into consideration that many factors influence behavior, including other gene polymorphisms or the complex neurotransmitter interactions in different brain areas [96–101]. For instance, functional brain imaging revealed an additive effect of *COMT* Met158 and *5-HTTLPR* S alleles on the response of the amygdale, hippocampal and limbic cortical areas to unpleasant stimuli, suggesting that persons with those alleles may show a lowered resilience against an anxiety mood [101]. An interesting study of Ro et al. (2018) indicates the differences in the expression of glucagon-like peptide 1 and 2 receptors (GLP-!R, GLP-2R) in patients suffering from mood disorders in comparison to healthy controls, with a greater susceptibility connected to higher BMI values. Both GLP-1R and GLP-2R are implicated in neuroprotection and the antidepressant effect [102]. Moreover, it has been found that a lower expression of the leptin receptor in the hippocampus and hypothalamus may have a significant impact on obesity and comorbid depression. Researchers found that obese individuals or those exposed to chronic unpredictable mild stress showed a diminished expression of the leptin receptor [103]. Nonetheless, mood disorders are complex in their nature and constitute a hard challenge for clinicians in their practice. Due to the growing problem of obesity, there is a need for creating more effective preventing programs that tackle the occurrence of affective disorders in this population. Hence, more studies focusing on the molecular basis of the pathogenesis and interplay between both disorders could bring a better understanding, which is essential for predicting the course and nature of the diseases.

4. Materials and Methods

4.1. Participants

The study was conducted on a population of 245 Caucasian people, who were diagnosed with primary obesity. Secondary causes of obesity were excluded in the Clinic of Endocrinology and Diabetology at the Collegium Medicum of the Nicolaus Copernicus in Bydgoszcz on the basis of a subjective and objective medical assessment, as well as on the basis of performed hormonal and metabolic tests. Significant physical diseases, addiction, substance abuse (e.g., cannabis misuse) or psychiatric and neurological illnesses were the excluding factors for participation in the study. All patients, after being given detailed information on the purpose and nature of the study, expressed written and informed consent of their participation. The study obtained the consent of the bioethical commission at the Nicolaus Copernicus University (No 533/ 2008, 15 Dec 2008).

4.2. Clinical Assessments and Measures

Building on the assessed anthropometric factors, the diagnosis of obesity was established. As a factor reflecting the amount of body fat, the BMI index was adopted. It was calculated as the ratio of weight (kg) to square of height (m).

4.3. Psychological Assessment

For the psychological assessment, we utilized the Temperament Evaluation of Memphis, Pisa, Paris and San Diego Autoquestionnaire (TEMPS-A) to perform an analysis of the dimensions of the affective temperament.

4.4. Genotyping

Genomic DNA was obtained from peripheral blood (5 mL) using the method developed by Lahiri and Schnabel (1993) [104]. The blood was collected on the EDTA medium and mixed, before being frozen in liquid nitrogen and stored at −80 °C prior to extraction. The polymorphisms of the *DAT1*, *COMT* and *DRD4* genes were determined using the polymerase chain reaction (PCR). The following primers were used: *DAT1* forward, 5′-TGTGGTGTAGGGAACGGCCTGAG-3′; *DAT1* reverse, 5′-CTTCCTGGAGGTCACGGCTCAAGG-3′; forward, 5′-AGCTCCAAGCGCGCTCACAG-3′; *COMT* reverse, 5′-CAAAGTGCGCATGCCCTCCC-3′; DRD4 forward: 5′-GCGACTACGTGGTCTACTCG-3′; and *DRD4* rewers: 5′-AGGACCCTCATGGCC TTGC-3′. The PCR products were then separated by agarose gel electrophoresis using O'RangeRuler™ 50 bp DNA Ladder (Fermentas) as a length marker (Figures 1–3).

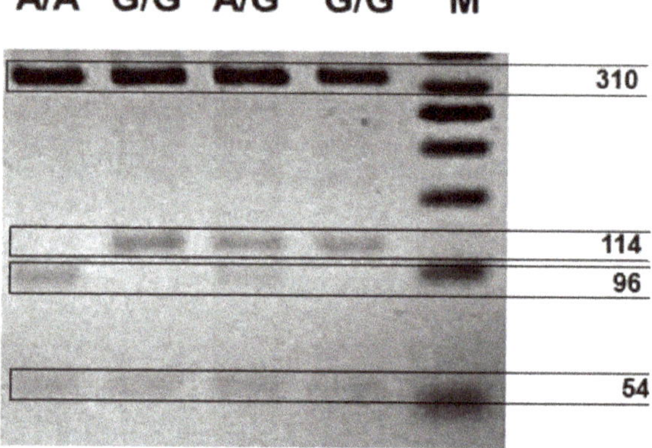

Figure 1. Photo of the digested *COMT* PCR products. The results are labeled by genotype: Met/Met (A/A), 96 bp only; Val/Met (A/G) 114 and 96 bp; and Val/Val (G/G), 114 bp only.

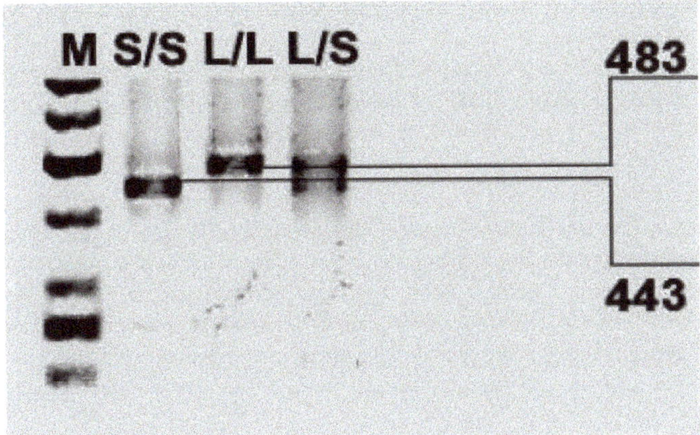

Figure 2. Photo of digested *DAT1* PCR products. The results are labeled by genotype: 10/10 (L/L) 483 bp only; 10/9 (L/S) 483 and 443 bp; and 9/9 (S/S) 443 bp only.

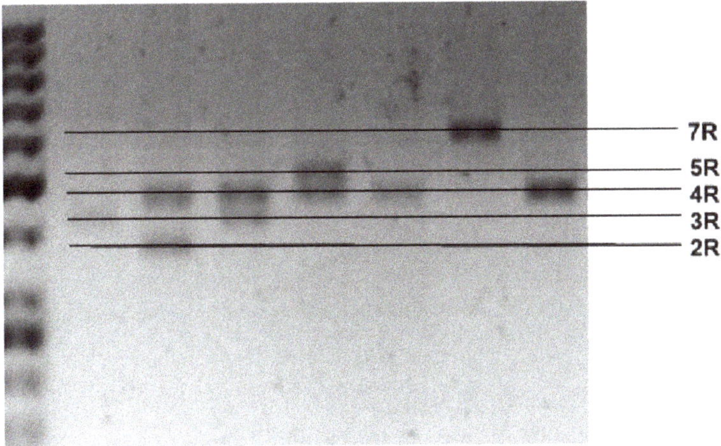

Figure 3. Photo of digested *DRD4* PCR products. Representative photo of separated *DRD4* PCR products depending on the genotype: LL—only 619 bp band (7R); S/S 379 bd (2R) or/and 427 bp (3R) or/and 523 bp (5R) band; L/S – 379 bd (2R) or 427 bp (3R) or 523 bp (5R) and 619 bp (7R) bands.

4.5. Statistical Analysis

Using the Shapiro-Wilk test, it was determined that the test group does not meet the normal distribution criteria. The statistical significance of the differences between the two groups was calculated using the Mann–Whitney U test, and for comparisons with three or more groups, the Kruskal–Wallis analysis of variance (ANOVA) was applied. The NIR Fisher test was used for post hoc analyses. Correlations between two quantitative variables were examined using the Spearman rank correlation test. To control for the effect of age and BMI, which both exhibit significant simple correlations with the dimensions of temperament, we analyzed the data with a partial Kendall regression (partial Kendall's Tau), the nonparametric technique that controls for one confounding [105].

An analysis of covariance (ANCOVA) was performed to examine the interaction effects. An effect size was determined using Cohen's d. The gathered data were analysed by means of StatSoft, Inc. (2017) using Statistica, version 13.0 software and the computer program "Utility Programs for Analysis of Genetic Linkage" (Copyright © 1988 J. Tot) was utilized to test for the goodness of fit to the

Hardy–Weinberg equilibrium. The distributions of all three analyzed genotypes were against the Hardy–Weinberg equilibrium.

Bonferroni corrections were used as multiple testing procedures. A logistic regression of data was performed to predict logit on TEMPS-A temperaments subscales (The Wald statistic in Logistic regression model).

5. Conclusions

To our knowledge, this is the first study analyzing the affective temperament in an obese population in the context of dopaminergic genes polymorphisms, including *COMT* Val158Met, *DRD4*, and *DAT1*. The results of our study indicate the connection between the irritable and cyclothymic dimensions in *COMT* heterozygotes only. We presume that the dopaminergic transmission modulated by these COMT gene polymorphisms may entail a significant expression of cyclothymic and irritable temperaments. This is a very interesting finding, giving rise to more sophisticated research in the future, utilizing neuroimaging studies.

6. Limitations

The main limitation of our study is the lack of a control group in order to gain more reliable results. Second, for the proper evaluation of the connection between the affective temperament and gene polymorphisms, our study group should be larger.

Author Contributions: I state that all authors have made significant contributions in regard to this research. A.B., J.P.-U. and M.B. conceived the idea for the study. J.P.-U. and M.B. contributed to the design of the research. M.B., J.P.-U., N.L. and A.K. were involved in data collection. M.B., N.L., and A.B. were involved in data analyze and interpretation. N.L. and M.B. wrote the manuscript and A.B. with A.T. and R.J. made correction and critically revised the paper. All authors agree to be accountable for all aspects of the work, as well as this manuscript was approved by all authors.

Funding: This research received no external funding.

Acknowledgments: The APC was funded by Nicolaus Copernicus University.

Conflicts of Interest: All authors declare no conflict of interests connected with this manuscript.

References

1. Eiber, R.; Vera, L.; Mirabel-Sarron, C.; Guelfi, J.-D. Self-esteem: A comparison study between eating disorders and social phobia. *L'Encéphale* **2003**, *29*, 35–41. [PubMed]
2. Madsen, S.A.; Grønbaek, H.; Olsen, H. Psychological aspects of obesity. *Ugeskr. Laeger* **2006**, *168*, 194–196. [PubMed]
3. Onyike, C.U.; Crum, R.M.; Lee, H.B.; Lyketsos, C.G.; Eaton, W.W. Is obesity associated with major depression? Results from the Third National Health and Nutrition Examination Survey. *Am. J. Epidemiol.* **2003**, *158*, 1139–1147. [CrossRef] [PubMed]
4. Pickering, R.P.; Grant, B.F.; Chou, S.P.; Compton, W.M. Are overweight, obesity, and extreme obesity associated with psychopathology? Results from the national epidemiologic survey on alcohol and related conditions. *J. Clin. Psychiatry* **2007**, *68*, 998–1009. [CrossRef] [PubMed]
5. Quek, Y.H.; Tam, W.W.S.; Zhang, M.W.B.; Ho, R.C.M. Exploring the association between childhood and adolescent obesity and depression: A meta-analysis. *Obes. Rev.* **2017**, *18*, 742–754. [PubMed]
6. Elmslie, J.L.; Silverstone, J.T.; Mann, J.I.; Williams, S.M.; Romans, S.E. Prevalence of overweight and obesity in bipolar patients. *J. Clin. Psychiatry* **2000**, *61*, 179–184. [CrossRef]
7. Wildes, J.E.; Marcus, M.D.; Fagiolini, A. Prevalence and correlates of eating disorder co-morbidity in patients with bipolar disorder. *Psychiatry Res.* **2008**, *161*, 51–58. [CrossRef] [PubMed]
8. Alciati, A.; D'Ambrosio, A.; Foschi, D.; Corsi, F.; Mellado, C.; Angst, J. Bipolar spectrum disorders in severely obese patients seeking surgical treatment. *J. Affect. Disord.* **2007**, *101*, 131–138. [CrossRef]
9. Heshmati, M.; Russo, S.J. Anhedonia and the brain reward circuitry in depression. *Curr. Behav. Neurosci. Rep.* **2015**, *2*, 146–153. [CrossRef] [PubMed]

10. Gatt, J.M.; Burton, K.L.; Williams, L.M.; Schofield, P.R. Specific and common genes implicated across major mental disorders: A review of meta-analysis studies. *J. Psychiatr. Res.* **2015**, *60*, 1–13. [CrossRef] [PubMed]

11. Nestler, E.J. Role of the brain's reward circuitry in depression: Transcriptional mechanisms. *Int. Rev. Neurobiol.* **2015**, *124*, 151–170. [PubMed]

12. Yokum, S.; Marti, N.C.; Smolen, A.; Stice, E. Relation of the multilocus genetic composite reflecting high dopamine signaling capacity to future increases in BMI. *Appetite* **2015**, *87*, 38–45. [CrossRef] [PubMed]

13. Serafini, G.; Muzio, C.; Piccinini, G.; Flouri, E.; Ferrigno, G.; Pompili, M.; Girardi, P.; Amore, M. Life adversities and suicidal behavior in young individuals: A systematic review. *Eur. Child Adolesc. Psychiatry* **2015**, *24*, 1423–1446. [CrossRef] [PubMed]

14. Buss, A.H.; Plomin, R. *Temperament: Early Developing Personality Traits*; Lawremce Erlbaum: Hillsdale, NJ, USA, 1984.

15. Kagan, J. *Galen's Prophecy: Temperament in Human Nature*; Basic Books: New York, NY, USA, 1994.

16. von Zerssen, D.; Akiskal, H.S. Personality factors in affective disorders: Historical developments and current issues with special reference to the concepts of temperament and character. *J. Affect. Disord.* **1998**, *51*, 1–5. [PubMed]

17. Cloninger, C.R.; Svrakic, D.M.; Przybeck, T.R. Can personality assessment predict future depression? A twelve-month follow-up of 631 subjects. *J. Affect. Disord.* **2006**, *92*, 35–44. [CrossRef] [PubMed]

18. Akiskal, H.S.; Akiskal, K.K. Cyclothymic, hyperthymic and depressive temperaments as subaffective variants of mood disorders. In *Annual Review*; Tasman, A., Riba, M.B., Eds.; American Psychatric Press: Washington, DC, USA, 1992; Volume 11, pp. 43–62.

19. Akiskal, H.S. The temperamental foundations of affective disorders. In *Interpersonal Factors in the Origin and Course of Affective Disorders*; Mundt, C., Hahlweg, K., Fiedler, P., Eds.; Gaskell: London, UK, 1996; pp. 3–30.

20. Akiskal, H.S.; Pinto, O. Soft bipolar spectrum: Footnotes to Kraepelin on the interface of hypomania, temperament and depression. In *Bipolar Disorders: 100 Years after Manic-Depressive Insanity*; Marneros, A., Angst, J., Eds.; Kluwer Academic: Dordrecht, The Netherlands, 2000; pp. 37–62.

21. Amann, B.; Mergl, R.; Torrent, C.; Perugi, G.; Padberg, F.; El-Gjamal, N.; Laakmann, G. Abnormal temperament in patients with morbid obesity seeking surgical treatment. *J. Affect. Disord.* **2009**, *118*, 155–160. [CrossRef]

22. Greenwood, T.A.; Badner, J.A.; Byerley, W.; Keck, P.E.; McElroy, S.L.; Remick, R.A.; Sadovnick, A.D.; Akiskal, H.S.; Kelsoe, J.R. Heritability and genome-wide SNP linkage analysis of temperament in bipolar disorder. *J. Affect. Disord.* **2013**, *150*, 1031–1040. [CrossRef]

23. Borkowska, A.; Bieliński, M.; Szczęsny, W.; Szwed, K.; Tomaszewska, M.; Kałwa, A.; Lesiewska, N.; Junik, R.; Gołębiewski, M.; Sikora, M.; et al. Effect of the 5-HTTLPR polymorphism on affective temperament, depression and body mass index in obesity. *J. Affect. Disord.* **2015**, *184*, 193–197. [CrossRef]

24. Roberts, R.E.; Kaplan, G.A.; Shema, S.J.; Strawbridge, W.J. Are the obese at greater risk for depression? *Am. J. Epidemiol.* **2000**, *152*, 163–170. [CrossRef]

25. Jantaratnotai, N.; Mosikanon, K.; Lee, Y.; McIntyre, R.S. The interface of depression and obesity. *Obes. Res. Clin. Pract.* **2017**, *11*, 1–10. [CrossRef]

26. Mannan, M.; Mamun, A.; Doi, S.; Clavarino, A. Prospective Associations between Depression and Obesity for Adolescent Males and Females—A Systematic Review and Meta-Analysis of Longitudinal Studies. *PLoS ONE* **2016**, *11*, e0157240. [CrossRef]

27. Zhao, Z.; Okusaga, O.O.; Quevedo, J.; Soares, J.C.; Teixeira, A.L. The potential association between obesity and bipolar disorder: A meta-analysis. *J. Affect. Disord.* **2016**, *202*, 120–123. [CrossRef]

28. ojko, D.; Buzuk, G.; Owecki, M.; Ruchała, M.; Rybakowski, J.K. Atypical features in depression: Association with obesity and bipolar disorder. *J. Affect. Disord.* **2015**, *185*, 76–80. [CrossRef]

29. Oniszczenko, W.; Dragan, W.; Chmura, A.; Lisik, W. Temperament as a risk factor for obesity and affective disorders in obese patients in a Polish sample. *Eat. Weight Disord.* **2015**, *20*, 233–239. [CrossRef]

30. Cameron, J.D.; Chaput, J.P.; Sjödin, A.M.; Goldfield, G.S. Brain on Fire: Incentive Salience, Hedonic Hot Spots, Dopamine, Obesity, and Other Hunger Games. *Annu. Rev. Nutr.* **2017**, *37*, 183–205. [CrossRef] [PubMed]

31. Naef, L.; Pitman, K.A.; Borgland, S.L. Mesolimbic dopamine and its neuromodulators in obesity and binge eating. *CNS Spectr.* **2015**, *20*, 574–583. [CrossRef] [PubMed]

32. Luo, S.X. Dopamine and Obesity: A Path for Translation? *Biol. Psychiatry* **2016**, *79*, e85–e86. [CrossRef] [PubMed]

33. Röttig, D.; Röttig, S.; Brieger, P.; Marneros, A. Temperament and personalityin bipolar I patients with and without mixed episodes. *J. Affect. Disord.* **2007**, *104*, 97–102. [CrossRef] [PubMed]

34. Tatlidil Yaylaci, E.; Kesebir, S.; Güngördü, Ö. The relationship between impulsivity and lipid levels in bipolar patients: Does temperament explain it? *Compr. Psychiatry* **2014**, *55*, 883–886. [CrossRef]

35. Bénard, M.; Camilleri, G.M.; Camilleri, G.M.; Etilé, F.; Méjean, C.; Bellisle, F.; Reach, G.; Hercberg, S.; Péneau, S. Association between Impulsivity and Weight Status in a General Population. *Nutrients* **2017**, *9*, 217. [CrossRef]

36. Piccinelli, M.; Wilkinson, G. Gender differences in depression: Critical review. *Br. J. Psychiatry* **2000**, *177*, 486–492. [CrossRef] [PubMed]

37. Fabricatore, A.N.; Wadden, T.A. Psychological aspects of obesity. *Clin. Dermatol.* **2004**, *22*, 332–337. [CrossRef] [PubMed]

38. Byrne, M.L.; O'Brien-Simpson, N.M.; Mitchell, S.A.; Allen, N.B. Adolescent-Onset Depression: Are Obesity and Inflammation Developmental Mechanisms or Outcomes? *Child Psychiatry Hum. Dev.* **2015**, *46*, 839–850. [CrossRef] [PubMed]

39. Cutolo, M.; Straub, R.H.; Bijlsma, J.W. Neuroendocrine–immune interactions in synovitis. *Nat. Rev. Rheumatol.* **2007**, *3*, 627–634. [CrossRef] [PubMed]

40. Rainville, J.R.; Tsyglakova, M.; Hodes, G.E. Deciphering sex differences in the immune system and depression. *Front. Neuroendocrinol.* **2018**, *50*, 67–90. [CrossRef] [PubMed]

41. Grinspoon, S.; Corcoran, C.; Stanley, T.; Baaj, A.; Basgoz, N.; Klibanski, A. Effects of hypogonadism and testosterone administration on depression indices in HIV-infected men. *J. Clin. Endocrinol. Metab.* **2000**, *85*, 60–65. [CrossRef]

42. Caserta, M.T.; Wyman, P.A.; Wang, H.; Moynihan, J.; O'Connor, T.G. Associations among depression, perceived self-efficacy, and immune function and health in preadolescent children. *Dev. Psychopathol.* **2011**, *23*, 1139–1147. [CrossRef]

43. Akiskal, H.S. Delineating irritable and hyperthymic variants of the cyclothymic temperament: Reassessing personality disorder constructs. *J. Pers. Disord.* **1992**, *6*, 326–342. [CrossRef]

44. Perugi, G.; Toni, C.; Maremmani, I.; Tusini, G.; Ramacciotti, S.; Madia, A.; Fornaro, M.; Akiskal, H.S. The influence of affective temperaments and psychopathological traits on the definition of bipolar disorder subtypes: A study on bipolar I Italian national sample. *J. Affect. Disord.* **2012**, *136*, 41–49. [CrossRef]

45. Cohen, S.; Doyle, W.J.; Turner, R.; Alper, C.M.; Skoner, D.P. Sociability and susceptibility to the common cold. *Psychol. Sci.* **2003**, *14*, 389–395. [CrossRef]

46. Brydon, L.; Walker, C.; Wawrzyniak, A.J.; Chart, H.; Steptoe, A. Dispositional optimism and stress-induced changes in immunity and negative mood. *Brain Behav. Immun.* **2009**, *23*, 810–816. [CrossRef]

47. Roy, B.; Diez-Roux, A.V.; Seeman, T.; Ranjit, N.; Shea, S.; Cushman, M. Association of optimism and pessimism with inflammation and hemostasis in the Multi-Ethnic Study of Atherosclerosis (MESA). *Psychosom. Med.* **2010**, *72*, 134–140. [CrossRef]

48. Anacker, C.; O'Donnell, K.J.; Meaney, M.J. Early life adversity and the epigenetic programming of hypothalamic-pituitary-adrenal function. *Dialogues Clin. Neurosci.* **2014**, *16*, 321–333.

49. Bartlett, A.A.; Singh, R.; Hunter, R.G. Anxiety and Epigenetics. *Adv. Exp. Med. Biol.* **2017**, *978*, 145–166.

50. Egger, G.; Liang, G.; Aparicio, A.; Jones, P.A. Epigenetics in human disease and prospects for epigenetic therapy. *Nature* **2004**, *429*, 457–463. [CrossRef]

51. Schroeder, M.; Hillemacher, T.; Bleich, S.; Frieling, H. The epigenetic code in depression: Implications for treatment. *Clin. Pharmacol. Ther.* **2012**, *91*, 310–314. [CrossRef]

52. Jankowska, A.M.; Millward, C.L.; Caldwell, C.W. The potential of DNA modifications as biomarkers and therapeutic targets in oncology. *Expert Rev. Mol. Diagn.* **2015**, *15*, 1325–1337. [CrossRef]

53. Shooshtari, P.; Huang, H.; Cotsapas, C. Integrative Genetic and Epigenetic Analysis Uncovers Regulatory Mechanisms of Autoimmune Disease. *Am. J. Hum. Genet.* **2017**, *101*, 75–86. [CrossRef]

54. Radtke, K.M.; Schauer, M.; Gunter, H.M.; Ruf-Leuschner, M.; Sill, J.; Meyer, A.; Elbert, T. Epigenetic modifications of the glucocorticoid receptor gene are associated with the vulnerability to psychopathology in childhood maltreatment. *Transl. Psychiatry* **2015**, *5*, e571. [CrossRef]

55. Farrell, C.; O'Keane, V. Epigenetics and the glucocorticoid receptor: A review of the implications in depression. *Psychiatry Res.* **2016**, *242*, 349–356. [CrossRef]

56. Dalton, V.S.; Kolshus, E.; McLoughlin, D.M. Epigenetics and depression: Return of the repressed. *J. Affect. Disord.* **2014**, *155*, 1–12. [CrossRef]

57. Dirven, B.C.J.; Homberg, J.R.; Kozicz, T.; Henckens, M.J.A.G. Epigenetic programming of the neuroendocrine stress response by adult life stress. *J. Mol. Endocrinol.* **2017**, *59*, R11–R31. [CrossRef]

58. Ancelin, M.L.; Scali, J.; Norton, J.; Ritchie, K.; Dupuy, A.M.; Chaudieu, I.; Ryan, J. Heterogeneity in HPA axis dysregulation and serotonergic vulnerability to depression. *Psychoneuroendocrinology* **2017**, *77*, 90–94. [CrossRef]

59. Evans, L.; Akiskal, H.S.; Keck, P.E., Jr.; McElroy, S.L.; Sadovnick, A.D.; Remick, R.A.; Kelsoe, J.R. Familiality of temperament in bipolar disorder: Support for a genetic spectrum. *J. Affect. Disord.* **2005**, *85*, 153–168. [CrossRef]

60. Kesebir, S.; Vahip, S.; Akdeniz, F.; Yüncü, Z.; Alkan, M.; Akiskal, H. Affective temperaments as measured by TEMPS-A in patients with bipolar I disorder and their first-degree relatives: A controlled study. *J. Affect. Disord.* **2005**, *85*, 127–133. [CrossRef]

61. Takeshima, M.; Oka, T. Comparative analysis of affective temperament in patients with difficult-to-treat and easy-to-treat major depression and bipolar disorder: Possible application in clinical settings. *Compr. Psychiatry* **2016**, *66*, 71–78. [CrossRef]

62. Serafini, G.; Geoffroy, P.A.; Aguglia, A.; Adavastro, G.; Canepa, G.; Pompili, M.; Amore, M. Irritable temperament and lifetime psychotic symptoms as predictors of anxiety symptoms in bipolar disorder. *Nord. J. Psychiatry* **2018**, *72*, 63–71. [CrossRef]

63. Signoretta, S.; Maremmani, I.; Liguori, A.; Perugi, G.; Akiskal, H.S. Affective temperament traits measured by TEMPS-I and emotional-behavioral problems in clinically-well children, adolescents, and young adults. *J. Affect. Disord.* **2005**, *85*, 169–180. [CrossRef]

64. Taylor, S. Association between COMT Val158Met and psychiatric disorders: A comprehensive meta-analysis. *Am. J. Med. Genet. B Neuropsychiatr. Genet.* **2018**, *177*, 199–210. [CrossRef]

65. Antypa, N.; Drago, A.; Serretti, A. The role of COMT gene variants in depression: Bridging neuropsychological, behavioral and clinical phenotypes. *Neurosci. Biobehav. Rev.* **2013**, *37*, 1597–1610. [CrossRef]

66. Bieliński, M.; Jaracz, M.; Lesiewska, N.; Tomaszewska, M.; Sikora, M.; Junik, R.; Kamińska, A.; Tretyn, A.; Borkowska, A. Association between COMT Val158Met and DAT1 polymorphisms and depressive symptoms in the obese population. *Neuropsychiatr. Dis. Treat.* **2017**, *13*, 2221–2229. [CrossRef]

67. Opmeer, E.M.; Kortekaas, R.; van Tol, M.J.; van der Wee, N.J.; Woudstra, S.; van Buchem, M.A.; Penninx, B.W.; Veltman, D.J.; Aleman, A. Influence of COMT val158met genotype on the depressed brain during emotional processing and working memory. *PLoS ONE* **2013**, *8*, e73290. [CrossRef]

68. Montirosso, R.; Provenzi, L.; Tavian, D.; Missaglia, S.; Raggi, M.E.; Borgatti, R. COMT(val158met) polymorphism is associated with behavioral response and physiologic reactivity to socio-emotional stress in 4-month-old infants. *Infant Behav. Dev.* **2016**, *45*, 71–82. [CrossRef]

69. Bornstein, S.R.; Schuppenies, A.; Wong, M.L.; Licinio, J. Approaching the shared biology of obesity and depression: The stress axis as the locus of gene-environment interactions. *Mol. Psychiatry* **2006**, *11*, 892–902. [CrossRef]

70. Luppino, F.S.; de Wit, L.M.; Bouvy, P.F.; Stijnen, T.; Cuijpers, P.; Penninx, B.W.; Zitman, F.G. Overweight, obesity, and depression: A systematic review and meta-analysis of longitudinal studies. *Arch. Gen. Psychiatry* **2010**, *67*, 220–229. [CrossRef]

71. Savitz, J.; van der Merwe, L.; Ramesar, R. Personality endophenotypes for bipolar affective disorder: A family-based genetic association analysis. *Genes Brain Behav.* **2008**, *7*, 869–876. [CrossRef]

72. Dávila, W.; Basterreche, N.; Arrue, A.; Zamalloa, M.I.; Gordo, E.; Dávila, R.; González-Torres, M.A.; Zumárraga, M. The influence of the Val158Met catechol-O-methyltransferase polymorphism on the personality traits of bipolar patients. *PLoS ONE* **2013**, *8*, e62900. [CrossRef]

73. Burdick, K.E.; Funke, B.; Goldberg, J.F.; Bates, J.A.; Jaeger, J.; Kucherlapati, R.; Malhotra, A.K. COMT genotype increases risk for bipolar I disorder and influences neurocognitive performance. *Bipolar Disord.* **2007**, *9*, 370–376. [CrossRef]

74. Enoch, M.A.; Xu, K.; Ferro, E.; Harris, C.R.; Goldman, D. Genetic origins of anxiety in women: A role for a functional catechol-O-methyltransferase polymorphism. *Psychiatr. Genet.* **2003**, *13*, 33–41. [CrossRef]

75. Massat, I.; Kocabas, N.A.; Crisafulli, C.; Chiesa, A.; Calati, R.; Linotte, S.; Kasper, S.; Fink, M.; Antonijevic, I.; Forray, C.; et al. COMT and age at onset in mood disorders: A replication and extension study. *Neurosci. Lett.* **2011**, *498*, 218–221. [CrossRef]

76. Baekken, P.M.; Skorpen, F.; Stordal, E.; Zwart, J.A.; Hagen, K. Depression and anxiety in relation to catechol-O-methyltransferase Val158Met genotype in the general population: The Nord-Trøndelag Health Study (HUNT). *BMC Psychiatry* **2008**, *8*, 48. [CrossRef]

77. Strakowski, S.M.; Sax, K.W.; McElroy, S.L.; Keck, P.E., Jr.; Hawkins, J.M.; West, S.A. Course of psychiatric and substance abuse syndromes co-occurring with bipolar disorder after a first psychiatric hospitalization. *J. Clin. Psychiatry* **1998**, *59*, 465–471. [CrossRef]

78. Golimbet, V.E.; Alfimova, M.V.; Gritsenko, I.K.; Ebstein, R.P. Relationship between dopamine system genes and extraversion and novelty seeking. *Neurosci. Behav. Physiol.* **2007**, *37*, 601–606. [CrossRef]

79. Hashimoto, R.; Noguchi, H.; Hori, H.; Ohi, K.; Yasuda, Y.; Takeda, M.; Kunugi, H.A. A possible association between the Val158Met polymorphism of the catechol-O-methyl transferase gene and the personality trait of harm avoidance in Japanese healthy subjects. *Neurosci. Lett.* **2007**, *428*, 17–20. [CrossRef]

80. Heck, A.; Lieb, R.; Ellgas, A.; Pfister, H.; Lucae, S.; Roeske, D.; Pütz, B.; Müller-Myhsok, B.; Uhr, M.; Holsboer, F.; et al. Investigation of 17 candidate genes for personality traits confirms effects of the HTR2A gene on novelty seeking. *Genes Brain Behav.* **2009**, *8*, 464–472. [CrossRef]

81. Tsai, S.J.; Hong, C.J.; Yu, Y.W.; Chen, T.J. Association study of catechol-O-methyltransferase gene and dopamine D4 receptor gene polymorphisms and personality traits in healthy young Chinese females. *Neuropsychobiology* **2004**, *50*, 153–156. [CrossRef]

82. Li, T.; Yu, S.; Du, J.; Chen, H.; Jiang, H.; Xu, K.; Fu, Y.; Wang, D.; Zhao, M. Role of novelty seeking personality traits as mediator of the association between COMT and onset age of drug use in Chinese heroin dependent patients. *PLoS ONE* **2011**, *6*, e22923. [CrossRef]

83. Kang, J.I.; Namkoong, K.; Kim, S.J. The association of 5-HTTLPR and DRD4 VNTR polymorphisms with affective temperamental traits in healthy volunteers. *J. Affect. Disord.* **2008**, *109*, 157–163. [CrossRef]

84. Akiskal, H.S.; Mendlowicz, M.V.; Jean-Louis, G.; Rapaport, M.H.; Kelsoe, J.R.; Gillin, J.C.; Smith, T.L. TEMPS-A: Validation of a short version of a self-rated instrument designed to measure variations in temperament. *J. Affect. Disord.* **2005**, *85*, 45–52. [CrossRef]

85. Erić, A.P.; Erić, I.; Ćurković, M.; Dodig-Ćurković, K.; Kralik, K.; Kovač, V.; Filaković, P. The temperament and character traits in patients with major depressive disorder and bipolar affective disorder with and without suicide attempt. *Psychiatr. Danub.* **2017**, *29*, 171–178. [CrossRef]

86. Parneix, M.; Pericaud, M.; Clement, J.P. Irritability associated with major depressive episodes: Its relationship with mood disorders and temperament. *Turk Psikiyatr. Derg.* **2014**, *25*, 106–113.

87. VanderBroek-Stice, L.; Stojek, M.K.; Beach, S.R.; vanDellen, M.R.; MacKillop, J. Multidimensional assessment of impulsivity in relation to obesity and food addiction. *Appetite* **2017**, *112*, 59–68. [CrossRef] [PubMed]

88. Gordon, J.A.; Hen, R. Genetic approaches to the study of anxiety. *Annu. Rev. Neurosci.* **2004**, *27*, 193–222. [CrossRef] [PubMed]

89. Depue, R.A.; Collins, P.F. Neurobiology of the structure of personality: Dopamine, facilitation of incentive motivation, and extraversion. *Behav. Brain Sci.* **1999**, *22*, 491–517. [CrossRef] [PubMed]

90. Ebstein, R.P.; Segman, R.; Benjamin, J.; Osher, Y.; Nemanov, L.; Belmaker, R.H. 5-HT2C (HTR2C) serotonin receptor gene polymorphism associated with the human personality trait of reward dependence: Interaction with dopamine D4 receptor (D4DR) and dopamine D3 receptor (D3DR) polymorphisms. *Am. J. Med. Genet.* **1997**, *74*, 65–72. [CrossRef]

91. Van Gestel, S.; Forsgren, T.; Claes, S.; Del-Favero, J.; Van Duijn, C.M.; Sluijs, S.; Nilsson, L.G.; Adolfsson, R.; Van Broeckhoven, C. Epistatic effect of genes from the dopamine and serotonin systems on the temperament traits of novelty seeking and harm avoidance. *Mol. Psychiatry* **2002**, *7*, 448–450. [CrossRef] [PubMed]

92. Jorm, A.F.; Prior, M.; Sanson, A.; Smart, D.; Zhang, Y.; Easteal, S. Association of a functional polymorphism of the serotonin transporter gene with anxiety-related temperament and behavior problems in children: A longitudinal study from infancy to the mid-teens. *Mol. Psychiatry* **2000**, *5*, 542–547. [CrossRef]

93. Kim, S.J.; Kim, Y.S.; Kim, C.H.; Lee, H.S. Lack of association between polymorphisms of the dopamine receptor D4 and dopamine transporter genes and personality traits in a Korean population. *Yonsei Med. J.* **2006**, *47*, 787–792. [CrossRef]

94. Congdon, E.; Lesch, K.P.; Canli, T. Analysis of DRD4 and DAT polymorphisms and behavioral inhibition in healthy adults: Implications for impulsivity. *Am. J. Med. Genet. B Neuropsychiatr. Genet.* **2008**, *147B*, 27–32. [CrossRef]

95. Maremmani, I.; Akiskal, H.; Signoretta, S.; Liguori, A.; Perugi, G.; Cloninger, C. The relationship of Kraepelian affective temperaments (asmeasured by TEMPS-I) to the tridimensional personality questionnaire (TPQ). *J. Affect. Disord.* **2005**, *85*, 17–27. [CrossRef]

96. Gonda, X.; Rihmer, Z.; Zsombok, T.; Bagdy, G.; Akiskal, K.K.; Akiskal, H.S. The 5HTTLPR polymorphism of the serotonin transporter gene is associated with affective temperaments as measured by TEMPS-A. *J. Affect. Disord.* **2006**, *91*, 125–131. [CrossRef] [PubMed]

97. Alexander, G.E.; DeLong, M.R.; Strick, P.L. Parallel organization of functionally segregated circuits linking basal ganglia and cortex. *Annu. Rev. Neurosci.* **1986**, *9*, 357–381. [CrossRef] [PubMed]

98. Benjamin, J.; Osher, Y.; Kotler, M.; Gritsenko, I.; Nemanov, L.; Belmaker, R.H.; Ebstein, R.P. Association between tridimensional personality questionnaire (TPQ) traits and three functional polymorphisms: Dopamine receptor D4 (DRD4), serotonin transporter promoter region (5-HTTLPR) and catechol O-methyltransferase (COMT). *Mol. Psychiatry* **2000**, *5*, 96–100. [CrossRef] [PubMed]

99. Smolka, M.N.; Bühler, M.; Schumann, G.; Klein, S.; Hu, X.Z.; Moayer, M.; Zimmer, A.; Wrase, J.; Flor, H.; Mann, K.; et al. Gene-gene effects on central processing of aversive stimuli. *Mol. Psychiatry* **2007**, *12*, 307–317. [CrossRef] [PubMed]

100. Drabant, E.M.; Hariri, A.R.; Meyer-Lindenberg, A.; Munoz, K.E.; Mattay, V.S.; Kolachana, B.S.; Egan, M.F.; Weinberger, D.R. Catechol O-methyltransferase val158met genotype and neural mechanisms related to affective arousal and regulation. *Arch. Gen. Psychiatry* **2006**, *63*, 1396–1406. [CrossRef] [PubMed]

101. Bagdy, G.; Juhasz, G.; Gonda, X. A new clinical evidence-based gene-environment interaction model of depression. *Neuropsychopharmacol. Hung.* **2012**, *14*, 213–220. [PubMed]

102. Mansur, R.B.; Fries, G.R.; Trevizol, A.P.; Subramaniapillai, M.; Lovshin, J.; Lin, K.; Vinberg, M.; Ho, R.C.; Brietzke, E.; McIntyre, R.S. The effect of body mass index on glucagon-like peptide receptor gene expression in the post mortem brain from individuals with mood and psychotic disorders. *Eur. Neuropsychopharmacol.* **2019**, *29*, 137–146. [CrossRef]

103. Yang, J.L.; Liu, X.; Jiang, H.; Pan, F.; Ho, C.S.; Ho, R.C. The Effects of High-fat-diet Combined with Chronic Unpredictable Mild Stress on Depression-like Behavior and Leptin/LepRb in Male Rats. *Sci. Rep.* **2016**, *6*, 35239. [CrossRef]

104. Lahiri, D.K.; Schnable, B. DNA isolation by a rapid method from human blood samples: Effects of MgCl2, EDTA, storage time, and temperature on DNA yield and quality. *Biochem. Genet.* **1993**, *1*, 321–328. [CrossRef]

105. Kaňková, Š.; Kodym, P.; Flegr, J. Direct evidence of Toxoplasma-induced changes in serum testosterone in mice. *Exp. Parasitol.* **2011**, *128*, 181–183. [CrossRef]

International Journal of
Molecular Sciences

Review

Genetic Overlap between General Cognitive Function and Schizophrenia: A Review of Cognitive GWASs

Kazutaka Ohi [1,2], Chika Sumiyoshi [3], Haruo Fujino [4], Yuka Yasuda [5], Hidenaga Yamamori [5], Michiko Fujimoto [6], Tomoko Shiino [5], Tomiki Sumiyoshi [7] and Ryota Hashimoto [5,8,*]

[1] Department of Neuropsychiatry, Kanazawa Medical University, Uchinada, Ishikawa 920-0293, Japan; ohi@kanazawa-med.ac.jp

[2] Medical Research Institute, Kanazawa Medical University, Ishikawa 920-0293, Japan

[3] Faculty of Human Development and Culture, Fukushima University, Fukushima 960-1296, Japan; sumiyoshi@educ.fukushima-u.ac.jp

[4] Graduate School of Education, Oita University, Oita 870-1192, Japan; fjinoh@hus.osaka-u.ac.jp

[5] Department of Pathology of Mental Diseases, National Institute of Mental Health, National Center of Neurology and Psychiatry, Tokyo 187-8553, Japan; yasuda@psy.med.osaka-u.ac.jp (Y.Y.); yamamori@psy.med.osaka-u.ac.jp (H.Y.); tshiino@ncnp.go.jp (T.S.)

[6] Department of Psychiatry, Osaka University Graduate School of Medicine, Suita, Osaka 565-0871, Japan; mfujimoto@psy.med.osaka-u.ac.jp

[7] Department of Preventive Interventions for Psychiatric Disorders, National Institute of Mental Health, National Center of Neurology and Psychiatry, Kodaira, Tokyo 187-8553, Japan; sumiyot@ncnp.go.jp

[8] Osaka University, Suita, Osaka 565-0871, Japan

*   Correspondence: ryotahashimoto55@ncnp.go.jp; Tel.: +81-42-346-1997; Fax: +81-42-346-2047

Received: 24 October 2018; Accepted: 26 November 2018; Published: 30 November 2018

Abstract: General cognitive (intelligence) function is substantially heritable, and is a major determinant of economic and health-related life outcomes. Cognitive impairments and intelligence decline are core features of schizophrenia which are evident before the onset of the illness. Genetic overlaps between cognitive impairments and the vulnerability for the illness have been suggested. Here, we review the literature on recent large-scale genome-wide association studies (GWASs) of general cognitive function and correlations between cognitive function and genetic susceptibility to schizophrenia. In the last decade, large-scale GWASs ($n > 30,000$) of general cognitive function and schizophrenia have demonstrated that substantial proportions of the heritability of the cognitive function and schizophrenia are explained by a polygenic component consisting of many common genetic variants with small effects. To date, GWASs have identified more than 100 loci linked to general cognitive function and 108 loci linked to schizophrenia. These genetic variants are mostly intronic or intergenic. Genes identified around these genetic variants are densely expressed in brain tissues. Schizophrenia-related genetic risks are consistently correlated with lower general cognitive function ($r_g = -0.20$) and higher educational attainment ($r_g = 0.08$). Cognitive functions are associated with many of the socioeconomic and health-related outcomes. Current treatment strategies largely fail to improve cognitive impairments of schizophrenia. Therefore, further study is needed to understand the molecular mechanisms underlying both cognition and schizophrenia.

Keywords: schizophrenia; general cognitive function; intelligence; GWAS; genetic correlation

---

1. Introduction

Cognitive functions play important roles in mental and physical well-beings. This is supported by observations that people with higher intelligence tend to have greater educational attainment, more professional jobs, higher incomes, and increased longevity [1,2]. Accordingly, impairments of cognitive functions result in social and occupational dysfunction which leads to poor life outcomes [3–8].

Cognitive disturbances are a core feature of schizophrenia—a psychiatric disorder with clinical and genetic heterogeneity [9,10]. Compared with healthy individuals, patients with schizophrenia demonstrate about a 1–2 standard deviation decline in performance on tests of several cognitive domains, including working, verbal and visual memories, processing speed, attention, social cognition, and intelligence [11–17]. Although the disorder is generally characterized by positive (e.g., hallucinations and delusions) and negative (blunted affect and withdrawal) symptoms, cognitive impairments should also be considered as an independent clinical dimension [18,19]. These impairments exist before the onset of illness and are worsened around it [20–22]. It has been suggested that cognitive deficits of schizophrenia may be resistant to treatment with antipsychotic drugs [23–26]. This indicates a need for clarifications of the mechanisms underlying these conditions.

Schizophrenia has a strong genetic basis with an estimated heritability of approximately 80% [27]. Cognitive functions such as general intelligence also have a genetic component ($h^2 = 0.33$–0.85) [28–32]. Despite the difference in heritability for intelligence between childhood ($h^2 = 0.45$) and adulthood ($h^2 = 0.80$), there is a high correlation between IQ levels in childhood and those in adulthood ($r_g = 0.89$) [33]. Relatives or twin siblings of patients with schizophrenia have also displayed impaired cognitive function to a lesser extent [9,34]. These findings suggest the contribution of genetic components to cognitive impairments in schizophrenia.

Genome-wide association studies (GWASs) that examine millions of genetic variants are a powerful tool to identify common variants responsible for susceptibility to common and complex diseases. The largest GWAS to date is the Psychiatric Genomics Consortium (PGC) using 36,989 patients with schizophrenia and 113,075 controls, which has identified 108 loci including genes and genetic variants related to schizophrenia [35]. Several consortia, such as the Cognitive Genomics Consortium (COGENT), Heart and Aging Research in Genomic Epidemiology Consortium (CHARGE), and UK Biobank (UKB), have performed GWASs to identify genetic loci related to cognitive function [36–43]. GWASs with fewer than 20,000 subjects did not find any significant loci [36,44–46]. These GWAS consortia used diverse assessment tools to represent targeted cognitive constructs in various samples, e.g., general cognitive function (g), Intelligence Quotient (IQ), fluid intelligence, etc., which could have been subject to phenotypic heterogeneity. By contrast, GWASs using samples from nearly 300,000 individuals successfully detected more than 100 genome-wide significant loci related to cognitive function [42,43]. In addition, part of the genetic correlation in the genetic effects identical between cognitive function and schizophrenia has been identified [38–42]. Therefore, cognitive functions have been proposed as an intermediate phenotype or biotype [9,15,47,48] to explain the mechanisms involved in the pathogenesis of schizophrenia.

In this article, we review the literature on recent large-scale GWASs of general cognitive function and genetic correlations between cognitive function and schizophrenia.

2. General Cognitive Function (g)

A number of tests have been used to measure various domains of cognitive functions. It is difficult to perform GWASs of cognitive functions uniformly because these cognitive tests vary among study cohorts. Twin and family studies show strong genetic correlations across diverse cognitive domains [49]. Under this circumstance, general cognitive function (g) is defined as a latent trait underlying shared variance across multiple subdomains of cognition [36,37,39,44,45]. To extract g, principal component analysis (PCA) is required on at least one cognitive measure across at least three domains, e.g., logical memory for verbal declarative memory, digit span for working memory, and digit symbol coding for processing speed. In other words, the first unrotated principal component of several distinct neuropsychological tests is obtained from the PCA. For example, an average of eight neuropsychological tests across COGENT cohorts were selected: digit span, digit symbol coding, verbal memory for words, visual memory, semantic fluency, word reading, verbal memory for stories, phonemic fluency, vocabulary, and the trail-making test [39]. The first principal component obtained accounted for approximately 40% of the variance in overall test performance. The *g* factors extracted from different cognitive tests were strongly correlated (>0.98) [50], supporting the universality of *g*.

Several cognitive GWASs have been performed using the *g* approach [15,36,37,39,44,45]. GWASs with fewer than 20,000 subjects did not find any genome-wide significant variants [15,36,44,45], while the GWASs with 35,298 [39] and 53,949 [37] subjects successfully identified two (*RP4-665J23.1* on 1p22.2 and *CENPO* on 2p23.3) and three (*MIR2113* on 6q16.1, *AKAP6/NPAS3* on 14q12 and *TOMM40/APOE* on 19q13.32) genome-wide significant loci, respectively (Table 1). However, neither these loci, nor the reproducibility of the findings, were consistent across studies.

3. Fluid Intelligence

Fluid-type intelligence requires swift thinking, relies relatively little on prior knowledge, and is often measured by unfamiliar and sometimes abstract materials [44]. By contrast, crystallized-type intelligence is typically assessed using tests such as those for acquired knowledge and vocabulary [44]. The discrepancy between fluid and crystallized intelligence becomes particularly noticeable in late adulthood—the age-related decline of fluid intelligence comes earlier and more rapidly [51,52].

To assess crystallized intelligence, either the National Adult Reading Test or the WAIS vocabulary subtest is used. Fluid intelligence, which may be equivalent to *g*, is assessed using PCA of data from several cognitive tests, such as logical memory, verbal fluency, auditory verbal learning tests (AVLT), and subtests from the Wechsler Adult Intelligence Scale (WAIS)-III [44]. Fluid intelligence is also measured by the verbal–numerical reasoning (VNR) test [42]. This test uses 13 multiple-choice questions—six verbal and seven numerical—which are presented on a touchscreen computer in either an assessment center or a web-based format at home [38,40]. Scores are obtained from the number of questions answered correctly in two minutes. With this method, the GWAS in UKB (n = 36,035) detected three genome-wide significant loci, including several genes, e.g., *CYP2D6* and *NAGA* at 22q13.2, *FUT8* at 14q23.3 and *PDE1C* at 7p14.3 [38].

Because performance on the VNR is correlated with *g* [40,53], the level of power in recent GWASs has been increased through combinations of *g* and fluid intelligence [40–43]. The total sample sizes in these studies were approximately 80,000–300,000 (Table 1). For example, one of the recent GWASs with 269,867 subjects identified 205 genome-wide significant loci [43]. This GWAS also identified some overlapping loci (2p23.3, 6q16.1, 7p14.3, 14q12, 19q13.32, and 22q13.2) consistent with previous reports [37–39], although these loci did not fully include lead genetic variants. The sample size in GWASs is positively correlated with the number of genome-wide associated loci detected (Figure 1, r^2 = 0.92, p = 1.18 × 10^{-5}). Several of these loci overlapped with those associated with schizophrenia, such as 1p21, 1p34, 2q24, 2q33, 3p21, 3q22, 4q24, 5q21, 6p22, 7q22, 8q24, 11q25, 12q24, 14q12, 14q32, 16q22, and 22q13 [35,41,43].

Table 1. A summary of GWASs of general cognitive function.

Authors (year)	n	Phenotypes	Ethnics	Participants	Consortium	Age Range	GWS Loci	SNP Hits	GWS Gene
Ohi et al. (2015) [15]	411	g or IQ	Japanese	Psychiatric healthy subjects	Osaka University	18–66	0	0	NA
Davies et al. (2011) [44]	3511	g	Caucasian	Nonclinical healthy samples	CAGES, LBC1921, LBC1936, ABC1936, etc.	64.6–79.1 *	0	0	1
Lencz et al. (2014) [36]	5000	g	Caucasian	General population (epidemiologically representative cohorts or mentally healthy cohorts)	COGENT	15.9–69.5 *	0	0	NA
Benyamin et al. (2014) [45]	17,989	g or IQ	European	Children	CHIC	6–18	0	0	0
Kirkpatrick et al. (2014) [46]	7100	IQ	Caucasian	Community-based family study samples	MTFS, SIBS	11.8–43.3 *	0	0	0
Davies et al. (2015) [37]	53,949	g	European	Population-based cohorts	CHARGE	>45	3	13	1
Davies et al. (2016) [38]	36,035	Fluid intelligence (VNR)	White British	Touchscreen-based community-dwelling individuals	UKB	40–73	3	149	17
Trampush et al. (2017) [39]	35,298	g	European	General population	COGENT	8–96	2	7	7
Sniekers et al. (2017) [40]	78,308	g, IQ or Fluid (VNR)	European	Web-base and touchscreen-based community-dwelling individuals and population-based cohorts	UKB, CHIC, MTFS, etc.	8–78	18	336	47
Lam et al. (2017) [41]	107,207	g, IQ or Fluid (VNR)	European	Web-base and touchscreen-based community-dwelling individuals and population-based cohorts	COGENT, UKB, CHIC, etc.	8–96	28	469	73
Davies et al. (2018) [42]	300,486	g or Fluid (VNR)	European	Web-base and touchscreen-based community-dwelling individuals and population-based cohorts	CHARGE, COGENT, UKB	16–102	148	11,600	709
Savage et al. (2018) [43]	269,867	g, IQ or Fluid (VNR)	European	Epidemiological cohorts	COGENT, UKB, etc.	5–98	205	12,110	507

VNR, Verbal–numerical reasoning; CAGES, Cognitive Aging Genetics in England and Scotland; LBC1921, LBC1936, Lothian Birth Cohorts of 1921 and 1936; ABC1936, Aberdeen Birth Cohort 1936; COGENT, Cognitive Genomics Consortium; MTFS, Minnesota Twin Family Study; SIBS, Sibling Interaction & Behavior Study; CHARGE, Heart and Aging Research in Genomic Epidemiology consortium; UKB, UK Biobank; GWS, Genome-wide significant. * Mean age range among cohorts was indicated.

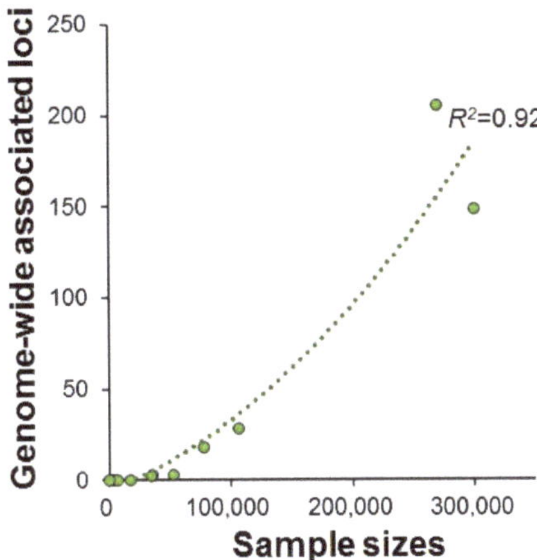

Figure 1. Relationship between sample sizes in GWASs of general cognitive function and genome-wide associated loci detected in each GWAS. Circles represent GWASs.

4. Educational Attainment

Educational attainment, represented by the number of years of education, is strongly influenced by genetic and environmental factors [54,55]. At least 20% of the variation among individuals is accounted for by genetic factors [54]. GWASs of educational attainment in 111,114 and 293,723 European individuals identified 14 genome-wide significant loci associated with the attainment of a college or university degree [38] and 74 loci associated with the number of years of schooling completed [55], respectively. Individuals with a higher level of intelligence tend to stay in school longer and attain higher qualifications than those with a lower level of intelligence. In addition, general cognitive ability (fluid intelligence) is correlated with educational attainment ($r_g > 0.70$) [38–41,43]. Therefore, educational attainment is useful as a proxy phenotype for general cognitive function in GWAS analyses. In fact, several loci, such as 1p31.1, 2q11.2, 3p21.31, 6q16.1, and 13q21.1, in a GWAS of educational attainment overlapped with those of general cognitive function.

5. Genes and Functions Related to General Cognitive Function

The genetic variants related to general cognitive function were mostly intronic or intergenic. The genes identified around these genetic variants were densely expressed in the brain [42,56], specifically striatal medium spiny neurons and hippocampal pyramidal CA1 neurons [43]. Common gene functions linked to general cognitive function were determined in gene-set analyses in some GWASs [40,42,43]. These functions include neurogenesis, regulation of nervous system development, neuronal differentiation, and regulation of cell development. Functions such as neuron projection and regulation of synaptic structure/activity were also associated with general cognitive function. As pathways related to these functions have been implicated in the pathophysiology for general cognitive function, these findings suggest that brain-expressed genes contribute to general cognitive function via neurodevelopmental processes in specific brain cells.

Smeland et al. (2017) extensively investigated shared genetic loci of the GWAS by conditional false discovery rate analysis and identified 21 genomic loci jointly influencing cognitive functions and vulnerability to schizophrenia [56]. Of the 21 loci, 18 showed a negative correlation between risk of schizophrenia and cognitive performance. The locus most strongly shared was detected on 22q13.2 that contains *TCF20*, *CYP2D6*, and *NAGA*. In addition, this locus was shown to have quantitative trait locus (eQTL). *NAGA* encodes lysosomal enzymes that modify glycoconjugates, and *CYP2D6* encodes cytochrome P450 enzymes that metabolize a broad range of drugs [56]. Other loci, including *KCNJ3*,

GNL3 and *STRC*, were also identified as eQTLs. Although these genes shared by two phenotypes are not localized in specific pathways, they may provide potential drug targets for improving cognitive impairments in patients with schizophrenia.

6. Polygenic Risk Score Analysis and Genetic Correlation between General Cognitive Function and Schizophrenia

Polygenic overlaps between alleles of general cognitive function and schizophrenia risk have been examined [36,57]. On the basis of the polygenic risk scores (PRS) derived from GWASs, a set of alleles associated with lower general cognitive function predicted an increased risk of vulnerability to schizophrenia. Conversely, polygenic alleles associated with schizophrenia-related risks predicted lower cognitive functions—particularly general cognitive function, performance IQ, attention, and working memory [36,57–63]. Thus, greater PRS related to risks for schizophrenia were associated with a greater decline in IQ after childhood in the general population [58]. So far, most studies on cognitive functions have used general population [36,57,58,60–63], and have not been specific to patients with schizophrenia [59,64].

Linkage disequilibrium score regression (LDSC) analysis estimates genetic variant correlations (r_g) from GWASs and is a powerful tool for investigating genetic architectures of common traits and diseases [65]. Studies using this method have consistently reported negative correlation between general cognitive function and schizophrenia-related risks, with r_g of approximately −0.2 (Figure 2) [39–43]. Specifically, higher educational attainment is associated with lower schizophrenia risk [66], whereas lower educational attainment predicts worse premorbid function and poorer outcomes [66]. These correlations would be reasonable in view of positive correlations between educational attainment and general cognitive function (Figure 3). However, recent studies found a positive correlation between educational attainment and schizophrenia (Figure 2) [55,67]. This discrepancy may be explained by at least two disease subtypes, i.e., patients with high intelligence, and those with cognitive impairments [68].

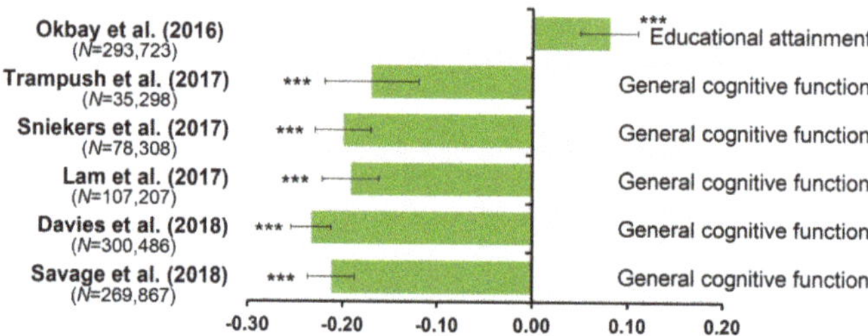

Figure 2. Genetic correlations (r_g) of educational attainment or general cognitive function with schizophrenia. Error bars indicate the SE of r_g. *** $p < 0.001$.

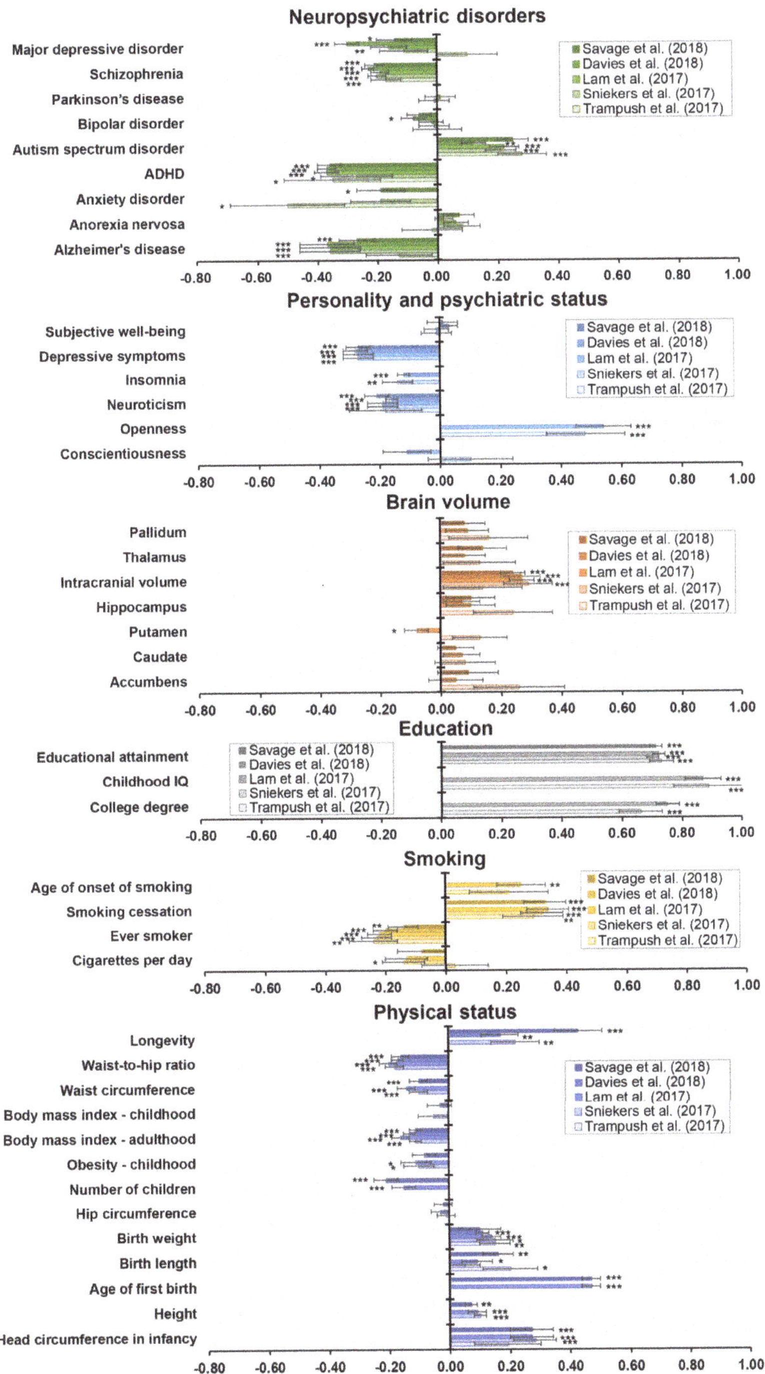

Figure 3. Genetic correlations (r_g) between general cognitive function and several phenotypes. Error bars indicate the SE of r_g. * $p < 0.05$, ** $p < 0.01$, *** $p < 0.001$.

7. Genetic Correlations between General Cognitive Function and Socioeconomic and Health-Related Outcomes

Cognitive function has been shown to be correlated with socioeconomic and health-related outcomes as well as neuropsychiatric disorders, as evidenced by LDSC analysis (Figure 3) [39–43]. Educational attainments provide the most robust correlations among other phenotypes. Specifically, better cognitive function was associated with a lower risk of several neuropsychiatric disorders, including schizophrenia, major depressive disorder, bipolar disorder, attention-deficit hyperactivity disorder, anxiety disorder, and Alzheimer's disease. By contrast, a higher risk of autism spectrum disorder was related to better cognitive function. As individuals around adolescence were included in correlational analyses (Table 1), the timing of cognitive assessment, i.e., before or after onset of the illness, may have obscured the results from these analyses.

Better cognitive function was associated with lower levels of neuroticism, depressive symptoms, and insomnia (Figure 3). Physical factors contributing were smaller waist-to-hip ratio and waist circumference, smaller volume of putamen, fewer numbers of cigarettes per day, less likelihood of having ever smoked, and lower body mass index in adulthood. Other factors affecting cognition included fewer children, higher levels of openness, age of onset of smoking and smoking cessation, larger intracranial volume, larger head circumference in infancy, height, birth length and weight, higher age of first birth, and greater longevity. These findings indicate that general cognitive function is related to socioeconomic and health-related outcomes in addition to neuropsychiatric disorders.

8. Intelligence Decline in Schizophrenia

Intelligence decline is conceptualized as intra-individual difference in intellectual performance between different time points [18,47,48,69]. Thus, it may be calculated by subtracting estimated premorbid IQ, as measured by the Adult Reading Test, and the present IQ, as measured by the WAIS. For the purpose of brief assessment, we have recently developed the WAIS-Short Form consisting of the Similarities and Symbol Search subtests [70]. Because clinical trials targeting cognitive impairment of schizophrenia have mostly yielded negative results, we suggest that patients without intelligence decline be excluded from participation. To date, no large-scale GWAS for intelligence decline in patients with schizophrenia has been performed, and further studies are needed.

The degree of intelligence decline in patients with schizophrenia is typically classified into three intellectual levels [18,23,47,69,71–77]:

(a)   Deteriorated group: patients with a difference of 10 points or more between premorbid IQ and present IQ;
(b)   Preserved group: patients with a difference of less than 10 points between premorbid IQ and present IQ (premorbid IQ above 90);
(c)   Compromised group: patients with a difference of less than 10 points between premorbid IQ and present IQ (premorbid IQ below 90).

The compromised IQ subgroup includes patients who have intellectual disability. Although cognitive impairments are a core feature of schizophrenia, approximately 30% of patients are classified into the preserved IQ subgroup [47].

So far, GWAS, PRS, or LDSC analysis has not been performed based on the above classification (deteriorated, preserved, and compromised IQ) in patients with schizophrenia. As the current diagnostic criteria for schizophrenia is independent of cognitive traits and genetic architectures, GWASs based on intelligence decline subgroups may reveal novel genetic variants specific to cognitive impairments. Caution is needed in interpreting data from IQ measures, as they are subject to non-specific consequences of schizophrenia, effects of medication, and cognitive decline preceding the onset of illness. Additionally, IQ scores by themselves cannot describe specific cognitive domains that are relatively more affected than others in individual patients.

9. Conclusions

In this paper, we reviewed the literature of recent large-scale GWASs targeting general cognitive function, a phenotype that captures shared variations in performance on tests of several cognitive domains. Studies on polygenic correlations between cognitive function and schizophrenia were also addressed. In the last decade, large-scale GWASs have identified more than 100 loci linked to general cognitive function and schizophrenia. Genetic variants identified are mostly intronic or intergenic, and genes around them are densely expressed in brain tissues. Substantial proportions of the heritability of these phenotypes are explained by polygenic architectures consisting of many genetic variants with small effects. General cognitive function has been reported to be genetically correlated with socioeconomic and health-related outcomes, as well as neuropsychiatric disorders. In particular, lower general cognitive function has been consistently correlated with schizophrenia risks. Current treatment strategies largely fail to improve cognitive impairments of schizophrenia. In order to progress, further study is needed to understand the shared pathogenesis for general cognitive function in relation to the illness.

Author Contributions: Conception and design: K.O. and R.H.; K.O., C.S., H.F., Y.Y., H.Y., M.F., T.S. (Tomoko Shiino), T.S. (Tomiki Sumiyoshi) and R.H. contributed to the analysis of publications, drafting of the manuscript and critical revision of the content.

Funding: This work was supported by Grants-in-Aid for Scientific Research (B) (25293250, 16H05375) and (C) (17K10321) and Young Scientists (B) (16K19784) from the Japan Society for the Promotion of Science (JSPS); the Grant for Assist KAKEN from Kanazawa Medical University (K2018-16); the YOKOYAMA Foundation for Clinical Pharmacology (YRY-1807), the Health and Labour Sciences Research Grants for Comprehensive Research on Persons with Disabilities from the Japan Agency for Medical Research and Development (AMED); a grant for Brain Mapping by Integrated Neurotechnologies for Disease Studies (Brain/MINDS) (AMED); and Intramural Research Grant for Neurological and Psychiatric Disorders of NCNP (29-1, 30-1, 30-8). The funders had no role in the study design, data collection and analysis, decision to publish, or preparation of the manuscript.

Conflicts of Interest: The authors declare no conflicts of interest.

Abbreviations

GWASs	Genome-Wide Association Studies
PGC	Psychiatric Genomics Consortium
COGENT	Cognitive Genomics Consortium
CHARGE	Heart and Aging Research in Genomic Epidemiology consortium
UKB	UK Biobank
IQ	Intelligence Quotient
PCA	Principal Component Analysis
WAIS	Wechsler Adult Intelligence Scale
AVLT	Auditory verbal learning Test
VNR	Verbal–Numerical Reasoning
LDSC	Linkage Disequilibrium Score Regression

References

1. Gottfredson, S.L. Why *g* matters: The complexity of everyday life. *Intelligence* **1997**, *24*, 79–132.
2. Batty, G.D.; Deary, I.J.; Gottfredson, L.S. Premorbid (early life) IQ and later mortality risk: Systematic review. *Ann. Epidemiol.* **2007**, *17*, 278–288.
3. Kahn, R.S.; Keefe, R.S. Schizophrenia is a cognitive illness: Time for a change in focus. *JAMA Psychiatry* **2013**, *70*, 1107–1112.
4. Green, M.F.; Kern, R.S.; Braff, D.L.; Mintz, J. Neurocognitive deficits and functional outcome in schizophrenia: Are we measuring the "right stuff"? *Schizophr. Bull.* **2000**, *26*, 119–136.
5. Green, M.F. What are the functional consequences of neurocognitive deficits in schizophrenia? *Am. J. Psychiatry* **1996**, *153*, 321–330.

6. Sawada, K.; Kanehara, A.; Sakakibara, E.; Eguchi, S.; Tada, M.; Satomura, Y.; Suga, M.; Koike, S.; Kasai, K. Identifying neurocognitive markers for outcome prediction of global functioning in individuals with first-episode and ultra-high-risk for psychosis. *Psychiatry Clin. Neurosci.* **2017**, *71*, 318–327.

7. Fujino, H.; Sumiyoshi, C.; Sumiyoshi, T.; Yasuda, Y.; Yamamori, H.; Ohi, K.; Fujimoto, M.; Hashimoto, R.; Takeda, M.; Imura, O. Predicting employment status and subjective quality of life in patients with schizophrenia. *Schizophr. Res. Cogn.* **2016**, *3*, 20–25.

8. Sumiyoshi, C.; Harvey, P.D.; Takaki, M.; Okahisa, Y.; Sato, T.; Sora, I.; Nuechterlein, K.H.; Subotnik, K.L.; Sumiyoshi, T. Factors predicting work outcome in Japanese patients with schizophrenia: Role of multiple functioning levels. *Schizophr. Res. Cogn.* **2015**, *2*, 105–112.

9. Ohi, K.; Shimada, T.; Nemoto, K.; Kataoka, Y.; Yasuyama, T.; Kimura, K.; Okubo, H.; Uehara, T.; Kawasaki, Y. Cognitive clustering in schizophrenia patients, their first-degree relatives and healthy subjects is associated with anterior cingulate cortex volume. *Neuroimage Clin.* **2017**, *16*, 248–256.

10. Simeone, J.C.; Ward, A.J.; Rotella, P.; Collins, J.; Windisch, R. An evaluation of variation in published estimates of schizophrenia prevalence from 1990 horizontal line 2013: A systematic literature review. *BMC Psychiatry* **2015**, *15*, 193.

11. Fujino, H.; Sumiyoshi, C.; Sumiyoshi, T.; Yasuda, Y.; Yamamori, H.; Ohi, K.; Fujimoto, M.; Umeda-Yano, S.; Higuchi, A.; Hibi, Y.; et al. Performance on the Wechsler Adult Intelligence Scale-III in Japanese patients with schizophrenia. *Psychiatry Clin. Neurosci.* **2014**, *68*, 534–541.

12. Ohi, K.; Hashimoto, R.; Yasuda, Y.; Fukumoto, M.; Nemoto, K.; Ohnishi, T.; Yamamori, H.; Takahashi, H.; Iike, N.; Kamino, K.; et al. The AKT1 gene is associated with attention and brain morphology in schizophrenia. *World J. Biol. Psychiatry* **2013**, *14*, 100–113.

13. Fukumoto, M.; Hashimoto, R.; Ohi, K.; Yasuda, Y.; Yamamori, H.; Umeda-Yano, S.; Iwase, M.; Kazui, H.; Takeda, M. Relation between remission status and attention in patients with schizophrenia. *Psychiatry Clin. Neurosci.* **2014**, *68*, 234–241.

14. Ohi, K.; Hashimoto, R.; Yasuda, Y.; Fukumoto, M.; Yamamori, H.; Umeda-Yano, S.; Fujimoto, M.; Iwase, M.; Kazui, H.; Takeda, M. Influence of the NRGN gene on intellectual ability in schizophrenia. *J. Hum. Genet.* **2013**, *58*, 700–705.

15. Ohi, K.; Hashimoto, R.; Ikeda, M.; Yamamori, H.; Yasuda, Y.; Fujimoto, M.; Umeda-Yano, S.; Fukunaga, M.; Fujino, H.; Watanabe, Y.; et al. Glutamate Networks Implicate Cognitive Impairments in Schizophrenia: Genome-Wide Association Studies of 52 Cognitive Phenotypes. *Schizophr. Bull.* **2015**, *41*, 909–918.

16. Horiguchi, M.; Ohi, K.; Hashimoto, R.; Hao, Q.; Yasuda, Y.; Yamamori, H.; Fujimoto, M.; Umeda-Yano, S.; Takeda, M.; Ichinose, H. Functional polymorphism (C-824T) of the tyrosine hydroxylase gene affects IQ in schizophrenia. *Psychiatry Clin. Neurosci.* **2014**, *68*, 456–462.

17. Hashimoto, R.; Noguchi, H.; Hori, H.; Ohi, K.; Yasuda, Y.; Takeda, M.; Kunugi, H. Association between the dysbindin gene (DTNBP1) and cognitive functions in Japanese subjects. *Psychiatry Clin. Neurosci.* **2009**, *63*, 550–556.

18. Badcock, J.C.; Dragovic, M.; Waters, F.A.; Jablensky, A. Dimensions of intelligence in schizophrenia: Evidence from patients with preserved, deteriorated and compromised intellect. *J. Psychiatr. Res.* **2005**, *39*, 11–19.

19. Ikebuchi, E.; Sato, S.; Yamaguchi, S.; Shimodaira, M.; Taneda, A.; Hatsuse, N.; Watanabe, Y.; Sakata, M.; Satake, N.; Nishio, M.; et al. Does improvement of cognitive functioning by cognitive remediation therapy effect work outcomes in severe mental illness? A secondary analysis of a randomized controlled trial. *Psychiatry Clin. Neurosci.* **2017**, *71*, 301–308.

20. Kremen, W.S.; Vinogradov, S.; Poole, J.H.; Schaefer, C.A.; Deicken, R.F.; Factor-Litvak, P.; Brown, A.S. Cognitive decline in schizophrenia from childhood to midlife: A 33-year longitudinal birth cohort study. *Schizophr. Res.* **2010**, *118*, 1–5.

21. Meier, M.H.; Caspi, A.; Reichenberg, A.; Keefe, R.S.; Fisher, H.L.; Harrington, H.; Houts, R.; Poulton, R.; Moffitt, T.E. Neuropsychological decline in schizophrenia from the premorbid to the postonset period: Evidence from a population-representative longitudinal study. *Am. J. Psychiatry* **2014**, *171*, 91–101.

22. Sheitman, B.B.; Murray, M.G.; Snyder, J.A.; Silva, S.; Goldman, R.; Chakos, M.; Volavka, J.; Lieberman, J.A. IQ scores of treatment-resistant schizophrenia patients before and after the onset of the illness. *Schizophr. Res.* **2000**, *46*, 203–207.

23. Leeson, V.C.; Sharma, P.; Harrison, M.; Ron, M.A.; Barnes, T.R.; Joyce, E.M. IQ trajectory, cognitive reserve, and clinical outcome following a first episode of psychosis: A 3-year longitudinal study. *Schizophr. Bull.* **2011**, *37*, 768–777.

24. Bilder, R.M.; Goldman, R.S.; Robinson, D.; Reiter, G.; Bell, L.; Bates, J.A.; Pappadopulos, E.; Willson, D.F.; Alvir, J.M.; Woerner, M.G.; et al. Neuropsychology of first-episode schizophrenia: Initial characterization and clinical correlates. *Am. J. Psychiatry* **2000**, *157*, 549–559.

25. Hill, S.K.; Schuepbach, D.; Herbener, E.S.; Keshavan, M.S.; Sweeney, J.A. Pretreatment and longitudinal studies of neuropsychological deficits in antipsychotic-naive patients with schizophrenia. *Schizophr. Res.* **2004**, *68*, 49–63.

26. Hoff, A.L.; Sakuma, M.; Wieneke, M.; Horon, R.; Kushner, M.; DeLisi, L.E. Longitudinal neuropsychological follow-up study of patients with first-episode schizophrenia. *Am. J. Psychiatry* **1999**, *156*, 1336–1341.

27. Sullivan, P.F.; Kendler, K.S.; Neale, M.C. Schizophrenia as a complex trait: Evidence from a meta-analysis of twin studies. *Arch. Gen. Psychiatry* **2003**, *60*, 1187–1192.

28. Swagerman, S.C.; de Geus, E.J.; Kan, K.J.; van Bergen, E.; Nieuwboer, H.A.; Koenis, M.M.; Hulshoff Pol, H.E.; Gur, R.E.; Gur, R.C.; Boomsma, D.I. The Computerized Neurocognitive Battery: Validation, aging effects, and heritability across cognitive domains. *Neuropsychology* **2016**, *30*, 53–64.

29. Husted, J.A.; Lim, S.; Chow, E.W.; Greenwood, C.; Bassett, A.S. Heritability of neurocognitive traits in familial schizophrenia. *Am. J. Med. Genet. B Neuropsychiatr. Genet.* **2009**, *150B*, 845–853.

30. Berrettini, W.H. Genetic bases for endophenotypes in psychiatric disorders. *Dialogues Clin. Neurosci.* **2005**, *7*, 95–101.

31. Chen, W.J.; Liu, S.K.; Chang, C.J.; Lien, Y.J.; Chang, Y.H.; Hwu, H.G. Sustained attention deficit and schizotypal personality features in nonpsychotic relatives of schizophrenic patients. *Am. J. Psychiatry* **1998**, *155*, 1214–1220.

32. Posthuma, D.; de Geus, E.J.; Boomsma, D.I. Perceptual speed and IQ are associated through common genetic factors. *Behav. Genet.* **2001**, *31*, 593–602.

33. Polderman, T.J.; Benyamin, B.; de Leeuw, C.A.; Sullivan, P.F.; van Bochoven, A.; Visscher, P.M.; Posthuma, D. Meta-analysis of the heritability of human traits based on fifty years of twin studies. *Nat. Genet.* **2015**, *47*, 702–709.

34. Toulopoulou, T.; Goldberg, T.E.; Mesa, I.R.; Picchioni, M.; Rijsdijk, F.; Stahl, D.; Cherny, S.S.; Sham, P.; Faraone, S.V.; Tsuang, M.; et al. Impaired intellect and memory: A missing link between genetic risk and schizophrenia? *Arch. Gen. Psychiatry* **2010**, *67*, 905–913.

35. Ripke, S.; Neale, B.M.; Corvin, A.; Walters, J.T.; Farh, K.H.; Holmans, P.A.; Lee, P.; Bulik-Sullivan, B.; Collier, D.A.; Huang, H. Biological insights from 108 schizophrenia-associated genetic loci. *Nature* **2014**, *511*, 421–427.

36. Lencz, T.; Knowles, E.; Davies, G.; Guha, S.; Liewald, D.C.; Starr, J.M.; Djurovic, S.; Melle, I.; Sundet, K.; Christoforou, A.; et al. Molecular genetic evidence for overlap between general cognitive ability and risk for schizophrenia: A report from the Cognitive Genomics consorTium (COGENT). *Mol. Psychiatry* **2014**, *19*, 168–174.

37. Davies, G.; Armstrong, N.; Bis, J.C.; Bressler, J.; Chouraki, V.; Giddaluru, S.; Hofer, E.; Ibrahim-Verbaas, C.A.; Kirin, M.; Lahti, J.; et al. Genetic contributions to variation in general cognitive function: A meta-analysis of genome-wide association studies in the CHARGE consortium (N = 53949). *Mol. Psychiatry* **2015**, *20*, 183–192.

38. Davies, G.; Marioni, R.E.; Liewald, D.C.; Hill, W.D.; Hagenaars, S.P.; Harris, S.E.; Ritchie, S.J.; Luciano, M.; Fawns-Ritchie, C.; Lyall, D.; et al. Genome-wide association study of cognitive functions and educational attainment in UK Biobank (*N* = 112 151). *Mol. Psychiatry* **2016**, *21*, 758–767.

39. Trampush, J.W.; Yang, M.L.; Yu, J.; Knowles, E.; Davies, G.; Liewald, D.C.; Starr, J.M.; Djurovic, S.; Melle, I.; Sundet, K.; et al. GWAS meta-analysis reveals novel loci and genetic correlates for general cognitive function: A report from the COGENT consortium. *Mol. Psychiatry* **2017**, *22*, 336–345.

40. Sniekers, S.; Stringer, S.; Watanabe, K.; Jansen, P.R.; Coleman, J.R.I.; Krapohl, E.; Taskesen, E.; Hammerschlag, A.R.; Okbay, A.; Zabaneh, D.; et al. Genome-wide association meta-analysis of 78,308 individuals identifies new loci and genes influencing human intelligence. *Nat. Genet.* **2017**, *49*, 1107–1112.

41. Lam, M.; Trampush, J.W.; Yu, J.; Knowles, E.; Davies, G.; Liewald, D.C.; Starr, J.M.; Djurovic, S.; Melle, I.; Sundet, K.; et al. Large-Scale Cognitive GWAS Meta-Analysis Reveals Tissue-Specific Neural Expression and Potential Nootropic Drug Targets. *Cell Rep.* **2017**, *21*, 2597–2613.

42. Davies, G.; Lam, M.; Harris, S.E.; Trampush, J.W.; Luciano, M.; Hill, W.D.; Hagenaars, S.P.; Ritchie, S.J.; Marioni, R.E.; Fawns-Ritchie, C.; et al. Study of 300,486 individuals identifies 148 independent genetic loci influencing general cognitive function. *Nat. Commun.* **2018**, *9*, 2098.

43. Savage, J.E.; Jansen, P.R.; Stringer, S.; Watanabe, K.; Bryois, J.; de Leeuw, C.A.; Nagel, M.; Awasthi, S.; Barr, P.B.; Coleman, J.R.I.; et al. Genome-wide association meta-analysis in 269,867 individuals identifies new genetic and functional links to intelligence. *Nat. Genet.* **2018**, *50*, 912–919.

44. Davies, G.; Tenesa, A.; Payton, A.; Yang, J.; Harris, S.E.; Liewald, D.; Ke, X.; Le Hellard, S.; Christoforou, A.; Luciano, M.; et al. Genome-wide association studies establish that human intelligence is highly heritable and polygenic. *Mol. Psychiatry* **2011**, *16*, 996–1005.

45. Benyamin, B.; Pourcain, B.; Davis, O.S.; Davies, G.; Hansell, N.K.; Brion, M.J.; Kirkpatrick, R.M.; Cents, R.A.; Franic, S.; Miller, M.B.; et al. Childhood intelligence is heritable, highly polygenic and associated with FNBP1L. *Mol. Psychiatry* **2014**, *19*, 253–258.

46. Kirkpatrick, R.M.; McGue, M.; Iacono, W.G.; Miller, M.B.; Basu, S. Results of a "GWAS plus:" general cognitive ability is substantially heritable and massively polygenic. *PLoS ONE* **2014**, *9*, e112390.

47. Ohi, K.; Sumiyoshi, C.; Fujino, H.; Yasuda, Y.; Yamamori, H.; Fujimoto, M.; Sumiyoshi, T.; Hashimoto, R. A Brief Assessment of Intelligence Decline in Schizophrenia As Represented by the Difference between Current and Premorbid Intellectual Quotient. *Front. Psychiatry* **2017**, *8*, 293.

48. Hashimoto, R.; Ikeda, M.; Ohi, K.; Yasuda, Y.; Yamamori, H.; Fukumoto, M.; Umeda-Yano, S.; Dickinson, D.; Aleksic, B.; Iwase, M.; et al. Genome-wide association study of cognitive decline in schizophrenia. *Am. J. Psychiatry* **2013**, *170*, 683–684.

49. Plomin, R.; Kovas, Y. Generalist genes and learning disabilities. *Psychol. Bull.* **2005**, *131*, 592–617.

50. Johnson, W.; Nijenhuis, J.T.; Bouchard, T.J., Jr. Still just 1 g: Consistent results from five test batteries. *Intelligence* **2008**, *36*, 81–95.

51. Salthouse, T.A. Localizing age-related individual differences in a hierarchical structure. *Intelligence* **2004**, *32*.

52. Craik, F.I.; Bialystok, E. Cognition through the lifespan: Mechanisms of change. *Trends Cogn. Sci.* **2006**, *10*, 131–138.

53. Deary, I.J.; Penke, L.; Johnson, W. The neuroscience of human intelligence differences. *Nat. Rev. Neurosci.* **2010**, *11*, 201–211.

54. Rietveld, C.A.; Medland, S.E.; Derringer, J.; Yang, J.; Esko, T.; Martin, N.W.; Westra, H.J.; Shakhbazov, K.; Abdellaoui, A.; Agrawal, A.; et al. GWAS of 126,559 individuals identifies genetic variants associated with educational attainment. *Science* **2013**, *340*, 1467–1471.

55. Okbay, A.; Beauchamp, J.P.; Fontana, M.A.; Lee, J.J.; Pers, T.H.; Rietveld, C.A.; Turley, P.; Chen, G.B.; Emilsson, V.; Meddens, S.F.; et al. Genome-wide association study identifies 74 loci associated with educational attainment. *Nature* **2016**, *533*, 539–542.

56. Smeland, O.B.; Frei, O.; Kauppi, K.; Hill, W.D.; Li, W.; Wang, Y.; Krull, F.; Bettella, F.; Eriksen, J.A.; Witoelar, A.; et al. Identification of Genetic Loci Jointly Influencing Schizophrenia Risk and the Cognitive Traits of Verbal-Numerical Reasoning, Reaction Time, and General Cognitive Function. *JAMA Psychiatry* **2017**, *74*, 1065–1075.

57. Hubbard, L.; Tansey, K.E.; Rai, D.; Jones, P.; Ripke, S.; Chambert, K.D.; Moran, J.L.; McCarroll, S.A.; Linden, D.E.; Owen, M.J.; et al. Evidence of Common Genetic Overlap Between Schizophrenia and Cognition. *Schizophr. Bull.* **2016**, *42*, 832–842.

58. McIntosh, A.M.; Gow, A.; Luciano, M.; Davies, G.; Liewald, D.C.; Harris, S.E.; Corley, J.; Hall, J.; Starr, J.M.; Porteous, D.J.; et al. Polygenic risk for schizophrenia is associated with cognitive change between childhood and old age. *Biol. Psychiatry* **2013**, *73*, 938–943.

59. Nakahara, S.; Medland, S.; Turner, J.A.; Calhoun, V.D.; Lim, K.O.; Mueller, B.A.; Bustillo, J.R.; O'Leary, D.S.; Vaidya, J.G.; McEwen, S.; et al. Polygenic risk score, genome-wide association, and gene set analyses of cognitive domain deficits in schizophrenia. *Schizophr. Res.* **2018**, *201*, 393–399. [CrossRef]

60. Liebers, D.T.; Pirooznia, M.; Seiffudin, F.; Musliner, K.L.; Zandi, P.P.; Goes, F.S. Polygenic Risk of Schizophrenia and Cognition in a Population-Based Survey of Older Adults. *Schizophr. Bull.* **2016**, *42*, 984–991.

61. Hatzimanolis, A.; Bhatnagar, P.; Moes, A.; Wang, R.; Roussos, P.; Bitsios, P.; Stefanis, C.N.; Pulver, A.E.; Arking, D.E.; Smyrnis, N.; et al. Common genetic variation and schizophrenia polygenic risk influence neurocognitive performance in young adulthood. *Am. J. Med. Genet. Part B Neuropsychiatr. Genet.* **2015**, *168B*, 392–401.

62. Germine, L.; Robinson, E.B.; Smoller, J.W.; Calkins, M.E.; Moore, T.M.; Hakonarson, H.; Daly, M.J.; Lee, P.H.; Holmes, A.J.; Buckner, R.L.; et al. Association between polygenic risk for schizophrenia, neurocognition and social cognition across development. *Transl. Psychiatry* **2016**, *6*, e924.

63. Hagenaars, S.P.; Harris, S.E.; Davies, G.; Hill, W.D.; Liewald, D.C.; Ritchie, S.J.; Marioni, R.E.; Fawns-Ritchie, C.; Cullen, B.; Malik, R.; et al. Shared genetic aetiology between cognitive functions and physical and mental health in UK Biobank (*N* = 112 151) and 24 GWAS consortia. *Mol. Psychiatry* **2016**, *21*, 1624–1632.

64. Van Scheltinga, A.F.; Bakker, S.C.; van Haren, N.E.; Derks, E.M.; Buizer-Voskamp, J.E.; Cahn, W.; Ripke, S.; Ophoff, R.A.; Kahn, R.S. Schizophrenia genetic variants are not associated with intelligence. *Psychol. Med.* **2013**, *43*, 2563–2570.

65. Bulik-Sullivan, B.K.; Loh, P.R.; Finucane, H.K.; Ripke, S.; Yang, J.; Patterson, N.; Daly, M.J.; Price, A.L.; Neale, B.M. LD Score regression distinguishes confounding from polygenicity in genome-wide association studies. *Nat. Genet.* **2015**, *47*, 291–295.

66. Swanson, C.L., Jr.; Gur, R.C.; Bilker, W.; Petty, R.G.; Gur, R.E. Premorbid educational attainment in schizophrenia: Association with symptoms, functioning, and neurobehavioral measures. *Biol. Psychiatry* **1998**, *44*, 739–747.

67. Power, R.A.; Steinberg, S.; Bjornsdottir, G.; Rietveld, C.A.; Abdellaoui, A.; Nivard, M.M.; Johannesson, M.; Galesloot, T.E.; Hottenga, J.J.; Willemsen, G.; et al. Polygenic risk scores for schizophrenia and bipolar disorder predict creativity. *Nat. Neurosci.* **2015**, *18*, 953–955.

68. Bansal, V.; Mitjans, M.; Burik, C.A.P.; Karlsson Linner, R.; Okbay, A.; Rietveld, C.A.; Begemann, M.; Bonn, S.; Ripke, S.; de Vlaming, R.; et al. Genome-wide association study results for educational attainment aid in identifying genetic heterogeneity of schizophrenia. *bioRxiv* **2018**. [CrossRef]

69. Fujino, H.; Sumiyoshi, C.; Yasuda, Y.; Yamamori, H.; Fujimoto, M.; Fukunaga, M.; Miura, K.; Takebayashi, Y.; Okada, N.; Isomura, S.; et al. Estimated cognitive decline in patients with schizophrenia: A multicenter study. *Psychiatry Clin. Neurosci.* **2017**, *71*, 294–300.

70. Sumiyoshi, C.; Fujino, H.; Sumiyoshi, T.; Yasuda, Y.; Yamamori, H.; Ohi, K.; Fujimoto, M.; Takeda, M.; Hashimoto, R. Usefulness of the Wechsler Intelligence Scale short form for assessing functional outcomes in patients with schizophrenia. *Psychiatry Res.* **2016**, *245*, 371–378.

71. Weickert, T.W.; Goldberg, T.E.; Gold, J.M.; Bigelow, L.B.; Egan, M.F.; Weinberger, D.R. Cognitive impairments in patients with schizophrenia displaying preserved and compromised intellect. *Arch. Gen. Psychiatry* **2000**, *57*, 907–913.

72. Kremen, W.S.; Seidman, L.J.; Faraone, S.V.; Tsuang, M.T. IQ decline in cross-sectional studies of schizophrenia: Methodology and interpretation. *Psychiatry Res.* **2008**, *158*, 181–194.

73. Potter, A.I.; Nestor, P.G. IQ subtypes in schizophrenia: Distinct symptom and neuropsychological profiles. *J. Nerv. Ment. Dis.* **2010**, *198*, 580–585.

74. Mercado, C.L.; Johannesen, J.K.; Bell, M.D. Thought disorder severity in compromised, deteriorated, and preserved intellectual course of schizophrenia. *J. Nerv. Ment. Dis.* **2011**, *199*, 111–116.

75. Ammari, N.; Heinrichs, R.W.; Pinnock, F.; Miles, A.A.; Muharib, E.; McDermid Vaz, S. Preserved, deteriorated, and premorbidly impaired patterns of intellectual ability in schizophrenia. *Neuropsychology* **2014**, *28*, 353–358.

76. Wells, R.; Swaminathan, V.; Sundram, S.; Weinberg, D.; Bruggemann, J.; Jacomb, I.; Cropley, V.; Lenroot, R.; Pereira, A.M.; Zalesky, A.; et al. The impact of premorbid and current intellect in schizophrenia: Cognitive, symptom, and functional outcomes. *NPJ Schizophr.* **2015**, *1*, 15043.

77. Weinberg, D.; Lenroot, R.; Jacomb, I.; Allen, K.; Bruggemann, J.; Wells, R.; Balzan, R.; Liu, D.; Galletly, C.; Catts, S.V.; et al. Cognitive Subtypes of Schizophrenia Characterized by Differential Brain Volumetric Reductions and Cognitive Decline. *JAMA Psychiatry* **2016**, *73*, 1251–1259.

Review

Alterations of Expression of the Serotonin 5-HT4 Receptor in Brain Disorders

Heike Rebholz [1],*, Eitan Friedman [1,2] and Julia Castello [1,2]

[1] Department of Molecular, Cellular and Biomedical Sciences, CUNY School of Medicine, New York, NY 10031, USA; Friedman@med.cuny.edu (E.F.); julia.csaval@gmail.com (J.C.)

[2] Ph.D. Programs in Biochemistry and Biology, The Graduate Center, City University of New York, New York, NY 10031, USA

* Correspondence: heikerebholz@gmail.com; Tel.: +1-212-650-8283

Received: 14 October 2018; Accepted: 6 November 2018; Published: 13 November 2018

Abstract: The serotonin 4 receptor, $5\text{-HT}_4\text{R}$, represents one of seven different serotonin receptor families and is implicated in a variety of physiological functions and their pathophysiological variants, such as mood and depression or anxiety, food intake and obesity or anorexia, or memory and memory loss in Alzheimer's disease. Its central nervous system expression pattern in the forebrain, in particular in caudate putamen, the hippocampus and to lesser extent in the cortex, predispose it for a role in executive function and reward-related actions. In rodents, regional overexpression or knockdown in the prefrontal cortex or the nucleus accumbens of $5\text{-HT}_4\text{R}$ was shown to impact mood and depression-like phenotypes, food intake and hypophagia; however, whether expression changes are causally involved in the etiology of such disorders is not clear. In this context, more data are emerging, especially based on PET technology and the use of ligand tracers that demonstrate altered $5\text{-HT}_4\text{R}$ expression in brain disorders in humans, confirming data stemming from post-mortem tissue and preclinical animal models. In this review, we would like to present the current knowledge of $5\text{-HT}_4\text{R}$ expression in brain regions relevant to mood/depression, reward and executive function with a focus on $5\text{-HT}_4\text{R}$ expression changes in brain disorders or caused by drug treatment, at both the transcript and protein levels.

Keywords: serotonin; 5-HT 4 receptor; 5-HT4R; depression; mood disorder; expression; Alzheimer's disease; cognition; Parkinson's disease

---

1. Introduction

5-HT receptors are composed of 7 families (5-HT_{1-7} receptors), comprising 14 structurally and pharmacologically distinct 5-HT receptor subtypes [1]. All receptors are G-protein-coupled, with the exception of the $5\text{-HT}_3\text{R}$ that belongs to the superfamily of ligand-gated ion channels. Members of all 7 receptor families are expressed in the brain: 5-HT_1 receptors are $G\alpha_{i/0}$-coupled and two receptors of this family, $5\text{-HT}_{1a}\text{R}$ and $5\text{-HT}_{1b}\text{R}$, have an important function as somatodendritic autoreceptors expressed on neurons of the raphe nuclei that produce 5-HT, but they are also expressed as postsynaptic heteroreceptors in several brain areas [2]. The three members of the $G\alpha_{q/11}$-coupled $5\text{-HT}_2\text{R}$ family have well defined roles in the periphery such as in the vascular system and muscle contraction; however, their function in the brain is not well understood A potential link between a $5\text{-HT}_{2C}\text{R}$ allele and vulnerability to affective disorders has been reported, and a number of antipsychotics have inverse agonist activity at 5-HT_{2C} receptors [3]. $5\text{-HT}_{2C}\text{R}$ KO mice are highly obese and suffer from epilepsies [4]. $5\text{-HT}_{2a}\text{R}$ mediates the hallucinatory and psychotic action of psychedelic drugs such as LSD or psylocibin [5]. Human brain $G\alpha_{i/0}$-coupled $5\text{-HT}_5\text{R}$ expression is localized to the cerebral cortex, hippocampus, cerebellum, and a role in mood and major depression was postulated, using pharmacological tools and knockout mice [6]. $5\text{-HT}_6\text{Rs}$ are postsynaptic $G_{\alpha s}$-coupled receptors strongly expressed in the striatum, nucleus accumbens and cortex, and moderately in the hippocampus, amygdala, and hypothalamus. They control among others central cholinergic function [6]. The $G\alpha_s$-coupled 5HT$_7$R is mainly expressed in the limbic system, and a potential role in sleep, circadian rhythmic activity and mood has been suggested [6]. Among all 5-HT receptors, the type 3 receptor is the only ligand-gated ion channel receptor triggering rapid depolarization via the opening of non-selective cation channels. 5-HT$_3$R expression in the forebrain is low, but higher levels are present in the hippocampus and amygdala [6].

5-HT$_4$R was initially identified in cultured mouse colliculi cells and guinea pig brain using a functional cAMP stimulation assay [7]. In 1995, its cloning was reported [8]. Two different splicing variants, a short one, found in the striatum and a long one in the whole brain [8] were initially described, while others found the short form also present more universally in the brain [9].

Expression in the brain is greatest in the basal ganglia, the hippocampal formation and the cortex, as shown in human and rat brain [10,11]. 5-HT$_4$R is also widely distributed in the body. Outside the CNS, it is found along the gastrointestinal tract (esophagus, ileum and colon) [12]. It is also present in the bladder, the heart and the adrenal glands. 5-HT$_4$ receptors are well known for their peripheral effects on the gastrointestinal tract, and are targets in the treatment of dyspepsia, gastroesophageal reflux disease, gastroparesis or irritable bowel syndrome [13,14]. Serotonin affects heart contractility through 5-HT$_4$R which is expressed in the human and pig atrium and ventricle, while, interestingly, in the rat it is only expressed in the atrium [15]. 5-HT$_4$R activation leads to the contraction of heart but also to tachycardia and arrhythmia [15]. The cardiac contractile effects of 5-HT$_4$R are restricted to human and pig atria and are absent from a large number of laboratory animals, such as rat, guinea pig, rabbit and frog [16]. 5-HT$_4$R was also shown to be overexpressed in the cortex of the adrenal gland of a subtype of Cushing syndrome patients, a condition caused by cortisone hyper-production [17].

Compared to other serotonin receptors, the gene encoding 5-HT$_4$R (*htr4*) is large and its architecture is complicated, with 38 exons spaced over 700 kb [18]. As a G-protein coupled receptor, 5-HT$_4$R signals through both G protein-dependent and G protein-independent pathways. The major G protein engaged by 5-HT$_4$R signaling is $G\alpha_s$, leading to the activation of the cAMP/PKA pathway [19]. The G-protein independent non-canonical pathway activates Src and ERK kinases, leading to pERK1/2 phosphorylation [20].

5-HT$_4$R KO mice develop normally, with no differences in body weight, metabolism, social behavior, or sleep pattern [21]. However, when stressed they exhibit reduced hypophagia [21] and re-expression of 5-HT4R in the medial prefrontal cortex rescues this phenotype [22]. In mice, 5-HT$_4$R was also shown to link appetite and feeding to addiction-related behaviors since 5-HT$_4$R activation in the nucleus accumbens provokes anorexia and hyperactivity, concurrently upregulating a gene induced by cocaine and amphetamine (CART) while knockdown thereof inhibits MDMA-induced hyperactivity [23].

One of the earliest functions attributed to 5-HT$_4$R in rodents is its excitatory effect on acetylcholine release in the frontal cortex and the hippocampus [24,25] which was linked to its role in enhancing memory and cognition [26–29]. For example, a two-week treatment with 5-HT$_4$R partial agonist RS67333 improved memory in the object recognition test in mice [30]. Olfactory associative learning was enhanced by another partial agonist (SL65.0155) in rats [31]. Other paradigms assessing social memory, autoshaping and spatial and place learning, showed a memory enhancing effect of 5-HT$_4$R stimulation [29,32,33]. Conversely, receptor antagonists induced a consistent deficit in (olfactory) associative memory formation [34,35], and weakened passive avoidance memory [36]. Paralleling these behavioral changes are structural plasticity effects of potentiated learning-induced dendritic spine growth in the hippocampus in mice, an effect which is abolished by 5-HT$_4$R inhibition [37].

5-HT$_4$ receptors have also been found to modulate GABA and dopamine release [18,26]. Serotonin depolarizes globus pallidus neurons, increases their firing rate and alters GABA release in a 5-HT$_4$R-dependent manner involving pre- and postsynaptic mechanisms [38]. In guinea pig, the 5-HT$_4$R agonist BIMU-8 increased GABA release from hippocampus indirectly via cholinergic muscarinic receptors [39]. 5-HT$_4$ receptors exert excitatory control on DA release in the striatum, while a receptor antagonist blocks this effect [40]. In freely moving rats, the 5-HT$_4$ antagonist GR125487 significantly reduced the nigrostriatal haloperidol-induced but not basal DA outflow without affecting the mesoaccumbal DA release, indicating that 5-HT$_4$R exerts facilitatory control under activated conditions [41]. This finding is also important in the context of Parkinson's disease where the substantia nigra is selectively vulnerable to degeneration compared to the VTA, leading to a depletion in striatal dopamine. In a rat model of PD, the 5-HT$_4$R agonist prucalopride selectively enhanced L-DOPA-stimulated DA release in the rat SNr and the PFC but not in the hippocampus or the striatum [42].

5-HT$_4$R also impacts global serotonergic tone. 5-HT$_4$R KO mice have diminished tissue levels of 5-HT and its main metabolite, 5-HIAA, increased serotonin transporter (SERT) at the protein and transcript levels, as well as decreased 5-HT$_{1A}$R binding sites [43]. 5-HT$_4$R is a component of a feedback loop projecting from the PFC to the dorsal raphe nuclei (DRN). More specifically, in mice, systemic 5-HT$_4$R stimulation or overexpression of 5-HT$_4$R in the mPFC increased the firing rate of DRN neurons, thus creating a positive feedback PFC-DRN loop involving 5-HT$_4$R activation in cortical projections neurons, glutamate release in the DRN and enhanced DRN firing [44–46].

5-HT$_4$R is a major candidate in mediating antidepressant drug action. As early as 1997, a role of 5-HT$_4$R in anxiety-like behavior was described in rats [47]. More recently, this topic has received more interest, possibly due to the need to identify novel, fast acting antidepressant drugs. Indeed, it was described in rodents that subchronic (3 days) treatment with 5-HT$_4$R agonist yields behavioral as well as biochemical responses in the hippocampus (CREB phosphorylation, neurogenesis) that are comparable to responses to treatment with SSRIs over 3 weeks [48], possibly through its action in the above mentioned PFC-DRN feedback loop [44–46].

These findings clearly indicate that 5-HT$_4$R is a major regulator of the homeostasis of several neurotransmitter systems, implying a role in brain disorders such as Alzheimer's, Huntington's, Parkinson's diseases or Major Depressive Disorder. Our review aims at summarizing the current knowledge of 5-HT$_4$R expression in the brain. We also want to present knowledge on cell-type specific expression, which has not yet been studied extensively, partly due to the lack of immunohistochemistry-competent antibodies as well as resolution limits of binding experiments in brain slices with radioactive antagonists.

2. Promoter Studies and Transcript Variants

Surprisingly little is known about the transcriptional regulation of the *htr4* gene across tissues. The human 5-HT$_4$ receptor gene is located on chromosome 5 (5q31–q33) and contains five exons and eight alternatively spliced cassettes that code for the internal and C-terminal splice variants [16]. Human *htr4* mRNA is transcribed from a very complex gene encompassing 38 exons spanning over 700 kb [18], and multiple C-terminal isoforms are expressed in specific tissues in the CNS. To date, it is not known how 5-HT$_4$ receptor expression is regulated in the brain, and so far we have only partial knowledge about the promoter, derived from human atrial tissue and placenta [16,49]. In the heart, the major transcription start site of the *htr4* gene is located at −3185 bp upstream of the first start codon [16]. In placenta, the 5'-UTR is even longer, spanning over 5100 bp upstream from the translation start site [49]. The different 5'-UTRs upstream of the translation initiation codon are interesting since they may hold an additional key to understand region and cell-type specific regulation of protein expression.

The human promoter lacks TATA and CAAT canonical motifs, but contains several transcription factors binding sites. Transient transfection assays with human 5-HT$_4$ receptor promoter-luciferase constructs identified an approx. 1.2 kb fragment of 5'-non-transcribed sequence as promoter in human cell lines but not monkey COS-7 cells [16] indicating that there is a tissue-specific expression of yet unknown transcription factors. We found in mouse brain that there is a region-specific negative transcriptional regulation of *htr4* exerted by the kinase CK2. Examination of conditional mouse knockouts of CK2 in the hippocampus, striatum and the cortex indicated an upregulation of 5-HT$_4$R mRNA selectively in the cortex [46]. Furthermore, in luciferase assays, using a 4 kb element upstream of the mouse gene fused to luciferase cDNA, expression was promoted when CK2 was inhibited or knocked down in human HeLa cells but not in Hek293 or monkey COS-7 cells, again underlining the importance of tissue-specific transcription factors.

Instead of the TATA box, Maillet et al. described the presence of a sequence in the human gene (TTCACTTT) that can function as a core promoter sequence similarly to the TATA box [16]. For other species, no promoter studies were performed.

There are differential transcription initiation sites in different tissues such as human heart and placenta while for the brain no such data are yet available. While the transcription initiation start site does not affect the protein-coding region, it may alter the transcription efficiency and the expression pattern of 5-HT$_4$R. It is hypothesized that such a long 5'-UTR reduces RNA translation and leads to low levels of expressed transcripts by causing premature initiation at a wrong ATG and preventing the ribosome from reaching the correct start codon [16].

Taken together, in particular in the human brain, there is a lack of data about the 5'-UTR, the promoter and the transcription factors that are active at the promoter for *htr4*.

3. SNPs in Non-Coding Regions

In addition to the 5'-UTRs, isoforms can also vary in the 3'-UTR. These 3'-UTRs are targets for post-transcriptional regulation by non-coding RNAs such as miRNAs. Within the 3'-UTR of the 5-HT$_4$R (b) and (i) isoforms from the GI tract from humans with irritable bowel syndrome (IBS), a single nucleotide polymorphism, termed 5-HT$_4$R (b_2) was found to be predominantly present in a subtype of IBS patients. This isoform lacks two of the three miRNA binding sites for miR-16 family/miR-103/107 and, compared to the full length 5-HT$_4$R (b) isoform, its expression yielded higher 5-HT$_4$R protein levels. It was further shown that miR-16 and miR-103 are responsible for the downregulation of the transcript in vitro which is impaired in the 5-HT$_4$R (b_2) mutant [50].

Another miRNA, *Let-7a*, was also postulated to have the potential to regulate 5-HT$_4$R [51].

Several genome wide association studies (GWAS) and meta-analyses have associated twelve intronic SNPs in the non-coding region of human *htr4* with pulmonary function [52,53]. The same SNPs have been associated with the clinical phenotypes of airflow obstruction and COPD and asthma [53,54]. A SNP in a non-coding region could affect transcriptional regulation or generate a splicing signal. In this context, the pulmonary function of 5-HT$_4$R KO mice was found to exhibit higher baseline lung resistance, confirming a role of 5-HT$_4$R in airway diseases [55]. No mechanistic studies have yet been performed to understand the impact of the described SNP on transcription and splicing.

4. Isoforms and Alternative Splicing

In contrast to promoter-dependent transcriptional initiation sites which will still yield the same transcript but alter expression levels, splicing affects the protein sequence.

Since the first publication in 1995 which described a short and a long isoform, several other isoforms were discovered: There are at least 11 human 5-HT$_4$ receptor splice variants (a–i,n) [18,56–59]. All splice variants differ at the C-terminus with the exception of 5-HT$_4$R (h) which is an internal splice variant with an insert in the 2nd extracellular loop [60], (Figure 1) and the (n) isoform which lacks the C-terminal exon [61].

Human 5-HT$_4$ receptor isoforms (a–i and n) are highly expressed in the central nervous system [18,56,61]. Isoform (b) is the most abundant form in the CNS and periphery, and is expressed in the caudate nucleus, putamen, amygdala, pituitary gland, and small intestine. Isoform (a) is highly expressed in the amygdala, hippocampus, nucleus accumbens, and caudate nucleus and at lower levels in the small intestine, the atrium, and pituitary gland. Isoform (c) is highly expressed in the pituitary gland and small intestine and to a lesser degree in the caudate nucleus, hippocampus, and putamen. Isoform (d) is not present in the CNS but is found in the small intestine [18,61,62]. Isoform (g) seems to be highly expressed in the hypothalamus and cortex [63]. The (n) variant, which lacks the alternatively spliced C-terminal exon, is abundantly expressed in human peripheral tissues and brain regions involved in mood disorders (frontal cortex, hippocampus) [61].

Mice are currently thought to have five [64] and rats four isoforms, with the fourth, (c1) isoform expressed in the gastrointestinal tract [59,63]. In rat brain, no significant difference in expression between the long and short variants has been found by ISH [65]. The C-terminal sequences will determine the baseline activity (with the shorter isoforms being more active) or the ability to recruit binding partners such as β-arrestins and GRKs, sorting nexins or the NHERF PDZ adaptor protein [19,66,67] This will affect internalization kinetics which are different between isoforms [68]. Finally, isoforms can differ in their G protein coupling, since the 5-HT$_4$R (b) isoform can couple via Gα_i as well as Gα_s [69].

To date, no specific isoform has been linked to a brain disorder; however, an interaction cannot be excluded since such studies have not been performed and would be very challenging. Most human studies using PET technology or radioactive labeling are based on ligands which cannot distinguish between isoforms. Quantitative RT-PCR was used to detect different isoforms and their expression in the rodent brain; however, no studies in disease models have employed this approach. The fact that mice or rats do not express the same isoforms than humans suggests that fine tuning of 5-HT$_4$R signaling through a differential expression of longer or short, more active versus less active isoforms, may occur in different species.

```
(a):     STTTINGSTHVLRYTVLHRGHHQELEKLPIHNDPESLESCF
(b):     STTTINGSTHVLRDAVECGGQWESQCHPPATSPLVAAQPSDT
(c)(l):STTTINGSTHVLSSGTETDRKKLWNKEEKIDQTIQMPKRKRKKKASLSYEDLILLGRKSCFREGK
(c)(s):STTTINGSTHVLSSGTETDRRNFGIRKRRLTKPS
(d):     STTTINGSTHVLRF
(e):     STTTINGSTHVLSFPLLFCNRPVPV
(f):     STTTINGSTHVLSPVPV
(g):     STTTINGSTHVLSGCSPVSSFLLLFCNRPVPV
(i):     STTTINGSTHVLRTDFLFDRDILARYWTKPARAGPFSGTLSIRCLTARKPVLGDAVECGGQWESQCHPPATSPLVAAQPSDT
(n):     STTTINGSTHVLR
```

Figure 1. Alignment of C-termini of isoforms found in human tissue: green: leucine 358, the last amino acid common to all variants. For the c isoform, a short and a long one were described. Yellow: S/T cluster necessary for b-arrestin dependent receptor endocytosis.

5. Post-Translational Regulation

5.1. Phosphorylation

The amount of membrane-localized and active GPCR is a result of the ratio between receptor endocytosis and recycling. Endocytosis is initiated through (S/T) phosphorylation of GPCRs in their intracellular domains by G protein-coupled receptor kinases (GRKs) and second messenger kinases such as PKA or PKC [70]. Binding of arrestins to GRK-phosphorylated receptors results in receptor desensitization [71] and internalization [72–76].

Fourteen phosphosites in the 3rd intracellular loop and in the C-terminal tail of 5-HT$_4$R that was heterologously expressed in retinal rod cells of the mouse were identified [77]; however, the identity of the kinases has not been determined. Neither has it been tested whether the phosphorylation of these sites is activity dependent.

In Hek293 cells, it was shown that GRK2 phosphorylates and desensitizes 5-HT$_4$R resulting in downregulation of the cAMP/PKA pathway, while GRK5-mediated 5-HT$_4$R phosphorylation resulted in reduced inhibition of ERK phosphorylation [19,78].

5.2. Palmitoylation

Palmitoylation is a lipid modification in which a cysteine SH group undergoes esterification with a palmitoyl group, generating an anchor to the lipid bilayer of the plasma membrane. This modification is readily reversible and, similar to phosphorylation/dephosphorylation, allows for rapid regulation of protein function, affecting GCPR endocytosis, phosphorylation, desensitization and ultimately cellular signaling. Biochemical studies in insect (Sf9) and mammalian cells (Cos7) showed that several

5-HT receptors (5-HT$_{1a}$R, 5-HT$_{1b}$R, 5-HT$_4$R and 5-HT$_7$R) are palmitoylated in their C-terminal tails. The mouse 5-HT$_4$R (a) variant is palmitoylated at 3 highly conserved cysteine sites and at a C-terminal cysteine that is variant-specific. Palmitoylation near or close to protein-protein interaction motifs will affect the binding properties of the receptor, impact on constitutive activity or internalization via β-arrestin-2 [79,80].

5.3. Glycosylation

Only one study describes two putative N–linked glycosylation sites that conform to the consensus sequence N–X–S/T (X being any amino acid but proline) for glycosylation. These are located on the extracellular side of 5-HT$_4$R, one at the N-terminus and one in the 2nd extracellular loop [77].

What is clearly missing in our understanding of all described post-translational modifications are data generated from physiologically expressed 5-HT$_4$R such as in mouse brain, a comparison between brain regions and an analysis in response to drug treatment or of brain disease models. Finally, the functionality of each of these modifications should be addressed, in particular on their effect on protein stability and receptor homeostasis.

6. Basal Expression

6.1. Transcript Level

5-HT$_4$R transcript expression in rodents mainly stems from in situ hybridization (ISH) experiments: In rat brain slices, ISH probes showed strong expression in the basal ganglia (caudate putamen, ventral striatum], olfactory tubercle, medial habenula and hippocampal formation while none was detected in globus pallidus and substantia nigra [65]. Similarly, human postmortem brains showed highest levels of 5-HT$_4$ receptor mRNA in caudate nucleus, putamen, nucleus accumbens, and the hippocampal formation but none in globus pallidus and substantia nigra [10].

A brain-wide comprehensive appraisal of cell-specific expression is still warranted; however, some evidence has been published: Dual-label in situ hybridization for 5-HT$_4$R and neuronal markers suggests expression in basal forebrain GABAergic parvalbumin synthesizing and glutamatergic cells and in glutamatergic pyramidal neurons in the medial prefrontal cortex and hippocampus of rat and guinea pig (CA1, CA3) [62,81]. 5-HT$_4$R mRNA is present in 60% of rat PFC pyramidal neurons of the frontal cortex as assessed by single cell mRNA/cDNA profiling [82,83].

In rat hippocampal slices, the 5-HT$_4$R agonist, cisapride, leads to increased hippocampal pyramidal cell activity and serotonin release, indirectly indicating that 5-HT$_4$R is expressed in these cells [84].

6.2. Protein Level

Our knowledge on 5-HT$_4$R protein expression stems to a large degree from radioactive ligand binding studies which for the most part has mirrored results of ISH studies. Indeed, a large number of radioligands exist that are specific to 5-HT$_4$R.

High densities of [3H]-GR 113808 or [125I]-SB 207710 binding sites are present in the ventral and dorsal striatum, substantia nigra, globus pallidus and ventral pallidum, interpeduncular nucleus, islands of Calleja, and olfactory tubercle in guinea pig, mouse and rat brain, lower densities are found in the hippocampus, septal region, neocortex, amygdala and colliculi as well as habenular and several thalamic and hypothalamic nuclei [65,85–87]. [125I]-SB 207710 binding in the caudate putamen shows a rostrocaudal and mediolateral increasing gradient of receptor densities, paralleling that observed for mRNA localization [65].

Kainic acid injection into the caudate-putamen of rats to destroy GABAergic striatal projection neurons resulted in a dramatic decrease of radioactive ligand binding, suggesting that 5-HT$_4$R is expressed in these neurons [11]. Similarly, 6-OHDA-lesion of dopaminergic neurons did not lead to a reduction in radioactive ligand binding but only to increased binding in the caudate putamen and globus pallidus. This allows the conclusion that 5-HT$_4$R expression does not occur in DA neurons of the SN [11]. These studies were confirmed by comparing ISH and radioligand labeling data: The presence of mRNA in the rat caudate putamen and its absence in substantia nigra pars compacta and the globus pallidus suggests again that receptors found in binding studies in the caudate putamen and globus pallidus are synthesized by striatonigral and striatopallidal cells [62]. Comparison of mRNA distribution with receptor distribution as visualized with [125I]-SB 207710 further indicates that $5\text{-}HT_4$ receptors are localized somatodendritically (e.g., in caudate putamen) and on axon terminals (e.g., in substantia nigra and globus pallidus) [65,88].

Transgenic Bac-GFP mice where GFP is expressed under the $5\text{-}HT_4R$ promoter are enabling a highly detailed look at protein expression in individual cells and confirm moderate to strong expression in the olfactory bulb, cerebral cortex, subicular cortex, hippocampus, striatum, globus pallidus, midbrain, pons medulla, cerebellum and weak expression in the piriform cortex, basal forebrain and the thalamus (Gensat Founder AU103). Dual immunohistochemical analysis showed expression of GFP in GABAergic spiny projection neurons but not in striatal interneurons [89]. Another transgenic mouse line where the β-galactosidase gene was knocked-in at the *htr4* gene locus shows LacZ localization in mature but not immature granule cells as suggested by staining with the neural marker, NeuN, and calbindin (mature granule cell marker) [90]. Another study confirmed that $5\text{-}HT_4$ receptors are expressed in efferent GABAergic neurons of the nucleus accumbens projecting to the lateral hypothalamus [23].

Species-specific differences of $5\text{-}HT_4R$ protein expression were found between mouse/rat and guinea pig in the globus pallidus, substantia nigra and interpeduncular nucleus [87].

Using [3H]-prucalopride and [3H]-GR116712 or [125I]-SB 207710 in binding studies of human post-mortem brain slices, the highest densities were found in the basal ganglia (caudate nucleus, putamen, nucleus accumbens, globus pallidus, substantia nigra). Moderate to low densities were detected in the hippocampal formation and in the cortical mantle [10]. Additionally, using the labeled antagonist GR 113808, expression in the human amygdala was reported [91]. In the neocortex, the binding showed a distinct lamination pattern with high levels in superficial layers and a band displaying lower levels in deep cortical layers [92]. Membrane binding studies with [3H]-GR 113808 resulted in highest binding in the human caudate nucleus, followed by substantial densities in the lenticular nucleus, the substantia nigra, the hippocampus and frontal cortex, whereas no binding could be detected in the cerebellum [82]

The expression data from all species studied are summarized in Table 1.

7. Changes in Expression in Brain Disorders and Changes Induced by Drug Treatment

In the healthy population there is a baseline difference of $5\text{-}HT_4R$ protein expression between sexes. Women show lower $5\text{-}HT_4R$ binding (by 13%) in the limbic system and the difference was most pronounced in the amygdala, which is highly involved in the processing and memorizing of emotions [93].

Studies using [3H]-GR 113808 in the rat have revealed that during development, prenatal expression is low, with the exception of the brainstem, indicating that $5\text{-}HT_4R$ is largely dispensable in development. Interestingly, the synchronous appearance of $5\text{-}HT_4$ receptors and cholinergic markers validates the notion of $5\text{-}HT_4R$-mediated control over acetylcholine release [94].

With age, $5\text{-}HT_4R$ expression goes down as older humans present lower $5\text{-}HT_4R$ binding [93].

Table 2 assembles data on expression changes in disease or that are pharmacologically induced.

Table 1. Compilation of studies of expression of the 5-HT4 receptor in human, mouse and rat brain in the basal/healthy state.

species	Tissue	Cell Type	Transcript/Protein	Method	Reference
human	caudate nucleus, putamen, nucleus accumbens, globus pallidus, substantia nigra. Lesser densities in hippocampus and cortex		protein	[3H]-R116712 and [3H]-pruclapride binding	Bonaventure et al., 2000, [10]
	caudate nucleus, putamen, nucleus accumbens, and in the hippocampal formation but not in globus pallidus and substantia nigra		mRNA	in situ hybridization	Bonaventure et al., 2000, [10]
	caudate nucleus, putamen, nucleus accumbens, globus pallidus, substantia nigra, hippocampus (CA1, subiculum), neocortex		protein	[125I-]SB 20771 binding	Varnas et al., 2003, [92],
human, calf, guinea pig	caudate nucleus, lenticular nucleus, substantia nigra, hippocampus, frontal cortex		protein	[3H]-GR 113808 binding to membrane preparations	Domenech et al., 1994, [82]
	caudate nucleus, lateral pallidum, putamen, medial pallidum, temporal cortex, hippocampus, amygdala, frontal cortex, cerebellar cortex		protein	[3H]-GR 113808 binding	Reynolds et al., 1995, [95]
rat	islands of Calleja, olfactory tubercle, ventral pallidum, fundus striati, amygdala, habenula and septo-hippocampal system. striatum, substantia nigra (lateralis), interpeduncular nucleus, superior colliculus		protein	[3H]-GR 113808 binding	Waeber et al., 1996, [94]
	caudate putamen, ventral striatum, medial habenula and hippocampus		mRNA	in situ hybridization	Vilaró et al., 1996, [65]
	prefrontal cortex	60% of pryamidal glutamatergic neurons	mRNA, protein	indirect through stimulation	Feng et al., 2001, [80]
	basal forebrain, hippocampus, cortex	GABAergic, glutamatergic and parvalbumin-containing neurons, hippocampal and cortical glutamatergic neurons	mRNA	in situ hybridization	Penas-Cazorla et al., 2015, [78]
rat, guinea pig	striatum, globus pallidaus, hippocampus, substantia nigra, olfactory tubercle		protein	[3H]-GR 113808 binding	Grossman et al., 1993, [82]
	striatum, hippocampus	striatal GABAergic projection neurons, projection from dentate granule cells to field CA3, habenulo-interpeduncular pathway, somatodendritically and axonally	mRNA, protein	[125I]-SB 207710 binding, in situ hybridization	Vilaro et al., 2005, [62]
mouse	striatum	GABAergic projection neurons but not dopaminergic neurons	protein	Kainic acid lesions and [3H]-GR 113808 binding	Compan et al., 1996, [11]
	striatum	GABAergic projection neurons	protein	immunohistochemistry	Egeland et al., 2011, [89]
	dentate gyrus	mature granule cells	protein	LacZ-IR staining	Imoto et al., 2015, [90]

Table 2. Compilation of studies of expression of the 5-HT4 receptor in human, mouse and rat brain in the disease state, in disease models or after drug treatment.

Species	Tissue	Condition/Treatment	Direction of Change	Transcript/Protein	Method	Reference
human	frontal cortex, caudate nucleus	suicide victims	up	protein	antagonist binding	Rosel et al., 2004, [96]
		association with bipolar disorder		SNPs	sequencing of PCR products	Ohtsuki et al.,2002, [97]
	hippocampus, frontal cortex	Alzheimer's disease	down	protein	[3H]-GR113808 binding	Reynolds et al.,1995, [95]
	putamen	Huntington's disease	down	protein	[3H]-GR113808 binding	Reynolds et al., 1995, [95]
	hippocampus	cognition, episodic memory, recall	negative correlation	protein	PET, [11C]-SB207145	Haahr et al., 2013, [98]
	nucleus accumbens, ventral pallidum, orbitofrontal cortex,hippocampus	body mass index, obesity	positive correlation	protein	PET, [11C]-SB207145	Haahr et al., 2012, [99]
rat	striatum, subthalamic nucleus, hippocampus	lesion of serotonergic nuclei	up in rostral dorsal, ventral striatum, substantia nigra, hippocampus	protein	[3H]-GR113808 binding	Compan et al., 1996, [11]
	striatum (caudate putamen, globus pallidus	lesion of DA neurons	up	protein	[3H]-GR113808 binding	Compan et al., 1996, [11]
	hippocampus, lateral globus pallidus	Flinders sensitive line (depression model)	down	protein	[3H]-SB207145 binding	Licht et al., 2009, [100]
	hippocampus, hypothalamus, caudate putamen, nucleus accumbens, Globus pallidus	21 days paroxetine (SSRI)	down after SSRI	protein	[3H]-SB207145 binding	Licht et al., 2009, [100]
	hippocampus, hypothalamus	4 days of 5-HT depletion	up after 5-HT depletion	protein	[3H]-SB207145 binding	Licht et al., 2009, [100]
	hippocampus (CA1), striatum	21 days of fluoxetine (SSRI)	down		[3H]-GR113808 binding	Vidal et al., 2009, [101]
	15 regions incl.hippocampus	learning: autoshaping test for food retrieval	upregulated in most regions	protein	[3H]-GR113808 binding	Manuel-Apolinar et al., 2005, [102]
	caudate putamen, nucleus accumbens	rat models of obesity	up	protein	[3H]SB207145 binding	Ratner et al., 2012, [103]
	hippocampus	maternal deprivation, unpredictable stress	down	mRNA, protein	qPCR and Western botting	Bai et al., 2014, [51]
mouse	striatum	6-OHDA lesion model of Parkinson's disease	down in D2 MSNs	mRNA	Affymetrix GeneChip microarray	Heiman et al., 2014, [104]
	ventral hippocampus	bulbectomy	up in ventral hippocampus, down in olfactory tubercles	protein	[3H]-SB207145 binding	Licht et al., 2010, [97]
	caudal putamen	GR (+/−) mice	up	protein	[3H]-SB207145 binding	Licht et al., 2010, [97]
	midbrain raphe nuclei and VTA	social defeat	up after defeat	mRNA	RNA seq	Kudryavtseva et al., 2017, [105]
	prefrontal cortex	restraint stress	up after restraint	mRNA	qPCR	Jean et al., 2017, [22]

7.1. Depression and Anxiety

The understanding of the roles that 5-HT$_4$ receptors play in mood disorders mainly comes from preclinical studies. Several rodent models of depression and anxiety, such as bulbectomy, glucocorticoid receptor heterozygous mice, social defeat stress or exposure to prenatal stress, all indicated changes in 5-HT$_4$R expression: In mice, the experience of social defeat led to 5-HT$_4$R mRNA up-regulation in the midbrain raphe nuclei and the VTA, as determined by RNA seq [105]. Similarly, restraint stress induced hypophagia and increased 5-HT$_4$R mRNA levels in the medial prefrontal cortex [22]. In contrast, maternal stress led to a reduction of all mouse 5-HT$_4$R variants on the mRNA level as assessed by qPCR, with the strongest difference observed for the (b) variant, while chemically induced 5-HT depletion in the embryo only affected the expression of the (b) variant in the embryonic telencephalon [106].

After bulbectomy, 5-HT$_4$R protein binding was increased in the rat ventral hippocampus and olfactory tubercles but unchanged in the dorsal hippocampus, frontal and caudal caudate putamen. 5-HT transporter (SERT) binding was unchanged in the hippocampus and caudate putamen and slightly down in lateral septum and globus pallidus [97]. GR(+/−) mice had increased 5-HT$_4$R binding in the caudal caudate putamen and the olfactory tubercles, decreased SERT binding in the frontal caudate putamen but no changes for 5-HT$_4$R and SERT in the hippocampus [97]. In contrast, in the Flinders Sensitive Line, a rat model of depression, 5-HT$_4$R binding was decreased in the dorsal and ventral hippocampus [100].

A 3-week long treatment regimen with the SSRI fluoxetine decreased the density of 5-HT$_4$ receptor binding in the CA1 field of hippocampus as well as in several areas of the striatum in rats [101]. In contrast, 5-HT$_4$R in layer 5 of the cerebral cortex was shown to be selectively upregulated after fluoxetine treatment in p11-GFP bacTRAP mice [107]. Interestingly, when 5-HT$_4$R expression was quantified by qPCR on whole cortical lysate no difference in response to fluoxetine treatment was detected, while a 16-fold upregulation in the deep cortical layers was found after TRAP purification. This study clearly demonstrates that methods of purification and enrichment are necessary to achieve a resolution that is sufficient to characterize the dynamics of 5-HT$_4$R expression. Given that chronic fluoxetine in mice lead to a specific upregulation in layer 5 of the cortex [107], it is clear that research into expression changes needs to be approached with techniques achieving high resolution since global expression changes might be counterweighed by cell-type and subregion-specific compensatory changes.

Data generated in humans with [^{11}C]-SB 207145 brain PET imaging suggest that 5-HT$_4$R is involved in the neurobiological mechanism underlying familial risk for depression, and that lower striatal but not cortical 5-HT$_4$ receptor binding is associated with an increased risk for developing major depressive disorder [108]. However, in the caudate nucleus, the relationship between 5-HT$_4$R and suicide risk was inverse: Postmortem studies found increased 5-HT$_4$ receptor binding in the caudate nucleus and frontal cortex of depressed suicide victims [96]. Polymorphisms of the *htr4* gene were found to correlate with major depression and/or bipolar disorders [109].

A PET study showed a global reduction in cerebral 5-HT$_4$R binding in healthy volunteers after a 3 week treatment with fluoxetine [110], pointing towards an inverse correlation of global 5-HT$_4$R binding and synaptic serotonin levels, or an activity-induced downregulation response.

In summary, there is strong evidence regarding the involvement of 5-HT$_4$R in the etiology and expression of depression; however, different preclinical models of depression and anxiety and binding studies in humans show different responses in 5-HT$_4$R expression in different brain regions that need to be further addressed.

7.2. Food Intake and Obesity

High levels of 5-HT$_4$R are observed in obese humans [99] and in overfed rats in the caudate putamen and the nucleus accumbens shell [103]. Injection of 5-HT$_4$R agonist into the nucleus accumbens reduces the drive to eat while injection of 5-HT$_4$R antagonist or knockdown in the nucleus accumbens induces hyperphagia in fed mice [111]. These data suggest that changes in 5-HT$_4$R expression may play a role in eating disorders. Indeed, PET studies showed a correlation between the body mass index and 5-HT$_4$R protein in the nucleus accumbens, ventral pallidum, the orbitofrontal cortex and hippocampus [99]. Furthermore, the density of 5-HT$_4$ receptors was found to be decreased in the temporal cortex of Alzheimer's disease patients who also suffer from hyperphagia [112].

7.3. Memory and Alzheimer's Disease

A role for 5-HT$_4$R in Alzheimer's disease has been described: The receptor was linked to APP processing and β-amyloid generation in rodent models of Alzheimer's disease. Chronic administration of 5-HT$_4$R agonists reduced β-amyloid pathology through the promotion of non-amyloidogenic cleavage of the precursor of Aβ and the consequent promotion of the neurotrophic protein, sAPPα, thereby alleviating AD pathology as well as reducing plaque load [113,114]. In a transgenic Alzheimer's mouse model, stimulation of 5-HT$_4$R exerted pro-cognitive effects, which resulted in enhanced learning through increasing acetylcholine levels [24,113,115,116]. This body of work is largely based on the use of 5-HT$_4$R pharmacological tools and shows that 5-HT$_4$R stimulation enhanced performance on memory tasks in rodents while receptor antagonists induced worsening of the performance on these tasks.

During memory consolidation in a food retrieval learning paradigm, 5-HT$_4$ radioligand binding showed an upregulation in olfactory lobule, caudate putamen, fundus striatum, hippocampus (CA2) and several cortical regions of young adult animals. In contrast, some but not all tested regions of older rats (hippocampal CA2 and CA3 areas, and frontal, parietal, and temporal cortex) expressed reduced 5-HT$_4$ receptor density [102] pointing towards age-dependent regulation of 5-HT$_4$R expression.

In humans, PET studies with [11C]-SB207145 as tracer and an episodic memory verbal learning test, resulted in an unexpected negative correlation of 5-HT$_4$R and memory function in healthy young volunteers. Thus, in humans, unlike what was hypothesized based on rodent studies, fewer hippocampal 5-HT$_4$Rs are representative of a better episodic memory function [98]. In newly diagnosed Alzheimer's disease patients, 5-HT$_4$R binding was positively correlated to β-amyloid burden and negatively to cognitive performance (MMSE score) suggesting that cerebral 5-HT$_4$R is upregulated during preclinical stage, possibly as compensatory effect to decreased levels of interstitial 5-HT [117].

No preclinical studies exist to date that show changes in 5-HT$_4$R expression in mouse models of Alzheimer's disease. In humans, [^{3}H]-GR 113808 labeling of post mortem brain tissue showed decreased 5-HT$_4$-receptor expression in the hippocampus and prefrontal cortex in patients with Alzheimer's disease [95]. However, another study contradicts these findings revealing no changes in 5-HT$_4$R density in Alzheimer's disease in frontal and temporal cortices [118].

Thus, to corroborate the relation between 5-HT$_4$R expression and memory function in humans, in healthy and disease states, further studies are warranted.

7.4. Schizophrenia

Limited evidence indicates that 5-HT$_4$R polymorphisms could predispose to schizophrenia [119] and attention deficit hyperactivity disorder (ADHD) [120].

7.5. Parkinson's Disease

Expression of 5-HT$_4$R was found to be altered in rodent models of PD. Depletion of dopamine neurons by 6-OHDA leads to increased 5-HT$_4$R receptor binding in the caudal caudate-putamen and globus pallidus (+93%) [11]. In contrast, in 6-OHDA lesioned mice, 5-HT$_4$R mRNA was reduced (4-fold) while L-DOPA treatment doubled the 5-HT$_4$R expression in the D2-SPNs. In D1-SPNs, changes only occurred after L-DOPA treatment (2-fold) [104]. For technical reasons in this study, no comparison of the total expression levels in D1- and D2-SPNs could be made. However, these findings are very interesting since they suggest a potential role for 5-HT$_4$R in L-DOPA induced dyskinesia. In post-mortem studies of PD subjects, 5-HT$_4$R binding in putamen and substantia nigra was found to be unaltered [91].

The small number of patients ($N = 6$), and the non-discrimination of medication, treatment duration and disease severity does, in our opinion, not allow a conclusive statement.

Future work involving spatially restricted deletions of 5-HT$_4$ receptors or local administration of pharmacological ligands is necessary to more precisely determine the cellular and circuit-based mechanisms by which 5-HT$_4$ receptors influence behavior.

8. Other Proteins Affecting 5-HT4R Signaling

8.1. SERT (5-HTT)

It is not surprising that genetic alteration of the serotonin transporter gene (5-HTT) has implications in mood disorders: For example, mice overexpressing SERT (OE) or with SERT depletion (KO) present anxiolytic-like or more anxious behaviors, respectively, when compared to WT littermates [121,122]. At the molecular level, in the homozygous SERT KO mice, the activity of the 5-HT$_{1A}$ autoreceptor is decreased [123,124] while 5-HT$_{2A}$ receptor function is enhanced [125–127]. Protein levels of 5-HT$_4$R are altered in the SERT KO and SERT OE mice. Precisely, autoradiography studies with [3H]-SB 207145 radioligand show increased 5-HT$_4$ receptor binding in the SERT OE mice in all brain regions but the amygdala. Inversely, in the SERT KO mice, 5-HT$_4$R binding is decreased in all regions studied. This is consistent with studies providing evidence that chronic treatment with SSRIs in healthy individuals decreased 5-HT4R binding as seen in PET imaging [110]. Studies in rodents replicate this result of decreased 5-HT$_4$R-dependent activation of adenylate cyclase and reduced electrophysiological activity in the hippocampus [128]. In a similar fashion, mice overexpressing 5-HT$_4$R in the mPFC exhibit stress-induced hypophagia and a corresponding 5-HT$_4$R-dependent downregulation of SERT and 5-HT$_{1A}$ transcripts. Oppositely, siRNA mediated knockdown of 5-HT$_4$R in the mPFC induces hyperphagia [22].

These studies are important because they highlight that altered 5-HT concentration is most likely responsible for changes in 5-HT$_4$R receptor binding as a compensatory mechanism; they also highlight the bi-directionality of this process, since exogenous alterations in 5-HT$_4$R levels induce changes in 5-HT availability, negatively regulating the expression of SERT as well as serotonin receptors.

8.2. Adaptor Protein p11

S100 calcium effector protein p11 (S100A10), a depression marker protein, has been identified in a yeast-based screening system as a binding partner to 5-HT$_4$R, with greater affinity to 5-HT$_4$R than to other serotonin receptors, such as 5-HT$_{1B}$ and 5-HT$_{1D}$ receptors [129]. p11 co-localizes with 5-HT$_4$R in brain regions that play an important role in major depressive disorder like cingulate cortex, hippocampus, amygdala and striatum as seen by in situ hybridization and immunohistochemistry using the transgenic bac-GFP mice where GFP is expressed under the 5-HT$_4$R promoter. p11 KO mice show reduced 5-HT$_4$R protein in radioligand binding assays, are behaviorally less sensitive to antidepressant treatment and do not respond to 5-HT$_4$R agonist. As binding partner of 5-HT$_4$R and adaptor protein for many other GPCRs, p11 recruits 5-HT$_4$R to the site of its action, the plasma membrane [129].

8.3. CK2

CK2 is a constitutively active and ubiquitously expressed kinase. Recently, CK2 has been identified as a negative regulator of the 5-HT$_4$R [46]. Knockdown or inhibition of CK2 in vitro elevates 5-HT$_4$R receptor-dependent cAMP generation and increases receptor localization at the plasma membrane in monkey COS7 cells. Interestingly, in the mouse brain, mRNA upregulation of the 5-HT$_4$R is specific to the PFC. Virally-mediated focal knockdown of CK2 or overexpression of 5-HT$_4$R in the mPFC generates an anti-depressed and anxiolytic-like phenotype that is similar to the phenotypes observed with CK2 knockout in the forebrain driven by Emx1-Cre or Drd1a-Cre. In addition, such conditional CK2 KO mice are more responsive to antidepressant drugs and 5-HT$_4$R agonist (RS 67333) treatment [46].

8.4. Testosterone

Several studies describe the relationship between sex hormones and serotonin in mood-related disorders. The prevalence of major depressive disorder is 1.7 times higher in women than in men [130]. Several studies correlate depressive episodes with hormonal changes especially in the menstrual cycle in women although the exact mechanism by which this happens is not clear [130]. In men, it has been found that plasma testosterone negatively correlates with brain 5-HT$_4$R binding in humans throughout the brain [131]. Higher levels of testosterone lead to increased serotonergic signaling but whether testosterone directly regulates levels through steroid hormone receptors co-localized with 5-HT$_4$R or by an indirect mechanism (e.g., increased of serotonergic tonus through other targets) to decrease expression of 5-HT$_4$R needs to be further examined.

8.5. Nav1.7

Nav1.7 is a voltage-gated sodium channel required for nociceptive neuronal activation. While humans lacking Nav1.7 and genetic KO mice show absence of pain, a pharmacological antagonist of this channel failed to decrease pain sensitivity, indicating that receptor signaling mediated activation of nociceptive neurons might not be the only mechanism involved in pain alleviation. For example, loss of Nav1.7 coincides with upregulation of met-enkephalin, an endogenous opioid peptide in sensory neurons, increasing opioid activity and anti-nociceptive signaling. In addition, Nav 1.7 KO mice present reduced levels of 5-HT$_4$R in dorsal root ganglia [132]. Both effects, i.e., changes in enkephalin and 5-HT$_4$R expression and signaling, take place in peripheral nociceptive neurons and together contribute to the analgesic effect [133].

9. Conclusions

It is clear that changes in 5-HT$_4$R expression correlate with several disease states. In order to clarify whether these changes are also causative or involved in the etiology of disease, the expression needs to be assessed on a cellular level in preclinical models. While 5-HT$_4$R overexpression in rodents, for example, through virus injection, is truly helpful in delineating the role of 5-HT$_4$R in certain brain regions and cell types, these experiments have the disadvantage of introducing the gene under an exogenous promoter thus leading to non-physiological levels of expression and lacking the opportunity to study transcriptional regulation. Thus, it is preferable to study transgenic mice in which a labeled version of the receptor is expressed under its endogenous promoter such as the transgenic mouse line where the β-galactosidase gene is knocked in at the *htr4* gene locus [90], enabling unambiguous cell identification or cell-type specific purifications and quantification methods. Human PET or post mortem studies are important to verify hypotheses but may not allow the resolution needed.

Another aspect that has to be taken into consideration is the fact that splicing variants differ between species. The factors responsible for these differences are unknown but may be important in understanding human pathologies. To bridge this knowledge gap, it would be interesting to generate, through streamlining the gene architecture by engineering/deleting of splicing sites, mice which expressing specific (human) variants only and to determine whether this will affect 5-HT$_4$R-dependent phenotypes (e.g., electrophysiological properties, neurotransmitter release, receptor homeostasis, behavior and biochemical signaling cascades). Once this has been established, we will be in a better position to develop more suitable 5-HT$_4$R mouse models to study human disease.

Conflicts of Interest: The authors declare no conflict of interest.

References

1. Palacios, J.M. Serotonin receptors in brain revisited. *Brain Res* **1645**. [CrossRef] [PubMed]
2. Riad, M.; Garcia, S.; Watkins, K.C.; Jodoin, N.; Doucet, E.; Langlois, X.; el Mestikawy, S.; Hamon, M.; Descarries, L. Somatodendritic localization of 5-ht1a and preterminal axonal localization of 5-ht1b serotonin receptors in adult rat brain. *J. Comp. Neurol.* **2000**, *417*, 181–194. [CrossRef]

3. Lerer, B.; Macciardi, F.; Segman, R.H.; Adolfsson, R.; Blackwood, D.; Blairy, S.; Del Favero, J.; Dikeos, D.G.; Kaneva, R.; Lilli, R.; et al. Variability of 5-ht2c receptor cys23ser polymorphism among european populations and vulnerability to affective disorder. *Mol. Psychiatry* **2001**, *6*, 579–585. [CrossRef] [PubMed]

4. Tecott, L.H.; Sun, L.M.; Akana, S.F.; Strack, A.M.; Lowenstein, D.H.; Dallman, M.F.; Julius, D. Eating disorder and epilepsy in mice lacking 5-ht2c serotonin receptors. *Nature* **1995**, *374*, 542–546. [CrossRef] [PubMed]

5. Vollenweider, F.X.; Kometer, M. The neurobiology of psychedelic drugs: Implications for the treatment of mood disorders. *Nat. Rev. Neurosci.* **2010**, *11*, 642–651. [CrossRef] [PubMed]

6. Hannon, J.; Hoyer, D. Molecular biology of 5-ht receptors. *Behav. Brain Res.* **2008**, *195*, 198–213. [CrossRef] [PubMed]

7. Dumuis, A.; Bouhelal, R.; Sebben, M.; Cory, R.; Bockaert, J. A nonclassical 5-hydroxytryptamine receptor positively coupled with adenylate cyclase in the central nervous system. *Mol. Pharmacol.* **1988**, *34*, 880–887. [PubMed]

8. Gerald, C.; Adham, N.; Kao, H.T.; Olsen, M.A.; Laz, T.M.; Schechter, L.E.; Bard, J.A.; Vaysse, P.J.; Hartig, P.R.; Branchek, T.A.; et al. The 5-ht4 receptor: Molecular cloning and pharmacological characterization of two splice variants. *EMBO J.* **1995**, *14*, 2806–2815. [CrossRef] [PubMed]

9. Claeysen, S.; Sebben, M.; Journot, L.; Bockaert, J.; Dumuis, A. Cloning, expression and pharmacology of the mouse 5-ht(4l) receptor. *FEBS Lett.* **1996**, *398*, 19–25. [CrossRef]

10. Bonaventure, P.; Hall, H.; Gommeren, W.; Cras, P.; Langlois, X.; Jurzak, M.; Leysen, J.E. Mapping of serotonin 5-ht(4) receptor mrna and ligand binding sites in the post-mortem human brain. *Synapse* **2000**, *36*, 35–46. [CrossRef]

11. Compan, V.; Daszuta, A.; Salin, P.; Sebben, M.; Bockaert, J.; Dumuis, A. Lesion study of the distribution of serotonin 5-ht4 receptors in rat basal ganglia and hippocampus. *Eur. J. Neurosci.* **1996**, *8*, 2591–2598. [CrossRef] [PubMed]

12. Tonini, M.; Pace, F. Drugs acting on serotonin receptors for the treatment of functional gi disorders. *Dig. Dis.* **2006**, *24*, 59–69. [CrossRef] [PubMed]

13. Hegde, S.S.; Wong, A.G.; Perry, M.R.; Ku, P.; Moy, T.M.; Loeb, M.; Eglen, R.M. 5-ht4 receptor mediated stimulation of gastric emptying in rats. *Naunyn-Schmiedebergs Arch. Pharmacol.* **1995**, *351*, 589–595. [CrossRef] [PubMed]

14. Callahan, M.J. Irritable bowel syndrome neuropharmacology. A review of approved and investigational compounds. *J. Clin. Gastroenterol.* **2002**, *35*, S58–S67. [CrossRef] [PubMed]

15. Qvigstad, E.; Brattelid, T.; Sjaastad, I.; Andressen, K.W.; Krobert, K.A.; Birkeland, J.A.; Sejersted, O.M.; Kaumann, A.J.; Skomedal, T.; Osnes, J.B.; et al. Appearance of a ventricular 5-ht4 receptor-mediated inotropic response to serotonin in heart failure. *Cardiovasc. Res.* **2005**, *65*, 869–878. [CrossRef] [PubMed]

16. Maillet, M.; Gastineau, M.; Bochet, P.; Asselin-Labat, M.L.; Morel, E.; Laverriere, J.N.; Lompre, A.M.; Fischmeister, R.; Lezoualc'h, F. Functional studies of the 5′-untranslated region of human 5-ht4 receptor mrna. *Biochem. J.* **2005**, *387*, 463–471. [CrossRef] [PubMed]

17. Cartier, D.; Lihrmann, I.; Parmentier, F.; Bastard, C.; Bertherat, J.; Caron, P.; Kuhn, J.M.; Lacroix, A.; Tabarin, A.; Young, J.; et al. Overexpression of serotonin4 receptors in cisapride-responsive adrenocorticotropin-independent bilateral macronodular adrenal hyperplasia causing cushing's syndrome. *J. Clin. Endocrinol. Metab.* **2003**, *88*, 248–254. [CrossRef] [PubMed]

18. Bockaert, J.; Claeysen, S.; Compan, V.; Dumuis, A. 5-ht4 receptors. *Curr. Drug Targets CNS Neurol. Disord.* **2004**, *3*, 39–51. [CrossRef] [PubMed]

19. Barthet, G.; Gaven, F.; Framery, B.; Shinjo, K.; Nakamura, T.; Claeysen, S.; Bockaert, J.; Dumuis, A. Uncoupling and endocytosis of 5-hydroxytryptamine 4 receptors. Distinct molecular events with different grk2 requirements. *J. Biol. Chem.* **2005**, *280*, 27924–27934. [CrossRef] [PubMed]

20. Barthet, G.; Framery, B.; Gaven, F.; Pellissier, L.; Reiter, E.; Claeysen, S.; Bockaert, J.; Dumuis, A. 5-hydroxytryptamine 4 receptor activation of the extracellular signal-regulated kinase pathway depends on src activation but not on g protein or beta-arrestin signaling. *Mol. Biol. Cell* **2007**, *18*, 1979–1991. [CrossRef] [PubMed]

21. Compan, V.; Zhou, M.; Grailhe, R.; Gazzara, R.A.; Martin, R.; Gingrich, J.; Dumuis, A.; Brunner, D.; Bockaert, J.; Hen, R. Attenuated response to stress and novelty and hypersensitivity to seizures in 5-ht4 receptor knock-out mice. *J. Neurosci.* **2004**, *24*, 412–419. [CrossRef] [PubMed]

22. Jean, A.; Laurent, L.; Delaunay, S.; Doly, S.; Dusticier, N.; Linden, D.; Neve, R.; Maroteaux, L.; Nieoullon, A.; Compan, V. Adaptive control of dorsal raphe by 5-ht4 in the prefrontal cortex prevents persistent hypophagia following stress. *Cell Rep.* **2017**, *21*, 901–909. [CrossRef] [PubMed]

23. Jean, A.; Laurent, L.; Bockaert, J.; Charnay, Y.; Dusticier, N.; Nieoullon, A.; Barrot, M.; Neve, R.; Compan, V. The nucleus accumbens 5-htr(4)-cart pathway ties anorexia to hyperactivity. *Transl. Psychiatry* **2012**, *2*, e203. [CrossRef] [PubMed]

24. Consolo, S.; Arnaboldi, S.; Giorgi, S.; Russi, G.; Ladinsky, H. 5-ht4 receptor stimulation facilitates acetylcholine release in rat frontal cortex. *Neuroreport* **1994**, *5*, 1230–1232. [CrossRef] [PubMed]

25. Siniscalchi, A.; Badini, I.; Beani, L.; Bianchi, C. 5-ht4 receptor modulation of acetylcholine outflow in guinea pig brain slices. *Neuroreport* **1999**, *10*, 547–551. [CrossRef] [PubMed]

26. Bijak, M.; Misgeld, U. Effects of serotonin through serotonin1a and serotonin4 receptors on inhibition in the guinea-pig dentate gyrus in vitro. *Neuroscience* **1997**, *78*, 1017–1026. [CrossRef]

27. Meneses, A.; Hong, E. Effects of 5-ht4 receptor agonists and antagonists in learning. *Pharmacol. Biochem. Behav.* **1997**, *56*, 347–351. [CrossRef]

28. Fontana, D.J.; Daniels, S.E.; Wong, E.H.; Clark, R.D.; Eglen, R.M. The effects of novel, selective 5-hydroxytryptamine (5-ht)4 receptor ligands in rat spatial navigation. *Neuropharmacology* **1997**, *36*, 689–696. [CrossRef]

29. Mohler, E.G.; Shacham, S.; Noiman, S.; Lezoualc'h, F.; Robert, S.; Gastineau, M.; Rutkowski, J.; Marantz, Y.; Dumuis, A.; Bockaert, J.; et al. Vrx-03011, a novel 5-ht4 agonist, enhances memory and hippocampal acetylcholine efflux. *Neuropharmacology* **2007**, *53*, 563–573. [CrossRef] [PubMed]

30. Quiedeville, A.; Boulouard, M.; Hamidouche, K.; Da Silva Costa-Aze, V.; Nee, G.; Rochais, C.; Dallemagne, P.; Fabis, F.; Freret, T.; Bouet, V. Chronic activation of 5-ht4 receptors or blockade of 5-ht6 receptors improve memory performances. *Behav. Brain Res.* **2015**, *293*, 10–17. [CrossRef] [PubMed]

31. Marchetti, E.; Jacquet, M.; Escoffier, G.; Miglioratti, M.; Dumuis, A.; Bockaert, J.; Roman, F.S. Enhancement of reference memory in aged rats by specific activation of 5-ht(4) receptors using an olfactory associative discrimination task. *Brain Res.* **2011**, *1405*, 49–56. [CrossRef] [PubMed]

32. Letty, S.; Child, R.; Dumuis, A.; Pantaloni, A.; Bockaert, J.; Rondouin, G. 5-ht4 receptors improve social olfactory memory in the rat. *Neuropharmacology* **1997**, *36*, 681–687. [CrossRef]

33. Lelong, V.; Dauphin, F.; Boulouard, M. Rs 67333 and d-cycloserine accelerate learning acquisition in the rat. *Neuropharmacology* **2001**, *41*, 517–522. [CrossRef]

34. Lamirault, L.; Simon, H. Enhancement of place and object recognition memory in young adult and old rats by rs 67333, a partial agonist of 5-ht4 receptors. *Neuropharmacology* **2001**, *41*, 844–853. [CrossRef]

35. Marchetti, E.; Dumuis, A.; Bockaert, J.; Soumireu-Mourat, B.; Roman, F.S. Differential modulation of the 5-ht(4) receptor agonists and antagonist on rat learning and memory. *Neuropharmacology* **2000**, *39*, 2017–2027. [CrossRef]

36. Galeotti, N.; Ghelardini, C.; Bartolini, A. Role of 5-ht4 receptors in the mouse passive avoidance test. *J. Pharmacol. Exp. Ther.* **1998**, *286*, 1115–1121. [PubMed]

37. Restivo, L.; Roman, F.; Dumuis, A.; Bockaert, J.; Marchetti, E.; Ammassari-Teule, M. The promnesic effect of g-protein-coupled 5-ht4 receptors activation is mediated by a potentiation of learning-induced spine growth in the mouse hippocampus. *Neuropsychopharmacology* **2008**, *33*, 2427–2434. [CrossRef] [PubMed]

38. Chen, L.; Yung, K.K.; Chan, Y.S.; Yung, W.H. 5-ht excites globus pallidus neurons by multiple receptor mechanisms. *Neuroscience* **2008**, *151*, 439–451. [CrossRef] [PubMed]

39. Bianchi, C.; Rodi, D.; Marino, S.; Beani, L.; Siniscalchi, A. Dual effects of 5-ht4 receptor activation on gaba release from guinea pig hippocampal slices. *Neuroreport* **2002**, *13*, 2177–2180. [CrossRef] [PubMed]

40. Bonhomme, N.; De Deurwaerdere, P.; Le Moal, M.; Spampinato, U. Evidence for 5-ht4 receptor subtype involvement in the enhancement of striatal dopamine release induced by serotonin: A microdialysis study in the halothane-anesthetized rat. *Neuropharmacology* **1995**, *34*, 269–279. [CrossRef]

41. Lucas, G.; Di Matteo, V.; De Deurwaerdere, P.; Porras, G.; Martin-Ruiz, R.; Artigas, F.; Esposito, E.; Spampinato, U. Neurochemical and electrophysiological evidence that 5-ht4 receptors exert a state-dependent facilitatory control in vivo on nigrostriatal, but not mesoaccumbal, dopaminergic function. *Eur. J. Neurosci.* **2001**, *13*, 889–898. [CrossRef] [PubMed]

42. Navailles, S.; Di Giovanni, G.; De Deurwaerdere, P. The 5-ht4 agonist prucalopride stimulates l-dopa-induced dopamine release in restricted brain regions of the hemiparkinsonian rat in vivo. *CNS Neurosci. Ther.* **2015**, *21*, 745–747. [CrossRef] [PubMed]

43. Conductier, G.; Dusticier, N.; Lucas, G.; Cote, F.; Debonnel, G.; Daszuta, A.; Dumuis, A.; Nieoullon, A.; Hen, R.; Bockaert, J.; et al. Adaptive changes in serotonin neurons of the raphe nuclei in 5-ht(4) receptor knock-out mouse. *Eur. J. Neurosci.* **2006**, *24*, 1053–1062. [CrossRef] [PubMed]

44. Lucas, G.; Compan, V.; Charnay, Y.; Neve, R.L.; Nestler, E.J.; Bockaert, J.; Barrot, M.; Debonnel, G. Frontocortical 5-ht4 receptors exert positive feedback on serotonergic activity: Viral transfections, subacute and chronic treatments with 5-ht4 agonists. *Biol. Psychiatry* **2005**, *57*, 918–925. [CrossRef] [PubMed]

45. Covington, H.E., 3rd; Lobo, M.K.; Maze, I.; Vialou, V.; Hyman, J.M.; Zaman, S.; LaPlant, Q.; Mouzon, E.; Ghose, S.; Tamminga, C.A.; et al. Antidepressant effect of optogenetic stimulation of the medial prefrontal cortex. *J. Neurosci.* **2010**, *30*, 16082–16090. [CrossRef] [PubMed]

46. Castello, J.; LeFrancois, B.; Flajolet, M.; Greengard, P.; Friedman, E.; Rebholz, H. Ck2 regulates 5-ht4 receptor signaling and modulates depressive-like behavior. *Mol. Psychiatry* **2018**, *23*, 872–882. [CrossRef] [PubMed]

47. Kennett, G.A.; Bright, F.; Trail, B.; Blackburn, T.P.; Sanger, G.J. Anxiolytic-like actions of the selective 5-ht4 receptor antagonists sb 204070a and sb 207266a in rats. *Neuropharmacology* **1997**, *36*, 707–712. [CrossRef]

48. Lucas, G.; Rymar, V.V.; Du, J.; Mnie-Filali, O.; Bisgaard, C.; Manta, S.; Lambas-Senas, L.; Wiborg, O.; Haddjeri, N.; Pineyro, G.; et al. Serotonin(4) (5-ht(4)) receptor agonists are putative antidepressants with a rapid onset of action. *Neuron* **2007**, *55*, 712–725. [CrossRef] [PubMed]

49. Hiroi, T.; Hayashi-Kobayashi, N.; Nagumo, S.; Ino, M.; Okawa, Y.; Aoba, A.; Matsui, H. Identification and characterization of the human serotonin-4 receptor gene promoter. *Biochem. Biophys. Res. Commun.* **2001**, *289*, 337–344. [CrossRef] [PubMed]

50. Wohlfarth, C.; Schmitteckert, S.; Hartle, J.D.; Houghton, L.A.; Dweep, H.; Fortea, M.; Assadi, G.; Braun, A.; Mederer, T.; Pohner, S.; et al. Mir-16 and mir-103 impact 5-ht4 receptor signalling and correlate with symptom profile in irritable bowel syndrome. *Sci. Rep.* **2017**, *7*, 14680. [CrossRef] [PubMed]

51. Bai, M.; Zhu, X.Z.; Zhang, Y.; Zhang, S.; Zhang, L.; Xue, L.; Zhong, M.; Zhang, X. Anhedonia was associated with the dysregulation of hippocampal htr4 and microrna let-7a in rats. *Physiol. Behav.* **2014**, *129*, 135–141. [CrossRef] [PubMed]

52. Hancock, D.B.; Eijgelsheim, M.; Wilk, J.B.; Gharib, S.A.; Loehr, L.R.; Marciante, K.D.; Franceschini, N.; van Durme, Y.M.; Chen, T.H.; Barr, R.G.; et al. Meta-analyses of genome-wide association studies identify multiple loci associated with pulmonary function. *Nat. Genet.* **2010**, *42*, 45–52. [CrossRef] [PubMed]

53. Wilk, J.B.; Shrine, N.R.; Loehr, L.R.; Zhao, J.H.; Manichaikul, A.; Lopez, L.M.; Smith, A.V.; Heckbert, S.R.; Smolonska, J.; Tang, W.; et al. Genome-wide association studies identify chrna5/3 and htr4 in the development of airflow obstruction. *Am. J. Respir. Crit. Care Med.* **2012**, *186*, 622–632. [CrossRef] [PubMed]

54. Soler Artigas, M.; Loth, D.W.; Wain, L.V.; Gharib, S.A.; Obeidat, M.; Tang, W.; Zhai, G.; Zhao, J.H.; Smith, A.V.; Huffman, J.E.; et al. Genome-wide association and large-scale follow up identifies 16 new loci influencing lung function. *Nat. Genet.* **2011**, *43*, 1082–1090. [CrossRef] [PubMed]

55. House, J.S.; Li, H.; DeGraff, L.M.; Flake, G.; Zeldin, D.C.; London, S.J. Genetic variation in htr4 and lung function: Gwas follow-up in mouse. *FASEB J.* **2015**, *29*, 323–335. [CrossRef] [PubMed]

56. Blondel, O.; Gastineau, M.; Langlois, M.; Fischmeister, R. The 5-ht4 receptor antagonist ml10375 inhibits the constitutive activity of human 5-ht4(c) receptor. *Br. J. Pharmacol.* **1998**, *125*, 595–597. [CrossRef] [PubMed]

57. Brattelid, T.; Kvingedal, A.M.; Krobert, K.A.; Andressen, K.W.; Bach, T.; Hystad, M.E.; Kaumann, A.J.; Levy, F.O. Cloning, pharmacological characterisation and tissue distribution of a novel 5-ht4 receptor splice variant, 5-ht4(i). *Naunyn-Schmiedebergs Arch. Pharmacol.* **2004**, *369*, 616–628. [CrossRef] [PubMed]

58. Coupar, I.M.; Desmond, P.V.; Irving, H.R. Human 5-ht(4) and 5-ht(7) receptor splice variants: Are they important? *Curr. Neuropharmacol.* **2007**, *5*, 224–231. [CrossRef] [PubMed]

59. Ray, A.M.; Kelsell, R.E.; Houp, J.A.; Kelly, F.M.; Medhurst, A.D.; Cox, H.M.; Calver, A.R. Identification of a novel 5-ht(4) receptor splice variant (r5-ht(4c1)) and preliminary characterisation of specific 5-ht(4a) and 5-ht(4b) receptor antibodies. *Eur. J. Pharmacol.* **2009**, *604*, 1–11. [CrossRef] [PubMed]

60. Bender, E.; Pindon, A.; van Oers, I.; Zhang, Y.B.; Gommeren, W.; Verhasselt, P.; Jurzak, M.; Leysen, J.; Luyten, W. Structure of the human serotonin 5-ht4 receptor gene and cloning of a novel 5-ht4 splice variant. *J. Neurochem.* **2000**, *74*, 478–489. [CrossRef] [PubMed]

61. Vilaro, M.T.; Domenech, T.; Palacios, J.M.; Mengod, G. Cloning and characterization of a novel human 5-ht4 receptor variant that lacks the alternatively spliced carboxy terminal exon. Rt-pcr distribution in human brain and periphery of multiple 5-ht4 receptor variants. *Neuropharmacology* **2002**, *42*, 60–73. [CrossRef]

62. Vilaro, M.T.; Cortes, R.; Mengod, G. Serotonin 5-ht4 receptors and their mrnas in rat and guinea pig brain: Distribution and effects of neurotoxic lesions. *J. Comp. Neurol.* **2005**, *484*, 418–439. [CrossRef] [PubMed]

63. Claeysen, S.; Faye, P.; Sebben, M.; Taviaux, S.; Bockaert, J.; Dumuis, A. 5-ht4 receptors: Cloning and expression of new splice variants. *Ann. N. Y. Acad. Sci.* **1998**, *861*, 49–56. [CrossRef] [PubMed]

64. Kaizuka, T.; Hayashi, T. Comparative analysis of palmitoylation sites of serotonin (5-ht) receptors in vertebrates. *Neuropsychopharmacol. Rep.* **2018**, *38*, 75–85. [CrossRef] [PubMed]

65. Vilaro, M.T.; Cortes, R.; Gerald, C.; Branchek, T.A.; Palacios, J.M.; Mengod, G. Localization of 5-ht4 receptor mrna in rat brain by in situ hybridization histochemistry. *Mol. Brain Res.* **1996**, *43*, 356–360. [CrossRef]

66. Claeysen, S.; Sebben, M.; Becamel, C.; Bockaert, J.; Dumuis, A. Novel brain-specific 5-ht4 receptor splice variants show marked constitutive activity: Role of the c-terminal intracellular domain. *Mol. Pharmacol.* **1999**, *55*, 910–920. [PubMed]

67. Joubert, L.; Hanson, B.; Barthet, G.; Sebben, M.; Claeysen, S.; Hong, W.; Marin, P.; Dumuis, A.; Bockaert, J. New sorting nexin (snx27) and nherf specifically interact with the 5-ht4a receptor splice variant: Roles in receptor targeting. *J. Cell Sci.* **2004**, *117*, 5367–5379. [CrossRef] [PubMed]

68. Pindon, A.; Van Hecke, G.; Josson, K.; Van Gompel, P.; Lesage, A.; Leysen, J.E.; Jurzak, M. Internalization of human 5-ht4a and 5-ht4b receptors is splice variant dependent. *Biosci. Rep.* **2004**, *24*, 215–223. [CrossRef] [PubMed]

69. Pindon, A.; van Hecke, G.; van Gompel, P.; Lesage, A.S.; Leysen, J.E.; Jurzak, M. Differences in signal transduction of two 5-ht4 receptor splice variants: Compound specificity and dual coupling with galphas- and galphai/o-proteins. *Mol. Pharmacol.* **2002**, *61*, 85–96. [CrossRef] [PubMed]

70. Benovic, J.L.; Strasser, R.H.; Caron, M.G.; Lefkowitz, R.J. Beta-adrenergic receptor kinase: Identification of a novel protein kinase that phosphorylates the agonist-occupied form of the receptor. *Proc. Natl. Acad. Sci. USA* **1986**, *83*, 2797–2801. [CrossRef] [PubMed]

71. Lohse, M.J.; Benovic, J.L.; Codina, J.; Caron, M.G.; Lefkowitz, R.J. Beta-arrestin: A protein that regulates beta-adrenergic receptor function. *Science* **1990**, *248*, 1547–1550. [CrossRef] [PubMed]

72. Krupnick, J.G.; Benovic, J.L. The role of receptor kinases and arrestins in g protein-coupled receptor regulation. *Annu. Rev. Pharmacol. Toxicol.* **1998**, *38*, 289–319. [CrossRef] [PubMed]

73. Pitcher, J.A.; Freedman, N.J.; Lefkowitz, R.J. G protein-coupled receptor kinases. *Annu. Rev. Biochem.* **1998**, *67*, 653–692. [CrossRef] [PubMed]

74. Stadel, J.M.; Strulovici, B.; Nambi, P.; Lavin, T.N.; Briggs, M.M.; Caron, M.G.; Lefkowitz, R.J. Desensitization of the beta-adrenergic receptor of frog erythrocytes. Recovery and characterization of the down-regulated receptors in sequestered vesicles. *J. Biol. Chem.* **1983**, *258*, 3032–3038. [PubMed]

75. Ferguson, S.S. Evolving concepts in g protein-coupled receptor endocytosis: The role in receptor desensitization and signaling. *Pharmacol. Rev.* **2001**, *53*, 1–24. [PubMed]

76. Willets, J.M.; Mistry, R.; Nahorski, S.R.; Challiss, R.A. Specificity of g protein-coupled receptor kinase 6-mediated phosphorylation and regulation of single-cell m3 muscarinic acetylcholine receptor signaling. *Mol. Pharmacol.* **2003**, *64*, 1059–1068. [CrossRef] [PubMed]

77. Salom, D.; Wang, B.; Dong, Z.; Sun, W.; Padayatti, P.; Jordan, S.; Salon, J.A.; Palczewski, K. Post-translational modifications of the serotonin type 4 receptor heterologously expressed in mouse rod cells. *Biochemistry* **2012**, *51*, 214–224. [CrossRef] [PubMed]

78. Barthet, G.; Carrat, G.; Cassier, E.; Barker, B.; Gaven, F.; Pillot, M.; Framery, B.; Pellissier, L.P.; Augier, J.; Kang, D.S.; et al. Beta-arrestin1 phosphorylation by grk5 regulates g protein-independent 5-ht4 receptor signalling. *EMBO J.* **2009**, *28*, 2706–2718. [CrossRef] [PubMed]

79. Ponimaskin, E.; Dumuis, A.; Gaven, F.; Barthet, G.; Heine, M.; Glebov, K.; Richter, D.W.; Oppermann, M. Palmitoylation of the 5-hydroxytryptamine4a receptor regulates receptor phosphorylation, desensitization, and beta-arrestin-mediated endocytosis. *Mol. Pharmacol.* **2005**, *67*, 1434–1443. [CrossRef] [PubMed]

80. Ponimaskin, E.G.; Heine, M.; Joubert, L.; Sebben, M.; Bickmeyer, U.; Richter, D.W.; Dumuis, A. The 5-hydroxytryptamine(4a) receptor is palmitoylated at two different sites, and acylation is critically involved in regulation of receptor constitutive activity. *J. Biol. Chem.* **2002**, *277*, 2534–2546. [CrossRef] [PubMed]

81. Penas-Cazorla, R.; Vilaro, M.T. Serotonin 5-ht4 receptors and forebrain cholinergic system: Receptor expression in identified cell populations. *Brain Struct. Funct.* **2015**, *220*, 3413–3434. [CrossRef] [PubMed]

82. Domenech, T.; Beleta, J.; Fernandez, A.G.; Gristwood, R.W.; Cruz Sanchez, F.; Tolosa, E.; Palacios, J.M. Identification and characterization of serotonin 5-ht4 receptor binding sites in human brain: Comparison with other mammalian species. *Mol. Brain Res.* **1994**, *21*, 176–180. [CrossRef]

83. Feng, J.; Cai, X.; Zhao, J.; Yan, Z. Serotonin receptors modulate gaba(a) receptor channels through activation of anchored protein kinase c in prefrontal cortical neurons. *J. Neurosci.* **2001**, *21*, 6502–6511. [CrossRef] [PubMed]

84. Roychowdhury, S.; Haas, H.; Anderson, E.G. 5-ht1a and 5-ht4 receptor colocalization on hippocampal pyramidal cells. *Neuropharmacology* **1994**, *33*, 551–557. [CrossRef]

85. Grossman, C.J.; Kilpatrick, G.J.; Bunce, K.T. Development of a radioligand binding assay for 5-ht4 receptors in guinea-pig and rat brain. *Br. J. Pharmacol.* **1993**, *109*, 618–624. [CrossRef] [PubMed]

86. Waeber, C.; Sebben, M.; Grossman, C.; Javoy-Agid, F.; Bockaert, J.; Dumuis, A. [3h]-gr113808 labels 5-ht4 receptors in the human and guinea-pig brain. *Neuroreport* **1993**, *4*, 1239–1242. [CrossRef] [PubMed]

87. Waeber, C.; Sebben, M.; Nieoullon, A.; Bockaert, J.; Dumuis, A. Regional distribution and ontogeny of 5-ht4 binding sites in rodent brain. *Neuropharmacology* **1994**, *33*, 527–541. [CrossRef]

88. Mengod, G.; Vilaro, M.T.; Raurich, A.; Lopez-Gimenez, J.F.; Cortes, R.; Palacios, J.M. 5-ht receptors in mammalian brain: Receptor autoradiography and in situ hybridization studies of new ligands and newly identified receptors. *Histochem. J.* **1996**, *28*, 747–758. [CrossRef] [PubMed]

89. Egeland, M.; Warner-Schmidt, J.; Greengard, P.; Svenningsson, P. Co-expression of serotonin 5-ht(1b) and 5-ht(4) receptors in p11 containing cells in cerebral cortex, hippocampus, caudate-putamen and cerebellum. *Neuropharmacology* **2011**, *61*, 442–450. [CrossRef] [PubMed]

90. Imoto, Y.; Kira, T.; Sukeno, M.; Nishitani, N.; Nagayasu, K.; Nakagawa, T.; Kaneko, S.; Kobayashi, K.; Segi-Nishida, E. Role of the 5-ht4 receptor in chronic fluoxetine treatment-induced neurogenic activity and granule cell dematuration in the dentate gyrus. *Mol. Brain* **2015**, *8*, 29. [CrossRef] [PubMed]

91. Wong, E.H.; Reynolds, G.P.; Bonhaus, D.W.; Hsu, S.; Eglen, R.M. Characterization of [3h]gr 113808 binding to 5-ht4 receptors in brain tissues from patients with neurodegenerative disorders. *Behav. Brain Res.* **1996**, *73*, 249–252. [CrossRef]

92. Varnas, K.; Halldin, C.; Pike, V.W.; Hall, H. Distribution of 5-ht4 receptors in the postmortem human brain–an autoradiographic study using [125i]sb 207710. *Eur. Neuropsychopharmacol.* **2003**, *13*, 228–234. [CrossRef]

93. Madsen, K.; Haahr, M.T.; Marner, L.; Keller, S.H.; Baare, W.F.; Svarer, C.; Hasselbalch, S.G.; Knudsen, G.M. Age and sex effects on 5-ht(4) receptors in the human brain: A [(11)c]sb207145 pet study. *J. Cereb. Blood Flow Metab.* **2011**, *31*, 1475–1481. [CrossRef] [PubMed]

94. Waeber, C.; Sebben, M.; Bockaert, J.; Dumuis, A. Regional distribution and ontogeny of 5-ht4 binding sites in rat brain. *Behav. Brain Res.* **1995**, *73*, 259–262. [CrossRef]

95. Reynolds, G.P.; Mason, S.L.; Meldrum, A.; De Keczer, S.; Parnes, H.; Eglen, R.M.; Wong, E.H. 5-hydroxytryptamine (5-ht)4 receptors in post mortem human brain tissue: Distribution, pharmacology and effects of neurodegenerative diseases. *Br. J. Pharmacol.* **1995**, *114*, 993–998. [CrossRef] [PubMed]

96. Rosel, P.; Arranz, B.; Urretavizcaya, M.; Oros, M.; San, L.; Navarro, M.A. Altered 5-ht2a and 5-ht4 postsynaptic receptors and their intracellular signalling systems ip3 and camp in brains from depressed violent suicide victims. *Neuropsychobiology* **2004**, *49*, 189–195. [CrossRef] [PubMed]

97. Licht, C.L.; Kirkegaard, L.; Zueger, M.; Chourbaji, S.; Gass, P.; Aznar, S.; Knudsen, G.M. Changes in 5-ht4 receptor and 5-ht transporter binding in olfactory bulbectomized and glucocorticoid receptor heterozygous mice. *Neurochem. Int.* **2010**, *56*, 603–610. [CrossRef] [PubMed]

98. Haahr, M.E.; Fisher, P.; Holst, K.; Madsen, K.; Jensen, C.G.; Marner, L.; Lehel, S.; Baare, W.; Knudsen, G.; Hasselbalch, S. The 5-ht4 receptor levels in hippocampus correlates inversely with memory test performance in humans. *Hum. Brain Mapp.* **2013**, *34*, 3066–3074. [CrossRef] [PubMed]

99. Haahr, M.E.; Rasmussen, P.M.; Madsen, K.; Marner, L.; Ratner, C.; Gillings, N.; Baare, W.F.; Knudsen, G.M. Obesity is associated with high serotonin 4 receptor availability in the brain reward circuitry. *Neuroimage* **2012**, *61*, 884–888. [CrossRef] [PubMed]

100. Licht, C.L.; Marcussen, A.B.; Wegener, G.; Overstreet, D.H.; Aznar, S.; Knudsen, G.M. The brain 5-ht4 receptor binding is down-regulated in the flinders sensitive line depression model and in response to paroxetine administration. *J. Neurochem.* **2009**, *109*, 1363–1374. [CrossRef] [PubMed]

101. Vidal, R.; Valdizan, E.M.; Mostany, R.; Pazos, A.; Castro, E. Long-term treatment with fluoxetine induces desensitization of 5-ht4 receptor-dependent signalling and functionality in rat brain. *J. Neurochem.* **2009**, *110*, 1120–1127. [CrossRef] [PubMed]

102. Manuel-Apolinar, L.; Rocha, L.; Pascoe, D.; Castillo, E.; Castillo, C.; Meneses, A. Modifications of 5-ht4 receptor expression in rat brain during memory consolidation. *Brain Res.* **2005**, *1042*, 73–81. [CrossRef] [PubMed]

103. Ratner, C.; Ettrup, A.; Bueter, M.; Haahr, M.E.; Compan, V.; le Roux, C.W.; Levin, B.; Hansen, H.H.; Knudsen, G.M. Cerebral markers of the serotonergic system in rat models of obesity and after roux-en-y gastric bypass. *Obesity* **2012**, *20*, 2133–2141. [CrossRef] [PubMed]

104. Heiman, M.; Heilbut, A.; Francardo, V.; Kulicke, R.; Fenster, R.J.; Kolaczyk, E.D.; Mesirov, J.P.; Surmeier, D.J.; Cenci, M.A.; Greengard, P. Molecular adaptations of striatal spiny projection neurons during levodopa-induced dyskinesia. *Proc. Natl. Acad. Sci. USA* **2014**, *111*, 4578–4583. [CrossRef] [PubMed]

105. Kudryavtseva, N.N.; Smagin, D.A.; Kovalenko, I.L.; Galyamina, A.G.; Vishnivetskaya, G.B.; Babenko, V.N.; Orlov, Y.L. serotonergic genes in the development of anxiety/depression-like state and pathology of aggressive behavior in male mice: Rna-seq data. *Mol. Biol* **2017**, *51*, 288–300. [CrossRef]

106. Chen, A.; Kelley, L.D.; Janusonis, S. Effects of prenatal stress and monoaminergic perturbations on the expression of serotonin 5-ht(4) and adrenergic beta(2) receptors in the embryonic mouse telencephalon. *Brain Res.* **2012**, *1459*, 27–34. [CrossRef] [PubMed]

107. Schmidt, E.F.; Warner-Schmidt, J.L.; Otopalik, B.G.; Pickett, S.B.; Greengard, P.; Heintz, N. Identification of the cortical neurons that mediate antidepressant responses. *Cell* **2012**, *149*, 1152–1163. [CrossRef] [PubMed]

108. Madsen, K.; Torstensen, E.; Holst, K.K.; Haahr, M.E.; Knorr, U.; Frokjaer, V.G.; Brandt-Larsen, M.; Iversen, P.; Fisher, P.M.; Knudsen, G.M. Familial risk for major depression is associated with lower striatal 5-ht(4) receptor binding. *Int. J. Neuropsychopharmacol.* **2014**, *18*. [CrossRef]

109. Ohtsuki, T.; Ishiguro, H.; Detera-Wadleigh, S.D.; Toyota, T.; Shimizu, H.; Yamada, K.; Yoshitsugu, K.; Hattori, E.; Yoshikawa, T.; Arinami, T. Association between serotonin 4 receptor gene polymorphisms and bipolar disorder in japanese case-control samples and the nimh genetics initiative bipolar pedigrees. *Mol. Psychiatry* **2002**, *7*, 954–961. [CrossRef] [PubMed]

110. Haahr, M.E.; Fisher, P.M.; Jensen, C.G.; Frokjaer, V.G.; Mahon, B.M.; Madsen, K.; Baare, W.F.; Lehel, S.; Norremolle, A.; Rabiner, E.A.; et al. Central 5-ht4 receptor binding as biomarker of serotonergic tonus in humans: A [11c]sb207145 pet study. *Mol. Psychiatry* **2014**, *19*, 427–432. [CrossRef] [PubMed]

111. Jean, A.; Conductier, G.; Manrique, C.; Bouras, C.; Berta, P.; Hen, R.; Charnay, Y.; Bockaert, J.; Compan, V. Anorexia induced by activation of serotonin 5-ht4 receptors is mediated by increases in cart in the nucleus accumbens. *Proc. Natl. Acad. Sci. USA* **2007**, *104*, 16335–16340. [CrossRef] [PubMed]

112. Tsang, S.W.; Keene, J.; Hope, T.; Spence, I.; Francis, P.T.; Wong, P.T.; Chen, C.P.; Lai, M.K. A serotoninergic basis for hyperphagic eating changes in Alzheimer's disease. *J. Neurol. Sci.* **2010**, *288*, 151–155. [CrossRef] [PubMed]

113. Baranger, K.; Giannoni, P.; Girard, S.D.; Girot, S.; Gaven, F.; Stephan, D.; Migliorati, M.; Khrestchatisky, M.; Bockaert, J.; Marchetti-Gauthier, E.; et al. Chronic treatments with a 5-ht4 receptor agonist decrease amyloid pathology in the entorhinal cortex and learning and memory deficits in the 5xfad mouse model of Alzheimer's disease. *Neuropharmacology* **2017**, *126*, 128–141. [CrossRef] [PubMed]

114. Giannoni, P.; Gaven, F.; de Bundel, D.; Baranger, K.; Marchetti-Gauthier, E.; Roman, F.S.; Valjent, E.; Marin, P.; Bockaert, J.; Rivera, S.; et al. Early administration of rs 67333, a specific 5-ht4 receptor agonist, prevents amyloidogenesis and behavioral deficits in the 5xfad mouse model of Alzheimer's disease. *Front. Aging Neurosci.* **2013**, *5*, 96. [CrossRef] [PubMed]

115. Bockaert, J.; Claeysen, S.; Compan, V.; Dumuis, A. 5-ht(4) receptors, a place in the sun: Act two. *Curr. Opin. Pharmacol.* **2011**, *11*, 87–93. [CrossRef] [PubMed]

116. Brodney, M.A.; Johnson, D.E.; Sawant-Basak, A.; Coffman, K.J.; Drummond, E.M.; Hudson, E.L.; Fisher, K.E.; Noguchi, H.; Waizumi, N.; McDowell, L.L.; et al. Identification of multiple 5-ht(4) partial agonist clinical candidates for the treatment of Alzheimer's disease. *J. Med. Chem.* **2012**, *55*, 9240–9254. [CrossRef] [PubMed]

117. Madsen, K.; Neumann, W.J.; Holst, K.; Marner, L.; Haahr, M.T.; Lehel, S.; Knudsen, G.M.; Hasselbalch, S.G. Cerebral serotonin 4 receptors and amyloid-beta in early Alzheimer's disease. *J. Alzheimers Dis.* **2011**, *26*, 457–466. [CrossRef] [PubMed]

118. Lai, M.K.; Tsang, S.W.; Francis, P.T.; Esiri, M.M.; Hope, T.; Lai, O.F.; Spence, I.; Chen, C.P. [3h]gr113808 binding to serotonin 5-ht(4) receptors in the postmortem neocortex of alzheimer disease: A clinicopathological study. *J. Neural Transm.* **2003**, *110*, 779–788. [PubMed]

119. Suzuki, T.; Iwata, N.; Kitamura, Y.; Kitajima, T.; Yamanouchi, Y.; Ikeda, M.; Nishiyama, T.; Kamatani, N.; Ozaki, N. Association of a haplotype in the serotonin 5-ht4 receptor gene (htr4) with japanese schizophrenia. *Am. J. Med. Genet. B Neuropsychiatr. Genet.* **2003**, *121B*, 7–13. [CrossRef] [PubMed]

120. Li, J.; Wang, Y.; Zhou, R.; Wang, B.; Zhang, H.; Yang, L.; Faraone, S.V. Association of attention-deficit/hyperactivity disorder with serotonin 4 receptor gene polymorphisms in han chinese subjects. *Neurosci. Lett.* **2006**, *401*, 6–9. [CrossRef] [PubMed]

121. Holmes, A.; Murphy, D.L.; Crawley, J.N. Reduced aggression in mice lacking the serotonin transporter. *Psychopharmacology* **2002**, *161*, 160–167. [CrossRef] [PubMed]

122. Jennings, K.A.; Loder, M.K.; Sheward, W.J.; Pei, Q.; Deacon, R.M.; Benson, M.A.; Olverman, H.J.; Hastie, N.D.; Harmar, A.J.; Shen, S.; et al. Increased expression of the 5-ht transporter confers a low-anxiety phenotype linked to decreased 5-ht transmission. *J. Neurosci.* **2006**, *26*, 8955–8964. [CrossRef] [PubMed]

123. Mannoury la Cour, C.; Boni, C.; Hanoun, N.; Lesch, K.P.; Hamon, M.; Lanfumey, L. Functional consequences of 5-ht transporter gene disruption on 5-ht(1a) receptor-mediated regulation of dorsal raphe and hippocampal cell activity. *J. Neurosci.* **2001**, *21*, 2178–2185. [CrossRef] [PubMed]

124. Gobbi, G.; Murphy, D.L.; Lesch, K.; Blier, P. Modifications of the serotonergic system in mice lacking serotonin transporters: An in vivo electrophysiological study. *J. Pharmacol. Exp. Ther.* **2001**, *296*, 987–995. [PubMed]

125. Basselin, M.; Fox, M.A.; Chang, L.; Bell, J.M.; Greenstein, D.; Chen, M.; Murphy, D.L.; Rapoport, S.I. Imaging elevated brain arachidonic acid signaling in unanesthetized serotonin transporter (5-htt)-deficient mice. *Neuropsychopharmacology* **2009**, *34*, 1695–1709. [CrossRef] [PubMed]

126. Qu, Y.; Villacreses, N.; Murphy, D.L.; Rapoport, S.I. 5-ht2a/2c receptor signaling via phospholipase a2 and arachidonic acid is attenuated in mice lacking the serotonin reuptake transporter. *Psychopharmacology* **2005**, *180*, 12–20. [CrossRef] [PubMed]

127. Jennings, K.A.; Sheward, W.J.; Harmar, A.J.; Sharp, T. Evidence that genetic variation in 5-ht transporter expression is linked to changes in 5-ht2a receptor function. *Neuropharmacology* **2008**, *54*, 776–783. [CrossRef] [PubMed]

128. Vidal, R.; Valdizan, E.M.; Vilaro, M.T.; Pazos, A.; Castro, E. Reduced signal transduction by 5-ht4 receptors after long-term venlafaxine treatment in rats. *Br. J. Pharmacol.* **2010**, *161*, 695–706. [CrossRef] [PubMed]

129. Warner-Schmidt, J.L.; Flajolet, M.; Maller, A.; Chen, E.Y.; Qi, H.; Svenningsson, P.; Greengard, P. Role of p11 in cellular and behavioral effects of 5-ht4 receptor stimulation. *J. Neurosci.* **2009**, *29*, 1937–1946. [CrossRef] [PubMed]

130. Albert, P.R. Why is depression more prevalent in women? *J. Psychiatry Neurosci.* **2015**, *40*, 219–221. [CrossRef] [PubMed]

131. Perfalk, E.; Cunha-Bang, S.D.; Holst, K.K.; Keller, S.; Svarer, C.; Knudsen, G.M.; Frokjaer, V.G. Testosterone levels in healthy men correlate negatively with serotonin 4 receptor binding. *Psychoneuroendocrinology* **2017**, *81*, 22–28. [CrossRef] [PubMed]

132. Minett, M.S.; Pereira, V.; Sikandar, S.; Matsuyama, A.; Lolignier, S.; Kanellopoulos, A.H.; Mancini, F.; Iannetti, G.D.; Bogdanov, Y.D.; Santana-Varela, S.; et al. Endogenous opioids contribute to insensitivity to pain in humans and mice lacking sodium channel nav1.7. *Nat. Commun.* **2015**, *6*, 8967. [CrossRef] [PubMed]

133. Isensee, J.; Krahe, L.; Moeller, K.; Pereira, V.; Sexton, J.E.; Sun, X.; Emery, E.; Wood, J.N.; Hucho, T. Synergistic regulation of serotonin and opioid signaling contributes to pain insensitivity in nav1.7 knockout mice. *Sci. Signal.* **2017**, *10*. [CrossRef] [PubMed]

MDPI
St. Alban-Anlage 66
4052 Basel
Switzerland
Tel. +41 61 683 77 34
Fax +41 61 302 89 18
www.mdpi.com

International Journal of Molecular Sciences Editorial Office
E-mail: ijms@mdpi.com
www.mdpi.com/journal/ijms